Lecture Notes in Artificial Intelligence 8443

Subseries of Lecture Notes in Computer Science

Vincent S. Tseng Tu Bao Ho
Zhi-Hua Zhou Arbee L.P. Chen
Hung-Yu Kao (Eds.)

Advances in Knowledge Discovery and Data Mining

18th Pacific-Asia Conference, PAKDD 2014
Tainan, Taiwan, May 13-16, 2014
Proceedings, Part I

Springer

Volume Editors

Vincent S. Tseng
National Cheng Kung University, Tainan, Taiwan, R.O.C.
E-mail: tsengsm@mail.ncku.edu.tw

Tu Bao Ho
Japan Advanced Institute of Science and Technology, Nomi, Ishikawa, Japan
E-mail: bao@jaist.ac.jp

Zhi-Hua Zhou
Nanjing University, China
E-mail: zhouzh@nju.edu.cn

Arbee L.P. Chen
National Chengchi University, Taipei, Taiwan, R.O.C.
E-mail: alpchen@cs.nccu.edu.tw

Hung-Yu Kao
National Cheng Kung University, Tainan, Taiwan, R.O.C.
E-mail: hykao@mail.ncku.edu.tw

ISSN 0302-9743 e-ISSN 1611-3349
ISBN 978-3-319-06607-3 e-ISBN 978-3-319-06608-0
DOI 10.1007/978-3-319-06608-0
Springer Cham Heidelberg New York Dordrecht London

Library of Congress Control Number: 2014936624

LNCS Sublibrary: SL 7 – Artificial Intelligence

Typesetting: Camera-ready by author, data conversion by Scientific Publishing Services, Chennai, India

Printed on acid-free paper

Springer is part of Springer Science+Business Media (www.springer.com)

Preface

PAKDD 2014 was the 18th conference of the Pacific Asia Conference series on Knowledge Discovery and Data Mining. The conference was held in Tainan, Taiwan, during May 13–16, 2014. Since its inception in 1997, the PAKDD conference series has been a leading international conference in the areas of data mining and knowledge discovery. It provides an inviting and inspiring forum for researchers and practitioners, from both academia and industry, to share new ideas, original research results, and practical experience. The 18th edition continues the great tradition with three world-class keynote speeches, a wonderful technical program, a handful of high quality tutorials and workshops, and a data mining competition.

The PAKDD 2014 conference received 371 submissions to the technical program, involving more than 980 authors in total. Each submitted paper underwent a rigorous double-blind review process and was reviewed by at least three Program Committee (PC) members as well as one senior PC member. Based on the extensive and thorough discussions by the reviewers, the senior PC members made recommendations. The Program Co-chairs went through each of the senior PC members' recommendations, as well as the submitted papers and reviews, to come up with the final selection. Overall, 100 papers were accepted in the technical program among 371 submissions, yielding a 27% acceptance rate. 40 of which (10.8%) had full presentations and 60 of which (16.2%) had short presentations. The technical program consisted of 21 sessions, covering the general fields of data mining and KDD extensively. We thank all reviewers (Senior PC, PC and external invitees) for their great efforts in reviewing the papers in a timely fashion. Without their hard work, we would not have been able to see such a high-quality program.

The conference program this year included three keynote talks by world-renowned data mining experts, namely, Professor Vipin Kumar from the University of Minnesota (*Understanding Climate Change: Opportunities and Challenges for Data Driven Research*); Professor Ming-Syan Chen from the National Taiwan University (*On Information Extraction for Social Networks*); Professor Jian Pei from the Simon Fraser University (*Being a Happy Dwarf in the Age of Big Data*). The program also included 12 workshops, which covered a number of exciting and fast growing hot topics. We also had 7 very timely and educational tutorials, covering the hot topics of social networks and media, pattern mining, big data, biomedical and health informatics mining and crowdsourcing. PAKDD 2014 also organized a data mining competition for those who wanted to lay their hands on mining interesting real-world datasets.

Putting together a conference on a scale like PAKDD 2014 requires tremendous efforts from the organizing team as well as financial support from the sponsors. We would like to express our special thanks to our honorary chairs,

Hiroshi Motoda and Philip S. Yu, for providing valuable advice and kind support. We thank Wen-Chih Peng, Haixun Wang, and James Bailey for organizing the workshop program. We also thank Mi-Yen Yeh, Guandong Xu and Seung-Won Hwang for organizing the tutorial program. As well, we thank Shou-De Lin, Nitesh Chawla and Hung-Yi Lo for organizing the data mining competition. We also thank Hung-Yu Kao for preparing the conference proceedings. Finally, we owe a big thank you to the great team of publicity co-chairs, local arrangement co-chairs, sponsorship chair and helpers. They ensured the conference attracted many local and international participants, and the conference program proceeded smoothly.

We would like to express our gratitude to all sponsors for their generous sponsorship and support. Special thanks are given to AFOSR/AOARD (Air Force Office of Scientific Research/Asian Office of Aerospace Research and Development) for their support to the success of the conference. We also wish to thank the PAKDD Steering Committee for offering the student travel support grant.

Finally, we hope you found the conference a fruitful experience and trust you had an enjoyable stay in Tainan, Taiwan.

May 2014

Vincent S. Tseng
Tu Bao Ho
Zhi-Hua Zhou
Arbee L.P. Chen
Hung-Yu Kao

Organization

Honorary Co-chairs

Hiroshi Motoda	Osaka University, Japan
Philip S. Yu	University of Illinois at Chicago, USA

General Co-chairs

Zhi-Hua Zhou	Nanjing University, China
Arbee L.P. Chen	National Chengchi University, Taiwan

Program Committee Co-chairs

Vincent S. Tseng	National Cheng Kung University, Taiwan
Tu Bao Ho	JAIST, Japan

Workshop Co-chairs

Wen-Chih Peng	National Chiao Tung University, Taiwan
Haixun Wang	Google Inc., USA
James Bailey	University of Melbourne, Australia

Tutorial Co-chairs

Mi-Yen Yeh	Academia Sinica, Taiwan
Guandong Xu	University of Technology Sydney, Australia
Seung-Won Hwang	POSTECH, Korea

Publicity Co-chairs

Takashi Washio	Osaka University, Japan
Tzung-Pei Hong	National University of Kaohsiung, Taiwan
Yu Zheng	Microsoft Research Asia, China
George Karypis	University of Minnesota, USA

Proceedings Chair

Hung-Yu Kao	National Cheng Kung University, Taiwan

Contest Co-chairs

Shou-De Lin	National Taiwan University, Taiwan
Nitesh Chawla	University of Notre Dame, USA
Hung-Yi Lo	Shih-Chien University, Taiwan

Local Arrangements Co-chairs

Jen-Wei Huang	National Cheng Kung University, Taiwan
Kun-Ta Chuang	National Cheng Kung University, Taiwan
Chuang-Kang Ting	National Chung Cheng University, Taiwan
Ja-Hwung Su	Kainan University, Taiwan

Sponsorship Chair

Yue-Shi Lee	Ming Chuan University, Taiwan

Registration Co-chairs

Hsuan-Tien Lin	National Taiwan University, Taiwan
Chien-Feng Huang	National University of Kaohsiung, Taiwan

Steering Committee

Chairs

Graham Williams	Australian Taxation Office, Australia
Tu Bao Ho (Co-Chair)	Japan Advanced Institute of Science and Technology, Japan

Life Members

Hiroshi Motoda	AFOSR/AOARD and Osaka University, Japan (Since 1997)
Rao Kotagiri	University of Melbourne, Australia (Since 1997)
Ning Zhong	Maebashi Institute of Technology, Japan (Since 1999)
Masaru Kitsuregawa	Tokyo University, Japan (Since 2000)
David Cheung	University of Hong Kong, China (Since 2001)
Graham Williams (Treasurer)	Australian National University, Australia (Since 2001)
Ming-Syan Chen	National Taiwan University, Taiwan (Since 2002)
Kyu-Young Whang	Korea Advanced Institute of Science & Technology, Korea (Since 2003)

Members

Huan Liu	Arizona State University, USA (Since 1998)
Chengqi Zhang	University of Technology Sydney, Australia (Since 2004)
Tu Bao Ho	Japan Advanced Institute of Science and Technology, Japan (Since 2005)
Ee-Peng Lim	Singapore Management University, Singapore (Since 2006)
Jaideep Srivastava	University of Minnesota, USA (Since 2006)
Zhi-Hua Zhou	Nanjing University, China (Since 2007)
Takashi Washio	Institute of Scientific and Industrial Research, Osaka University, Japan (Since 2008)
Thanaruk Theeramunkong	Thammasat University, Thailand (Since 2009)
P. Krishna Reddy	International Institute of Information Technology, Hyderabad (IIIT-H), India (Since 2010)
Joshua Z. Huang	Shenzhen Institutes of Advanced Technology, Chinese Academy of Sciences, China (Since 2011)
Longbing Cao	Advanced Analytics Institute, University of Technology Sydney, Australia (Since 2013)
Jian Pei	School of Computing Science, Simon Fraser University, Canada (Since 2013)
Myra Spiliopoulou	Information Systems, Otto-von-Guericke-University Magdeburg, Germany (Since 2013)

Senior Program Committee Members

James Bailey	University of Melbourne, Australia
Michael Berthold	University of Konstanz, Germany
Longbing Cao	University of Technology Sydney, Australia
Sanjay Chawla	University of Sydney, Australia
Lei Chen	Hong Kong University of Science and Technology, Hong Kong
Ming-Syan Chen	National Taiwan University, Taiwan
Peter Christen	The Australian National University, Australia
Ian Davidson	UC Davis, USA
Wei Fan	IBM T.J. Watson Research Center, USA
Bart Goethals	University of Antwerp, Belgium
Xiaohua Hu	Drexel University, USA
Ming Hua	Facebook, USA
Joshua Huang	Shenzhen Institutes of Advanced Technology, Chinese Academy of Sciences, China

George Karypis	University of Minnesota, USA
Hisashi Kashima	University of Tokyo, Japan
Shonali Krishnaswamy	Institute for Infocomm Research, Singapore
Jiuyong Li	University of South Australia, Australia
Ee-Peng Lim	Singapore Management University, Singapore
Chih-Jen Lin	National Taiwan University, Taiwan
Charles Ling	The University of Western Ontario, Canada
Huan Liu	Arizona State University, USA
Jiming Liu	Hong Kong Baptist University, Hong Kong
Nikos Mamoulis	University of Hong Kong, Hong Kong
Wee Keong Ng	Nanyang Technological University, Singapore
Jian Pei	Simon Fraser University, Canada
Wen-Chih Peng	National Chiao Tung University, Taiwan
P. Krishna Reddy	International Institute of Information Technology, Hyderabad (IIIT-H), India
Dou Shen	Baidu, China
Kyuseok Shim	Seoul National University, Korea
Myra Spiliopoulou	Otto-von-Guericke University Magdeburg, Germany
Jaideep Srivastava	University of Minnesota, USA
Masashi Sugiyama	Tokyo Institute of Technology, Japan
Dacheng Tao	University of Technology Sydney, Australia
Thanaruk Theeramunkong	Thammasat University, Thailand
Hanghang Tong	CUNY City College, USA
Shusaku Tsumoto	Shimane University, Japan
Haixun Wang	Google, USA
Jianyong Wang	Tsinghua University, China
Wei Wang	University of California at Los Angeles, USA
Takashi Washio	Osaka University, Japan
Ji-Rong Wen	Microsoft Research Asia, China
Xindong Wu	University of Vermont, USA
Xing Xie	Microsoft Research Asia, China
Hui Xiong	Rutgers Univesity, USA
Takahira Yamaguchi	Keio University, Japan
Xifeng Yan	UC Santa Barbara, USA
Jieping Ye	Arizona State University, USA
Jeffrey Xu Yu	The Chinese University of Hong Kong, Hong Kong
Osmar Zaiane	University of Alberta, Canada
Chengqi Zhang	University of Technology Sydney, Australia
Yanchun Zhang	Victoria University, Australia
Yu Zheng	Microsoft Research Asia, China
Ning Zhong	Maebashi Institute of Technology, Japan
Xiaofang Zhou	The University of Queensland, Australia

Program Committee Members

Shafiq Alam	University of Auckland, New Zealand
Aijun An	York University, Canada
Hideo Bannai	Kyushu University, Japan
Gustavo Batista	University of Sao Paulo, Brazil
Bettina Berendt	Katholieke Universiteit Leuven, The Netherlands
Chiranjib Bhattachar	Indian Institute of Science, India
Jiang Bian	Microsoft Research, China
Marut Buranarach	National Electronics and Computer Technology Center, Thailand
Krisztian Buza	University of Warsaw, Poland
Mary Elaine Califf	Illinois State University, USA
Rui Camacho	Universidade do Porto, Portugal
K. Selcuk Candan	Arizona State University, USA
Tru Cao	Ho Chi Minh City University of Technology, Vietnam
James Caverlee	Texas A&M University, USA
Keith Chan	The Hong Kong Polytechnic University, Hong Kong
Chia-Hui Chang	National Central University, Taiwan
Muhammad Cheema	Monash University, Australia
Chun-Hao Chen	Tamkang University, Taiwan
Enhong Chen	University of Science and Technology of China, China
Jake Chen	Indiana University-Purdue University Indianapolis, USA
Ling Chen	University of Technology Sydney, Australia
Meng Chang Chen	Academia Sinica, Taiwan
Shu-Ching Chen	Florida International University, USA
Songcan Chen	Nanjing University of Aeronautics and Astronautics, China
Yi-Ping Phoebe Chen	La Trobe University, Australia
Zheng Chen	Microsoft Research Asia, China
Zhiyuan Chen	University of Maryland Baltimore County, USA
Yiu-ming Cheung	Hong Kong Baptist University, Hong Kong
Silvia Chiusano	Politecnico di Torino, Italy
Kun-Ta Chuang	National Cheng Kung University, Taiwan
Bruno Cremilleux	Universite de Caen, France
Bin Cui	Peking University, China
Alfredo Cuzzocrea	ICAR-CNR and University of Calabria, Italy
Bing Tian Dai	Singapore Management University, Singapore
Dao-Qing Dai	Sun Yat-Sen University, China

Anne Denton	North Dakota State University, USA
Bolin Ding	Microsoft Research, USA
Wei Ding	University of Massachusetts Boston, USA
Guozhu Dong	Wright State University, USA
Dejing Dou	University of Oregon, USA
Haimonti Dutta	Columbia University, USA
Vladimir Estivill-Castro	Griffith University, Australia
Philippe Fournier-Viger	University of Moncton, Canada
Dragan Gamberger	Rudjer Boskovic Institute, Croatia
Cong Gao	Nanyang Technological University, Singapore
Jun Gao	Peking University, China
Junbin Gao	Charles Sturt University, Australia
Yong Guan	Iowa State University, USA
Ravi Gupta	Bioinformatics Head SciGenom Labs, India
Sung-Ho Ha	Kyungpook National University, Korea
Michael Hahsler	Southern Methodist University, USA
Yi Han	National Defence Technology University, China
Choochart Haruechaiyasak	National Electronics and Computer Technology Center, Thailand
Jing He	Victoria University, Australia
Jingrui He	IBM Research, USA
Robert Hilderman	University of Regina, Canada
Shoji Hirano	Shimane University, Japan
Chin-Kuan Ho	Multimedia University, Malaysia
Jaakko Hollmen	Aalto University, Finland
Tzung-Pei Hong	National Univesity of Kaohsiung, Taiwan
Kuo-Wei Hsu	National Chengchi University, Taiwan
Wynne Hsu	National University of Singapore, Singapore
Jun Huan	University of Kansas, USA
Jen-Wei Huang	National Cheng Kung University, Taiwan
Eyke Huellermeier	University of Marburg, Germany
Daisuke Ikeda	Kyshu University, Japan
Akihiro Inokuchi	Osaka University, Japan
Sanjay Jain	National University of Singapore, Singapore
Toshihiro Kamishima	National Institute of Advanced Industrial Science and Technology, Japan
Murat Kantarcioglu	University of Texas at Dallas, USA
Ben Kao	The University of Hong Kong, China
Hung-Yu Kao	National Cheng Kung University, Taiwan
Panagiotis Karras	Rutgers University, USA
Hiroyuki Kawano	Nanzan University, Japan
John Keane	The University of Manchester, UK
Latifur Khan	University of Texas at Dallas, USA
Hiroyuki Kitagawa	University of Tsukuba, Japan

Irena Koprinska	University of Sydney, Australia
Walter Kosters	Universiteit Leiden, The Netherlands
Marzena Kryszkiewicz	Warsaw University of Technology, Poland
James Kwok	Hong Kong University of Science and Technology, China
Wai Lam	The Chinese University of Hong Kong, Hong Kong
Wang-Chien Lee	Pennsylvania State University, USA
Yue-Shi Lee	Ming Chuan University, Taiwan
Yuh-Jye Lee	University of Science and Technology, Taiwan
Philippe Lenca	Telecom Bretagne, France
Carson K. Leung	University of Manitoba, Canada
Chengkai Li	The University of Texas at Arlington, USA
Chun-hung Li	Hong Kong Baptist University, Hong Kong
Gang Li	Deakin University, Australia
Jinyan Li	University of Technology Sydney, Australia
Ming Li	Nanjing University, China
Tao Li	Florida International University,USA
Xiaoli Li	Institute for Infocomm Research, Singapore
Xue Li	The University of Queensland, Australia
Xuelong Li	Chinese Academy of Sciences, China
Yidong Li	Beijing Jiaotong Univeristy, China
Zhenhui Li	Pennsylvania State University, USA
Grace Lin	Institute of Information Industry, Taiwan
Hsuan-Tien Lin	National Taiwan University, Taiwan
Shou-De Lin	National Taiwan University, Taiwan
Fei Liu	Bosch Research, USA
Qingshan Liu	NLPR Institute of Automation Chinese Academy of Science, China
David Lo	Singapore Management University, Singapore
Woong-Kee Loh	Sungkyul University, South Korea
Chang-Tien Lu	Virginia Polytechnic Institute and State University, USA
Hua Lu	Aalborg University, Denmark
Jun Luo	Hua Wei Noah's Ark Lab, Hong Kong
Ping Luo	Institute of Computing Technology, Chinese Academy of Sciences, China
Shuai Ma	Beihang University, China
Marco Maggini	Università degli Studi di Siena, Italy
Luong Chi Mai	Inst. of Information Technology, Vietnam Academy of Science and Technology, Vietnam
Bradley Malin	Vanderbilt University, USA
Hiroshi Mamitsuka	Kyoto University, Japan
Giuseppe Manco	Università della Calabria, Italy
David Martens	University of Antwerp, Belgium

Florent Masseglia	Inria, France
Tao Mei	Microsoft Research Asia, China
Xiaofeng Meng	Renmin University of China, China
Nguyen Le Minh	JAIST, Japan
Pabitra Mitra	Indian Institute of Technology Kharagpur, India
Mohamed Mokbel	University of Minnesota, USA
Yang-Sae Moon	Kangwon National University, Korea
Yasuhiko Morimoto	Hiroshima University, Japan
J. Nath	Indian Insitute of Technology, India
Richi Nayak	Queensland University of Technologies, Australia
See-Kiong Ng	Institute for Infocomm Research, A*STAR, Singapore
Wilfred Ng	Hong Kong University of Science and Technology, Hong Kong
Ngoc Thanh Nguyen	Wroclaw University of Technology, Poland
Xuan Vinh Nguyen	University of Melbourne, Australia
Tadashi Nomoto	National Institute of Japanese Literature, Japan
Masayuki Numao	Osaka University, Japan
Manabu Okumura	Japan Advanced Institute of Science and Technology, Japan
Salvatore Orlando	University of Venice, Italy
Jia-Yu Pan	Google, USA
Dhaval Patel	Indian Institute of Technology, Roorkee, India
Yonghong Peng	University of Bradford, UK
Jean-Marc Petit	INSA Lyon, France
Clifton Phua	SAS Institute, Singapore
Dinh Phung	Deakin University, Australia
Vincenzo Piuri	Università degli Studi di Milano, Italy
Oriol Pujol	University of Barcelona, Spain
Weining Qian	East China Normal University, China
Chedy Raissi	Inria, France
Chandan Reddy	Wayne State University, USA
Patricia Riddle	University of Auckland, New Zealand
Hong Shen	Adelaide University, Australia
Jialie Shen	Singapore Management University, Singapore
Yi-Dong Shen	Chinese Academy of Sciences, China
Masashi Shimbo	Nara Institute of Science and Technology, Japan
Lisa Singh	Georgetown University, USA
Andrzej Skowron	University of Warsaw, Poland
Min Song	New Jersey Institute of Technology, USA
Mingli Song	Zhejiang University, China

Aixin Sun	Nanyang Technological University, Singapore
Yizhou Sun	Northeastern University, USA
Thepchai Supnithi	National Electronics and Computer Technology Center, Thailand
David Taniar	Monash University, Australia
Xiaohui (Daniel) Tao	The University of Southern Queensland, Australia
Tamir Tassa	The Open University, Israel
Srikanta Tirthapura	Iowa State University, USA
Ivor Tsang	Nanyang Technological University, Singapore
Jeffrey Ullman	Stanford University, USA
Sasiporn Usanavasin	SIIT, Thammasat University, Thailand
Marian Vajtersic	University of Salzburg, Austria
Kitsana Waiyamai	Kasetsart University, Thailand
Hui Wang	University of Ulster, UK
Jason Wang	New Jersey Science and Technology University, USA
Lipo Wang	Nanyang Technological University, Singapore
Xiang Wang	IBM TJ Watson, USA
Xin Wang	University of Calgary, Canada
Chih-Ping Wei	National Taiwan University, Taiwan
Raymond Chi-Wing Wong	Hong Kong University of Science and Technology, Hong Kong
Jian Wu	Zhejiang University, China
Junjie Wu	Beihang University, China
Xintao Wu	University of North Carolina at Charlotte, USA
Guandong Xu	University of Technology Sydney, Australia
Takehisa Yairi	University of Tokyo, Japan
Seiji Yamada	National Institute of Informatics, Japan
Christopher Yang	Drexel University, USA
De-Nian Yang	Academia Sinica, Taiwan
Min Yao	Zhejiang University, China
Mi-Yen Yeh	Academia Sinica, Taiwan
Dit-Yan Yeung	Hong Kong University of Science and Technology, China
Jian Yin	Hong Kong University of Science and Technology, China
Xiaowei Ying	Bank of America, USA
Jin Soung Yoo	IUPU, USA
Tetsuya Yoshida	Hokkaido University, Japan
Clement Yu	University of Illinois at Chicago, USA
Aidong Zhang	State University of New York at Buffalo, USA
Bo Zhang	Tsinghua University, China
Daoqiang Zhang	Nanjing University of Aeronautics and Astronautics, China

Du Zhang	California State University, USA
Harry Zhang	University of New Brunswick, Canada
Junping Zhang	Fudan University, China
Mengjie Zhang	Victoria University of Wellington, New Zealand
Min-Ling Zhang	Southeast University, China
Shichao Zhang	University of Technology, Australia
Wenjie Zhang	University of New South Wales, Australia
Ying Zhang	University of New South Wales, Australia
Zhongfei Zhang	Binghamton University, USA
Zili Zhang	Deakin University, Australia
Peixiang Zhao	Florida State University, USA
Bin Zhou	University of Maryland Baltimore County, USA
Jiayu Zhou	Samsung Research America, USA
Shuigeng Zhou	Fudan University, China
Wenjun Zhou	University of Tennessse - Knoxville, USA
Feida Zhu	Singapore Management University, Singapore
Xingquan Zhu	Florida Atlantic University, USA
Albrecht Zimmerman	Katholieke Universiteit Leuven, Belgium

Table of Contents – Part I

Pattern Mining

Social Network and Social Media

Classification

Graph and Network Mining

Applications

Privacy Preserving

Table of Contents – Part II

Recommendation

Feature Selection and Reduction

Machine Learning

Temporal and Spatial Data

Novel Algorithms

Clustering

Biomedical Data Mining

Stream Mining

Outlier and Anomaly Detection

Multi-Sources Mining

Unstructured Data and Text Mining

MalSpot: Multi² Malicious Network Behavior Patterns Analysis

Hing-Hao Mao[1], Chung-Jung Wu[1], Evangelos E. Papalexakis[2], Christos Faloutsos[2], Kuo-Chen Lee[1], and Tien-Cheu Kao[1]

[1] Institute for Information Industry, Taipei, Taiwan
{chmao,cklonger,kclee,tckao}@iii.org.tw

[2] Cargegie Mellon University, Pittsburgh, PA, USA
{epapalex,christos}@cs.cmu.edu

Abstract. What are the patterns that typical network attackers exhibit? For a given malicious network behaviors, are its attacks spread uniformly over time? In this work, we develop MALSPOT, multi-resolution and multi-linear (Multi^2) network analysis system in order to discover such malicious patterns, so that we can use them later for attack detection, when attacks are concurrent with legitimate traffic. We designed and deployed MALSPOT, which employs multi-linear analysis with different time resolutions, running on top of MapReduce (Hadoop), and we identify patterns across attackers, attacked institutions and variation of time scales. We collect over a terabyte of proven malicious traces (along with benign ones), from the Taiwanese government security operation center (G-SOC) , during the entire year of 2012. We showcase the effectiveness of MALSPOT, by discovering interesting patterns and anomalies on this enormous dataset. We observed static and time-evolving patterns, that a vast majority of the known malicious behavior seem to follow.

Keywords: multi-resolution, tensor, anmoaly detection, multi-linear, uncorrelated levels.

1 Introduction

In today's wide interconnected world, malicious network attacks have a long incubation period, and as a result, existing state-of-the-art information security/data analysis mechanisms find it very challenging to compete against those attacks in a timely manner. Information security monitoring enterprises have limited information and, hence, fail to see the big picture of the attacks that are being orchestrated. However, due to it's immense scale, the Internet provides us with a large variety of data, both structured (e.g. logs), as well as unstructured, that can be used in aid of information security analysis. Thus, today's Internet's scale calls for big data analysis techniques. The main focus of the present work is to investigate large-scale and stealthy malware behaviour. Our analysis is based on considerable number of logs from real security information event management systems.

V.S. Tseng et al. (Eds.): PAKDD 2014, Part I, LNAI 8443, pp. 1–14, 2014.

Detection of stealthy attacks is particularly challenging, since, statistically, they are hard to distinguish from normal connections. Furthermore, obtaining attack data (e.g. through a network sniffer or a honeynet) poses challenges in its own right. Dainotti et al.[1] employ a horizontal scan of the entire IPv4 address space, in order to detect attacks created by the *Sality* botnet. Chen et al.[8] propose a scalable network forensic mechanism for stealthy, self-propagating attack detection; However, both these state-of-the-art methods are specialized in terms of the attacks they target. Therefore, there is still need for a tool that is able to identify attacks without specific assumptions of their characteristics or their behaviour/propagation pattern.

Key to the discovery of malicious patterns is the summarization of the network behaviour, characteristics, and propagation of a connection. Therefore, we need a systematic and scalable approach which is able to effectively summarize large heterogeneous pieces of data that represent different aspects of the network. Such tools can be drawn from time evolving graph mining and tensor analysis literature. In particular tensors or multi-dimensional arrays have appeared in numerous interesting application including clustering, trend & anomaly detection [6], and network forensics [3]. In [13] the authors propose GigaTensor, a scalable, distributed algorithm for large scale tensor decomposition. In this work, we leverage GigaTensor to the end of stealthy malware detection, without assuming prior knowledge on the malware's behaviour.

A more formal definition of the problem at hand is as follows

Problem 1. Attack Patterns Discovery in MalSpot

- **Given**: (1) intrusion detection system (IDS) event logs, recording ⟨ event_name, timestamp, target_ip ⟩ (2) Honeynet firewall logs, recording ⟨ source_ip, target_ip, timestamp ⟩
- **Find**: (1) the suspicious and common patterns in all three modes/aspects of the data, (2) provide an intuitive visualization of the above patterns, and (3) scale up in millions of nodes in our network.

Guided by the format of the data at hand, we propose MalSpot which choses to formulate the problem as multi-linear solution as well as tensor analysis. Additionally, we propose to experiment with the granularity of the time window in our data; hence, we propose a *multi-resolution* approach, which will be able to identify different types of anomalies, in *uncorrelated* levels of temporal granularity.

Definition 1 (Uncorrelated Levels). *Two different levels of temporal granularity are called* uncorrelated, *if the network behavior in those levels, for a particular node, or set of nodes, is significantly different. For instance, a set of nodes may experience different patterns in network traffic in an hourly scale, as opposed to a daily scale.*

By leveraging multi-level, multi-linear analysis of the aforementioned data, we are able to conduct scalable and efficient anomaly detection. Our main contributions are the following:

- **Design** of the MalSpot system: leveraging **hadoop**, our proposed system can scale up to arbitrarily big datasets, and it is scales near linearly with the network trace size (number of non-zero entries).
- **Discoveries:** we report that attacks come in different flavours: there are, for example, attacks that are particularly short in time and stop showing after a period of time, while other attacks are more persistent, or focus only on a specific port.

The benefit from using MalSpot as opposed to standard techniques is the fact that by doing so, we are able to detect correlations of entities participating in a heterogeneous network for a very long term, and additionally detect multi-aspect correlations of entities (e.g. using the port number as a dimension), the comparison is as shown in Table 1.

Table 1. Qualitative analysis of commercial SIEM event analysis packages, as compared to our proposed method.

	Time Granularity	Anomaly Detection	Pattern Discovery
MalSpot	*Second to Year*	✓	✓
Splunk	Second to Day	x	x
ArcSight (HP)	Second to Min	x	x

The rest of this paper is organized as follows: related work is outlined in Section 2. We first elaborate on MalSpot, in Section 3, we then provide experimental studies in Section 4. Finally, Section 5 concludes the discussion and highlights future directions.

2 Related Work

For handling huge collections of time-evolving events, [9] proposes a multi-resolution clustering scheme for time series data using k-means clustering and progressively renes the clusters. In order to discover the streaming pattern in multiple time-series, [11] propose SPIRIT (Streaming Pattern dIscoveRy in multIple Time-series) which can incrementally find correlations and hidden variables by means of using principal components analysis (PCA) and singular value decomposition (SVD) to summarize the key trends in the entire stream collection. In [10] the authors propose TriMine which consider multiple features to provide hidden topics modeling and forecast future events. In [4] the authors apply SVD to compress sensor data sequences by decomposing them into local patterns and weight variables.

Tensors and tensor decompositions have been extensively used in a variety of fields, including but not limited to Data Mining, Chemometrics, Signal Processing and Psychology. A concise review of tensor decompositions in the literature can be found in [5]. In this particular work, we focus on the so called

CP/PARAFAC decomposition, however, [5] provides an overview of the entire variety of decompositions that have been introduced.

In the immediate field of interest, anomaly detection, there has been a fair amount of tensor applications. In particular, [2] develop a decomposition model that is suitable for stream data mining and anomaly detection. The authors of [6] introduce a scalable anomaly detection framework using tensors. In [12] the authors perform anomaly detection in a (source IP, destination IP, port number) dataset, and in [7] the authors operate on (source IP, destination IP, port number, timestamp) dataset. Finally, in [3] the authors propose a framework for anomaly detection in multi-aspect data that is based in tensor decompositions.

In terms of scalable tensor decompositions, [12] proposes a fast, sparse and approximate method that scales very well mostly in multicore environments. In [13], the authors propose a MapReduce, scalable and distributed algorithm for CP/PARAFAC; this suits better our purpose, since our data resides in a distributed file system.

3 MalSpot

In this section, we describe the MalSpot, an approach for finding the pattern in huge data from scalable design. The MalSpot has two modes, i.e., single-resolution mode and multi-resolution mode based on multi-linear analysis process. The notations are shown in Table 2.

Initially, given the data description, provided a few lines above, we are able to form three mode tensors, whose non-zero entries correspond to the non-zero entries of the network logs. Since the logs record counts of events, and due to high data skew, it is often the case that a few set of connections will outnumber the rest of the connections, in terms of counts. To that end, we have two choices, in order to alleviate this issue: We may, either, make our data binary, where the tensor, we may take the logarithm of the counts, so that we compress very big values.

Tensor Formulation of Our Problem

In order to form a tensor out of the data that we posses, we create a tensor entry for each (i, j, k) triple of, say (source IP, target IP, timestamp) that exists in our data log. The choice for the value for each (i, j, k) varies: we can have the raw counts of connections, we can compress that value (by taking its logarithm), or we can simply indicate that such a triplet exists in our log, by setting that value to 1.

Tensor decomposition leverages multi-linear algebra in order to analyze such high-order data. The canonical polyadic (CP) or PARAFAC decomposition we employ can be seen as a generalization of the Singular Value Decomposition (SVD) for matrices. CP/PARAFAC decomposes a tensor to the weighted sum of outer products of mode-specific vectors for a 3-order tensor. Formally, for an M-mode tensor $\underline{\mathbf{X}}$ of size $\{I_1 \times I_2 \times \cdots \times I_M\}$, its CP/PARAFAC decomposition of rank R yields $\underline{\mathbf{X}} \approx \sum_{r=1}^{R} \lambda(\mathbf{a}_r^{(1)} \circ \ldots \circ \mathbf{a}_r^{(M)}) = \sum_{r=1}^{R} \prod_{m=1}^{M} \mathbf{a}_r^{(m)}$ where $\circ$ denotes the outer product, and $\prod$ is in the sense of vector outer product multiplication (and not in the traditional multiplication operation).

Table 2. Table of Symbols

Notations	Definitions and Descriptions
D	raw data from different types of information security logs with three different kinds of features
x_1,x_2,x_3	the three features defined in data D
$\mathbf{A},\mathbf{B},\mathbf{C}$	tensor factor matrices, associated with x_1, x_2 and x_3
$\underline{\mathbf{X}}$	the 3-mode tensor
$\mathbf{C}$	covariance matrix for measuring the prioritizing of investigation from clusters
R	the rank for tensor decomposition
k	the number of clusters given for the malicious patterns clustering
λ	the threshold of the top-n selected result
k-means($\mathbf{M}$,k)	scalable k-means algorithms given matrix $\mathbf{M}$
GigaTensor(χ)	scalable tensor decomposition [13]
$\oplus$	string concatenation
$Cov(G^+, G^*)$	covariance measures between clustering groups G^+ and G^*

3.1 Network Malicious Behavior Decomposition

Given the data description, provided a few lines above, we are able to form three mode tensors, whose non-zero entries correspond to the non-zero entries of the network logs. Since the logs record counts of events, and due to high data skew, it is often the case that a few set of connections will outnumber the rest of the connections, in terms of counts. To that end, we have two choices, in order to alleviate this issue: We may, either, make our data binary, where the tensor, instead of counts, stores 1 or 0, depending on whether a specific triplet exists in the logs, or, we may take the logarithm of the counts, so that we compress very big values.

Introduction to Tensors

We start by introducing a few definitions. A tensor is essentially a multi-dimensional extension of a matrix; more precisely, a n-mode tensor is a structure that is indexed by n indices. For instance, a 1-mode tensor is a vector, a 2-mode tensor is a matrix, and a 3-mode tensor is a cubic structure. In this work, we focus on three-mode tensors, however, given the appropriate data, we can readily extend our techniques to higher modes.

Tensor Formulation of Our Problem

In order to form a tensor out of the data that we posses, we create a tensor entry for each (i, j, k) triple of, say (source IP, target IP, timestamp) that exists in our data log. The choice for the value for each (i, j, k) varies: we can have the raw counts of connections, we can compress that value (by taking its logarithm), or we can simply indicate that such a triplet exists in our log, by setting that value to 1.

The $\underline{\mathbf{X}}$ is no longer approximate if R is equal to rank($\underline{\mathbf{X}}$), however, we often want to decompose $\underline{\mathbf{X}}$ to $R \ll$ rank($\underline{\mathbf{X}}$), so that we force similar patterns to map to the same low rank basis. Forcing R to be small is *key* to our application, because in this way, we force connections behaving similarly to map to the same low rank subspace.

3.2 Single Resolution Mode in MalSpot

We obtain suspicious patterns via scalable tensor decomposition, as mentioned in previously. In order to find out the groups of similar patterns, we use a scalable k-means implementation, in MapReduce, so that we cluster different malicious patterns, produced by the tensor decomposition. For the cluster, we choose to use the cosine similarity (or rather its inverse) as a distance measure. The cosine distance we used in this study is shown as $similarity(\mathbf{p},\mathbf{q}) = cos(\theta) = \frac{\mathbf{p}\cdot\mathbf{q}}{\|\mathbf{p}\|\|\mathbf{q}\|}$ where $\mathbf{p}$ and $\mathbf{q}$ are pairs of columns of the factor matrices $\mathbf{A}$, $\mathbf{B}$ or $\mathbf{C}$, produced by the decomposition.

Algorithm 1: MALSPOT algorithm (single-resolution mode)

Input: Dataset D with x_1,x_2 and x_3, with size of l, m and n, and a decomposition rank R, clustering number k

Output: Prioritized malicious patterns groups G

/* step 1: Tensor Construction */

MapReduce:

key $\leftarrow x_1 \oplus x_2 \oplus x_3$

Map $\langle$ *key* , 1 $\rangle \leftarrow D$

Tensor $\underline{\mathbf{X}} \leftarrow$ Reduce $\langle$ *key* , *count* $\rangle$, i=1 to $|D|$

/* step 2: Decomposition */

$\langle \mathbf{A}_{[l\times R]}, \mathbf{B}_{[m\times R]}, \mathbf{C}_{[n\times R]} \rangle \leftarrow GigaTensor(\underline{\mathbf{X}})$

/* step 3: Clustering and Ranking */

$\mathbf{a}_i \in$ columns of $\mathbf{A}, i = 1...R \leftarrow$ k-means($\mathbf{A}, k$)

$\mathbf{b}_i \in$ columns of $\mathbf{B}, i = 1...R \leftarrow$ k-means($\mathbf{B}, k$)

$\mathbf{c}_i \in$ columns of $\mathbf{C}, i = 1...R \leftarrow$ k-means($\mathbf{C}, k$)

$\{G_A \mid \{G_A^1,...,G_A^k\}\} \leftarrow$ PrioritizeCov($\mathbf{a}$,λ)

$\{G_B \mid \{G_B^1,...,G_B^k\}\} \leftarrow$ PrioritizeCov($\mathbf{b}$,λ)

$\{G_C \mid \{G_C^1,...,G_C^k\}\} \leftarrow$ PrioritizeCov($\mathbf{c}$,λ)

After clustering, we obtain different groups of connections, as summarized by decomposing $\underline{\mathbf{X}}$. Prioritization of each group is helpful for recommending groups of anomalous connections to domain experts, for further inspection. The covariance distance between two clusters G^+ and G^* as $Cov(G^+, G^*) = \frac{\sum(G_i^+ - \overline{G^+})(G_i^* - \overline{G^*})}{n-1}$. We use the covariance matrix $\mathbf{C}_{[k\times k]}$ to store all-pairs of clusters ; we then rank the groups according to $\sum_{i=1}^{k} \mathbf{C}(:,i)$ and choose the top-k^* to show to a domain expert, for further inspection.

3.3 Multi Resolution Mode in Malspot

As opposed to the single-resolution mode MALSPOT, the multi-resolution version takes significant changes among different temporal granularities into account, in order to identify pieces of the data that bear the *uncorrelated levels* characteristic. In order to introduce the multi-resolution mode of MALSPOT, let $T_1, T_2, \ldots, T_n$ be the different time granularities (i.e., hourly, daily, weekly and so on). For each T_i, there are k clusters denoted by G_i^j ($j = 1$ to k). In order to identify uncorrelated levels among these different temporal levels of resolution, we use the adjacency matreix of G_i, which we denote by A_i. A detailed outline of the procedure we follow is shown in Algorithm 2.

Algorithm 2: MALSPOT algorithm - (multi-resolution mode)

Input: For each time scale T_i we have a set of k clusters $G_i = \{G_i^j \mid j = 1 \text{ to } k\}$ where $\bigcup_{j=1}^{k} G_i^j = D$ and $G_i^m \bigcap G_i^n = \emptyset \, \forall m \neq n$

Output: data $\mathbf{N}_{i,i+1}$ that change clusters from T_i to T_{i+1} and $I_{i,i+1}(a)$ the degree of similarity between $F_i(a)$ and $F_{i+1}(a)$.

```
/* step 1: Generating Adjacency matrices A_i = {a_mn} of G_i */
Calculate a_mn according to (1)
/* step 2: Generating reduced adjacency matrices A_i^reduce = {a'_mn} */
Calculate a'_mn according to (2)
/* step 3: Find data N_{i,i+1} that change clusters from T_i to T_{i+1} */
A_i^diff = A_{i+1}^reduce - A_i^reduce
Find the minimum number of lines L = {(p,l)} where l represents row number
  or column number that cross the nonzero rows or columns of A_i^diff, p represents
  nonzero row or column entry of I.
N_{i,i+1} <- {p|(p,l) ∈ L for some l}
/* step 4: Calculate the probability of migration of each datum */
Calculate I_{i,i+1}(a) ∀i and ∀a according to (4)
```

In Algorithm 2, we have a set of k clusters $G_i = \{G_i^j \mid j = 1 \text{ to } k\}$ where $\bigcup_{j=1}^{k} G_i^j = D$, which are generated iteratively from different temporal resolutions. In the first step, we use the adjacency matrices $\mathbf{A}_i = \{a_{mn}\}$ in order to record whether a member of a cluster *changes* its cluster assignment between different resolutions, as shown in Eq. 1

$$a_{mn} \leftarrow \begin{cases} 1, & \begin{cases} \text{a.} & \text{if } m < n \text{ and data } m \text{ and } n \text{ are in the same cluster} \\ \text{b.} & \text{if } m = n \text{ and data } m \text{ itself forms a cluster} \end{cases} \\ 0, & \text{otherwise} \end{cases} \quad (1)$$

In order to reduce the computational complexity, we summarize the the reduced adjacency matrices as $\mathbf{A}_i^{\text{reduce}} = \{a'_{mn}\}$, by employing the characteristics of transitivity among clusters as shown in Eq. 2.

$$a'_{mn} \leftarrow \begin{cases} 1, & \text{if } a_{mn} = 1 \text{ and } a_{pn} = 0 \quad \forall p > m \\ 0, & \text{otherwise} \end{cases} \quad (2)$$

Let F_i be a function that maps the data in time scale T_i to its cluster. That is, $F_i : D \mapsto G_i$ s.t. $F_i(a) = G_i^j$ iff $a \in G_i^j$. For any $a \in D$, denote $C_{i \to i+1}(a) = F_i(a) \bigcap N_{i,i+1}$ (C denotes *change*), $C_{i \leftarrow i+1}(a) = F_{i+1}(a) \bigcap N_{i,i+1}$, $R_{i \to i+1}(a) = F_i(a) \backslash N_{i,i+1} = F_i(a) - O_{i \to i+1}(a)$ (R denotes that the assignment remain the same), $R_{i \leftarrow i+1}(a) = F_{i+1}(a) \backslash N_{i,i+1}$, the conditions considering change or no change are shown in Eq. 3.

$$S_{i,i+1}(a) \leftarrow \begin{cases} C_{i \to i+1}(a) \bigcap C_{i \leftarrow i+1}(a), & \text{if } a \in N_{i,i+1} \\ R_{i \to i+1}(a) \bigcap R_{i \leftarrow i+1}(a), & \text{if } a \notin N_{i,i+1} \end{cases} \tag{3}$$

$I_{i,i+1}(a)$ is the degree of similarity between $F_i(a)$ and $F_{i+1}(a)$ is as shown in Equation (4).

$$I_{i,i+1}(a) = \frac{\sum_{a \in S_{i,i+1}(a)} p_{i,i+1}(a)}{\sum_{a \in F_{i+1}(a)} p_{i,i+1}(a)} \tag{4}$$

where the $p_{i,i+1}(a)$ is given as in Eq. 5.

$$p_{i,i+1}(a) \leftarrow |S_{i,i+1}(a)| - 1 \tag{5}$$

4 Experiments

In this section we show the effectiveness of MalSpot in detecting anomalous behavior in diverse settings of information security logs, i.e., Security Operations Center (SOC) event logs and Honeynet firewall logs. We design all the experiments in order to answer the following questions:

- **Q1: Malicious pattern discovery:** how can MalSpot effectively identify malicious events in a variety of sites or for a long term. Especially, how effective is MalSpot in detecting random scanning, and hit-list scan behavior?
- **Q2: Providing insights to domain experts:** what is an intuitive and concise way of presenting the detected anomalies to domain experts and network administrators, so that they can, in turn, validate our methodology, as well as further investigate a set of attacks.

4.1 Dataset and Environment

As we mention in the problem definition, we analyze data coming from three different sources. We use three different types and sources of datasets in our experiments to demonstrate MalSpot performances. These two datasets are 1) a Honeynet dataset, 2) an intrusion detection system (IDS) events dataset from Taiwan government; we summarize the datasets in Table 3. The Honeynet dataset was collected especially for the purposes of this work, using the Honeynet project system from 10 distributed sites in Taiwan[1]. The IDS events dataset is

[1] Honeynet project, `http://www.Honeynet.org/`

collected from the Security Operation Center of the Taiwanese government, for the entire year of 2012. This dataset includes the IDS triggered alerts from 61 government institutes with 4,081 types of events. These types of events can be categorized into 39 classes, Of which 33 classes of attack is defined by Snort IDS, the other 6 classes are custom by us (including: blacklist, high threat malware behavior and so on). In total, the SOC dataset contains 828,069,066 events, the dataset details could be shown as Table 3. .

Table 3. Datasets harvested & analyzed

Dataset	Description	Dimension	Nonzeros
Honeynet	Gathered from 10 distributed Honeynet sensors in Jan. 2013	368K x 64K x 31	3243K
G-SOC (Type)	Taiwan official institutes IDS events in 2012	[8187, 361, 52, 12] x 4081 x 61	800M+
G-SOC(Cat)	Taiwan official institutes IDS events in 2012	[8187, 361, 52, 12] x 39 x 61	1742K

We leveraged the scalability of GigaTensor[13] in 16 nodes of a Hadoop cluster where each machine has 2 quad-core Intel 2.83 GHz CPU, 8 GB RAM, and 2 Terabytes disk. The Apache Mahout(Scalable Machine Learning and Data Mining)[2] version 0.7 is used for supporting clustering algorithms.

4.2 Analysis of the Honeynet Data

For this dataset, we set $R = 5$ as the low rank of the tensor decomposition; after decomposing the tensor, we obtain three factor matrices each representing one of the three modes of our data: source IP, target IP and timestamp respectively. Each column of those factor matrices corresponds to one out of the R latent groups, in our low rank representation of the data. Based on this low rank embedding of the data, we compare pairs of columns for each factor matrix, in order to detect outliers. For example, given the factor matrix **A** that corresponds to the source IP, if we plot, say, columns 1 and 2 against each other, we will see a scatterplot that contains one point for each source IP; given this scatterplot, we are able to detect the outliers. We henceforth refer to the 'score' for each source IP (or any other entity associated to a particular mode of a tensor), as expressed by the values of the columns of the corresponding factor matrix, as TensorScore. We show our most outstanding results in Figure 1.

According to the scatterplots obtained from tensor analysis, in Figure 1(a)-(c), we may observe the relative attackers' relations according to different directions of the TensorScore. In Figure 1(a), we found two outliers (denoted as A and C) out of three clusters. After further inspection of the participating of the attacks, we found out that the attackers in group A are focused on port port 110

[2] Apache Mahout, `http://mahout.apache.org/`

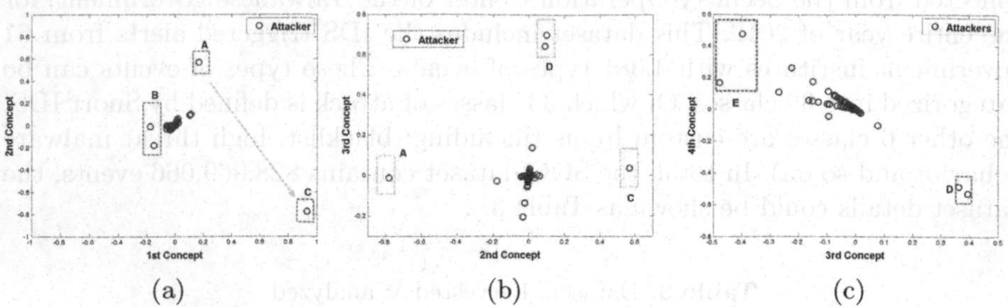

Fig. 1. Scatter plot in Honeynet result from both attackers and victims views. (a) In 1st-2nd concept of source IP view, we observe three different cases, case A is POP3 (port 110) brute force attacks, and case C is port scanning in port 25. Case B contains a lot of instances but cannot be separated in this plot ; (b) In $2^{nd}3^{rd}$ concept sof source IP view, case C and D appear in this plot which is medium scale of scanning behavior, using ports 22, 23, 135 and 445; (c) In 4th-5th concepts of source IP view, case E appears which represents another scanning behavior.

and perform POP3 probing. The attackers in group C attempted to use ports 50 79 aiming to perform a large scale port scan. Both attack groups (A and C) appear only on a single day (January 10th and January 25th respectively). In Figure 1(b) and (c), we are able to discover a new set of anomalies, as we choose a different couple of latent factor vectors to obtain the TensorScore from. In attack group D, we were able to identify an attacker who attempted to trigger 14,652 connections to 236 target Honeynet system IP addresses, with a duration of 8 days. We present the attacks that belong to group E; those attacks focus on a particular Microsoft Network security vulnerability that is associated with ports 139 and 445.

4.3 IDS Event Result

The IDS events consist of an entire year's worth of data, collected by the G-SOC of Taiwan in 2012. The single resolution mode of MalSpot use the day scale granularity to analyze these logs. In Fig. 2(a), we illustrate the two groups that MalSpot was able to spot in the IDS event logs. The first group is associated with the Web and native IDS event rules, whereas the second group is related to the blacklist-based event rules.

We proceed to the second step of our analysis, by setting $k = 5$ and cluster the tensor decomposition latent groups as shown in Fig. 2(b). This post-analysis, forces hosts with similar characteristics to end up in the same cluster. For instance, cluster 1 contains a vast number of hosts that are related to a large scale government institute. In clusters 4 and 5, we mostly observe service-oriented information systems tend.

In order to evaluate the multi-resolution nature of MalSpot, we select 4 different time resolutions, i.e., hour, day, week and month, and seek to identify the uncorrelated levels. The result is shown as Fig. 4. In fact, hosts A, B and C

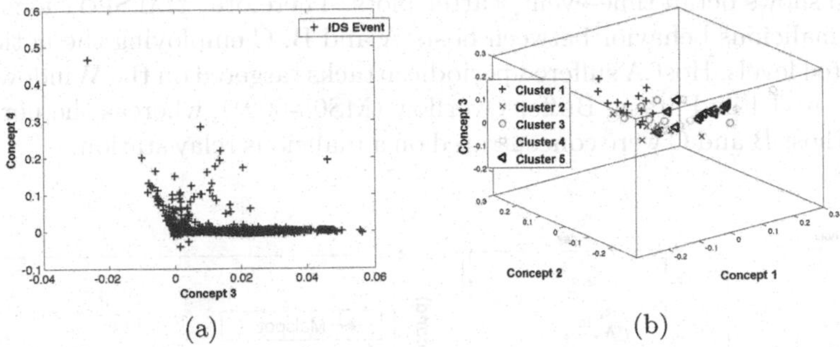

Fig. 2. IDS alert events scatterplot: (a) In 3rd-4th concept of event view (IDS alerts), we observe three different cases, part of alerts are triggered more often and part of them triggered rarely; (b) from the organization's view, we can see 5 groups are clustered together.

are grouped as the same cluster in both hourly and daily levels. MALSPOT is able to select the uncorrelated levels for B and C in weekly and monthly granularities, respectively.

In Fig. 3, we use event classes with 4 different temporal resolutions to identify uncorrelated levels of activity, for various institutes. We identify that institute A has an uncorrelated level of activity between weekly and daily granularity, as opposed to institutes B and C. From further investigation, institute A has suffered from a "system-call detect" event class during the uncorrelated time period (e.g., B and C have a uniform distribution of activity during the entire year, but A is skewed towards early 2012). Additionally, MALSPOT offers huge savings in computational time in order to detect the aforementioned attacks, when compared to competing methods.

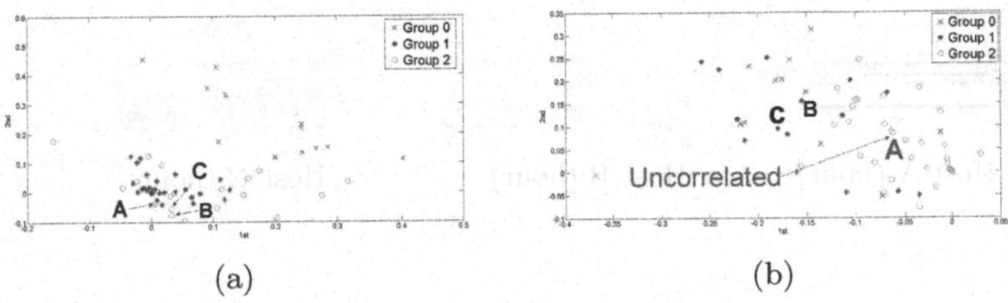

Fig. 3. Scatter plot of different time resolutions of G-SOC(cat) dataset. Each point in the scatter-plot denotes an institution. (a) day resolution and (b) week resolution.

Based on the Fig. 4 (a), we plot the scatter-plots from the hour and week (X-axis) versus the triggered event types (Y-axis). We observed a critical difference between institute A and institutes B, C with respect to attack periodicity. Institutes B and C suffered so-called "WEB-MISC TOP10.dll access" attacks [3] while A did

[3] This event is generated when an attempt is made to exploit a buffer overflow in the Trend Micro InterScan eManager.

not. Fig. 5 shows detail time-event scatter plots. Therefore, MALSPOT identifies potential malicious behavior between hosts A and B, C employing the notion of uncorrelated levels. Host A suffered periodic attacks targeted on the Windows OS (Windows ANI File Parsing Buffer Overflow (MS05-002)), whereas the periodic attacks of host B and C were concentrated on a malicious relay station.

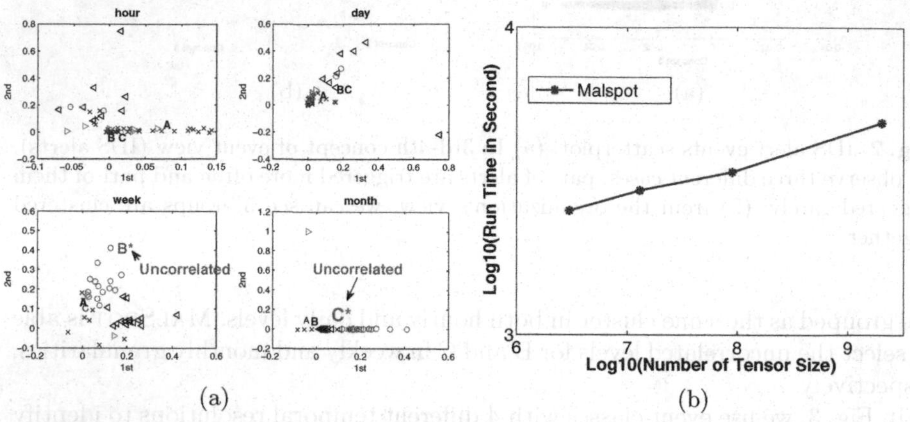

(a) (b)

Fig. 4. (a) Scatter plot of different time resolutions. Each point in scatterplot denotes an institution. (b) The scalability of MALSPOT, as the input size grows.

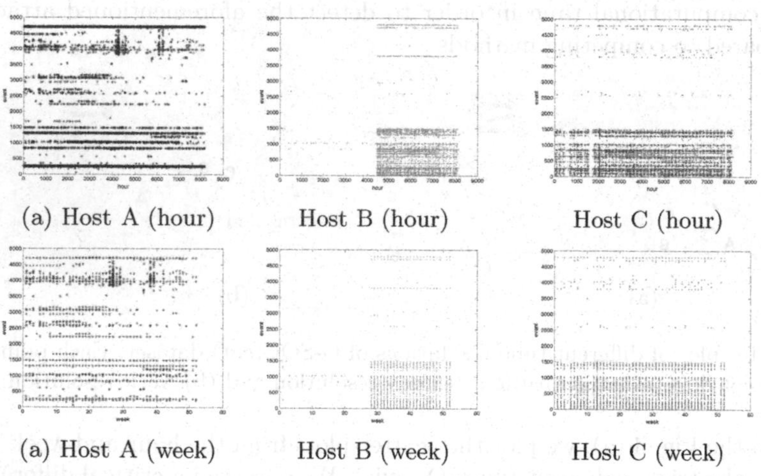

(a) Host A (hour) Host B (hour) Host C (hour)

(a) Host A (week) Host B (week) Host C (week)

Fig. 5. Scatter plot of different time resolutions. (X-axis is time scale and Y-axis is event types). We can see the difference in the distribution of scatterplots.

5 Conclusion

In this work, we develop a big data analytics system in order to discover malicious patterns in a variety of network/malware propagation settings, so that we can further use them for attack detection and prevention, when attacks are concurrent with legitimate traffic. By conducting large-scale information security data analysis, our proposed method, MALSPOT, can easily identify the patterns in massive IDS logs, spamming delivery logs, and Honeynet firewall logs, pertaining to long-term and stealthy attack behavior.

The contributions of this work are the following:

- **Design** of the MALSPOT system: based on `hadoop`, it can scale up to arbitrary-size datasets, and it is nearly linear as the log trace size grows.
- **Discoveries:** We report very interesting attack patterns, and positively identified attacks, as detected by MALSPOT.
- **Scalability:** regardless of data scale or data source variety, MALSPOT is able to detect attacks efficiently and effectively.

Acknowledgement. This material is based upon work supported by the National Science Foundation under Grants No. IIS-1247489 and CNS-1314632 Research was sponsored by the Defense Threat Reduction Agency and was accomplished under contract No. HDTRA1-10-1-0120. Also, sponsored by the Army Research Laboratory and was accomplished under Cooperative Agreement Number W911NF-09-2-0053. This work is also partially supported by a Google Focused Research Award. Any opinions, findings, and conclusions or recommendations expressed in this material are those of the author(s) and do not necessarily reflect the views of the funding parties.

References

1. Dainotti, A.: Analysis of a "/0" stealth scan from a botnet. In: IMC 2012 (2012)
2. Sun, J., Papadimitriou, S., Yu, P.S.: Window-based tensor analysis on high-dimensional and multi-aspect streams. In: ICDM, pp. 1076–1080 (2006)
3. Maruhashi, K., Guo, F., Faloutsos, C.: Multiaspectforensics: Pattern mining on large-scale heterogeneous networks with tensor analysis. In: ASONAM 2011 (2011)
4. Kishino, Y., Sakurai, Y., Yanagisawa, Y., Suyama, T., Naya, F.: Svd-based hierarchical data gathering for environmental monitoring. In: Proceedings of the 2013 ACM Conference on Pervasive and Ubiquitous Computing Adjunct Publication, UbiComp 2013 Adjunct, pp. 9–12. ACM, New York (2013)
5. Kolda, T.G., Bader, B.W.: Tensor decompositions and applications. SIAM Review 51(3), 455–500 (2009)
6. Kolda, T., Sun, J.: Scalable tensor decompositions for multi-aspect data mining. In: ICDM (2008)
7. Koutra, D., Papalexakis, E.E., Faloutsos, C.: Tensorsplat: Spotting latent anomalies in time. In: 2012 16th Panhellenic Conference on Informatics (PCI), pp. 144–149. IEEE (2012)

8. Chen, L.-M., Chen, M.-C., Liao, W., Sun, Y.S.: A scalable network forensics mechanism for stealthy self-propagating attacks. Computer Communications (2013)
9. Lin, J., Vlachos, M., Keogh, E., Gunopulos, D.: Iterative incremental clustering of time series. In: Bertino, E., Christodoulakis, S., Plexousakis, D., Christophides, V., Koubarakis, M., Böhm, K. (eds.) EDBT 2004. LNCS, vol. 2992, pp. 106–122. Springer, Heidelberg (2004)
10. Matsubara, Y., Sakurai, Y., Faloutsos, C., Iwata, T., Yoshikawa, M.: Fast mining and forecasting of complex time-stamped events. In: Proceedings of the 18th ACM SIGKDD International Conference on Knowledge Discovery and Data Mining, KDD 2012, pp. 271–279. ACM, New York (2012)
11. Papadimitriou, S., Yu, P.: Optimal multi-scale patterns in time series streams. In: Proceedings of the 2006 ACM SIGMOD International Conference on Management of Data, pp. 647–658. ACM, New York (2006)
12. Papalexakis, E.E., Faloutsos, C., Sidiropoulos, N.D.: ParCube: Sparse parallelizable tensor decompositions. In: Flach, P.A., De Bie, T., Cristianini, N. (eds.) ECML PKDD 2012, Part I. LNCS, vol. 7523, pp. 521–536. Springer, Heidelberg (2012)
13. Kang, U., Papalexakis, E., Harpale, A., Faloutsos, C.: Gigatensor: scaling tensor analysis up by 100 times - algorithms and discoveries. In: Proceedings of the 18th ACM SIGKDD International Conference on Knowledge Discovery and Data Mining, KDD 2012, pp. 316–324. ACM, New York (2012)

Extracting Diverse Patterns with Unbalanced Concept Hierarchy

M. Kumara Swamy, P. Krishna Reddy, and Somya Srivastava

Centre of Data Engineering
International Institute of Information Technology-Hyderabad (IIIT-H)
Gachibowli, Hyderabad, India - 500032
kumaraswamy@research.iiit.ac.in, pkreddy@iiit.ac.in, somya@amazon.com

Abstract. The process of frequent pattern extraction finds interesting information about the association among the items in a transactional database. The notion of *support* is employed to extract the frequent patterns. Normally, in a given domain, a set of items can be grouped into a category and a pattern may contain the items which belong to multiple categories. In several applications, it may be useful to distinguish between the pattern having items belonging to multiple categories and the pattern having items belonging to one or a few categories. The notion of diversity captures the extent the items in the pattern belong to multiple categories. The items and the categories form a concept hierarchy. In the literature, an approach has been proposed to rank the patterns by considering the balanced concept hierarchy. In a real life scenario, the concept hierarchies are normally unbalanced. In this paper, we propose a general approach to calculate the rank based on the diversity, called *drank*, by considering the unbalanced concept hierarchy. The experiment results show that the patterns ordered based on *drank* are different from the patterns ordered based on *support*, and the proposed approach could assign the *drank* to different kinds of unbalanced patterns.

Keywords: data mining, association rules, frequent patterns, diversity, diverse rank, interestingness, concept hierarchy, algorithms.

1 Introduction

In the field of data mining, the process of frequent pattern mining has been widely studied [1]. The related concepts of frequent pattern mining are as follows [2]. Let $I = \{i_1, i_2, \cdots, i_n\}$ be the set of n items and D be the database of m transactions. Each transaction is identified with unique identifier and contains n items. Let $X \subseteq I$ be a set of items, referred to as an item set or a *pattern*. A pattern that contains k items is a k-item pattern. A transaction T is said to contain X if and only if $X \subseteq T$. The *frequency* or *support* of a pattern X in D, denoted as $f(X)$, is the number of transactions in D containing X. The support X, denoted as $S(X)$, is the ratio of its frequency to the $|D|$ i.e., $S(X) = \frac{f(X)}{|D|}$. The pattern X is frequent if its support is not less than the user-defined minimum support threshold, i.e., $S(X) \geq minSup$.

V.S. Tseng et al. (Eds.): PAKDD 2014, Part I, LNAI 8443, pp. 15–27, 2014.

The techniques to enumerate frequent patterns generates large number of patterns which could be uninteresting to the user. Research efforts are on to discover interesting frequent patterns based on constraints and/or user-interest by using various interestingness measures such as closed [3], maximal [4], top-k [5], pattern-length [6] and cost (utility) [7].

Normally, in a given domain, a set of items can be grouped into a category and a pattern may contain the items which belong to multiple categories. In several applications, it may be useful to distinguish between the pattern having items belonging to multiple categories and the pattern having items belonging to the one or a few categories. The existing frequent pattern extraction approaches do not distinguish the patterns based on the diversity. The notion of diversity captures the extent of items in the pattern belong to multiple categories. The items and the categories form concept hierarchy. In [8], an effort has been made to rank the patterns based on diversity by considering balanced concept hierarchy. However, in real life scenarios, the concept hierarchies are unbalanced. In this paper, we have proposed an approach to assign the diverse rank, called *drank*, to patterns by considering unbalanced concept hierarchy. The proposed approach is a general approach which can be applied to calculate *drank* value by considering both balanced and unbalanced concept hierarchies. Experiments on the real-world data set show that patterns ordered based on *drank* are different from the patterns ordered based on *support*, and the proposed approach could assign the *drank* to different kinds of unbalanced patterns.

In the literature, the concept hierarchies have been used to discover the generalized association rules in [9] and multiple-level association rules in [10]. In [11], a keyword suggestion approach based on the concept hierarchy has been proposed to facilitate the user's web search. The notion of *diversity* has been widely exploited in the literature to assess the interestingness of summaries [12],[13],[14]. In [15], an effort has been made to extend the *diversity-based* measures to assess the interestingness of the data sets using the diverse association rules. The diversity is defined as the variation in the items' frequencies. Such a method cannot be directly applied to rank the patterns based on the diversity. Moreover, the work in [15] has focused on comparing the data sets using diverse association rules. In this paper, we developed a framework to compute the diversity of patterns by analyzing the categories of items.

The rest of the paper is organized as follows. In the next section, we explain about concept hierarchy and diversity of pattern. In section 3, we explain the approach to computing the *drank* of a pattern by considering balanced concept hierarchy. In section 4, we present the proposed approach. In section 5, we present experimental results. The last section contains summary and conclusions.

2 About Concept Hierarchy and Diversity of Patterns

The notion of concept hierarchy plays the main role in assigning the rank to a pattern based on the diversity. In this section, we explain about concept hierarchy and the basic idea employed in the proposed approach to calculate the diversity.

2.1 Concept Hierarchy

A pattern contains data items. A concept hierarchy is a tree in which the data items are organized in an hierarchical manner. In this tree, all the leaf nodes represent the *items*, the internal nodes represent the *categories* and the top node represents the *root*. The *root* could be a virtual node.

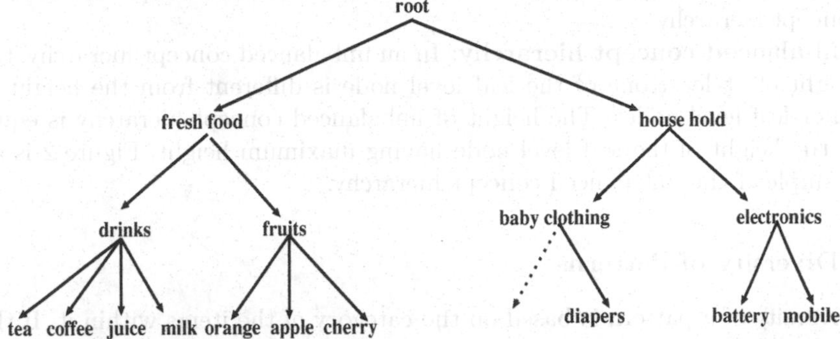

Fig. 1. An example of balanced concept hierarchy

Let C be a concept hierarchy. A node in C may be an item, category or root. The height of *root* node is 0. Let n be a node in C. The height of n, is denoted as $h(n)$, is equal to the number of edges on the path from *root* to n.

Figure 1 represents a concept hierarchy. In this, the items *orange*, *apple* and *cherry* are mapped to the category *fruits*. Similarly, the categories *drinks* and *fruits* are mapped to the category *fresh food*. Finally, the categories *fresh food* and *house hold* are mapped to *root*.

The concept hierarchy having height h has the same number of levels. The items at the given height are said to be at the same level. In C, all the lower-level nodes, except the *root*, are mapped to the immediate higher level nodes. In this

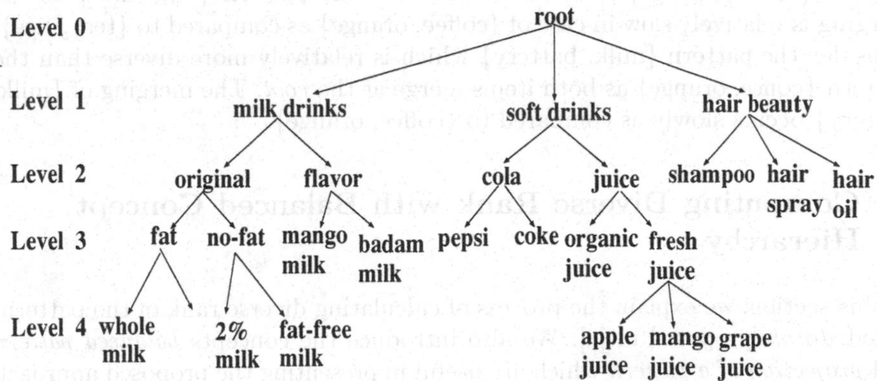

Fig. 2. An example of unbalanced concept hierarchy

paper, we consider the concept hierarchies in which a lower level node is mapped to only one higher level node.

The concept hierarchies can be balanced or unbalanced.

- **Balanced concept hierarchy:** In balanced concept hierarchy, the height of all leaf level nodes is the same. The height of balanced concept hierarchy is equal to the height of a leaf level item. Figure 1 is an example of balanced concept hierarchy.
- **Unbalanced concept hierarchy:** In an unbalanced concept hierarchy, the height of at least one of the leaf level node is different from the height of other leaf level nodes. The height of unbalanced concept hierarchy is equal to the height of the leaf level node having maximum height. Figure 2 is an example of the unbalanced concept hierarchy.

2.2 Diversity of Patterns

The diversity of a pattern is based on the category of the items within it. If the items of a pattern are mapped to the same/few categories in a concept hierarchy, we consider that the pattern has low diversity. Relatively, if the items are mapped to multiple categories, we consider that the pattern has more diversity. We have developed an approach to assign the diversity for a given pattern based on the merging behavior in the corresponding concept hierarchy. If the pattern merges into few higher level categories quickly, it has low diversity. Otherwise, if the pattern merges into one or a few high level categories slowly, it has relatively high diversity value.

As an example, consider the concept hierarchy in Figure 1. For the pattern {tea, juice}, the items *tea* and *juice* are mapped to the next level category *drinks.* In this case, the merging occurs quickly. For the pattern {coffee, orange}, the items *coffee* is mapped to category *drinks* and item *orange* maps to the category *fruits.* Further, both the categories *drinks* and *fruits* are mapped to the category *fresh food*, and the category *fresh food* in turn maps to *root*. We say that the pattern {coffee, orange} is more diverse than the pattern {tea, juice} as the merging is relatively slow in case of {coffee, orange} as compared to {tea, juice}. Consider the pattern {milk, battery} which is relatively more diverse than the pattern {coffee, orange} as both items merge at the *root*. The merging of {milk, battery} occurs slowly as compared to {coffee, orange}.

3 Computing Diverse Rank with Balanced Concept Hierarchy

In this section, we explain the process of calculating diverse rank of the pattern, called *drank*, proposed in [8]. We also introduce the concepts *balanced pattern* and *projection of a pattern* which are useful in presenting the proposed approach in the next section.

Definition 1. ***Balanced Pattern (BP):*** *Consider a pattern $Y = \{i_1, i_2, \cdots, i_n\}$ with 'n' items and a concept hierarchy of height 'h'. The pattern Y is called balanced pattern, if the height of all the items in Y is equal to 'h'.*

Definition 2. ***Projection of Balanced Concept Hierarchy for*** Y $(P(Y/C))$: *Let Y be BP and C be balanced concept hierarchy. The P(Y/C) is the projection of C for Y which contains the portion of C. All the nodes and edges exists in the paths of the items of Y to the root, along with the items and the root, are included in P(Y/C). The projection P(Y/C) is a tree which represents a concept hierarchy concerning to the pattern Y.*

Given two patterns of the same length, different merging behavior can be realized, if we observe how the items in the pattern are mapped to higher level nodes. That is, one pattern may quickly merge to few higher level items within few levels and the other pattern may merge to few higher level items by crossing more number of levels. By capturing the process of merging, we define the notion of diverse rank (*drank*). So, *drank*(Y) is calculated by capturing how the items are merged from leaf-level to *root* in P(Y/C). It can be observed that a given pattern maps from the leaf level to the *root* level through a merging process by crossing intermediate levels. At a given level, several lower level items/categories are merged into the corresponding higher level categories.

Two notions are employed to compute the diversity of a BP: *Merging Factor (MF)* and *Level Factor (LF)*.

We explain about MF after presenting the notion of generalized pattern.

Definition 3. ***Generalized Pattern (GP(Y, l, P(Y/C))):*** *Let Y be a pattern, 'h' be the height of P(Y/C) and 'l' be an integer. The $GP(Y, l, P(Y/C))$ indicates the GP of Y at level 'l' in P(Y/C). Assume that the $GP(Y, l+1, P(Y/C))$ is given. The $GP(Y, l, P(Y/C))$ is calculated based on the GP of Y at level $(l+1)$. The $GP(Y, l, P(Y/C))$ is obtained by replacing every item at level $(l+1)$ in $GP(Y, l+1, P(Y/C))$ with its corresponding parent at the level 'l' with duplicates removed, if any.*

The notion of merging factor at level l is defined as follows.

Merging factor (MF(Y, l, P(Y/C))): Let Y be BP and l be the height. The merging factor indicates how the items of a pattern merge from the level $l+1$ to the level l $(0 \leq l < h)$. If there is no change, the MF(Y,l) is 1. If all items merges to one node, the MF(Y,l) value equals to 0. So, the MF value at the level l is denoted by MF(Y,l, P(Y/C)) which is equal to the ratio of the number of nodes in (GP(Y, l, P(Y/C)-1) to the number of nodes in (GP(Y, l+1, P(Y/C)-1).

$$MF(Y, l, P(Y/C)) = \frac{|GP(Y,\ l,\ P(Y/C))| - 1}{|GP(Y,\ l+1,\ P(Y/C))| - 1} \tag{1}$$

We now define the notion of level factor to determine the contribution of nodes at the given level.

Level Factor (LF(l,P(Y/C)): For a given P(Y/C), h be the height of $P(Y/C)$ $\neq \{0, 1\}$. Let l be such that $1 \leq l \leq (h-1)$. The LF value of P(Y/C) height l

indicates the contribution of nodes at l to *drank*. We can assign equal, linear or exponential weights to each level. Here, we provide a formula which assigns the weight to the level such that the weight is in proportion the level number.

$$LF(l, P(Y/C)) = \frac{2 * (h - l)}{h * (h - 1)} \tag{2}$$

Diverse rank of a pattern Y: The *drank* of BF Y for a given C, is calculated by summing up the product of MF and LF from the leaf level to the *root* of $P(Y/C)$. The formula is as follows.

$$drank(Y,C) = \sum_{l=h-1}^{l=0} MF(Y, l, P(Y/C)) * LF(l, P(Y/C)) \tag{3}$$

where, Y is BP, h is height of P(Y/C).

4 Computing Diverse Rank with Unbalanced Concept Hierarchy

In this section, we explain the approach to assign the *drank* to unbalanced pattern. The term unbalanced pattern is defined as follows.

Definition 4. ***Unbalanced Pattern (UP)**: Consider a pattern Y and an unbalanced concept hierarchy U of height 'h'. A pattern is called unbalanced pattern, if the height of at least one of the item in Y is less than 'h'.*

The notion of unbalanced-ness depends on how the heights of the nodes in the concept hierarchy are distributed. It can be noted that we consider a pattern as unbalanced pattern, if the height of at least one item is less than the height of unbalanced concept hierarchy. Suppose, all the items of a pattern are at the height, say k. The pattern X is unbalanced, if k is less than the height of concept hierarchy.

The basic idea to compute *drank* of UP is as follows. We first convert the unbalanced concept hierarchy to balanced concept hierarchy called, "extended unbalanced concept hierarchy" by adding dummy nodes and edges. We calculate the *drank* of UP with Equation 3 by considering the "extended unbalanced concept hierarchy". Next, we reduce the *drank* in accordance with the number of dummy nodes and edges. So, the *drank* of UP is relative to the *drank* of the same pattern computed by considering all of its items are at the leaf level of the extended unbalanced concept hierarchy.

Given UP and the corresponding unbalanced concept hierarchy U, the following steps should be followed to calculate the *drank* of UP.

(i) Convert the U to the corresponding extended U.
(ii) Compute the effect of the dummy nodes and edges.
(iii) Compute the *drank*.

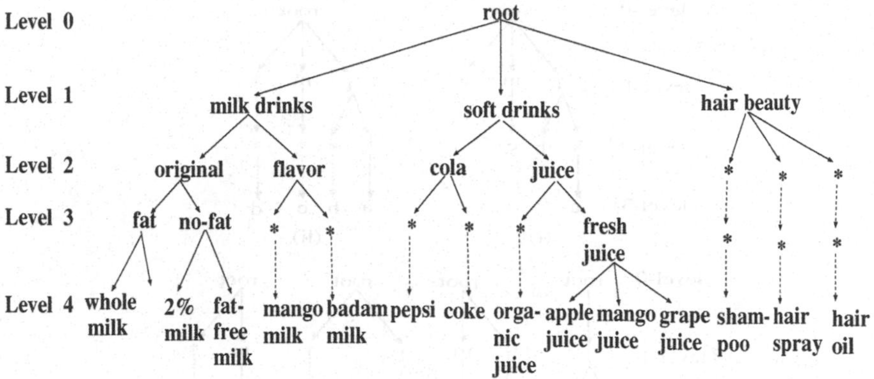

Fig. 3. Extended Unbalanced Concept Hierarchy for the Figure 2

(i) Convert the Unbalanced Concept Hierarchy to Extended Unbalanced Concept Hierarchy
We define the extended unbalanced concept hierarchy as follows.

Definition 5. ***Extended Unbalanced Concept Hierarchy (E)***: *For a given unbalanced concept hierarchy U with height 'h', we convert U into extended U, say E, by adding dummy nodes and edges such that the height of each leaf level item is equal to 'h'.*

Figure 3 shows the extended unbalanced concept hierarchy of Figure 2. In Figure 3, '$*$' indicates the dummy node and dotted line indicates the dummy edge.

We define the projection of extended unbalanced concept hierarchy for Y as follows.

Definition 6. ***Projection of Extended Unbalanced Concept Hierarchy of*** Y ***(P(Y/E))***: *Let Y be UP, U be unbalanced concept hierarchy, and E be the corresponding extended unbalanced concept hierarchy of U. The projection of E for the unbalanced pattern Y is $P(Y/E)$. The $P(Y/E)$ contains the portion of U which includes all the paths of the items of Y from the root.*

It can be noted that, in addition to real nodes/edges, $P(Y/E)$ may contain dummy nodes/edges.

As an example, consider the unbalanced concept hierarchy shown in Figure 4(i). In this figure, the items a, b, c, and d are located at different levels. We find the longest path $\langle root, l, k, a\rangle$ in the unbalanced concept hierarchy. The additional dummy nodes and edges are added such that all the items are at the height h. This extended unbalanced concept hierarchy is shown in Figure 4(ii). The projections of the patterns {a, b}, {b, c}, {b, d}, and {c, d} are shown in Figures 4(iii), 4(iv), 4(v), and 4(vi) respectively.

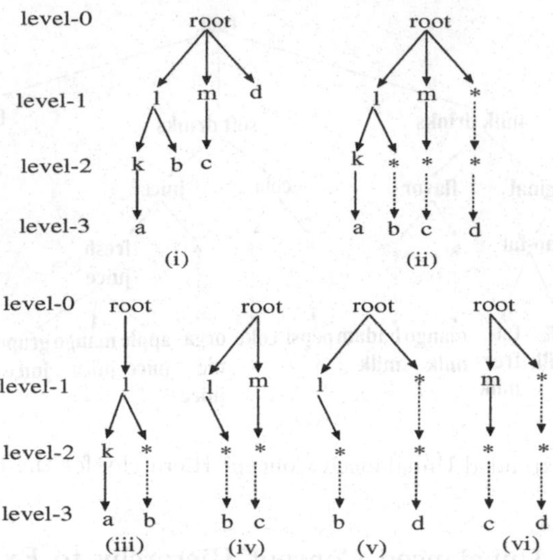

Fig. 4. (i) Unbalanced Concept Hierarchy, (ii) Extended Unbalanced Concept Hierarchy of (i). The projection of Extended Unbalanced Concept Hierarchies for the patterns {a, b}, {b, c}, {b, d}, and {c, d} are shown in (iii), (iv), (v), and (vi) respectively.

(ii) Compute the Effect of the Dummy Nodes and Edges

We define the notion of adjustment factor to compute the effect of the dummy nodes and edges.

Adjustment factor (AF): We define the *AF* at the given level. The *Adjustment Factor (AF)* at level l helps in reducing the *drank* by measuring the contribution of dummy edges/nodes relative to the original edges/nodes at the level l. The *AF* for a pattern Y at a level l should depend on the ratio of number of real edges formed with the children of the real nodes in $P(Y/E)$ versus total number of edges formed with the children of real and dummy nodes at l in $P(Y/E)$. The value of *AF* at a given height should lie between 0 and 1. If the number of real edges is equals to zero, *AF* is zero. If the pattern at the given level does not contain dummy nodes/edges, the value of *AF* becomes 1. Note that the *AF* value is not defined at the leaf level nodes as children do not exist. The *AF* for Y at height l is denoted as $AF(Y, l, P(Y/E))$ and is calculated by the following formula.

$$AF(Y, l, P(Y/U)) = \frac{\#\ of\ Real\ Edges\ of\ UP(Y, l, P(Y/E))}{\#\ of\ Total\ Edges\ of\ UP(Y, l, P(Y/E))} \tag{4}$$

where *numerator* is the number of edges formed with the children of the real nodes and *denominator* is the number of edges formed with the children of both real and dummy nodes at the level l in $P(Y/E)$.

Consider the frequent pattern $Y = \{whole\ milk,\ pepsi,\ coke,\ shampoo\}$ in Figure 3. The level of the item *whole milk* is 4. As l is between $(0, h)$, we calculate the number of edges at level 3. At the level 3, the number of real edges in $P(Y/E)$ is 1 and the total number edges including real and dummy edges in $P(Y/E)$ is 4, i.e., $AF(Y, 4, P(Y/U)) = \frac{1}{4} = 0.25$. Similarly, the AF value at level 2, level 1, level and 0.75, 1, and 1 respectively.

(iii) Computing the *drank* of UP

The *drank* of UP is a function of MF, AF and LF.

Definition 7. ***Diverse rank of a frequent pattern Y (drank(Y)):*** *Let Y be the pattern and U be the unbalanced concept hierarchy of height 'h'. The drank of Y, denoted by drank(Y), is given by the following equation.*

$$drank(Y, U) = \sum_{l=h-1}^{l=0} [MF(Y, l, P(Y/E)) * AF(Y, l, P(Y/E))] * LF(l, P(Y/E)) \tag{5}$$

where, h is the height of the $P(P/E)$, E is the extended unbalanced concept hierarchy, $MF(Y, l, P(Y/E))$ is the MF of Y at level l, $LF(l, P(Y/E))$ is the LF at level l and $AF(Y, l, P(Y/E))$ is the AF of Y at level l.

It can be noted that Equation 5 can be used for computing *drank* of the patterns with both balanced and unbalanced concept hierarchy as the values of AF becomes 1 at all levels in case of balanced concept hierarchy.

5 Experiment Results

For conducting experiments, we have considered the groceries data set which contains 30 days of point-of-sale transaction data from a typical local grocery outlet. The data set contains 9,835 transactions and 168 items. The average transaction size in the data set is 4.4. The maximum and minimum transaction size is 32 and 1 respectively. To generate a concept hierarchy for the items, a web based Grocery API provided by Tesco [20] (a United Kingdom Grocery Chain Store) is used. Some of the items of the transactional data that were not listed in the concept hierarchy of Tesco are added manually by consulting the domain experts. There are 220 items (excluding internal nodes) available in the concept hierarchy of Tesco. We have removed the items which do not exist in the transactional data. The distribution of remaining 168 items at different levels in the concept hierarchy are shown in Table 1.

Top Diverse Patterns: In Table 3, we present the list of the top 3-item patterns ordered by *drank*. In this table, the first column shows the pattern, the second column shows the *support* of the pattern, the third column shows the *drank* of UP, the fourth column shows the *drank* of UP with extended concept hierarchy (E), and the final column shows the difference from *drank* of UP and the *drank* of UP with E.

Top Frequent Patterns: Table 4 contains the list of top 3-item patterns ordered by *support* value.

Table 1. Distribution of items in unbalanced concept hierarchy

Level No.	No. of items
0	0
1	1
2	29
3	104
4	34

Table 2. Distribution of items in the simulated concept hierarchy

Level No.	No. of items
4	5
5	14
6	12
7	36
8	17
9	25
10	12
11	40
12	7

From these two tables, we can observe that there are no common patterns between them. The results show that the pattern having the highest *drank* value may not be the patterns with the highest *support*. Similarly, the patterns with the highest *support* may not have the highest value of *drank*. So, the patterns ordered by *drank* indicates a different kind of knowledge which could not be extracted by *support*.

Table 3. Patterns ordered by *drank*

Top 10 3-item diverse patterns	Support (%)	drank of UP	drank of UP with E	Diff.
{rolls-buns, soda, sausages}	1.0	1.00	1.0	0.00
{soda, rolls-buns, other vegetables}	1.0	1.00	1.0	0.00
{rolls-buns, soda, shopping bags}	0.6	1.00	1.0	0.00
{soda, whole milk, shopping bags}	0.7	0.89	1.0	0.11
{rolls-buns, whole milk, newspapers}	0.8	0.89	1.0	0.11
{rolls-buns, bottled water, other vegetables}	0.7	0.89	1.0	0.11
{rolls-buns, bottled water, yogurt}	0.7	0.89	1.0	0.11
{rolls-buns, soda, whole milk}	0.6	0.89	1.0	0.11
{rolls-buns, soda, yogurt}	0.9	0.89	1.0	0.11
{rolls-buns, bottled water, whole milk}	0.9	0.89	1.0	0.11

The *drank* value of UP is obtained after reducing the effect of dummy nodes/ edges from E. It can be noted that the *drank* of UP with E indicates the value of diversity without considering AF. So, the value in the final column of Table 3 and Table 4 indicates the degree of unbalanced-ness. If the value is low, the UP is less imbalanced and if the value is high, the UP is highly imbalanced. However, in Table 3 and Table 4, the values in the last column are very low. This is due to the fact that the concept hierarchy is relatively more balanced.

Influence of Adjustment Factor: In this experiment, we change the concept hierarchy such that it becomes very unbalanced. For this, we increase the level of some items. A random number from the list {1, 2, 4, 4, 6, 8, 8} is chosen to

Table 4. Patterns ordered by *support*

Top 10 frequent patterns	**Support (%)**	**drank of UP**	**drank of UP with E**	**Diff.**
{whole milk, other vegetables, root vegetables}	2.3	0.29	0.33	0.04
{yogurt, whole milk, other vegetables}	2.2	0.31	0.33	0.02
{rolls-buns, whole milk, other vegetables}	1.8	0.67	0.75	0.08
{whole milk, tropical fruit, other vegetables}	1.7	0.44	0.50	0.06
{rolls-buns, yogurt, whole milk}	1.6	0.78	0.83	0.05
{yogurt, whole milk, root vegetables}	1.5	0.31	0.33	0.02
{yogurt, whole milk, tropical fruit}	1.5	0.31	0.33	0.02
{whipped sour cream, whole milk, other vegetables}	1.5	0.31	0.33	0.02
{whole milk, pip fruit, other vegetables}	1.4	0.44	0.50	0.06
{soda, whole milk, other vegetables}	1.4	0.67	0.75	0.08

add the number of edges to increase the height of the items. The height of the simulated concept hierarchy is 12 and the distribution of items are shown in the Table 2.

Table 5 provides the details of *drank* of UP by considering the simulated concept hierarchy. In this table, the first column shows the pattern, the second column shows the *drank* of UP, the third column shows the *drank* of UP with E, and the last column shows difference between the *drank* of UP and the *drank* of UP with E. The values in the last column show that the notion of AF, along with MF and LF, helps in computing the *drank* for all kinds of patterns including less unbalanced patterns to high unbalanced patterns.

Table 5. Patterns by considering the simulated concept hierarchy

Patterns	**drank of UP**	**drank of UP with E**	**Diff.**
{whole milk, beef, root vegetables}	0.705	0.866	0.161
{whole milk, root vegetables, frozen vegetables}	0.750	0.933	0.183
{rolls-buns, whole milk, pork}	0.634	0.933	0.299
{pastry yogurt, other vegetables}	0.416	0.804	0.388
{yogurt, pip fruit, other vegetables}	0.454	0.866	0.412
{rolls-buns, whole milk, pip fruit}	0.434	0.933	0.499
{whole milk, yogurt, other vegetables}	0.036	0.805	0.769
{rolls-buns, whole milk, yogurt}	0.045	0.938	0.893
{rolls-buns, whole milk, other vegetables}	0.032	0.933	0.901

6 Summary and Conclusions

Finding interesting patterns is one of the issues in frequent pattern mining. Several interestingness measures have been proposed to extract the subset of frequent patterns according to the application's requirements. In this paper, we have proposed a new interestingness measure to rank the patterns based on diversity. We have proposed a general approach to assign the *drank* to the patterns by considering unbalanced concept hierarchy. For computing the *drank* of a pattern, the unbalanced concept hierarchy is being converted into balanced concept hierarchy by adding dummy nodes and edges. The notion of *adjustment factor* is proposed to remove the effect of the dummy nodes and edges. The *drank* is calculated using the notions of *merging factor*, *level factor* and *adjustment factor*. The experiments on the real world data set show that the patterns with high *drank* are different from the patterns with high *support*. Also, the proposed approach could assign the *drank* to patterns having different degrees of unbalanced-ness.

As a part of future work, we are planning to refine the approach by considering all types of unbalanced hierarchies. We are also planning to investigate how the notion of diversity influences the performance of frequent pattern based clustering, classification and recommendation algorithms.

References

1. Han, J., Cheng, H., Xin, D., Yan, X.: Frequent pattern mining: current status and future directions. Data Min. Knowl. Discov. 15(1), 55–86 (2007)
2. Agrawal, R., Srikant, R.: Fast algorithms for mining association rules. In: 20th Intl. Conf. on VLDB, pp. 487–499 (1994)
3. Zaki, M.J., Hsiao, C.-J.: Efficient algorithms for mining closed itemsets and their lattice structure. IEEE TKDE 17(4), 462–478 (2005)
4. Hu, T., Sung, S.Y., Xiong, H., Fu, Q.: Discovery of maximum length frequent itemsets. Information Sciences 178(1), 69–87 (2008)
5. Minh, Q.T., Oyanagi, S., Yamazaki, K.: Mining the K-most interesting frequent patterns sequentially. In: Corchado, E., Yin, H., Botti, V., Fyfe, C. (eds.) IDEAL 2006. LNCS, vol. 4224, pp. 620–628. Springer, Heidelberg (2006)
6. Wang, J., Han, J., Lu, Y., Tzvetkov, P.: TFP: an efficient algorithm for mining top-k frequent closed itemsets. IEEE TKDE 17(5), 652–663 (2005)
7. Hu, J., Mojsilovic, A.: High-utility pattern mining: A method for discovery of high-utility item sets. Pattern Recogn. 40(11), 3317–3324 (2007)
8. Somya, S., Uday Kiran, R., Krishna Reddy, P.: Discovering Diverse-Frequent Patterns in Transactional Databases. In: International Conference on Management of Data (COMAD 2011), Bangalore, India, pp. 69–78 (2011)
9. Srikant, R., Agrawal, R.: Mining generalized association rules. In: VLDB, Zurich, Switzerland, pp. 407–419 (1995)
10. Han, J., Fu, Y.: Mining multiple-level association rules in large databases. IEEE TKDE 11(5), 798–805 (1999)
11. Chen, Y., Xue, G.-R., Yu, Y.: Advertising keyword suggestion based on concept hierarchy. In: WSDM 2008, pp. 251–260. ACM, USA (2008)
12. Geng, L., Hamilton, H.J.: Interestingness measures for data mining: a survey. ACM Comput. Surv. 38(3), 1–32 (2006)
13. Hilderman, R.J., Hamilton, H.J.: Knowledge Discovery and Measures of Interest. Kluwer Academic Publishers, Norwell (2001)

14. Zbidi, N., Faiz, S., Limam, M.: On mining summaries by objective measures of interestingness. Machine Learning 62, 175–198 (2006)
15. Huebner, R.A.: Diversity-based interestingness measures for association rule mining. In: ASBBS 2009, Las Vegas (2009)
16. Brin, S., Motwani, R., Silverstein, C.: Beyond market baskets: Generalizing association rules to correlations. SIGMOD Rec. 26(2), 265–276 (1997)
17. Liu, B., Hsu, W., Mun, L.-F., Lee, H.-Y.: Finding interesting patterns using user expectations. IEEE TKDE 11(6), 817–832 (1999)
18. McGarry, K.: A survey of interestingness measures for knowledge discovery. Knowl. Eng. Rev. 20, 39–61 (2005)
19. Omiecinski, E.: Alternative interest measures for mining associations in databases. IEEE TKDE 15(1), 57–69 (2003)
20. Tesco: Grocery api (2013), https://secure.techfortesco.com/tescoapiweb/

Efficiently Depth-First Minimal Pattern Mining

Arnaud Soulet[1] and François Rioult[2]

[1] Université François Rabelais Tours, LI
3 Place Jean Jaurès,
F-41029 Blois, France
arnaud.soulet@univ-tours.fr

[2] Université de Caen, GREYC
Campus Côte de Nacre,
F-14032 Caen Cédex, France
francois.rioult@unicaen.fr

Abstract. Condensed representations have been studied extensively for 15 years. In particular, the maximal patterns of the equivalence classes have received much attention with very general proposals. In contrast, the minimal patterns remained in the shadows in particular because of their difficult extraction. In this paper, we present a generic framework for minimal patterns mining by introducing the concept of minimizable set system. This framework addresses various languages such as itemsets or strings, and at the same time, different metrics such as frequency. For instance, the free and the essential patterns are naturally handled by our approach, just as the minimal strings. Then, for any minimizable set system, we introduce a fast minimality check that is easy to incorporate in a depth-first search algorithm for mining the minimal patterns. We demonstrate that it is polynomial-delay and polynomial-space. Experiments on traditional benchmarks complete our study.

Keywords: Pattern mining, condensed representation, minimal pattern.

1 Introduction

Minimality is an essential concept of pattern mining. Given a function f and a language $\mathcal{L}$, a minimal pattern X is one of the smallest pattern with respect to the set inclusion in $\mathcal{L}$ satisfying the property $f(X)$. Interestingly, the whole set of minimal patterns forms a condensed representation of $\mathcal{L}$ adequate to f: it is possible to retrieve $f(Y)$ for any pattern of Y in $\mathcal{L}$. Typically, the set of free itemsets [1] (also called generators or key itemsets [2]) is a condensed representation of all itemsets (here, f and $\mathcal{L}$ are respectively the frequency and the itemset language). Of course, it is often more efficient to extract minimal patterns rather than all patterns because they are less numerous. In addition, minimal patterns have a lot of useful applications including higher KDD tasks: producing the most relevant association rules [3], building classifiers [4] or generating minimal traversals [5]. Minimality has been studied in the case of different functions (like frequency [6] and condensable functions [7]) and different languages (e.g., itemsets [1] and sequences [8]). Although the minimality has obvious advantages [9],

V.S. Tseng et al. (Eds.): PAKDD 2014, Part I, LNAI 8443, pp. 28–39, 2014.

very few studies are related to the minimality while maximality (i.e., closed patterns) has been widely studied. In particular, to the best of our knowledge, there is no framework as general as those proposed for maximality [10].

We think that a current major drawback of minimal patterns lies in their inefficient extraction. This low efficiency comes mainly from the fact that most existing algorithms use a levelwise approach [1,7,11] (i.e., breadth-first search/generate and test method). As they store all candidates in memory during the generation phase, the extraction may fail due to memory lack. To tackle this memory pitfall, it seems preferable to adopt a depth-first traversal which often consumes less memory and is still very fast. However, check whether the minimality is satisfied or not is very difficult in a depth-first traversal. In the case of frequency with itemsets, the best way for evaluating the minimality for a pattern (saying *abc*) is to compare its frequency with that of all its direct subsets (here, *ab*, *ac* and *bc*). But, when the pattern *abc* is achieved by a depth-first traversal, only frequencies of *a* and *ab* have previously been calculated. As the frequency of *ac* and *bc* are unknown, it is impossible to check whether the frequency of *abc* is strictly less than that of *ac* and *bc*. To cope with this problem, [11,12] have adopted a different traversal with reordered items. For instance, when the itemset *abc* is reached by this new traversal, *c*, *b*, *bc*, *a*, *ac* and *bc* were previously scanned and their frequency are known for checking whether *abc* is minimal. Unfortunately, such a method requires to store all the patterns in memory (here, *c*, *b*, *bc* and so on) using a trie [11] or an hash table [12]. For this reason, existing DFS proposals [11,12] do not solve the low memory consumption issue as expected.

Contributions. The main goal of this paper is to present a generic and efficient framework for minimal pattern mining by providing a depth-first search algorithm. We introduce the notion of *minimizable set system* which is at the core of the definition of this framework. This latter covers a broad spectrum of minimal patterns including all the languages and measures investigated in [7,10]. Fast minimality checking in a depth-first traversal is achieved thanks to the notion of *critical objects* which depends on the minimizable set system. Based on this new technique, we propose the DEFME algorithm. It mines the minimal patterns for any minimizable set system using a depth-first search algorithm. To the best of our knowledge, this is the first algorithm that enumerates minimal patterns in polynomial delay and in linear space with respect to the dataset.

The outline of this paper is as follows. In Section 2, we propose our generic framework for minimal pattern mining based on set systems. We introduce our fast minimality checking method in Section 3 and we indicate how to use it by sketching the DEFME algorithm. Section 4 provides experimental results. In Section 5, we discuss some related work in light of our framework.

2 Minimizable Set System Framework

2.1 Basic Definitions

A *set system* $(\mathcal{F}, E)$ is a collection $\mathcal{F}$ of subsets of a *ground set* E (i.e. $\mathcal{F}$ is a subset of the power set of E). A member of $\mathcal{F}$ is called a *feasible set*. A *strongly accessible* set system $(\mathcal{F}, E)$ is a set system where for every feasible sets X, Y satisfying $X \subset Y$, there is an element $e \in Y \setminus X$ such that $Xe \in \mathcal{F}$[1]. Obviously, itemsets fits this framework with the set system $(2^{\mathcal{I}}, \mathcal{I})$ where $\mathcal{I}$ is the set of items. $(2^{\mathcal{I}}, \mathcal{I})$ is even strongly accessible. But the notion of set system allows considering more sophisticated languages. For instance, it is easy to build a family set $\mathcal{F}_S$ denoting the collection of substrings of $S = abracadabra$ by encoding each substring $s_{k+1}s_{k+2}\ldots s_{k+n}$ by a set $\{(s_{k+1}, 1), (s_{k+2}, 2), \ldots, (s_{k+n}, n)\}$. The set sytem $(\mathcal{F}_S, E_S = \bigcup \mathcal{F}_S)$ is also strongly accessible. The set system formalism has already been used to describe pattern mining problems (see for instance [10]).

Intuitively, a pattern always describes a set of objects. This set of objects is obtained from the pattern by means of a *cover operator* formalized as follows:

Definition 1 (Cover operator). *Given a set of objects $\mathcal{O}$, a cover operator $cov : 2^E \to 2^{\mathcal{O}}$ is a function satisfying $cov(X \cup Y) = cov(X) \cap cov(Y)$ for every $X \in 2^E$ and $Y \in 2^E$.*

This definition indicates that the coverage of the union of two patterns is exactly the intersection of their two covers. For itemsets, a natural cover operator is the extensive function of an itemset X that returns the set of tuple identifiers supported by X: $cov_{\mathcal{I}}(X) = \{o \in \mathcal{O} \mid X \subseteq o\}$. But, in general, the cover is not the final aim: the cardinality of $cov_{\mathcal{I}}(X)$ corresponds to the frequency of X. In the context of strings, the index list of a string X also define a cover operator: $cov_S(X) = \{i \mid \forall (s_j, j) \in X,\ (s_j, j+i) \in S\}$. Continuing our example with the string $S =$ *abracadabra*, it is not difficult to compute the index lists $cov_S(\{(a,1)\}) = \{0,3,5,7,10\}$ and $cov_S(\{(b,2)\}) = \{0,7\}$ and then, to verify $cov_S(\{(a,1),(b,2)\}) = cov_S(\{(a,1)\}) \cap cov_S(\{(b,2)\}) = \{0,7\}$.

For some languages, the same pattern is described by several distinct sets and then it is necessary to have a canonical form. For example, consider the set $\{(a,1),(b,2),(r,3)\}$ corresponding to the string *abr*. Its suffix $\{(b,2),(r,3)\}$ encodes the same string *br* as $\{(b,1),(r,2)\}$. The latter is the canonical form of the string *br*. To retrieve the canonical form of a pattern, we introduce the notion of canonical operator:

Definition 2 (Canonical operator). *Given two set systems $(\mathcal{F}, E)$ and $(\mathcal{G}, E)$, a canonical operator $\phi : \mathcal{F} \cup \mathcal{G} \to \mathcal{F}$ is a function satisfying (i) $X \subset Y \Rightarrow \phi(X) \subset \phi(Y)$ and (ii) $X \in \mathcal{F} \Rightarrow \phi(X) = X$ for all sets $X, Y \in \mathcal{G}$.*

In this definition, the property (i) ensures us that the canonical forms of two comparable sets with respect to the inclusion remain comparable. The property (ii) means that the set system $(\mathcal{F}, E)$ includes all canonical forms. Continuing our example about strings, it is not difficult to see that $\phi_S : \{(s_k, k), (s_{k+1}, k+1), \ldots, (s_{k+n}, n)\} \mapsto \{(s_k, 1), (s_{k+1}, 2), \ldots, (s_{k+n}, n-k+1)\}$ satisfies the two desired properties (i) and (ii). For instance, $\phi_S(\{(b,2),(r,3)\})$ returns the canonical form of the string $\{(b,2),(r,3)\}$ which is $\{(b,1),(r,2)\}$.

[1] We use the notation Xe instead of $X \cup \{e\}$.

2.2 Minimizable Set System

Rather than considering an entire set system, it is wise to select a smaller part that provides the same information (w.r.t. a cover operator). For this, it is necessary that this set system plus the cover operator form a *minimizable* set system:

Definition 3 (Minimizable set system). *A minimizable set system is a tuple* $\langle(\mathcal{F}, E), \mathcal{G}, cov, \phi\rangle$ *where:*

- $(\mathcal{F}, E)$ *is a finite, strongly accessible set system. A feasible set in* $\mathcal{F}$ *is called a pattern.*
- $(\mathcal{G}, E)$ *is a finite, strongly accessible set system satisfying for every feasible set* $X, Y \in \mathcal{F}$ *such that* $X \subseteq Y$ *and element* $e \in E$, $X\setminus\{e\} \in \mathcal{G} \Rightarrow Y\setminus\{e\} \in \mathcal{G}$. *A feasible set in* $\mathcal{G}$ *is called a generalization.*
- $cov : 2^E \to 2^{\mathcal{O}}$ *is a cover operator.*
- $\phi : \mathcal{F} \cup \mathcal{G} \to \mathcal{F}$ *is a canonical operator such that for every feasible set* $X \in \mathcal{G}$, *it implies* $cov(\phi(X)) = cov(X)$.

Let us now illustrate the role of $\mathcal{G}$ compared to $\mathcal{F}$ in the case of strings. In fact, $\mathcal{G}_S$ gathers all the suffixes of any pattern of $\mathcal{F}_S$. Typically, $\{(b, 2), (r, 3)\} \in \mathcal{G}_S$ is a generalization of $\{(a, 1), (b, 2), (r, 3)\} \in \mathcal{F}_S$. As said above, $\{(b, 2), (r, 3)\}$ has an equivalent form in $\mathcal{F}_S$: $\phi_S(\{(b, 2), (r, 3)\}) = \{(b, 1), (r, 2)\}$. By convention, we extend the definition of cov_S to $\mathcal{G}_S$ by considering that $cov_S(\phi_S(X)) = cov_S(X)$. In addition, it is not difficult to see that $\mathcal{G}_S$ satisfies the desired property with respect to $\mathcal{F}_S$: for every feasible set $X, Y \in \mathcal{F}_S$ such that $X \subseteq Y$ and element $e \in E_S$, $X \setminus \{e\} \in \mathcal{G}_S \Rightarrow Y \setminus \{e\} \in \mathcal{G}_S$. Indeed, if $X \setminus \{e\}$ is a suffix of X, it means that e is the first letter. If we consider a specialization of X and we again remove the first letter, we also obtain a suffix belonging to $\mathcal{G}_S$. Therefore, $\langle(\mathcal{F}_S, E_S), \mathcal{G}_S, cov_S, \phi_S\rangle$ is a minimizable set system.

Obviously, a minimizable set system can be reduced to a system of smaller cardinality of which the patterns are called the *minimal patterns*:

Definition 4 (Minimal pattern). *A pattern* X *is minimal for* $\langle(\mathcal{F}, E), \mathcal{G}, cov, \phi\rangle$ *iff* $X \in \mathcal{F}$ *and for every generalization* $Y \in \mathcal{G}$ *such that* $Y \subset X$, $cov(Y) \neq cov(X)$. $\mathcal{M}(\mathcal{S})$ *denotes the set of all minimal patterns.*

Definition 4 means that a pattern is minimal whenever its cover differs from that of any generalization. For example, for the cover operator cov_S, the minimal patterns have a *strictly* smaller cover than their generalizations. The string *ab* is not minimal due to its suffix *b* because $cov_S(\{(b, 2)\}) = cov_S(\{(a, 1), (b, 2)\}) = \{0, 7\}$. For our running example, the whole collection of minimal strings is $\mathcal{M}(\mathcal{S}_S) = \{a, b, r, c, d, ca, ra, da\}$.

Given a minimizable set system $\mathcal{S} = \langle(\mathcal{F}, E), \mathcal{G}, cov, \phi\rangle$**, the minimal pattern mining problem consists in enumerating all the minimal patterns for** $\mathcal{S}$**.**

3 Enumerating the Minimal Patterns

This section aims at effectively mining all the minimal patterns in a depth-first search manner (Section 3.3). To do this, we rely on two key ideas: the pruning of the search space (Section 3.1) and the fast minimality checking (Section 3.2).

Before, it is important to recall that the minimal patterns are sufficient to induce the cover of any pattern. From now, we consider a minimizable set system $\mathcal{S} = \langle(\mathcal{F}, E), \mathcal{G}, cov, \phi\rangle$. The minimal patterns $\mathcal{M}(\mathcal{S})$ is a lossless representation of all patterns of $\mathcal{F}$ in the sense we can find the cover of any pattern.

Theorem 1 (Condensed representation). *The set of minimal patterns is a concise representation of $\mathcal{F}$ adequate to cov: for any pattern $X \in \mathcal{F}$, there exists $Y \subseteq X$ such that $\phi(Y) \in \mathcal{M}(\mathcal{S})$ and $cov(\phi(Y)) = cov(X)$.*

Theorem 1 means that $\mathcal{M}(\mathcal{S})$ is really a condensed representation of $\mathcal{S}$ because the minimal pattern mining enables us to infer the cover of any pattern in $\mathcal{S}$. For instance, the cover of the non-minimal pattern $\{(a,1),(b,2)\}$ equals to that of the minimal pattern $\phi(\{(b,2)\}) = \{(b,1)\}$: $cov_{\mathcal{S}}(\{(a,1),(b,2)\}) = cov_{\mathcal{S}}(\{(b,1)\}) = \{0,7\}$. It is preferable to extract $\mathcal{M}(\mathcal{S})$ instead of $\mathcal{S}$ because its size is lower (and, in general, much lower) than the total number of patterns.

3.1 Search Space Pruning

The first problem we face is fairly classical. Given a minimizable set system $\mathcal{S} = \langle(\mathcal{F}, E), \mathcal{G}, cov, \phi\rangle$, the number of patterns $|\mathcal{F}|$ is huge in general (in the worst case, it reaches $2^{|E|}$ patterns). So, it is absolutely necessary not to completely scan the search space for focusing on the minimal patterns. Effective techniques can be used to prune the search space due to the downward closure of $\mathcal{M}(\mathcal{S})$:

Theorem 2 (Independence system). *If a pattern X is minimal for $\mathcal{S}$, then any pattern $Y \in \mathcal{F}$ satisfying $Y \subseteq X$ is also minimal for $\mathcal{S}$.*

The proof of this theorem strongly relies on a key lemma saying that a non-minimal pattern has a direct generalization having the same cover (proofs are omitted due to lack of space):

Lemma 1. *If X is not mininal, there exists $e \in X$ such that $X \setminus \{e\} \in \mathcal{G}$ and $cov(X) = cov(X \setminus \{e\})$.*

For instance, as the string *da* is minimal, the substrings *d* and *a* are also minimal. More interestingly, as *ab* is not minimal, the string *abr* is not minimal. It means that the string *ab* is a cut-off point in the search space. In practice, anti-monotone pruning is recognized as a very powerful tool whatever the traversal of the search space (level by level or in depth).

3.2 Fast Minimality Checking

The main difficulty in extracting the minimal patterns is to test whether a pattern is minimal or not. As we mentioned earlier, this is particularly difficult in a depth-first traversal because all subsets have not yet been enumerated. Indeed, depth-first approaches only have access to the first parent branch contrary to levelwise approaches. To overcome this difficulty, we introduce the concept of *critical objects* inspired from critical edges in case of minimal traversals [13]. Intuitively, the critical objects of an element e for a pattern X are objects that are not covered by X due to the element e. We now give a formal definition of the critical objects derived from any cover operator:

Definition 5 (Critical objects). *For a pattern X, the critical objects of an element $e \in X$, denoted by $\widehat{cov}(X, e)$ is the set of objects that belongs to the cover of X without e and not to the cover of e: $\widehat{cov}(X, e) = cov(X \setminus e) \setminus cov(e)$.*

Let us illustrate the critical objects with our running example. For $\{(a, 1), (b, 2)\}$, the critical objects $\widehat{cov}(ab, a)$ of the element $(a, 1)$ correspond to $\emptyset$ $(= \{0, 7\} \setminus \{0, 3, 5, 7, 10\})$. It means that the addition of a to b has no impact on the cover of ab. At the opposite, for the same pattern, the critical objects of $(b, 2)$ are $\{3, 5, 10\}$ $(= \{0, 3, 5, 7, 10\} \setminus \{0, 7\})$. It is due to the element b that ab does not cover the objects $\{3, 5, 10\}$.

The critical objects are central in our proposition for the following reasons: 1) the critical objects easily characterize the minimal patterns; and 2) the critical objects can efficiently be computed in a depth-first search algorithm.

The converse of Lemma 1 says that a pattern is minimal if its cover differs from that of its generalization. We can reformulate this definition thanks to the notion of critical objects as follows:

Property 1 (Minimality). $X \in \mathcal{F}$ is minimal if $\forall e \in X$ such that $X \setminus e \in \mathcal{G}$, $\widehat{cov}(X, e) \neq \emptyset$.

Typically, as b is a generalization of the string ab and at the same time, $\widehat{cov}(ab, a)$ is empty, ab is not minimal. Property 1 means that checking whether a candidate X is minimal only requires to know the critical objects of all the elements in X. Unlike the usual definition, no information is required on the subsets. Therefore, the critical objects allow us to design a depth-first algorithm if (and only if) computing the critical objects does not also require information on the subsets.

In a depth-first traversal, we want to update the critical objects of an element e for the pattern X when a new element e' is added to X. In such case, we now show that the critical objects can efficiently be computed by intersecting the old set of critical objects $\widehat{cov}(X, e)$ with the cover of the new element e':

Property 2. The following equality holds for any pattern $X \in \mathcal{F}$ and any two elements $e, e' \in E$: $\widehat{cov}(Xe', e) = \widehat{cov}(X, e) \cap cov(e')$.

For instance, Definition 5 gives $\widehat{cov}_S(a, a) = \{1, 2, 4, 6, 8, 9\}$. As $cov_S(b) = \{0, 7\}$, we obtain that $\widehat{cov}_S(ab, a) = \widehat{cov}_S(a, a) \cap cov_S(b) = \{1, 2, 4, 6, 8, 9\} \cap \{0, 7\} = \emptyset$. Interestingly, Property 2 allows us to compute the critical objects of any element included in a pattern X having information on a single branch. This is an ideal situation for a depth-first search algorithm.

3.3 Algorithm DeFMe

The algorithm DeFMe takes as inputs the current pattern and the current tail (the list of the remaining items to be checked) and it returns all the minimal patterns containing X (based on *tail*). More precisely, Line 1 checks whether X is minimal or not. If X is minimal, it is output (Line 2). Lines 3-14 explores the subtree containing X based on the tail. For each element e where Xe is a pattern of $\mathcal{F}$ (Line 4) (Property 1), the branch is built with all the necessary information. Line 7 updates the cover and Lines 8-11 updates the critical objects using Property 2. Finally, the function DeFMe is recursively called at Line 12 with the updated tail (Line 5).

Algorithm 1. DeFMe(X, *tail*)

Input: X is a pattern, *tail* is the set of the remaining items to be used in order to generate the candidates. Initial values: $X = \emptyset, tail = E$.
Output: polynomially incrementally outputs the minimal patterns.

```
1: if ∀e ∈ X, cov^(X,e) ≠ ∅ then
2:   print X
3:   for all e ∈ tail do
4:     if Xe ∈ F then
5:       tail := tail \ {e}
6:       Y := Xe
7:       cov(Y) := cov(X) ∩ cov(e)
8:       cov^(Y,e) := cov(X) \ cov(e)
9:       for all e' ∈ X do
10:        cov^(Y,e') := cov^(X,e') ∩ cov(e)
11:      end for
12:      DeFMe(Y, tail)
13:    end if
14:  end for
15: end if
```

Theorems 3 and 4 demonstrate that the algorithm DeFMe has an efficient behavior both in space and time. This efficiency mainly stems from the inexpensive handling of covers/critical objects as explained by the following property:

Property 3. The following inequality holds for any pattern $X \in \mathcal{F}$:

$$|cov(X)| + \sum_{e \in X} |\widehat{cov}(X, e)| \leq |cov(\emptyset)|$$

Property 3 means that for a pattern, the storage of its cover plus that of all the critical objects is upper bounded by the number of objects (i.e., $|cov(\emptyset)|$). Thus, it is straightforward to deduce the memory space required by the algorithm:

Theorem 3 (Polynomial-space complexity). *$\mathcal{M}(\mathcal{S})$ is enumerable in $O(|cov(\emptyset)| \times m)$ space where m is the maximal size of a feasible set in $\mathcal{F}$.*

In practice, the used memory space is very limited because m is small. In addition, the amount of time between each output pattern is polynomial:

Theorem 4 (Polynomial-delay complexity). *$\mathcal{M}(\mathcal{S})$ is enumerable in $O(|E|^2 \times |cov(\emptyset)|)$ time per minimal pattern.*

It is not difficult to see that between two output patterns, DEFME requires a polynomial number of operations assuming that the membership oracle is computable in polytime (Line 4). Indeed, the computation of the cover and that of the critical objects (Lines 7-11) is linear with the number of objects due to Property 3; the loop in Line 3 does not exceed $|E|$ iterations and finally, the number of consecutive backtracks is at most $|E|$.

4 Experimental Study

The aim of our experiments is to quantify the benefit brought by DEFME both on effectiveness and conciseness. We show its effectiveness with the problem of free itemset mining for which several prototypes already exist in the literature. Then we instantiate DEFME to extract the collection of minimal strings and compare its size with that of closed strings. All tests were performed on a 2.2 GHz Opteron processor with Linux operating system and 200 GB of RAM memory.

4.1 Free Itemset Mining

We designed a prototype of DEFME for itemset mining as a proof of concept and we compared it with two other prototypes: ACMINER based on a levelwise algorithm [1] and NDI[2] based on a depth-first traversal with reordered items [11]. For this purpose, we conducted experiments on benchmarks coming from the FIMI repository and the 2004 PKDD Discovery Challenge[3]. The first three columns of Table 1 give the characteristics of these datasets. The fourth column gives the used minimal support threshold. The next three columns report the running times and finally, the last three columns indicate the memory consumption.

The best performances are highlighted in bold in Table 1 for both time and space. ACMINER is by far the slowest prototype. Its levelwise approach is particularly penalized by the large amount of used memory. Except on the genomic datasets 74x822 and 90x27679, the running times of NDI clearly outperform those of DEFME. As a piece of information, Figure 1 details, for various minsup thresholds, the speed of DEFME. It plots the number of minimal patterns it extracted for each second of computing time.

Concerning memory consumption, DEFME is (as expected) the most efficient algorithm. In certain cases, the increase of the storage memory would not be sufficient to treat the most difficult datasets. Here, ACMINER and NDI are

[2] As this prototype mines non-derivable itemsets, it enable us to compute free patterns when the depth parameter is set to 1.

[3] `fimi.ua.ac.be/data/` and `lisp.vse.cz/challenge/ecmlpkdd2004/`

Table 1. Characteristics of benchmarks and results about free itemset mining

				time (s)			memory (kB)		
dataset	objects	items	minsup	ACMINER	NDI	DEFME	ACMINER	NDI	DEFME
74x822	74	822	88%	fail	fail	**45**	fail	fail	**3,328**
90x27679	90	27,679	91%	fail	fail	**79**	fail	fail	**13,352**
chess	3,196	75	22%	6,623	**187**	192	3,914,588	1,684,540	**8,744**
connect	67,557	129	7%	34,943	**115**	4,873	2,087,216	1,181,296	**174,680**
pumsb	49,046	2,113	51%	70,014	**212**	548	7,236,812	1,818,500	**118,240**
pumsb*	49,046	2,088	5%	21,267	**202**	4,600	5,175,752	2,523,384	**170,632**

not suitable to process genomic datasets even with 200GB of RAM memory and relatively high thresholds. More precisely, Figure 1 plots the ratio between NDI's and DEFME's memory use for various minsup thresholds. It is easy to notice that this ratio quickly leads NDI to go out of memory. DEFME works with bounded memory and then is not minsup limited.

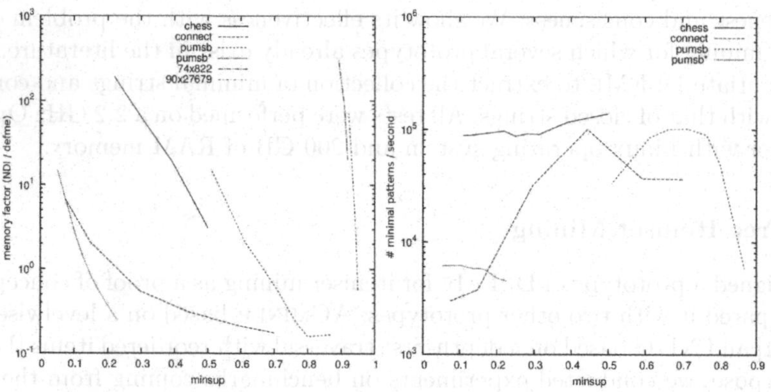

Fig. 1. Ratio of mining speed (left) and memory use (right) of NDI by DEFME

4.2 Minimal String Mining

In this section, we adopt the formalism of strings stemming from our running example. We compared our algorithm for minimal string mining with the MAXMOTIF prototype provided by Takeaki Uno that mines closed strings [10]. Our goal is to compare the size of condensed representations based on minimal strings with those based on all strings and all closed strings. We do not report the execution times because MAXMOTIF developed in Java is much slower than DEFME (developed in C++). Experiments are conducted on two datasets: chromosom[4] and msnbc coming from the UCI ML repository (www.ics.uci.edu/~mlearn).

Figure 2 and 3 report the number of strings/minimal strings/closed strings mined in chromosom and msnbc. Of course, whatever the collection of patterns,

[4] This dataset is provided with MAXMOTIF: research.nii.ac.jp/~uno/codes.htm

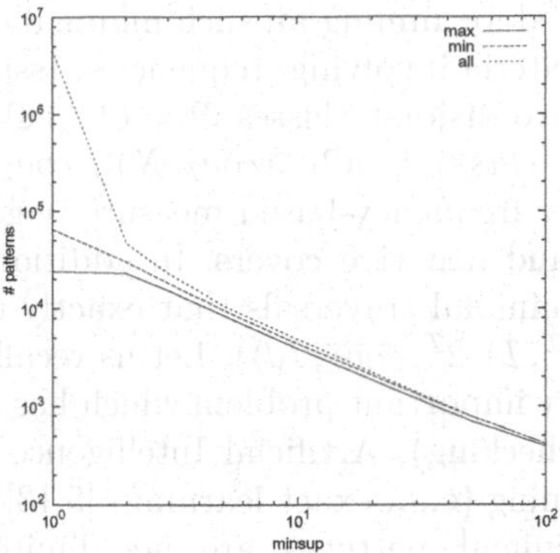

Fig. 2. Number of patterns in chromosom

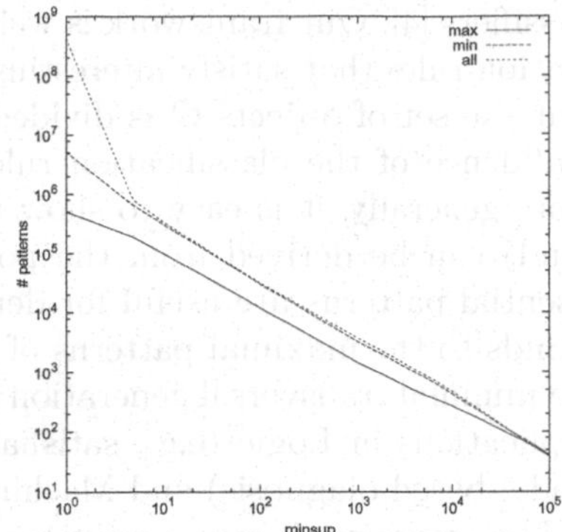

Fig. 3. Number of patterns in msnbc

the number of patterns increases with the decrease of the minimal frequency threshold. Interestingly, the two condensed representations become particularly useful when the frequency threshold is very small. Clearly the number of minimal strings is greater than the number of closed strings, but the gap is not as important as it is the case with free and closed itemsets.

5 Related Work

The collection of minimal patterns is a kind of condensed representations. Let us recall that a condensed representation of the frequent patterns is a set of patterns that can regenerate all the patterns that are frequent with their frequency. The success of the condensed representations stems from their undeniable benefit to reduce the number of mined patterns by eliminating redundancies. A large number of condensed representations have been proposed in literature [6,14]: closed itemsets [2], free itemsets [1], essential itemsets [15], Non-Derivable Itemsets [11], itemsets with negation [16] and so on. Two ideas are at the core of the condensed representations: the closure operator [14] that builds equivalence classes and the principle of inclusion-exclusion. As the inclusion-exclusion principle only works for the frequency, this paper exclusively focuses on minimal patterns considering equivalence classes. In particular, as indicated above the system $\mathcal{S}_\mathcal{I} = \langle(2^\mathcal{I}, \mathcal{I}), 2^\mathcal{I}, cov_\mathcal{I}, Id\rangle$ is minimizable and $\mathcal{M}(\mathcal{S}_\mathcal{I})$ corresponds exactly to the free itemsets (or generators). The frequency of each itemset is computed using the cardinality of the cover. Replace the cover operator $cov_\mathcal{I}$ by $\overline{cov_\mathcal{I}} : X \mapsto \{o \in \mathcal{O} \mid X \cap o = \emptyset\}$ leads to a new minimizable set system $\langle(2^\mathcal{I}, \mathcal{I}), 2^\mathcal{I}, \overline{cov_\mathcal{I}}, Id\rangle$ of which minimal patterns are essential itemsets [15]. The disjunctive frequency of an itemset X is $|\mathcal{O}| - |\overline{cov_\mathcal{I}}(X)|$.

Minimal pattern mining has a lot of applications and their use is not limited to obtain frequent patterns more efficiently. Their properties are useful for higher KDD tasks. For instance, minimal patterns are used in conjunction of closed patterns to produce non-redundant [3] or informative rules [2]. The sequential rules also benefit from minimality [17]. It is also possible to exploit the minimal patterns for mining the classification rules that are the key elements of associative classifiers [4]. Our framework is well-adapted for mining all such minimal classification rules that satisfy interestingness criteria involving frequencies.Assuming that the set of objects $\mathcal{O}$ is divided into two disjoint classes $\mathcal{O} = \mathcal{O}_1 \cup \mathcal{O}_2$, the confidence of the classification rule $X \rightarrow class_1$ is $|\mathcal{O}_1 \cap cov_{\mathcal{I}}(X)|/|cov_{\mathcal{I}}(X)|$. More generally, it is easy to show that any frequency-based measure (e.g., lift, bond) can be derived from the positive and negative covers. In addition, the essential patterns are useful for deriving minimal traversals that exactly corresponds to the maximal patterns of $\mathcal{M}(\langle(2^{\mathcal{I}}, \mathcal{I}), 2^{\mathcal{I}}, \overline{cov_{\mathcal{I}}}, Id\rangle)$. Let us recall that the minimal transversal generation is a very important problem which has many applications in Logic (e.g., satisfiability checking), Artificial Intelligence (e.g., model-based diagnosis) and Machine Learning (e.g., exact learning) [5,13].

The condensed representations of minimal patterns are not limited to frequency-based measures or itemsets. Indeed, it is also possible to mine the minimal patterns adequate to classical aggregate functions like min, max or sum [7]. Minizable set systems are also well-adapted for such measures. For instance, let us consider the function $cov_{min}(X) = \{val(i) | \exists i \in \mathcal{I}, val(i) \leq min(X.val)\}$ that returns all the possible values of val less than $min(X.val)$. This function is a cover operator and $\langle(2^{\mathcal{I}}, \mathcal{I}), 2^{\mathcal{I}}, cov_{\min}, Id\rangle$ is even a minimizable set system. The minimal patterns adequate to min correspond to the minimal patterns of the previous set system. Furthermore, the value $min(X.val)$ could be obtained as follows $\max(cov_{min}(X))$. A similar approach enables us to deal with max and sum. In parallel, several studies have extended the notion of generators to address other languages such as sequences [8,18], negative itemsets [19], graphs [20]. Unfortunately no work proposes a generic framework to extend the condensed representations based on minimality to a broad spectrum of languages as it was done with closed patterns [10]. For instance, [1,2,11,12] only address itemsets or [8,18] focus exclusively on sequences. In this paper, we have made the connection between the set systems and only two languages: itemsets and strings due to space limitation. Numerous other languages can be represented using this set system framework. In particular, all the languages depicted by [10] are suitable.

6 Conclusion

By proposing the new notion of *minimizable set system*, this paper extended the paradigm of minimal patterns to a broad spectrum of functions and languages. This framework encompasses the current methods since the existing condensed representations (e.g., free or essential itemsets) fit to specific cases of our framework. Besides, DefMe efficiently mines such minimal patterns even in difficult datasets, which are intractable by state-of-the-art algorithms. Experiments also showed on strings that the sizes of the minimal patterns are smaller than the total number of patterns.

Of course, we think that there is still room to improve our implementation even if it is difficult to find a compromise between generic method and speed. We especially want to test the ability of the minimal patterns for generating minimal classification rules with new types of data, such as strings. Similarly, it would be interesting to build associative classifiers from minimal patterns.

Acknowledgments. This article has been partially funded by the Hybride project (ANR-11-BS02-0002).

References

1. Boulicaut, J.-F., Bykowski, A., Rigotti, C.: Approximation of frequency queries by means of free-sets. In: Zighed, D.A., Komorowski, J., Żytkow, J.M. (eds.) PKDD 2000. LNCS (LNAI), vol. 1910, pp. 75–85. Springer, Heidelberg (2000)
2. Pasquier, N., Bastide, Y., Taouil, R., Lakhal, L.: Efficient mining of association rules using closed itemset lattices. Inf. Syst. 24(1), 25–46 (1999)
3. Zaki, M.J.: Generating non-redundant association rules. In: KDD, pp. 34–43 (2000)
4. Liu, B., Hsu, W., Ma, Y.: Integrating classification and association rule mining. In: KDD, pp. 80–86 (1998)
5. Eiter, T., Gottlob, G.: Hypergraph transversal computation and related problems in logic and AI. In: Flesca, S., Greco, S., Leone, N., Ianni, G. (eds.) JELIA 2002. LNCS (LNAI), vol. 2424, pp. 549–564. Springer, Heidelberg (2002)
6. Calders, T., Rigotti, C., Boulicaut, J.-F.: A survey on condensed representations for frequent sets. In: Boulicaut, J.-F., De Raedt, L., Mannila, H. (eds.) Constraint-Based Mining. LNCS (LNAI), vol. 3848, pp. 64–80. Springer, Heidelberg (2006)
7. Soulet, A., Crémilleux, B.: Adequate condensed representations of patterns. Data Min. Knowl. Discov. 17(1), 94–110 (2008)
8. Lo, D., Khoo, S.C., Li, J.: Mining and ranking generators of sequential patterns. In: SDM, pp. 553–564. SIAM (2008)
9. Li, J., Li, H., Wong, L., Pei, J., Dong, G.: Minimum description length principle: Generators are preferable to closed patterns. In: AAAI, pp. 409–414 (2006)
10. Arimura, H., Uno, T.: Polynomial-delay and polynomial-space algorithms for mining closed sequences, graphs, and pictures in accessible set systems. In: SDM, pp. 1087–1098. SIAM (2009)
11. Calders, T., Goethals, B.: Depth-first non-derivable itemset mining. In: SDM, pp. 250–261 (2005)
12. Liu, G., Li, J., Wong, L.: A new concise representation of frequent itemsets using generators and a positive border. Knowl. Inf. Syst. 17(1), 35–56 (2008)
13. Murakami, K., Uno, T.: Efficient algorithms for dualizing large-scale hypergraphs. In: ALENEX, pp. 1–13 (2013)
14. Hamrouni, T.: Key roles of closed sets and minimal generators in concise representations of frequent patterns. Intell. Data Anal. 16(4), 581–631 (2012)
15. Casali, A., Cicchetti, R., Lakhal, L.: Essential patterns: A perfect cover of frequent patterns. In: Tjoa, A.M., Trujillo, J. (eds.) DaWaK 2005. LNCS, vol. 3589, pp. 428–437. Springer, Heidelberg (2005)
16. Kryszkiewicz, M.: Generalized disjunction-free representation of frequent patterns with negation. J. Exp. Theor. Artif. Intell. 17(1-2), 63–82 (2005)
17. Lo, D., Khoo, S.C., Wong, L.: Non-redundant sequential rules - theory and algorithm. Inf. Syst. 34(4-5), 438–453 (2009)
18. Gao, C., Wang, J., He, Y., Zhou, L.: Efficient mining of frequent sequence generators. In: WWW, pp. 1051–1052. ACM (2008)
19. Gasmi, G., Yahia, S.B., Nguifo, E.M., Bouker, S.: Extraction of association rules based on literalsets. In: Song, I.-Y., Eder, J., Nguyen, T.M. (eds.) DaWaK 2007. LNCS, vol. 4654, pp. 293–302. Springer, Heidelberg (2007)
20. Zeng, Z., Wang, J., Zhang, J., Zhou, L.: FOGGER: an algorithm for graph generator discovery. In: EDBT, pp. 517–528 (2009)

Fast Vertical Mining of Sequential Patterns Using Co-occurrence Information

Philippe Fournier-Viger[1], Antonio Gomariz[2], Manuel Campos[2], and Rincy Thomas[3]

[1] Dept. of Computer Science, University of Moncton, Canada
[2] Dept. of Information and Communication Engineering, University of Murcia, Spain
[3] Dept. of Computer Science, SCT, Bhopal, India
philippe.fournier-viger@umoncton.ca, {agomariz,manuelcampos}@um.es, rinc_thomas@rediffmail.com

Abstract. Sequential pattern mining algorithms using a vertical representation are the most efficient for mining sequential patterns in dense or long sequences, and have excellent overall performance. The vertical representation allows generating patterns and calculating their supports without performing costly database scans. However, a crucial performance bottleneck of vertical algorithms is that they use a generate-candidate-and-test approach that can generate a large amount of infrequent candidates.To address this issue, we propose pruning candidates based on the study of item co-occurrences. We present a new structure named CMAP (Co-occurence MAP) for storing co-occurrence information. We explain how CMAP can be used to prune candidates in three state-of-the-art vertical algorithms, namely SPADE, SPAM and ClaSP. An extensive experimental study with six real-life datasets shows that (1) co-occurrence-based pruning is effective, (2) CMAP is very compact and that (3) the resulting algorithms outperform state-of-the-art algorithms for mining sequential patterns (GSP, PrefixSpan, SPADE and SPAM) and closed sequential patterns (ClaSP and CloSpan).

Keywords: sequential pattern mining, vertical database format, candidate pruning.

1 Introduction

Mining useful patterns in sequential data is a challenging task. Many studies have been proposed for mining interesting patterns in sequence databases [9]. Sequential pattern mining is probably the most popular research topic among them. A subsequence is called *sequential pattern* or *frequent sequence* if it frequently appears in a sequence database, and its frequency is no less than a user-specified minimum support threshold *minsup* [1]. Sequential pattern mining plays an important role in data mining and is essential to a wide range of applications such as the analysis of web click-streams, program executions, medical data, biological data and e-learning data [9]. Several efficient algorithms have been proposed

V.S. Tseng et al. (Eds.): PAKDD 2014, Part I, LNAI 8443, pp. 40–52, 2014.

for sequential pattern mining such as ClaSP [7], CloSpan [12], GSP [11], PrefixSpan [10], SPADE [13] and SPAM [3]. Sequential pattern mining algorithms can be categorized as using a *horizontal database format* (e.g. CloSpan, GSP and PrefixSpan) or a *vertical database format* (e.g. ClaSP, SPADE, SPAM). Using the vertical format provides the advantage of generating patterns and calculating their supports without performing costly database scans [3,7,13]. This allows vertical algorithms to perform better on datasets having dense or long sequences than algorithms using the horizontal format, and to have excellent overall performance [2,3,7]. However, a crucial performance bottleneck of vertical algorithms is that they use a generate-candidate-and-test approach, which can generate a large amount of patterns that do not appear in the input database or are infrequent. An important research questions that arises is: Could we design an effective candidate pruning method for vertical mining algorithms to improve mining performance? Answering this question is challenging. It requires designing a candidate pruning mechanism (1) that is effective at pruning candidates and (2) that has a small runtime and memory cost. Moreover, the mechanism should preferably be generic. i.e. applicable to any vertical mining algorithms.

In this paper, we present a solution to this issue based on the study of item co-occurrences. Our contribution is threefold. First, to store item co-occurrence information, we introduce a new data structure named *Co-occurrence MAP* (CMAP). CMAP is a small and compact structure, which can be built with a single database scan.

Second, we propose a generic candidate pruning mechanism for vertical sequential pattern mining algorithms based on the CMAP data structure. We describe how the pruning mechanism is integrated in three state-of-the-art algorithms ClaSP, SPADE and SPAM. We name the resulting algorithms CM-ClaSP, CM-SPADE and CM-SPAM.

Third, we perform a wide experimental evaluation on six real-life datasets. We compare the performance of CM-ClaSP, CM-SPADE and CM-SPAM with state-of-the-art algorithms for mining sequential patterns (GSP, PrefixSpan, SPADE and SPAM) and closed sequential patterns (ClaSP and CloSpan). Results show that the modified algorithms (1) prune a large amount of candidates, (2) and are up to eight times faster than the corresponding original algorithms and (3) that CM-ClaSP and CM-SPADE have respectively the best performance for sequential pattern mining and closed sequential pattern mining.

The rest of the paper is organized as follows. Section 2 defines the problem of sequential pattern mining and reviews the main characteristics of ClaSP, SPADE and SPAM. Section 3 describes the CMAP structure, the pruning mechanism, and how it is integrated in ClaSP, SPADE and SPAM. Section 4 presents the experimental study. Finally, Section 5 presents the conclusion.

2 Problem Definition and Related Work

Definition 1 (sequence database). Let $I = \{i_1, i_2, ..., i_l\}$ be a set of items (symbols). An *itemset* $I_x = \{i_1, i_2, ..., i_m\} \subseteq I$ is an unordered set of distinct items. The *lexicographical order* $\succ_{lex}$ is defined as any total order on I. Without loss of generality, it is assumed in the following that all itemsets are ordered according to $\succ_{lex}$. A *sequence* is an ordered list of itemsets $s = \langle I_1, I_2, ..., I_n \rangle$ such that $I_k \subseteq I$ $(1 \leq k \leq n)$. A *sequence database SDB* is a list of sequences $SDB = \langle s_1, s_2, ..., s_p \rangle$ having sequence identifiers (SIDs) $1, 2...p$. **Example.** A sequence database is shown in Fig. 1 (left). It contains four sequences having the SIDs 1, 2, 3 and 4. Each single letter represents an item. Items between curly brackets represent an itemset. The first sequence $\langle \{a, b\}, \{c\}, \{f, g\}, \{g\}, \{e\} \rangle$ contains five itemsets. It indicates that items a and b occurred at the same time, were followed by c, then f, g and lastly e.

SID	Sequences
1	⟨{*a*, *b*},{*c*},{*f*, *g*},{*g*},{*e*}⟩
2	⟨{*a*, *d*},{*c*},{*b*},{*a*, *b*, *e*, *f*}⟩
3	⟨{*a*},{*b*},{*f*},{*e*}⟩
4	⟨{*b*},{*f*, *g*}⟩

ID	Pattern	Support
p1	⟨{*a*},{*f*}⟩	3
p2	⟨{*a*},{c}{*f*}⟩	2
p3	⟨{*b*},{f,g}⟩	2
p4	⟨{*g*},{*e*}⟩	2
p5	⟨{*c*},{f}⟩	2
p6...	⟨{*b*}⟩	4

Fig. 1. A sequence database (left) and some sequential patterns found (right)

Definition 2 (sequence containment). A sequence $s_a = \langle A_1, A_2, ..., A_n \rangle$ is said to be *contained in* a sequence $s_b = \langle B_1, B_2, ..., B_m \rangle$ iff there exist integers $1 \leq i_1 < i_2 < ... < i_n \leq m$ such that $A_1 \subseteq B_{i1}, A_2 \subseteq B_{i2}, ..., A_n \subseteq B_{in}$ (denoted as $s_a \sqsubseteq s_b$). **Example.** Sequence 4 in Fig. 1 (left) is contained in Sequence 1.

Definition 3 (prefix). A sequence $s_a = \langle A_1, A_2, ..., A_n \rangle$ is a *prefix* of a sequence $s_b = \langle B_1, B_2, ..., B_m \rangle, \forall n < m$, iff $A_1 = B_1, A_2 = B_2, ..., A_{n-1} = B_{n-1}$ and the first $|A_n|$ items of B_n according to $\succ_{lex}$ are equal to A_n.

Definition 4 (support). The *support* of a sequence s_a in a sequence database SDB is defined as the number of sequences $s \in SDB$ such that $s_a \sqsubseteq s$ and is denoted by $sup_{SDB}(s_a)$.

Definition 5 (sequential pattern mining). Let *minsup* be a threshold set by the user and SDB be a sequence database. A sequence s is a *sequential pattern* and is deemed *frequent* iff $sup_{SDB}(s) \geq minsup$. The *problem of mining sequential patterns* is to discover all sequential patterns [11]. **Example.** Fig. 1 (right) shows 6 of the 29 sequential patterns found in the database of Fig. 1 (left) for $minsup = 2$.

Definition 6 (closed sequential pattern mining). A sequential pattern s_a is said to be *closed* if there is no other sequential pattern s_b, such that s_b is a superpattern of s_a, $s_a \sqsubseteq s_b$, and their supports are equal. The problem of *closed sequential patterns* is to discover the set of closed sequential patterns, which is a compact summarization of all sequential patterns [7,12].

Definition 7 (horizontal database format). A *sequence database in horizontal format* is a database where each entry is a sequence. **Example.** Fig. 1 (left) shows an horizontal sequence database.

Definition 8 (vertical database format). A *sequence database in vertical format* is a database where each entry represents an item and indicates the list of sequences where the item appears and the position(s) where it appears [13]. **Example.** Fig. 2 shows the vertical representation of the database of Fig. 1 (left).

From the vertical representation, a structure named *IdList* [13] can be associated with each pattern. IdLists allow calculating the support of a pattern quickly by making join operations with IdLists of smaller patterns. To discover sequential patterns using IdLists, a single database scan is required to create IdLists of patterns containing a single items, since IdList of larger patterns are obtained by performing the aforementioned join operation (see [13] for details). Several works proposed alternative representations for IdLists to save time in join operations, being the bitset representation the most efficient one [3,2].

a	
SID	**Itemsets**
1	1
2	1,4
3	1
4	

b	
SID	**Itemsets**
1	1
2	3,4
3	2
4	1

c	
SID	**Itemsets**
1	2
2	2
3	
4	

d	
SID	**Itemsets**
1	
2	1
3	
4	

e	
SID	**Itemsets**
1	5
2	4
3	4
4	

f	
SID	**Itemsets**
1	3
2	4
3	3
4	2

g	
SID	**Itemsets**
1	3,4
2	
3	
4	2

Fig. 2. The vertical representation of the example database shown in Figure 1(left)

The horizontal format is used by Apriori-based algorithms (e.g. GSP) and pattern-growth algorithms (e.g. CloSpan and PrefixSpan). The two main algorithms using the vertical database format are SPADE and SPAM. Other algorithms are variations such as bitSPADE [2] and ClaSP [7]. SPADE and SPAM differ mainly by their candidate generation process, which we review thereafter.

Candidate Generation in SPAM. The pseudocode of SPAM is shown in Fig. 3. SPAM take as input a sequence database SDB and the *minsup* threshold. SPAM first scans the input database SDB once to construct the vertical representation of the database $V(SDB)$ and the set of frequent items F_1. For each frequent item $s \in F_1$, SPAM calls the SEARCH procedure with $\langle s \rangle$, F_1, $\{e \in F_1 | e \succ_{lex} s\}$, and *minsup*. The SEARCH procedure outputs the pattern $\langle\{s\}\rangle$ and recursively explore candidate patterns starting with the prefix $\langle\{s\}\rangle$. The SEARCH procedure takes as parameters a sequential pattern *pat* and two sets of items to be appended to *pat* to generate candidates. The first set S_n represents items to be appended to *pat* by *s*-extension. The *s*-extension of a sequential pattern $\langle I_1, I_2, ...I_h\rangle$ with an item x is defined as $\langle I_1, I_2, ...I_h, \{x\}\rangle$. The second set S_i represents items to be appended to *pat* by *i*-extension. The *i*-extension of a sequential pattern $\langle I_1, I_2, ...I_h\rangle$ with an item x is defined as $\langle I_1, I_2, ...I_h \cup \{x\}\rangle$. For each candidate *pat* generated by an extension, SPAM calculate its support to determine if it is frequent. This is done by making a join operation (see [3] for details) and counting the number of sequences where the pattern appears. The IdList representation used by SPAM is based on bitmaps to get faster operations [3]. If the pattern *pat* is frequent, it is then used in a recursive call to SEARCH to generate patterns starting with the prefix *pat*. Note that in the recursive call, only items that resulted in a frequent pattern by extension of *pat* are considered for extending *pat*. SPAM prunes the search space by not extending infrequent patterns. This can be done due to the property that an infrequent sequential pattern cannot be extended to form a frequent pattern [1].

```
SPAM(SDB, minsup)
Scan SDB to create V(SDB) and identify F1, the list of frequent items.
FOR each item s ∈ F1,
      SEARCH(⟨s⟩, F1, {e ∈ F1 | e ≻lex s}, minsup).
```

```
SEARCH(pat, Sn, In, minsup)
Output pattern pat.
Stemp := Itemp := ∅
FOR each item j ∈ Sn,
      IF the s-extension of pat is frequent THEN Stemp := Stemp ∪{i}.
FOR each item j∈ Stemp,
      SEARCH(the s-extension of pat with j, Stemp , {e ∈ Stemp | e ≻lex j}, minsup).
FOR each item j ∈ In,
      IF the i-extension of pat is frequent THEN Itemp := Itemp ∪{i}.
FOR each item j ∈ Itemp,
      SEARCH(i-extension of pat with j, Stemp , {e ∈ Itemp | e ≻lex j}, minsup).
```

Fig. 3. The pseudocode of SPAM

Candidate Generation in SPADE. The pseudocode of SPADE is shown in Fig. 4. The SPADE procedure takes as input a sequence database SDB and the *minsup* threshold. SPADE first constructs the vertical database $V(SDB)$ and identifies the set of frequent sequential patterns F_1 containing frequent items. Then, SPADE calls the ENUMERATE procedure with the equivalence class of size 0. An *equivalence class* of size k is defined as the set of all frequent patterns containing k items sharing the same prefix of $k - 1$ items. There is only an equivalence class of size 0 and it is composed of F_1. The ENUMERATE procedure receives an equivalence class F as parameter. Each member A_i of the equivalence class is output as a frequent sequential pattern. Then, a set T_i, representing the equivalence class of all frequent extensions of A_i is initialized to the empty set. Then, for each pattern $A_j \in F$ such that $i \succ_{lex} j$, the pattern A_i is merged with A_j to form larger pattern(s). For each such pattern r, the support of r is calculated by performing a join operation between IdLists of A_i and A_j. If the cardinality of the resulting IdList is no less than *minsup*, it means that r is a frequent sequential pattern. It is thus added to T_i. Finally, after all pattern A_j have been compared with A_i, the set T_i contains the whole equivalence class of patterns starting with the prefix A_i. The procedure ENUMERATE is then called with T_i to discover larger sequential patterns having A_i as prefix. When all loops terminate, all frequent sequential patterns have been output (see [13] for the proof that this procedure is correct and complete).

SPADE and SPAM are very efficient for datasets having dense or long sequences and have excellent overall performance since performing join operations to calculate the support of candidates does not require scanning the original database unlike algorithms using the horizontal format. For example, the well-known PrefixSpan algorithm, which uses the horizontal format, performs a database projection for each item of each frequent sequential pattern, in the worst case, which is extremely costly. The main performance bottleneck of vertical mining algorithms is that they use a generate-candidate-and-test approach and therefore spend lot of time evaluating patterns that do not appear in the input database or are infrequent. In the next section, we present a novel method based on the study of item co-occurrence information to prune candidates generated by vertical mining algorithms to increase their performance.

```
SPADE(SDB, minsup)
    Scan SDB to create V(SDB) and identify F1 the list of frequent items.
    ENUMERATE(F1).

ENUMERATE(an equivalence class F)
    FOR each pattern Ai ∈ F
        Output Ai.
        Ti := ∅.
        FOR each pattern Aj ∈ F, with j ≥ i
            R = MergePatterns(Ai, Aj)
            FOR each pattern r ∈ R
                IF sup(R) ≥ minsup THEN
                    Ti := Ti ∪{R};
        ENUMERATE(Ti)
```

Fig. 4. The pseudocode of SPADE

3 Co-occurrence Pruning

In this section, we introduce our approach, consisting of a data structure for storing co-occurrence information, and its properties for candidate pruning for vertical sequential pattern mining. Then, we describe how the data structure is integrated in three state-of-the-art vertical mining algorithms, namely ClaSP, SPADE and SPAM.

3.1 The Co-occurrence Map

Definition 9. An item k is said to *succeed by i-extension* to an item j in a sequence $\langle I_1, I_2, ..., I_n \rangle$ iff $j, k \in I_x$ for an integer x such that $1 \leq x \leq n$ and $k \succ_{lex} j$.

Definition 10. An item k is said to *succeed by s-extension* to an item j in a sequence $\langle I_1, I_2, ..., I_n \rangle$ iff $j \in I_v$ and $k \in I_w$ for some integers v and w such that $1 \leq v < w \leq n$.

Definition 11. A *Co-occurrence MAP* (CMAP) is a structure mapping each item $k \in I$ to a set of items succeeding it. We define two CMAPs named $CMAP_i$ and $CMAP_s$. $CMAP_i$ maps each item k to the set $cm_i(k)$ of all items $j \in I$ succeeding k by i-extension in no less than *minsup* sequences of SDB. $CMAP_s$ maps each item k to the set $cm_s(k)$ of all items $j \in I$ succeedings k by s-extension in no less than *minsup* sequences of SDB. **Example.** The CMAP structures built for the sequence database of Fig. 1(left) are shown in Table 1, being $CMAP_i$ on the left part and $CMAP_s$ on the right part. Both tables have been created considering a *minsup* of two sequences. For instance, for the item f, we can see that it is associated with an item, $cm_i(f) = \{g\}$, in $CMAP_i$, whereas it is associated with two items, $cm_s(f) = \{e, g\}$, in $CMAP_s$. This indicates that both items e and g succeed to f by s-extension and only item g does the same for i-extension, being all of them in no less than *minsup* sequences.

Table 1. $CMAP_i$ and $CMAP_s$ for the database of Fig. 1 and $minsup = 2$

$CMAP_i$	
item	is succeeded by (i-extension)
a	$\{b\}$
b	$\emptyset$
c	$\emptyset$
e	$\emptyset$
f	$\{g\}$
g	$\emptyset$

$CMAP_s$	
item	is succeeded by (s-extension)
a	$\{b, c, e, f\}$
b	$\{e, f, g\}$
c	$\{e, f\}$
e	$\emptyset$
f	$\{e, g\}$
g	$\emptyset$

Size Optimization. Let $n = |I|$ be the number of items in SDB. To implement a CMAP, a simple solution is to use an $n \times n$ matrix (two-dimensional array) M where each row (column) correspond to a distinct item and such that each entry $m_{j,k} \in M$ represents the number of sequences where the item k succeed to the item i by i-extension or s-extension. The size of a CMAP would then be $O(n^2)$. However, the size of CMAP can be reduced using the following strategy. It can be observed that each item is succeeded by only a small subset of all items for most datasets. Thus, few items succeed another one by extension, and thus, a CMAP may potentially waste large amount of memory for empty entries if we consider them by means of a $n \times n$ matrix. For this reason, in our implementations we instead implemented each CMAP as a hash table of hash sets, where an hashset corresponding to an item k only contains the items that succeed to k in at least *minsup* sequences.

3.2 Co-occurrence-Based Pruning

The CMAP structure can be used for pruning candidates generated by vertical sequential pattern mining algorithms based on the following properties.

Property 1 (pruning an i-extension). Let be a frequent sequential pattern A and an item k. If there exists an item j in the last itemset of A such that k belongs to $cm_i(j)$, then the i-extension of A with k is infrequent. **Proof.** If an item k does not appear in $cm_i(j)$, then k succeed to j by i-extension in less than *minsup* sequences in the database SDB. It is thus clear that appending k by i-extension to a pattern A containing j in its last itemset will not result in a frequent pattern. □

Property 2 (pruning an s-extension). Let be a frequent sequential pattern A and an item k. If there exists an item $j \in A$ such that the item k belongs to $cm_s(j)$, then the s-extension of A with k is infrequent. **Proof.** If an item k does not appear in $cm_s(j)$, then k succeeds to j by s-extension in less than *minsup* sequences from the sequence database SDB. It is thus clear that appending j by s-extension to a pattern A containing k will not result in a frequent pattern. □

Property 3 (pruning a prefix). The previous properties can be generalized to prune all patterns starting with a given prefix. Let be a frequent sequential pattern A and an item k. If there exists an item $j \in A$ (equivalently j in the last itemset of A) such that there is an item $k \in cm_s(j)$ (equivalently in $cm_i(j)$), then all supersequences B having A as prefix and where k succeeds j by s-extension (equivalently i-extension to the last itemset) in A in B are infrequent. **Proof.** If an item k does not appear in $cm_s(j)$ (equivalently $cm_i(j)$), therefore k succeeds to j in less than *minsup* sequences by s-extension (equivalently i-extension to the last itemset) in the database SDB. It is thus clear that no frequent pattern containing j (equivalently j in the last itemset) can be formed such that k is appended by s-extension (equivalently by i-extension to the last itemset). □

3.3 Integrating Co-occurrence Pruning in Vertical Mining

Integration in SPADE. The integration in SPADE is done as follows. In the ENUMERATE procedure, consider a pattern r obtained by merging two patterns $A_i = P \cup x$ and $A_j = P \cup y$, being P a common prefix for A_i and A_j. Let y be the item that is appended to A_i to generate r. If r is an i-extension, we use the $CMAP_i$ structure, otherwise, if r is an s-extension, we use $CMAP_s$. If the last item a of r does not have an item $x \in cm_i(a)$ (equivalently in $cm_s(a)$), then the pattern r is infrequent and r can be immediately discarded, avoiding the join operation to calculate the support of r. This pruning strategy is correct based on Properties 1, 2 and 3.

Note that to perform the pruning in SPADE, we do not have to check if items of the prefix P are succeeded by the item $y \in A_j$. This is because the items of P are also in A_j. Therefore, checking the extension of P by y was already done, and it is not necessary to do it again.

Integration in SPAM. The CMAP structures are used in the SEARCH procedure as follows. Let a sequential pattern *pat* being considered for *s*-extension ($x \in S_n$) or *i*-extension ($x \in S_i$) with an item x (line 3). If the last item a in *pat* does not have an item $x \in cm_s(a)$ (equivalently cm_i), then the pattern resulting from the extension of *pat* with x will be infrequent and thus the join operation of x with *pat* to count the support of the resulting pattern does not need to be performed (by Property 1 and 2). Furthermore, the item x should not be considered for generating any pattern by *s*-extension (*i*-extension) having *pat* as prefix (by Property 3). Therefore x should not be added to the variable S_{temp} (I_{temp}) that is passed to the recursive call to the SEARCH procedure.

Note that to perform the pruning in SPAM, we do not have to check for extensions of *pat* with x for all the items since such items, except for the last one, have already been checked for extension in previous steps.

Integration in ClaSP. We have also integrated co-occurrence pruning in ClaSP [7], a state of the art algorithm for closed sequential pattern mining. The integration in ClaSP is not described here since it is done as in SPAM since ClaSP is based on SPAM.

4 Experimental Evaluation

We performed experiments to assess the performance of the proposed algorithms. Experiments were performed on a computer with a third generation Core i5 processor running Windows 7 and 5 GB of free RAM. We compared the performance of the modified algorithms (CM-ClaSP, CM-SPADE, CM-SPAM) with state-of-the-art algorithms for sequential pattern mining (GSP, PrefixSpan, SPADE and SPAM) and closed sequential pattern mining (ClaSP and CloSpan). All algorithms were implemented in Java. Note that for SPADE algorithms, we use the version proposed in [2] that implement IdLists by means of bitmaps. All memory measurements were done using the Java API. Experiments were carried on six real-life datasets having varied characteristics and representing four different types of data (web click stream, text from books, sign language utterances and protein sequences). Those datasets are *Leviathan*, *Sign*, *Snake*, *FIFA*, *BMS* and *Kosarak10k*. Table 2 summarizes their characteristics. The source code of all algorithms and datasets used in our experiments can be downloaded from `http://goo.gl/hDtdt`.

The experiments consisted of running all the algorithms on each dataset while decreasing the *minsup* threshold until algorithms became too long to execute, ran out of memory or a clear winner was observed. For each dataset, we recorded the execution time, the percentage of candidate pruned by the proposed algorithms and the total size of CMAPs. The comparison of execution times is shown in Fig. 5. The percentage of candidates pruned by the proposed algorithms is shown in Table 3.

Effectiveness of Candidate Pruning. CM-ClaSP, CM-SPADE and CM-SPAM are generally from about 2 to 8 times faster than the corresponding original algorithms (ClaSP, SPADE and SPAM). This shows that co-occurrence

Table 2. Dataset characteristics

dataset	sequence count	distinct item count	avg. seq. length (items)	type of data
Leviathan	5834	9025	33.81 (std= 18.6)	book
Sign	730	267	51.99 (std = 12.3)	language utterances
Snake	163	20	60 (std = 0.59)	protein sequences
FIFA	20450	2990	34.74 (std = 24.08)	web click stream
BMS	59601	497	2.51 (std = 4.85)	web click stream
Kosarak10k	10000	10094	8.14 (std = 22)	web click stream

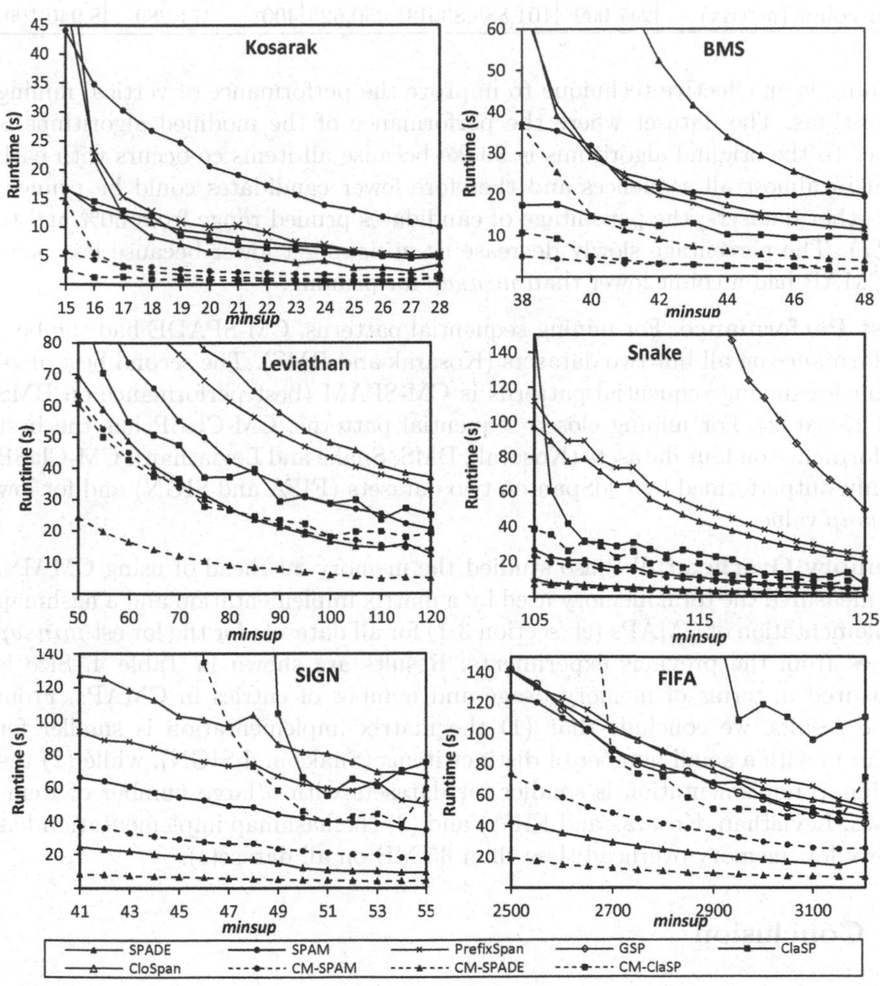

Fig. 5. Execution times

Table 3. Candidate reduction

	BMS	Kosarak	Leviathan	Snake	Sign	Fifa
CM-SPAM	78 to 93 %	94 to 98 %	50 to 51 %	28%	63 %	61 to 68 %
CM-SPADE	75 to 76 %	98 %	50 %	25 to 26 %	69 %	63 to 69 %
CM-ClaSP	79 to 93%	75 %	50 to 52 %	18 %	63 %	67 to 68 %

Table 4. CMAP implementations comparison

	BMS	Kosarak	Leviathan	Snake	Sign	Fifa
minsup	38	16	60	105	43	2500
CMAP Size (hashmap)	**0.5 MB**	**33.1 MB**	**15 MB**	64 KB	3.19 MB	**0.4 MB**
CMAP Size (matrix)	0.9 MB	388 MB	310 MB	**1.7 KB**	**0.2 MB**	34.1 MB
Pair count (hashmap)	50,885	58,772	41,677	144	17,887	2,500
Pair count (matrix)	247,009	101,888,836	81,450,625	400	71,289	8,940,100

pruning is an effective technique to improve the performance of vertical mining algorithms. The dataset where the performance of the modified algorithms is closer to the original algorithms is Snake because all items co-occurs with each item in almost all sequences and therefore fewer candidates could be pruned. For other datasets, the percentage of candidates pruned range from 50% and to 98 %). The percentage slowly decrease as *minsup* get lower because less pairs in CMAP had a count lower than *minsup* for pruning.

Best Performance. For mining sequential patterns, CM-SPADE had the best performance on all but two datasets (Kosarak and BMS). The second best algorithm for mining sequential patterns is CM-SPAM (best performance on BMS and Kosarak). For mining closed sequential patterns, CM-ClaSP has the best performance on four datasets (Kosarak, BMS, Snake and Leviathan). CM-ClaSP is only outperformed by CloSpan on two datasets (FIFA and SIGN) and for low *minsup* values.

Memory Overhead. We also studied the memory overhead of using CMAPs. We measured the total memory used by a matrix implementation and a hashmap implementation of CMAPs (cf. section 3.1) for all datasets for the lowest *minsup* values from the previous experiments. Results are shown in Table 4. Size is measured in terms of memory usage and number of entries in CMAPs. From these results, we conclude that (1) the matrix implementation is smaller for datasets with a small number of distinct items (Snake and SIGN), while (2) the hashmap implementation is smaller for datasets with a large number of items (BMS, Leviathan, Kosarak and FIFA) and (3) the hashmap implementation has a very low memory overhead (less than 35 MB on all datasets).

5 Conclusion

Sequential pattern mining algorithms using the vertical format are very efficient because they can calculate the support of candidate patterns by avoiding costly database scans. However, the main performance bottleneck of vertical mining algorithms is that they usually spend lot of time evaluating candidates that do not appear in the input database or are infrequent. To address this problem, we presented a novel data structure named CMAP for storing co-occurrence information. We have explained how CMAPs can be used for pruning candidates generated by vertical mining algorithms. We have shown how to integrate CMAPs in three state-of-the-art vertical algorithms. We have performed an extensive experimental study on six real-life datasets to compare the performance of the modified algorithms (CM-ClaSP, CM-SPADE and CM-SAPM) with state-of-the-art algorithms (ClaSP, CloSpan, GSP, PrefixSpan, SPADE and SPAM). Results show that the modified algorithms (1) prune a large amount of candidates, (2) are up to 8 times faster than the corresponding original algorithms and (3) that CM-SPADE and CM-ClaSP have respectively the best performance for mining sequential patterns and closed sequential patterns.

The source code of all algorithms and datasets used in our experiments can be downloaded from `http://goo.gl/hDtdt`.

For future work, we plan to develop additional optimizations and also to integrate them in sequential rule mining [5], top-k sequential pattern mining [4] and maximal sequential pattern mining [6].

Acknowledgement. This work is partially financed by a National Science and Engineering Research Council (NSERC) of Canada research grant, a PhD grant from the Seneca Foundation (Regional Agency for Science and Technology of the Region de Murcia), and by the Spanish Office for Science and Innovation through project TIN2009-14372-C03-01 and PlanE, and the European Union by means of the European Regional Development Fund (ERDF, FEDER).

References

1. Agrawal, R., Ramakrishnan, S.: Mining sequential patterns. In: Proc. 11th Intern. Conf. Data Engineering, pp. 3–14. IEEE (1995)
2. Aseervatham, S., Osmani, A., Viennet, E.: bitSPADE: A Lattice-based Sequential Pattern Mining Algorithm Using Bitmap Representation. In: Proc. 6th Intern. Conf. Data Mining, pp. 792–797. IEEE (2006)
3. Ayres, J., Flannick, J., Gehrke, J., Yiu, T.: Sequential pattern mining using a bitmap representation. In: Proc. 8th ACM SIGKDD Intern. Conf. Knowledge Discovery and Data Mining, pp. 429–435. ACM (2002)
4. Fournier-Viger, P., Gomariz, A., Gueniche, T., Mwamikazi, E., Thomas, R.: TKS: Efficient Mining of Top-K Sequential Patterns. In: Motoda, H., Wu, Z., Cao, L., Zaiane, O., Yao, M., Wang, W. (eds.) ADMA 2013, Part I. LNCS, vol. 8346, pp. 109–120. Springer, Heidelberg (2013)
5. Fournier-Viger, P., Nkambou, R., Tseng, V.S.: RuleGrowth: Mining Sequential Rules Common to Several Sequences by Pattern-Growth. In: Proc. ACM 26th Symposium on Applied Computing, pp. 954–959 (2011)
6. Fournier-Viger, P., Wu, C.-W., Tseng, V.S.: Mining Maximal Sequential Patterns without Candidate Maintenance. In: Motoda, H., Wu, Z., Cao, L., Zaiane, O., Yao, M., Wang, W. (eds.) ADMA 2013, Part I. LNCS, vol. 8346, pp. 169–180. Springer, Heidelberg (2013)

7. Gomariz, A., Campos, M., Marin, R., Goethals, B.: ClaSP: An Efficient Algorithm for Mining Frequent Closed Sequences. In: Pei, J., Tseng, V.S., Cao, L., Motoda, H., Xu, G. (eds.) PAKDD 2013, Part I. LNCS, vol. 7818, pp. 50–61. Springer, Heidelberg (2013)
8. Han, J., Kamber, M.: Data Mining: Concepts and Techniques, 2nd edn. Morgan Kaufmann, San Francisco (2006)
9. Mabroukeh, N.R., Ezeife, C.I.: A taxonomy of sequential pattern mining algorithms. ACM Computing Surveys 43(1), 1–41 (2010)
10. Pei, J., Han, J., Mortazavi-Asl, B., Wang, J., Pinto, H., Chen, Q., Dayal, U., Hsu, M.: Mining sequential patterns by pattern-growth: the PrefixSpan approach. IEEE Trans. Knowledge Data Engineering 16(11), 1424–1440 (2004)
11. Srikant, R., Agrawal, R.: Mining Sequential Patterns: Generalizations and Performance Improvements. In: Apers, P.M.G., Bouzeghoub, M., Gardarin, G. (eds.) EDBT 1996. LNCS, vol. 1057, pp. 3–17. Springer, Heidelberg (1996)
12. Yan, X., Han, J., Afshar, R.: CloSpan: Mining closed sequential patterns in large datasets. In: Proc. 3rd SIAM Intern. Conf. on Data Mining, pp. 166–177 (2003)
13. Zaki, M.J.: SPADE: An efficient algorithm for mining frequent sequences. Machine Learning 42(1), 31–60 (2001)

Balancing the Analysis of Frequent Patterns

Arnaud Giacometti, Dominique H. Li, and Arnaud Soulet

Université François Rabelais Tours, LI EA 6300
3 Place Jean Jaurès, F-41029 Blois, France
firstname.lastname@univ-tours.fr

Abstract. A main challenge in pattern mining is to focus the discovery on high-quality patterns. One popular solution is to compute a numerical score on how well each discovered pattern describes the data. The best rating patterns are then the most analyzed by the data expert. In this paper, we evaluate the quality of discovered patterns by anticipating of how user analyzes them. We show that the examination of frequent patterns with the notion of support led to an unbalanced analysis of the dataset. Certain transactions are indeed completely ignored. Hence, we propose the notion of balanced support that weights the transactions to let each of them receive user specified attention. We also develop an algorithm ABSOLUTE for calculating these weights leading to evaluate the quality of patterns. Our experiments on frequent itemsets validate its effectiveness and show the relevance of the balanced support.

Keywords: Pattern mining, stochastic model, interestingness measure.

1 Introduction

For twenty years, the pattern mining algorithms have gained performance and now arrive to quickly extract patterns from large amounts of data. However, evaluate and ensure the quality of extracted patterns remains a very open issue. In general, a pattern is considered to be relevant if it deviates from what was expected from a knowledge model. In the literature, there are two broad categories of knowledge models [1,2,3]: user-driven and data-driven ones. *User-driven* approaches discover interesting patterns with subjective models based on user oriented information, such as domain knowledge, beliefs or preferences. *Data-driven* approaches discover interesting patterns with objective models based on the statistical properties applied to data, such as frequency of patterns. Most often these methods neglect how the user will analyze the collection of patterns. In this paper, we present a novel approach, named *analysis-driven*, to evaluate discovered patterns by foreseeing how the collection will be analyzed.

Before presenting in depth our motivations, we first recall the context of frequent itemset mining [4]. Let $\mathcal{I}$ be a set of distinct literals called *items*, an itemset corresponds to a non-null subset of $\mathcal{I}$. A transactional dataset is a multi-set of itemsets, named *transactions*. Table 1 (a) presents such a dataset $\mathcal{D}$ where 4 transactions $t_1, \ldots, t_4$ are described by 4 items $A, \ldots, D$. The *support* of an

V.S. Tseng et al. (Eds.): PAKDD 2014, Part I, LNAI 8443, pp. 53–64, 2014.

itemset X in a dataset $\mathcal{D}$ is the fraction of transactions containing X. An itemset is *frequent* if it appears in a database with support no less than a user specified threshold. For instance, Table 1 (b) shows the frequent itemsets of $\mathcal{D}$ (with 0.25 as minimal support) that are ranked according to the support.

Table 1. An unbalanced analysis of a dataset by frequent patterns

(a) Dataset $\mathcal{D}$

TID	Items
t_1	AB
t_2	AC
t_3	BC
t_4	D

(b) The mined pattern set P

PID	Pattern	Support
p_1	A	0.50
p_2	B	0.50
p_3	C	0.50
p_4	D	0.25
p_5	AB	0.25
p_6	AC	0.25
p_7	BC	0.25

(c) Analysis resulting from the scoring

		Reality	Expected
TID	Length	Proportion Π	Preference ρ
t_1	1.25	0.3125	> 0.25
t_2	1.25	0.3125	> 0.25
t_3	1.25	0.3125	> 0.25
t_4	0.25	0.0625	< 0.25
Sum:	4	1	1

In a post-analysis process based on the scoring of patterns, we assume that an analyst examines each pattern with a diligence proportional to an interestingness measure (here, the support) for analyzing the dataset (we will justify our proposal of analysis model in Section 3). Hence, it is not difficult to see that the attention paid to A is twice that for AB since their support is respectively 0.5 and 0.25. We also assume that the analysis proportion of each transaction is proportional to the sum of the time spent on each pattern covering it. The transaction t_1 covered by A, B and AB is then analyzed for a length of 1.25 $(= 0.5 + 0.5 + 0.25)$. Therefore, according to this observation, the individual analysis of t_1, t_2, or t_3 in the analysis is 5 times greater than that of t_4 as shown by the third column of Table 1 (c). Assuming that the user considers all transactions equally interesting (see the preference vector ρ in the fourth colmun), we say that such an analysis is *unbalanced.* It means that some transactions are understudied while others are overstudied! For us, a good way to balance the analysis is to increase the score (here the support) of the pattern D. Its score increase will also increase the analysis proportion of t_4 and decrease that of the other three transactions. More interestingly, we think that D is peculiar and deserves a higher valuation because it describes a singular transaction that is described by no other pattern.

In this work, we seek to balance the analysis of the dataset by proposing a new interestingness measure to rate the patterns. More precisely, a preference vector $\rho : \mathcal{D} \rightarrow (0, 1]$ indicates the intensity of his/her interest for each transaction such that $\sum_{t \in \mathcal{D}} \rho(t) = 1$. Indeed, the user does not always devote the same interest in all transactions as in our example above. Sometimes he/she prefers to focus on rare data as it is often the case in fraud detection or with biomedical data. An analysis is balanced when each transaction t is studied with an acuity $\Pi(t, \mathcal{M})$ corresponding to that specified by a user-preference vector $\rho(t)$: $\Pi(t, \mathcal{M}) = \rho(t)$ where $\Pi(t, \mathcal{M})$ is the analysis proportion of the transaction t given an analysis model $\mathcal{M}$ that simulates analysis sessions conducted to understand the data. To the best of our knowledge, we propose the first model, called *scoring analysis model*, to simulate the sessions of analyzing pattern sets according to an interestingness measure. Its strength is to rely on a stochastic model successfully used in Information Retrieval [5,6] while integrating the preferences given by the user. As main contribution, we then introduce the *balanced support* that induces balanced analysis under our model. This measure removes the equalized axiom of support saying that every transaction has the same weight in its calculation. It gives a higher weight to the most singular transactions in the calculation of the balanced support. For instance, in Table 1, the transaction t_4 will be weighted 5 times more than others so that it receives the same attention as other transactions. We also develop an algorithm, ABSOLUTE, to compute the balanced support. Our experiments show its effectiveness for balancing frequent itemsets and compare the balanced support with the traditional support.

The rest of this paper is organized as follows. Section 2 introduces the basic notations. In Section 3, we introduce the scoring analysis model allowing us to simulate the behavior of an analyst. Under this model, we propose in Section 4 the balanced support and the algorithm ABSOLUTE to compute it. Section 5 presents experiments demonstrating its efficiency and the interest of the balanced support. Section 6 reviews some related work. We conclude in Section 7.

2 Preliminaries

For the sake of clarity, we illustrate our definitions with the notion of itemsets but, our problem is not limited to a particular type of pattern. We consider a *language* $\mathcal{L}$ and a dataset $\mathcal{D}$ that is a multiset of $\mathcal{L}$ (or another language). A *specialization relation* $\preceq$ is a partial order relation on $\mathcal{L}$ [7]. Given a specialization relation $\preceq$ on $\mathcal{L}$, $l \preceq l'$ means that l is more general than l', and l' is more specific than l. For instance, A is more general than AB w.r.t $\subseteq$.

Given two posets $(\mathcal{L}_1, \preceq_1)$ and $(\mathcal{L}_2, \preceq_2)$, a binary relation $\lhd \subseteq \mathcal{L}_1 \times \mathcal{L}_2$ is a *cover relation* iff for any $l_1 \lhd l_2$, we have $l_1' \lhd l_2$ (resp. $l_1 \lhd l_2'$) for any pattern $l_1' \preceq_1 l_1$ (resp. $l_2 \preceq_2 l_2'$). The relation $l_1 \lhd l_2$ means that l_1 covers l_2, and l_2 is covered by l_1. The cover relation is useful to relate different languages together (e.g., for linking patterns to data). Note that a specialization relation on $\mathcal{L}$ is also a cover relation on $\mathcal{L} \times \mathcal{L}$. For instance, the set inclusion is used for determinating which patterns of P cover a transaction of $\mathcal{D}$. Given two pattern sets $L \subseteq \mathcal{L}$, $L' \subseteq \mathcal{L}'$ and a cover relation $\lhd \subseteq \mathcal{L} \times \mathcal{L}'$, the *covered patterns* of L' by $l \in L$ is the set of patterns of L' covered by the pattern l: $L'_{\rhd l} = \{l' \in L' | l \lhd l'\}$. Dually, the *covering patterns* of L for $l' \in L'$ is the set of patterns of L covering the pattern l': $L_{\lhd l'} = \{l \in L | l \lhd l'\}$. With Table 1, we obtain that $\mathcal{D}_{\supseteq A} = \{t_1, t_2\}$ and $P_{\subseteq t_1} = \{A, B, AB\}$.

Pattern discovery takes advantage of interestingness measures to evaluate the relevancy of a pattern. The *support* of a pattern φ in the dataset $\mathcal{D}$ can be considered as the proportion of transactions covered by φ [4]: $Supp(\varphi, \mathcal{D}) = |\mathcal{D}_{\rhd\varphi}|/|\mathcal{D}|$. A pattern is said to be *frequent* when its support exceeds a user-specified minimal threshold. For instance, with a minimal threshold 0.25, the pattern A is frequent because $Supp(A, \mathcal{D}) = |\{t_1, t_2\}|/4$ (≥ 0.25). Thereafter, any function $f : \mathcal{L} \to \mathbb{R}$ is extended to any pattern set $P \subseteq \mathcal{L}$ by considering $\widetilde{f}(P) = \sum_{\varphi \in P} f(\varphi)$. For instance, $\widetilde{Supp}(P, \mathcal{D})$ corresponds to $\sum_{\varphi \in P} Supp(\varphi, \mathcal{D}) = 2.5$ with $P = \{A, B, C, D, AB, AC, BC\}$ in Table 1.

3 Simulating Analysis Using a Scoring of Patterns

3.1 Scoring Analysis Model

In this section, we propose the *scoring analysis model* to simulate an analyst faced with a set of scored patterns. This model generates sessions by randomly picking patterns taking into account the scoring of patterns. More precisely, the "simulated analyst" randomly draws a pattern by favoring those with the highest measure, and then studies each transaction covered by this pattern during a constant period weighted by its preference vector. Indeed, it is important to benefit from these preferences for better approximating the user behavior. After each pattern analysis, the session can be interrupted (if the analyst is satisfied, no longer has time to pursue, etc.) or continued (if the analyst is dissatisfied, wants more information, etc). This interruption of the session of analysis can be modeled by a halting probability. We now formalize this model:

Definition 1 (Scoring analysis model). *Let $\mathcal{D}$ be a dataset, $P \subseteq \mathcal{L}$ a pattern set, $m : \mathcal{L} \to [0, 1]$ an interestingness measure and ρ a preference vector.*

The scoring analysis model with a halting probability $\alpha \in (0, 1)$ and a unit length $\delta > 0$, denoted by $\mathcal{S}_{\rho,\alpha,\delta}$, generates sessions with the following process:

1. *Pick (with replacement) a pattern φ of P with probability distribution $p(\gamma) = m(\gamma)/\widetilde{m}(P)$ (where $\gamma \in P$).*
2. *Study each transaction $t \in \mathcal{D}$ covered by φ during a length $\delta \times \rho(t)$.*
3. *Stop the session with probability α or then, continue at Step 1.*

Basically, Step 1 favors the analysis of patterns having the highest measure (with replacement because the end-user can re-analyze a pattern in the light of another). Step 2 takes into account the user-preferences for the analysis of transactions. Simulating a data expert by randomly picking patterns may seem strange and unrealistic at first. However, this mechanism has been successfully used in other high-level tasks such as web browsing [5] and text analysis [6]. We think that the strength of our stochastic model is to describe the average behavior of users. By analogy with the random surfer model, each pattern would be a web page. The web pages would then be completely interconnected where each link is weighted by the support of the destination page. In this context, the probability α would correspond to the probability of interrupting navigation.

3.2 Analysis Proportion of a Transaction under $\mathcal{S}_{\rho,\alpha,\delta}$

Starting from the scoring analysis model, we desire to derive the analysis proportion of each transaction.

Theorem 1 (Analysis proportion). *The analysis proportion $\Pi(t, \mathcal{S}_{\rho,\alpha,\delta})$ of the transaction t is:*

$$\Pi(t, \mathcal{S}_{\rho,\alpha,\delta}) = \frac{\widetilde{m}(P_{\triangleleft t}) \times \rho(t)}{\sum_{t' \in \mathcal{D}} \widetilde{m}(P_{\triangleleft t'}) \times \rho(t')}$$

Theorem 1 (proofs are omitted due to lack of space) means that the analysis proportion of a pattern is independent of the parameters α and δ. Therefore, in the following, $\Pi(t, \mathcal{S}_{\rho,\alpha,\delta})$ is simply denoted $\Pi(t, \mathcal{S}_\rho)$. Let us consider the analysis proportion of each transaction of Table 1 in light of Theorem 1 using a uniform preference vector. As the itemsets A, B and AB cover t_1, we obtain that $\widetilde{Supp}(P_{\triangleleft t_1}, \mathcal{D}) = 0.5 + 0.5 + 0.25 = 1.25$. The same result is obtained for t_2 and t_3 and similarly, $\widetilde{Supp}(P_{\triangleleft t_4}, \mathcal{D}) = 0.25$. Finally, $\Pi(t_1, \mathcal{S}_\rho) = \widetilde{Supp}(P_{\triangleleft t_1}, \mathcal{D}) / \sum_{t' \in \mathcal{D}} \widetilde{Supp}(P_{\triangleleft t'}, \mathcal{D}) = 1.25/(3 \times 1.25 + 0.25) = 5/16 = 0.3125$ and $\Pi(t_4, \mathcal{S}_\rho) = 0.25/4 = 1/16 = 0.0625$. It means that under the scoring analysis model, the transaction t_4 will be less analyzed than the transactions t_1, t_2 or t_3 as indicated in Table 1 (c).

3.3 Balanced Analysis under $\mathcal{S}_{\rho,\alpha,\delta}$

We now deduce what a balanced analysis with respect to ρ is under the scoring analysis model:

Property 1 (Balanced analysis). The analysis of $\mathcal{D}$ by the pattern set P with m is balanced with respect to ρ under the scoring analysis model iff for any transaction $t \in \mathcal{D}$, the following relations holds:

$$\widetilde{m}(P_{\triangleleft t}) = \frac{1}{|\mathcal{D}|} \times \sum_{t' \in \mathcal{D}} \widetilde{m}(P_{\triangleleft t'})$$

The crucial observation highlighted by Property 1 is that the balance of an analysis is independent of the preference vector specified by the user, under the scoring analysis model. Indeed, the preference vector ρ involved in the right side of the equation $\Pi(t, \mathcal{M}) = \rho(t)$ is canceled by the one appearing in the analysis proportion (see Theorem 1). Consequently, if the analysis of a dataset $\mathcal{D}$ by a pattern set P with a measure m is balanced with respect to a given preference vector ρ, then it is also balanced with respect to any other preference vector. However, note that the analysis length of a transaction will take into account the considered preference vector.

Let us compute whether the analysis of the dataset $\mathcal{D}$ by the pattern set P with the measure *Supp* (see Table 1) is balanced under the model $\mathcal{S}_\rho$ using Property 1. First, the transaction t_1 is too much studied because $\widetilde{Supp}(P_{\triangleleft t_1}, \mathcal{D}) = \widetilde{Supp}(\{A, B, AB\}, \mathcal{D}) = 1.25$ and $1/|\mathcal{D}| \times \sum_{t' \in \mathcal{D}} \widetilde{Supp}(P_{\triangleleft t'}, \mathcal{D}) = 1/4 \times (3 \times 1.25 + 0.25) = 1$. Conversely, as $\widetilde{Supp}(P_{\triangleleft t_4}, \mathcal{D}) = \widetilde{Supp}(\{D\}, \mathcal{D}) = 0.25$ (< 1), the transaction t_4 is not studied enough. In Section 5, we observe that the use of frequent patterns with the support for the analysis of datasets coming from the UCI repository always leads to an unbalanced analysis.

4 Balancing the Analysis of Patterns

4.1 Axiomatization of Support

Under the scoring analysis model, we aim at balancing the analysis of the dataset by proposing a new interestingness measure that satisfies the equation of Property 1. At this stage, the right question is "what characteristics should satisfy this measure?" Unfortunately we found that the support does not lead to balanced analysis. However, this extremely popular measure is both intuitive for experts and useful in many applications. Moreover, it is an essential atomic element to build many other interestingness measures. For all these reasons, we desire a measure that leads to balanced analysis while maintaining the fundamental properties of the support. To achieve this goal, we first dissect the support by means of its axiomatization (we only focus on the support measure and not on the frequent itemset mining as proposed in [8]).

Property 2 (Support axioms). The support is the only interestingness measure m that simultaneously satisfies the three below axioms for any dataset $\mathcal{D}$:

1. **Normalized:** If a pattern φ covers no transaction (resp. all transactions), then its value $m(\varphi)$ is equal to 0 (resp. 1).
2. **Cumulative:** If patterns φ_1 and φ_2 cover respectively the set of transactions T_1 and T_2 such that $T_1 \cap T_2 = \emptyset$, then the value $m(\varphi)$ of a pattern φ covering exactly $T_1 \cup T_2$ is $m(\varphi_1) + m(\varphi_2)$.
3. **Equalized:** If two patterns cover the same number of transactions, then they have the same value for m.

Clearly the first axiom does not constitute the keystone of support, since similar normalizations are widely used by other measures (e.g., confidence or J-measure). Furthermore, it has no impact on the fact that an analysis is balanced or not, since Step 1 of scoring analysis model performs another normalization. Conversely, we believe that the other two axioms (not verified by other measures) are the main characteristics of the support. If we do not find reason to reconsider the cumulative axiom, we think the third is not fair. Ideally, an interestingness measure should favor the patterns covering the least covered transactions as explained in the introduction. Thus, the value of a measure should not only depend on the number of transactions covered but also on the *singularity* of these transactions. To this end, we propose to retain the first two axioms and to substitute the equalized axiom by the axiom of balance: *a measure of interest must lead to the balanced analysis of the dataset by the pattern set.*

4.2 Balanced Support

We first introduce a relaxation of the support by removing the constraint due to the equalized axiom:

Definition 2 (Weighted support). *Given a function $w : \mathcal{D} \rightarrow \Re^+$, the weighted support of a pattern φ in the dataset $\mathcal{D}$ is defined as: $Supp_w(\varphi, \mathcal{D}) = \sum_{t \in \mathcal{D}_{\triangleright\varphi}} w(t) / \sum_{t \in \mathcal{D}} w(t)$.*

It is not difficult to see that the weighted support satisfies the normalized axiom and the cumulative axiom. Now it only remains to choose the right vector w to get a balanced analysis. A naive idea would be to use the preference vector ρ to weight the support. That does not work in the general case: $Supp_{\rho_u}$ (where $\rho_u : t \mapsto 1/|\mathcal{D}|$) corresponds exactly to the support *Supp* which lead to an unbalanced analysis as shown by our running example (see Section 3.3) or observed in experimental study (see Section 5). In fact, to find the right weights, it is necessary to solve the equation of Property 1 by using the definition of the weighted support. Then, the weighted support induced by these weights defines the *optimal balanced support*:

Definition 3 (Optimal balanced support). *If it exists, the optimal balanced support of a pattern $\varphi \in P$ in the dataset $\mathcal{D}$ with the pattern set P, denoted by $\mathrm{BS}^*(\varphi, \mathcal{D}, P)$, is the weighted support where the weight w satisfies the following equation for all transactions $t \in \mathcal{D}$:*

$$\widetilde{Supp_w}(P_{\lhd t}, \mathcal{D}) = \frac{1}{|\mathcal{D}|} \times \sum_{t' \in \mathcal{D}} \widetilde{Supp_w}(P_{\lhd t'}, \mathcal{D})$$

Interestingly, this definition underlines that the whole set of mined patterns P is necessary to compute the optimal balanced support of any individual pattern. Let us illustrate the above equation with the example given by Table 1. With the weight w_{bal} where $t_1, t_2, t_3 \mapsto 1/8$ et $t_4 \mapsto 5/8$, we obtain $Supp_{w_{bal}}(A, \mathcal{D}) = Supp_{w_{bal}}(B, \mathcal{D}) = 1/8 + 1/8 = 2/8$; $Supp_{w_{bal}}(AB, \mathcal{D}) = 1/8$ and $Supp_{w_{bal}}(D, \mathcal{D}) = 5/8$. Then, we can check that the equation of Definition 3 is satisfied $\widetilde{Supp_{w_{bal}}}(P_{\lhd t_1}, \mathcal{D}) = \widetilde{Supp_{w_{bal}}}(\{A, B, AB\}, \mathcal{D}) = 2/8 + 2/8 + 1/8 = 5/8$ (similar for t_2 and t_3) and $\widetilde{Supp_{w_{bal}}}(P_{\lhd t_4}, \mathcal{D}) = \widetilde{Supp_{w_{bal}}}(\{D\}, \mathcal{D}) = 5/8$. In other words, $Supp_{w_{bal}}$ corresponds exactly to the optimal balanced support $\mathrm{BS}^*(\varphi, \mathcal{D}, P)$.

Theorem 2. *The optimal balanced support (if it exists) is the single interestingness measure that satisfies the normalized and cumulative axioms, and that leads to a balanced analysis.*

Theorem 2 achieves our main goal as stated in introduction. However, the equation of Definition 3 can admit no solution and then the optimal balanced support is not defined. For instance, it is impossible to adjust the weighted support for balancing the analysis of $\mathcal{D} = \{A, B, AB\}$ by $P = \{A, B, AB\}$. Indeed, whatever the weighted support, the transaction AB is still more analyzed than the other two since it is covered by all patterns. So, the next section proposes an algorithm to approximate the optimal balanced support by minimizing the deviation between $\widetilde{Supp_w}(P_{\lhd t}, \mathcal{D})$ and $\sum_{t' \in \mathcal{D}} \widetilde{Supp_w}(P_{\lhd t'}, \mathcal{D})/|\mathcal{D}|$.

4.3 Approximating the Balanced Support

ABSOLUTE (for an anagram of balanced support) returns the weights w such that the analysis $\mathcal{S}_{\rho,\alpha,\delta}(\mathcal{D}, P, Supp_w)$ is balanced as better as possible. Its input parameters consist in a pattern set P, a dataset $\mathcal{D}$ and a threshold ϵ. The latter is the maximal difference expected between two weight vectors stemming from consecutive iterations before terminating the algorithm. The weights outputted by ABSOLUTE enable us to define the *(approximated) balanced support* $\mathrm{BS}(\varphi, \mathcal{D}, P)$.

Algorithm 1. ABSOLUTE

Input: a dataset $\mathcal{D}$, a set of patterns P, a difference threshold ϵ
Output: a weight vector that balances the support
1: **for all** $t \in \mathcal{D}$ **do**
2: $\quad w_0[t] \leftarrow 1/|\mathcal{D}|$
3: **end for**
4: $i \leftarrow 0$
5: **repeat**
6: $\quad W \leftarrow 0$
7: $\quad$ **for all** $t \in \mathcal{D}$ **do**
8: $\quad\quad w_{i+1}[t] \leftarrow w_i[t] \times \frac{\frac{1}{|\mathcal{D}|} \times \sum_{t' \in \mathcal{D}} \widetilde{Supp_w}(P_{\lhd t'}, \mathcal{D})}{\widetilde{Supp_{w_i}}(P_{\lhd t}, \mathcal{D})}$ *// Correct the weight of t*
9: $\quad\quad W \leftarrow W + w_{i+1}[t]$
10: $\quad$ **end for**
11: $\quad diff \leftarrow 0$
12: $\quad$ **for all** $t \in \mathcal{D}$ **do**
13: $\quad\quad w_{i+1}[t] \leftarrow w_{i+1}[t]/W$ *// Normalize the weight of t*
14: $\quad\quad diff \leftarrow diff + |w_{i+1}[t] - w_i[t]|$ *// Update diff*
15: $\quad$ **end for**
16: $\quad i \leftarrow i + 1$
17: **until** $diff/|\mathcal{D}| < \epsilon$
18: **return** w_i

Note that in Algorithm 1, w_i are symbol tables where the keys are transactions. Lines 1-3 initialize all the weigths with $1/|\mathcal{D}|$. The main loop (Lines 5-17) adjusts the weigths until the sum of differences between w_{i+1} and w_i is less than ϵ. More precisely, Lines 7-10 correct the weight of each transaction. Using Definition 3, Line 8 computes the new weight $w_{i+1}[t]$ by multiplying the previous weight $w_i[t]$ by the ratio between the average coverage (i.e., a constant $1/|\mathcal{D}| \times \sum_{t' \in \mathcal{D}} \widetilde{Supp_{w_i}}(P_{\lhd t'}, \mathcal{D})$ shared by all transactions) and the coverage of t (i.e., $\widetilde{Supp_{w_i}}(P_{\lhd t}, \mathcal{D})$). For instance, if the coverage of t is below the average coverage, the ratio is above 1 and the new weight is stronger. Thus, it increases the support of all the patterns covering this transaction. This operation therefore operates a local balance for each transaction. Nevertheless, there is also a global modification since a normalization is performed on these weights at Line 13 (where W is computed Line 9). Line 14 updates *diff* (initialized Line 11) accumulating the difference between w_{i+1} and w_i for all the transactions. Finally, Line 18 returns the last weights that correspond to a balanced analysis.

5 Experimental Evaluation

This section evaluates the effectiveness of the algorithm for balancing the analysis and to compare the quality of the balanced support with respect to the usual one. All experiments reported below were conducted with a difference threshold $\epsilon = 10^{-5}$ on datasets coming from the UCI Machine Learning Repository (`www.ics.uci.edu/~mlearn/MLRepository.html`). Given a minimal support threshold *minsupp*, we select all the frequent itemsets for P. Increasing the weight of singular transactions does not cause the extraction of random noise patterns because the final patterns are selected from the collection of frequent patterns. For simplicity, we use the uniform preference vector $\rho_u : t \mapsto 1/|\mathcal{D}|$ for ρ.

Table 2. ABSOLUTE on UCI benchmarks for frequent itemsets (*minsupp* = 0.05)

Dataset	$\|P\|$	# of iter.	$D_{KL}{}^{\rho_u}_{Supp}$	$D_{KL}{}^{\rho_u}_{BS}$	Gain
abalone	2,527	17	0.180	0.021	8.72
anneal	25,766	24	0.339	0.013	**26.2**
austral	20,386	29	0.076	0.006	**12.6**
breast	2,226	41	0.145	0.006	**22.5**
cleve	11,661	35	0.172	0.012	**14.7**
cmc	2,789	23	0.091	0.002	**50.8**
crx	34,619	24	0.122	0.011	**10.9**
german	124,517	17	0.172	0.029	5.96
glass	3,146	52	0.084	0.005	**17.7**
heart	16,859	37	0.116	0.014	7.95
hepatic	511,071	13	0.568	0.040	**14.1**
horse	17,084	19	0.275	0.017	**15.9**
lymph	275,278	34	0.138	0.016	8.34
page	3,190	42	0.054	0.004	**14.8**
vehicle	187,449	24	0.636	0.020	**31.3**
wine	12,656	46	0.154	0.009	**17.8**
zoo	586,579	34	0.353	0.019	**18.2**

Efficiency of ABSOLUTE Table 2 (columns 2-3) presents the number of patterns and the number of iterations required by ABSOLUTE for balancing all the frequent patterns. Note that we do not provide running times because they are very low. Indeed, the worst case is the balancing time for all the frequent patterns on zoo, but it does not exceed 16 seconds performed on a 2.5 GHz Xeon processor with the Linux operating system and 2 GB of RAM memory (ABSOLUTE is implemented in C++). Table 2 shows that the number of iterations varies between 13 and 52. No simple relationship was found between the number of iterations and the features of datasets.

Table 2 also reports the Kullback-Leibler divergence for support and BS (columns 4-6). Let us recall that Kullback-Leibler divergence defined by $D_{KL}(P||Q) = \sum_i P(i) \times \log \frac{P(i)}{Q(i)}$ measures the difference between two probability distributions P and Q [9]. For any transaction t, we fix $P(t) = \rho_u(t)$ as reference and $Q(t) = \Pi(t, \mathcal{M}_{\rho_u})$ as model. Table 2 shows that ABSOLUTE reaches its goal since the Kullback-Leibler divergence is always significantly reduced by benefiting from the balanced support. This divergence is at least divided by 5 and it is even divided by more than 10 in 13 datasets. The average gain is 17.56 for frequent itemsets. Similar experiments conducted on collections of free and closed itemsets [10] gave respectively an average gain of 12.36 and 11.94.

Effectiveness of Balanced Support. We desire to quantify the number of non-correlated patterns (i.e., the number of extracted patterns that are spurious) with a usual/balanced support. Unfortunately, the pattern discovery process is unsupervised and the (ir)relevant patterns are unknown. We tackle this issue

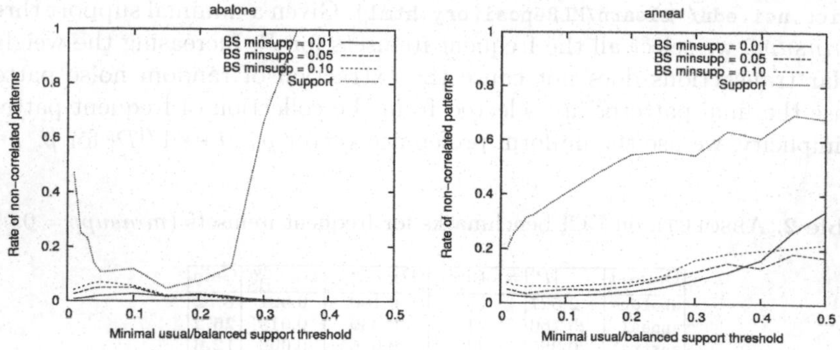

Fig. 1. Estimating the number of non-correlated patterns for *Supp* and BS

by using an experimental protocol inspired by [11]. The idea is to make the assumption that a pattern is non-correlated if this pattern is also extracted (by the same method) in a random dataset $\mathcal{D}^*$ having the same characteristics as $\mathcal{D}$ (i.e., the same dimensions and the same support for each item).

Figure 1 depicts the ratio of non-correlated patterns (averaged from 10 random datasets $\mathcal{D}^*$) for `abalone` and `anneal` for frequent itemsets with a minimal usual/balanced support varying between 0 and 0.5. This ratio is the number of non-correlated patterns divided by the total number of patterns. For the balanced support, we use three collections of frequent patterns P obtained with $minsupp = 0.01/0.05/0.10$ independently of the second threshold applied to balanced support. Given a minimal threshold (see horizontal axis), the ratio of non-correlated patterns for *Supp* is always higher than that of BS and most of times, with a significant difference. Interestingly, the change of $minsupp$ for the collection of patterns has a marginal impact on the ratio of non-correlated patterns. Recall that balanced support only differs from the traditional one by replacing the equalized axiom by the axiom of balance (see Section 4.1). So it is this axiom that enables our measure to keep out uncorrelated patterns. More generally, this experience justifies the interest of a balanced analysis and even the usefulness of the scoring analysis model for simulating an analysis.

6 Related Work

As mentioned in the introduction, many interestingness measures have been proposed for evaluating the pattern interest as alternative to the support [2,12,3]. They can be categorized into two sets [1]: user-driven measures and data-driven ones. Among the data-driven approaches, the statistical models are often based on the null hypothesis. A pattern is interesting if it covers more transactions than what was expected. Some models simply require the frequency of items forming the itemset [12], others rely on its subsets [13,14] or even, patterns already extracted [15]. These methods consider that all transactions have the same weight. However, in practice, the user tends to attach more importance to information that describes the least common facts. Thus, the most singular transactions should have an important weight in the evaluation of patterns that describe such transactions. In this sense, this paper proposes another alternative resting on the integration of the analysis method into the metric. To the best of our knowledge, this way has not yet been explored in the literature. A major and original consequence of our approach lies in the fact that each transaction contributes with a different weight in the balanced support and this weight depends on the entire extracted collection.

However, the problem of unbalance induced by a pattern set is indirectly addressed by several approaches removing the patterns that describe transactions covered by other patterns. For instance, the condensed representations [10] which remove redundant patterns, often decrease the unbalanced of the analysis. But, empirical experiments have shown that the unbalance remains important (see Section 5). In the same way, global models based on patterns [16,17,18] favor balanced analyses of the dataset. Indeed, one goal of these approaches is to describe all the data by choosing the smallest set of patterns. The overlap between the coverings of the different patterns is very reduced (ideally each transaction should be described by a unique pattern as it is the case with a decision tree). Unfortunately, relevant patterns may be removed from such models. Our approach balances the analysis of the dataset by preserving the whole set of patterns to avoid losing information.

Rather than modifying the collection of mined patterns, it would be possible to modify the initial dataset in order to satisfy user preferences. Sampling methods [19,20] are widely used in machine learning and data mining in particular to correct a problem of unbalance between classes. There is no reason that the change of the dataset with a usual sampling method leads to a balanced analysis. We think that our approach is complementary to those of sampling.

7 Conclusion

In this paper, we introduce the scoring analysis model for simulating analysis sessions of a dataset by means of a pattern set. Under this model, we define the balanced support that induces a balanced analysis of the dataset for any user-specified preference vector. We propose the algorithm ABSOLUTE to iteratively calculate transaction weights leading to the balanced support. This new interestingness measure strongly balances the analysis and in parallel, it enables us to filter-out non-correlated patterns. The originality of our work is to show that the integration of the analysis method to drive the data mining is profitable.

In future work, we are interested in examining our approach on real-world data for better understanding the semantic of the balanced support: what are the patterns which balanced support is much higher than traditional support? What are domains and datasets where the balanced support is most appropriate? Dually, we must also study the properties of the weights resulting from ABSOLUTE that could be interesting to identify the outliers. Furthermore, the prospects of using the scoring analysis model are manifold. For instance, this model could be used to balance other measures of interest like the confidence.

References

1. Freitas, A.A.: Are we really discovering "interesting" knowledge from data. Expert Update (the BCS-SGAI Magazine) 9, 41–47 (2006)
2. McGarry, K.: A survey of interestingness measures for knowledge discovery. Knowledge Eng. Review 20(1), 39–61 (2005)
3. Geng, L., Hamilton, H.J.: Interestingness measures for data mining: A survey. ACM Comput. Surv. 38(3) (2006)
4. Agrawal, R., Srikant, R.: Fast algorithms for mining association rules in large databases. In: VLDB, pp. 487–499. Morgan Kaufmann (1994)
5. Brin, S., Page, L.: The anatomy of a large-scale hypertextual web search engine. Computer Networks 30(1-7), 107–117 (1998)
6. Mihalcea, R., Tarau, P.: Textrank: Bringing order into text. In: EMNLP, pp. 404–411. ACL (2004)
7. Mannila, H., Toivonen, H.: Levelwise search and borders of theories in knowledge discovery. Data Min. Knowl. Discov. 1(3), 241–258 (1997)
8. Calders, T., Paredaens, J.: Axiomatization of frequent itemsets. Theor. Comput. Sci. 290(1), 669–693 (2003)
9. Kullback, S., Leibler, R.A.: On information and sufficiency. Annals of Mathematical Statistics 22, 49–86 (1951)
10. Calders, T., Rigotti, C., Boulicaut, J.F.: A survey on condensed representations for frequent sets. In: Boulicaut, J.-F., De Raedt, L., Mannila, H. (eds.) Constraint-Based Mining. LNCS (LNAI), vol. 3848, pp. 64–80. Springer, Heidelberg (2006)
11. Gionis, A., Mannila, H., Mielikäinen, T., Tsaparas, P.: Assessing data mining results via swap randomization. TKDD 1(3) (2007)
12. Omiecinski, E.: Alternative interest measures for mining associations in databases. IEEE Trans. Knowl. Data Eng. 15(1), 57–69 (2003)
13. Tatti, N.: Probably the best itemsets. In: Rao, B., Krishnapuram, B., Tomkins, A., Yang, Q. (eds.) KDD, pp. 293–302. ACM (2010)
14. Webb, G.I.: Self-sufficient itemsets: An approach to screening potentially interesting associations between items. TKDD 4(1) (2010)
15. Mampaey, M., Vreeken, J., Tatti, N.: Summarizing data succinctly with the most informative itemsets. TKDD 6(4), 16 (2012)
16. Bringmann, B., Zimmermann, A.: The chosen few: On identifying valuable patterns. In: ICDM, pp. 63–72. IEEE Computer Society (2007)
17. Fürnkranz, J., Knobbe, A.: Guest editorial: Global modeling using local patterns. Data Min. Knowl. Discov. 21(1), 1–8 (2010)
18. Gamberger, D., Lavrac, N.: Expert-guided subgroup discovery: Methodology and application. J. Artif. Intell. Res. (JAIR) 17, 501–527 (2002)
19. Chawla, N.V.: Data mining for imbalanced datasets: An overview. In: Data Mining and Knowledge Discovery Handbook, pp. 875–886. Springer (2010)
20. Liu, H., Motoda, H.: On issues of instance selection. Data Min. Knowl. Discov. 6(2), 115–130 (2002)

Balanced Seed Selection for Budgeted Influence Maximization in Social Networks

Shuo Han[1,2], Fuzhen Zhuang[1], Qing He[1], and Zhongzhi Shi[1]

[1] Key Laboratory of Intelligent Information Processing, Institute of Computing Technology, Chinese Academy of Sciences, Beijing 100190, China

[2] University of Chinese Academy of Sciences, Beijing 100049, China

{hans,zhuangfz,heq,shizz}@ics.ict.ac.cn

Abstract. Given a budget and a network where different nodes have different costs to be selected, the budgeted influence maximization is to select seeds on budget so that the number of final influenced nodes can be maximized. In this paper, we propose three strategies to solve this problem. First, Billboard strategy chooses the most influential nodes as the seeds. Second, Handbill strategy chooses the most cost-effective nodes as the seeds. Finally, Combination strategy chooses the "best" seeds from two "better" seed sets obtained from the former two strategies. Experiments show that Billboard strategy and Handbill strategy can obtain good solution efficiently. Combination strategy is the best algorithm or matches the best algorithm in terms of both accuracy and efficiency, and it is more balanced than the state-of-the-art algorithms.

Keywords: Budgeted Influence Maximization, Information Propagation, Social Networks.

1 Introduction

Influence maximization is a hot topic for viral marketing and has been heavily studied in the previous literature [5,3,4]. The traditional problem statement is to find a k-node set of seeds that propagate influence so that the number of resulting influenced nodes can be maximized. This definition is proposed by Kempe at al. [5] and implies an assumption that each node has an uniform cost to be chosen. Following his work, most of the existing works comply with the same assumption and focus on the k-node influence maximization problem [11,3,4]. However, this assumption does not accord with most real-world scenarios. For example, in the domain of online advertising service, different web sites have different advertising prices. If a company promotes its product by online advertisement, how to choose the web sites on budget? Spend much money on some few famous portal sites or choose less popular web sites to add the number of advertisements? Obviously, this problem is different from k-node influence maximization.

The problem statement of influence maximization can be extended to a generalized form. Given a social network where nodes may have different costs to be selected, an influence diffusion model and a budget, influence maximization is to find a seed set within the budget that maximizes the number of final influenced nodes. Nguyen et al. [10] call this problem budgeted influence maximization(BIM). In mathematics, k-node influence maximization and budgeted influence maximization have two different mathematical abstractions [8,6]. And it has been proved that the algorithms for the

V.S. Tseng et al. (Eds.): PAKDD 2014, Part I, LNAI 8443, pp. 65–77, 2014.

former will no longer produce satisfactory solutions for the latter [6]. Thus, most of the existing approaches for k-node influence maximization do not suit for the BIM problem.

The other challenge to influence maximization is the efficiency. A common method to this problem is Greedy algorithm [5], which does well in accuracy but has a bad performance in efficiency. Some researchers adopt different ideas to address this problem efficiently [3,2,4]. However, either they improve the efficiency at the cost of effectiveness, or the proposed approach can only address k-node influence maximization problem. To our knowledge, there are only two existing studies focusing on budgeted influence maximization [7,10], and both of them are based on Greedy algorithm. Although these studies concentrate on improving Greedy algorithm to reduce the runtime, they still can not overcome its intrinsic flaw of expensive computation.

In this paper, we tackle the problem of budgeted influence maximization and aim to propose an approach that has good performance in both accuracy and efficiency. We first analyze real networks empirically, including defining node roles and studying seed selection heuristics, which is the foundation of the proposed algorithm. Then, we advance three strategies to address this problem. The first one is Billboard strategy that chooses the most influential nodes as the seeds. The second one is Handbill strategy that chooses the most cost-effectvie nodes as the seeds. And the third one, called Combination strategy, uses Simulated Annealing to choose the best combination of nodes from the two nonoverlapping sets resulting from the previous two strategies. Experiments show that Billboard strategy and Handbill strategy can solve this problem efficiently and obtain good accuracies. Combination strategy is the best algorithm or matches the best algorithm in terms of both accuracy and efficiency, and is more balanced than the state-of-the-art algorithms.

2 Problem Statement and Preliminaries

2.1 Problem Statement

A social network can be modeled as a graph $G = (V, E)$, where vertices v represent individuals and edges E represent the relationship between two individuals, each vertice $v \in V$ has a cost $c(v)$ denoting the expense when it is selected. A diffusion model describes the spread of an information through the social network G. In this paper, we adopt Independent Cascade(IC) model for simulating influence propagation, which is a classical information propagation model and is widely used in the previous influence maximization research [5,7,4]. Given a network G, a diffusion model and a budget B, budgeted influence maximization is to find a seed set $S \in V$ such that subjecting to the budget constraint $\sum_{v \in S} c(v) \leq B$, the number of final influenced individuals $\sigma(S)$ can be maximized. The important notations used in this paper are declared in Table 1.

Table 1. Important Notations

Notation	Description
$G = (V, E)$	A social network with vertex set V and edge set E
B	The total budget
$N(v)$	The out-neighbors of node v
$c(v)$	The cost of node v
$ce(v)$	The cost-efficient value of node v to influence propagate
$\sigma(S)$	The final influence of seed set S

2.2 Billboards and Handbills

There are two natural ideas commonly used to select seeds for the BIM problem in the real-life scenario. In this paper, we call them Billboard strategy and Handbill strategy. Billboard strategy is to choose the most influential nodes as seeds in the network. We call these nodes billboards, because they are like billboard advertisements that are located in high traffic area and can create the most impactful visibility to people. Handbill strategy is to choose the low-cost nodes as seeds. We call them handbills, because handbills are inexpensive to produce but can be distributed to a large number of individuals.

The two strategies select seeds from two different kinds of node candidates. They are billboard nodes and handbill nodes. In this paper, we distinguish them according to the degree. Since most real networks have scale-free property [1], we could adopt Pareto principle to distinguish between billboards and handbills, that is, the top 20% of nodes with the largest degree can be defined as billboards and the remainders are handbills. Fig.1 is a degree Pareto chart for a citation network (called HEP-PH network and its description is given in Section 4.1). From the chart, the degree distribution has a long tail and fits Pareto's law. The border degree for this network is 19, that is, the nodes whose degree is larger than 19 are billboards and the others are handbills.

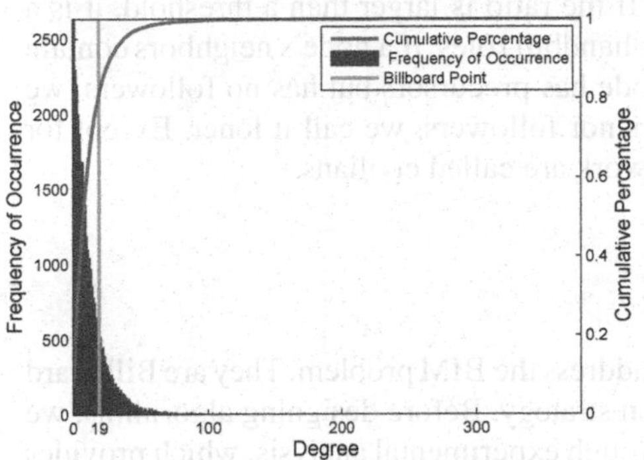

Fig. 1. Degree Distribution (HEP-PH)

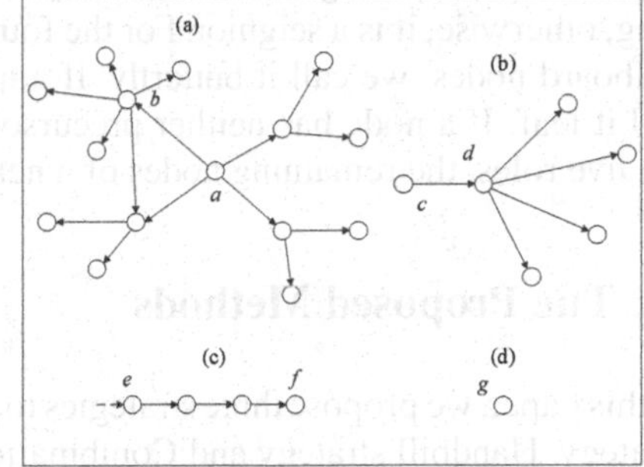

Fig. 2. Basic Structures of Social Network

2.3 Node Roles

A node role is a characterization of the part it plays in a network structure [11]. Nodes with different degrees or in different locations can be assigned different roles for describing their abilities to influence propagation. In our study, we first abstract four typical topologies from real networks, which are the basic structures to compose a network. They are respectively multi-cluster, single cluster, chain and loner, shown in Fig.2. Observing the nodes in these topologies, we define six roles to characterize different kinds of nodes, and they can cover all the nodes in the network. The definitions are follows.

- **King.** King is a global hub node in a multi-cluster topology, providing connections to many local communities. It has a wide influence by two characteristics. Firstly, it is a high-degree node that can influence many nodes directly. Secondly, many of its neighbors are also high-degree nodes. The influence can further propagate by the influential neighbors indirectly. In Fig.2, node a is a king node.

- **Seignior.** Seignior is a hub node in a local community. It is also a high-degree node. However, comparing with the king, it does not have that many influential neighbors and its influence is limited to a local region. In Fig.2, node b is a seignior.
- **Butterfly.** We call a node butterfly from the word "butterfly effect" meaning that a low-degree node can have wide influence by activating its influential neighbors. For example, in Twitter network, once a grassroot's tweet is forwarded by a celebrity, it will be popular explosively. In Fig.2, node c is a butterfly.
- **Leaf.** Leaf is the end of a communication chain. In Fig.2, node f is a leaf node. It only accepts information from spreaders, but can not further transmit it to others.
- **Loner.** Loner is an isolated node that has no connections with others, shown as node g in Fig.2.
- **Civilian.** Except for the above five types of nodes, the remaining nodes of a network can be categorized as civilians. Civilian nodes have no particular characteristics for influence propagation. In Fig.2, node e is a civilian.

Based on the above description, we distinguish the roles by quantification. We define kings and seigniors as billboard nodes, which have the top 20% largest degree in the network, and the other four roles are handbill nodes. We distinguish kings from seigniors by calculating the ratio of the sum of a node's neighbors' degree to its own degree and comparing it with a threshold. If the ratio is larger than a threshold, it is a king, otherwise, it is a seignior. For the four handbill roles, if a node's neighbors contain billboard nodes, we call it butterfly. If a node has precursors but has no followers, we call it leaf. If a node has neither precursors nor followers, we call it loner. Except for the five roles, the remaining nodes of a network are called civilians.

3 The Proposed Methods

In this paper, we propose three strategies to address the BIM problem. They are Billboard strategy, Handbill strategy and Combination strategy. Before designing algorithms, we first investigate seed selection heuristics through experimental analysis, which provides the basis for the proposed algorithms.

3.1 Seed Selection Heuristics

An commonly used method for the BIM problem is Modified Greedy algorithm [7,10]. In this section, we compute the seed set with Modified Greedy algorithm and obtain some heuristics from the seeds, which are the foundation of the proposed algorithms.

Firstly, we calculate the proportion of each role in a real network according to the role definition. The statistical result of a citation network (HEP-PH) is shown in Table 2.

Table 2. Proportion of Each Role in the HEP-PH Network

Role	King	Seignior	Butterfly	Leaf	Loner	Civilian
Proportion	0.0398	0.1652	0.3369	0.0721	0	0.3860

Then, we compute the seeds with Modified Greedy algorithm and calculate the proportion of each role in the seed set. The basic idea of Modified Greedy algorithm is to compute the solutions by using two different heuristics and return the better one as the result. Consequently, we can get two seed sets and each set is computed by one heuristic. The first heuristic is the basic greedy heuristic:

$$v_k = \underset{v \in V \setminus S_{k-1}}{\operatorname{argmax}} \; \sigma(S_{k-1} \cup v) - \sigma(S_{k-1}), \tag{1}$$

where v_k is the target seed in step k, S_{k-1} is the seed set in step $k-1$, and $\sigma(S_{k-1})$ is the influence of seed set S_{k-1}. We perform the Greedy algorithm on HEP-PH network and calculate the proportion of each role in the seed set. The result is shown in Table 3.

The second heuristic takes cost into account and choose the most cost-effective nodes as the seeds.

$$v_k = \underset{v \in V \setminus S_{k-1}}{\operatorname{argmax}} \; \frac{\sigma(S_{k-1} \cup v) - \sigma(S_{k-1})}{c(v)}, \tag{2}$$

where $c(v)$ is the cost of node v. The proportion of each role in the seeds computed by this heuristic is shown as Table 4.

Table 3. Proportion of Each Role in the Seed Set with Formula (1)

Budget	King	Seignior	Butterfly	Leaf	Loner	Civilian
1000	0.7273	0.0909	0.1818	0	0	0
3000	0.7857	0.1071	0.0714	0	0	0.0357
5000	0.8235	0.1373	0.0392	0	0	0

Table 4. Proportion of Each Role in the Seed Set with Formula (2)

Budget	King	Seignior	Butterfly	Leaf	Loner	Civilian
1000	0.1667	0.0513	0.6026	0	0	0.1795
3000	0.1216	0.0378	0.6622	0	0	0.1784
5000	0.1961	0.0784	0.6118	0	0	0.1137

From the above calculations, we can obtain two heuristic methods. The first one is role heuristic. Comparing Table 3 with Table 2, we could find that although there are very few king nodes in the network, they account for a large percentage in the seed set obtained by basic greedy heuristic. The basic greedy heuristic chooses the most influential seeds in the network, which is in accord with Billboard strategy. Then, the approach of Billboard strategy should distinguish kings from seigniors. Comparing Table 4 with Table 2, we could find that when we take cost into account, the butterfly node has more advantage than the other roles in influence propagation. Since, Handbill strategy is to choose the most cost-effective nodes as the seeds, the approach of Handbill strategy should distinguish butterfly from other roles.

The second heuristic method is distance heuristic. We measure the distance between each pair of the selected seeds, that is the length of the shortest path form a seed to another one. The average distance of HEP-PH network is larger than 2. Thus, in the proposed algorithms, we do not choose the nodes whose neighbors have already existed in the seed set, which can avoid the overlap of the seeds' influence.

In our study, we also do the same analyses on three other networks (their descriptions are declared in Section 4.1) and obtain similar conclusions. Due to the space limitation, we do not show the analysis results here.

3.2 Billboard Strategy

Billboard strategy is the idea that chooses the most influential nodes in the network as the seeds. For example, companies put advertisements on major media and products are endorsed by celebrities, both of the marketing actions are Billboard strategy.

We have defined that Billboard strategy chooses seeds from billboard nodes, whose roles are kings and seigniors. And according to the role heuristic, the proposed approach of Billboard strategy should distinguish kings from seigniors. The distinguishing characteristic of kings from seigniors is that not only king nodes themselves but also most of their neighbors are high-degree nodes. It implies that when we evaluate the influence of a node, we need to consider its neighbors' abilities to influence propagation. Then, evaluating a node's influence can be converted as calculating the expected number of influenced nodes in its two-hop area.

Theorem 1. *Given a social network $G = (V, E)$ and the IC model with a small propagation probability p, let $N(u) = \{v|v \in V, e_{u,v} \in E\}$ be the out-neighbors of node u, $outD(v)$ be the out-degree of node v. The expected number of influenced nodes in seed u's two-hop area is estimated by*

$$1 + \sum_{v \in N(u)} (1 + outD(v) \cdot p) \cdot p - \sum_{\exists v_i, v_j \in N(u), e_{v_i,v_j} \in E} p^3. \tag{3}$$

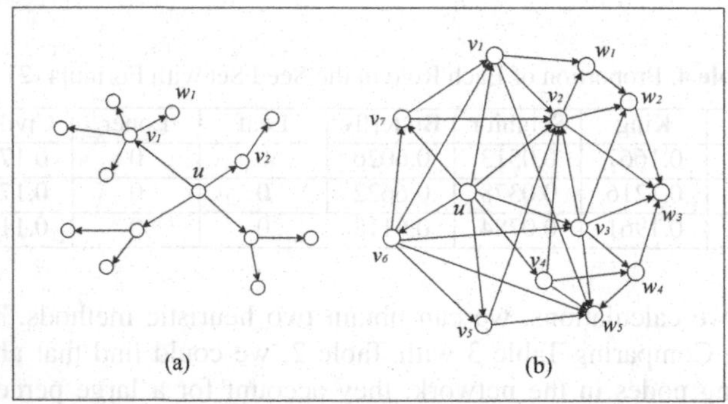

Fig. 3. The Topology of a Seed's Two-hop Area

Proof. Firstly, We consider a simple situation that there are no connections between the nodes in a seed's two-hop area, shown as Fig.3(a). Suppose node u is the seed, node $v_i \in N(u)$ is the seed's neighbor, and node $w_j \in \{w|w \in N(v_i), v_i \in N(u)\}$ is the seed's neighbor's neighbor. The probability that node v_i is influenced by seed u is p and the probability that node w_j is influenced by seed u is p^2. Then, the expected number of influenced nodes in seed u's two-hop area can be defined as:

$$1 + \sum_{v \in N(u)} (1 + outD(v) \cdot p) \cdot p, \tag{4}$$

where $outD(v)$ is the out-degree of node v. In this definition, 1 means node u itself, that is sure to be activated, and $\sum_{v \in N(u)} (1 + outD(v) \cdot p) \cdot p$ means the number of the potentially influenced nodes in its neighbors and neighbors' neighbors.

However, real-world networks usually have many tightly-knit groups that are characterized by a high density of ties [12], shown as Fig.3(b). In this situation, Formula (4) can not estimate a seed's influence accurately. For example, suppose that there are m edges connecting from node $v_i (i \neq 2)$ to node v_2. In the correct calculation, the probability p_{u,v_2} that node v_2 is influenced by node u is:

$$\begin{aligned} p_{u,v_2} &= 1-(1-p)(1-p^2)^m \\ &= 1-(1-p)[1-mp^2+C_m^2 p^4+o(p^4)] \\ &= p+mp^2-mp^3+o(p^3). \end{aligned}$$

Suppose that there are n edges connecting from v_i to node w_5. In the correct calculation, the probability p_{u,w_5} that node w_5 is influenced by node u is:

$$\begin{aligned} p_{u,w_5} &= 1-(1-p^2)^n \\ &= 1-[1-np^2+C_n^2 p^4+o(p^4)] \\ &= np^2+o(p^3). \end{aligned}$$

Since the propagation probability p is usually very small, we can ignore $o(p^3)$.

However, in formula (4), we calculate the probability p_{u,v_2} as $p+mp^2$, since the probability that node v_2 is directly influenced by seed u is p and indirectly influenced by the m neighbors is mp^2. And we calculate the probability p_{u,w_5} as np^2, since node w_5 is indirectly influenced by the n neighbors. Comparing this calculation with the above derivation, for each edge $e \in \{e_{v_i,v_j} | e_{v_i,v_j} \in E, v_i \in N(u), v_j \in N(u)\}$, Formula (4) should minus p^3. Then, we can get Formula (3).

From the above, the algorithm of Billboard strategy can be stated as follows. We only take billboards that are the top 20% nodes with the largest degree in the network as the candidates for seed selection, and calculate their abilities to influence propagation by Formula (3). We in turn select the next best candidate with the largest ability value as the seed until the budget is exhaust. Based on the distance heuristic, when choosing a seed, we would judge whether its neighbors have already existed in the seed set. If not yet, we will choose it, otherwise, we will ignore it and take the next one.

3.3 Handbill Strategy

Handbill strategy is the idea that chooses the most cost-effective nodes as the seeds. In the real world, advertisers, limiting each location's cost to increase the advertising locations, can also broaden the awareness of product.

We have defined that Handbill strategy chooses seeds from handbill nodes, whose roles are butterfly, leaf, loner and civilian. And according to the role heuristic, the proposed approach of Handbill strategy should prioritize butterfly nodes. The distinguishing characteristic of butterfly nodes from other handbill roles is that there are some high-degree nodes existing in their neighbors. They can indirectly influence more nodes by their high-degree neighbors. Then, Formula (3) can also evaluate a node's influence for Handbill strategy. Moreover, since Handbill strategy is sensitive to cost, we divide the influence of a node by its cost and evaluate a node's cost-effective value as:

$$ce(u) = (1+\sum_{v \in N(u)} (1+outD(v) \cdot p) \cdot p - \sum_{\exists v_i, v_j \in N(u), e_{v_i,v_j} \in E} p^3)/c(u), \quad (5)$$

where $ce(u)$ is the cost-effective value of node u, $c(u)$ is the cost of node u, $N(u)$ is the out-neighbors of node u and $outD(v)$ is the out-degree of node v.

From the above, the algorithm of Handbill strategy can be stated as follows. We only take handbills that are the nodes with the bottom 80% largest degree in the network as the candidates for seed selection, and calculate their cost-effective value by Formula (5). We in turn select the next best candidate with the largest cost-effective value as the seed until the budget is exhaust. Based on the distance heuristic, when choosing a seed, we would judge whether its neighbors have already existed in the seed set. If not yet, we will choose it, otherwise, we will ignore it and take the next one.

3.4 Combination Strategy

After performing the above two strategies, we could get two nonoverlapping seed sets. One is billboard set, where the seeds have great influence. The other one is handbill set, where the seeds are cost-effective. Then, we proposed Combination strategy to select the "best" seeds from the two "better" seed sets and obtain a compositive solution. The proposed approach is based on Simulated Annealing algorithm [9], which taking influence maximization as the objective, searches an approximate solution in the two set of nodes.

We first give a brief introduction of the procedure of SA algorithm as follows.

(1) The algorithm firstly initializes an initial state S_0, an initial temperature T_0, an annealing schedule $T(t)$ and an objective function $E(S_t)$.

(2) At each step t, it produces a new state S_t' from the neighbors of the current state S_t. It probabilistically decides between moving the system to state S_t' or staying in state S_t. If $E(S_t') \geq E(S_t)$, the system moves to the new state S_t' with the probability 1; otherwise, the system does this move with a probability of $exp(-(E(S_t') - E(S_t))/T_t)$.

(3) The algorithm iteratively does step (2), until the system reaches a good enough state, or until the temperature T decreases to 0.

The Combination strategy is outlined in Algorithm 1. In this strategy, we evaluate the influence of a node by Formula(3) and define the objective function $E(S)$ as the sum of the influence of each node in the seed set. We set the initial seed set as billboard seed set. The algorithm has two levels of node replacements. The first one is billboard replacement(lines 3-10). In this level, we reduce a billboard $b \in S$ and add the equal cost of handbills $h_i \notin S$ to produce a neighbor set. Then, we judge whether to accept the new set(lines 4-10). If accepted, the algorithm comes into the second level of replacement: handbill replacement(lines 12-20). In the second level, we randomly replace a handbill $h \in S$ with the equal cost of other handbills $h_i \notin S$ to produce a neighbor set, and judge whether to accept the new set(lines 14-20). We repeat this replacement for q times. The billboard replacement is the outer iteration. We in turn try to replace each billboard. When all the billboard replacements have been executed, the algorithm is finished.

4 Experiments

4.1 Datasets

We use four real-world networks in our experiments. The first one is HEP-PH citation network, where nodes are papers and an directed edge means one paper cites another. The second one is an Email network, which records one day of email communication in a school, where nodes are email addresses and edges are communication records. The third one is a P2P network, where nodes represent hosts and edges represent the

Algorithm 1. Combination Strategy

Input: Graph $G = (V, E)$, budget B, billboard seed set S_B, handbill seed set S_H, initial temperature T_0, temperature drop ΔT, objective function $E(S)$ and the number of loop q.
Output: The final seed set S.

$t \leftarrow 0, T_t \leftarrow T_0, S \leftarrow S_B, |S_B| = k$;
for $i \leftarrow 1$ to k **do** $flag \leftarrow$ false;
  create a neighbor set S'; /*replace a billboard $S_B(i)$ with the equal cost of handbills*/
  calculate the influence difference $\Delta E \leftarrow E(S') - E(S)$;
  if $\Delta E \geq 0$ **then**
    $S \leftarrow S'$, $flag \leftarrow$ true;
  else
    create a random number $\xi \in U(0, 1)$;
    if $exp(\Delta E/T_t) > \xi$ **then**
      $S \leftarrow S'$, $flag \leftarrow$ true;
  if $flag$ is true **then**
    while $q > 0$ **do** $q \leftarrow q - 1$
      create a neighbor set S'; /*replace a handbill with the equal cost of other handbills*/
      calculate the influence difference $\Delta E \leftarrow E(S') - E(S)$;
      if $\Delta E \geq 0$ **then**
        $S \leftarrow S'$;
      else
        create a random number $\xi \in U(0, 1)$;
        if $exp(\Delta E/T_t) > \xi$ **then**
          $S \leftarrow S'$;
  $T_t \leftarrow T_t - \Delta T$
return S;

connections between two hosts. And the last one is a Web network, where nodes are web pages and edges are links. Some of them are used in recent influence maximization research [3,4,10]. The basic statistics of the datasets are given in Table 5.

Table 5. Statistics of Datasets

Dataset	HEP-PH	Email	P2P	Web
Nodes	34,546	27,018	62,586	148,468
Edges	421,578	66,012	147,892	356,294

4.2 Experiment Setup

In the experiment, we adopt simulation method [5] to compute the influence propagation of the resulting seed set. Given a seed set, we repeat the simulations for R=10,000 times and take the average. We use IC model to simulate the influence propagation and set the propagation probability to be 0.1. In Combination strategy, we set T_0=1,000,000 and ΔT=1,000. All the experiments are performed on a server with Intel(R) Core i7-4770K CPU, 32G memory. All the codes are written in Java.

We evaluate the accuracy and efficiency of the three proposed strategies with three state-of-the-art algorithms. The first baseline algorithm is CELF Greedy algorithm(CELF G) [7], which is the basic greedy algorithm with CELF speed optimization. The second one is Modified Greedy algorithm(Modified G) [7], which has been introduced in Section 3.1. The third one is DegreeDiscountIC algorithm(DDIC) [3]. It is a fast algorithm for influence maximization problem, and we modify it to make it suit for BIM problem.

4.3 Experiment Results

Accuracy When Varying Budget. In this experiment, we evaluate the accuracy of each algorithm by varying the budget. In the real world, the advertising price of a web site is always relevant to its popularity. Then we define the cost of a node u as $outD(u) \cdot p + 1$, where $outD(u)$ is the out-degree of node u and p is the propagation probability.

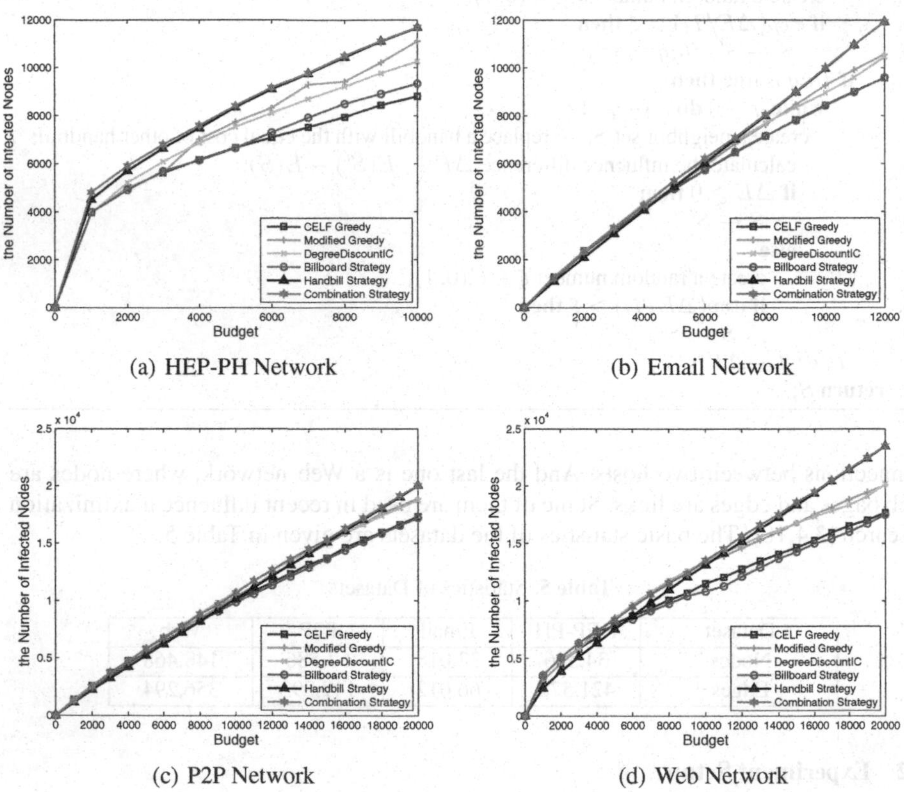

(a) HEP-PH Network (b) Email Network

(c) P2P Network (d) Web Network

Fig. 4. Influence Propagation of Different Algorithms as Budget is Varied

The experiment results are shown in Fig. 4. For HEP-PH network, Combination strategy has the best accuracy. It outperforms the second best algorithm Handbill strategy by 1.19% averagely, but this advantage reduces with the increase of budget. Combination strategy and Handbill strategy are superior to Billboard strategy obviously, which implies it is better to choose cost-effective nodes rather than influential nodes as seeds for HEP-PH network. Modified G and DDIC have similar accuracies. They are respectively inferior to Combination strategy by 10.36% and 13.3% averagely. And CELF G has the worst performance in accuracy.

For Email network, when the budget $B \in [0, 5, 000]$, Billboard strategy is better than Handbill strategy in accuracy. But when $B \in (5, 000, 12, 000]$, the latter overtakes the former. When $B \in [0, 2, 000]$, Modified G and CELF G have the best accuracies and outperform Combination strategy by 2.7%. When $B \in (2, 000, 4, 000]$, DDIC has the best accuracy and outperforms Combination strategy by 1.45%. When $B \in (4, 000, 12, 000]$, Combination strategy becomes the best algorithm in accuracy.

For P2P network, when the budget $B \in [0, 8, 000]$, Billboard strategy has a better performance than Handbill strategy in accuracy. But when $B \in (8, 000, 20, 000]$, the latter overtakes the former. When $B \in [0, 4, 000]$, Modified G and CELF G have the best accuracies and outperform Combination strategy by 2.6% averagely. When $B \in (4, 000, 10, 000]$, DDIC overtakes Greedy algorithm to be the best algorithm and slightly outperforms Combination strategy by 1.15%. When $B \in (10, 000, 20, 000]$, Combination strategy becomes the best algorithm in accuracy. It outperforms DDIC by 3.03% and outperforms Modified G by 3.71% averagely.

For Web network, when the budget $B \in [0, 5, 000]$, Billboard strategy is better than Handbill strategy in accuracy. But when $B \in (5, 000, 20, 000]$, the latter overtakes the former. When $B \in [0, 4, 000]$, Combination strategy has the best accuracy and outperforms the second best algorithm Modified G by 2.18%. When $B \in (4, 000, 9, 000]$, DDIC overtakes Combination strategy and becomes the best algorithm in accuracy. It slightly outperforms Combination strategy by 1.72%. When $B \in (9, 000, 20, 000]$, Combination strategy once again becomes the best algorithm. It outperforms Modified G by 10.85% and outperforms DDIC by 10.89% averagely.

Accuracy with Different Cost Definitions. This experiment is to evaluate the accuracy of each algorithm with different cost definitions. We respectively define the cost of a node u as $outD(u) + 1$ and $outD(u) \cdot c + 1$, where $outD(u)$ is the out-degree of node u, p is the propagation probability and c is a random number in $(0, 1]$. Due to the space limitation, we only show the results of HEP-PH network in Fig.5.

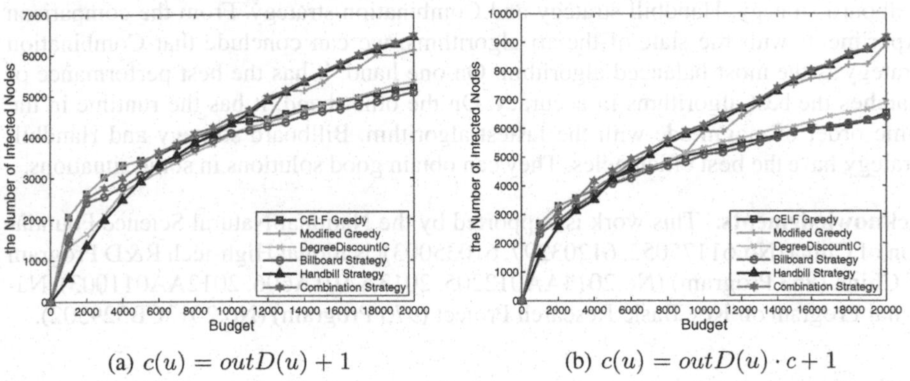

(a) $c(u) = outD(u) + 1$ (b) $c(u) = outD(u) \cdot c + 1$

Fig. 5. Influence Propagation of Different Algorithms as Cost is Varied

With these two cost definitions, when the budget is small, Billboard strategy always has a better accuracy than Handbill strategy. But with the increase of budget, Handbill strategy overtakes the former and even outperforms most other algorithms. Combination strategy has an excellent and stable performance in accuracy. When the budget is small, it may fall behind with the best algorithm slightly. However, as the budget increases, it narrows the gap rapidly and becomes the best algorithm, which outperforms Handbill strategy slightly and outperforms other algorithms obviously. Modified G has a good accuracy when the budget is small. But it is inferior to Combination strategy and Handbill strategy when the budget is large. We can find that the curve of Modified G have some drops, which implies the Greedy algorithm falls into local optimums. CELF G performs well when the budget is small, but it can not maintain the advantage when the budget is increasing. Comparing with other algorithms, DDIC has a low accuracy.

Efficiency We compare the runtimes of the six algorithms, when the budget is 5,000 and the propagation probability is 0.1 on the four networks.

From Table 6, we observe that Billboard strategy, Handbill strategy, Combination algorithm and DDIC have the runtime in the same order of magnitude. And Combination algorithm is slightly slower than the other three. By contrast, CELF G has a much longer runtime and it is slower than the fastest algorithm by three orders of magnitude. Modified G has the longest runtime, it is even slower than CELF G by 2 times at least.

Table 6. Runtime(seconds) of Each Algorithm on Different Networks

Network	Billboard	Handbill	Combination	CELF G	Modified G	DDIC
HEP-PH	5.89	8.23	23.21	4.95×10^3	1.23×10^4	6.85
Email	4.91	6.96	19.78	2.27×10^3	7.63×10^3	5.08
P2P	6.32	9.61	27.76	5.05×10^3	1.26×10^4	7.53
Web	11.03	13.49	40.25	7.36×10^3	2.09×10^4	13.42

5 Conclusions

Unlike most of the existing works on k-node influence maximization, this paper focuses on budgeted influence maximization. We address the BIM problem by three strategies: Billboard strategy, Handbill strategy and Combination strategy. From the comparison experiments with the state-of-the-art algorithms, we can conclude that Combination strategy is the most balanced algorithm. On one hand, it has the best performance or matches the best algorithms in accuracy. On the other hand, it has the runtime in the same order of magnitude with the fastest algorithm. Billboard strategy and Handbill strategy have the best efficiencies. They can obtain good solutions in some situations.

Acknowledgments. This work is supported by the National Natural Science Foundation of China (No.61175052, 61203297, 61035003), National High-tech R&D Program of China (863 Program) (No.2014AA012205, 2013AA01A606, 2012AA011003), National Program on Key Basic Research Project (973 Program) (No.2013CB329502).

References

1. Bollobás, B., Riordan, O., Spencer, J., Tusnády, G., et al.: The degree sequence of a scale-free random graph process. Random Structures & Algorithms 18(3), 279–290 (2001)
2. Chen, W., Wang, C., Wang, Y.: Scalable influence maximization for prevalent viral marketing in large-scale social networks. In: Proceedings of the 16th ACM SIGKDD International Conference on Knowledge Discovery and Data Mining, pp. 1029–1038. ACM (2010)

3. Chen, W., Wang, Y., Yang, S.: Efficient influence maximization in social networks. In: Proceedings of the 15th ACM SIGKDD International Conference on Knowledge Discovery and Data Mining, pp. 199–208. ACM (2009)
4. Jiang, Q., Song, G., Cong, G., Wang, Y., Si, W., Xie, K.: Simulated annealing based influence maximization in social networks. In: AAAI (2011)
5. Kempe, D., Kleinberg, J., Tardos, É.: Maximizing the spread of influence through a social network. In: Proceedings of the Ninth ACM SIGKDD International Conference on Knowledge Discovery and Data Mining, pp. 137–146. ACM (2003)
6. Khuller, S., Moss, A., Naor, J.S.: The budgeted maximum coverage problem. Information Processing Letters 70(1), 39–45 (1999)
7. Leskovec, J., Krause, A., Guestrin, C., Faloutsos, C., VanBriesen, J., Glance, N.: Cost-effective outbreak detection in networks. In: Proceedings of the 13th ACM SIGKDD International Conference on Knowledge Discovery and Data Mining, pp. 420–429. ACM (2007)
8. Megiddo, N., Zemel, E., Hakimi, S.L.: The maximum coverage location problem. SIAM Journal on Algebraic Discrete Methods 4(2), 253–261 (1983)
9. Metropolis, N., Rosenbluth, A.W., Rosenbluth, M.N., Teller, A.H., Teller, E.: Equation of state calculations by fast computing machines. The Journal of Chemical Physics 21, 1087 (1953)
10. Nguyen, H., Zheng, R.: On budgeted influence maximization in social networks. IEEE Journal on Selected Areas in Communications 31(6), 1084–1094 (2013)
11. Scripps, J., Tan, P.N., Esfahanian, A.H.: Node roles and community structure in networks. In: Proceedings of the 9th WebKDD and 1st SNA-KDD 2007 Workshop on Web Mining and Social Network Analysis, pp. 26–35. ACM (2007)
12. Watts, D.J., Strogatz, S.H.: Collective dynamics of small-world networks. Nature 393(6684), 440–442 (1998)

Persistent Community Detection in Dynamic Social Networks

Siyuan Liu[1], Shuhui Wang[2], and Ramayya Krishnan[1]

[1] Heinz College, Carnegie Mellon University, Pittsburgh, PA, 15213, USA
{siyuan,rk2x}@cmu.edu

[2] Key Lab of Intell. Info. Process., Inst. of Comput. Tech., CAS, Beijing, 100190, China
wangshuhui@ict.ac.cn

Abstract. While community detection is an active area of research in social network analysis, little effort has been devoted to community detection using time-evolving social network data. We propose an algorithm, Persistent Community Detection (PCD), to identify those communities that exhibit persistent behavior over time, for usage in such settings. Our motivation is to distinguish between steady-state network activity, and impermanent behavior such as cascades caused by a noteworthy event. The results of extensive empirical experiments on real-life big social networks data show that our algorithm performs much better than a set of baseline methods, including two alternative models and the state-of-the-art.

Keywords: Community detection, persistent behavior, social networks.

1 Introduction

Much effort has been devoted to the development of community detection algorithms [7,2,3,5], which can be used to identify clusters of nodes in social network data whose connections exhibit similar tendencies. Such clusterings may be intended for use as a predictive feature, or as a crude summary of network structure. In empirical studies, these algorithms often produce clusters that agree with general intuition about the network that is being studied, corresponding closely with known affiliations or genres held by the network nodes.

Here we develop a community detection model for *time-evolving* network data, and use this model to analyze a real-world call network. This data set is challenging to analyze, in part because of its large size (3.6 million users), and more importantly, because its structure also appears to change over time. To illustrate that the network is time-varying, Figure 1 shows a Q-Q plot of the degree distribution for calls made between 7-8AM and 8-9AM on a single day. The plot suggests that 7-8AM exhibits higher call density, and also a more heavy-tailed distribution (i.e., the largest values of the degree distribution grow more extreme). However, it is unclear to what extent the change in layout is due to time-varying structure (as opposed to being an artifact of the visualization process), and more importantly, how to quantify our observations.

In light of these concerns, we take a statistical model-based approach. Statistical modeling of dynamic network structure is challenging, and still nascent as a field of research. Sociological theories for dynamic networks are not as well developed as for

V.S. Tseng et al. (Eds.): PAKDD 2014, Part I, LNAI 8443, pp. 78–89, 2014.

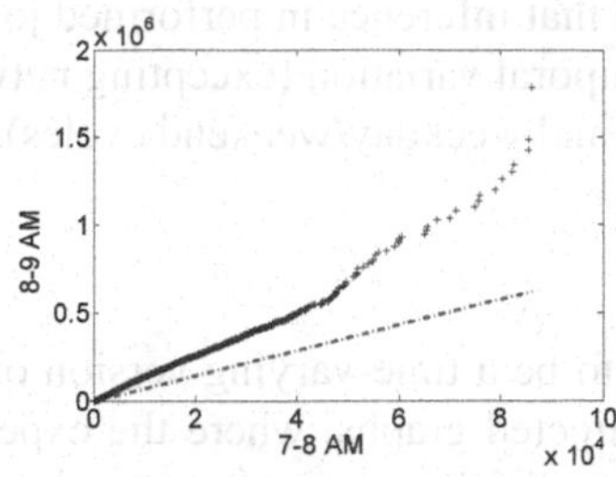

Fig. 1. Dynamics in mobile phone social networks on March 3rd, 2008. Q-Q plot of call volumes from 7-8 AM and 8-9 AM.

static networks, and hence less guidance is available for modeling. Perhaps due to this, existing time-varying models are typically designed to model many potential types of behavior [14]. Here we take a different approach. Our model is designed to detect only a single type of behavior; specifically, we find communities that exhibit *persistent* levels of communication over time. Our motivation is to distinguish between steady-state activity and impermanent behavior, such as cascades caused by a noteworthy event. We feel that persistence is the simplest type of dynamic behavior, making it a logical next step from the static setting.

Our contributions are summarized as follows.

1. We formally define a new network structure, *persistent community*, which exhibits persistent behavior/ structure over time.
2. We propose a novel algorithm to detect persistent community by a time and degree corrected blockmodel. We also provide inference of the model.
3. We conduct extensive empirical experiments on real-life big social networks data. Interesting findings and discussions are provided.

The rest of the paper is organized as follows. Section 2 surveys the previous work on community detection in dynamic networks. Section 3 proposes our algorithm. The empirical experiment results are reported in Section 4. At last, we conclude our work and give the future research directions in Section 5.

2 Related Work

The literature on static community detection is very large. Various extensions to the basic community model have been proposed, such as overlapping or mixed community membership [1,13], degree-corrections which allow for heterogeneity within communities [6,16], and community detection from trajectories [12]. In particular, without degree-correction, maximum likelihood methods often group the nodes according to their degree (i.e. their number of neighbors) [6]. As such behavior is typically undesirable, we will also include degree-corrections in our model.

Recent attention has been paid to community detection in dynamic social networks. Existing approaches, such as [7,8,4], generally detect communities separately for each time slot, and then determine correspondences by various methods [15,18,17]. However, such approaches often result in community structures with high temporal variation

[10,9]. Our approach differs in that inference in performed jointly across time, so as to find communities with low temporal variation (excepting network-wide fluctuations in call volume, such as night/day and weekday/weekend cycles).

3 Methodology

Our model can be considered to be a time-varying version of degree-corrected blockmodel in [6], specialized to directed graphs, where the expected call volumes within each community are assumed to all follow a single network-wide trajectory over time.

3.1 Time and Degree-Corrected Blockmodel

General Model. We assume a network of N nodes over T time periods, where nodes are free to leave and re-enter the network. Let $\mathcal{I}_t \subset \{1,\ldots,N\}, t = 1,\ldots,T$ denote the subset of nodes which are present in the network at time t. Let K denote the number of communities, which determines the model order. Let $A^{(t)} \in \mathbb{N}^{N\times N}$ denote a matrix of call counts at time $t = 1,\ldots,T$; i.e., for $i \neq j$, $A_{ij}^{(t)}$ denotes the number of calls from node i to node j at time t. Our model is that the elements of $A^{(1)},\ldots,A^{(T)}$ are independent Poisson random variables, whose parameters are jointly parameterized (with explanation of all parameters to follow):

$$A_{ij}^{(t)} \sim \text{Pois}\left(\lambda_{ij}^{(t)}\right)$$

$$\lambda_{ij}^{(t)} = \begin{cases} \theta_i^{(t)} \phi_j^{(t)} \mu^{(t)} \omega_{g_i g_j}^{(t)} & \text{if } i \in \mathcal{I}_t, j \in \mathcal{I}_t \\ 0 & \text{otherwise} \end{cases}$$

We see that the expected number of calls λ_{ij} for each dyad (i, j) is a function of parameters g, ω, θ, ϕ, and μ. We now describe each parameter, its allowable values, and its function:

1. The vector $g \in \{1,\ldots,K\}^N$ assigns each node to a latent community in $1,\ldots,K$.
2. The matrix $\omega^{(t)} \in \mathbb{R}^{K\times K}$ gives the expected total call volume between each community at time $t = 1,\ldots,T$. In other words, $\omega_{ab}^{(t)}$ is the expected call volume from community a to community b at time t. To enforce persistence, ω is restricted to satisfy the following constraint:

   $$\omega_{aa}^{(t)} = \omega_{aa}^{(t')} \qquad t,t' \in 1,\ldots,T,\ a \in 1,\ldots,K. \tag{1}$$

   As a result, intra-community call volumes are modeled as being constant over time (up to the network-wide effect of μ, which we discuss shortly), while inter-community call volumes may follow arbitrary trajectories over time.
3. The vector $\theta^{(t)} \in [0,1]^N$ controls the out-degree for each node at time $t = 1,\ldots,T$. Nodes whose element in $\theta^{(t)}$ is high will have higher expected outgoing call volumes than whose with low values in $\theta^{(t)}$. This allows for heterogeneity within communities. For identifiability, θ is restricted to satisfy the following constraint:

   $$\sum_{i\in\mathcal{I}_t, g_i=a} \theta_i^{(t)} = 1 \qquad t = 1,\ldots,T,\ a = 1,\ldots,K$$

The effect of this constraint is that $\omega_{ab}^{(t)}$ determines the total number of calls from community a to community b, while θ determines the proportion of calls emanating from each node in community a.

4. The vector $\phi^{(t)} \in [0,1]^N$ controls the in-degree for each node at time $t = 1, \ldots, T$, but is otherwise analogous to θ. A similar constraint is also enforced:

$$\sum_{i \in \mathcal{I}_t, g_i = a} \phi_i^{(t)} = 1 \qquad t = 1, \ldots, T, \; a = 1, \ldots, K.$$

5. The scalar $\mu^{(t)} \in [0,1]$ for $t = 1, \ldots, T$, modifies the total network call volume as a function of t. This allows for network-wide trends to be modelled, such as day/night or weekday/weekend cycles. For identifiability, μ is restricted to satisfy the following constraint:

$$\sum_{t=1}^{T} \mu^{(t)} = 1.$$

The effect of this constraint is that $T\omega_{aa}^{(1)}$ determines the total number of calls within community a, while μ determines the proportion of those calls occurring within each time slot.

Discussion. While self-calls are disallowed in a phone network, we note that our model assigns nonzero probability to positive values of $A_{ii}^{(t)}$. This is a simplification which decouples estimation of $\theta^{(1:T)}, \phi^{(1:T)}, \mu^{(1:T)}$ and $\omega^{(1:T)}$, leading to analytically tractable expressions for the parameter estimates. Self-calls predicted under the model should be disregarded as a modeling artifact. As the number of predicted self-calls will be a vanishing fraction of the total call volume, the effect will be negligible.

In the data, there exist pairs of individuals with extremely high call volumes, exceeding an average of 10 calls to each other per day. Such pairs are very sparse in the data ($< 1\%$ of all dyads), and do not seem to conform to the idea of community-based calling behavior. It is unlikely that the Poisson-based community model will explain these dyads. As such, we have opted to treat these dyads as outliers, and remove them before estimating the model parameters. Our interpretation is that the data is best described by the community based model, plus a small set of dyads whose high call volumes distinguish them from the overall network.

We note that θ and ϕ involve large numbers of parameters, as they are allowed to vary over time. A simpler model, in which θ and ϕ are constant over time, was considered. However, formulas for parameter inference become significantly more complicated in this case, unless $\mathcal{I}_t$ is also constant over time. If $\mathcal{I}_t$ is constant over time, so that nodes cannot enter and leave the network, then the equations to be presented in Section 3.2 may be used with only slight modification if θ and ϕ are held constant over time.

3.2 Inference

We will estimate g and $\{\theta^{(t)}, \phi^{(t)}, \mu^{(t)}, \omega^{(t)}\}_{t=1}^{T}$ by maximum likelihood. We show here that given g, the maxmizing values of the remaining parameters can be found analytically, so that the maximizing the likelihood consists of a search over all community assignments in $\{1,\dots,K\}^N$. While this exact maximization is computationally intractable, heuristic methods seem to give good results in practice, and we use a greedy search method described in [6] with multiple restarts.

We now derive formula for the remaining parameter estimates given g. The joint distribution of $A^{(1:T)} \equiv (A^{(1)},\dots,A^{(T)})$ (or equivalently the likelihood) is given by the product of Poisson distributions

$$L(\theta,\phi,\omega,\mu,g;A^{(1:T)}) = \prod_{t=1}^{T}\prod_{i,j\in\mathcal{I}_t} \frac{\left(\theta_i^{(t)}\phi_j^{(t)}\mu^{(t)}\omega_{g_ig_j}^{(t)}\right)^{A_{ij}^{(t)}}}{A_{ij}!} \times \exp\left(-\theta_i^{(t)}\phi_j^{(t)}\mu^{(t)}\omega_{g_ig_j}^{(t)}\right).$$

This expression can be simplified using the following intermediate terms. Given g and $A^{(1:T)}$, for all i,j,t let:

$$d_{i\cdot}^{(t)} = \sum_{j\in\mathcal{I}_t} A_{ij}^{(t)} \qquad d_{\cdot i}^{(t)} = \sum_{j\in\mathcal{I}_t} A_{ji}^{(t)},$$

$$m_{ab}^{(t)} = \sum_{i,j\in\mathcal{I}_t} A_{ij}^{(t)} 1\{g_i = a, g_j = b\},$$

$$m_{aa}^{(\cdot)} = \sum_{t=1}^{T} m_{aa}^{(t)} \qquad m_{\cdot\cdot}^{(t)} = \sum_{a=1}^{K} m_{aa}^{(t)}.$$

In words, $d_{i\cdot}^{(t)}$ and $d_{\cdot i}^{(t)}$ are the out-degree and in-degree of node i at time t; $m_{ab}^{(t)}$ is the call volume between communities a and b at time t; $m_{aa}^{(\cdot)}$ is the total call volume within community a over all time, and $m_{\cdot\cdot}^{(t)}$ is the total intra-community call volume (versus inter-community call volume) at time t. Using these terms, the likelihood L can be written as

$$L(\theta,\phi,\omega,\mu,g;A^{(1:T)}) = \frac{1}{\prod_{t,i,j} A_{ij}!}\prod_{t=1}^{T}\prod_{i=1}^{N}\left(\left[\theta_i^{(t)}\right]^{d_{i\cdot}^{(t)}}\left[\phi_i^{(t)}\right]^{d_{\cdot i}^{(t)}}\right) \times \prod_{t=1}^{T}\prod_{a,b=1}^{K}\left(\left[\mu^{(t)}\omega_{ab}^{(t)}\right]^{m_{ab}^{(t)}}\exp\left(-\mu^{(t)}\omega_{ab}^{(t)}\right)\right),$$

where we have used the constraints that $\sum_{i\in\mathcal{I}_t,g_i=a}\theta_i^{(t)}$ and $\sum_{i\in\mathcal{I}_t,g_i=a}\phi_i^{(t)} = 1$. The function $\ell \equiv \log L$ is given by

$$\ell(\theta,\phi,\mu,g) = \sum_{t=1}^{T}\sum_{i=1}^{N}\left(d_{i\cdot}^{(t)}\log\theta_i^{(t)} + d_{\cdot i}^{(t)}\log\phi_i^{(t)}\right) + \sum_{t=1}^{T}\sum_{a,b=1}^{K}\left(m_{ab}^{(t)}\log\left[\mu^{(t)}\omega_{ab}^{(t)}\right] - \mu^{(t)}\omega_{ab}^{(t)}\right).$$

We observe that ℓ can be grouped into terms which can be separately maximized,

$$\begin{aligned}\ell(\theta,\phi,\mu,g) &= \sum_{t=1}^{T}\sum_{i=1}^{N}\left(d_{i\cdot}^{(t)}\log\theta_i^{(t)} + d_{\cdot i}^{(t)}\log\phi_i^{(t)}\right)\\ &+\sum_{t=1}^{T}\sum_{a=1}^{K}\left(m_{aa}^{(t)}\log\omega_{aa}^{(t)} - \mu^{(t)}\omega_{aa}^{(t)}\right)\\ &+\sum_{t=1}^{T}\sum_{a=1}^{K} m_{aa}^{(t)}\log\mu^{(t)}\\ &+\sum_{t=1}^{K}\sum_{a\neq b}\left(m_{ab}^{(t)}\log\left[\mu^{(t)}\omega_{ab}^{(t)}\right] - \mu^{(t)}\omega_{ab}^{(t)}\right).\end{aligned}$$

For fixed g, the maximizing value of the other parameters $\{\theta^{(t)},\phi^{(t)},\mu^{(t)},\omega^{(t)}\}_{t=1}^{T}$ can be analytically shown to satisfy for $t = 1,\dots,T$:

$$\begin{aligned}\mu^{(t)} &= \frac{m_{\cdot\cdot}^{(t)}}{\sum_{\tau=1}^{T} m_{\cdot\cdot}^{(t)}}\\ \omega_{ab}^{(t)} &= \frac{m_{ab}^{(t)}}{\mu^{(t)}} && a\neq b\\ \omega_{aa}^{(t)} &= m_{aa}^{(\cdot)} && a = 1,\dots,K\\ \theta_i^{(t)} &= \frac{d_{i\cdot}^{(t)}}{\sum_{i\in\mathcal{I}_t} d_{i\cdot}1\{g_i = a\}} && i = 1,\dots,N\\ \phi_i^{(t)} &= \frac{d_{\cdot i}^{(t)}}{\sum_{i\in\mathcal{I}_t} d_{\cdot i}1\{g_i = a\}} && i = 1,\dots,N\end{aligned}$$

Subsitution of the optimal values yields a function of g:

$$\begin{aligned}\ell(g) &= \sum_{t=1}^{T}\sum_{a=1}^{K}\left[H\left(\{d_{i\cdot}^{(t)}\}_{i\in\mathcal{I}_t,g_i=a}\right) + H\left(\{d_{\cdot i}^{(t)}\}_{i\in\mathcal{I}_t,g_i=a}\right)\right]\\ &+\sum_{t=1}^{T}\sum_{a\neq b} h(m_{ab}^{(t)}) + \sum_{a=1}^{K} h(m_{aa}^{(\cdot)}) + H\left(\{m_{\cdot\cdot}^{(t)}\}_{t=1}^{T}\right),\end{aligned} \tag{2}$$

where the mapping H is given by $H(\{x_i\}_{i=1}^{k}) = \sum_{i=1}^{k} x_i \log\frac{x_i}{\sum_{j=1}^{k} x_j}$ for $x\in\mathbb{R}_+^k$, and the mapping h is given by $h(x) = x\log x - x$ for $x\in\mathbb{R}_+$.

We estimate the model parameters by optimizing $\ell(g)$ over all group assignments $g\in\{1,\dots,K\}^N$. While it is intractable to find a global maximum, a local maximum can be found using the method described in [6]. After multiple restarts, the highest scoring local optima was chosen for the parameter estimate.

4 Empirical Experiment Results

4.1 Description of Data and Fitting Procedure

The call records correspond to all mobile calls involving a particular service provider, with origin and destination within a particular city region, in the year 2008. The city area is roughly 8700 km^2, is covered by 5120 base stations, and serves 3.6 million mobile phone users. The data set is roughly 1 TB in size (more than 10 billion phone call records). The data set was prepared by the service provider, for the purpose of data mining to improve service and marketing capabilities. It contains call metadata (such as phone number, date of call, instant location of call), linked with customer profile information.

Using the call metadata, a set of directed adjacency matrices $\left(A^{(1)}, \ldots, A^{(365)}\right)$ was created, with nodes corresponding to customers (i.e., phone numbers), and edge weights corresponding to the number of calls between the sender and receiver on each day of the year. Community labels g were chosen to maximize ℓ as given by Eq.(2), using the algorithm described in Section 3.2, with the number of groups K chosen to be 800.

4.2 Out of Sample Prediction

To test the model, out of sample prediction was conducted, by randomly withholding 5% of the dyads from the fitting procedure. After fitting, the probability of connection according to the model (i.e., $P\left(A_{ij}^{(t)} > 0\right)$) was used to predict which of the withheld dyads had non-zero call volume. The model was highly predictive of the withheld dyads, with precision is 0.74±0.05 and recall is 0.53±0.05.

4.3 Description of Model Fit

To describe the fitted model, we give the following statistics. From the CDF giving the fraction of customers belong to the k largest communities, for $k = 1, \ldots, 800$, it shows that the majority of customers are concentrated in the 10 largest communities by the algorithm. Figure 2 (a) shows the fitted intra-community call densities, i.e., the call volume divided by the number of dyads in each community. The figure shows that the network is quite dense, with 30% of users in the largest community, in which every pair of members experienced an average of 0.008 phone calls over the course of a year. Figure 2 (b) plots the call densities (i.e., normalizing by the square of the community size). Figure 2 (c) shows the fitted inter-community call volumes $\omega_{ab}^{(t)}$, for $a \neq b$ and $t = 1, \ldots, T$, as a quantile-plot. Based on Figure 2 (b) and Figure 2 (c), we see that the inter- and intra- community parameters follow different different distributions. Figure 3 (a) and 3 (b) shows the fitted degree corrections θ and ϕ as quantile plots. Figure 4 shows the fitted time-corrections $\mu^{(1)}, \ldots, \mu^{(T)}$. We note that larger values of μ occur on holidays and weekends.

To further understand the inferred communities, we compared the community labels g with groupings produced by various customer covariates included in the customer profile information:

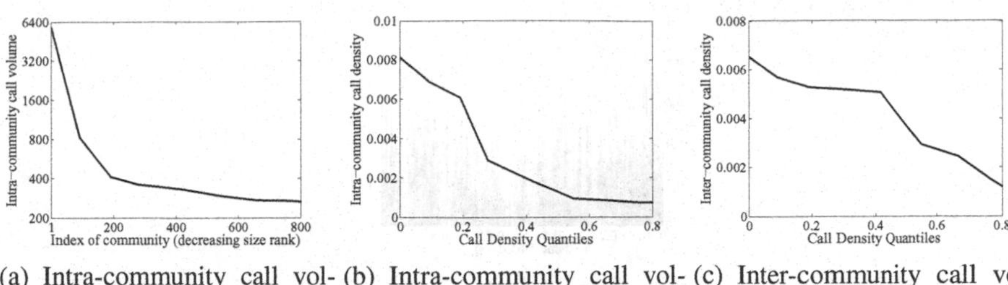

(a) Intra-community call volume distribution (b) Intra-community call volume density distribution (c) Inter-community call volume density distribution

Fig. 2. (a) Call volumes for within-community communication, (b) Call densities for within-community communication, (c) Call densities for between-community communication

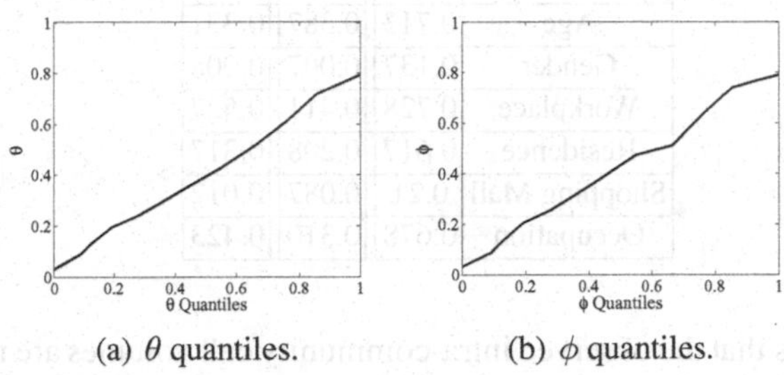

(a) θ quantiles. (b) ϕ quantiles.

Fig. 3. Fitted values of in-degree and out-degree corrections θ and ϕ, in ranked order

- Age: the age of the customer, grouped by increments of six months.
- Gender: the gender of the customers.
- Workplace: the geographic region containing the registered workplace address of the customer. In our record, we have almost ten thousand workplace address and half a million customers provide this information.
- Residence: the geographic region containing the registered home address of the customer. In our record, we have almost ten thousand workplace address and half a million customers provide this information.
- Shopping mall: the shopping mall location. In our record, we have fifty shopping mall locations. On the other hand, based on the location information of each call, we are able to localize each customer, as reported in [11].
- Occupation: The occupational category reported by the customer. In our record, one hundred thousand customers have this information.

Table 1 (second column) reports the Jacaard similarity between the inferred community labels g and the customer covariates [1]. We find that the model has high similarity with Age, Workplace, and Residence, and Occupation, but not with Gender and Shopping Mall.

Figure 5 (a) shows the correlation coefficient for the intra-community and inter-community call volumes, for time lags varying from 10 to 40 weeks.

[1] $J(A, B) = \frac{|A \cap B|}{|A \cup B|}$, where A and B are two label sets.

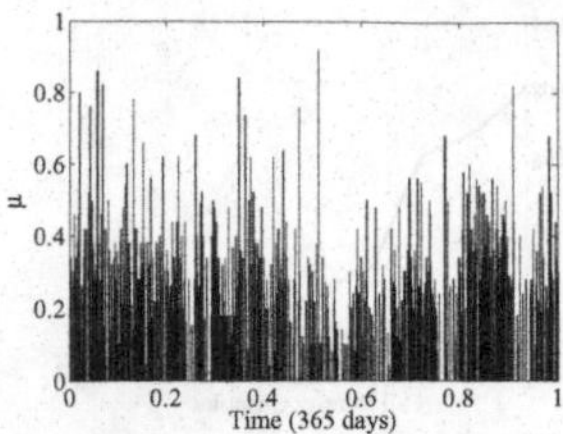

Fig. 4. Time-corrections $\mu^{(t)}$ as a function of time

Table 1. Community Detection Evaluation on PCD

Covariate	PCD	A1	A2
Age	0.713	0.387	0.331
Gender	0.137	0.007	0.008
Workplace	0.728	0.411	0.527
Residence	0.617	0.208	0.317
Shopping Mall	0.21	0.087	0.012
Occupation	0.678	0.310	0.423

The figure shows that the observed intra-community call volumes are more persistent over time compared to the inter-community call volumes, suggesting that the method is successful in finding communities with persistent intra-community call volumes.

4.4 Work v.s. Leisure Groupings

To differentiate work and leisure interactions, the data was separated into weekdays and weekends, and then g was fit separately by Eq.(2) on the two scenarios. As shown in Table 2, we found that the weekday groupings corresponded more closely to (place of employment, or some other covariate), which the weekend groups were more closely aligned with family relationships (which are recorded in the data set).

We also notice that the usage of persistence constraints had a larger effect in the weekend groups; this suggests that the social/weekend groups are less visible in the data (i.e., a "weaker signal"), causing the model regularization to have greater effect.

Table 2. Weekday grouping v.s. weekend grouping (Jacaard Similarity)

Covariate	Weekday	Weekend
Age	0.731	0.702
Gender	0.152	0.120
Workplace	0.801	0.568
Residence	0.578	0.817
Shopping Mall	0.124	0.453
Occupation	0.831	0.542

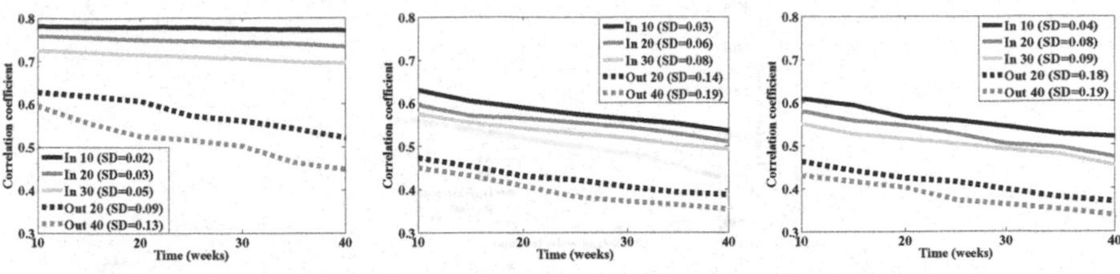

(a) Persistence evaluation of our PCD algorithm. (b) Persistence evaluation of the alternative model 1. (c) Persistence evaluation of the alternative model 2.

Fig. 5. Persistence evaluation. Within-group call volumes are persistent over time, up to network-wide trends. In contrast, the call volumes in the off-diagonal plots are much lower, and do not follow network-wide trends, suggesting that communication between groups was more sporadic. The persistence of the communities detected by alternative model 1 and 2 are not good as the result of PCD.

4.5 Comparison with Other Community Detection Algorithms

The results of our method were compared against several other algorithms. Specifically:

- A1: The persistence constraint $\omega_{aa}^{(t)} = \omega_{aa}^{(t')}$ given by Eq. 1 is removed when fitting the model. As a result, the intra-community expected call volumes are no longer constrained to follow any particular trajectory over time.
- A2: A static degree-corrected blockmodel, as described in [6], is fit to the static matrix $A = \sum_{t=1}^{T} A^{(t)}$.
- DSBM: A bayesian approach [17] for detecting communities in dynamic social networks.

Under the algorithm A1, as shown in Table 1 (third column), the groupings differed significantly compared to those found by our proposed method, and did not correspond as well to observed covariates, as described in Table 1 (second column). Figure 5 (b) shows the average correlation coefficients for the intra- and inter-community call volumes. We observe that the coefficients are lower compared to our proposed algorithm, suggesting that the call volumes are less persistent over time.

Under the algorithm A2, similar findings resulted, as shown in Table 1 (fourth column) and Figure 5 (c). It is interesting that the similarity results of A1 and A2 are different and can be interpreted by the methods we use. For A1, we release the persistence constraint for intra-community connection, while for A2, we use a static model. For Age and Gender, A1 and A2 give similar results, but for Workplace, Residence and Occupation, A2 can give much better similarity result than A1, while for Shopping Mall, A1 is better. It means in working places and resident locations, people prefer to make connections within communities, while in shopping mall locations, the dynamics of social connection is much stronger and impacted by time of day and day of week. We can further interpret the result as that human social behavior is not only impacted by real life behaviors, but also the time of day and day of week.

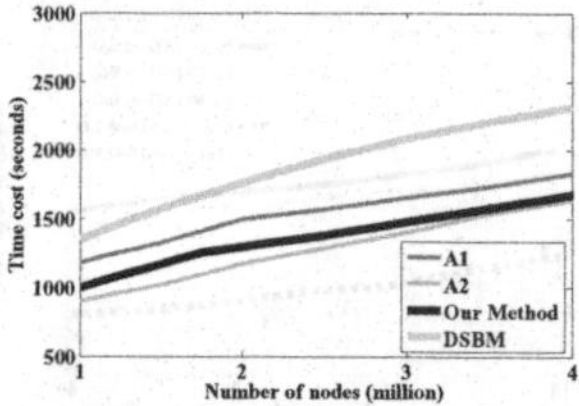

Fig. 6. Efficiency evaluation of different algorithms. The running time cost of our algorithm (PCD) is much lower than two other baseline algorithms.

4.6 Computational Runtime

In Figure 6, we report computational runtimes for the different algorithms. The results show that our method runs much faster than two other baseline methods. For a fixed number of random restarts, the runtime scales nearly linearly for the graph sizes considered here. All algorithms were conducted on a standard server (Linux), with four Intel Core Quad CPUs, Q9550 2.83 GHz and 32 GB main memory.

5 Conclusion and Future Work

In this paper, we studied an interesting but challenging problem, persistent community detection in evolving social graphs. Extensive empirical experiment results show that our proposed method performs much better than a set of baseline methods, in merits of persistence in time series analysis, consistency in social graph structure and efficiency in algorithm running time cost.

In the future, we are going to apply our method to online social networks (e.g., Facebook and Twitter), and then we would like to compare the persistent community in mobile social networks with the persistent community in online social networks.

Acknowledgments. This research was supported by the Singapore National Research Foundation under its International Research Centre @ Singapore Funding Initiative and administered by the IDM Programme Office, Media Development Authority (MDA) and the Pinnacle Lab at Singapore Management University. Shuhui Wang was supported in part by National Basic Research Program of China (973 Program): 2012CB316400, and National Natural Science Foundation of China: 61303160. The authors also thank David Choi for valuable discussions and support regarding this work.

References

1. Anandkumar, A., Ge, R., Hsu, D., Kakade, S.M.: A tensor spectral approach to learning mixed membership community models. CoRR, –1–1 (2013)
2. Barbieri, N., Bonchi, F., Manco, G.: Cascade-based community detection. In: WSDM 2013, pp. 33–42 (2013)
3. D'Amore, R.J.: Expertise community detection. In: SIGIR 2004, pp. 498–499 (2004)

4. Drugan, O.V., Plagemann, T., Munthe-Kaas, E.: Detecting communities in sparse manets. IEEE/ACM Trans. Netw, 1434–1447 (2011)
5. Fortunato, S.: Community detection in graphs. CoRR, –1–1 (2009)
6. Karrer, B., Newman, M.: Stochastic blockmodels and community structure in networks. Physical Review E 83(1), 016107 (2011)
7. Leskovec, J., Lang, K.J., Mahoney, M.W.: Empirical comparison of algorithms for network community detection. In: WWW 2010, pp. 631–640 (2010)
8. Lin, W., Kong, X., Yu, P.S., Wu, Q., Jia, Y., Li, C.: Community detection in incomplete information networks. In: Proc. of WWW 2012 (2012)
9. Lin, Y.-R., Chi, Y., Zhu, S., Sundaram, H., Tseng, B.L.: Facetnet: a framework for analyzing communities and their evolutions in dynamic networks. In: WWW 2008, pp. 685–694 (2008)
10. Lin, Y.-R., Chi, Y., Zhu, S., Sundaram, H., Tseng, B.L.: Analyzing communities and their evolutions in dynamic social networks. TKDD, –1–1 (2009)
11. Liu, S., Kang, L., Chen, L., Ni, L.M.: Distributed incomplete pattern matching via a novel weighted bloom filter. In: ICDCS 2012, pp. 122–131 (2012)
12. Liu, S., Wang, S., Jeyarajah, K., Misra, A., Krishnan, R.: TODMIS: Mining communities from trajectories. In: ACM CIKM (2013)
13. Nguyen, N.P., Dinh, T.N., Tokala, S., Thai, M.T.: Overlapping communities in dynamic networks: their detection and mobile applications. In: MOBICOM 2011, pp. 85–96 (2011)
14. Skyrms, B., Pemantle, R.: A dynamic model of social network formation. Proceedings of the National Academy of Sciences of the United States of America 97(16), 9340–9346 (2000)
15. Tang, L., Wang, X., Liu, H.: Community detection via heterogeneous interaction analysis. Data Min. Knowl. Discov., 1–33 (2012)
16. Yan, X., Jensen, J.E., Krzakala, F., Moore, C., Shalizi, C.R., Zdeborov, L., Zhang, P., Zhu, Y.: Model selection for degree-corrected block models. CoRR, –1–1 (2012)
17. Yang, T., Chi, Y., Zhu, S., Gong, Y., Jin, R.: Detecting communities and their evolutions in dynamic social networks - a bayesian approach. Machine Learning, 157–189 (2011)
18. Zhang, Y., Wang, J., Wang, Y., Zhou, L.: Parallel community detection on large networks with propinquity dynamics. In: KDD 2009, pp. 997–1006 (2009)

Detecting and Analyzing Influenza Epidemics with Social Media in China*

Fang Zhang[1], Jun Luo[1,2], Chao Li[1], Xin Wang[3], and Zhongying Zhao[1]

[1] Shenzhen Institutes of Advanced Technology
Chinese Academy of Sciences, Shenzhen, China
[2] Huawei Noah's Ark Lab, Hong Kong, China
[3] Department of Geomatics Engineering, University of Calgary, Canada
{fang.zhang,jun.luo,chao.li1,zy.zhao}@siat.ac.cn,
xcwang@ucalgary.ca

Abstract. In recent years, social media has become important and omnipresent for social network and information sharing. Researchers and scientists have begun to mine social media data to predict varieties of social, economic, health and entertainment related real-world phenomena. In this paper, we exhibit how social media data can be used to detect and analyze real-world phenomena with several data mining techniques. Specifically, we use posts from TencentWeibo to detect influenza and analyze influenza trends. We build a support vector machine (SVM) based classifier to classify influenza posts. In addition, we use association rule mining to extract strongly associated features as additional features of posts to overcome the limitation of 140 words for posts. We also use sentimental analysis to classify the reposts without feature and uncommented reposts. The experimental results show that by combining those techniques, we can improve the precision and recall by at least ten percent. Finally, we analyze the spatial and temporal patterns for positive influenza posts and tell when and where influenza epidemic is more likely to occur.

Keywords: Influenza Epidemics, Social Media, Data Mining.

1 Introduction

Influenza is a severe disease and seasonally spreads around the world in epidemics, causing over 3 million yearly cases of severe illness and about 250,000 to 500,000 yearly death[1] Global attention has been drawn to this issue from both medical and technical perspectives. However, influenza is unable to be detected under the traditional surveillance system both effectively and efficiently, thus making the disease monitoring a challenging topic.

In recent years, social media, for instance, Facebook, Twitter, MySpace and TencenWeibo, has become a popular platform among people on which they create, share, and

* This research is partially funded by the NSF of China (Grant No. 11271351 and 61303167) and the Basic Research Program of Shenzhen (Grant No. JCYJ20130401170306838 and JC201105190934A). Xin Wang's research is partially funded by the NSERC Discovery Grant.

[1] http://en.wikipedia.org/wiki/Influenza

V.S. Tseng et al. (Eds.): PAKDD 2014, Part I, LNAI 8443, pp. 90–101, 2014.

propagate information. Social media gains ascendancy over traditional media because of its better performance in stability, fast propagation and efficient resource utilization. Therefore, it is gradually replacing traditional medias and grows in fast pace as the platform of useful information sharing. Recent work has demonstrated that prediction of varieties of phenomena can be made by using social media data. These phenomena include disease transmission [18], movie box-office revenues [4], and even elections [20]. In this paper we illustrate how social media can be used to detect and analyze influenza epidemics in China. Specifically, we consider the task of detecting and analyzing influenza by utilizing the posts from TencentWeibo, one of the most popular social networks with more than 500 million users in China.

We first extract influenza-like posts from our TencentWeibo data corpus. The most common influenza symptoms are chills, fever, runny nose, sore throat, headache, coughing, fatigue and discomfort. Although, these symptoms as keywords can be utilized to determine whether a post is an influenza-like post, inaccurate, ambiguous or keywords related posts might still disturb the collection of the real influenza-like posts such as (all posts and words are translated from Chinese to English in this paper):

- One should have more water when catching flu.
- Avian flu is under epidemics this spring.
- Jesus, fevering, have I got cold?

These posts all mention the word of "flu" or flu symptoms. Nevertheless that does not mean that the posters have been affected by influenza. We consider these posts (news, advices or suspicion) as negative influenza posts. Our goal is to detect positive influenza posts and analysis influenza epidemics in China with TencentWeibo data. As discussed above, it is necessary to extract positive influenza posts from the whole dataset to get more accurate results. In this paper, we propose a machine learning based classifier to filter out negative influenza posts with 0.900 precision and 0.913 recall.

Next, after classification, TencentWeibo data is analyzed and processed from the perspective of time and space respectively. From the perspective of time we can find out which place is more likely for influenza outbreak and from the perspective of space, we can discover when is more likely for influenza outbreak in one city or a certain province in China.

This paper is organized as follows: related works are presented in next section. In Section 3, a short introduction to TencentWeibo and the characteristics of our dataset are provided. In Section 4, several data mining techniques which are used in our research are introduced. In Section 5, evaluation of our model is shown. In Section 6, our model is applied in detecting and analyzing influenza epidemics in China. We conclude and give the future work in Section 7.

2 Related Works

In recent years, scientists have been using social media data or other information to detect influenza epidemics and to provide earlier influenza warnings.

Espino et al. [7] proposed a public health early warning system by utilizing data from telephone triage (TT) which is a public service to give advice to users via telephone in 2003. They obtained TT data from a healthcare call center services and software company. By investigating the relationship between the number of telephone calls and influenza epidemics, then reported a signification correlation.

Magruder [13] utilized the amount of over-the-counter (OTC) drug sales to build a possible early warning indicator of human disease like influenza. Influenza patients requirement for anti-influenza drugs makes this approach reasonable. They reported the magnitude of correlations between clinical data and some OTC sales data and then measured the time lead after controlling for day-of-week effects and some holiday effects.

Ginsberg et al. [8] built a system, utilizing Google web search queries, to generate more comprehensive models for use in influenza surveillance. Their approach demonstrated high precision, obtaining an average correlation of 0.97 with the CDC-observed influenza-like illness (ILI) percentage.

Lampos el al. [11] proposed a regression model, by applying Balasso, the bootstrapped version of Lasso, for tracking the prevalence of ILI in part of UK using the contents of Twitter. Compared to the actual HPA's ILI rates, their model achieved high accuracy.

Aramaki el al. [3] proposed a system to detect influenza epidemics. First, the system extracts influenza related tweets via Twitter API. Next, a support vector machine (SVM) based classifier was used to extract tweets that mention actual influenza patients. Their approach was feasible with 0.89 correlation to the gold standard.

However, these previous approaches ignored some major characteristics of posts that may impede the classification. First, all posts and reposts have 140 word limitation. That could cause limited features we can use in SVM. Second, the reposts could be no comments or the comments without features. We propose words association rules and sentiment analysis to overcome those problems and improve the classification precision and recall.

3 Dataset

3.1 TencentWeibo Dataset

Launched in April, 2010 by Tencent Holding Limited, TencentWeibo is a Chinese micro-blogging (weibo) website, which is extremely popular around China, consisting of more than half a billion of users (0.54 billion users by Dec, 2012[2]). Like Twitter, each user of TencentWeibo has a set of followers, and from this point TencentWeibo can be considered as a social network. Users can upload and share with its followers photos, videos and text within a 140 word limit, known as posts like tweets in Twitter, that typically consist of personal information about the users. The posts composed by one user are displayed on the user's profile page, so that its followers can either just read, comment or repost the same content and post to their own pages. For one user, it is also possible to send a direct message to another user. A repost, called retweet in Twitter, is a post made by one user that is forwarded by another user. Reposts are useful and fast for information spreading, like photos, videos, text and links through TencentWeibo community. Due to its huge amount of users and prevalence, TencentWeibo is

[2] http://it.21cn.com/itnews/a/2013/0121/11/20247640.shtml

increasingly used by a number of companies and organizations to advertise products and disseminate information. Mining TencentWeibo data to make the future prediction on some social phenomena has become an innovative approach in China.

3.2 Dataset Characteristics

The data we used in our experiments was obtained by downloading the posts from TencentWeibo.com with TencentWeibo Search API. We used "flu" and its common influenza symptoms, such as fever, runny nose, as keywords to ensure that the posts we obtained were influenza related. We obtained 2.59 million influenza related posts over a period of six months from Nov, 2012 to May, 2013. Most of these posts contain location information which indicates the poster's living city. By accurately classifying influenza posts from this data set, we can analyze spatial and temporal patterns of influenza epidemics.

3.3 Label Rules

Three annotators are responsible for assigning positive or negative label to every post in both training dataset and test dataset. One post is labeled as a positive only when it meet one of the following requirements.

- Post indicates the poster has influenza. Since each post has one attribute showing the city name which indicates where the post is sent, we can use this information to do spatial analysis of the positive posts.
- The post mentions other person (relative or friend) has influenza and also mentions the location of the other person. For example, one post says "My poor brother got fever in Beijing". Then we annotate it as positive post and the count of influenza case in Beijing will be increased by one. Otherwise, if there is no indication of location, then the post is annotated as negative post since it has no use to our analysis.
- For reposts, if one repost has no comment, we consider the reposter is consented with the original poster, thus the repost is annotated as the original post's label. If the repost has comment, we label it according to the previous two rules.

Each of these three annotators individually labeled a post x as negative (-1) or positive (+1) influenza-like post described as y_1, y_2, and y_3. Each post was given the final label by the following function $L = \sum_{i=1}^{3} y_i, y_i \in \{+1, -1\}$, where the positive value of L indicates a positive influenza post, while the negative value of L indicates a negative influenza post.

4 Methodology

We build a support vector machine (SVM) [5, 12, 14, 16] based classifier to classify influenza posts with the help of association rule mining and sentiment analysis. SVMs are well-known supervised learning models used in machine learning, particularly for text classification and regression analysis. In terms of linear SVM, the training data set **D** with points is defined as below.

$$\mathbf{D} = \{(\boldsymbol{x}_i, y_i) | \boldsymbol{x}_i \in \mathbf{R}^{\mathbf{p}}, y_i \in \{+1, -1\}\}_{i=0}^{n} \quad (1)$$

Where $\overrightarrow{x_i}$ is a p-dimensional real vector and y_i is the label of the point $\overrightarrow{x_i}$, indicating to which class $\overrightarrow{x_i}$ belongs. And the classification function of linear SVM is:

$$f(\boldsymbol{x}) = sign(\sum_i \alpha_i y_i \boldsymbol{x}_i^T \boldsymbol{x} + b) \quad (2)$$

Where the value of $f(\boldsymbol{x})$ indicates the point's class, α_i is Lagrange multiplier and b is the intercept.

In the rest of this section, several techniques that we utilized in our model will be illustrated.

4.1 Association Rule Mining

Most TencentWeibo posts are short texts with significant characteristics of short length and few features. If potential strongly associated features can be added to the original texts, making longer length and more diversified features, classification performance will be improved. In data mining, association rule learning [1, 2, 9, 21] is a popular and well researched method for discovering interesting relations between variables in large databases. For these reasons, association rule learning is applied to find strong rules in our data.

According to the original definition by Agrawal et al. [1] the problem of association rule mining in our research is defined as: Let $I = \{i_1, i_2, ..., i_n\}$ be a set of n texts features called items. Let $D = \{t_1, t_2, ..., t_n\}$ be a set of posts called the database. In a given database D, an association rule is similar to a form of A⇒B where A, B∈I and A⋂B=∅, the sets of items A and B are respectively called antecedent and consequent of the rule. An easy attempt of differentiation of strong rules is calculating its support and confidence, and thus, mining frequent patterns is the key to obtain strong association rules.

Methods like Apriori [2] can be used for mining association rules and frequency patterns. Apriori is not an efficient algorithm as it needs to find all the candidate itemsets and to repeatedly scan the data base during the process. However, in our research, since frequent patterns with 2 features, such as {cold, runny nose}, are needed and the candidate sets with more than 2 features ($k > 2$) are avoided, Apriori algorithm becomes efficient and is applied in our research.

Frequent patterns with a given minimum thresholds on support and confidence are regarded as strong association rules which then will be utilized to extend short posts to improve classification performance. For example, if "cold"⇒"runny nose" is a strong association rule in our data base, then word token "runny nose" will be added into the texts of the posts which contain word token "cold" as a feature.

4.2 Sentiment Analysis

Sentiment analysis [6, 15, 17, 19] refers to the application of natural language processing (NLP), computational linguistics, and identification plus extraction of subjective information over text analysis[3] Generally speaking, sentiment analysis is designed for

[3] http://en.wikipedia.org/wiki/Sentiment_analysis

acquiring the attitude of the corresponding author or lecturer upon his or her contextual works on a comprehensive level. The basics of sentiment analysis is to categorize the absolute standing or meaning of the given material words based on the opinion delivered into three classification positive, negative, or neutral.

In our research, an important component of sentiment analysis which focuses on the automatic identification for whether a repost contains positive or negative opinion about influenza is to identify the emotion expressed in the posts if the according poster has infected with influenza.

In our dataset, each post that is downloaded by keywords is one of the following three kinds:

- Posts with features after feature selection.
- Commented reposts with an original post, but no features in comment.
- Uncommented reposts with an origin post.

For the first kind of posts, the SVM classifier directly classifies them. As to the rest two kinds, we separate the reposts r into two parts, comment part c and original post part o. Take "Fortunately, I didn't. || @ someone: I got flu" as an example. "Fortunately, I didn't. "is part c and "I got flu." is part o. After feature selection, however, this post has no features. The SVM randomly classifies it as positive influenza post or negative one; nevertheless, the poster definitely has not got influenza. In this situation, sentiment analysis is needed to improve the SVM's precision. Our approach can be described as:

$$L(r) = \begin{cases} s(c) \times f(o), f(o) = +1 \\ f(c), f(o) = -1 \end{cases}, \tag{3}$$

$$s(c) = \begin{cases} +1, \textit{no negative word in c} \\ -1, \textit{has negative word in c} \end{cases} \tag{4}$$

where $f(o)$ is defined in formula 2, $s(c)$ indicates whether comment part c has negative attitude to the origin part o (we regard reposters have positive attitude towards the original post if the according reposts have no comment on the original posts), and the value of $L(r)$ indicates the repost's class.

5 Experiments

We collected about 2.59 million posts posted within the time period from Nov 2012 to May 2013, using TencentWeibo Search API. We separated those posts into three groups.

Training data consists of 4092 posts which were randomly selected by computer and annotated by 3 annotators. Then these posts were used for the purpose of SVM classifier training.

Test data consists of 2500 posts randomly selected by computer and annotated by 3 annotators like training data. These data were used to evaluate the SVM based classifier.

Experiment data are the rest of the posts collected. They were used in experiments of influenza epidemics detection and analysis. Those posts were separated into six groups by month within the time range that we studied, from Nov 2012 to April 2013.

Table 1. Examples of positive and negative weighted significant features of our SVM classifier

Positive Weighted Feature	Weight	Negative Weighted Feature	Weight
have cold	1713.39	fatigue	132.99
feel ill	385.70	faint	86.98
fevering	146.85	healthy	56.43
runny nose	85.06	question	47.41
very	82.26	share	46.23
sore throat	48.61	later	43.00
serious	48.22	little	41.64
rhinobyon	40.01	lack	39.99
headache	38.67	sneeze	37.04
seemingly	37.43	nervous	36.01

We first applied feature selection for SVM training because of three reasons: (1) To improve the efficiency of training and testing process. (2) To delete noisy features. (3) To improve classification precision.

We calculated Chi-squared value for every word token that appeared in the training data set. As SVM features, top 1,000 word tokens ranked by Chi-squared value were utilized. Before word segmentation and vectorization, punctuation and special characters were striped, mentions of user names (the"@" tag), reposts (the "||" tag) and expressions (the "/" tag) were removed, and all other language characters were ignored. Table 1 lists examples of significant features we used as SVM features.

Besides precision and recall, $F_1 = \dfrac{2 * precision * recall}{precision + recall}$ which is a weighted harmonic mean that trades off precision versus recall was utilized to evaluate our classifier.

In our experiment, the kernel of SVMs is linear. The evaluation of this SVM classifier on test set showed 0.79 precision, 0.80 recall and 0.795 F_1. There are two reasons to cause relative low precision and recall:

- A TencentWeibo post consists text with a 140 word limit and most of the posts are short texts with only several words. Not even a single feature is contained in some processed short posts after word segmentation and vectorization. Our SVM classifier's scheme of random labeling them leads to relative low precision and recall.
- In terms of reposts, reposts without comment are also qualified for the condition above, as reposts with comment always indicate the reposter's attitude on the original post. However a SVM classifier is not capable of analyzing poster's sentiment on this kind of post.

We handled the first case by applying rule mining to extend word-segmented posts to obtain more features before vectorization. Based on our training data, we learned some word association rules as shown in Table 2 with the threshold of 0.01 minimum support and 0.6 minimum confidence. However, the performance of the SVM classifier with 0.797 precision, 0.804 recall and 0.800 F_1did not improve too much.

We then utilized sentiment analysis to improve the evaluation of classification. First, we collected thousands of emotional-related words and put them into 2 groups

Table 2. Examples of word associations rules with support and confidence

Feature 1	Feature 2	Support	Confidence
feeling ill	having cold	0.085	0.67
sore throat	having cold	0.011	0.60
runny nose	having cold	0.017	0.65
sneezing	runny nose	0.013	0.65
serious	having cold	0.012	0.69
question	having solved	0.011	0.65
lack	voting	0.011	0.65
bad	having cold	0.015	0.67

Table 3. Examples of both emotional negative and emotional positive words

Negative	no	don't think so	deny	don't agree	disappoint	abhorrent	annoyed	angry	insane	bad
Positive	yes	I think so	accept	agree	gladness	amused	happy	glad	smart	good

(emotional negative and emotional positive). Table 3 lists examples of both emotional negative and emotional positive words. We then applied formula 3 and 4 to classify reposts with an original post.

As shown in Figure 1, when considering word associations and sentiment of posters, the classification performance substantially improves, achieving up to 0.900 precision, 0.913 recall and 0.905 F_1.

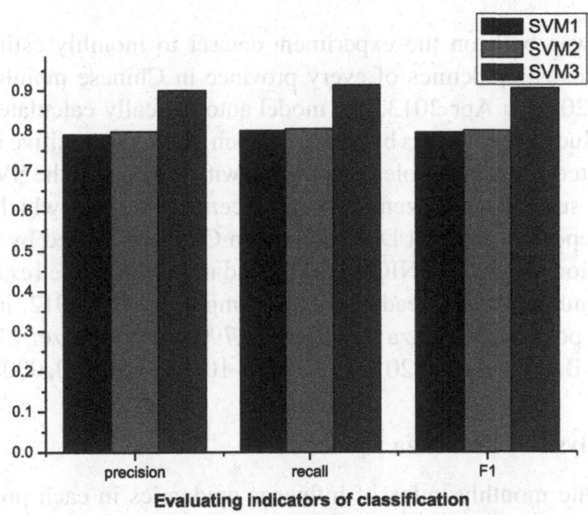

Fig. 1. Summary of evaluation results. SVM1 represents original SVM classifier, SVM2 represents the SVM based classifier with association rule mining, and SVM3 represents the SVM based classifier with association rule mining and sentiment analysis.

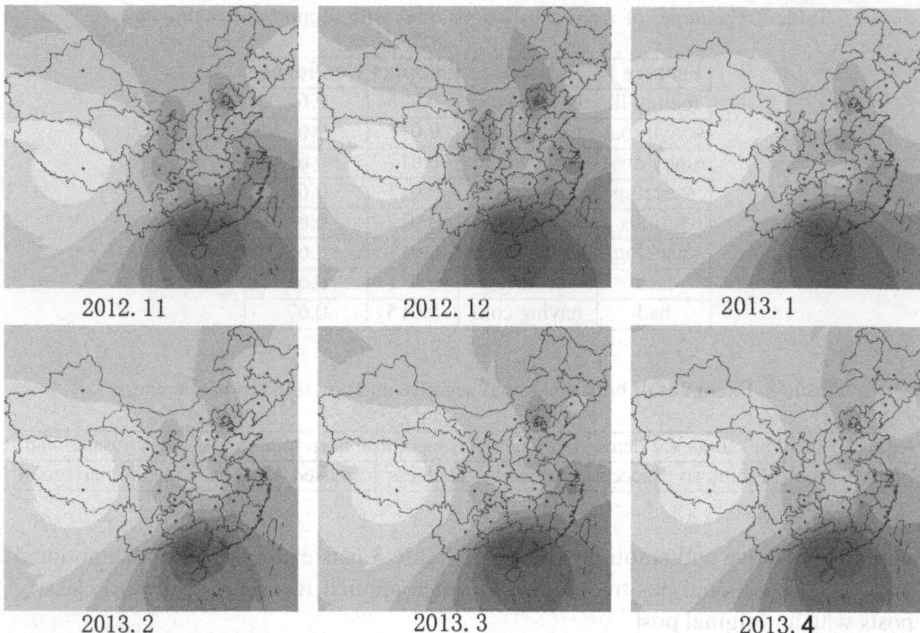

Fig. 2. Spatial analysis of influenza epidemics in China from Nov 2012 to Apr 2013. The darker colored area indicates a higher influenza index and the lighter colored means a lower index.

6 Spatial and Temporal Analysis of Influenza Epidemics

A simple model was built on the experiment dataset to monthly estimate the average extent of influenza epidemics of every province in Chinese mainland. For every month from Nov 2012 to Apr 2013, our model automatically calculates the index of each provinces influenza epidemics by summing each provinces positive influenza posts which were extracted from the whole posts corpus with the help of the SVM based classifier, dividing the sum by net citizen scale of the certain province which was obtained from Statistical Report on Internet Development in China published by China Internet Network Information Center (CNNIC) in 2012, and multiplying the result with 10,000 to make the final number more readable. For example, in Nov 2012, in our database Beijing has 6,916 positive influenza posts and 13,790,000 net citizens. Therefore, the influenza index of Beijing in Nov 2012 is $6,916 \times 10,000/13,790,000 = 5.015$.

6.1 Spatial Analysis of Influenza Epidemics

After computing the monthly index of influenza epidemics in each province, we assign that value to the capital city of the corresponding province. The we use Kriging [10]which is a geostatistical estimator that infers the value of a random field at an unobserved location to spatially interpolate influenza epidemics index of the whole map. Figure 2 represents the distribution of influenza epidemics in China from Nov 2012 to April 2013. From Figure 2, we obtain some conclusions as below:

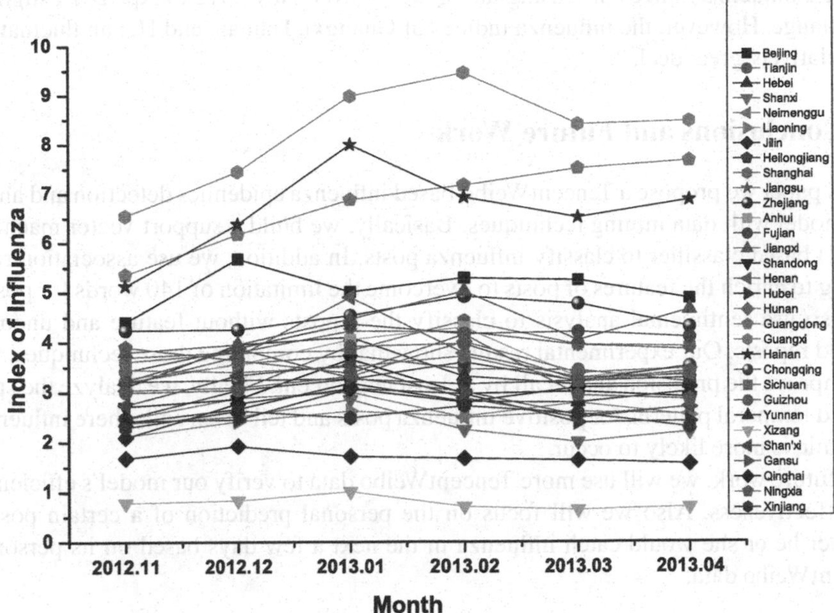

Fig. 3. Time analysis of influenza epidemics in China from Nov 2012 to Apr 2013. Different kinds of dots represent different provinces and each dot shows the influenza index of a given province in the given month.

- With the drop of temperature from Nov 2012 to Dec 2012, high influenza-index regions are increasing and low influenza-index regions are reducing. On the contrary, with the rise of temperature from Dec 2012 to Apr 2013, high influenza-index regions are reducing and low influenza-index regions are increasing.
- From Nov 2012 to Apr 2013, southeast coastal provinces including Guangdong, Guangxi, and Hainan have higher influenza indices. While west provinces including Xinjiang, Qinghai, and Tibet have relatively lower influenza indices.
- Areas such as Beijing-Tianjin, Chengdu-Chongqing, Yangtze River Delta, and Pearl River Delta with bigger fluid population and more density of population have higher influenza indices. Areas such as Xinjiang, Xizang (Tibet), Qinghai, and Gansu where density of population is far smaller than the areas mentioned above have lower influenza indices.

6.2 Temporal Analysis of Influenza Epidemics

Figure 3 represents the influenza indices of each province in China mainland from Nov, 2012 to Apr, 2013. Some conclusions can be obtained by observing Figure 3:

- From Nov 2012 to Apr 2013, Guangxi, Guangdong, and Hainan have relatively higher influenza indices than other provinces. Xizang and Qinghai contrarily have relatively lower influenza indices.

- The influenza indices of Xizang and Qinghai from Nov 2012 to Apr 2013 slightly change. However, the influenza indices of Guangxi, Hainan, and Henan fluctuate a relatively great deal.

7 Conclusions and Future Work

In this paper we propose a TencentWeibo based influenza epidemics detection and analysis model with data mining techniques. Basically, we build a support vector machine (SVM) based classifier to classify influenza posts. In addition, we use association rule mining to enrich the features of posts to overcome the limitation of 140 words for posts. We also use sentimental analysis to classify the reposts without feature and uncommented reposts. Our experimental results show that by combining those techniques, we can improve the precision and recall by at least ten percent. Finally, we analyze the spatial and temporal patterns for positive influenza posts and tell when and where influenza epidemic is more likely to occur.

In future work, we will use more TencentWeibo data to verify our model's efficiency and effectiveness. Also we will focus on the personal prediction of a certain poster whether he or she would catch influenza in the next a few days based on its personal TencentWeibo data.

References

1. Agrawal, R., Imieliński, T., Swami, A.: Mining association rules between sets of items in large databases. ACM SIGMOD Record 22, 207–216 (1993)
2. Agrawal, R., Srikant, R., et al.: Fast algorithms for mining association rules. In: Proc. 20th Int. Conf. Very Large Data Bases, VLDB, vol. 1215, pp. 487–499 (1994)
3. Aramaki, E., Maskawa, S., Morita, M.: Twitter catches the flu: Detecting influenza epidemics using twitter. In: Proceedings of the Conference on Empirical Methods in Natural Language Processing, pp. 1568–1576. Association for Computational Linguistics (2011)
4. Asur, S., Huberman, B.A.: Predicting the future with social media. In: 2010 IEEE/WIC/ACM International Conference on Web Intelligence and Intelligent Agent Technology (WI-IAT), vol. 1, pp. 492–499. IEEE (2010)
5. Cortes, C., Vapnik, V.: Support-vector networks. Machine Learning 20(3), 273–297 (1995)
6. de Haaff, M.: Sentiment analysis, hard but worth it!, customerthink (2010), http://www.customerthink.com/blog/sentiment_analysis_hard_but_worth_it
7. Espino, J.U., Hogan, W.R., Wagner, M.M.: Telephone triage: a timely data source for surveillance of influenza-like diseases. In: AMIA Annual Symposium Proceedings, vol. 2003, p. 215. American Medical Informatics Association (2003)
8. Ginsberg, J., Mohebbi, M.H., Patel, R.S., Brammer, L., Smolinski, M.S., Brilliant, L.: Detecting influenza epidemics using search engine query data. Nature 457(7232), 1012–1014 (2008)
9. Hipp, J., Güntzer, U., Nakhaeizadeh, G.: Algorithms for association rule mining - general survey and comparison. ACM SIGKDD Explorations Newsletter 2(1), 58–64 (2000)
10. Krige, D.G.: A statistical approach to some mine valuation and allied problems on the Witwatersrand. PhD thesis, University of the Witwatersrand (1951)
11. Lampos, V., De Bie, T., Cristianini, N.: Flu detector - tracking epidemics on twitter. In: Balcázar, J.L., Bonchi, F., Gionis, A., Sebag, M. (eds.) ECML PKDD 2010, Part III. LNCS, vol. 6323, pp. 599–602. Springer, Heidelberg (2010)

12. Lin, C.-J.: A guide to support vector machines, Department of Computer Science, National Taiwan University (2006)
13. Magruder, S.: Evaluation of over-the-counter pharmaceutical sales as a possible early warning indicator of human disease. Johns Hopkins University APL Technical Digest 24, 349–353 (2003)
14. Manning, C.D., Raghavan, P., Schütze, H.: Introduction to information retrieval, vol. 1. Cambridge University Press, Cambridge (2008)
15. Melville, P., Gryc, W., Lawrence, R.D.: Sentiment analysis of blogs by combining lexical knowledge with text classification. In: Proceedings of the 15th ACM SIGKDD International Conference on Knowledge Discovery and Data Mining, pp. 1275–1284. ACM (2009)
16. Meyer, D., Leisch, F., Hornik, K.: The support vector machine under test. Neurocomputing 55(1), 169–186 (2003)
17. Pang, B., Lee, L., Vaithyanathan, S.: Thumbs up?: sentiment classification using machine learning techniques. In: Proceedings of the ACL 2002 Conference on Empirical Methods in Natural Language Processing, vol. 10, pp. 79–86. Association for Computational Linguistics (2002)
18. Sadilek, A., Kautz, H., Silenzio, V.: Predicting disease transmission from geo-tagged microblog data. In: Twenty-Sixth AAAI Conference on Artificial Intelligence, p. 11 (2012)
19. Thelwall, M., Buckley, K., Paltoglou, G., Cai, D., Kappas, A.: Sentiment strength detection in short informal text. Journal of the American Society for Information Science and Technology 61(12), 2544–2558 (2010)
20. Tumasjan, A., Sprenger, T.O., Sandner, P.G., Welpe, I.M.: Predicting elections with twitter: What 140 characters reveal about political sentiment. In: Proceedings of the Fourth International AAAI Conference on Weblogs and Social Media, pp. 178–185 (2010)
21. Witten, I.H., Frank, E.: Data Mining: Practical machine learning tools and techniques. Morgan Kaufmann (2005)

Analyzing Location Predictability on Location-Based Social Networks

Defu Lian[1,2], Yin Zhu[3], Xing Xie[2], and Enhong Chen[1]

[1] University of Science and Technology of China
[2] Microsoft Research
[3] Hong Kong University of Science and Technology
[4] liandefu@mail.ustc.edu.cn, yinz@cse.ust.hk, xingx@microsoft.com
cheneh@ustc.edu.cn

Abstract. With the growing popularity of location-based social networks, vast amount of user check-in histories have been accumulated. Based on such historical data, predicting a user's next check-in place is of much interest recently. There is, however, little study on the limit of predictability of this task and its correlation with users' demographics. These studies can give deeper insight to the prediction task and bring valuable insights to the design of new prediction algorithms. In this paper, we carry out a thorough study on the limit of check-in location predictability, i.e., to what extent the next locations are predictable, in the presence of special properties of check-in traces. Specifically, we begin with estimating the entropy of an individual check-in trace and then leverage Fano's inequality to transform it to predictability. Extensive analysis has then been performed on two large-scale check-in datasets from Jiepang and Gowalla with 36M and 6M check-ins, respectively. As a result, we find 25% and 38% potential predictability respectively. Finally, the correlation analysis between predictability and users' demographics has been performed. The results show that the demographics, such as gender and age, are significantly correlated with location predictability.

Keywords: Location predictability, entropy, LBSN.

1 Introduction

With the proliferation of smart phones and the development of positioning technologies, users can obtain location information more easily than ever before. This development has triggered a new kind of social network service - location-based social networks (LBSNs). In a LBSN, people can not only track and share location-related information of an individual, but also leverage collaborative social knowledge learned from them. "Check-in" is such user-generated and location-related information, being used to represent the process of announcing and sharing users' current locations in LBSNs.

In this paper, we are interested in predicting users' future check-in locations based on their location histories accumulated in LBSNs. In particular, we attempt to determine at which Point Of Interest (POI), such as a clothing store or

V.S. Tseng et al. (Eds.): PAKDD 2014, Part I, LNAI 8443, pp. 102–113, 2014.

a western restaurant, a user will check in next. Though this problem has been recently investigated [1,2,3], there is little study on the limit of predictability, i.e., to what degree the next check-in locations are predictable, and its correlation with users' demographics. We believe this study will bring valuable insights to the design of prediction algorithms and help to understand users' behavior from both social and physical perspectives.

The limit of location predictability was first studied on cell tower location sequences in [4]. The authors discovered a 93% potential predictability in human mobility. However, a check-in trace is quite different from a cell tower trace in the following three aspects. First, check-in is a proactive behavior comparing to the passive recording of cell tower traces. In other words, a user might not check in at boring places where he has actually been but may check in at locations where there is no visiting behavior. Therefore, check-in locations are usually discontinuous, and many important mobility patterns could have been lost. Second, the spatial granularity of check-in locations is much finer than cell tower locations (e.g., a point location versus an area of one square kilometers). Thus there are more candidate locations to choose for check-in so that it is much more difficult to predict next check-in location. Last but not least, users are equipped with rich profile information and social relationships, since their check-ins are usually shared on different social networks. This would be helpful for developing more accurate algorithms. In our work, we analyze the problem of check-in location prediction in the presence of these characteristics.

To study the limit of location predictability, we begin with estimating the entropy of an individual check-in trace by first considering an individual check-in trace as a sample of underlying stochastic processes and then calculating the entropy of stochastic processes. We then leverage Fano's inequality [5] to transform the estimated entropy into the limit of predictability for each user. The limit of check-in location predictability is measured for each user on two large-scale check-in datasets from Jiepang and Gowalla with 36M and 6M check-ins, respectively. As a result, we find 25% and 38% potential predictability on these two datasets, respectively.

However, according to our observation, the variance of location predictability among population is large. It implies there is large diversity of human mobility patterns among population. To better understand such large diversity, we can study the difference in predictability of users with different demographics. Particularly, we will perform correlation analysis between predictability and demographics. This task can be more easily done than ever before since users have been already equipped with rich profile information on social networks, including gender, age, social relationship and so on. By conducting case studies on these check-in datasets, we show that the demographics including users' gender, age and influence (measured as the number of followers) as well as the repetitiveness of check-ins (measured as the ratio of the number of check-ins to locations) are significantly correlated with location predictability. More specifically, the mobility of students is higher predictable since their activity areas are usually constrained around the campus and their mobility patterns tend to be more regular; the users with high social influence are hard to predict because they don't usually repeat to check in at those familiar locations. In this case, it is evident that incorporating demographics into the prediction task could be beneficial.

2 Related Work

The limit of location predictability was first studied in [4] on cell tower data, on which they derived an upper bound of predictability from the entropy of the individual location sequence and found a 93% potential predictability in human mobility. They also studied on a lower bound, Regularity, which predicted the next location as the most frequently visited location at given hour of week. Following them, in [6], the authors studied the predictability from mobile sensor data and also found a high potential predictability on mobile sensor data. And in [7] the authors investigated the scaling effects on the predictability using the high resolution GPS trajectories and derived another equivalent statistical quantities to the predictability. In their conclusion, they stated that high predictability was still present at the very high spatial/temporal resolutions. Although all these work focused on the analysis of the limit of predictability, their mobility data differs from ours in the following two major aspects. First, the check-in trace is more discontinuous since a user might not check in many places which he has actually visited. Second, the spatial granularity of check-ins is even finer than physical locations since there might be many different semantic locations in the same physical location. These properties lower the check-in location predictability and only achieve a 25%-40% potential predictability.

As for the correlation between mobility patterns and demographics, in [8], the authors analyzed mobility patterns based on the travel diaries of hundreds of volunteers and figured out people with different occupation had distinct mobility patterns. In particular, students and employees were more tending to move among those frequented locations than retirees. From the perspective of predictability, it seems that students and employees were more easily predicted. The direct correlation between predictability and users' demographics was also studied in [4] on cell tower traces logged by hundreds of volunteers with some demographics. They concluded that there were no significant gender- or age-based differences. Different from theirs, we perform analysis on check-in traces of hundreds of thousands of users on social networks, where users are usually equipped with rich profile information. The results of analysis show that user's demographics including age, gender, social influence and so on, are significantly correlated with location predictability.

3 Check-in Datasets

We perform our analysis on two check-in datasets. The first check-in dataset is from Jiepang, which is a Chinese location-based social network similar to Foursquare. For the sake of protecting privacy, in these LBSNs, users' historical check-ins are not shown to strangers. Thus we cannot directly obtain users' check-ins from these LBSNs without becoming their friends. However, users may share their check-ins as tweets on other social networking platforms, such as Weibo and Twitter. For example, Jiepang check-ins are synchronized on Weibo as a particular type of tweets (called location tweets). Thus these check-ins can be crawled from these social networking platforms via their open APIs. Some check-in datasets were also crawled in this way [2,3].

We crawled 36,143,085 Jiepang check-ins at 1,000,457 POIs from 454,375 users via the Weibo API from March. 2011 to March. 2013, where each user has 80 check-ins and check in at 47 POIs on average. If we distribute these check-ins into their date, we find that each user only make 1.5 check-ins each day on average. If we distribute these POIs into 3 km^2 regions, each region owns 13 POIs on average and up to 13,068 POIs in the maximal case. In addition, users on Weibo may fill their profile information more precisely so we also crawled these data, including age, gender, and social relationship as well as tags.

The other check-in dataset, used in [9] and crawled from Gowalla from Feb. 2009 to Oct. 2010, contains 6,423,854 check-ins at 1,280,969 POIs from 107,092 users, where each user has 60 check-ins and check in at 37 POIs on average. If we distribute these check-ins into their date, we find that each user only make 2.1 check-ins each day on average. If we distribute these POIs into 3 km^2 regions, each region owns 7 POIs on average and up to 3,940 POIs in the maximal case.

Based on the above statistics, it is easily observed that the frequency of check-ins is significantly smaller than calling or messaging (SMS) and that location density on check-in datasets is significantly higher than cell towers since each cell tower covers a 3-km^2 perception area on average.

In order to guarantee that the entropy of location sequence is well estimated, we only reserve those users with more than 50 check-ins. As a result, 144,053 and 27,693 users are then kept on Jiepang and Gowalla, respectively. All remaining users on Jiepang have gender information while 53,377 out of them have age information. Moreover, they have 3.9 tags and 15 followees on average. Based on the filtered datasets, we perform extensive analysis after presenting the limits of predictability and then compare them with cell tower traces.

4 Location Predictability

Assume we predict the nth check-in location L_n for user u, given her past location sequence of length $n-1$, $h_{n-1} = \{l_1, l_2, ..., l_{n-1}\}$. From the probabilistic perspective, we need to model the probability distribution of L_n given h_{n-1}, i.e., $P(L_n|h_{n-1})$. In the context of prediction, we choose the location $\hat{l}$ with the maximum probability

$$\hat{l} = \arg\max_{l} P(L_n = l|h_{n-1}). \tag{1}$$

Intuitively, if the distribution of $P(L_n|h_{n-1})$ is flat, the prediction $\hat{l}$ with the maximum probability has a low likelihood of being correct; if the distribution peaks at location $\hat{l}$ significantly, then the prediction can be made with high confidence. Thus the probability at $\hat{l}$ (denoted as $\pi(h_{n-1})$) contains the full predictive power including the potential long range dependency. Summing $\pi(h_{n-1})$ over all possible sequences of length $n-1$, the predictability at the nth location is defined as

$$\Pi(n) \equiv \sum_{h_{n-1}} P(h_{n-1})\pi(h_{n-1}), \tag{2}$$

where $P(h_{n-1})$ is the probability of observing h_{n-1}.

After averaging the predictability over all time indices and taking the limit, each user's predictability Π is defined as

$$\Pi \equiv \lim_{n\to\infty} \frac{1}{n} \sum_{i=1}^{n} \Pi(i) \tag{3}$$

The limit that Π can reach is estimated by first calculating the entropy of a user's check-in location sequence using non-parametric approaches, and then transforms the estimated entropy into the limit of Π using Fano's inequality [5].

4.1 Entropy of Check-in Location Sequence

The history of check-in locations of a user can be considered as one sample path of its underlying stochastic process, e.g., Markov process. Therefore, the entropy estimation of the location history is equivalent to deriving the entropy rate of the stochastic process. According to the definition of entropy, the entropy rate of a stationary stochastic process $\mathcal{L} = \{L_i\}$ is defined as,

$$S \equiv \lim_{n\to\infty} \frac{1}{n} \sum_{i=1}^{n} S(L_i|H_{i-1}) \tag{4}$$

where $S(L_i|H_{i-1})$ is the conditional entropy of L_i given H_{i-1}, which is a random variable corresponding to h_{i-1} (i.e., the past location sequence of length $i-1$). If the stochastic process lacks any long range temporal correlations, i.e., $P(L_i|h_{i-1}) = P(L_i)$, its entropy is $S^{unc} = -\sum_{l=1}^{N} P(l)\log_2 P(l)$, where $P(l)$ is the probability of being at location l and N is the number of locations. In this case, the user moves around N locations according to previous visiting frequency, which we named as the MostFreq algorithm. Another special entropy of interest is the random entropy $S^{rand} = \log_2 N$, obtained when $P(l) = \frac{1}{N}$. In this case, the user moves around N locations randomly, which we named as the Random algorithm. It is obvious that $0 \le S \le S^{unc} \le S^{rand} < \infty$.

One practical way of calculating the entropy of the user's location history is to fix a underlying stochastic process model and then estimate its parameters, e.g., transition probability of first-order Markov process, and finally derive the entropy rate. This method follows a parametric way and somewhat over-specific. From the non-parametric perspective it can also be achieved to use an estimator based on Lempel-Ziv data compression [10]. This method doesn't assume the stochastic process model and thus is more general. In [10], the authors discussed three kinds of LZ estimators and proved that they can converge to the real entropy of a time series when the length of observation sequence is approaching infinity. They applied them to calculate the entropy of English texts (number of bits storage). One estimator for a time series with n steps is defined as follows:

$$S \approx \frac{\ln n}{\frac{1}{n}\sum_{i=1}^{n} \Lambda_i^i} \tag{5}$$

where Λ_i^i is the length of the shortest substring starting at position i which doesn't previously appear from position 1 to $i-1$.

To get the real entropy for any user, her location must be recorded continuously (e.g., hourly). However, the cell tower traces used in [4] only contain locations when a person uses her phone, e.g., she sends a short text message or makes a call, and thus exhibits discontinuity and bursting characteristics in temporal dimension. To handle bursting, the authors merged locations within the same hour. To deal with discontinuity, they first studied the relationship between the entropy of discontinuous location history and the degree of discontinuity, and then extrapolated the entropy where the degree of discontinuity was zero. However, in addition to discontinuity and bursting, check-ins are at the granularity of POIs instead of regions in the cell tower traces. POIs are physical coordinates with semantic labels so that it is possible for different POIs to share the same physical coordinates. Thus POIs are even finer-grained than physical coordinates. For instance, shops in the same building share the physical location. As the check-in POIs within the same hour may have different semantic labels, it is difficult to merge them so that the subsequent extrapolation cannot be applied. Instead, we can simply use the entropy calculated from the check-in history. This is reasonable to some extent since the benefit of extrapolation results from imputing unseen locations while imputing unseen check-in location is more difficult than imputing physical locations.

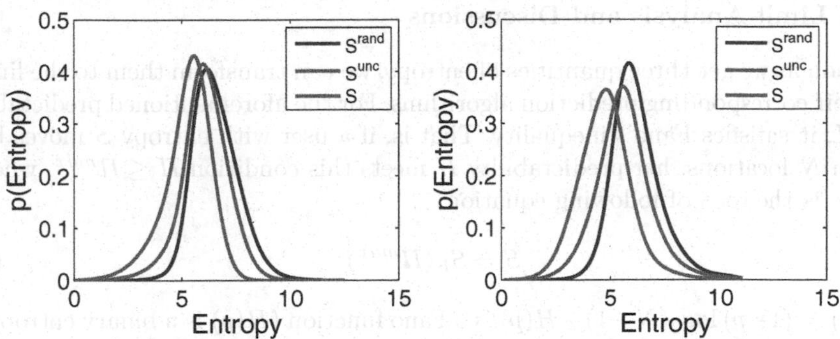

Fig. 1. The distribution of S, S^{unc} and S^{rand} across user population on Jiepang (left) and Gowalla (right)

Next we measure three entropy quantities S, S^{unc} and S^{rand} for each user, and show their probability distribution across users on Jiepang and Gowalla separately in Figure 1. Compared to the results obtained from cell tower traces, there is no big gap between $P(S)$ and $P(S^{rand})$ on both check-in datasets. Indeed, $P(S^{rand})$ peaks around 5.8 on both check-in datasets, indicating if we assume users moved randomly, their next locations can be found on average in any of $2^{5.8}$=56 locations. However, $P(S)$ peaks around 5.59 and 4.83 on Jiepang and Gowalla, respectively. In other words, if we make a prediction with the help of their past history, we reduce less than 2 bits of uncertainty and must choose among $2^{5.59}$=48 and $2^{4.83}$=28 locations on average, respectively, which is much larger than the corresponding number ($2^{0.8}$=1.74) obtained from cell tower traces. Therefore, the prediction of check-in location is more difficult than the cell tower location. To get deep understanding on the difficulty of prediction on LBSNs, we compile statistics on what percentage of transition across locations will repeat. The result indicates that there are only 3.4% and 6.5% repetitive transitions across locations on Jiepang and Gowalla, respectively. This may be because users' proactive check-in behaviour renders checking in at locations without actual visit and missing check-ins at locations where they often go. To continue analyzing Figure 1, we observe that the difference between $P(S^{rand})$ and $P(S)$ on the Jiepang check-in dataset is smaller than on the Gowalla check-in dataset, which means that check-in location prediction on Jiepang is more difficult. This is in line with the previous results that there are larger repetitive transitions across locations on Gowalla than on Jiepang. Comparing $P(S^{unc})$ with $P(S^{rand})$, there is only a small gap on the Jiepang check-in dataset, which indicates that a large number of locations are checked in only once since the average times of users' check-in at POIs is less than 2. The extra part of $P(S)$ over $P(S^{unc})$ can be explained by the temporal correlation between locations in the location sequence and thus helps us to understand the effect of the sequential patterns. Due to their small gap, we could foresee the limited benefit from sequential patterns in the check-in traces.

4.2 Limit Analysis and Discussions

As soon as we get three quantities of entropy, we can transform them to the limit of their corresponding prediction algorithms. For the aforementioned predictability Π, it satisfies Fano's inequality. That is, if a user with entropy S moves between N locations, her predictability Π meets this condition $\Pi \leq \Pi^{max}$, where Π^{max} is the root of following equation

$$S = S_F(\Pi^{max}) \tag{6}$$

$S_F(p) = (1-p)\log_2(N-1) + H(p)$ is a Fano function ($H(p)$ is a binary entropy), which is concave and monotonically decreases with p when $p \in [\frac{1}{N}, 1)$. Therefore, the satisfaction of $\Pi \leq \Pi^{max}$ only requires $S_F(\Pi) \geq S = S_F(\Pi^{max})$ since it is easily verified that $\Pi \geq \frac{1}{N}$ and $\Pi^{max} \geq \frac{1}{N}$. Based on the concavity and monotonicity properties of $S_F(p)$ as well as Jensen's inequality, $S_F(\Pi) \geq S$ is equivalent to $S_F(\pi(h_{n-1})) \geq S(L_n|h_{n-1})$. The latter inequality is simply the well-known Fano's inequality when the probability of error prediction is $1 - \pi(h_{n-1})$ so that $\Pi \leq \Pi^{max}$ is proved.

Since $\Pi^{max} \geq \frac{1}{N}$, Eq (6) has a unique root. We can leverage any root finding algorithm, e.g., Newton's method to find the solution of Π^{max}. Similarly from S^{rand} and S^{unc}, we can determine Π^{rand} and Π^{unc}, which are limits of Random and MostFreq, respectively.

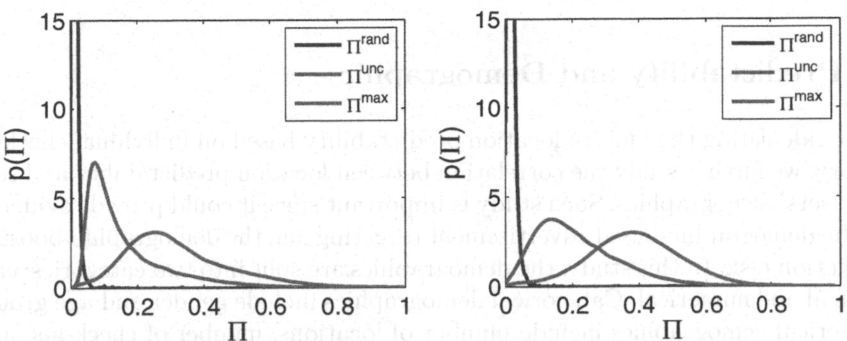

Fig. 2. The distribution of Π^{max}, Π^{unc}, Π^{rand} on Jiepang (left) and Gowalla (right)

After determining Π^{max}, Π^{unc} and Π^{rand} for each user using Eq (6), we demonstrate their distribution over all users in Figure 2. We can see that Π^{max} on the Jiepang and Gowalla check-in datasets peaks around 0.25 and 0.38, respectively, which means that *no matter* how good the algorithm is, its accuracy can not be larger than 25% and 38%, respectively when using their individual check-in history. In addition, we can see that the variance of Π^{max} is larger than that obtained from cell tower traces and thus the predictability in terms of check-in location varies more from person to person. In other words, some people show high regularity while others do not. Π^{unc} shows a similar trend to Π but is left shifted and more narrowly distributed. Thus Π^{unc} shows more universal patterns across user population than Π^{max}. This observation agrees with the phenomena that users check in at the most locations only once.

Although there is 25%-38% potential predictability on the check-in data, there are two assumptions behind. The first is not to impute unseen check-in locations. The reason of making this assumption is that it is difficult to impute unseen check-in locations in practice. The second assumption is not to resort to other information, such as check-ins of friends and similar users. This assumption mainly results from the dominated effect of individual check-in history according to our observation and the conclusion in [1,2]. In the future, we can relax the second assumption to consider these information so as to get a higher limit of predictability. Although the potential predictability is much lower than that of cell tower traces, it is still a theoretically tight bound, which is difficult to achieve in practice. In order to approach this bound, according to the definition of entropy, we should leverage all orders of sequential patterns, from 0 order (MostFreq) to possibly highest (the number of all check-ins) order. In practice, significant high-order sequential patterns are difficult to discover due to limited check-in history, so the combination of 0 order and other low order sequential patterns can be a possible way of designing prediction algorithm [3]. Additionally, for the sake of approaching this bound, according to Fano's inequality, the probability of error $1 - \pi(h_{n-1})$ should be distributed as uniformly as possible over all locations except the most possible one $\hat{l}$. To achieve this, we can introduce other information, such as the time of next check-in, to increase $\pi(h_{n-1})$.

5 Predictability and Demographic

After calculating the limit of location predictability based on individual check-in history, we further study the correlation between location predictability and several users' demographics. Such study is important since it could provide evidence for the demographic-based advertisement targeting and the demographic-boosted prediction task. In this study, the demographics are split into two categories: categorical and numerical. Categorical demographics include gender and age group. Numerical demographics include number of locations, number of check-ins, and number of followers. The age information is numerical in practice, but we quantized it into the following ordinal groups, i.e., "<19", "19-23", "24-28", "29-33", "34-38", ">38" for better visualization. Since the ages of most users are distributed within [19, 38], users both younger than 19 and older than 38 are aggregated into separate groups.

For categorical variables, we perform analysis of variance (ANOVA) test for the statistical correlation between a demographic measure and user's predictability. For numerical variables, since we don't know the concrete form of their correlation, we calculate a non-parametric correlation, i.e., Kendall rank correlation coefficient, and perform non-parametric hypothesis test, i.e., tau test, to see the significance of their correlation.

Table 1. Analysis of variance (ANOVA) for testing the correlation between gender as well as age and predictability. F is F-statistics and p is the p-value of statistical test.

	Gender		Age	
	F	p	F	p
Predictability	4584	<1e-10	260.1	<1e-10

Before performing ANOVA test, we first check the assumption of ANOVA test on the categorical variables and predictability. The result of testing shows the assumption can be satisfied with a p-value smaller than 1e-10. The result of ANOVA test is shown in Table 1. From this table, we observe that the correlation of predictability with the categorical demographics including gender and age group is significant. And gender is more correlated with predictability, which indicates that male users and female users show different degree of predictability. In order to see how these categorical variables are correlated with predictability, we draw the box plot of predictability with respect to these categorical variables and show them in Figure 3. From them, we have the following observations: 1) male users show higher regularity than female users. According to the statistics of the categories of POIs, the four most frequent check-in categories from female users are residential, coffee shop, shopping mall and chaffy dish while the four most frequented check-in categories made by male users are residential, airport, office building and subway station. Thus this observation sounds reasonable. 2) young users (age<24) are easier to predict. This is because these young users are mainly students at school so that their check-ins are constrained around their campus. According to the statistics of the POI categories, three most frequent check-in categories by these young users are teaching building, dormitory and campus. According to the analysis to the tags of each user, 10% of these young users are tagged by "*students*" while only 1% of elder users are such tagged.

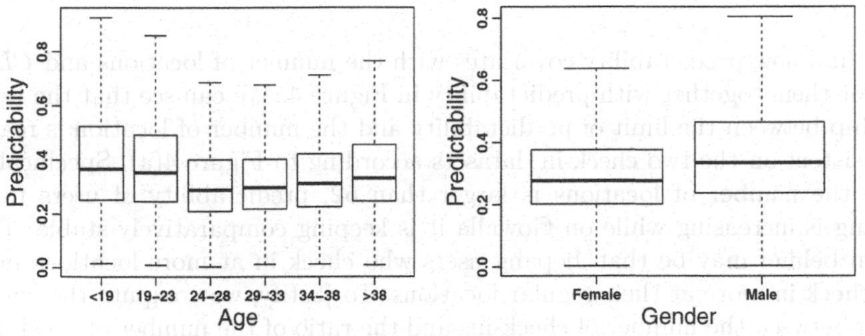

Fig. 3. Box plot of predictability with respect to gender and age

Next we study the correlation between numerical variables and predictability and show the results of tau test in Table 2. From this table, we can answer the following two questions. 1) Are users with larger social influence harder to predict? We measure social influence as the number of followers (#F) in this paper. Users with more followers are more cautious about their reputation so that they will not check in at those boring locations, such as home, subway station. According to this table (last column), the answer to this question is yes and thus the immediately preceding explanation is justified. 2) which of these three factors, the number of locations (#L), the number of check-ins (#C) and the ratio of the number of check-ins to the number of locations which we name as *CLR*, are consistently and significantly correlated to predictability? From this table, we see that the correlation between predictability and the number of locations is significant but not consistent on Jiepang (positively correlated) and Gowalla (negatively correlated). More discussions are provided in Figure 4 and in the subsequent paragraphs. As for the number of check-ins and its ratio *CLR* to the number of locations, they are both significantly correlated to the limit of location predictability on both datasets from the perspectives of statistics testing. However, *CLR* is more strongly correlated with location predictability than the number of locations. The principle reason is that the larger number of check-ins doesn't necessarily indicate more repetitive patterns since some users like to check in at many neighbour locations which they didn't visit in practice.

Table 2. Kendall rank correlation test between numerical profile variables and predictability. Z means the Z-statistics in τ test of Kendall rank correlation, τ is the correlation coefficient and p is the p-value. The cells with bold font indicate negative correlation and the cell with bold italic font shows insignificant correlation.

	Dataset	Stat	#L	#C	CLR	#F
UB	Jiepang	Z	42.6	184.5	516.5	**-10.3**
		p	<1e-10	<1e-10	<1e-10	**<1e-10**
		τ	0.075	0.325	0.907	**-0.018**
	Gowalla	Z	**-56.7**	24.4	212.1	
		p	**<1e-10**	<1e-10	<1e-10	
		τ	**-0.228**	0.098	0.850	

To find how predictability covariates with the number of locations and *CLR*, we plot them together with predictability in Figure 4. We can see that the relationship between the limit of predictability and the number of location is really inconsistent on the two check-in datasets according to Figure 4(a). Specifically, when the number of locations is larger than 52, predictability of users from Jiepang is increasing while on Gowalla it is keeping comparatively stable. The reason behind may be that Jiepang users who check in at more locations may also check in more at their regular locations. To justify, we compute the correlation between the number of check-ins and the ratio of the number of check-ins to the number of locations (*CLR*). We find that on Jiepang there exists positive Kendall rank correlation ($\tau = 0.155$) between them while on Gowalla they are negatively correlated ($\tau = -0.208$). This means that when checking in at more locations, the average number of check-ins at locations is increasing on Jiepang. This implies that these users also check in more at these familiar locations. However, a contrast trend is observed on Gowalla. Thus, the correlation between predictability and the number of locations on check-ins seems incompatible with the result discovered in [4], that regularity was inversely proportional

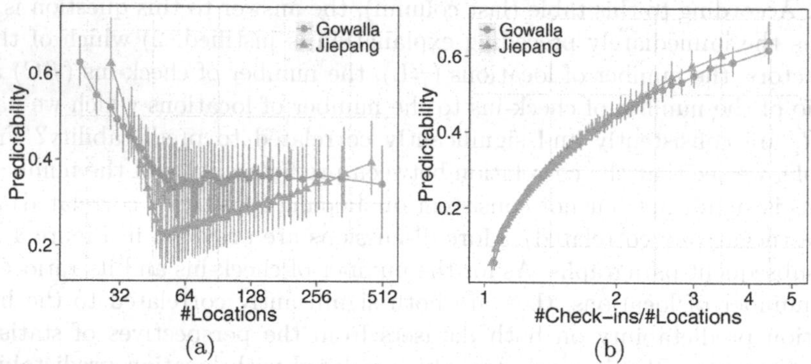

Fig. 4. Predictability with respect to some continuous profile variables. Continuous profile variables are quantized into 50 groups by its uniform quantiles. For each group, we show the median, 25% and 75% quantile of predictability in the error bar plot.

to $N^{-\frac{1}{4}}$, where N is the number of locations. However, according to Figure 4(b), the correlation between the limit of location predictability and CLR is consistently positive on both datasets. This indicates larger CLR could imply more repetitive patterns so that users' behaviour can be more accurately predicted.

6 Conclusion and Future Work

We have analyzed check-in location predictability on two large scale check-in datasets from Jiepang and Gowalla, and found 25% and 38% potential predictability respectively. Then we have studied the correlation between location predictability and users' demographics. The results show that the check-in behaviour of the male users and the young students are more higher predicted. By comparing the correlation between location predictability and the number of locations on two check-in datasets, we have not observed the universal correlation between them. In other words, the number of locations is not directly correlated to location predictability. However, the number of check-ins and its ratio to the number of locations was significantly and positively correlated to predictability, but the degree of the ratio's correlation to predictability is stronger.

In the future, we will extend our predictability analysis from a single user to user groups and study location predictability in the presence of both real friendship on social network and virtual friendship based on location visiting history. This analysis can be helpful to predict check-ins at novel locations where users have never visited before.

References

1. Chang, J., Sun, E.: Location3: How users share and respond to location-based data on social. In: Proc. of ICWSM 2011 (2011)
2. Noulas, A., Scellato, S., Lathia, N., Mascolo, C.: Mining user mobility features for next place prediction in location-based services. In: Proc. of ICDM 2012, pp. 1038–1043. IEEE (2012)
3. Gao, H., Tang, J., Liu, H.: Exploring social-historical ties on location-based social networks. In: Proc. of ICWSM 2012 (2012)
4. Song, C., Qu, Z., Blumm, N., Barabási, A.: Limits of predictability in human mobility. Science 327(5968), 1018–1021 (2010)
5. Fano, R.: Transmission of information: a statistical theory of communications. M.I.T. Press (1961)
6. Jensen, B., Larsen, J., Jensen, K., Larsen, J., Hansen, L.: Estimating human predictability from mobile sensor data. In: IEEE International Workshop on Machine Learning for Signal Processing (MLSP), pp. 196–201. IEEE (2010)
7. Lin, M., Hsu, W., Lee, Z.: Predictability of individuals' mobility with high-resolution positioning data. In: Proc. of Ubicomp 2012, pp. 381–390. ACM (2012)
8. Yan, X.Y., Han, X.P., Wang, B.H., Zhou, T.: Diversity of individual mobility patterns and emergence of aggregated scaling laws. Scientific Reports 3 (2013)
9. Cho, E., Myers, S., Leskovec, J.: Friendship and mobility: user movement in location-based social networks. In: Proc. of KDD 2011, pp. 1082–1090 (2011)
10. Kontoyiannis, I., Algoet, P., Suhov, Y., Wyner, A.: Nonparametric entropy estimation for stationary processes and random fields, with applications to English text. IEEE Transactions on Information Theory 44(3), 1319–1327 (1998)

ReadBehavior: Reading Probabilities Modeling of Tweets via the Users' Retweeting Behaviors

Jianguang Du, Dandan Song, Lejian Liao, Xin Li, Li Liu, Guoqiang Li, Guanguo Gao, and Guiying Wu

Beijing Engineering Research Center of High Volume Language Information Processing & Cloud Computing Applications, Beijing Key Laboratory of Intelligent Information Technology, School of Computer Science & Technology, Beijing Institute of Technology, Beijing, China, 100081
{dujianguang,sdd,liaolj,xinli,3120120376,lgqsj}@bit.edu.cn,
{guanguogao1005,wuguiying1989}@gmail.com

Abstract. Along with *twitter*'s tremendous growth, studying users' behaviors, such as retweeting behavior, have become an interesting research issue. In literature, researchers usually assumed that the *twitter* user could catch up with all the tweets posted by his/her friends. This is untrue most of the time. Intuitively, modeling the reading probability of each tweet is of practical importance in various applications, such as social influence analysis. In this paper, we propose a *ReadBehavior* model to measure the probability that a user reads a specific tweet. The model is based on the user's retweeting behaviors and the correlation between the tweets' posting time and retweeting time. To illustrate the effectiveness of our proposed model, we develop a *PageRank*-like algorithm to find influential users. The experimental results show that the algorithm based on *ReadBehavior* outperforms other related algorithms, which indicates the effectiveness of the proposed model.

1 Introduction

Micro-blog services, such as *Twitter*, have grown rapidly in recent years. It has more than 500 million registered users and 200 million active users. More than 400 million tweets are posted per day[1]. On *twitter*, the user whose tweets are followed is called *friend*, while the user who is following is called *follower*.

Followers may get lots of tweets from their *friends*[2]. One scenario is that the user has too many active friends, who post tweets very frequently. In this scenario, he/she is unable to read all the tweets. Previous works, such as on identifying influential users [1], assumed that *followers* could read all the tweets on their micro-blog spaces. However, it is actually untrue most of the time. Another work [2] systematically investigated the underlying mechanism of the retweeting

[1] http://expandedramblings.com/index.php/march-2013-by-the-numbers-a-few-amazing-twitter-stats/.

[2] Without causing misunderstanding, we will use "tweets" to denote "tweets from users' *friends*" in the rest of this paper.

V.S. Tseng et al. (Eds.): PAKDD 2014, Part I, LNAI 8443, pp. 114–125, 2014.

behaviors, and divided the tweets into three categories: retweet, ignore[3], and miss. Nonetheless, it is not obvious to detect whether the tweets are missed or ignored in some situations, even with the help of users' login time.

In this paper, we propose a *ReadBehavior* model to measure the reading probability of each tweet, where reading probability means the probability of a tweet that is read by a user. The model is based on the user's retweeting behavior, which is the correlation between the tweet's original posting time and the corresponding retweeting time. *ReadBehavior* is of practical importance and can be beneficial for various applications, such as social influence analysis.

To illustrate the effectiveness of *ReadBehavior*, we develop an **I**mproved **P**age**R**ank (*IPR*) algorithm, which is an extension of the *PageRank* [3] algorithm. Both *IPR* and *PageRank* find influential users on the whole network. *PageRank* assumes that all the tweets are read by the user, whereas *IPR* calculates the reading probability of tweets, and then estimates the number of tweets that are read.

Experiments are then conducted to evaluate the performance of our proposed model. The results show that *IPR* outperforms other related algorithms, such as *PageRank*, *FollowersNum*, and *TweetsNum*. Consequently, the results also verify the effectiveness of *ReadBehavior*.

The main contributions of this paper include:

1. We propose a *ReadBehavior* model to measure the reading probability of each tweet posted to a user according to his/her retweeting behaviors. To the best of our knowledge, it is the first model to measure the reading probability of tweets.
2. Based on our proposed model, we develop an *IPR* algorithm to find influential users. The experimental results show the advantages and effectiveness of the proposed model.

The rest of this paper is organized as follows. A review of related work is given in section 2. Section 3 presents the proposed *ReadBehavior* model in detail. Following that, in Section 4, we present *IPR*, which is based on *ReadBehavior*. Then, experimental results are presented in Section 5. Finally, Section 6 concludes this paper.

2 Related Work

There have been quite some studies on micro-blog services, especially on *Twitter*, e.g., identifying influential users. The Web Ecology project [4] measured influential users for a 10-day period. This work performed a comparison of three measures of influence - retweet, reply, and mention. Cha et al. [5] used mentions, retweets, and the number of followers as influence measurement. Kwak et al. [6] used *PageRank* of the network constructed by *followers*. Bakshy et al. [7] used the information cascades. Recently, researchers also studied social

[3] "ignore" means that *followers* read the tweets but do not choose to retweet them.

influence from the topics perspective [8,9,10,11,12]. Romero et al. [1] designed the influence-passivity algorithm to measure influential users. However, all these methods assumed that *followers* could read all the tweets. Whereas in reality, users may miss some of the tweets.

Other works studied the retweeting behaviors. Boyd et al. [13] treated retweeting behaviors as conversations inside Twitter, and studied the basic issues about retweet. Hong et al. [14] studied the problem of predicting popular tweets according to the future retweets. Petrovic et al. [15] explored tweets features to predict whether a tweet would be retweeted. Benevenuto et al. [16] analyzed the user workloads based on users' behaviors. Uysal and Feng [17,18] proposed a tweet ranking method to help users to catch up with valuable tweets based on the retweet history. Compared with our model, they always ignored the tweets time (tweets' posting time), which we treat as an important factor.

There were some works that mentioned the tweets time. Yang et al. [2] studied the underlying mechanism of the retweeting behaviors. They divided the tweets into three categories - retweet, ignore, and miss, and then classified the retweeting delay into short-term intervals and long-term intervals. However, they did not mention how to deal with the situation when it was not clear whether a tweet was missed or ignored. Dabeer et al. [19] proposed response probability that bear the similarity with susceptibility using the tweets time. However, as far as we know, very few researchers have studied the tweets time in the view of measuring the reading probability.

3 The Proposed Model

To measure the reading probabilities of the tweets on users' micro-blog spaces, we propose the *ReadBehavior* model according to users' retweeting behaviors. In this section, we will describe the model in detail.

3.1 *Terminology*, *Assumption*, and *Fact*

When users reading tweets on their micro-blog spaces, the timeline lists the tweets in reverse chronological order. Once they find an interesting tweet, they will retweet it. This is denoted as the retweeting behavior.

For the simplicity of the analysis, we have the following *terms*, *assumptions*, and *facts*. We should make it clear that our assumptions are based on users' general reading habits. Although we do not think about special cases, the experimental results in Section 5 still show the advantages of our model.

Term. *Check* means the user's reading behavior. *Check period* means a period when the user is continuously reading the tweets, and *check time* means the start time of the *check period*.

Assumption 1. Users read the tweets from top to bottom, i.e., they read from the latest one to the earliest one. Although the mobile *Twitter* client starts from the last tweet that was shown to the user, he/she must refresh to get new tweets with the latest one on the top.

Assumption 2. If users encounter a tweet which they have already read, they will not read the tweets below it.

Assumption 3. Once users read an interesting tweet, they will retweet it, i.e., they will not read back to retweet a tweet.

Assumption 4. Assume that time $t_i^{'}$ and time $t_j^{'}$ are two adjacent retweets time during one *check period*, and t_i and t_j are the corresponding tweets time, respectively. If $t_i < t_j$, then $t_i^{'} > t_j^{'}$, and the user must read the tweets posted between t_i and t_j. Because we think he/she is continuously reading the tweets.

Fact 1. Given a tweet posted at time t_i, and it was retweeted at time $t_i^{'}$. The *check time* of this reading behavior must be between t_i and $t_i^{'}$.

Fact 2. If a user retweets a tweet, he/she must have read the tweet.

Fact 3. Tweets, which are posted after the *check time*, are not able to be read by the user in this *check period*. If users refresh to get new tweets, we treat it as a new *check period*.

3.2 *ReadBehavior* Model

Two Retweeting Behaviors. In the proposed *ReadBehavior* model, we study the retweets time and the corresponding original tweets time. For example, assume that user A was reading the tweets. When he/she found an interesting tweet C_i which was posted at time t_i, he/she retweeted it at time $t_i^{'}$ ($t_i < t_i^{'}$). After that, he/she may continue reading or become idle. Assume that there also exists another retweeting behavior at time $t_j^{'}$, and the original tweet C_j was posted at time t_j ($t_j < t_j^{'}$). There are three intuitive scenarios considering the time sequence if $t_i < t_j$:

Scenario 1: $t_i < t_j < t_j^{'} < t_i^{'}$

Scenario 2: $t_i < t_i^{'} < t_j < t_j^{'}$

Scenario 3: $t_i < t_j < t_i^{'} < t_j^{'}$

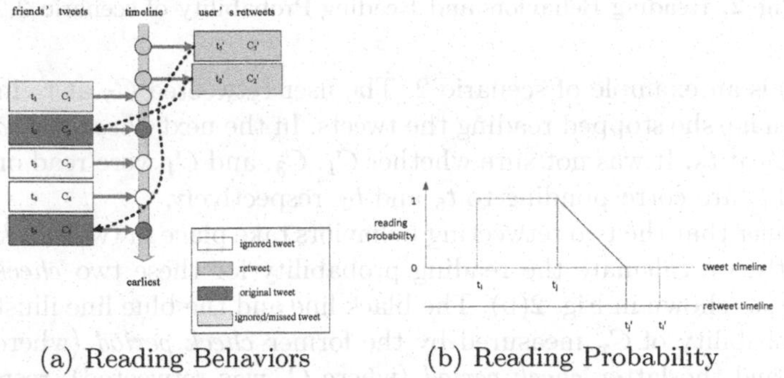

(a) Reading Behaviors (b) Reading Probability

Fig. 1. Reading Behaviors and Reading Probability of scenario 1

For scenario 1, we show an intuitive example in Fig. 1(a). When the user was reading the tweets, he/she retweeted C_2 at $t_2^{'}$. Then he/she continued reading the tweets, and retweeted C_5 at $t_5^{'}$. As the user was continuously reading the tweets, he/she read the tweets posted between t_2 and t_5. It was not sure whether C_1 was read or missed. Here t_i and t_j are corresponding to t_5 and t_2, respectively.

The reading probability is illustrated by Fig. 1(b). According to *Fact 2*, the two retweeting behaviors take place in one *check period*. The reading probability of every tweet posted between t_i and t_j (including t_i and t_j) is 1 according to *Assumption 4*. The *check time* is between t_j and $t_j^{'}$ according to *Fact 1*. Similarly, the reading probabilities of tweets newer than $C_j^{'}$ are 0 according to *Fact 3*. Because the reading probabilities of tweets posted between t_j and $t_j^{'}$ are descending, we assume that the reading probabilities are linearly descending. So given t_x which is between t_i and $t_i^{'}$, then the reading probability of C_x in this *check period* is measured by:

$$P(C_x) = \begin{cases} 1, & t_i \leqslant t_x \leqslant t_j \\ \frac{t_j' - t_x}{t_j' - t_j}, & t_j < t_x < t_j' \\ 0, & t_j' \leqslant t_x \leqslant t_i' \end{cases} \tag{1}$$

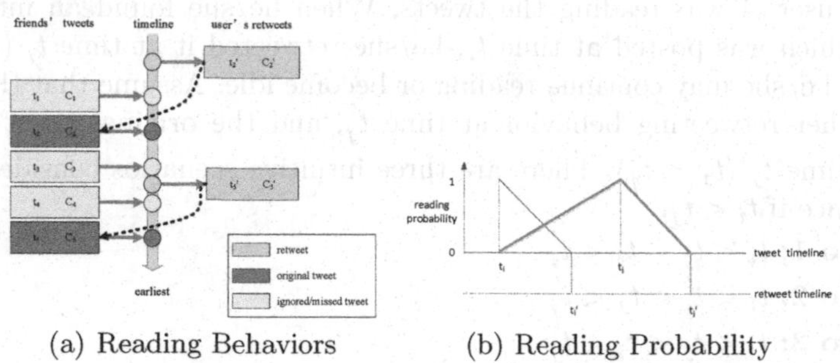

(a) Reading Behaviors (b) Reading Probability

Fig. 2. Reading Behaviors and Reading Probability of scenario 2

Fig. 2(a) is an example of scenario 2. The user retweeted C_5 at $t_5^{'}$ in a *check period*. Then he/she stopped reading the tweets. In the next *check period*, he/she retweeted C_2 at $t_2^{'}$. It was not sure whether C_1, C_3, and C_4 were read or missed. Here t_i and t_j are corresponding to t_5 and t_2, respectively.

We can infer that the two retweeting behaviors take place in two *check periods*. Consequently, we calculate the reading probability for these two *check periods* respectively, as shown in Fig. 2(b). The black line and the blue line illustrate the reading probability of C_x measured by the former *check period* (where C_i was retweeted) and the latter *check period* (where C_j was retweeted), respectively. For the former *check period*, similar to the analysis of scenario 1, the reading probabilities of tweets posted at t_i and $t_i^{'}$ (if exist) are 1 and 0, respectively. If $t_i < t_x < t_i^{'}$, the reading probability of C_x is $\frac{t_i^{'} - t_x}{t_i^{'} - t_i}$. For the latter *check period*, the reading probabilities of tweets posted before $t_j^{'}$ are 0 according to *Fact 3*. If t_x

is between t_j and $t_j^{'}$, the reading probability of C_x is $\frac{t_j^{'}-t_x}{t_j^{'}-t_j}$. And if t_x is between t_i and t_j, the reading probability of C_x is $\frac{t_x-t_i}{t_j-t_i}$. The tweets posted earlier than C_i are not read according to *Assumption 2*. Now we take both the *check periods* into account. When the reading probability of C_x can be measured by two *check periods*, we will choose the higher one.

To sum up, the reading probability of C_x posted between t_i and $t_j^{'}$ is measured by:

$$P(C_x) = \begin{cases} \max(\frac{t_i^{'}-t_x}{t_i^{'}-t_i}, \frac{t_x-t_i}{t_j-t_i}), & t_i \leqslant t_x < t_j \\ \frac{t_j^{'}-t_x}{t_j^{'}-t_j}, & t_j \leqslant t_x \leqslant t_j^{'} \end{cases} \tag{2}$$

Fig. 3(a) is an example of scenario 3. Similar to scenario 2, C_6 was retweeted at $t_6^{'}$ in the first *check period*. C_4 was retweeted at $t_4^{'}$ in the second *check period*. Here t_i and t_j are corresponding to t_6 and t_4, respectively.

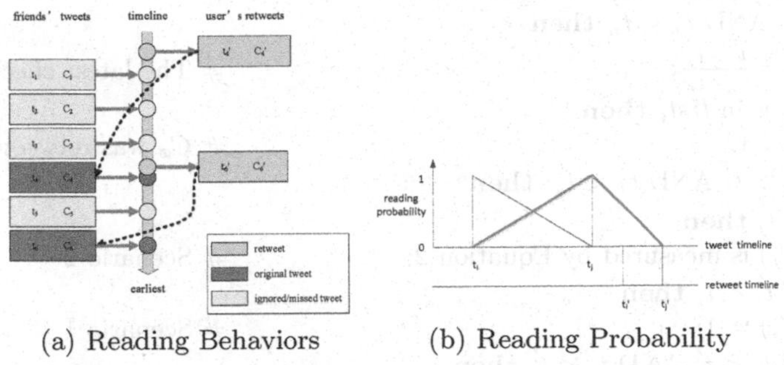

(a) Reading Behaviors (b) Reading Probability

Fig. 3. Reading Behaviors and Reading Probability of scenario 3

The main difference of this scenario from scenario 2 is that the retweet time $t_i^{'}$ is just after the tweet time t_j. It means that the first *check time* is between t_j and $t_i^{'}$ (just after t_j), as shown in Fig. 3(b). Then similar to scenario 2, the reading probability of C_x posted between t_i and $t_j^{'}$ is measured by:

$$P(C_x) = \begin{cases} \max(\frac{t_j-t_x}{t_j-t_i}, \frac{t_x-t_i}{t_j-t_i}), & t_i \leqslant t_x < t_j \\ \frac{t_j^{'}-t_x}{t_j^{'}-t_j}, & t_j \leqslant t_x \leqslant t_j^{'} \end{cases} \tag{3}$$

For tweets which are earlier than both of the original tweets, the reading probabilities of them can be measured by earlier *check periods*, which will be discussed in the following section.

Three and More Retweeting Behaviors. The above three scenarios are the essential situations. Now we expand them to three and more continuous retweeting behaviors, which are the combination of the three scenarios. We can summarize that if t_x is between t_i and t_j, the reading probability of C_x can be measured by one of the three scenarios, otherwise, the reading probability of C_x can be measured by additional boundary conditions.

Formally, assume the retweets time sequence is $(t_1^{'}, t_2^{'}, ..., t_{n-1}^{'}, t_n^{'})$, and the corresponding tweets time sequence is $(t_1, t_2, ..., t_{n-1}, t_n)$[4]. $list_t$ is the time list of the tweets that are retweeted, t_{start} is the start time of the dataset and t_{end} is the end time of the dataset. If C_x is between two adjacent retweeting behaviors, t_i and t_j are the former tweet time and the latter tweet time, respectively, then the reading probability of C_x is measured by Algorithm 1.

Algorithm 1. Reading probability of tweet C_x posted at time t_x

Input:
The time of a tweet posted by user A's friends;
The set of user A's retweets time and the corresponding tweets time;
Output:
The reading probability of the tweet;

if $t_x > t_n$ AND $t_x < t_n^{'}$ **then**
  $P(C_x) = \frac{t_n^{'} - t_x}{t_n^{'} - t_n}$; # The latest check period
else if t_x is in $list_t$ **then**
  $P(C_x) = 1$; # C_x was retweeted
else if $t_x > t_i$ AND $t_x < t_j^{'}$ **then**
  if $t_i^{'} < t_j$ **then**
    $P(C_x)$ is measured by Equation 2; # Scenario 2
  else if $t_j^{'} < t_i^{'}$ **then**
    $P(C_x) = 1$; # Scenario 1
  else if $t_j < t_i^{'}$ AND $t_i^{'} < t_j^{'}$ **then**
    $P(C_x)$ is measured by Equation 3; # Scenario 3
  end if
else if $t_x \leqslant t_1$ AND $t_x \geqslant t_{start}$ **then**
  $P(C_x) = \frac{t_x - t_{start}}{t_1 - t_{start}}$; # The earliest check period
else
  $P(C_x) = 0$;
end if
return $P(C_x)$;

With the reading probability of each tweet, we estimate the number of tweets read by a user. Assume that user B is a friend of A, and the number of tweets B posted is n, then the number of tweets read by A from B is measured by:

$$Num_{BA} = \sum_{x=1}^{n} P(C_x) \tag{4}$$

[4] Though the tweets time sequence is in the order of time, the corresponding retweets time sequence may not be.

4 The Improved *PageRank* Algorithm

In this section, we will describe the *IPR* algorithm with our proposed *ReadBehavior* model. First of all, we have the following definition:

Definition 1. $\boldsymbol{G = (N, E, W)}$ is a directed graph. N is the set of nodes. E is the set of arcs. W is the set of arc weights. If user i has ever retweeted from user j, then there is a directed arc (i, j) between them.

Based on the proposed model, we develop an *IPR* algorithm, which extends from the well-known *PageRank* algorithm. *PageRank* [3] was used by researchers to find influential users in social media. It took both the pairwise influence and the link structure into account. The main difference between *IPR* and *PageRank* is the calculation of the arc weight in Definition 1. The arc weight of *PageRank* is measured by:

$$w_{ij} = \frac{S_{ij}}{Q_j}, \tag{5}$$

where S_{ij} is the number of tweets that i has retweeted from j, and Q_j is the number of tweets posted by j.

While the arc weight of *IPR* is measured by:

$$w_{ij} = \frac{S_{ij}}{Num_{ij}}, \tag{6}$$

where S_{ij} is the number of tweets posted by j and retweeted by i, and Num_{ij} is the estimated number of tweets read by i from j according to Equation 4.

With the directed graph G, both *PageRank* and *IPR* iteratively compute to find influential users. Since G is weighted, the random surfer probability from i to j is measured by $\frac{w_{ij}}{\sum_{k:(i,k)\in E} w_{ik}}$, where E is the set of arcs in G.

5 Experiments

To evaluate the effectiveness of our proposed model, we present the experiments on a large-scale *Twitter* dataset in this section.

5.1 *Twitter* Dataset

A set of *Twitter* data from [6] is prepared for this study. The dataset contains a continuous stream of about 132 million tweets. In order to compare with the work in [1], we also use the tweets with URLs. We get about 3.38 million users posting at least one URL. In the following descriptions, we will use "tweets" to represent "tweets with URLs" in short. Similar to [1], we choose to concern users who have at least 7 tweets with URLs, and exclude the invalid users whose user ID cannot be accessed with the usernames. After that, we get 497,782 users with 17,592,586 tweets. We then strike out the isolated users (users that never retweet or are retweeted by others). Finally, we get 74,813 users who post 3,150,334 tweets. Among them, 83,356 pairs of users have retweeting relationships, and the number of retweets is 103,774.

5.2 Comparison of *IPR* with Other Related Algorithms

In this section, we study the effectiveness of *IPR* in finding influential users. Several related comparison algorithms are conducted, which include:

- ***FollowerNum***, which measures the influence of users by the number of *followers*. This measurement is widely used in many *Twitter* services.
- ***TweetsNum***, which measures the influence of users by the number of tweets posted by them.
- ***PageRank***, which measures the influence of users taking both link structure and pairwise influence into account. Nevertheless, unlike *IPR*, the arc weight is measured by Equation 5.

Evaluation Method. All the four algorithms can find influential users. However, there is no existing method to directly compare their performances. We use a cross-validation method [20] to compare these algorithms. The method is described as below:

Given four algorithms A, B, C, and D, and the sets of Top-K influential users discovered by them are I_A, I_B, I_C, and I_D, respectively. The criterion set is denoted as:

$$I_2 = (I_A \cap I_B) \cup (I_A \cap I_C) \cup (I_A \cap I_D) \cup (I_B \cap I_C) \cup (I_B \cap I_D) \cup (I_C \cap I_D) \quad (7)$$

Then, the precision of algorithm X is:

$$P_X = \frac{|I_X \cap I_2|}{|I_X|} \quad (8)$$

The recall is:

$$R_X = \frac{|I_X \cap I_2|}{|I_2|} \quad (9)$$

A better algorithm will get a higher precision score and a higher recall score. Because the best algorithm should have the greatest contribution to the criterion set.

Performance and Analysis. We compare the precision, recall, and F measure of Top-K ($K = 200, 250, 300, 350, 400, 450, 500$) influential users of the above four algorithms. Fig. 4 shows the results.

We see from Fig. 4(a) that the precisions of all the algorithms increase when the parameter K increases. In addition, as K increases, the precisions of *IPR* and *PageRank* significantly increase, whereas the precisions of *FollowersNum* and *TweetsNum* only increase little. This is because both *IPR* and *PageRank* have greater contributions to the criterion set in Equation 7. The greater contributions cause the numerator of Equation 8 to grow at faster rate comparing with the denominator. We also find that the precisions of all the algorithms are low. The reason is that the Top-K influential users discovered by these algorithms are rarely correlated with each other. However, as K increases, the correlation

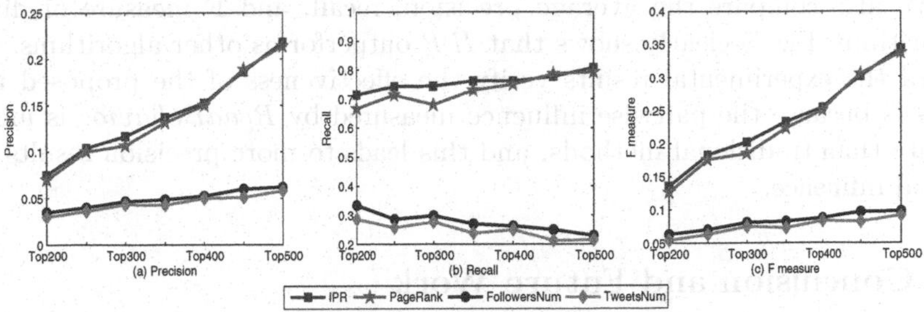

Fig. 4. Precision, Recall, and F measure of different algorithms in finding influential users

between *IPR* and *PageRank* becomes stronger. This also indicates the significant performance increasing of *IPR* and *PageRank*.

Fig. 4(b) shows the recalls of different algorithms. Surprisingly, as K increases, the recalls of *FollowersNum* and *TweetsNum* decrease. We note that the decrease in recall can be attributed to two main factors: first, both algorithms have little contribution to the criterion set, and second, both algorithms do not correlate with other algorithms. So these two factors cause the value of the numerator of Equation 9 to be small. We also observe that the recalls of *IPR* and *PageRank* are really high, which indicates that most elements of the criterion set are consist of the influential users discovered by the two algorithms.

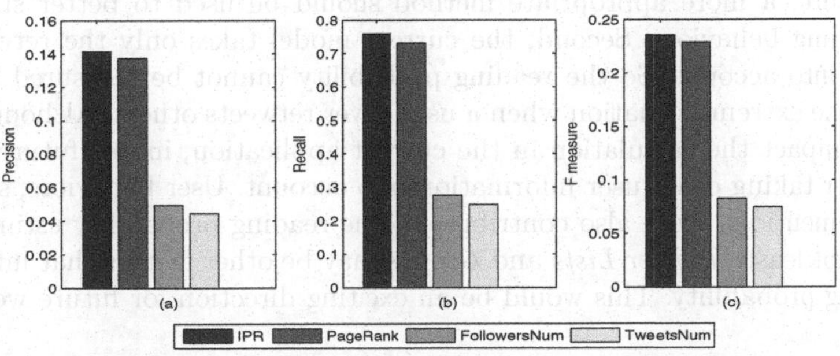

Fig. 5. The average of Precision, Recall, and F measure of different algorithms

By combining precision and recall, we get a comprehensive measurement-F measure-of those algorithms, as shown in Fig. 4(c). It is clear that *IPR* outperforms other algorithms in most cases except $K = 450$. When checking in detail, we find that the influential users (discovered by *IPR*), whose ranks are between 400 and 450, post and retweet small number of tweets containing URLs. Moreover, we find that *IPR* and *PageRank* significantly outperform *FollowersNum* and *TweetsNum* in all of the measurements, which suggests that both the number of followers and the number of tweets may not be good measures of influence.

We also compare the average precision, recall, and F measure of different algorithms. Fig. 5 clearly shows that *IPR* outperforms other algorithms.

All the experimental results verify the effectiveness of the proposed model. This is because the pairwise influence measured by *ReadBehavior* is more accurate than traditional methods, and this leads to more precision results of the global influence.

6 Conclusion and Future Work

The number of messages on the micro-blog space is large, and it is almost impossible for users to catch up with all the messages. Motivated by the fact, this paper focuses on proposing a *ReadBehavior* model to measure the reading probabilities of tweets posted by a user's friends according to his/her retweeting behaviors. To the best of our knowledge, this work is the first to measure the reading probabilities of tweets. To illustrate the effectiveness of our proposed model, an **I**mproved **P**age**R**ank (*IPR*) algorithm is proposed to find influential users on the whole network. We find from the experimental results that *IPR* outperforms its competitors, which indicate the effectiveness of our model. We should emphasize that the model not only can be used to measure influential users, but also can be used to a lot of applications which capture the reading behaviors of users.

Nevertheless, as the first attempt, it still has spaces for improvement. First, as a preliminary model, the linear approximation is used due to its simpleness. In the future, a more appropriate method should be used to better simulate users' reading behaviors. Second, the current model takes only the retweeting behaviors into account. So the reading probability cannot be measured by our model in the extreme situation when a user never retweets others. Although this does not impact the calculation in the current application, in our future work, we consider taking other user information into account. User behaviors, such as reply and mention, could also contribute to the reading probability estimation. Last but not least, *Twitter Lists* and *Groups* may be other factors that influence the reading probability. This would be an exciting direction for future work.

Acknowledgments. This work is funded by the National Program on Key Basic Research Project (973 Program, Grant No. 2013CB329605), National Natural Science Foundation of China (NSFC, Grant Nos. 60873237, 61300178 and 61003168), and Beijing Higher Education Young Elite Teacher Project (Grant No. YETP1198).

References

1. Romero, D.M., Galuba, W., Asur, S., Huberman, B.A.: Influence and passivity in social media. In: Gunopulos, D., Hofmann, T., Malerba, D., Vazirgiannis, M. (eds.) ECML PKDD 2011, Part III. LNCS, vol. 6913, pp. 18–33. Springer, Heidelberg (2011)

2. Yang, Z., Guo, J., Cai, K., Tang, J., Li, J., Zhang, L., Su, Z.: Understanding retweeting behaviors in social networks. In: CIKM, pp. 1633–1636. ACM (2010)
3. Page, L., Brin, S., Motwani, R., Winograd, T.: The pagerank citation ranking: bringing order to the web (1999)
4. Leavitt, A., Burchard, E., Fisher, D., Gilbert, S.: The Influentials: New Approaches for Analyzing Influence on Twitter. Webecology Project (September 2009)
5. Cha, M., Haddadi, H., Benevenuto, F., Gummadi, P.K.: Measuring user influence in twitter: The million follower fallacy. In: ICWSM, vol. 10, pp. 10–17 (2010)
6. Kwak, H., Lee, C., Park, H., Moon, S.: What is Twitter, a social network or a news media? In: WWW 2010: Proceedings of the 19th International Conference on World Wide Web, pp. 591–600. ACM, New York (2010)
7. Bakshy, E., Hofman, J.M., Mason, W.A., Watts, D.J.: Everyone's an influencer: quantifying influence on twitter. In: Proceedings of the Fourth ACM International Conference on Web Search and Data Mining, pp. 65–74. ACM (2011)
8. Liu, L., Tang, J., Han, J., Jiang, M., Yang, S.: Mining topic-level influence in heterogeneous networks. In: CIKM, pp. 199–208. ACM (2010)
9. Lin, C.X., Mei, Q., Han, J., Jiang, Y., Danilevsky, M.: The joint inference of topic diffusion and evolution in social communities. In: Proceedings of the 11th International Conference on Data Mining (ICDM), pp. 378–387. IEEE (2011)
10. Tang, J., Sun, J., Wang, C., Yang, Z.: Social influence analysis in large-scale networks. In: Proceedings of the 15th ACM SIGKDD International Conference on Knowledge Discovery and Data Mining, pp. 807–816. ACM (2009)
11. Weng, J., Lim, E.P., Jiang, J., He, Q.: Twitterrank: finding topic-sensitive influential twitterers. In: Proceedings of the Third ACM International Conference on Web Search and Data Mining, pp. 261–270. ACM (2010)
12. Macskassy, S.A., Michelson, M.: Why do people retweet? anti-homophily wins the day. In: Proceedings of the Fifth International AAAI Conference on Weblogs and Social Media, pp. 209–216 (2011)
13. Boyd, D., Golder, S., Lotan, G.: Tweet, tweet, retweet: Conversational aspects of retweeting on twitter. In: 2010 43rd Hawaii International Conference on System Sciences (HICSS), pp. 1–10. IEEE (2010)
14. Hong, L., Dan, O., Davison, B.D.: Predicting popular messages in twitter. In: Proceedings of the 20th International Conference Companion on World Wide Web, pp. 57–58. ACM (2011)
15. Petrovic, S., Osborne, M., Lavrenko, V.: Rt to win! predicting message propagation in twitter. In: Prof. of AAAI on Weblogs and Social Media (2011)
16. Benevenuto, F., Rodrigues, T., Cha, M., Almeida, V.: Characterizing user behavior in online social networks. In: Proceedings of the 9th ACM SIGCOMM Conference on Internet Measurement Conference, pp. 49–62. ACM (2009)
17. Uysal, I., Croft, W.B.: User oriented tweet ranking: a filtering approach to microblogs. In: Proceedings of the 20th ACM International Conference on Information and Knowledge Management, pp. 2261–2264. ACM (2011)
18. Feng, W., Wang, J.: Retweet or not?: personalized tweet re-ranking. In: Proceedings of the Sixth ACM International Conference on Web Search and Data Mining, pp. 577–586. ACM (2013)
19. Dabeer, O., Mehendale, P., Karnik, A., Saroop, A.: Timing tweets to increase effectiveness of information campaigns. In: Proc. ICWSM (2011)
20. Zhaoyun, D., Yan, J., Bin, Z., Yi, H.: Mining topical influencers based on the multi-relational network in micro-blogging sites. Communications, China 10(1), 93–104 (2013)

Inferring Strange Behavior from Connectivity Pattern in Social Networks

Meng Jiang[1], Peng Cui[1], Alex Beutel[2], Christos Faloutsos[2], and Shiqiang Yang[1]

[1] Department of Computer Science and Technology, Tsinghua University, Beijing, China
jm06@mails.tsinghua.edu.cn, {cuip,yangshq}@tsinghua.edu.cn
[2] Computer Science Department, Carnegie Mellon University, PA, USA
{abeutel,christos}@cs.cmu.edu

Abstract. Given a multimillion-node social network, how can we summarize connectivity pattern from the data, and how can we find unexpected user behavior? In this paper we study a complete graph from a large who-follows-whom network and spot lockstep behavior that large groups of followers connect to the same groups of followees. Our first contribution is that we study strange patterns on the adjacency matrix and in the spectral subspaces with respect to several flavors of lockstep. We discover that (a) the lockstep behavior on the graph shapes dense "block" in its adjacency matrix and creates "ray" in spectral subspaces, and (b) partially overlapping of the behavior shapes "staircase" in the matrix and creates "pearl" in the subspaces. The second contribution is that we provide a fast algorithm, using the discovery as a guide for practitioners, to detect users who offer the lockstep behavior. We demonstrate that our approach is effective on both synthetic and real data.

1 Introduction

Given a large social network, how can we catch strange user behaviors, and how can we find intriguing and unexpected connectivity patterns? While the strange behaviors have been documented across services ranging from telecommunication fraud [1] to deceptive Ebay's reviews [2] to ill-gotten Facebook's page-likes [3], we study here a complete graph of more than *117 million* users and *3.33 billion* edges in a popular microblogging service Tencent Weibo (Jan. 2011). Several recent studies have used social graph data to characterize connectivity patterns, with a focus on understanding the community structure [4–6] and the cluster property [7, 8]. However, no analysis was presented to demonstrate what strange connectivity pattern we can infer strange behavior from and how.

In this paper, we investigate *lockstep behavior* pattern on Weibo's "who-follows-whom" graph, that is, groups of followers acting together, consistently following the same group of followees, often with little other activity. Therefore, though the followees are not popular, they could have a large number of followers. We study different types of lockstep behavior, characterize connectivity patterns in the adjacency matrix of the graph, and examine the associated patterns in

V.S. Tseng et al. (Eds.): PAKDD 2014, Part I, LNAI 8443, pp. 126–138, 2014.

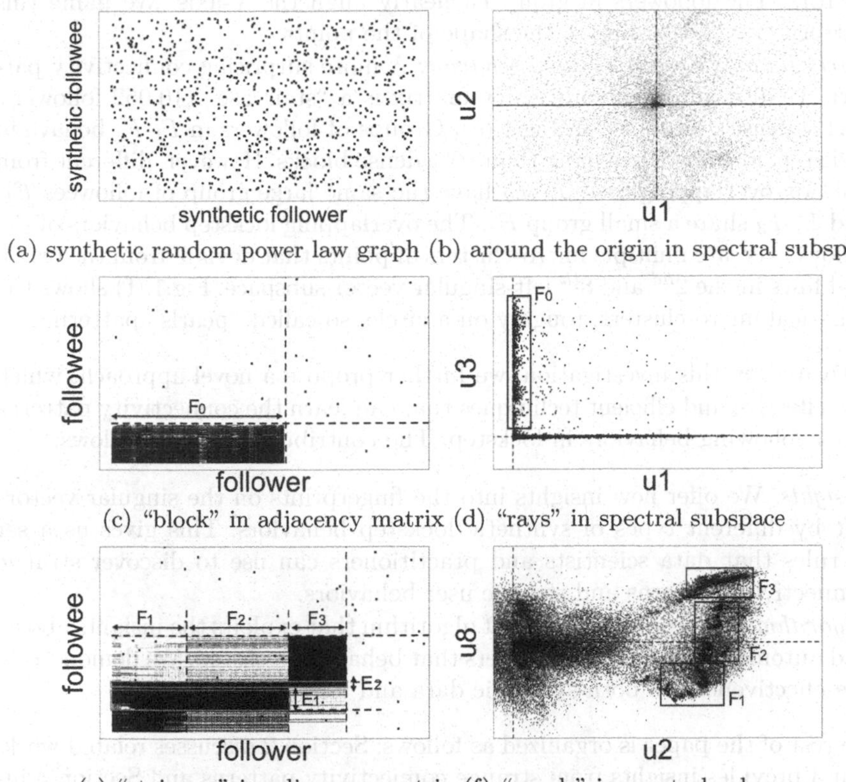

(a) synthetic random power law graph (b) around the origin in spectral subspace

(c) "block" in adjacency matrix (d) "rays" in spectral subspace

(e) "staircase" in adjacency matrix (f) "pearls" in spectral subspace

Fig. 1. Lockstep behavior shows interesting connectivity patterns and spectral patterns: On synthetic graph, followers are around the origin in all spectral subspaces. On WEIBO, non-overlapping lockstep behaviors of followers in group F_0 shape a dense "block" in adjacency matrix and create "rays" in spectral subspace. Overlapping lockstep behaviors of followers in group F_1-F_3 create a "staircase" and "pearls".

spectral subspaces. Fig.1.(a,c,e) plot connections in the matrix, in which a black point shows the follower on the X-axis connecting to the followee on the Y-axis. Fig.1.(b,d,f) plot each follower node by its values in a pair of the left-singular vectors of the adjacency matrix. These figures visualize the spectral subspaces, and the dashed lines are X- and Y-axis. Specifically, we show that

- *No lockstep behavior:* According to the Chung-Lu model [9], we generate a random power law graph where no lockstep behavior exists. The adjacency matrix in Fig.1.(a) has no large, dense components. We study every 2-dimensional spectral subspace of this synthetic graph and observe that follower nodes are around the original point, as shown in Fig.1.(b).
- *Non-overlapping lockstep behavior:* On WEIBO, there is a group of followers in F_0 connecting to the same group of followees. Thus, the adjacency matrix shows a large, dense "block" (83,208 followers, 81.3% dense) in Fig.1.(c). Fig.1.(d) plots the spectral subspace formed by the 1st and 3rd left-singular

vectors. The followers in group F_0 neatly align the Y-axis. We name this pattern "ray" according to the shape of the points.
- *Partially overlapping lockstep behavior:* A more surprising connectivity pattern we discover in the adjacency matrix is a "staircase" (10,052 followers, 43.1% dense), as shown in Fig.1.(e). Groups of followers in F_1-F_3 behave in lockstep, forming three more than 89% dense blocks. However, different from the non-overlapping case, F_1-F_2 have the same large group of followees E_1, and F_1-F_3 share a small group E_2. The overlapping lockstep behaviors of the followers create multiple micro-clusters of points that deviate from the origin and lines in the 2nd and 8th left-singular vector subspace. Fig.1.(f) shows the spherical micro-clusters, roughly on a circle, so called "pearls" pattern.

Motivated by this investigation, we further propose a novel approach, which include effective and efficient techniques that can learn the connectivity patterns and infer following behaviors in lockstep. The contributions are as follows:

- *Insights:* We offer new insights into the fingerprints on the singular vectors left by different types of synthetic lockstep behaviors. This gives us a set of rules that data scientists and practitioners can use to discover strange connectivity patterns and strange user behaviors.
- *Algorithm:* We propose an efficient algorithm that exploits the insights above, and automatically find the followers that behave in lockstep. We demonstrate the effectiveness on both synthetic data and a real social graph.

The rest of the paper is organized as follows: Section 2 discusses related work. Section 3 provides insights from strange connectivity patterns and Section 4 introduces our algorithm inferring lockstep behaviors. We give experimental results in Section 5 and conclude in Section 6.

2 Related Work

A great deal of work has been devoted to mining connectivity patterns. For finding social communities, Leskovec *et al.* [4] capture the intuition of a cluster as set of users with better internal connectivity than external connectivity. Clauset *et al.* [10] and Wakita *et al.* [11] infer community structure from network topology by optimizing the modularity. It is desirable that user of a community have a dense internal links and small number of links connected to users of other communities. For graph clustering and partitioning, Ng *et al.* [12] present a spectral clustering algorithm using eigenvectors of matrices derived from the data. Huang *et al.* [13] devise a spectral bi-partitioning algorithm using the second eigenvector of the normalized Laplacian matrix.

The properties of spectral subspaces have recently received much attention. Prakash *et al.* [14] show that the singular vectors of mobile call graphs, when plotted against each other, have separate lines along specific axes, which is associated with the presence of tightly-knit communities. The authors propose SpoKEn to chip the communities embeded in the graphs. Ying *et al.* [15] suggest that the lines formed by nodes in well-structured communities are not necessarily axes aligned. Wu *et al.* [16] give theoretical studies to explain the existence of orthogonal lines in the spectral subspaces.

However, none of the above approaches provided a guide for practitioners to understand real settings, namely, non-overlapping and partially overlapping lockstep behaviors, with an explanation for the strange spectral patterns we observe ("staircase" and "pearl"), and strange connectivity patterns.

3 Guide for Lockstep Behavior Inference

In this section, we first introduce how to plot spectral subspaces. We then study different types of lockstep behavior, show the connectivity patterns. and give a list of rules on which type of behavior the spectral patterns represent.

3.1 Spectral-subspace Plot

The concept of "spectral-subspace plot" is fundamental. The intuition behind it is that it is a visualization tool to help us see strange patterns. Let A be the $N \times N$ adjacency matrix of our social graph. Each user can be envisioned as an N-dimensional point; a spectral-subspace plot is a projection of those points in N dimensions, into a suitable 2-dimensional subspace. Specifically, the subspace is spanned by two singular vectors.

More formally, the k-truncated singular value decomposition (SVD) is a factorization of the form $A = U\Sigma V^T$, where Σ is a $k \times k$ diagonal matrix with the first k singular values, and U and V are orthonormal matrices of dimensions $N \times k$. U and V contain as their columns the left- and right- singular vectors, respectively. Let $u_{n,i}$ be the (n,i) entry of matrix U, and similarly, $v_{n,i}$ is the entry of matrix V. The score $u_{n,i}$ is the coordinate of n-th follower on the i-th left-singular vector. Thus, we define (i,j)-left-spectral-subspace plot as the scatter plot of the points $(u_{n,i}, u_{n,j})$, for $n = 1, \ldots, N$. This plot is exactly the projection of all N followers on the i-th and j-th left-singular vectors. We have the symmetric definition for the N users as followees: (i,j)-right-spectral-subspace plot is the scatter plot of the points $(v_{n,i}, v_{n,j})$, for $n = 1, \ldots, N$. Clearly, it is easy to visualize such 2-dimensional plots; if used carefully, the plots can reveal a lot of information about the adjacency matrix, as we will show shortly.

As we had shown in Fig.1.(a-b), normally, given a random power law graph, we would expect to find a cloud of points around the origin in all the spectral subspaces. However, we find strange shapes ("ray" and "pearl") in some left-spectral-subspace plots of WEIBO data. The question we want to answer here is: *What kind of user behavior could cause "rays" and "pearls" in spectral subspaces?*

The short answer is different types of lockstep behavior. We explain below in more detail what type of lockstep behavior generates such the odd patterns.

3.2 "Ray" for Non-overlapping Lockstep Behavior

In order to enumerate all the types of lockstep behavior, we introduce concepts of "camouflage" and "fame". If a group of followers F had monetary incentives

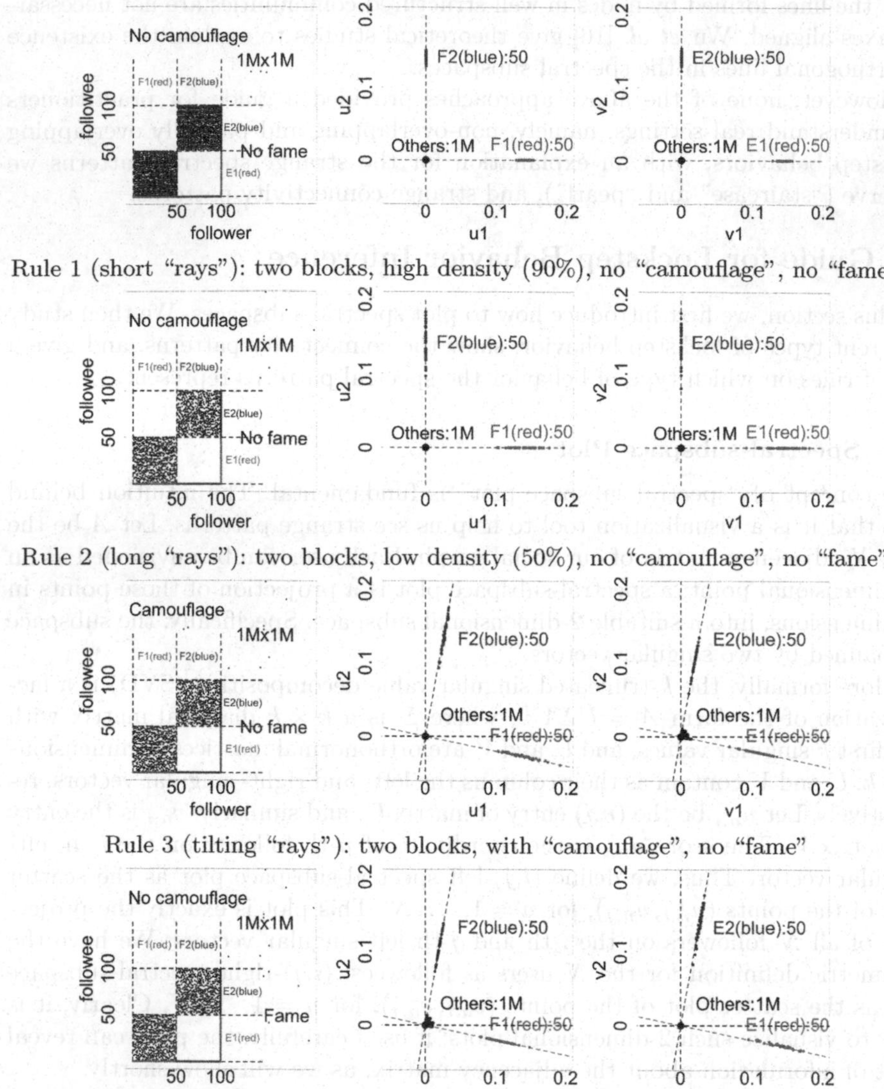

Fig. 2. Rule 1-3 ("rays"): non-overlapping blocks in adjacency matrix

to follow the same group of followees E in lockstep, they could follow additional followees who are not in E, which is called "camouflage" that helps look normal. Similarly, the group of followees E could have additional followers who are not in F, which we succinctly call "fame".

With these concepts, we can now study users' lockstep behavior with synthetic datasets. We first generate a $1M \times 1M$ random power law graph and then inject two groups of followers that separately operate in lockstep. In detail, we create

50 new followers in group F_1 to consistently follow 50 followees in group E_1. Similarly, we create another new follower group F_2 to follow a followee group E_2. Thus, if we plot black dots for non-zero entries in the adjacency matrix in the left side of Fig.2, we spot two 50×50 non-overlapping, dense blocks. Properties of the non-overlapping lockstep behavior are discussed as follows:

- *Density:* High, if a new follower connects to 90% of the related followee group; low, if the ratio is as small as 50%.
- *Camouflage:* With camouflage, if the follower connects to 0.1% of other followees; no camouflage, if he follows only the new followees and no one else.
- *Fame:* With fame, if a new followee is also followed by 0.1% of other followers; no fame, if the followee is followed by no one else.

The spectral subspaces formed by left- and right-singular vectors are plotted in the middle and right of Fig.2, respectively. We spot footprints left in these plots by the different types of non-overlapping lockstep behavior and summarize the following rules:

- *Rule 1 (short "rays"):* If the lockstep behavior of followers is compact on the graph, the adjacency matrix contains one or more non-overlapping blocks of high density like 90%. The spectral-subspace plots show short rays: a set of points that densely fall along a line that goes through the origin.
- *Rule 2 (long "rays"):* If a group of followers and a group of followees are consistently but loosely connected, the adjacency matrix contains blocks of low density like 50%. The plots show long rays: the rays stretch into lines aligned with the axes and elongate towards the origin.
- *Rule 3 (tilting "rays"):* If the follower group has "camouflage" or the followee group has "fame", the adjacency matrix shows sparse external connections outside the blocks. Different from Rule 1-2, a more messy set of rays come out of the origin at different angles, called tilting rays.

In summary, we find that non-overlapping lockstep behavior creates rays on the spectral-subspace plots: as the density decreases, the rays elongate; as the followers add camouflage or the followees add fame, the rays tilt.

3.3 "Pearl" for Partially Overlapping Lockstep Behavior

If a group of followers consistently follows their related group of followees, and partially connect to other groups of followees, we say they have partially overlapping lockstep behavior.

Here we inject the random power law graph with three follower groups F_i, for $i = 1, \ldots, 3$, and five followee groups E_i, for $i = 1, \ldots, 5$. Each follower group has 1,000 fans and each followee group has 10 idols. Followers in F_1 connect to followees in E_1-E_3; followers in F_2 connect to followees in E_2-E_4; and followers in F_3 connect to followees in E_3-E_5; Fig.3.(a) plots the adjacency matrix and (b) plots the left- and right-spectral subspaces. We summarize a new rule here.

– *Rule 4 ("pearls"):* Overlapping lockstep behavior creates "staircase" in the matrix, that is, multiple dense blocks that are overlapping due to followers from each block also connecting to some followees in some other blocks. The spectral-subspace plots show "pearls" as a set of points that form spherical-like high density regions within roughly a same radius from the origin, reminiscent of pearls in a necklace.

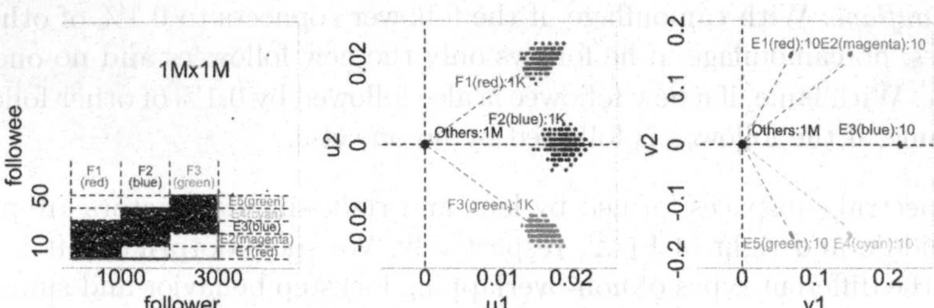

(a) adjacency matrix (b) left-spectral subspace plot (c) right-spectral subspace plot

Fig. 3. Rule 4 ("pearls"): a "staircase" of three partially overlapping blocks

In our case, Fig.3.(b) shows "pearls" of three clusters, each having 1,000 followers in groups from F_1 to F_3. Fig.3.(c) shows five clusters, each having 10 followees in E_1 to E_5. If the follower groups share some followees, or followee groups have the same followers, their clusters are close on these plots.

With the insights into patterns on spectral-subspace plots (Rule 1-4), it is now easy for a practitioner to predict connectivity patterns in the adjacency matrix and infer different types of lockstep behavior.

4 Lockstep Behavior Inference Algorithm

Our lockstep behavior inference algorithm has two steps:

– *Seed selection:* Following Rule 1-4 in Sect.3, select nodes as seeds of followers that behave in lockstep, simiply called "lockstep" followers.
– *"Lockstep" propagation:* Propagate "lockstep" score between followers and followees, and thus catch the lockstep behaviors.

4.1 Seed Selection

The algorithm can start with any kind of seeds, even randomly selected ones. However, careful selection of seeds obviously accelerates the response time. Fig.4 shows how we conduct the seed selection.

First, generate a range of spectral-subspace plots. We compute the top k left-singular vectors $u_1, \ldots, u_k$, and plot all the follower points in the subspace formed by each pair of the singular vectors. For example, Fig.4.(a) shows "rays" and "pearls" in (1,3)- and (2,8)-left-spectral-subspace plot, respectively.

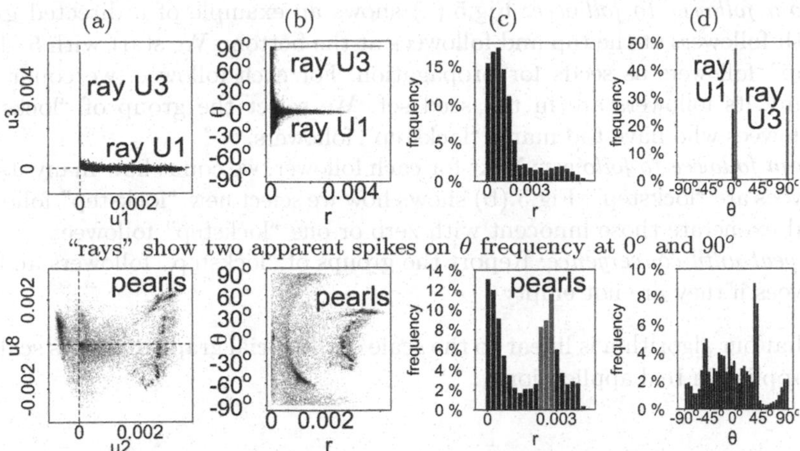

"pearls" show a spike on r frequency at a much-greater-than-zero value

Fig. 4. Find "rays" and "pearls": (a) spectral-subspace plot (b) hough transform (c) bin plot of perpendicular distance r frequency (d) bin plot of rotation angle θ frequency

Second, use the points as input to Hough transform and plot them in polar coordinates (r,θ), where r is the perpendicular distance and θ is the rotation angle. As shown in Fig.4.(b), for "rays", it shows two straight lines at $\theta = 0^o$ and $\theta = 90^o$; for "pearls", it shows a set of micro-clusters at some big r values.

Third, divide r and θ axes into bins and plot node frequencies in each bin. Therefore, for "rays", the θ-bin plot shows two apparent spikes at 0° and 90°; for "pearls", the r-bin plot shows a single spike apart from $r = 0$. With median filtering, we can detect the spikes and then catch the related nodes as seeds.

Notice that if there is no lockstep behavior, no dense block in the adjacency matrix, the spectral-subspace plots show a cloud of points around the origin, as shown in Fig.1.(a-b). The node frequency of angle θ should be almost a constant, and the node frequency of distance r should decrease smoothly with the value increasing. The r- and θ-bin plots are omitted for saving space.

4.2 "Lockstep" Propagation

We now interpret how we start with the seeds and refine a group of followers and followees with lockstep behavior. The "lockstep" value of a followee is defined as the percentage of the seeds or "lockstep" followers who are its followers. Similarly, the "lockstep" value of a follower is defined as the percentage of the "lockstep" followees who are its followees. We need a threshold to decide which users are new "lockstep" followers/followees and here we use 0.8 as default.

The algorithm recursively propagates this value from followers to followees, and vice versa, like what Belief Propagation method does. In more detail, we explain the steps as follows.

– *From follower to followee:* Fig.5.(a) shows an example of a directed graph with followers at the top and followees at the bottom. We start with 5 "lockstep" followers as seeds for propagation. For each followee, we count how many its followers are in the seed set. We select the group of "lockstep" followees who have too many "lockstep" followers.
– *From followee to follower:* Next for each follower, we count how many its followees are "lockstep". Fig.5.(b) shows how we select new "lockstep" followers and exonerate those innocent with zero or one "lockstep" followee.
– *Repeat until convergence:* Report the groups of "lockstep" followers and followees if they are not empty.

Note that our algorithm is linear to the scale of the social graph and thus scalable to be applied in real applications.

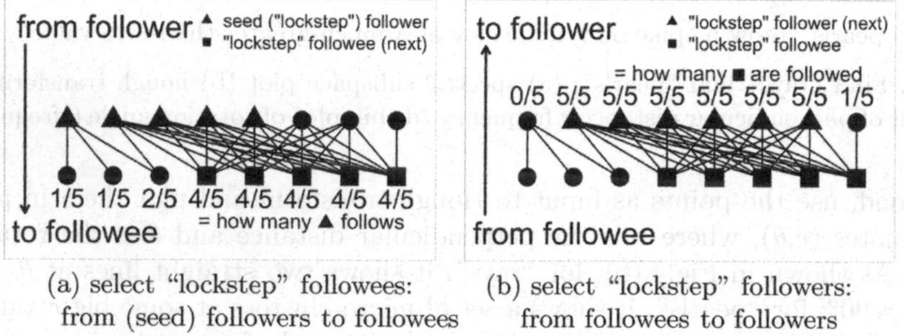

(a) select "lockstep" followees: from (seed) followers to followees (b) select "lockstep" followers: from followees to followers

Fig. 5. Find lockstep behavior of users by propagating "lockstep" value: select followers (followees) who have too many "lockstep" followees (followers)

5 Experimental Results

In this section we present our empirical evaluation, first on a large, real-world graph, and then on synthetic graphs where the ground truth is known.

5.1 Real-world Graph

We operate our algorithm on the 100-million-node social graph WEIBO. Table 1 report the statistics of strange connectivity patterns that we find on the network.

– *"Blocks" and "staircase":* With the proposed rules and algorithm, we catch a dense block with the "ray" pattern and a staircase of three overlapping blocks with the "pearl" pattern on spectral-subspace plots. Fig.1.(c,e) have show the adjacency matrix and their sets of followers F_0 and F_1-F_3.
– *High density, small "camouflage" and small "fame":* The density of every block is greater than 80%, while the density of the "staircase" is only 43%. It proves that the staircase consists of partially overlapping blocks. The camouflage, that is the connectivity between "lockstep" followers and other followees, is as small as 0.2% dense. The fame is smaller than 2%.

Table 1. Statistics of connectivity patterns formed by groups of users with lockstep behavior: The density of the "block" and blocks in "staircase" is greater than 80%, while the WEIBO followers have little "camouflage" and followees have little "fame"

	"ray" F_0	"pearl" F_1	"pearl" F_2	"pearl" F_3	"pearl" Total
Num. seeds	100	1,239	107	990	—
Size of block	$83,208 \times 30$	$3,188 \times 135$	$7,210 \times 79$	$2,457 \times 148$	$10,052 \times 270$
Density	81.3%	91.3%	92.6%	89.1%	43.1%
Camouflage	0.14%	0.06%	0.10%	0.05%	0.07%
Fame	0.05%	1.93%	1.94%	1.72%	1.73%
Out-degree	231±109	310±7	312±7	304±5	310±7
In-degree	2.0±1.4	9±6	10±6	17±13	12±9

The above numbers validate the existence of non-overlapping and partially overlapping lockstep behavior and also the effectiveness of our method. Further, we give additional evidence of the similar personalities of the "lockstep" followers.

- *Strange profiles:* The login-names of 10,787 accounts from the "lockstep" users are like "a#####" (# is a digital number, for example, "a27217"). Their self-declared dates of birth are in lockstep the Jan. 1st. They were probably created by a script, as opposed to natural users.
- *Small in-degree values of followers:* The average in-degree value of followers in the single "block" is as small as 2.0, while that of followers in the "staircase" is smaller than 20. The "lockstep" followers actively connect to their followees but they have little reputation themselves.
- *Similar out-degree values of followers:* The out-degree values of "lockstep" followers in the "staircase" are similarly around 300. In Fig.6, we plot the out-degree distribution of the graph in log-log scale and spot two spikes, which means abnormally high frequency of nodes who have around 300 followees. After we remove the "lockstep" followers, we find out that the spikes disappear and the distribution becomes smoother.

For the last point, we want to say, most graphs exhibit smooth degree distributions, often obeying a heavy-tailed distribution (power law, lognormal, etc). Deviations from smoothness are strange: Border *et al.* [17] said that in the case of the web graph, the spikes were due to link farms. Thus, if the removal of some "lockstep" users makes the degree plots smoother, then we have one more reason to believe that indeed those users were strange.

5.2 Synthetic Data

Here we want to validate the effectiveness of Rule 3 (tilting "rays") and 4 ("pearls"). We inject a group of followers and followees operating in lockstep on a 1-million-node random power law graph. The goal is to predict who are the injected nodes. We adopt *Accuracy* to qualify the performance, which is the ratio of correct predictions.

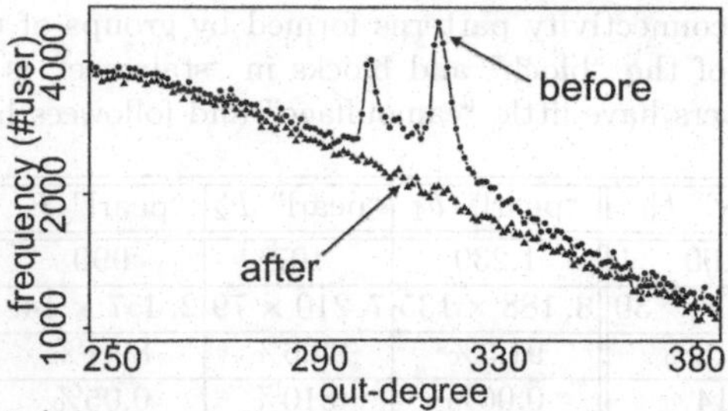

Fig. 6. The out-degree distribution (in log-log scale) becomes smoother after the removal of "lockstep" followers. The followers have similar out-degree values, i.e., similar numbers of followees from the same group.

First, we add camouflage to the followers, i.e., we increase the density of connections between the followers and other followees on the graph from 0 to 0.01. We compare the performance of different versions of our algorithm: one considers Rule 3 when it selects seeds from spectral-subspace plots, and the other does not. Rule 3 says when the followers have camouflage, the rays tilt. Fig.7.(a) shows that both accuracy values decrease with the camouflage increasing, and the algorithm that considers Rule 3 performs much better.

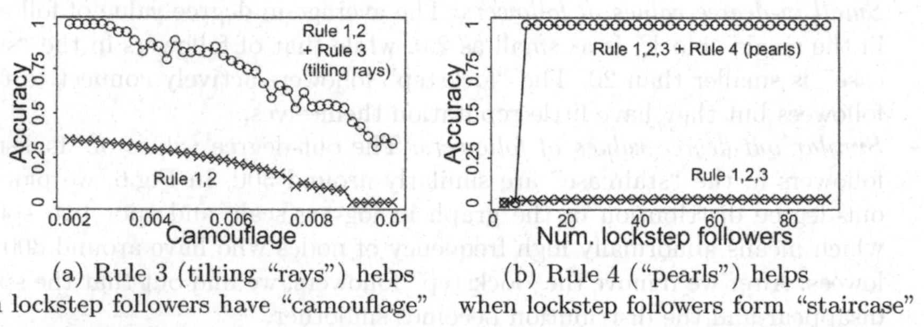

(a) Rule 3 (tilting "rays") helps when lockstep followers have "camouflage" (b) Rule 4 ("pearls") helps when lockstep followers form "staircase"

Fig. 7. Effectiveness of Rule 3-4. If the accuracy is higher, the performance is better.

Second, we inject partially overlapping lockstep behavior. In other words, we put a "staircase" in the adjacency matrix. We change the size of the staircase, i.e., the number of followers. One of the algorithms compared here considers Rule 4 and the other does not. Rule 4 says when there is a staircase, some spectral-subspace plots have "pearls". Fig.7.(b) shows that our algorithm that fully considers all the rules is sensitive to the number of "lockstep" followers. When it is bigger than 7, which is big enough for the behaviors to show footprints in the eigenspaces, we can catch over 95% of the followers, while the version that does not consider Rule 4 fails to predict them.

6 Conclusion

In this paper, we have proposed a novel method to infer users' lockstep behaviors from connectivity patterns on large "who-follows-whom" social graphs. We offer new understanding into the plots of spectral subspaces. The suspicious "ray" and "pearl" patterns are created by different types of lockstep behaviors. Using the insights, we design a fast algorithm to detect such behavior patterns. We demonstrate the effectiveness of our method on both a large real-world graph and synthetic data with injected lockstep behaviors.

Acknowledgement. This work is supported by National Natural Science Foundation of China, No. 61370022, No. 61003097, No. 60933013, and No. 61210008; International Science and Technology Cooperation Program of China, No. 2013DFG12870; National Program on Key Basic Research Project, No. 2011CB302206; NExT Research Center funded by MDA, Singapore, WBS:R-252-300-001-490. Thanks for the support of National Science Foundation, No. CNS-1314632; Army Research Laboratory under Cooperative Agreement Number W911NF-09-2-0053; U.S. Army Research Office (ARO) and Defense Advanced Research Projects Agency (DARPA), No. W911NF-11-C-0088; and the National Science Foundation Graduate Research Fellowship, Grant No. DGE-1252522.

Any opinions, findings and conclusions or recommendations expressed in this material are those of the author(s) and do not necessarily reflect the views of the National Science Foundation, DARPA, or other funding parties. The U.S. Government is authorized to reproduce and distribute reprints for Government purposes notwithstanding any copyright notation here on.

References

1. Becker, R.A., Volinsky, C., Wilks, A.R.: Fraud detection in telecommunications: History and lessons learned. Technometrics 52(1) (2010)
2. Chau, D.H., Pandit, S., Faloutsos, C.: Detecting fraudulent personalities in networks of online auctioneers. In: Fürnkranz, J., Scheffer, T., Spiliopoulou, M. (eds.) PKDD 2006. LNCS (LNAI), vol. 4213, pp. 103–114. Springer, Heidelberg (2006)
3. Beutel, A., Xu, W., Guruswami, V., Palow, C., Faloutsos, C.: CopyCatch: stopping group attacks by spotting lockstep behavior in social networks. In: Proceedings of the 22nd International Conference on World Wide Web, pp. 119–130 (2013)
4. Leskovec, J., Kevin, J.L., Dasgupta, A., Mahoney, M.W.: Statistical properties of community structure in large social and information networks. In: Proceedings of the 17th International Conference on World Wide Web, pp. 695–704 (2008)
5. Fortunato, S.: Community detection in graphs. Physics Reports 486(3), 75–174 (2010)
6. Chen, J., Saad, Y.: Dense subgraph extraction with application to community detection. IEEE Transactions on Knowledge and Data Engineering 24(7), 1216–1230 (2012)
7. Zha, H., He, X., Ding, C., Simon, H., Gu, M.: Bipartite graph partitioning and data clustering. In: Proceedings of the Tenth International Conference on Information and Knowledge Management, pp. 25–32 (2001)

8. Günnemann, S., Boden, B., Färber, I., Seidl, T.: Efficient Mining of Combined Subspace and Subgraph Clusters in Graphs with Feature Vectors. In: Pei, J., Tseng, V.S., Cao, L., Motoda, H., Xu, G. (eds.) PAKDD 2013, Part I. LNCS, vol. 7818, pp. 261–275. Springer, Heidelberg (2013)
9. Chung, F., Lu, L.: The average distances in random graphs with given expected degrees. Proceedings of the National Academy of Sciences 99(25), 15879–15882 (2002)
10. Clauset, A., Newman, M.E., Moore, C.: Finding community structure in very large networks. Physical Review E 70(6), 066111 (2004)
11. Wakita, K., Tsurumi, T.: Finding community structure in mega-scale social networks. In: Proceedings of the 16th International Conference on World Wide Web, pp. 1275–1276 (2007)
12. Ng, A.Y., Jordan, M.I., Weiss, Y.: On spectral clustering: Analysis and an algorithm. In: Advances in Neural Information Processing Systems, vol. 2, pp. 849–856 (2002)
13. Huang, L., Yan, D., Taft, N., Jordan, M.I.: Spectral clustering with perturbed data. In: Advances in Neural Information Processing Systems, pp. 705–712 (2008)
14. Prakash, B.A., Sridharan, A., Seshadri, M., Machiraju, S., Faloutsos, C.: Eigen-Spokes: Surprising patterns and scalable community chipping in large graphs. In: Zaki, M.J., Yu, J.X., Ravindran, B., Pudi, V. (eds.) PAKDD 2010. LNCS, vol. 6119, pp. 435–448. Springer, Heidelberg (2010)
15. Ying, X., Wu, X.: On Randomness Measures for Social Networks. In: SIAM International Conference on Data Mining, vol. 9, pp. 709–720 (2009)
16. Wu, L., Ying, X., Wu, X., Zhou, Z.: Line orthogonality in adjacency eigenspace with application to community partition. In: Proceedings of the 22nd International Joint Conference on Artificial Intelligence, pp. 2349–2354 (2011)
17. Broder, A., Kumar, R., Maghoul, F., Raghavan, P., Rajagopalan, S., Stata, R., Tomkins, A., Wiener, J.: Graph structure in the web. Computer Networks 33(1), 309–320 (2000)

Inferring Attitude in Online Social Networks Based on Quadratic Correlation

Cong Wang* and Andrei A. Bulatov

Simon Fraser University Burnaby, B.C. Canada
{cwa9,abulatov}@sfu.ca

Abstract. The structure of an online social network in most cases cannot be described just by links between its members. We study online social networks, in which members may have certain attitude, positive or negative, toward each other, and so the network consists of a mixture of both positive and negative relationships. Our goal is to predict the sign of a given relationship based on the evidences provided in the current snapshot of the network. More precisely, using machine learning techniques we develop a model that after being trained on a particular network predicts the sign of an unknown or hidden link. The model uses relationships and influences from peers as evidences for the guess, however, the set of peers used is not predefined but rather learned during the training process. We use quadratic correlation between peer members to train the predictor. The model is tested on popular online datasets such as Epinions, Slashdot, and Wikipedia. In many cases it shows almost perfect prediction accuracy. Moreover, our model can also be efficiently updated as the underlying social network evolves.

Keywords: Signed Networks, machine learning, quadratic optimization.

1 Introduction

Online social networks provide a convenient and ready to use model of relationships between individuals. Relationships representing a wide range of social interactions in online communities are useful for understanding individual attitude and behaviour as a part of a larger society.

While the bulk of research in the structure on social networks tries to analyze a network using the topology of links (relationships) in the network [21], relationships between members of a network are much richer, and this additional information can be used in many areas of social networks analysis. In this paper we consider signed social networks, which consist of a mixture of both positive and negative relationships. This type of networks has attracted attention of researchers in different fields [6,10,17]. This framework is also quite natural in recommender systems [3]where we can exploit similarities as well as dissimilarities between users and products.

* We'd like to thank Dr.Jian Pei for his valuable suggestions and feedbacks on our work.

V.S. Tseng et al. (Eds.): PAKDD 2014, Part I, LNAI 8443, pp. 139–150, 2014.

Over the last several years there has been a substantial amount of work done studying signed networks, see, e.g. [14,7,8,5,12,15,23,24]. Some of the studies focused on a specific online network, such as Epinions [8,18], where users can express trust or distrust to others, a technology news site Slashdot [12,13], whose users can declare others 'friends' or 'foes', and voting results for adminship of Wikipedia [5]. Others develop a general model that fits several different networks [7,14]. We build upon these works and attempt to combine the best in the two approaches by designing a general model that nevertheless can be tuned up for specific networks.

Edge Sign Prediction. Following Guha et al. [8] and Kleinberg et al. [14], [17] we consider a signed network as a directed (or undirected) graph, every edge of which has a sign, either positive to indicate friendship, support, approval, or negative to indicate enmity, opposition, disagreement. In the edge sign prediction problem, given a snapshot of the signed network, the goal is to predict the sign of a given link using the information provided in the snapshot. Thus, the edge sign problem is similar to the much studied link prediction problem [16,11], only we need to predict the sign of a link rather than the link itself.

Several different approaches have been taken to tackle this problem. Kunegis et al. [4] studied the friends and foes on Slashdot using network characteristics such as clustering coefficient, centrality and PageRank; Guha et al. [8] used propagation algorithms based on exponentiating the adjacency matrix to study how trust and distrust propagate in Epinion. Later Kleinberg et al. [14] took a machine learning approach to identify features, such as local relationship patterns and degree of nodes, and their relative weight, to build a general model predicting the sign of a given link. They train their predictor on some dataset, to learn the weights of these features by logistic regression.

Our Contribution. In this paper we also take the machine learning approach, only instead of focusing on a particular network or building a general model across different networks, we build a model that is unique to each individual network, yet can be trained on different networks. We suggest several new features into both the modeling of signed networks and the method of processing the model.

The basic assumption of our model is that users' attitude can be determined by the opinions of their peers in the network (compare to the balance and status theories [6] from social psychology, see [14]). Intuitively, peer opinions are guesses from peers on the sign of the link from a source to target node. We assume that peer opinions are only partially known, some of them are hidden. We introduce two new components into the model: set of trusted peers and influence.

Not all peer opinions are equally reliable, and we therefore choose a set of trusted peers whose opinions are important in determining the user's action. The set of trusted peers is one of the features our algorithm learns during the training phase. The algorithm forms a set of trusted peers for each individual node. The optimal composition of such a set is not quite trivial, because even trusted peers may disagree, and sometimes it is beneficial to have trusted peers who disagree.

Thus, the set of trusted peers of a node has to form a wide knowledge base on other nodes in the network.

While peer opinions provide important information, this knowledge is sometimes incomplete. Relying solely on peer opinions implies that the attitude of a user would always agree with the attitude of a peer. However, it also matters is how this opinion correlates with the opinion of the user we are evaluating. To take this correlation into account we introduce another feature into the model, influence. Suppose the goal is to learn the sign of the link between user A and user B, and C is a peer of A. Then if A tends to disagree with C, then positive attitude of C towards B should be taken as indication that A's attitude towards B is less likely to be positive. The opinion of C is then considered to be the product of his attitude towards B and his influence on A. Usually, influence is not given in the snapshot of the network and has to be learned together with other unknown parameters. We experiment with different ways of defining peer opinion, and found that using relationships and influences together is more effective than using relationships alone.

To learn the weights of features providing the best accuracy we use the standard quadratic correlation technique from machine learning [9]. This method involves finding an optimum of a quadratic polynomial, and while being relatively computationally costly, tends to provide very good accuracy. To mitigate the cost of computation we use two approaches. Firstly, we apply several techniques to split the problem and avoid solving large quadratic problems. Secondly, we attempt to make the main algorithm independent on a specific tool of quadratic optimization so that this step consuming the bulk of processing time can be easily improved as better solvers appear.

2 Approach

In our method, we start with the underlying model of a network, then proceed to the machine learning formulation of the edge sign prediction problem, and finally describe the method to solve the resulting quadratic optimization problem.

2.1 Underlying Model

We are given a snapshot of the current state of a network. Such a snapshot is represented by a directed graph $G = (V, E)$, where nodes represent the members of the network and edges represent the links (relationships). Some of the links are signed to indicate positive or negative relationships. Let $s_{x,y}$ denote the sign of the relationship from x to y in the network. It may take two different values, $\{-1, 1\}$, indicating negative and positive relationships respectively.

To estimate the sign $s_{x,y}$ of a relationship from x to y, we collect peer opinions. In different versions of the model a peer can be any node of the network, or any node linked to x. Let $p^z_{x,y} \in \{-1, 0, 1\}$ denote the peer opinion of peer z on the sign $s_{x,y}$. When $p^z_{x,y} = 1$ or $p^z_{x,y} = -1$, it indicates that the z believes that $s_{x,y} = 1$ or $s_{x,y} = -1$ respectively. When $p^z_{x,y} = 0$ that means z does not have enough knowledge to make a valid estimation.

Another assumption made in our model is that not every peer can make a reliable estimation. Therefore we divide all peers of a node into two categories, and count the opinions only of the peers from the first category, trusted peers. The problem of how to select a set of trusted peers and use their opinions for the estimation will be addressed later. Let P_x denote the set of trusted peers of a. We estimate the sign $s_{x,y}$ of a relationship from x to y by collecting the opinions of peers $z \in P_x$. If the sum of the opinions is nonnegative, then we say $s_{x,y}$ should be 1, otherwise, it should be -1. This can be expressed as,

$$s_{x,y} = sign\left(\sum_{z \in P_x} p^z_{x,y}\right).$$

2.2 Machine Learning Approach

Our approach to selecting an optimal set of trusted peers is to consider the quadratic correlations between each pair of peers. The overall performance of a set of peers is determined by the sum of the individual performances of each of them together with the sum of their performance in pairs. The individual performance measures the accuracy of individual estimations, while the pairwise performance measures the degree of difference between the estimations of the pair of peers. We want to maximize the accuracy of each individual and the diversity of each pair at the same time.

Our goal is to use the information in G to build a predictor $S(x, y)$ that predicts the sign $s_{x,y}$ of an unknown relationship from x to y with high accuracy. Function $S(x, y)$ is defined as the sign of the sum of peer opinions as follows. Let

$$F_x(y) = \sum_{z \in P_x} p^z_{x,y} \tag{1}$$

be the sum of individual peer opinions. Then set

$$S(x,y) = \begin{cases} 1, & \text{if } F_x(y) \geq 0\ , \\ -1, & \text{if } F_x(y) < 0. \end{cases}$$

Since P_x is unknown, we introduce a new variable $w_{z,x} \in \{0, 1\}$ which indicates if a node $z \in V$ should be included into set P_x. Hence, we rewrite (1) using the characteristic function $w_{z,x}$ as,

$$F_x(y) = \sum_{z \in V} w_{z,x} p^z_{x,y}. \tag{2}$$

Quadratic optimization problem. We are now ready to set the machine learning problem. A training dataset (a subset of G) is given. Every entry of the training dataset is a known edge along with its sign. Let a training dataset be $D = \{(x_i, y_i, s_{x_i,y_i}) | i = 1, ..., M\}$. The goal is to minimize the objective function, finding the optimal weight vector $w = \{w_x | x \in V\}$, where $w_x = \{w_{z,x} | z \in V\}$.

We use machine learning methods [9] to train the predictor $S(x, y)$ and learn an optimal weight vector such that the objective function below is minimized.

$$w^{opt} = argmin_w \left(\sum_{(x,y,s_{x,y}) \in D} \left(\frac{1}{N} \sum_{z \in V} (w_{z,x} p^z_{x,y} - s_{x,y})^2 + \lambda |w| \right) \right).$$

Note that there will be more details on peer opinion terms $p^z_{x,y}$.

2.3 Peer Opinion Variants

As mentioned earlier, we are going to test our model using different peer opinion formulations. First, let $s'_{x,y}$ be extension of $s_{x,y}$ to edges with unknown sign and also to pairs of nodes that are not edges defined by

$$s'_{x,y} = \begin{cases} s_{x,y}, & \text{if } s_{x,y} \text{ exists,} \\ 0, & \text{otherwise.} \end{cases}$$

Simple-adjacent. The simplest option, later referred to as *Simple-adjacent*, is based on the given information, to formulate peer opinions using existing relationships from peers to the target node, that is, $p^z_{x,y} = s'_{z,y}$.

Standard-pq. In the *Standard-pq* option the influences $r_{z,x} \in \{-1, 0, 1\}$ associated with each pair of vertices z, x is an unknown parameter. A positive influence, $r_{z,x} = 1$, indicates that the attitude of z affects x positively, while a negative influence, $r_{z,x} = -1$ indicates that the attitude of z affects x negatively, and our expectation of $p^z_{x,y}$ based on z's opinion $s'_{z,y}$ has to be reversed, that is, $p^z_{x,y} = s'_{z,y} r_{z,x}$. Also, in the Standard-pq mode we consider nonadjacent nodes as potential peers to accommodate the problem of possible missing edges. Details of the Standard-pq mode is explained in the experiment section. Since the standard formulation gives us the best result in experiments, we use it throughout our discussion. Using the standard formulation, we rewrite Equation (2) as

$$F_x(y) = \sum_{z \in V} w_{z,x} s'_{z,y} r_{z,x}. \tag{3}$$

Standard-adjacent. Finally, in the *Standard-adjacent* option the peers of x are restricted to the neighbours of x. $F_x(y) = \sum_{z \in N(x)} w_{z,x} s'_{z,y} r_{z,x}$. The rest is defined in the same way as for the *Standard-pq* option.

2.4 Simplifying the Model

In our model, we are given a directed complete graph $G = (V, E)$. In (3), both $w_{z,x}$ and $r_{z,x}$ are unknown parameters. Since $r_{z,x} \in \{-1, 0, 1\}$, we can reduce the number of unknown parameters by considering all possible values of $r_{z,x}$, and rewriting $F_x(y)$ as $F_x(y) = \sum_{z\in V} w^+_{z,x}s'_{z,y} - w^-_{z,x}s'_{z,y}$,where $w_{z,x} = w^+_{z,x} + w^-_{z,x}$ for $w^+_{z,x}, w^-_{z,x} \in \{0,1\}$. If $w^+_{z,x} = 1$, then $z \in P_x$ and $r_{z,x} = 1$. Similarly, $w^-_{z,x} = 1$ indicates that $z \in P_x$ and $r_{z,x} = -1$. When both $w^-_{z,x} = 0$ and $w^+_{z,x} = 0$, then $z \notin P_x$. When $w^-_{z,x} = 1$ and $w^+_{z,x} = 1$, the two term cancel out each other. Moreover, the regularization term ensures such case will not happen as $w^-_{z,x} = w^+_{z,x} = 0$ is always better than $w^-_{z,x} = w^+_{z,x} = 1$. Although $r_{z,x}$ can take three possible values, there are only two terms in Equation (3) since when $r_{z,x} = 0$, the term is also zero regardless of the value of $s'_{z,y}$.

Now to minimize the objective function, we need to determine the optimal weight vector $w = \{w_x | x \in V\}$ where $w_x = \{w^+_{z,x}, w^-_{z,x} | z \in V\}$ such that

$$w^{opt} = argmin_w \left(\sum_{(x,y,s_{x,y})\in D} \left(\frac{1}{N} \sum_{z\in V} (w^+_{z,x} - w^-_{z,x}) s'_{z,y} - s_{x,y} \right)^2 + \lambda |w| \right). \tag{4}$$

From the definition, we know that w_x and w_y are independent for different nodes x and y. Instead of solving for w^{opt} directly, we can solve w^{opt}_x for each $x \in V$ separately, and then combine their values to get $w^{opt} = \{w^{opt}_x | x \in V\}$

$$w^{opt}_x = argmin_{w_x} \left(\sum_{(x,y,s_{x,y})\in D} \left(\frac{1}{N} \sum_{z\in V} (w^+_{z,x} - w^-_{z,x}) s'_{z,y} - s_{x,y} \right)^2 + \lambda |w_x| \right). \tag{5}$$

Now, instead of solving a QUBO of size $2n^2$, we could solve n QUBOs of size $2n$ separately which can be solved approximately by a heuristic solver. Another approach (similar to [20]) is to further simplify the problem, as it is still challenging to solve each of these size $2n$ QUBOs exactly.

Breaking down the problem. In order to find a good approximation of the optimal solution to the QUBO defined by (4), we could break it down to much smaller QUBOs. Given a subset $U \subset V$, let $w_{x,U} = \{w^+_{z,x}, w^-_{z,x} | z \in U\}$, and define a restricted optimization problem as follows

$$w^{opt}_{x,U} = argmin_{w_{x,U}} \left(\sum_{(x,y,s_{x,y})\in D} \left(\frac{1}{N} \sum_{z\in U} (w^+_{z,x} - w^-_{z,x}) s'_{z,y} - s_{x,y} \right)^2 + \lambda |w_{x,u}| \right). \tag{6}$$

The optimal solution w^{opt}_x can be approximated by combining w^{opt}_{x,U_i} for $V = \bigcup_{i=1}^m U_i$. Next, we describe a method to decompose V and to combine w^{opt}_{x,U_i}.

2.5 Method

The optimization problem defined by (4) is a quadratic unconstrained binary optimization problem which is NP-hard in general. To solve the problem, we use two approaches. First, we solve the problem for each individual node separately, as given by (5), using METSLib Tabu search heuristics. The clear setback of this method is that for large problems the tabu search does not find the best solution. Second, we apply a similar method as described in [20] to reduce the size of the problem dramatically. In order to do that the variables are first ordered according to their individual prediction error: For each data point $(a, u, s_{a,u})$ in the training dataset, we count e_{v_-}, the number of instances $p^v_{a,u} \neq s_{a,u}$ when $r_{v,a} = -1$, and e_{v_+}, the number of instances $p^v_{a,u} \neq s_{a,u}$ when $r_{v,a} = 1$ separately. Note that since $p^v_{a,u} = r_{v,a}s'_{v,u}$, we can compute this number; and that if u is not a neighbour of v then it contributes to both e_{v_-} and e_{v_+}. Then, we replace v by v_+ and v_- with individual prediction error, e_{v_+} and e_{v_-} respectively. The subset U is iteratively selected by picking the first d nodes in the list that are not yet considered. The value of d is an important parameter of the algorithm and is selected manually at the beginning of the algorithm. In the experiment section, we show how the prediction accuracy changes as the size of d changes. The sorting and selecting processes not only reduce the amount of computation, but also allow us to consider the relevant nodes first. The small subproblems are now solved by the brute-force method or Cplex. Algorithm 1 describes the method we use to solve each subproblem. Algorithm 2 uses Algorithm 1 as a subroutine and explains how the problem is broken down into subproblems and also how to combine the solutions of subproblems to obtain an approximate solution.

Algorithm 1. Learn the parameters for a subset

Require: training dataset: *TD*, validation dataset: *VD*, a subset of nodes: U
Ensure: values of w and $Z \subset U$ where Z is the set of trusted peers

```
Z = ∅, e_min = |TD|
for λ = λ_min to λ_max do
    Z_current = ∅
    solve the optimization w^opt = argmin_w(Σ_{i=1}^M (1/N Σ_{z∈U}(w_z^+ − w_z^-)s_{z,y} − s_{x,y})^2 +
    λ|w|)
    if w_z == 1 then
        Z_current = Z_current ∪ z
    end if
    Measure the validation error e_val on VD using Z_current
    if e_val < e_min then
        Z = Z_current, e_min = e_val
    end if
end for
```

2.6 Running Time

Although the QUBO problem is NP-hard in general, the proposed algorithm can be very efficient when using the right solver and right parameter d. Let $T(d)$ denote the time of solving a size d optimization problem defined by (6), and k_λ be the number of λ values tested. The running time of Algorithm 1 is $O(k_\lambda T(d))$. Let n_v denote the number of neighbours of a node v. By our definition, n_v is at most $\|V\|$. In Algorithm 2, Algorithm 1 is repeated at most $\frac{n_v}{d}$ times. Therefore,

Algorithm 2. Solve the optimization problem

Require: training dataset: TD, validation dataset: VD, The size of the subset: d
Ensure: values of w_z for $z \in V$, and the set of trusted peers Z
  $e_{old} = |TD|$, $e_{new} = |TD|$-1, $Z, Z_{current} = \emptyset$
  sort nodes of V by their individual prediction errors in increasing order
  U = the first d nodes in V
  while $e_{old} > e_{new}$ **do**
    $Z = Z_{current} \cup Z$
    $w_z, Z_{current}$ = Algorithm 1(TD, VD, U) e_{old}=e_{new}
    Measure the validation error e_{new} on vd using Z
    update U with the next d nodes in V
  end while

the running time of Algorithm 2 is $O(\frac{n_v k_\lambda}{d} T(d))$. Traditionally, the best λ for a model is determined before training stage through cross validations. In our model, we are building a personalized predictor for each node. Hence, we need to pick a λ for each node separately. During the training stage, we test $k_\lambda = 25$ different values in the range $[\lambda_{min} = .001,\ \lambda_{max} = 0.25]$, and use the λ which gives the lowest validation error. Since k_λ is a constant, the running time crucially depends on the efficiency of the QUBO solver. For example, in the experiments, when $d = 10$ and $T(d)$ is limited to 1 sec for METSLib-solver [2], the predictor for a node can be determined instantly. Yet, when $d = 10$, using Cplex-solver [1] would take a few seconds to minutes to determine the predictor for a node.

Since the main focus of our paper is on the prediction accuracy of the model, we do not measure and compare the running times for different solvers, and we keep our experiments on a standard set of solvers rather than some exclusive ones. Although the model is currently limited by the power of the software solvers, it has shown a good potential. Its performance will improve as better solutions are found by more efficient solvers. One of such solvers could be the quantum system which is rapidly developing at D-wave System. A recent work [19] which compares the performances of different software solvers with D-wave hardware on different combinatorial optimization problems shows promise.

3 Experiment

3.1 Datasets

We use three datasets borrowed from [14]. In order to make comparison possible the datasets are unchanged rather than updated to their current status. The dataset statistics is therefore also from [14] (see Table 1).

3.2 Parameters of Datasets

In our experiment, we split each dataset into two parts. We randomly pick one tenth of the dataset for testing. The remaining dataset is used for training.

Table 1. Basic statistics on the datasets

Dataset	Epinions	Slashdot	Wikipedia
Nodes	119217	82144	7118
Edges	841200	549202	103747
+1 edges	85.0%	77.4%	78.7%
-1 edges	15.0%	22.6%	21.2%

Table 2. Number of edges passing the threshold

Dataset	Epinions	Slashdot	Wikipedia
(p,q)=(15,20)	247725	25436	51372
embeddedness 25 ([14])	205796	21780	28287

The training dataset is split into two equal parts, half for training and half for validating during the training process.

When the dataset is sparse, it is hard to build good classifiers due to the lack of training and testing data. In order to get a better understanding of the performance of the model, edge embeddedness of an edge uv is introduced in [14,7] as the number of common neighbours (in the undirected sense) of u and v. Instead of testing the model over the entire dataset, they only consider the performance restricted to subsets of edges of different levels of minimum embeddedness. For example, Kleinberg et al. [14] restrict the analysis to edges with minimum embeddedness 25.

Similarly, we also introduce two parameters that restrict the analysis of our model. Instead of considering the edge embeddedness of a link, we consider the node embeddedness of the source and target nodes. The first parameter, p, controls which nodes are considered peers of a given node v. A node u is considered a peer of v if it has at least p common neighbours with v. When $p = 0$, we consider every node in the network as a peer of v. The second parameter, q, restricts the set of links whose sign we attempt to predict. We try to predict the sign $s_{u,v}$, only if u is connected to at least q peers of v.

In Table 3, we show the dependence of the prediction accuracy of our model on different values of p and q. The data in Table 3 is obtained using option *Standard-pq* with $d = 10$ solved by Cplex. As p and q grows, the performance of the model clearly improves. However, increasing the values of p and q severely restricts the set of nodes that can be processed. We choose somewhat optimal values of these parameters, $q = 20$ and $p = 15$ and use them in the rest of our experiments. It is also worth to notice that values $q = 20$ and $p = 15$ are less restrictive than edge embeddedness 25. As shown in Table 2, more edges pass the $q = 20$ and $p = 15$ threshold than the edge embeddedness 25 threshold.

3.3 Experimental Results

As explained before, our model depends on several parameters: internal parameters, such as, the peer opinion variant and the method of solving the QUBO, and external parameter, such as, balancing the dataset. We first make the comparison for different internal parameters using the original unbalanced datasets.

In Table 6, we compare the performance of our model using different peer opinion formulations. As shown in the table, standard formulations (*Standard-adjacent*,

Table 3. Prediction Accuracy for Different Values for p, q

Dataset	Epinions	Slashdot	Wiki
(p,q)=(10,0)	91.7%	84.2%	85.0%
(p,q)=(10,10)	92.8%	91.6%	86.5%
(p,q)=(10,20)	93.7%	93.9%	86.6%
(p,q)=(10,30)	95.6%	95.1%	87.6%
(p,q)=(15,0)	93.7%	87.7%	85.2%
(p,q)=(15,10)	95.8%	96.1%	86.3%
(p,q)=(15,20)	96.2%	97.9%	86.9%
(p,q)=(15,30)	96.3%	96.0%	88.5%
(p,q)=(20,0)	93.5%	87.4%	85.0%
(p,q)=(20,10)	96.2%	98.1%	86.8%
(p,q)=(20,20)	96.3%	98.6%	86.9%
(p,q)=(20,30)	96.5%	99.2%	89.0%

Table 4. Prediction Accuracy of Balanced Approach

Dataset	Epinions	Slashdot	Wiki
Standard-pd (average)	85.14 %	82.82%	62.58%
false negative	0.84%	0.50%	2.10%
false positive	28.89%	33.8%	72.6%
Standard-pq (balanced)	89.36%	85.37%	76.81%
false negative	6.20%	14.37%	22.5%
false positive	22.93%	15.35%	23.90%
All123 ([14])	93.42%	93.51%	80.21%
EIG ([8])	85.30%	N/A	N/A

Table 5. Prediction Accuracy of Different Values of d

Dataset	Epinions	Slashdot	Wiki
d=10 (exact)	96.36%	98.00%	87.69%
d=10 (cplex)	96.17%	97.31%	86.98%
d=25 (cplex)	96.20%	97.52%	85.60%
METSLib	93.03%	96.79%	86.22%

Table 6. Prediction Accuracy of Different Algorithms

Dataset	Epinions	Slashdot	Wiki
Standard-adj	96.59%	97.93%	87.29%
Standard-pq	96.36%	98.00%	87.69%
Simple-adj	95.93%	97.68%	87.03%
HOC-5 ([7])	90.80%	84.69%	86.05%
All123 ([14])	90-95%	90-95%	N/A
EIG ([8])	93.60%	N/A	N/A

Standard-pq) have better prediction accuracy then the simple formulation (*Simple-adjacent*), so it is useful to introduce influences. For Slashdot and Wikipedia, restricting peers to neighbours (*Standard-adjacent*) is not as effective as using the set of nodes with at least $p = 15$ common neighbours as peers (*Standard-pq*), although the difference is neglegible. Surprisingly, for Epinions, it is slightly better to only consider neighbours as peers. We compare the results with those of [7] (HOC-5) and [14] (All123). Unfortunately, Kleinberg et al. [14] provide only a (somewhat wide) range of the results their model produces on such datasets. Yet,even such partial results allow us to conclude that collecting opinions from trusted peers is an effective method to infer people's attitude.

In Table 5, we compare the performance of our model with different values of d. We expected the prediction accuracy to increase as the value of d increases. However, experimental result shows that it is not the case. Limited by the strength of each solver, accuracy of the algorithm is very sensitive to the quality of approximation. But still, we think the prediction accuracy should increase if we can find a better solution for the problem for larger value of d.

In [8] and [14] the authors use certain techniques to test their approaches on unbiased datasets. They use, however, different ways to balance the dataset and/or results. For instance, [8] does not change the dataset (Epinions), but, since the dataset is biased toward positive links, they find the error ratio separately for positive and negative links, and then average the results. More precisely, they test the method on a set of randomly sampled edges that naturally contains more positive edges. Then they record the error rate on all negative edges, sample randomly the same number of positive edges (from the test set), find the error rate on them, and report the mean of the two numbers.

The approach of [14] is different. Instead of balancing the results they balance the dataset itself. In order to do that they keep all the negative edges in the datasets, and then sample the same number of positive edges removing the rest of them. All the training and testing is done on the modified datasets.

Although we have reservations about both approaches, we tested our model in both settings. The results are shown in Table 4. Observe that since our approach is to train the predictor for a particular dataset rather than finding and tuning up general features as it is done in [8] and [14], and the test datasets are biased toward positive edges, it is natural to expect that predictions are biased toward positive edges as well. This is clearly seen from Table 4. We therefore think that average error rate does not properly reflect the performance of our algorithm.

In the case of balanced datasets our predictor does not produce biased results, again as expected. This, however, is the only case when its performance is worse than some of the previous results. One way to explain this is to note that density of the dataset is crucial for accurate predictions made by the quadratic correlation approach. Therefore we had to lower the embeddedness threshold used in this part of the experiment to $p = 5$, $q = 5$, while [14] still tests only edges of embeddedness at least 25.

4 Conclusion

We have investigated the link sign prediction problem in online social networks with a mixture of both positive and negative relationships. We have shown that a better prediction accuracy can be achieved using personalized features such as peer opinions. Although the improvement upon previous results is not significant, a nearly perfect prediction accuracy is hard to achieve. Another advantage of the model is that it accommodates the dynamic nature of online social networks by building a predictor for each individual nodes independently. This enables fast updates as the underlying network evolves over time.

References

1. IBM ILOG CPLEX Optimizer (2010), `http://goo.gl/uKIx6K`
2. METSLib (2011), `https://projects.coin-or.org/metslib`
3. Avesani, P., Massa, P., Tiella, R.: A trust-enhanced recommender system application: Moleskiing. In: SAC, pp. 1589–1593 (2005)
4. Brzozowski, M., Hogg, T., Szabó, G.: Friends and foes: ideological social networking. In: CHI, pp. 817–820 (2008)
5. Burke, M., Kraut, R.: Mopping up: modeling Wikipedia promotion decisions. In: CSCW, pp. 27–36 (2008)

6. Cartwright, D., Harary, F.: Structure balance: A generalization of Heider's theory. Psychological Review 63(5), 277–293 (1956)
7. Chiang, K.-Y., Natarajan, N., Tewari, A., Dhillon, I.: Exploiting longer cycles for link prediction in signed networks. In: CIKM, pp. 1157–1162 (2011)
8. Guha, R.V., Kumar, R., Raghavan, P., Tomkins, A.: Propagation of trust and distrust. In: WWW, pp. 403–412 (2004)
9. Guyon, I., Gunn, S., Nikravesh, M., Zadeh, L.A.: Feature extraction: foundations and applications. STUDFUZZ, vol. 207. Springer, Heidelberg (2006)
10. Harary, F.: On the notion of balance of a signed graph. Michigan Math. J. 2(2), 143–146 (1953)
11. Al Hasan, M., Chaoji, V., Salem, S., Zaki, M.: Link prediction using supervised learning. In: SDM, 2006 Workshop on Link Analysis, Counterterrorism and Security (2006)
12. Kunegis, J., Lommatzsch, A., Bauckhage, C.: The Slashdot Zoo: mining a social network with negative edges. In: WWW 2009, pp. 741–750 (2009)
13. Lampe, C., Johnston, E., Resnick, P.: Follow the reader: filtering comments on Slashdot. In: CHI, pp. 1253–1262 (2007)
14. Leskovec, J., Huttenlocher, D.P., Kleinberg, J.M.: Predicting positive and negative links in online social networks. In: WWW, pp. 641–650 (2010)
15. Li, Y., Chen, W., Wang, Y., Zhang, Z.-L.: Influence diffusion dynamics and influence maximization in social networks with friend and foe relationships. In: WSDM 2013, pp. 657–666 (2013)
16. Liben-Nowell, D., Kleinberg, J.: The link prediction problem for social networks. In: CIKM 2003, pp. 556–559 (2003)
17. Liben-Nowell, D., Kleinberg, J.: The link-prediction problem for social networks. J. Am. Soc. Inf. Sci. Technol. 58(7), 1019–1031 (2007)
18. Massa, P., Avesani, P.: Controversial users demand local trust metrics: An experimental study on epinions.com community. In: AAAI, pp. 121–126 (2005)
19. McGeoch, C., Wang, C.: Experimental evaluation of an adiabiatic quantum system for combinatorial optimization. In: CF 2013, pp. 23:1–23:11 (2013)
20. Neven, H., Denchev, V., Rose, G., Macready, W.: Training a large scale classifier with the quantum adiabatic algorithm. CoRR, abs/0912.0779 (2009)
21. Newman, M.E.J.: The structure and function of complex networks. SIAM Review 45(2), 167–256 (2003)
22. Sarwar, B., Karypis, G., Konstan, J., Riedl, J.: Analysis of recommendation algorithms for e-commerce. In: ACM Conf. on Electronic Commerce, pp. 158–167 (2000)
23. Symeonidis, P., Tiakas, E., Manolopoulos, Y.: Transitive node similarity for link prediction in networks with positive and negative links. In: RecSys 2010, pp. 183–190 (2010)
24. Wang, C., Satuluri, V., Parthasarathy, S.: Local probabilistic models for link prediction. In: ICDM 2007, pp. 322–331 (2007)

Hash-Based Stream LDA: Topic Modeling in Social Streams

Anton Slutsky, Xiaohua Hu, and Yuan An

College of Computing and Informatics, Drexel University, USA
{as3463,xh29,yuan.an}@drexel.edu

Abstract. We study the problem of topic modeling in continuous social media streams and propose a new generative probabilistic model called Hash-Based Stream LDA (HS-LDA), which is a generalization of the popular LDA approach. The model differs from LDA in that it exposes facilities to include inter-document similarity in topic modeling. The corresponding inference algorithm outlined in the paper relies on efficient estimation of document similarity with Locality Sensitive Hashing to retain the knowledge of past social discourse in a scalable way. The historical knowledge of previous messages is used in inference to improve quality of topic discovery. Performance of the new algorithm was evaluated against classical LDA approach as well as the stream-oriented On-line LDA and SparseLDA using data sets collected from the Twitter microblog system and an IRC chat community. Experimental results showed that HS-LDA outperformed other techniques by more than 12% for the Twitter dataset and by 21% for the IRC data in terms of average perplexity.

1 Introduction

In this paper we are motivated by the problem of topic discovery in social media. We recognize that topic discovery systems for online social discourse need to address a set of challenges associated with the scale of modern social media outlets such as Twitter, chat systems and others. To be useful, these systems must operate continuously for extended periods of time, as social conversations do not stop, produce output in a timely fashion to remain relevant and ensure high quality of output.

Commonly used data mining techniques handle the problem of social stream topic discovery by applying batching heuristics to process the never-ending stream of messages. Since retaining all messages is not feasible in practice, current topic modeling approaches improve quality of topic discovery by retaining globally applicable statistics such as topic-word counters, but fail to take advantage of document-level information as no technique has existed so far to retain such information in a scalable and meaningful way.

Therefore, in this work we propose a new generative probabilistic model called Hash-based Stream LDA (HS-LDA), which is a generalization of the popular Latent Dirichlet Allocations (LDA) [1]. The model improves upon previous works by introducing a theoretical framework that makes it possible to retain the knowledge of

V.S. Tseng et al. (Eds.): PAKDD 2014, Part I, LNAI 8443, pp. 151–162, 2014.

historical stream messages in a scalable way and use this knowledge to improve the quality of topic discovery in social streams. Further, an efficient inference mechanism for the HS-LDA model is outlined, which makes use of the scalable hashing algorithm called Locality Sensitive Hashing (LSH) [2]. We show that the HS-LDA model and the associated inference algorithm are well suited for topic discovery in streams by comparing the predictive power of the topic models inferred by HS-LDA with that of topics learned by applying the classical LDA, On-line LDA [3] and SparseLDA [4] approaches to stream data. Evaluation was performed using data collected from the Twitter microblog site and an IRC chat system. Our experiments showed that HS-LDA outperformed other techniques by more than 12% for the Twitter dataset and by 21% for the IRC data in terms of average perplexity.

The paper is organized as follows. In section 2, current state of the art of topic modeling and stream mining is discussed. Section 3 introduces the HS-LDA model, outlines an efficient inference algorithm and discusses its application to stream data. In section 4, comparison of performance of our method to that of other modeling approaches in terms of perplexity is presented. Section 5 concludes the paper and outlines future work.

2 Related Works

The seminal work on Latent Dirichlet Allocation (LDA) [1] provides basis for numerous extensions and generalizations in the field topic modeling. LDA considers document collections as bag-of-words assemblies that are generated by stochastic processes. To generate a document, a random process first selects a topic from a distribution over topics and then generates a word by sampling the associated topic-word distribution. Both the topic and the word distributions are governed by hidden (or latent) parameters.

The LDA framework is designed to operate on a fixed set of documents and cannot be applied to stream data directly as converting an unbounded number of documents to a finite collection is not possible. To overcome this challenge, many approaches limit the training scope by aggregating messages based on attributes such as authorship or hash tag annotations and training models based on these aggregates [5], [6, 7].

An interesting recent work by Want et al. introduced an efficient topic modeling technique called TM-LDA for stream data. This approach is based on the notion that if document topic model is known at time t, at time $t+1$ a new topic model can be predicted and an error can be computed by comparing the "old" and the "new" topic models. This error computation reduces the challenge of estimating topic models for new documents to a least-squares problem, which can be solved efficiently. Focusing on the popular Twitter micro-blog data, TM-LDA selects a set of individual authors and trains a separate model for each of the authors. To accomplish this, TM-LDA monitors Twitter for an extended period of time (a week's worth of data was collected in the original work) and then trains a model to be able to predict new messages.

The idea of using authorship to improve topic modeling quality is not unique to TM-LDA. A recent work by Xu et al. modified the well-known Author-Topic [8] model for Twitter data [6]. Xu et al. extended the insight of the Author-Topic model by taking advantage of additional features available in Twitter such as links, tags, etc.

Another way to approach topic modeling in streams is to apply LDA machinery to snapshots or buffers of documents of fixed size. Online Variational Inference for

LDA [9] is one such technique. The algorithm assembles mini-batches of documents at periodic intervals and uses Expectation Maximization (EM) algorithm to infer distribution parameters by holding topic proportions fixed in the E-step and then re-computing topic proportions as if the entire corpus consisted of document mini-batches repeated numerous times. Topic parameters are then adjusted using the weighted average of previous values of each topic proportion.

Another approach termed On-line LDA [3] considers the data stream as a sequence of time-sliced batches of documents. The approach processes each time-slice batch using the classical LDA sampling techniques, with the variation being that the corresponding collapsed Gibbs sampler initialization is augmented with the inclusion of topic-word counters from histories of pervious time-slice batches. The histories are maintained using a fixed-length sliding window and the contribution of each history to the current slice initialization is predicated upon a set of weights associated with each element in the sliding window.

In another work, Yao et al in [4] considered topic discovery in streaming documents and proposed the SparseLDA model. Noticing that the efficiency of sampling-based inference depends on how fast the sampling distribution can be evaluated for each token, their work enhanced the inference procedure in a way as to allow parts of computations used in sampling to be pre-computed, thus improving performance. Further, the sampling procedure proposed by Yao et al. restricted training to a fixed collection of training documents and then, for each test document, sampled topics using counts from the training data and test document only, ignoring the rest of the stream.

The explosion of micro-blog popularity has attracted much attention from outside of the topic modeling community. One particularly interesting application is the field of first story detection. Conceptually, first story detection is concerned with locating emergent clusters of similar stream messages, which are said to be indicative of particularly interesting and currently relevant stories. First story detection approaches require the ability to discover clusters of similar documents in near real-time fashion, which is difficult to accomplished using classic clustering tools since the computational complexity of commonly used clustering algorithms (hierarchical, partitioning, etc.) is quite high. Therefore, recent works on first story detection have seized upon the concept of Locality Sensitive Hashing (LSH) [2], which is an approach for identifying a datum neighborhood in constant time [10]. In [10], Petrovic et al use a combination of LSH and inverse index searching to show that clusters of similar documents may be identified in constant time with exceptional accuracy and low variability.

3 Hash-Based Stream LDA

As noted in the preceding survey of related works, many approaches to topic modeling in streams have been developed in recent years. A number of these approaches [3, 9] attempted to enhance quality by preserving various aspects of topic inference calculations and predicating topic learning upon past knowledge. Unfortunately, none of these techniques were successful in retaining the knowledge of stream documents relying instead on storing global structures such as topic-word multinomials. Hurdles for retaining document knowledge are two-pronged – 1) the number of documents in streams is unbounded making storage of individual document information not feasible, and 2) since previous documents do not get replayed in streams, retaining records of their presence directly may be meaningless for topic modeling.

Therefore, this section introduces the new Hash-Based Stream LDA (HS-LDA) model, which provides a mechanism for retaining document knowledge for stream modeling in a scalable and meaningful way. HS-LDA is a generative probabilistic model that describes a process for generating a document collection. Like LDA, in HS-LDA each document is viewed as a mixture of underlying topics and each word is generated by drawing from a topic-word distribution. HS-LDA departs from LDA by imagining that, in addition to words, the generative process also emits certain auxiliary objects that are not directly observable in data. In order to refer to these objects in an intuitive way, we reach out to the physical world for a descriptive analogy. We borrow from particle physics nomenclature and recall that, in physics, a neutrino is a nearly massless, uncharged particle that is detectable only through its interactions with other matter [11]. Since the auxiliary objects postulated by the HS-LDA model are similarly ethereal, we introduce the notion of *HS-LDA neutrinos* (or *pseudo-neutrinos* for short), as the analogy with the real particle seems appropriate.

Following the analogy, as the physical neutrinos are said to be classifiable into a collection of categories [11], the HS-LDA *pseudo-neutrinos* are also thought to belong to a fixed set of possible types (or flavors). The physics analogy is abandoned at this point, however, as HS-LDA makes no further claims as to the properties or nature of each flavor. The generative process is graphically outlined in Figure 2.

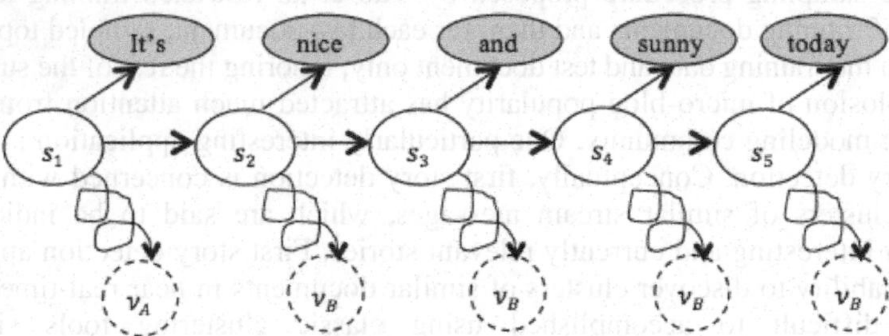

Fig. 1. Visualization of the HS-LDA generative process. Ovals $s_1,\ldots,s_5$ represent process states, shaded ovals represent word generation and dashed circles represent emissions of neutrinos ν of types A and B. Dashed circles surrounding neutrinos labels aim to emphasize the notion that neutrinos are assumed to be present but difficult to detect.

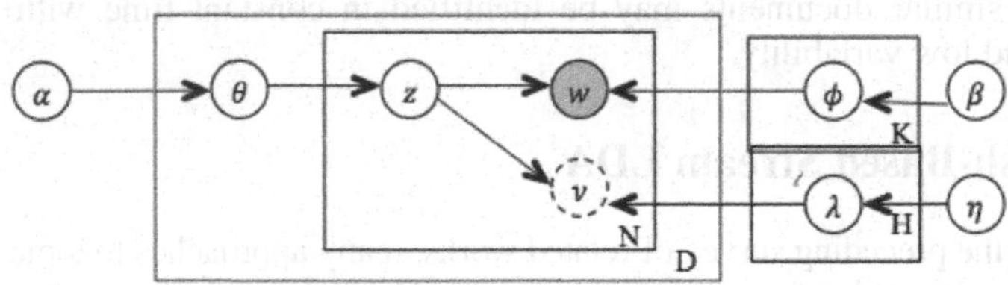

Fig. 2. Graphical model representation of HS-LDA. N is the number of words in a document, D is the number of documents, K is the number of topics and H is the number of pseudo-neutrino types. α, η and β are Dirichlet prior vectors that are assumed to be symmetrical in this paper. θ represents the vector multinomial over topics, ϕ is the multinomial over words, z is the topic draw, w stands for a word realization and ν is the emitted pseudo-neutrino. The clear circles represent hidden entities, shaded circles represent directly observable entities and the dashed circles stand for indirectly detectable ones.

In Figure 3, the generative process is outlined. There, words are generated in a way common to many LDA-type models by drawing from a distribution over words. Unlike other approaches, however, a pseudo-neutrino is also emitted by a draw from a multinomial distribution parameterized by a vector of topic-specific neutrino type proportions.

1. For each topic $k \in \{1, \ldots, K\}$:
2. Generate $\phi_k = \{\phi_{k,1}, \ldots, \phi_{k,V}\}^T \sim Dir(\cdot\,|\beta)$
3. Generate $\lambda_k = \{\lambda_{k,1}, \ldots, \lambda_{k,H}\}^T \sim Dir(\cdot\,|\eta)$
4. For each document d:
5. Generate $\theta^{(d)} \sim Dir(\cdot\,|\alpha)$
6. For each $i \in \{1, \ldots, N_d\}$
7. Generate $z_i \in \{1, \ldots, K\} \sim Mult(\cdot\,|\theta^{(d)})$
8. Generate $w_i \in \{1, \ldots, V\} \sim Mult(\cdot\,|\phi_{z_i})$
9. Generate $v_i \in \{1, \ldots, H\} \sim Mult(\cdot\,|\lambda_{z_i})$

Fig. 3. Generative process for HS-LDA: ϕ_k is a vector consisting of parameters for the multinomial distribution over words corresponding to kth topic, λ_k is a vector consisting of parameters for the multinomial distribution over neutrino types corresponding to kth topic, α is the Dirichlet document topic prior vector, β word prior vector, η is the neutrino type prior vector and N_d is the number or words in document d and K is the number of topics

It is important to note that if a user were to restrict the set of possible neutrino types to just a single type (say {"root"}), HS-LDA would become equivalent to LDA as all draws of type label assignments would be the same making the generative branch from z to v redundant. Therefore, HS-LDA is a generalization of Latent Dirichlet Allocations [1], which is important to note since the general nature of HS-LDA suggests that its insight can be applied to other models that extend LDA, of which there are many. Later sections will take advantage of this fact and show the experimental results of application of HS-LDA to other successful models.

3.1 Gibbs Sampling with HS-LDA

The generative probabilistic HS-LDA model describes the process of document collection creation. The hidden model parameters θ, ϕ and λ may be estimated using a Monte Carlo procedure, which is relatively easy to implement, does not require a lot of memory and produces output that is competitive with that of other more complicated and slower algorithms [3, 12]. The rest of the section describes the derivation of an efficient sampling algorithm used to infer models parameters with HS-LDA.

We start by framing the problem of topic discovery in terms of collections of D documents containing K topics expressed over W words and H pseudo-neutrino types. The task of learning topic models is to discover the makeup of θ, ϕ and λ, which can be estimated by evaluating the probability of a topic having observed both a word and a pseudo-neutrino. The posterior distribution is formally stated as $P(\boldsymbol{z}|\boldsymbol{w},\boldsymbol{v}) = \frac{P(\boldsymbol{w},\boldsymbol{z},\boldsymbol{v})}{\sum_{\boldsymbol{z}} P(\boldsymbol{w},\boldsymbol{z},\boldsymbol{v})}$

The joint distribution $P(\boldsymbol{w},\boldsymbol{v},\boldsymbol{z})$ can be computed by considering that Dirichlet priors α, β and η in the HS-LDA model are conjugate to θ,ϕ and λ respectively. Since $P(\boldsymbol{w},\boldsymbol{v},\boldsymbol{z}) = P(\boldsymbol{w}|\boldsymbol{v},\boldsymbol{z})P(\boldsymbol{v}|\boldsymbol{z})P(\boldsymbol{z})$ by the chain rule and since $\boldsymbol{w}$ and $\boldsymbol{v}$ are conditionally independent in our model (see Figure 2), $P(\boldsymbol{w}|\boldsymbol{v},\boldsymbol{z}) = P(\boldsymbol{w}|\boldsymbol{z})$ which simplifies the joint distribution as

$$P(\boldsymbol{w},\boldsymbol{v},\boldsymbol{z}) = P(\boldsymbol{w}|\boldsymbol{z})P(\boldsymbol{v}|\boldsymbol{z})P(\boldsymbol{z})$$

Observing that ϕ,λ and θ only appear in first, second and third terms respectively, each term may be evaluated separately. Integrating out ϕ,λ and θ in each term gives

$$P(\boldsymbol{w}|\boldsymbol{z}) = \left(\frac{\Gamma(W\beta)}{\Gamma(\beta)^W}\right)^K \prod_{j=1}^K \left(\frac{\prod_w \Gamma\left(n_j^{(w)}+\beta\right)}{\Gamma\left(n_j^{(\cdot)}+W\beta\right)}\right) \tag{1a}$$

$$P(\boldsymbol{v}|\boldsymbol{z}) = \left(\frac{\Gamma(H\eta)}{\Gamma(\eta)^H}\right)^K \prod_{j=1}^K \left(\frac{\prod_v \Gamma\left(n_j^{(v)}+\eta\right)}{\Gamma\left(n_j^{(\cdot)}+H\eta\right)}\right) \tag{1b}$$

$$P(\boldsymbol{z}) = \left(\frac{\Gamma(K\alpha)}{\Gamma(a)^K}\right)^D \prod_{d=1}^D \left(\frac{\prod_j \Gamma\left(n_j^{(d)}+\alpha\right)}{\Gamma\left(n_{\cdot}^{(d)}+K\alpha\right)}\right) \tag{1c}$$

where $n_j^{(w)}$ is the number of times word w has been assigned to topic j, $n_j^{(d)}$ is the number of time a word from document d has been assigned to topic j, $n_j^{(v)}$ is the number of times a neutrino of type v has been assigned to topic j, $n_j^{(\cdot)}$ and $n_{\cdot}^{(d)}$ are the total numbers of assignments in topic j and document d respectively. $\Gamma(\cdot)$ is the standard gamma function.

Since computing the exact distributions in Equations 1a-c is intractable [1, 12], we follow the pattern in other topic modeling approaches [3, 4, 6-8, 13] and estimate θ, ϕ and λ by relying on the Gibbs sampling procedure. The Gibbs procedure operates by iteratively sampling all variables from their distributions conditioned on their current values and data and updating variables for each new state. The full conditional distribution $P(z_i = j|\boldsymbol{z}_{-i},\boldsymbol{w},\boldsymbol{v})$ that is necessary for the Gibbs sampling algorithm is obtained by probabilistic argument [12] as well as by observing that first terms in each of the Equations 1a-c are constant and values of denominators and numerators of second terms are proportional to the arguments of their gamma functions. Therefore, the sampling equation is as follows:

$$P(z_i = j|\boldsymbol{z}_{-i},\boldsymbol{w},\boldsymbol{v}) \propto \frac{n_{-i,j}^{(w_i)}+\beta}{n_{-i,j}^{(\cdot)}+W\beta}\frac{n_{-i,j}^{(d)}+\alpha}{n_{-i,\cdot}^{(d)}+K\alpha}\frac{n_{-i,j}^{(v_i)}+\eta}{n_{-i,j}^{(\cdot)}+H\eta} \tag{2}$$

where, $n_{-i,j}^{(v_i)}$ is the count of times neutrino v_i has been assigned to topic j excluding current assignment and $n_{-i,j}^{(\cdot)}$ is the total number of topics j assignments in any document excluding current assignment. Reader may notice that denominators in the first and third product terms in Equation 2 have identical counters. That is because, in the HS-LDA model, the number of words is always exactly the same as the number of neutrino emissions by process construction.

The Gibbs sampling algorithm can be implemented in an on-line fashion by first initializing topic assignments to a random state and then using Equation 2 to assign words to topics. The algorithm operates by reconsidering data for a number of iterations during which new states of topic assignments are found using Equation 2. The algorithm is fast as the only information necessary to estimate the new state is the word, topic and neutrino counters, which can be cached and updated efficiently [12].

3.2 Pseudo-Neutrino Detection

The sampling algorithm outlined in the previous section estimates parameter values by relying on two detectable quantities – words and pseudo-neutrino emissions. To detect pseudo-neutrinos, which cannot be observed directly in text, we assumed a Gaussian distribution of pseudo-neutrinos in documents, as this distribution was common to many phenomena [14]. With this assumption, we could refer to all pseudo-neutrinos in a given document in a meaningful way by identifying the most common (or mean) neutrino type. That is, for $H \in \mathbb{Z}^{+}$ possible pseudo-neutrino types, we assumed that there existed a mean pseudo-neutrino type $1 \leq c_{\mu}^{d} \leq H$ for each document d. With that, a rough approximation vector of pseudo-neutrino assignments $h_d = \{h_{d,1}, \dots, h_{d,H}\}$ could be constructed for each document d of size N_d such that $h_{d,i} = \begin{cases} N_d, if\ i = c_{\mu}^{d} \\ 0, otherwise \end{cases}$.

Constructing the vector h_d as described in the previous paragraph suggested that a meaningful approximation of document pseudo-neutrinos could be found by identifying a representative (mean) neutrino type for each document. To locate the representative flavor, we noticed that pseudo-neutrino types essentially constituted a kind of vocabulary akin to that of words. With that, considering topics from conceptual point of view, intuitively, documents on the same topic would be close to one another in terms of similarity of their content regardless of the vocabulary used to express the content (e.g. for any language, documents about the 'World Cup' sporting event would contain text related to the even in that language). With that, since the number of pseudo-neutrino types was known, clustering documents into H clusters based on word similarity would approximate document-level (mean) neutrino types as cluster indices could be used as the neutrino type identifiers.

To implement this intuition in practice, we searched for a clustering strategy that would perform in a scalable way while at the same time ensuring that similar documents were likely to share a cluster. We realized that by restricting $H = 2^n$ for some positive integer n, it would be possible to make use of Locality Sensitive Hashing (LSH)[2].

LSH relies on existence of a set of hash functions $\mathcal{H}$ (referred to as a *function family*) for some *d*-dimensional coordinate space $\mathbb{R}^d$ where each hash function can be efficiently implemented with the help of Random Projections (RP) [15]. To use LSH, we start by defining a function space $f: \mathbb{R}^d \to \{0,1\}$ and constructing a function family $\mathcal{H} = \{f_1, \dots, f_{\log_2 H} | f_i \in f\}$. Each function $f_i \in \mathcal{H}$ is associated with a random projection vector $p_i^{\text{random}} \in \mathbb{R}^d$ with components that are selected at random from a Gaussian distribution $\mathcal{N}(0,1)$ [16]. Each random projection is used to compute a dot-product between it and any point $p \in \mathbb{R}^d$ allowing the mapping function to be constructed in the following way:

$$f_i(p) = \begin{cases} 1\ if\ p \cdot p_i^{\text{random}} \geq 0 \\ 0\ if\ p \cdot p_i^{\text{random}} < 0 \end{cases} \quad [2]$$

Then, for any $p \in \mathbb{R}^d$, LSH hash value is constructed by invoking each of the functions in $\mathcal{H}$ on p and concatenating output bits as a bit string. Treating the bit string as a binary number, a mapping function assigns p to a number between one and H as follows:

$$map(p) = ||_{i=1}^{|\mathcal{H}|} f_i(p)$$

Since the bit string generated by the above procedure is of finite size, the space of possible values is bound by $2^{|\mathcal{H}|}$. Recalling that $H = 2^n$ and $|\mathcal{H}| = \log_2 H = n$, function map can be used to map each point in $\mathbb{R}^d$ to a positive integer bound by H.

Further, since it is proven in [17] (proof omitted here) that $P(f_i(p) = f_i(q)) = 1 - \frac{\angle(p,q)}{\pi}$ holds for any function $f_i \in \mathcal{H}$ and all points $p, q \in \mathbb{R}^d$, the probability of LSH hash collision for two vectors increases with the decrease to the angle between them. Then, since the value of cosine of two vectors is directly related to the size of the angle

$$P(f_i(p) = f_i(q)) \propto \cos(\angle(p,q))$$

where $\angle$ is the angle between the two vectors in radians[1].

Therefore, since LSH hashing allowed for fast clustering of vectors in a way that preserved document similarity, LSH was used to approximate the mean pseudo-neutrino type by treating LSH hash value as the type identifier. To make use of LSH hashing in topic modeling, we restricted the size of the set H to be a power of two and rewrote the sampling equation (Equation 2) in terms of LSH hash family F of size $\log_2 H$ as:

$$P(z_i = j|\mathbf{z}_{-i}, \mathbf{w}, \boldsymbol{x}) \propto \frac{n_{-i,j}^{(w_i)}+\beta}{n_{-i,j}^{(\cdot)}+W\beta} \frac{n_{-i,j}^{(d)}+\alpha}{n_{-i,\cdot}^{(d)}+K\alpha} \frac{n_{-i,j}^{(h_d^F)}+\eta}{n_{-i,j}^{(\cdot)}+H\eta} \quad (3)$$

where h_d^F is the hash value of document d, $n_{-i,j}^{(h_d^F)}$ is the number of words from documents with hash value h_d^F assigned to topic j excluding current assignment and $n_{-i,j}^{(\cdot)}$ is the total number of words in any document assigned to topic j excluding current assignment. The sampling algorithm, then, proceeds as outlined in section 3.1 using Equation 3 to assign words to topics.

4 Evaluation

In order to validate the utility of our model, the approach was tested on two distinct data sets. Our first data set consisted of 1,000,000 English language messages collected from Twitter micro-blog site using its public sampling API over a period of one week. The second data set was comprised of 300,000 English language chatroom messages collected by connecting to the public *irc.freenode.net* public chat server and monitoring chat rooms with more than 150 chatters for the same one week period. Filtering of non-English texts was accomplished with the help of the open source *language-detection*[2] library.

The language models produced by our approach were compared to those learned by On-line LDA [3] and SparseLDA [4] as these models were designed to operate efficiently on stream data. In addition, to provide a common baseline, topic models

---

[1] Unusual angle operator used to avoid confusion with topic modeling notation.
[2] `https://code.google.com/p/language-detection/`

learned by HS-LDA were compared to those discovered by the classic LDA [1] algorithm. We did not evaluate our approach against TM-LDA as it required partitioning by author as well as a significant and static training sample to be collected prior to producing any output at all. These constraining requirements made TM-LDA unfit for continuous topic modeling application, which was the motivation of this work.

To compare language models, evaluation was performed using the perplexity measure over held-out subset of data $\overline{W} = \{\overline{w}_1, \ldots, \overline{w}_n\}$ given language model M and the training data calculating perplexity using the following formula:

$$perp_M(\overline{W}) = \exp\left(-\frac{1}{n}\sum_{i=1}^{n}\frac{1}{|\overline{w}_i|}\sum_{j=1}^{|\overline{w}_i|}\log\left(p_m(\overline{w}_{ij})\right)\right)$$

where $n = |\overline{W}|$, $\overline{w} \in \overline{W}$, $\overline{w}$ is the jth term in the ith string in the held-out collection and $p_M(w \in \overline{w})$ is the probability of term w as per the learned language model M. Further, to account for possible overfitting, our evaluation was validated using the 5-fold cross-validation.

4.1 Parameter Selection

As pointed out in earlier works [10, 18] Locality Sensitive Hashing is highly sensitive to choices of the hash family size. This choice governs the scatter within each hash bucket as chance of collision decreases with the increase of hash family size. Therefore, hash family size selection was approached from the point of view of estimating a reasonable number of buckets for the number of messages expected.

Considering the Twitter micro-blog service as being one of the most vibrant and popular social forums today, we experimented with the numbers of English language messages that could be downloaded over a given period. Recalling the industry-oriented motivation for this work and selecting one working week as the target period (timeframe common to the industry environment) the number of messages that could be gathered from Twitter's sampling service was empirically estimated to number in some millions. Realizing that if the number of hash family function was chosen to be high (ex.: $2^{20} = 1{,}048{,}576$) the algorithm could potentially map every message into an individual bucket, negating the entire insight of HS-LDA. With that, the reasonable number of hash functions for our experiments was chosen to be 17 ($2^{17} = 131{,}072$) as this number would allow for variability within each cluster while at the same time providing reasonable specificity.

4.2 Experimental Setup and Results

Having thus chosen the hash family size, HS-LDA was evaluated against LDA, On-line LDA and Sparse LDA using the two test datasets. For all models, the number of topics was chosen to be 100 and experimented with various hyperparameter settings. Results reported here were for hyperparameter values of $\alpha = 0.05$, $\beta = 0.05$ and $\eta = 1$ as these values produced best results for all models.

Figure 4 shows perplexity results for the two test datasets. In order to provide a readable graphic, the Simple Moving Average (SMA) smoothing technique was applied to raw results, setting the moving average window set to 10,000.

While cross-model comparison shows that HS-LDA approach outperformed other models in terms of perplexity, performance of LDA-type models could be sensitive to parameter choices [19]. Since some parameter choices could be more beneficial to performances of some frameworks and less so for others and since all models used for evaluation were derivative of the classic LDA model, we applied the insight of the HS-LDA approach to each test algorithm and conducted a pairwise comparison in terms of perplexity, thus controlling for model parameter sensitivity. Figures 5-6 show pairwise comparisons for each test model with the same approach augmented with HS-LDA (LDA/HS-LDA pairwise comparison is not reported as it can be found in Figure 4).

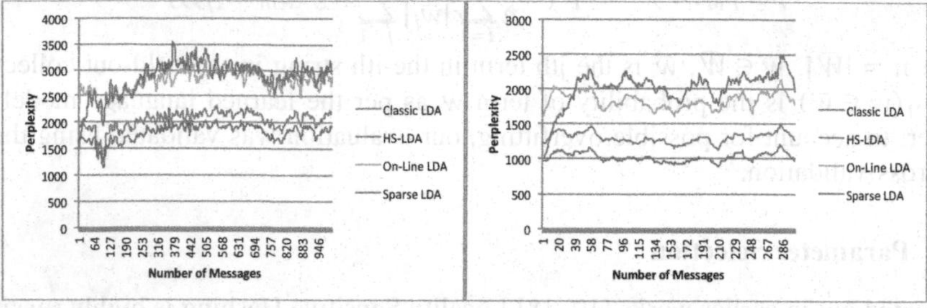

Fig. 4. Smoothed perplexity results for Twitter (left) and IRC (right) dataset

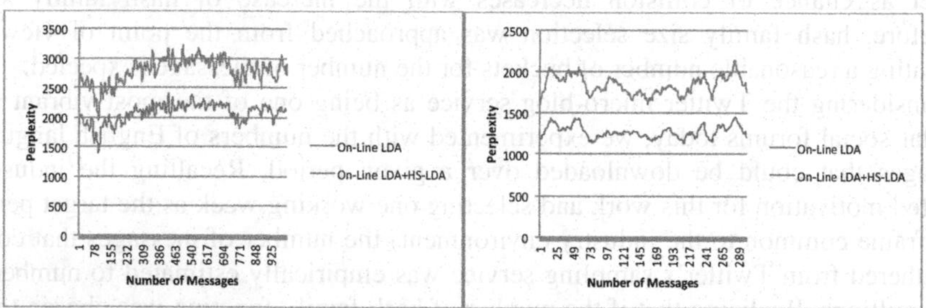

Fig. 5. Pairwise comparison of On-Line LDA and On-Line LDA augmented with HS-LDA for Twitter (left) and IRC (right) test sets

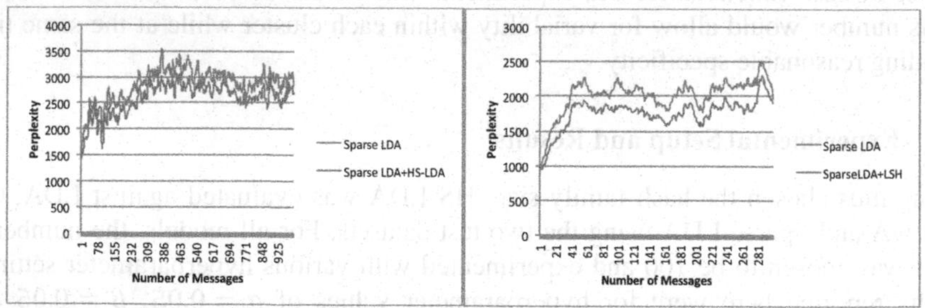

Fig. 6. Pairwise comparison of Sparse LDA and Sparse LDA augmented with HS-LDA for Twitter (left) and IRC (right) test sets

To summarize results in numerical way, average perplexities are reported for all tested models in Table 1. The purpose of this report is to identify the model with the highest predictive prowess as well as to quantify amount of improvement in terms of percentages.

Table 1. Average perplexity results for Twitter and IRC datasets

Model	Average Perplexity (Twitter)	Average Perplexity (IRC)
LDA	2044.42	1300.92
On-Line LDA	2773.99	1835.74
Sparse LDA	2860.27	1998.53
HS-LDA	**1803.67**	**1023.12**

In Table 1, HS-LDA outperformed other models by at least approximately 12% for the Twitter dataset and 21% for the IRC chatroom data. Significantly better predictive power of resulting topic models learned from the chatroom discourse may be explained by noting that chatrooms are often oriented towards particular themes, thus introducing loose structuring to social discourse. Such structuring does not exist in Twitter where the discourse is entirely unstructured, making the job of theme discovery more difficult.

5 Conclusions and Future Work

To improve the quality of topic models learned from social media streams, we introduced the new HS-LDA model for topic modeling, which was a generalization of the well-known LDA topic discovery technique. We experimented on large data sets collected from popular social media services and showed that our model outperformed other state-of-the-art stream topic modeling techniques in all cases. Further, we enhanced other topic modeling approaches with the insight of HS-LDA and showed that applying core notions of HS-LDA to other techniques improves their performance in terms of predictive power of resulting topic models.

While our results showed improvement in all cases where HS-LDA insight was used, combining HS-LDA with other models aimed at preserving global context did not immediately result in substantial performance gains. It seems, however, that such a combination has merit and we will continue this investigation in the future work.

Further, while this work was instrumental in moving towards the goal of constructing an industry-grade stream topic monitoring system, one of the major hurdles for constructing such a system with HS-LDA was the necessity to specify the number of topics. In our future work, we plan to investigate topic modeling approaches based on the popular Chinese Restaurant Process paradigm and will attempt to apply the insight of HS-LDA to dynamically discovered topic allocations.

References

1. Blei, D.M., Ng, A.Y., Jordan, M.I.: Latent dirichlet allocation. J. Mach. Learn. Res. 3, 993–1022 (2003)
2. Indyk, P., Motwani, R.: Approximate nearest neighbors: towards removing the curse of dimensionality. In: Proceedings of the Thirtieth Annual ACM Symposium on Theory of Computing, pp. 604–613. ACM (1998)
3. AlSumait, L., Barbará, D., Domeniconi, C.: On-line LDA: Adaptive Topic Models for Mining Text Streams with Applications to Topic Detection and Tracking. In: ICDM, pp. 3–12. IEEE Computer Society (2008)
4. Yao, L., Mimno, D., McCallum, A.: Efficient methods for topic model inference on streaming document collections. In: Proceedings of the 15th ACM SIGKDD International Conference on Knowledge Discovery and Data Mining, pp. 937–946. ACM (2009)
5. Hong, L., Davison, B.D.: Empirical study of topic modeling in Twitter. In: Proceedings of the First Workshop on Social Media Analytics, pp. 80–88. ACM (2010)
6. Xu, Z., Lu, R., Xiang, L., Yang, Q.: Discovering User Interest on Twitter with a Modified Author-Topic Model. In: 2011 IEEE/WIC/ACM International Conference on Web Intelligence and Intelligent Agent Technology (WI-IAT), pp. 422–429 (2011)
7. Wang, Y., Agichtein, E., Benzi, M.: TM-LDA: efficient online modeling of latent topic transitions in social media. In: Proceedings of the 18th ACM SIGKDD International Conference on Knowledge Discovery and Data Mining, pp. 123–131. ACM (2012)
8. Rosen-Zvi, M., Griffiths, T., Steyvers, M., Smyth, P.: The author-topic model for authors and documents. In: Proceedings of the 20th Conference on Uncertainty in Artificial Intelligence, pp. 487–494. AUAI Press (2004)
9. Hoffman, M.D., Blei, D.M., Bach, F.: Online learning for latent dirichlet allocation. In: NIPS (2010)
10. Petrović, S., Osborne, M., Lavrenko, V.: Streaming first story detection with application to Twitter. In: Human Language Technologies: The 2010 Annual Conference of the North American Chapter of the Association for Computational Linguistics, pp. 181–189. Association for Computational Linguistics (2010)
11. Wang, K.C.: A Suggestion on the Detection of the Neutrino. Phys. Rev. 61, 97 (1942)
12. Griffiths, T.L., Steyvers, M.: Finding scientific topics. Proceedings of the National Academy of Sciences 101, 5228–5235 (2004)
13. Kim, H., Sun, Y., Hockenmaier, J., Han, J.: ETM: Entity Topic Models for Mining Documents Associated with Entities. In: ICDM 2012, pp. 349–358 (2012)
14. Patel, J.K., Read, C.B.: Handbook of the normal distribution. Marcel Dekker Inc. (1996)
15. Bingham, E., Mannila, H.: Random projection in dimensionality reduction: applications to image and text data. In: Proceedings of the Seventh ACM SIGKDD International Conference on Knowledge Discovery and Data Mining, pp. 245–250. ACM (2001)
16. Slaney, M., Casey, M.: Locality-Sensitive Hashing for Finding Nearest Neighbors [Lecture Notes]. IEEE Signal Processing Magazine 25, 128–131 (2008)
17. Ravichandran, D., Pantel, P., Hovy, E.: Randomized algorithms and NLP: using locality sensitive hash function for high speed noun clustering. In: Proceedings of the 43rd Annual Meeting on Association for Computational Linguistics, pp. 622–629. Association for Computational Linguistics (2005)
18. Ture, F., Elsayed, T., Lin, J.: No free lunch: brute force vs. locality-sensitive hashing for cross-lingual pairwise similarity. In: Proceedings of the 34th International ACM SIGIR Conference on Research and Development in Information Retrieval, pp. 943–952. ACM (2011)
19. Panichella, A., Dit, B., Oliveto, R., Di Penta, M., Poshyvanyk, D., De Lucia, A.: How to effectively use topic models for software engineering tasks? an approach based on genetic algorithms. In: Proceedings of the 2013 International Conference on Software Engineering, pp. 522–531. IEEE Press (2013)

Two Sides of a Coin: Separating Personal Communication and Public Dissemination Accounts in Twitter

Peifeng Yin[1], Nilam Ram[2], Wang-Chien Lee[1], Conrad Tucker[3], Shashank Khandelwal[4], and Marcel Salathé[4]

[1] Department of Computer Science & Engineering, Pennsylvania State University
[2] Human Development and Psychology, Pennsylvania State University
[3] School of Engineering Design Technology, Pennsylvania State University
[4] Department of Biology, Pennsylvania State University
{pzy102,wlee}@cse.psu.edu,
{nur5,ctucker4,khandelwal,salathe}@psu.edu

Abstract. There are millions of accounts in Twitter. In this paper, we categorize twitter accounts into two types, namely *Personal Communication Account (PCA)* and *Public Dissemination Account (PDA)*. PCAs are accounts operated by individuals and are used to express that individual's thoughts and feelings. PDAs, on the other hand, refer to accounts owned by non-individuals such as companies, governments, etc. Generally, Tweets in PDA (i) disseminate a specific type of information (e.g., job openings, shopping deals, car accidents) rather than sharing an individual's personal life; and (ii) may be produced by non-human entities (e.g., bots). We aim to develop techniques for identifying PDAs so as to (i) facilitate social scientists to reduce "noise" in their study of human behaviors, and (ii) to index them for potential recommendation to users looking for specific types of information. Through analysis, we find these two types of accounts follow different temporal, spatial and textual patterns. Accordingly we develop probabilistic models based on these features to identify PDAs. We also conduct a series of experiments to evaluate those algorithms for cleaning the Twitter data stream.

1 Introduction

As Twitter[1] has grown, many different kinds of user accounts have emerged. At a general level, we roughly classify them into two categories: (i) Personal Communication Account (PCA) and (ii) Public Dissemination Account (PDA). PCAs are accounts that are usually operated by unique individuals and used for interpersonal communication (e.g., to share personal experiences and opinions). PDAs, in contrast, are typically linked to and operated by a company[2], a web

[1] http://twitter.com
[2] https://twitter.com/#!/citi

V.S. Tseng et al. (Eds.): PAKDD 2014, Part I, LNAI 8443, pp. 163–175, 2014.

site[3] or a program[4] and used to disseminate specific news and information (e.g., locations of car accidents, shopping deals, crimes).

Existence of PDA may cause problems when attempting to study human behavior using Twitter data. Recently, a Twitter-based analysis in *Science* suggested that changes in overall tweet sentiment over time (hours of the day) may be interpreted as evidence of biologically-based diurnal cycles in the mood patterns of humans [7]. However, an underlying assumption is that the tweets were produced by human individuals as part of their natural daily lives. If the data stream is a mixture of PCAs (humans) and PDAs (corporate/entity), the conclusion may be unwarranted. To illustrate it, In Figure 1 we plot the time evolution of average sentiment for 2,787 PCAs and 389 PDAs that were labelled manually. As can be seen, the daily diurnal cycle, i.e., the mood increases in the early morning and decrease later, is easily discernable in the overall data (Figure 1(a)) as similar to the previously published analysis. However, contrary to expectations, we find that the cycles are much less prominent in the PCA (human individual) sub-sample (Figure 1(b)) than in the PDA (corporate/entity) sub-sample (Figure 1(c)). The importance of separating the different account types for reaching accurate conclusions is clear.

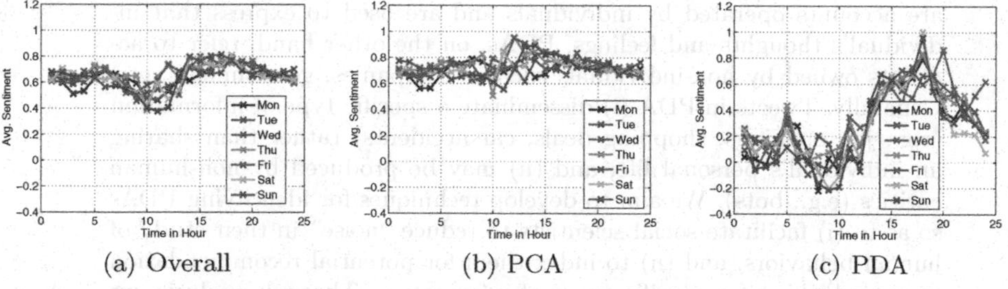

(a) Overall (b) PCA (c) PDA

Fig. 1. Sentiment change within a 24 hour time period across different account types

Beyond potentially adding noise to researchers' data streams, PDAs are potentially very useful for various types of data consumers. For instance, an individual looking for jobs may follow PDAs that publish job postings of a particular type (e.g., web development) in a particular geographic area (e.g., New York). Similarly, shoppers may follow PDAs that provide timely notification of nearby sale events and hot deals. In sum, as PDAs' tweets are often focused on a very specific topic and formatted in a uniform manner, they are relatively easy to process and may thus provide rich content for individuals, researchers, and the recommendation engines that support those populations.

The enormous size of the Twitter data stream makes it highly impractical to manually check the account type. In this paper we develop and test a variety

[3] `https://twitter.com/#!/WHERE`

[4] `https://twitter.com/#!/memcrime`

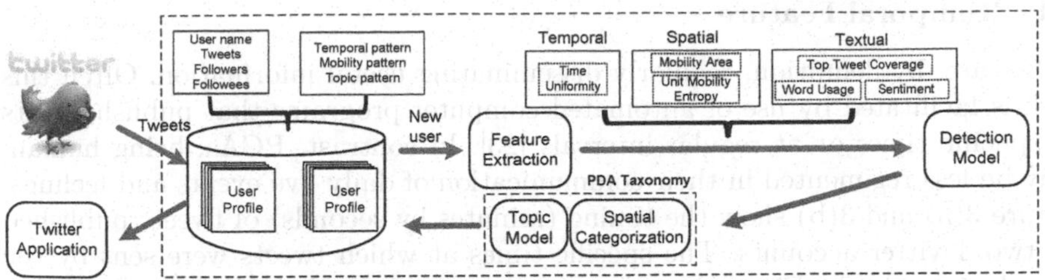

Fig. 2. The framework for PDA detection

of techniques for automatic classification of PDAs and PCAs using multiple temporal, spatial, and textual features of accounts' tweet publishing patterns.

Figure 2 gives an overview of the proposed framework for PDA detection. As shown, tweets are continuously sent to the database. Once a new user arrives, her profile of raw data is checked and different types of features are extracted. Specifically as illustrated in Figure 2, there are temporal, spatial and textual features (details are discussed in Section 2). With extracted features, a classification model is then employed to determine the account type. After a PDA is detected, the system checks its posted tweets to model its topic as well as categorizing its spatial characteristics. Finally, all extracted features and topic models are also saved in the database. Twitter applications, e.g., user recommendation, can then be built upon the knowledge mined in this framework.

The rest of the paper is organized as follows. Section 2 describes the feature extraction. Section 3 provides details of our models. Section 4 reports the evaluation of our model and shows some PDAs found using our model. Section 5 reviews relevant research and finally Section 6 concludes the whole paper.

2 Feature Extraction

In this section we discuss the extraction features used to identify PCAs and PDAs. The work makes use of an archive of geo-tagged tweets published between March 1, 2011 and January 18, 2012 [1]. During this time, 39,994,126 geo-tagged tweets (with latitude and longitude attached) were posted by 1,506,937 users. Ground truth classification data were generated by randomly selecting 5,000 accounts that published at least 200 tweets and manually labeling them as PCA, PDA, or unknown. Of the 5,000 randomly selected accounts 2,787 were PCAs, 389 PDAs, 0 spam accounts and 1,824 unknown accounts[5]. These data were then used to extract and analyze the temporal, spatial and textual features of PCAs and PDAs.

[5] In our manual search of the data we did not identify any spam accounts. They may not appear in the geo-tagged tweet stream, perhaps to maintain anonymity, or because Twitter had already detected and blocked the tweet content (in which case we would have labeled them as unknown).

2.1 Temporal Feature

PDAs are, by definition, regularly disseminating useful information. Often this task is facilitated by use of automated computer programs that publish tweets at specific times or at regular intervals [23]. In contrast, PCAs, being human, may be less regimented in their communication of daily live events and feelings. Figure 3(a) and 3(b) show the timing (minutes by seconds) of tweets published by two Twitter accounts. The specific times at which tweets were sent by the user depicted in the Figure 3(a) are spread relatively uniformly across the space. That is, the user does not appear to have a preference for specific minute-second combinations. In contrast, the user depicted in the right panel tweets at very specific times. While this program/bot- controlled PDA is not able to get the tweet out at exactly the same second each hour, the temporal distribution is clearly non-uniform.

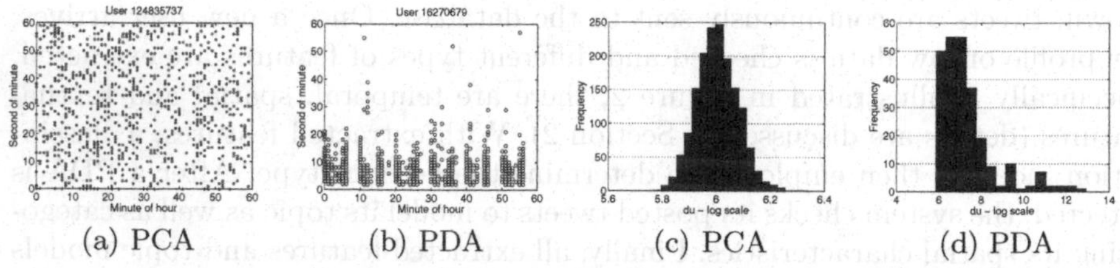

(a) PCA (b) PDA (c) PCA (d) PDA

Fig. 3. Time distribution of tweets published by PCA and PDA

Consider the two-dimensional time space shown in Figure 3(a) and 3(b) where the x-axis is the exact minute (0-59) within the hour that the tweet was published, and the y-axis is the exact second (0-59) of that minute. Tweets' timestamp information can be used to locate each tweet as a point in this space. For each account, we count the number of tweets within each section of the grid and compute the sum of the difference between the observed frequency and the expected uniform frequency to obtain a temporal feature. Formally, let g denote the total number of grids and each time stamp can be converted to a g-dimensional vector $\boldsymbol{x} = (x_1, \cdots, x_g)$, where $x_i\{0,1\}$ and $\sum_{i=1}^{g} x_i = 1$. This vector indicates which grid the time stamp belongs to. Suppose there are N tweets and the expected number of tweets falling in each grid should be N/g for a uniform distribution. We define a *time uniformity* metric du to measure the difference between the observed time distribution and a uniform one.

$$\boldsymbol{Y} = \sum_{i=1}^{N} \boldsymbol{x}_i - \frac{N}{g} \cdot \boldsymbol{I} \quad du = \frac{\boldsymbol{Y} \cdot \boldsymbol{Y}^T}{N/g} \tag{1}$$

where $\boldsymbol{I} = \underbrace{(1, \cdots, 1)}_{g}$ denotes a g-dimensional unit vector.

The lower value of du suggests a higher probability of uniform distribution. As can be seen in Figure 3(c) and 3(d) the distribution of du for PCAs satisfies a log-Gaussian distribution centered around 6, while the PDAs *du* are skewed from 6 upwards.

2.2 Spatial Feature

PCA's and PDA's tweets may also exhibit different spatial distributions. As people go about their daily lives, they often tend to move around within a limited area, periodically switch between previously visited locations (e.g., home and work), and are constrained by the physical parameters bounding how fast they can travel between locations [8,20,5,4]. In contrast, PDAs, by their very nature, are not constrained. Twitter APIs can be used to tweet from multiple locations simultaneously and/or purposively designate the geo-locations that should be attached to each tweet. Figure 4(a) and 4(b) show the footprints of geo-located tweets respectively published by a PCA and a PDA. It can be seen that the PCA tweets from a small area (303.0233 km^2) in New York, while the PDA tweets from all across the United States (about 9.6302×10^6 km^2). The narrow and sharp peak in Figure 4(c) indicates a PCA visits the same locations repeatedly. In contrast, the density distribution of a PDA in Figure 4(d) is much flatter, indicating that this account rarely tweets from the same locations.

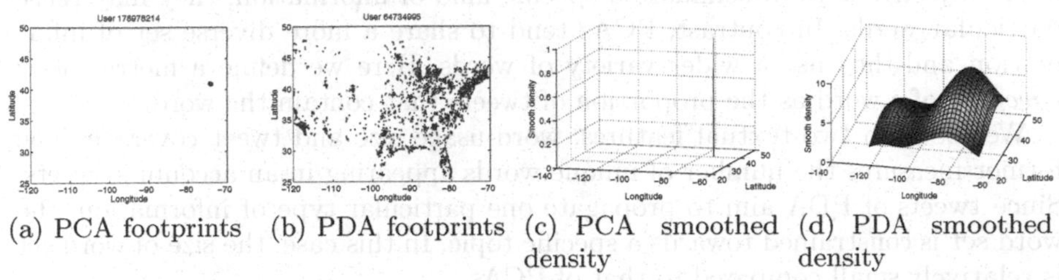

(a) PCA footprints (b) PDA footprints (c) PCA smoothed density (d) PDA smoothed density

Fig. 4. Spatial pattern of tweets published by a PCA and a PDA. The x-axis is the longitude and the y-axis is the latitude. In Figure 4(c) and Figure 4(d), the z-axis represents the smoothed frequency of visits.

We define two metrics, namely *Mobility Area (MA)* and *Unit Mobility Entropy (UME)* to capture the spatial features. MA is a measure of an account's mobility range. For a set of points in geographic space $\langle p_1, \cdots, p_n \rangle$, where $p_i = (x_i, y_i)$ consists of a longitude x_i and a latitude y_i, we can find a minimum bounding box $\langle p_{min} = (x_{min}, y_{min}), p_{max} = (x_{max}, y_{max}) \rangle$ that covers all points. MA is defined as the surface area of the bounding box in the earth.

$$\begin{aligned} MA = \texttt{Area}(p_{min}, p_{max}) &= \int_{x_{min}}^{x_{max}} \int_{y_{min}}^{y_{max}} R^2 \cdot \cos(x) dy dx \\ &= R^2 (y_{max} - y_{min})(\sin(x_{max}) - \sin(x_{min})) = R^2 \triangle y \triangle \sin(x) \end{aligned} \tag{2}$$

where R is constant representing the radius of the earth.

UME measures the diversity of spatial locations visited during a specific unit of time. A smaller value indicates a higher probability of revisiting the same location. Formally, given a unique set of locations $\langle p_1, \cdots, p_n \rangle$ that appear in one's tweets and a minimum bounding box (p_{min}, p_{max}) that covers these points, the UME is defined in Equation (3).

$$UME = \frac{\sum_{i=1} \frac{f_i}{\sum_{j=1}^{n} f_j} \log \frac{\sum_{j=1}^{n} f_j}{f_i}}{\Delta T} \tag{3}$$

where f_i represents the frequency of tweets that contains the geographical point p_i and ΔT is the time interval between the earliest tweet and the most recent one.

Furthermore, we can calculate the "moving" speed of the account by checking the time stamp and geo-coding of its successive tweets. For some PDAs, where account holders may publish tweets from multiple, distant locations within a short interval, moving speed may be quite large. In contrast, PCAs are bounded by the physical constraints on human mobility.

2.3 Textual Feature

Content of PDA's tweets may also differ from that of PCA's. Given that PDAs' main objective is to disseminate a specific kind of information, they may reuse particular words. In contrast, PCAs tend to share a more diverse set of information and thus use a wider variety of words. Here we define a metric *tweet coverage* of a word as the proportion of tweets that contain the word.

We focus on two textual features: word-usage size and tweet coverage. The former measures the number of unique words appearing in an account's tweets. Since tweets of PDA aim to propagate one particular type of information, the word set is constrained towards a specific topic. In this case, the size of word set is relatively small compared to that of PCAs.

Formally, let $W = \langle w_1, \cdots, w_n \rangle$ denote the global word set and f_i^u denote the number of user u's tweets that contain the word w_i. For N posted tweets of an account, the word-usage size of the user ws^u is defined in Equation (4).

$$ws^u = \sum_{i=1}^{n} \mathbf{1}_{f_i^u \neq 0} \tag{4}$$

Tweet coverage is the probability of a single word appearing in the tweet. Given a user u's N tweets, the tweet coverage for a word w_i can be computed by $\frac{f_i^u}{N}$. Particularly, we are focused on the mean of top-k ($k \leq n$) words (referred to as top-k mean) and tweet-coverage variance of all words (referred to as global variance) for that user. Suppose we sort the words in a non-ascending order based on their tweet frequency, i.e., $\forall i, j \in [1, ws^u]$, we have $f_i^u \leq f_j^u \Leftrightarrow i \geq j$. The top-$k$ mean μ_u and global variance σ_u^2 of tweet coverage for the user account u are defined in Equation (5).

$$\mu_u = \frac{\sum_{i=1}^{k} f_i^u}{N \cdot k} \quad \sigma_u^2 = \frac{\sum_{i=1}^{n} (\frac{f_i^u}{N} - \mu)^2}{ws^u} \tag{5}$$

Figure 5(a) and 5(b) show the words' tweet coverage of 1,000 randomly sampled tweets for a PCA and a PDA. It can be easily seen that the tweet coverage for a PCA is quite low and the maximal one is about 0.04, i.e., the most frequent word appears in 4% of her published tweets. In contrast, the PDA (Figure 5(b)) uses some words in almost every tweet. In this example the PDA is a corporate account that tweets jobs for mobile phone retail in different US cities. The words with 100% tweet coverage are *job*, *mobile* and *retail*.

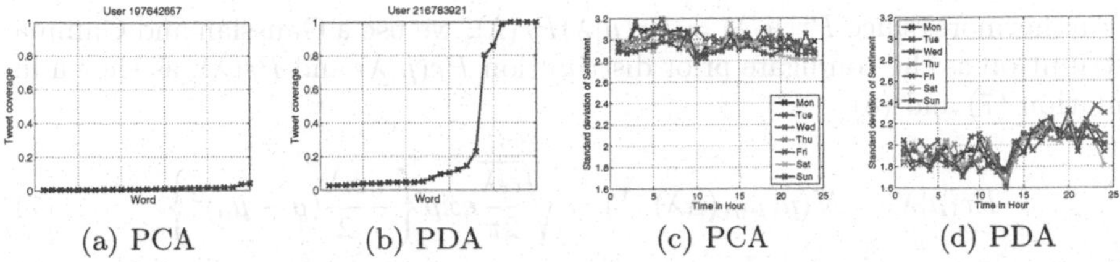

(a) PCA (b) PDA (c) PCA (d) PDA

Fig. 5. Tweet coverage and sentiment distribution for PCA and PDA

Moreover, since PCA's tweets are a reflection of their daily life, the sentiment of the tweets are more likely to fluctuate than that of PDAs. By adopting the lexicon for word-sentiment in existing works [16,17], we estimate a sentiment score for each tweet. Figure 5(c) and 5(d) show the standard deviation of PCA's and PDA's sentiment in Tweets covering a 24 hour time period. It can be seen that on average the PDA's tweets display less fluctuation of sentiment than the PCA's. Therefore, the deviation of sentiment is also extracted as a feature.

3 Detection Model

In this section we describe details of our detection model, including model development, parameter learning and detection function. Specifically to fit the temporal, spatial and textual features of PCAs, we propose a generative model that is adapted for stream training data. Classification of PDAs is solved by detecting the outliers of the fitted model.

3.1 Model Development

Without loss of generality, let $\mathbf{D} = \langle x_1, \cdots, x_n \rangle$ denote the values of extracted features. Here each element x_i represents the feature value of an account. The semantics depends on the feature types. For instance, x_i could indicate the log-value of du (see Equation (1) for definition) for a temporal feature, or ma (Equation (2)) for a spatial feature, or ws (Equation (4)) for a textual feature. Based on maximum-likelihood theory, to learn the model parameter μ, λ, we need to maximize the probability $Pr(\mu, \lambda|\mathbf{D})$. Using Bayesian inference, $Pr(\mu, \lambda|\mathbf{D}) = Pr(\mathbf{D}|\mu, \lambda)Pr(\mu, \lambda)$, where the $Pr(\mu, \lambda)$ is the *prior distribution*.

Under the assumption that the data $\mathbf{D}$ is generated by some Gaussian distribution $\mathcal{N}(\mu, \lambda^{-2})$, we can write the probability as in Equation (6).

$$
\begin{aligned}
Pr(\mathbf{D}|\mu,\lambda) &= \left(\frac{\lambda}{2\pi}\right)^{-\frac{n}{2}} exp\left\{-\frac{\lambda}{2}\sum_{i=1}^{n}(x_i-\mu)^2\right\} \\
&= \frac{1}{\sqrt{2\pi}}\left[\lambda^{1/2} exp\left(-\frac{\lambda\mu^2}{2}\right)\right]^n exp\left\{\lambda\mu\sum_{i=1}^{n}x_i - \frac{\lambda}{2}\sum_{i=1}^{n}x_i^2\right\}
\end{aligned}
\tag{6}
$$

Furthermore, since $Pr(\mu, \lambda) = Pr(\mu|\lambda)Pr(\lambda)$, we use a Gaussian and Gamma distribution as the conjugate prior distribution $Pr(\mu|\lambda)$ and $Pr(\lambda)$, as shown in Equation (7) and (8).

$$
Pr(\mu|\lambda) = \mathcal{N}(\mu; \mu_0, (\alpha\lambda)^{-1}) = \sqrt{\frac{\alpha\lambda}{2\pi}} exp\left\{-\frac{\alpha\lambda}{2}(\mu-\mu_0)^2\right\} \tag{7}
$$

where μ_0, α are prior distribution parameters for μ.

$$
Pr(\lambda) = \mathcal{G}(\lambda; a, b) = \frac{1}{(a-1)!} b^a \lambda^{a-1} exp(-b\lambda) \tag{8}
$$

Therefore, the prior distribution is represented by a product of a Gaussian and a Gamma distribution. Conversion to match the format of posterior distribution in Equation (6) gives us

$$
\begin{aligned}
Pr(\mu,\lambda) &= Pr(\mu|\lambda)Pr(\lambda) = \mathcal{N}(\mu; \mu_0, (\alpha\lambda)^{-1})\mathcal{G}(\lambda; a, b) \\
&\propto \left[\lambda^{1/2} exp\left(-\frac{\lambda\mu^2}{2}\right)\right]^{\alpha} exp\left\{\beta\lambda\mu - \gamma\lambda\right\}
\end{aligned}
\tag{9}
$$

Note that in Equation (9), we define for simplicity new parameters $\beta = \alpha\mu_0$, $\gamma = \frac{\alpha\mu_0^2}{2} + b$ to. Also, to maintain consistency with the posterior distribution, we constrain $a = \frac{\alpha+1}{2}$.

After unifying the posterior and prior distribution, we can represent the probability $Pr(\mu, \lambda|\mathbf{D})$ as below:

$$
\begin{aligned}
&Pr(\mu,\lambda|\mathbf{D}) = Pr(\mathbf{D}|\mu,\lambda)Pr(\mu|\lambda)Pr(\lambda) \propto \\
&\left[\lambda^{1/2} exp\left(-\frac{\lambda\mu^2}{2}\right)\right]^{n+\alpha} exp\left\{(\beta + \sum_{i=1}^{n}x_i)\lambda\mu - (\gamma + \frac{1}{2}\sum_{i=1}^{n}x_i^2)\lambda\right\}
\end{aligned}
\tag{10}
$$

3.2 Model Training for Stream Data

In previous subsection we unified the prior and posterior distribution in Equation (10). Now we can define the objective function and learn the parameters by maximizing it.

Let $\theta = \langle\alpha, \beta, \gamma\rangle$ denote the parameters for prior distribution and we define the objective function as the log of the probability $Pr(\mu, \lambda|\mathbf{D})$, i.e., $\mathfrak{L}(\mu, \lambda) =$

$\log Pr(\mu, \lambda|\mathbf{D}, \theta)$. By setting partial differential $\frac{\partial \mathfrak{L}}{\partial \mu}$ and $\frac{\partial \mathfrak{L}}{\partial \lambda}$ to 0, we can estimate the value for model parameters. Without loss of generality, suppose there are two sets of training samples coming in a stream, where $\mathbf{X} = \langle x_1, \cdots, x_{n_1} \rangle$ arrives first and it is followed by $\mathbf{Y} = \langle y_1, \cdots, y_{n_2} \rangle$. Also, let (μ_x, λ_x) denote the model parameters learned purely based on $\mathbf{X}$, the sequential learning process is then illustrated in Equation (11) and (12).

$$\mu = \frac{\beta + \sum_{i=1}^{n_1} x_i + \sum_{j=1}^{n_2} y_j}{\alpha + n_1 + n_2} = \mu_x + \frac{\sum_{j=1}^{n_2} (y_j - \mu_x)}{\alpha + n_1 + n_2} \tag{11}$$

$$\lambda^{-1} = \lambda_x^{-1} + \frac{\sum_{j=1}^{n_2} [(y_j - \mu_x)^2 - \lambda_x^{-1}]}{\alpha + n_1 + n_2} - \left[\frac{\sum_{j=1}^{n_2} (y - \mu_x)}{\alpha + n_1 + n_2} \right]^2 \tag{12}$$

From the Equation (11) and Equation (12) we can see good characteristics of the model. Suppose the model has been trained based on the dataset $\mathbf{X}$ and a new dataset $\mathbf{Y}$ comes. With such sequential learning equations, instead of re-training on the whole dataset, we can simply update the model parameters with the new dataset.

3.3 Detection Function

Given the extracted features of a target account, we use the trained model to compute the probability that this account is generated by the model. The higher the value is, the more likely the account is a PCA.

Formally, suppose there is an unknown account with features $\boldsymbol{u}_0 = \langle f_1, \cdots, f_6 \rangle$. The parameters of corresponding Gaussian distribution are denoted by $\boldsymbol{M} = \langle (\cdots, (\mu_i, \lambda_i^{-1}), \cdots \rangle$. The ranking score $\boldsymbol{S}_{u_0}$ is a vector of log-likelihood that the given feature vector is generated by the model M.

$$\begin{aligned} \boldsymbol{S}_{u_0} &= \texttt{Rank}(\boldsymbol{u}_0, \boldsymbol{M}) = \langle \cdots, \log \mathcal{N}(f_i; \mu_i, \lambda_i^{-1}), \cdots \rangle \\ \log \mathcal{N}(f_i; \mu_i, \lambda_i^{-1}) &= \left(\log \lambda_i - \frac{\lambda_i}{2} (f_i - \mu_i)^2 \right) + C \end{aligned} \tag{13}$$

where C is a constant that is independent of the target account u_0 and model parameters.

The final detection is a voting process. Given the threshold vector $\boldsymbol{\delta}$, the detection function will judge whether the given account is PDA in each feature dimension. If the number of votes exceeds a threshold v_δ, the function will output 1, indicating the account is classified as PDA.

$$\texttt{Detect}(u_0, \mathbf{M}, \delta) \begin{cases} 1, & \text{if } \sum \{\texttt{Rank}(u_0, \mathbf{M}) \leq \boldsymbol{\delta} ? 1 : 0\} \geq v_\delta \\ 0, & \text{otherwise} \end{cases} \tag{14}$$

3.4 Other Classifiers

Besides the proposed probabilistic model, we also examine the utility of other widely used classifiers in this framework. Particularly we examine Support Vector

Machine (SVM), K-Nearest Neighbor (KNN), Decision Tree and Naive Bayes. The latter two are both implemented by Weka [10].

The SVM we adopt is developed by LIBSVM [2]. Since the number of PDA is far smaller than that of PCA, we develop three variant SVM for imbalanced classification problem. The first one over-samples minor class and is denoted as *DUP-SVM*. The second one under-samples the major class and is referred to as *RED-SVM*. Finally we increase the misclassification cost of the minority class to 100 times than that of the majority class and is referred to as *Biased-SVM*.

4 Evaluation

This section we evaluate the classifiers in terms of both effectiveness measured by F_β and efficiency measured by training time. All evaluation are based on four-fold cross-validation and average performance is reported. Then we run our generative model on the unlabeled data to mine new PDAs. The data set we use for evaluation is a collection of manually labeled accounts, 389 PDAs and 2,787 PCAs. For F_β, we set $\beta = 0.25$ because precision is relatively more important than recall in our case.

4.1 Experiments

Figure 6(a) shows the impact of threshold on the performance of the generative model. Generally, small threshold can achieve high accuracy PDA detection but may miss many PDAs. On the other hand, big threshold may reduce missing rate but lead to false identification. Figure 6(b) shows the general comparison of different methods on PDA/PCA classification. Particularly we choose the feature threshold $\boldsymbol{\delta}$ and vote threshold v_δ with regarding to maximize PDA's and PCA's F-measure, which are respectively denoted as Mdf and Mcf. The tuning process is not shown in this paper due to the page limit. We can see that our probabilistic model is either better than or close to the best performance of other classifiers. We also use synthetic data to evaluate the efficiency. Figure 6(c) shows the result and it can be seen that the time cost of training our model increases slowest as the data set grows. Note that the time cost for KNN mainly comes from classification where all training data is scanned for each classification task. The experiment shows it takes KNN 1.0334 seconds to classify one account when the training data set is 100,000. For SVM, the time cost is 401.785 seconds for 20,000 training samples in our experiment.

4.2 Exploration

In this section we run our trained model on the unlabeled data to explore new possible PDAs. By the time the paper is written, we have detected 13,871 PDAs.

Table 1 shows some of the detected PDAs. In the table, some twitter account has such symbols as **XXX** and **YYY**. They mean there is a bunch of twitter accounts with similar naming rules, where **XXX** means the type of jobs while **YYY** stands for a particular area name, e.g., tmj_tx_intern is a PDA that tweets internship in Texas, sp_arizona tweets about deals and coupons in Arizona.

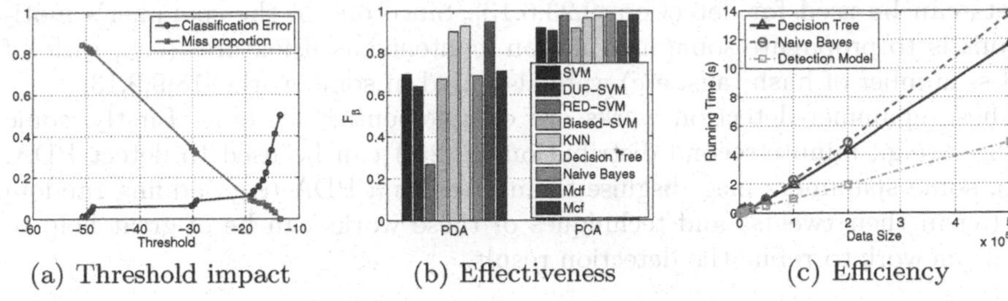

(a) Threshold impact (b) Effectiveness (c) Efficiency

Fig. 6. Experiment results

Table 1. Result of exploration

Topic	Key Words	PDA	Description
	tweetmyjobs	tmj_**XXX**_**YYY**	tweeting jobs in different areas
job	intern, internship	Get**XXX**Jobs	tweeting jobs
	job, jobs	Memphiscareers	tweeting jobs in Memphis
	theft, traffic	TotalTraffic**YYY**	real time traffic in **YYY** areas
traffic	accident, police	PinellasCo911	Fire/EMS 911 Dispatches for Pinellas County, Florida
	delay	HPD_scanner	police incidents in Houston
	healthcare, nursing	tmj_**YYY**_health	tweeting jobs of health
health	hospitality, medical	tmj_**YYY**_nursing	tweeting jobs of nursing
	rescue	tofire**YYY**	fire incidents in Toronto
	university, education	berkeleymedia	real time news in Berkeley
education	instructor, news	tmj_**YYY**_edu	educational jobs in **YYY** areas
	hall	_SchoolSpring	teaching and education jobs
	coupon, free	sp_**YYY**	deals and coupons in US
coupon	service, hotel	eatcheapnearu	best restaurants with discounts
	restaurant	KidsDineFree	restaurants providing free kid-meal

5 Related Work

Two areas are related, (i) human mobility modeling, and (ii) spam detection.

As more and more people use geo-enabled smart phones to share their locations via social media, there is a large number of studies on modeling individual's mobility pattern. Generally there are two categories: (i) predicting user's location [3,11,12], and (ii) modeling continuous moving behavior [4,5,14]. These works and our work are of mutual benefit to each other. On one hand, observations of these works serve as a guideline for us to design spatial features for our detection model. On the other hand, these existing works do not differentiate common user accounts and non-human accounts. Our work can facilitate them to reduce "noise" in the data.

Many works studied the methods to battle with spammer in Twitter. In [19,18,6,22,15,21], following/followee structure was exploited.Also, spammers are usually controlled by some program and thus the times tamp of their published tweets can be used for detection [9,23,6,13]. Since one of the spammer's motivations is to propagate some information, content-based features (e.g., ratio of URLs, number of hash tags, etc) were also used in some works [18,9,9,13].

These spammer-detection works are complementary to ours. Firstly, some features (e.g., minute-second distribution in [23]) can be used to detect PDA. Also, some spammers may disguise themselves as a PDA (e.g., adding random geo-tag in their tweets) and techniques of these works can be of great help to our framework to refine the detection result.

6 Conclusion and Future Work

The Twitter data stream is an immensely rich data resource to which many different types of entities are contributing. As such, effective use may require separation of tweets by account type. We identified types of accounts that may be especially interesting to researchers and information consumers, with specific concentration on *Public Dissemination Account (PDA)* and *Personal Communication Account (PCA)*. To separate PDAs from millions of PCAs in Twitter, we defined and extracted temporal, spatial and textual features of each account's tweets and compared our proposed probabilistic model to other conventional classifiers including SVM, KNN, Decision Tree and Naive Bayes. The experiment showed while the probabilistic model displays better or similar performance with these classifiers, it shows higher efficiency in training and is more adapted for stream data.

In future work we plan to strengthen our current system so that it can automatically build a detailed and dynamic taxonomy of PDAs, thus turning gold specks into nuggets that are more easily mined out of the Twitter stream.

References

1. Bodnar, T., Salathé, M.: Validating models for disease detection using twitter. In: WWW, pp. 699–702 (2013)
2. Chang, C.-C., Lin, C.-J.: LIBSVM: A library for support vector machines. ACM Tran. on IST 2, 27:1–27:27 (2011)
3. Cheng, Z., Caverlee, J., Lee, K.: You are where you tweet: a content-based approach to geo-locating twitter users. In: CIKM, pp. 759–768 (2010)
4. Cheng, Z., Caverlee, J., Lee, K., Sui, D.Z.: Exploring millions of footprints in location sharing services. In: ICWSM, pp. 81–88 (2011)
5. Cho, E., Myers, S.A., Leskovec, J.: Friendship and mobility: user movement in location-based social networks. In: KDD, pp. 1082–1090 (2011)
6. Chu, Z., Gianvecchio, S., Wang, H., Jajodia, S.: Who is tweeting on twitter: human, bot, or cyborg? In: ACSAC, pp. 21–30 (2010)
7. Golder, S.A., Macy, M.W.: Diurnal and seasonal mood vary with work, sleep, and daylength across diverse cultures. Science 333(6051), 1878–1881 (2011)
8. González, M.C., Hidalgo, C.A., Barabási, A.-L.: Understanding individual human mobility patterns. Nature 435, 779–782 (2008)
9. Grier, C., Thomas, K., Paxson, V., Zhang, M.: @spam: the underground on 140 characters or less. In: CCS, pp. 27–37 (2010)

10. Hall, M., Frank, E., Holmes, G., Pfahringer, B., Reutemann, P., Witten, I.H.: The weka data mining software: An update. SIGKDD Explorations 11(1), 10–18 (2009)
11. Hecht, B., Hong, L., Suh, B., Chi, E.H.: Tweets from justin bieber's heart: the dynamics of the location field in user profiles. In: CHI, pp. 237–246 (2011)
12. Kinsella, S., Murdock, V., O'Hare, N.: "i'm eating a sandwich in glasgow": modeling locations with tweets. In: SMUC, pp. 61–68 (2011)
13. Laboreiro, G., Sarmento, L., Oliveira, E.: Identifying automatic posting systems in microblogs. In: Antunes, L., Pinto, H.S. (eds.) EPIA 2011. LNCS, vol. 7026, pp. 634–648. Springer, Heidelberg (2011)
14. Noulas, A., Scellato, S., Mascolo, C., Pontil, M.: An empirical study of geographic user activity patterns in foursquare. In: ICWSM, pp. 570–573 (2011)
15. Song, J., Lee, S., Kim, J.: Spam filtering in twitter using sender-receiver relationship. In: Sommer, R., Balzarotti, D., Maier, G. (eds.) RAID 2011. LNCS, vol. 6961, pp. 301–317. Springer, Heidelberg (2011)
16. Taboada, M., Brooke, J., Tofiloski, M., Voll, K., Stede, M.: Lexicon-based methods for sentiment analysis. Computational Linguistics 37(2), 267–307 (2011)
17. Thelwall, M., Buckley, K., Paltoglou, G., Cai, D., Kappas, A.: Sentiment strength detection in short informal text. Journal of the American Society for Information Science and Technology 61(12), 2544–2558 (2011)
18. Lea, D.: Detecting spam bots in online social networking sites: A machine learning approach. In: Foresti, S., Jajodia, S. (eds.) Data and Applications Security and Privacy XXIV. LNCS, vol. 6166, pp. 335–342. Springer, Heidelberg (2010)
19. Wang, A.H.: Don't follow me - spam detection in twitter. In: SECRYPT, pp. 142–151 (2010)
20. Wang, D., Pedreschi, D., Song, C., Giannotti, F., Barabasi, A.-L.: Human mobility, social ties, and link prediction. In: KDD, pp. 1100–1108 (2011)
21. Yang, C., Harkreader, R.C., Gu, G.: Die free or live hard? Empirical evaluation and new design for fighting evolving twitter spammers. In: Sommer, R., Balzarotti, D., Maier, G. (eds.) RAID 2011. LNCS, vol. 6961, pp. 318–337. Springer, Heidelberg (2011)
22. Yardi, S., Romero, D.M., Schoenebeck, G., Boyd, D.: Detecting spam in a twitter network. First Monday 15(1) (2010)
23. Zhang, C.M., Paxson, V.: Detecting and analyzing automated activity on twitter. In: Spring, N., Riley, G.F. (eds.) PAM 2011. LNCS, vol. 6579, pp. 102–111. Springer, Heidelberg (2011)

Latent Features Based Prediction on New Users' Tastes

Ming-Chu Chen and Hung-Yu Kao

Department of Computer Science and Information Engineering,
National Cheng Kung University, Tainan, Taiwan, R.O.C.
mcchen@ikmlab.csie.ncku.edu.tw, hykao@mail.ncku.edu.tw

Abstract. Recommendation systems have become popular in recent years. A key challenge in such systems is how to effectively characterize new users' tastes — an issue that is generally known as the cold-start problem. New users judge the system by the ability to immediately provide them with what they consider interesting. A general method for solving the cold-start problem is to elicit information about new users by having them provide answers to interview questions. In this paper, we present Matrix Factorization K-Means (MFK), a novel method to solve the problem of interview question construction. MFK first learns the latent features of the user and the item through observed rating data and then determines the best interview questions based on the clusters of latent features. We can determine similar groups of users after obtaining the responses to the interview questions. Such recommendation systems can indicate new users' tastes according to their responses to the interview questions. In our experiments, we evaluate our methods using a public dataset for recommendations. The results show that our method leads to better performance than other baselines.

Keywords: Recommendation System, Collaborative Filtering, Cold Start.

1 Introduction

Recommendation systems have become increasingly popular in recent years and are widely used in e-commerce and in social networks such as Amazon[1], Netflix[2] and Facebook[3]. Amazon.com recommends specific products that customers may be interesting when they are shopping. Facebook recommends friends to users by analyzing the relationships among users. The goal of a recommendation system is to provide personalized recommendations for products that are aligned with users' tastes.

There are two types of users in recommendation systems: existing users and new users. Existing users are those who already have historical data attached to them, and new users are those who have not previously been evaluated in the recommendation system. However, we can also categorize all items as existing items or new items by considering whether the item has received ratings. Thus, there are four partitions in a recommendation system, consisting of two types of users and two types of items. Fig. 1 illustrates these partitions.

---

1 http://www.amazon.com
2 https://signup.netflix.com/global
3 https://www.facebook.com

V.S. Tseng et al. (Eds.): PAKDD 2014, Part I, LNAI 8443, pp. 176–187, 2014.

	Existing User	New User
Existing Item	1	3
New Item	2	4

Fig. 1. The four partitions in a recommendation system

Partition 1 consists of the recommendations of existing items to existing users and represents the standard situation in recommendation systems. Many *Collaborative Filtering* techniques, such as *k-nearest neighbors* (KNN) and *singular value decomposition* (SVD), work well in this situation. Partition 2 is a situation that includes recommendations of new items to existing users. *Content-based Filtering* works well in this category because it does not rely on the historical data of users. This approach can recommend items if the content information of the items is available [3], for example, the actors, genre and release year of a movie or the text in a book. Partition 3 is an important part of recommendation systems and represents the situation of recommendation of existing items to new users. To survive, recommendation systems always need to attract new users as customers. The system should indicate the taste of a new user within a short time. New users may continue to use the system if they find what they are looking for in the recommendations offered by the system. Therefore, the situation defined by Partition 3 is the focus of this paper. Partition 4 indicates recommendations of new items to new users, which is difficult because it represents a case in which there are no historical data for both items and users.

This situation with new users and new items is referred to as the *cold-start problem* [12]. A natural method for solving the cold-start problem to elicit new users' information by having them answer interview questions [9]. The system characterizes the users based on their responses to the questions, thus it can provide appropriate recommendations in the future. The interview process should not be time-consuming. Customers will become impatient and leave the system if they are presented with too many interview questions. Furthermore, the system should construct a rough profile for each new user by asking a limited number of interview questions. The decision tree method has been found to work well for the interview process [4, 5, 14]. In a decision tree, each node represents a question that is determined by certain measurements. They build the best decision tree in the interview process through the optimization of a certain loss function defined in the training step. However, some users who have similar tastes might have different paths because of their different responses to just one question. The decision tree method locally chooses the interview question and groups the users in the nodes. The method only considers the behavior of users in certain nodes. We argue that this method should be used to globally build the framework of the interview process.

2 Related Work

2.1 Collaborative Filtering

The term *Collaborative Filtering* was developed by the author of the first recommendation system, called Tasestry [6]. The system was designed to recommend documents to prevent users from becoming inundated by a huge stream of documents. In recent years, many studies of recommendation systems have focused on collaborative filtering approaches. This method can identify the new user-item association by analyzing the relationships between both users and items. The two primary types of collaborative filtering methods are memory-based [2] and model-based [13]. Memory-based methods compute the relationships between items and users. The item-item approach [8] predicts the rating of a user of an item based on ratings by the same user of neighboring items. Two items are considered to be neighbors if they receive similar ratings by users. Model-based methods are an alternative approach that characterizes users and items by rating data. This approach has become popular in recent studies of collaborative filtering because it can handle large datasets effectively and has good prediction performance. *Matrix Factorization* [7] is one of the most popular approaches in collaborative filtering; it uses a low-dimension vector to represent each user and item and learns the vectors by user-item rating data.

2.2 Cold-Start Collaborative Filtering

A recommendation system knows nothing about the new users because there is no information available. The most direct way to learn the preferences of new users is to ask them for ratings of items. Each interview question asks the user to rate items selected by the system according some criteria. Several studies have focused on the strategies for finding the best items through interview questions. The GroupLens team [9, 11] provides a balanced strategy that considers both the popularity and the entropy of movies based on the selection of interview questions. In a decision tree, each node is an interview question, and users are directed to one of the child nodes according by their responses at the parent node. These responses serve [5, 14] to learn a ternary decision tree based on the interview questions. The decision is created and optimized by analyzing the user rating data. All the tree nodes are items for the interview questions. Users have three choices—"like," "unlike," or "unknown"—for their responses to the items in each node on their path. The users are split into three separate groups in each node. Thus, the next question is determined by the user's response to the current question. Different responses to the current question result in different following questions. New users are characterized after answering the questions from the root node to the leaf nodes. The work in [1] presented a new similarity measure based on neural learning and also shown good results on Netflix and Movielens databases.

3 Our Approach

We first group the training users by observed the rating data. Because the real-world data are sparse, and no user provides ratings for all of the items, we group the users to find virtual users who have similar rating behaviors to the users in same group. When a new user comes into the system, she will be asked for responses to the interview questions. The system then calculates the similarity between the new user and the virtual users according to their responses. Thus, the system can predict the new users' tastes based on the similar virtual users. Our system's flowchart is presented in Fig. 2. We first apply the feature extraction model to determine the latent features of the users and the items, and then we use a question selection model to find the best interview questions by fitting the latent user features.

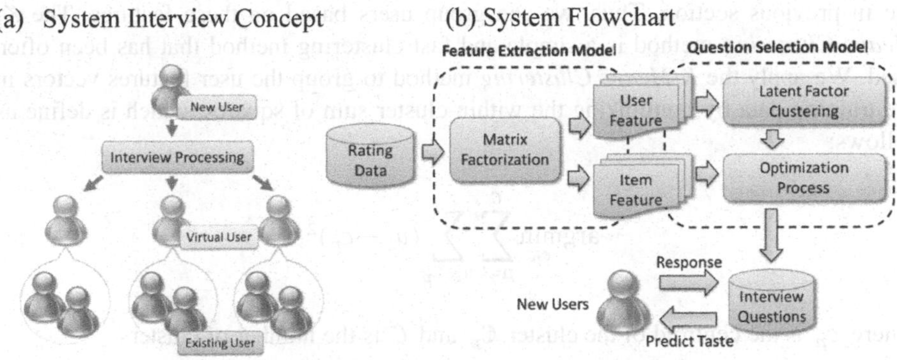

Fig. 2. The system interview concept and flowchart

3.1 Low-Rank Matrix Factorization

Matrix Factorization maps users and items to the latent factor space of dimensionality D. Each user i is associated with a vector $u_i \in \mathbb{R}^D$, and each item j is also associated with a vector $v_j \in \mathbb{R}^D$. For a given item j, the elements of v_j measure the extent of each factor. For a given user i, the elements of u_i measure the interest of each factor. The dot product, $u_i^T v_j$, captures the interaction between user i and item j and approximates the rating r_{ij}, which is user i's rating of item j. Thus, the rating matrix R can be approximated by the product of the user matrix $U \in \mathbb{R}^{D\times M}$ and item matrix $V \in \mathbb{R}^{D\times N}$, which contain M user feature vectors and N item feature vectors, respectively.

$$R \approx U^T V \tag{1}$$

This model is closely related to *singular value decomposition* (SVD), which is a well-known technique in information retrieval. The system learns the models by fitting the observed ratings to predict the unknown ratings. It should avoid overfitting the observed data by regularizing the learned parameters. Thus, the regularization terms are added to the following equation. The system learns the user and item feature matrix by minimizing the regularized squared error. Here, $\mathbb{O}$ is the set that contains all observed ratings and λ is a constant that controls the extent of the regularization.

$$\underset{U,V}{\operatorname{argmin}} \sum_{r_{ij} \in \mathbb{O}} (r_{ij} - u_i^T v_j)^2 + \lambda(\|u_i\|^2 + \|v_j\|^2) \tag{2}$$

3.2 Latent Factor Clustering

The goal of grouping users is to find several cliques that can represent different tastes of users. Each user has his or her own attributes in the features vector, as described in the in previous section. Thus, we can group users based on these features. The *K-Means Clustering* method is a simple and fast clustering method that has been often used. We apply the *K-Means Clustering* method to group the user features vectors in the training data by minimizing the within-cluster sum of squares, which is define as follows:

$$\underset{c_p}{\operatorname{argmin}} \sum_{p=1}^{C} \sum_{u_i \in \mathbb{C}_p} (u_i - c_p)^2 \tag{3}$$

where c_p is the centroid of the cluster $\mathbb{C}_p$ and C is the number of clusters.

3.3 Question Selection Process

We present an optimization process to ensure that the questions we select can identify the new users' tastes. We first define an objective function:

$$Err(\mathbb{Q}) = \sum_{r_{ij} \in \mathbb{O}} (r_{ij} - \hat{r}_{ij,\mathbb{Q}})^2 \tag{4}$$

where $\mathbb{Q}$ denotes the interview question set. We define the value $Err(\mathbb{Q})$ as the error if we use $\mathbb{Q}$ as the interview questions. Thus, we can calculate the error for each item in the training data. The $\hat{r}_{ij,\mathbb{Q}}$ denote the predicting value for the observed rating r_{ij} after asking the interview questions in $\mathbb{Q}$. We define $\hat{r}_{ij,\mathbb{Q}}$ as the following:

$$\hat{r}_{ij,\mathbb{Q}} = \begin{cases} \frac{\sum_{p=1}^{K} c_p}{K} \cdot v_j & \textit{if user i answers more than one question} \\ u_{avg} \cdot v_j & \textit{otherwise} \end{cases} \tag{5}$$

where K denotes the number of nearest neighbors and c_p is one of cluster centers that approaches the user u_i. We use the average of the choices of the cluster centers by the *K-Nearest Neighbor* (KNN) *method* to prevent overfitting. Next, we calculate the similarity of the answers to the current interview questions to find the k-nearest clusters to the users. The similarity of cluster center c_k and user u_i is defined as follows:

$$sim(c_p, u_i) = \sum_{v_j \in \mathbb{Q}_t} I_{ij}(c_p v_j - r_{ij})^2 \tag{6}$$

where $\mathbb{Q}_t$ is the current interview question set and v_j is one of the items in $\mathbb{Q}_t$. I_{ij} equals 1 if $r_{ij} \in \mathbb{O}$ and 0 otherwise. It is a constant that indicates whether the user u_i has a rating to item v_j. Finally, we select the interview question by considering the error of the current and previous questions. We choose an item to be an interview questions because it has minimum error when this item and the previous item we select are both interview questions. The following equation shows how we choose the interview questions:

$$q_h = \underset{j \in \mathbb{S}}{argmin}\{ Err(\mathbb{Q}_{q_h=j})\} \tag{7}$$

where q_h denotes the $\underline{h}$'th interview question and $\mathbb{Q}_{q_h=j}$ means item j is the $\underline{h}$'th interview question in $\mathbb{Q}$. $\mathbb{S}$ denotes a set that contains a number of items. The following section describes how we select the items in $\mathbb{S}$.

3.4 Question Selection Strategies

The popularity of an item indicates how familiar the users are with it. The more ratings the item receives, the more popular it is considered. Thus, we define these items as popular items. The advantage of choosing the most popular items to be interview questions is that new users usually provide answers. However, popular items are not effectively characterized by users. We cannot indicate the users' tastes by their responses to an item that nearly everyone liked. Users often give a high rating to such popular items. We cannot identify users' tastes by simply referencing the users' responses to popular items. The contention of the item indicates how widely the receiving ratings spread. The more widely the ratings spread, the more controversial the item is. Prior works [10] define the entropy of item to indicate how controversial it is, as defined by the following equation:

$$\text{Entropy}(v_j) = -\sum_s p_s \lg(p_s) \tag{8}$$

where p_s denotes the fraction of ratings of item v_j that equal s. For example, an item v_j has been rated by 1000 people, while there are a total of 2000 people in the data set. In the 1000 ratings data, there are 100 ratings equal to 1, 100 ratings equal to 2, 200 ratings equal to 3, 300 ratings equal to 4 and 300 ratings equal to 5. The $p_s, 1 \leq s \leq 5$ are 100/1000, 100/1000, 200/1000, 300/1000, 300/1000, respectively. The same amount of ratings for each score leads to the maximum entropy, while all ratings having the same score leads to the minimum entropy.

Popular items usually are not controversial, while items that have widely spread ratings are often not popular enough. [9] uses a balance strategy to consider both popularity and contention at the same time. It ranks the items by the product of entropy and the log of popularity. It takes the log of popularity because the number of ratings of items is an exponential distribution. Popularity will dominate the score if it does not take the log.

$$(\log Popularity(v_j)) \times \text{Entropy}(v_j) \tag{9}$$

4 Experiment

4.1 Data Set

We used the Movielens dataset for our experiment, which is a public dataset that can be downloaded from GroupLens[4] website. The data set contains a total of 1,000,209 rating data items from 6,040 users on 3,951 movies.

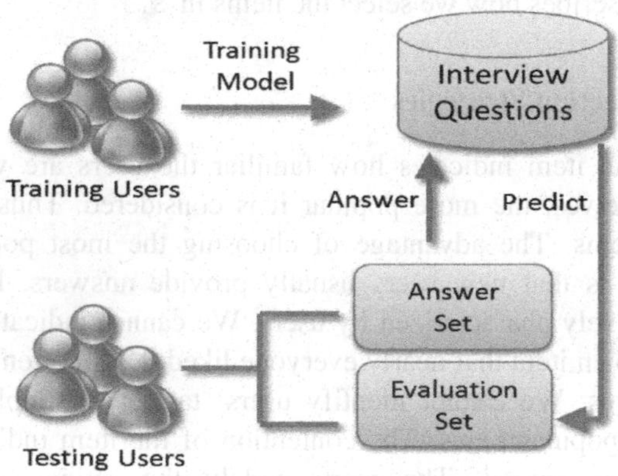

Fig. 3. System evaluation process

We split the dataset into two separate sets: the training set and the testing set. The training set contains 75% of all users, and the testing set contains 25% of the users. In the testing set, we split the users' ratings into an answer set and an evaluation set, which contain 75% and 25% of the ratings, respectively. The answer set is used to

---

4 http://www.grouplens.org/node/73/

generate the user response for the interview questions. Each tested user answers the interview question based on the ratings in the answer set and answers "Unknown" if there is no such rating in the answer set. The evaluation set is used to evaluate the performance after the interview process. The system characterizes each new user after the interview process and predicts the ratings in the evaluation set. Then, the RMSE mentioned previously is calculated to evaluate the performance.

4.2 Evaluation

Root mean square error (RMSE) measurement has been widely used to evaluate the performance of collaborative filtering algorithm. The RMSE formula is defined as follows:

$$\mathrm{RMSE} = \sqrt[2]{\frac{\sum_{r_{ij} \in T} (r_{ij} - \hat{r}_{ij})^2}{|T|}} \qquad (10)$$

where T is the set of test ratings, r_{ij} is the ground-truth rating of user i on movie j and $\hat{r}_{ij}$ is the value predicted by models. The smaller value of RMSE means better performance of the model because the prediction ratings are approaching the ground-truth rating.

4.3 Comparisons

Because MFK methods predict the rating by finding the neighbors of feature clusters according to the responses to interview questions, we compare two types of baseline to our approach. The first type of baseline is the method of predicting ratings. We use three baselines to predict the new users' ratings in the evaluation set. The second baseline is the strategy used to select the interview questions. We use different strategies to determine the interview questions and predict the ratings by finding the neighbor cluster according to the response.

- *Global Average:*
  We do not characterize each new user by the interview questions. Instead, we return only the global average rating (i.e., 3.58) to be the prediction of new users' ratings.

- *User Average:*
  For each tested user, we return rating $\hat{r}_i$, as defined in the following, to predict the observed rating in the evaluation set.

- *Item Average:*
  We predict the observed rating r_{ij} by the average received rating of item j.

Table 1 shows the performance of our approach compared with baselines. Note that $MFK_{Balance}$ in this case is the performance we show for seven interview questions, and 0.9583 is the predicted performance of users who answer zero to the seven interview questions.

Table 1. RMSE for different methods

Global Average	1.1072
User Average	1.0299
Item Average	0.9745
$MFK_{Balance}$	0.9583

The following strategies are our approaches, which combine the selection criteria and optimization processes we described in Section 3.

- $MFK_{Popularity}$.
  We rank all items by their popularity and include the top 100 items in the question set. Then, we determine the interview questions by using the optimizing process described in Section 3.3.

- $MFK_{Convention}$.
  We rank all items by their entropy score and include the top 100 items in the question set. Then, we determine the interview questions by using the optimizing process described in Section 3.3.

- $MFK_{Balance}$.
  We rank all items by the score defined in Equation (9) and include the top 100 items in the question set. Then, we determine the interview questions by using the optimizing process described in Section 3.3.

In Table 2, the performance of the popularity strategy shows a substantial improvement when we show seven questions. However, the entropy strategy does not improve when the number of shown questions increases. Fig. 4 shows that more than 90% of users do not give any response to interview questions that have high entropy scores. For almost all the interview questions, the system cannot indicate users' tastes correctly through the "Unknown" response. The balance method shows better performance than the popularity methods. This finding indicates that considering the entropy of an item is a way to improve the performance. In our approach, we consider both the popularity and the entropy of items, and we also implement an optimization process to ensure that we obtain the best performance compared with other strategies.

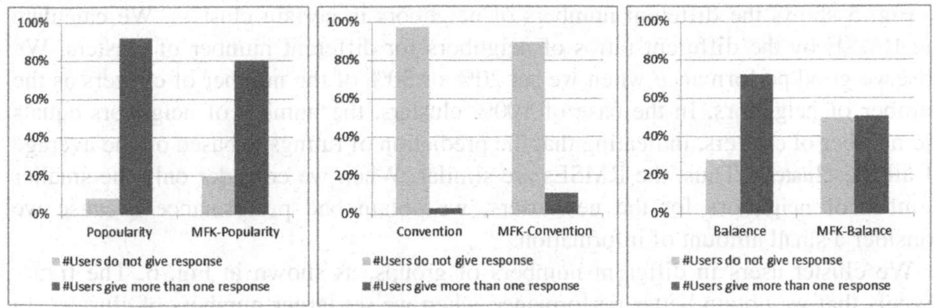

Fig. 4. The ratio of non-responses to users who provide more than one response

Table 2. Comparison with baselines

#Shown Questions	3	4	5	6	7
Random	0.9808	0.9812	0.9812	0.9812	0.9812
Popularity	0.9902	0.9847	0.9766	0.9722	0.9662
Convention	0.9833	0.9839	0.9833	0.9839	0.9837
Balance	0.9869	0.9825	0.9788	0.9671	0.9634
$MFK_{Popularity}$	0.9822	0.9820	0.9828	0.9743	0.9628
$MFK_{Convention}$	0.9839	0.9884	0.9892	0.9832	0.9892
$MFK_{Balance}$	0.9674	0.9614	0.9604	0.9591	0.9583

4.4 Impact of Parameters

Fig. 5 and Fig. 6 show the different numbers of neighbors in certain clusters. We will discuss the effect of the number of clusters and the number of user neighbors selected by their response to interview questions.

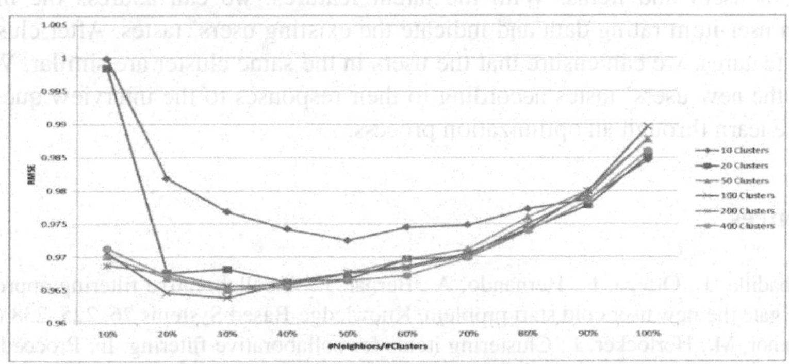

Fig. 5. RMSE in different numbers of neighbors with certain clusters

Fig. 5 shows the different numbers of neighbors in certain clusters. We calculate the RMSE by the different ratios of neighbors for different number of clusters. We observe good performance when we set 20% or 30% of the number of clusters as the number of neighbors. In the case of 100% clusters, the number of neighbors equals the number of clusters, indicating that the prediction of ratings is based on the average of all the clusters. Thus, the RMSEs are similar. When we consider only the smaller number of neighbors for the new users, we obtain bad performance because we consider a small amount of information.

We cluster users in different numbers of groups, as shown in Fig. 6. The figure reveals that we obtain better performance when we set larger numbers of clusters. In large numbers of clusters, users can find their neighbors more accurately because there are more different numerical responses to the interview questions.

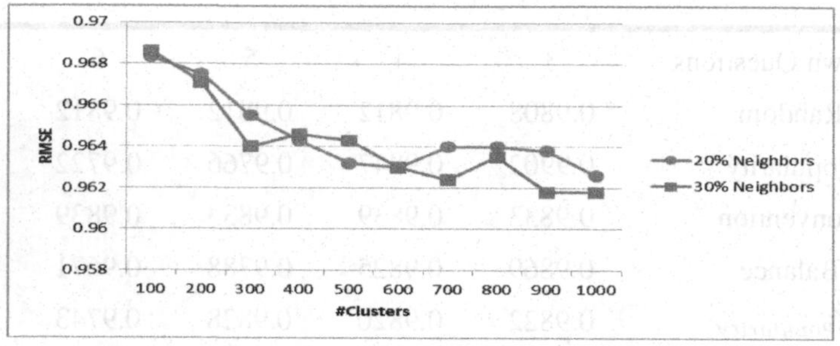

Fig. 6. RMSE in different numbers of clusters

5 Conclusions

In this paper, we proposed the MFK algorithm, which combines user feature selection and interview question optimization to address the new user problem. We can determine the best interview questions and extract the latent features by minimizing the objective functions. We first use the feature extraction model to extract the latent features of users and items. With the latent features, we can address the missing values in user-item rating data and indicate the existing users' tastes. After clustering the user features, we can ensure that the users in the same cluster are similar. We can indicate the new users' tastes according to their responses to the interview questions, which we learn through an optimization process.

References

[1] Bobadilla, J., Ortega, F., Hernando, A., Bernal, J.: A collaborative filtering approach to mitigate the new user cold start problem. Knowledge-Based Systems 26, 225–238 (2012)

[2] Connor, M., Herlocker, J.: Clustering items for collaborative filtering. In: Proceedings of the ACM SIGIR Workshop on Recommender Systems, SIGIR 1999 (1999)

[3] Gantner, Z., Drumond, L., Freudenthaler, C., Rendle, S., Schmidt-Thieme, L.: Learning Attribute-to-Feature Mappings for Cold-Start Recommendations. Presented at the Proceedings of the 2010 IEEE International Conference on Data Mining (2010)

[4] Golbandi, N., Koren, Y., Lempel, R.: On bootstrapping recommender systems. In: Proceedings of the 19th ACM International Conference on Information and Knowledge Management, Toronto, ON, Canada, pp. 1805–1808 (2010)

[5] Golbandi, N., Koren, Y., Lempel, R.: Adaptive bootstrapping of recommender systems using decision trees. In: Proceedings of the Fourth ACM International Conference on Web Search and Data Mining, Hong Kong, China, pp. 595–604 (2011)

[6] Goldberg, D., Nichols, D., Oki, B.M., Terry, D.: Using collaborative filtering to weave an information tapestry. Commun. ACM 35, 61–70 (1992)

[7] Koren, Y., Bell, R., Volinsky, C.: Matrix Factorization Techniques for Recommender Systems. Computer 42, 30–37 (2009)

[8] Linden, G., Smith, B., York, J.: Amazon.com Recommendations: Item-to-Item Collaborative Filtering. IEEE Internet Computing 7, 76–80 (2003)

[9] Rashid, A.M., Albert, I., Cosley, D., Lam, S.K., McNee, S.M., Konstan, J.A., et al.: Getting to know you: learning new user preferences in recommender systems. In: Proceedings of the 7th International Conference on Intelligent User Interfaces, San Francisco, California, USA, pp. 127–134 (2002)

[10] Rashid, A.M., Karypis, G., Riedl, J.: Learning preferences of new users in recommender systems: an information theoretic approach. SIGKDD Explor. Newsl. 10, 90–100 (2008)

[11] Resnick, P., Iacovou, N., Suchak, M., Bergstrom, P., Riedl, J.: GroupLens: an open architecture for collaborative filtering of netnews. In: Proceedings of the 1994 ACM Conference on Computer Supported Cooperative Work, Chapel Hill, North Carolina, United States, pp. 175–186 (1994)

[12] Schein, A.I., Popescul, A., Ungar, L.H., Pennock, D.M.: Methods and metrics for cold-start recommendations. In: Proceedings of the 25th Annual International ACM SIGIR Conference on Research and Development in Information Retrieval, Tampere, Finland, pp. 253–260 (2002)

[13] Su, X., Khoshgoftaar, T.M.: A survey of collaborative filtering techniques. Adv. in Artif. Intell. 2009, 2 (2009)

[14] Zhou, K., Yang, S.-H., Zha, H.: Functional matrix factorizations for cold-start recommendation. In: Proceedings of the 34th International ACM SIGIR Conference on Research and Development in Information Retrieval, Beijing, China, pp. 315–324 (2011)

Supervised Nonlinear Factorizations Excel In Semi-supervised Regression

Josif Grabocka[1], Erind Bedalli[2], and Lars Schmidt-Thieme[1]

[1] ISML Lab, University of Hildesheim
Samelsonplatz 22, 31141 Hildesheim, Germany
{josif,schmidt-thieme}@ismll.uni-hildesheim.de

[2] Department of Mathematics and Informatics, University of Elbasan
Rruga Rinia, Elbasan, Albania
erind.bedalli@uniel.edu.al

Abstract. Semi-supervised learning is an eminent domain of machine learning focusing on real-life problems where the labeled data instances are scarce. This paper innovatively extends existing factorization models into a supervised nonlinear factorization. The current state of the art methods for semi-supervised regression are based on supervised manifold regularization. In contrast, the latent data constructed by the proposed method jointly reconstructs both the observed predictors and target variables via generative-style nonlinear functions. Dual-form solutions of the nonlinear functions and a stochastic gradient descent technique which learns the low dimensionality data are introduced. The validity of our method is demonstrated in a series of experiments against five state-of-art baselines, clearly improving the prediction accuracy in eleven real-life data sets.

Keywords: Supervised Matrix Factorization, Nonlinear Dimensionality Reduction, Feature Exctraction.

1 Introduction

Regression is a core task of machine learning, aiming at identifying the relationship between a series of predictor variables and a special target variable (labeled instances) of interest [1]. Practitioners often face budget constraints in recording/measuring instances of the target variable, in particular due to the need for domain expertise [2]. On the other hand, the instances composed of predictor variables alone (unlabeled instances) appear in abundant amounts because they typically originate from less expensive automatic processes. Eventually the research community realized the potential of unlabeled instances as an important guidance in the learning process, establishing the rich domain of semi-supervised learning [2]. Semi-supervised learning is expressed on two flavors: regression and classification, depending on the metric used to evaluate the prediction of the target variable.

The principle of incorporating unlabeled instances relies heavily on exploring the geometric structure of unlabeled data, addressing the synchronization of the

V.S. Tseng et al. (Eds.): PAKDD 2014, Part I, LNAI 8443, pp. 188–199, 2014.

detected structural regularities against the positioning of the labeled instances. A stream of research focuses on the notion of clusters, where the predicted target values were influenced by connections to labeled instances through dense data regions [3,4]. The other prominent stream elaborates on the idea that data is closely encapsulated in a reduced dimensionality space, known as the manifold principle. Subsequently, the method of Manifold Regularization restricted the learning algorithm by imposing the manifold geometry via the addition of structural regularization penalty terms [5]. Discretized versions of the manifold regularization highlighted the structural understanding of data through elaborating the graph Laplacian regularization [6]. The extrapolating power of manifold regularization have been extended to involve second-order Hessian energy regularization [7], and parallel vector field regularization [8].

Throughout this study we introduce a semi-supervised regression model. The underlying foundation of our approach considers the observed data variables to be dependent on a smaller set of hidden/latent variables. The proposed method builds a low-rank representation of the data which can reconstruct both the predictor variables and the target variable via nonlinear functions. The target variable is utilized in guiding the reduction process, which in comparison to unsupervised methods, help filtering only those features which boosts the target prediction accuracy [9,10]. The proposed method operates by constructing latent nonlinear projections, opposing techniques guiding the reconstruction linearly [9]. Therefore we extend supervised matrix factorization into non-linear capabilities. The nonlinear matrix factorization belongs to the family of models known as Gaussian Process Latent Variable Modeling [11]. Our stance on nonlinear projections is further elaborated in Section 3.

The *modus operandi* of our paper is defined as a joint nonlinear reconstruction of the predictors and target variable by optimizing the regression quality over the training data. The nonlinear functions are defined as regression weights in a mapped data space, which are expressed and learned in the dual-form using the kernel theory. In addition, a stochastic gradient descent algorithm is introduced for updating the latent data based on the learned dual regression weights. In the context of semi-supervision our model can operate with very few labeled instances. Detailed explanation of the method and all necessary derivations are described in Section 4.

No previous paper has attempted to compare factorization approaches against the state of the art in manifold regularization, regarding semi-supervised regression problems. In order to demonstrate the superiority of the presented method we implemented and compared against five strong state-of-art methods. A battery of experiments over eleven real life datasets at varying number of labeled instances is conducted. Our method clearly outperforms five state-of-art baselines in the vast majority of the experiments as discussed in Section 5. The main contributions of this study are:

- Formulated a supervised nonlinear factorizations model
- Developed a learning algorithm in the dual formulation
- Conducted a throughout empirical analysis against the state of the art (manifold regularization)

2 Related Work

Even though a plethora of **regression** models have been proposed, yet Support Vector Machines (SVMs) are among the strongest general purpose learning models. A particular implementation of SVMs tailored for approximating square error loss is called Least Square SVMs (LS-SVM) [12], and is shown to perform equivalently to the epsilon loss regression SVMs [13]. This study empirically compares against LS-SVM, in order to demonstrate the additive gain of incorporating unlabeled information.

The **semi-supervised regression** research was boosted by the elaboration of the unlabeled instances' structure into the regression models. A major stream explored the cluster notion in utilizing high density unlabeled instances' regions for predicting the target values [3,4]. The other stream, called Manifold Regularization, assumes the data lie on a low-dimensional manifold and that the structure of the manifold should be respected in regressing target values of the unlabeled instances [5]. A discretized variant of the regularization was proposed to include the graph Laplacian representation of the unlabeled data as a penalty term [6]. The regularization of the manifold surfaces have been extended to involve second-order Hessian energy regularization [7], while a formalization of the vector field theory was employed in the so-called Parallel Field Regularization (PFR) [8]. In addition, a recent elaboration of surface smoothing included energy minimizations called total variation and Euler's elastica [14]. Another study attempts to discover eigenfunctions of the integral operator derived from both labeled and unlabeled instances [15], while efforts have extended to incorporate kernel theory to manifold regularization [16]. In this study we compare against three of the strongest baselines, the Laplacian regularization, the Hessian Regularization and the PFR regularization. In contrast to these existing approaches, our novel method explores hidden data structures via latent nonlinear reconstructions of both predictors and target variable.

Supervised Dimensionality Reduction involves label information as a guidance for dimensionality reduction. The Linear Discriminant Analysis is the pioneer of supervised decomposition [17]. SVMs were adjusted to high dimensional data through reducing the dimensionality via kernel matrix decomposition [18]. Generalized linear models [19] and Bayesian mixture modeling [10] have also been combined with supervised dimensionality reduction. Furthermore convolutional and sampling layers of convolutional networks are functioning as supervised decomposition [20]. The field of Gaussian Process Latent Variable Models (GPLVM) aims at detecting latent variables through a set of functions having a joint Gaussian distribution [21]. A similar model to ours has utilized GPLVM for pose estimation in images [22].

Due to its empirical success, matrix factorization has been employed in detecting latent features, while supervised matrix factorization is engineered to emphasize the target variable [9]. For the sake of clarity, methods that reduce the dimensionality in a nonlinear fashion such as the kernel PCA [23], or kernel non-negative matrix factorization [24], should not be confused with the proposed method, because such methods are unsupervised in terms of target variable. Our method offers novelty compared to state-of-art techniques in proposing joint nonlinear reconstruction of both predictors and target variables, in a semi-supervised fashion, from a minimalistic latent decomposition through dual-form nonlinearity.

3 Elaborated Principle

The majority of machine learning methods expect a target variable to be a consequence of, or directly related to, the predictor variables. This study operates over the hypothesis that both the predictors and the target variables are *observed* effects of other hidden *original* factors/variables which are not recorded/known. Our method extracts original variables which can jointly approximate both predictors and target variables in a nonlinear fashion. The current study claims that original variables contain less noise and therefore better predict the target variable, while empirical results of Section 5 demonstrate its validity.

Let us assume the unknown original data to be composed of D-many hidden variables in N training and N' testing instances and denoted as $Z \in \mathbb{R}^{(N+N')\times D}$. Assume we could observe M-many predictor variables $X \in \mathbb{R}^{(N+N')\times M}$ and one target variable $Y \in \mathbb{R}^{N}$, with the aim of accurately predicting the test targets $Y_t \in \mathbb{R}^{N'}$. Semi-supervised scenarios where $N' > N$ are taken into consideration. Our method learns the original variables Z and nonlinear functions $g_j, h \in \mathbb{R}^D \rightarrow \mathbb{R}$ which can *jointly* approximate X, Y. Equation 1 describes the idea, while we included natural Gaussian noise with variance σ_X, σ_Y in the process. We introduce a syntactic notation $\mathbb{N}_a^b = \{a, a+1, \ldots, b-1, b\}$.

$$X_{i,j} = g_j(Z_{i,:}) + \mathcal{N}(0, \sigma_X) \; ; \quad Y_i = h(Z_{i,:}) + \mathcal{N}(0, \sigma_Y) \qquad (1)$$
$$i \in \mathbb{N}_1^{N+N'} \; , \; j \in \mathbb{N}_1^M$$

4 Supervised Nonlinear Factorizations (SNF)

As aforementioned, our novelty relies on learning a latent low-rank representation Z from observed data X, Y, such that the predictor variables and the target variable are *jointly* reconstructible from the low-rank data via *nonlinear* functions. Nonlinearity is achieved by expanding the low-rank data Z to a (probably much) higher-dimensional space $\mathbb{R}^F$ space via a mapping $\psi : \mathbb{R}^D \rightarrow \mathbb{R}^F$. Linear hyperplanes $V \in \mathbb{R}^{F\times M}, W \in \mathbb{R}^F$ with bias terms $V^0 \in \mathbb{R}^M, W^0 \in \mathbb{R}$ can therefore approximate X, Y in the mapped space as described in Equation 2.

$$\hat{X}_{i,j} = \langle \psi(Z_{i,:}), V_{:,j} \rangle + V_j^0 \; ; \quad \hat{Y}_i = \langle \psi(Z_{i,:}), W \rangle + W^0 \qquad (2)$$
$$i \in \mathbb{N}_1^{N+N'} \; , \; j \in \mathbb{N}_1^M$$

4.1 Maximum Aposteriori Optimization

Consecutively the objective is to maximize the joint likelihood of the predictors X, target Y and the maximum aposteriori estimators V, V^0, W, W^0 as shown in Equation 3. The hyperplanes parameters incorporate normal priors $V \sim \mathcal{N}(0, \lambda_V^{-1}), W \sim \mathcal{N}(0, \lambda_W^{-1})$. The distribution of the observed variables is also assumed normal $X \sim \mathcal{N}(\langle \psi(Z), V \rangle, \sigma_X)$ and $Y \sim \mathcal{N}(\langle \psi(Z), W \rangle, \sigma_Y)$ and independently distributed. The logarithmic likelihood, depicted in Equation 4 converts the objective to a summation of terms.

$$\underset{\psi(Z),V,V^0,W,W^0}{\text{argmax}} \prod_{i=1}^{N+N'} p(X_{i,:}|\psi(Z_{i,:}), V, V^0)\, p(V) \prod_{l=1}^{N} p(Y_l|\psi(Z_{l,:}), W, W^0)\, p(W) \tag{3}$$

$$\underset{\psi(Z),V,V^0,W,W^0}{\text{argmax}} \sum_{i=1}^{N+N'} \log\left(p(X_{i,:}|\psi(Z_{i,:}), V, V^0)\right) + \sum_{l=1}^{N} \log\left(p(Y_l|\psi(Z_{l,:}), W, W^0)\right) + \sum_{j=1}^{M} \log\left(p(V_{:,j})\right) + \log\left(p(W)\right) \tag{4}$$

Inserting the normal probability into Equation 4 converts logarithmic likelihoods into L2 norms with Tikhonov regularization terms as shown in Equation 5 and the variance terms σ_X, σ_Y drop out as constants. An additional biased regularization term $\lambda_Z \langle Z, Z \rangle$ is included in order to help the latent data avoid over-fitting.

$$\underset{Z,V,V^0,W,W^0}{\text{argmin}} \left(\sum_{j=1}^{M} \langle \xi_{:,j}, \xi_{:,j} \rangle + \lambda_V \langle V_{:,j}, V_{:,j} \rangle \right) + \langle \phi, \phi \rangle + \lambda_W \langle W, W \rangle + \lambda_Z \langle Z, Z \rangle$$

$$\text{subject to: } \xi_{i,j} = X_{i,j} - \langle \psi(Z_{i,:}), V_{:,j} \rangle - V_j^0,\ i \in \mathbb{N}_1^{N+N'},\ j \in \mathbb{N}_1^M \tag{5}$$

$$\phi_l = Y_l - \langle \psi(Z_{l,:}), W \rangle - W^0, \quad l \in \mathbb{N}_1^N$$

Computing the $\psi(Z)$ directly is intractable, therefore we will derive the dual-form representation in the next Section 4.2, where the kernel trick will be utilized to compute Z in the original space $\mathbb{R}^D$.

4.2 Dual-Form Solution - Learning the Nonlinear Regression Weights

The optimization of Equation 5 is carried on in an alternated fashion. Hyperplane weights V, V^0, W, W^0 are converted to dual variables and then solved by keeping Z fixed, while in a second step Z is solved keeping the dual weights fixed. This section is dedicated to learning the nonlinear weights in the dual-form. Each of the M-many predictors loss terms from Equation 5, (one per each predictor variable $X_{:,j}$) can be learned isolated as described in the sub-objective function J_j of Equation 6. To facilitate forthcoming derivations we multiplied the objective function by $\frac{1}{2\lambda_V}$.

$$\underset{V_{:,j},V_j^0}{\text{argmin}}\ J_j = \frac{1}{2\lambda_V}\langle \xi_{:,j}, \xi_{:,j}\rangle + \frac{1}{2}\langle V_{:,j}, V_{:,j}\rangle \tag{6}$$
$$\xi_{i,j} = X_{i,j} - \langle \psi(Z_{i,:}), V_{:,j}\rangle - V_j^0$$

In order to optimize Equation 6, the equality conditions are added to the objective function through Lagrange multipliers $\alpha_{i,j}$. The inner minimization objective is solved by computing stationary solution points $V_{:,j}, V_j^0, \xi_{:,j}$ and eliminating out the first derivatives ($\frac{\partial L_j}{\partial V_{:,j}} = 0, \frac{\partial L_j}{\partial \xi_{:,j}} = 0, \frac{\partial L_j}{\partial V_j^0} = 0$) as shown in Equation 7.

$$\begin{aligned}
\underset{\alpha_{:,j},V_j^0}{\text{argmax}}\ \underset{V_{:,j},V_j^0,\xi_{:,j}}{\text{argmin}}\ L_j &= \frac{1}{2\lambda_V}\langle \xi_{:,j}, \xi_{:,j}\rangle + \frac{1}{2}\langle V_{:,j}, V_{:,j}\rangle \\
&+ \sum_{i=1}^{N+N'} \alpha_{i,j}\left(X_{i,j} - \langle \psi(Z_{i,:}), V_{:,j}\rangle - V_j^0 - \xi_{i,j}\right) \qquad (7)\\
\rightarrow\ V_{:,j} &= \sum_{i=1}^{N+N'} \alpha_{i,j}\psi(Z_{i,:}) \\
\xi_{:,j} &= \lambda_V \alpha_{:,j} \\
\sum_{i=1}^{N+N'} \alpha_{i,j} &= 0
\end{aligned}$$

Replacing the stationary point solution of $V_{:,}, \xi_{:}, j, V_j^0$ back into the objective function 7, we get rid of the variables $V_{:,j}, \xi_{:,j}$, yielding Equation 8.

$$\begin{aligned}
\underset{\alpha_{:,j},V_j^0}{\text{argmax}}\ & - \sum_{i=1,l=1}^{N+N^t} \alpha_{i,j}\alpha_{l,j}\langle \psi(Z_{i,:}), \psi(Z_{l,:})\rangle - \lambda_V\langle \alpha_{:,j}, \alpha_{:,j}\rangle \\
& + 2\sum_{i=1}^{N+N^t} \alpha_{i,j}\left(X_{i,j} - V_j^0\right) \qquad (8)
\end{aligned}$$

The solution of the dual maximization is given through eliminating the derivative of $\frac{\partial L}{\partial \alpha_{:,j}} = 0$ as presented in Equation 9. The kernel notation is introduced as $K_{i,l} = \langle \psi(Z_{i,:}), \psi(Z_{l,:})\rangle$.

$$2\lambda_V\alpha_{:,j} + 2K\cdot\alpha_{:,j} + 2\langle \mathbf{1}, V_j^0\rangle = 2X_{:,j} \ \rightarrow\ (K + \lambda_V I)\,\alpha_{:,j} + \langle \mathbf{1}, V_j^0\rangle = X_{:,j} \tag{9}$$

Combining Equation 9 and the constraint of Equation 8, the final nonlinear reconstruction solution is given through the closed-form formulation of $\alpha_{:,j}, V_j^0$ as depicted in Equation 10.

$$\begin{bmatrix} V_j^0 \\ \alpha_{:,j}\end{bmatrix} = \begin{bmatrix} 0 & \mathbf{1}^T \\ \mathbf{1} & K + \lambda_V I\end{bmatrix}^{-1}\begin{bmatrix} 0 \\ X_{:,j}\end{bmatrix} \tag{10}$$

Symmetrical to the predictors case, a dual-form maximization objective function is created and a *mot-a-mot* procedure like Section 4.2 can be trivially adopted in solving the nonlinear regression for the target variable. The derived solution is shown in Equation 11. Instead of using the symbol α we denote the dual-weights of the target regression dual problem using the symbol ω.

$$\begin{bmatrix} W^0 \\ \omega \end{bmatrix} = \begin{bmatrix} 0 & \mathbf{1}^T \\ \mathbf{1} & K^Y + \lambda_W I \end{bmatrix}^{-1} \begin{bmatrix} 0 \\ Y \end{bmatrix}; \tag{11}$$
$$\text{where } K^Y_{i,l} = \langle \psi(Z_{i,:}), \psi(Z_{l,:}) \rangle; \quad i,l \in \mathbb{N}_1^N$$

A prediction of the target value of a test instance $t \in \mathbb{N}_{N+1}^{N+N'}$ is conducted using the learned dual weights as shown in Equation 12.

$$\hat{Y}_t = \sum_{i=1}^{N} \omega_i \, K^Y(Z_{i,:}, Z_{t,:}) + W^0 \tag{12}$$

Stochastic Gradient Descent - Learning the Low Dimensionality Representation. A novel algorithm is applied to learn Z for optimizing Equation 8. Sub-losses composing only of $\alpha_{i,j}, \alpha_{l,j}$ are defined for all combinations $\forall i \in \mathbb{N}_1^{N+N^t}, \forall l \in \mathbb{N}_1^{N+N^t}$ and Z is updated in order to optimize each sub-loss in a stochastic gradient descent fashion as presented in Equation 13. The addition of penalty terms controlled by the hyper-parameter λ_Z which controls the regularization of Z as described in Equation 5. Our model called Supervised Nonlinear Factorizations (SNF) utilizes polynomial kernels with the derivatives needed for gradient descent represented in Equation 13.

$$Z_{i,k} \leftarrow Z_{i,k} + \eta \left(\alpha_{i,j} \alpha_{l,j} \frac{\partial K_{i,l}}{\partial Z_{i,k}} - \lambda_Z Z_{i,k} \right) \tag{13}$$
$$Z_{l,k} \leftarrow Z_{l,k} + \eta \left(\alpha_{i,j} \alpha_{l,j} \frac{\partial K_{i,l}}{\partial Z_{l,k}} - \lambda_Z Z_{l,k} \right)$$
$$K_{i,l} = (\langle Z_{i,:}, Z_{l,:} \rangle + 1)^d \;\rightarrow\; \frac{\partial K_{i,l}}{\partial Z_{r,k}} = d\,(\langle Z_{i,:}, Z_{l,:} \rangle + 1)^{d-1} \times \begin{cases} Z_{l,k} & \text{if } r = i \\ Z_{i,k} & \text{if } r = l \\ 0 & \text{else} \end{cases}$$

Algorithm 1 combines all the steps of the proposed method. During each epoch all predictors' non-linear weights are solved and the latent data Z is updated. The target model is updated multiple times after each predictor model to boost convergence. The learning algorithm makes use of two different learning rates in updating Z, one for the predictors' loss (η_X) and one for the target loss (η_Y).

Algorithm 1. Learn SNF

Require: Data $X \in \mathbb{R}^{(N+N^t)\times M}, Y \in \mathbb{R}^{N^l}$, Latent Dimension: D, Learn Rates: η_X, η_Y, Number of Iterations: NumIter, Regularization parameters: $\lambda_Z, \lambda_V, \lambda_W$

Randomly set: $Z \in R^{(N+N')\times D}, V^0 \in \mathbb{R}^M, W^0 \in \mathbb{R}, \alpha \in \mathbb{R}^{(N+N')\times M}, \omega \in \mathbb{R}^{N^l}$,

for $1 \ldots$ NumIter **do**

  for $j \in \{1 \ldots M\}$ **do**

    Compute $K_{i,l} = K(Z_{i,:}, Z_{l,:}),\ i, l \in \mathbb{N}_1^{N+N'}$

    Solve $\begin{bmatrix} V_j^0 \\ \alpha_{:,j} \end{bmatrix} = \begin{bmatrix} 0 & \mathbf{1}^T \\ \mathbf{1} & K + \lambda_V I \end{bmatrix}^{-1} \begin{bmatrix} 0 \\ X_{:,j} \end{bmatrix}$

    for $i \in \mathbb{N}_1^{N+N'}$, $l \in \mathbb{N}_1^{N+N'}$, $k \in \mathbb{N}_1^D$ **do**

      $Z_{i,k} \leftarrow Z_{i,k} + \eta_X \left(\alpha_{i,j}\alpha_{l,j} \frac{\partial K_{i,l}}{\partial Z_{i,k}} - \lambda_Z Z_{i,k} \right)$

      $Z_{l,k} \leftarrow Z_{l,k} + \eta_X \left(\alpha_{i,j}\alpha_{l,j} \frac{\partial K_{i,l}}{\partial Z_{l,k}} - \lambda_Z Z_{l,k} \right)$

    end for

    Compute $K_{i,l}^Y = K(Z_{i,:}, Z_{l,:}),\ i, l \in \mathbb{N}_1^N$

    Solve $\begin{bmatrix} W^0 \\ \omega \end{bmatrix} = \begin{bmatrix} 0 & \mathbf{1}^T \\ \mathbf{1} & K^Y + \lambda_W I \end{bmatrix}^{-1} \begin{bmatrix} 0 \\ Y \end{bmatrix}$

    for $i \in \mathbb{N}_1^{N+N'}$, $l \in \mathbb{N}_1^{N+N'}$, $k \in \mathbb{N}_1^D$ **do**

      $Z_{i,k} \leftarrow Z_{i,k} + \eta_Y \left(\omega_i \omega_l \frac{\partial K_{i,l}^Y}{\partial Z_{i,k}} - \lambda_Z Z_{i,k} \right)$

      $Z_{l,k} \leftarrow Z_{l,k} + \eta_Y \left(\omega_i \omega_l \frac{\partial K_{i,l}^Y}{\partial Z_{l,k}} - \lambda_Z Z_{l,k} \right)$

    end for

  end for

end for

return $Z, \alpha, V^0, \omega, W^0$

5 Empirical Results

Five strong state of the art baselines and empirical evidence over eleven datasets are mainly the outline of our experiments, which will be detailed in this section, together with the results and their interpretation.

5.1 Baselines

The proposed method Supervised Nonlinear Factorizatoins (SNF) is compared against the following five baselines:

- **Least Square Support Vector Machines (LS-SVM)** [12] is a strong general purpose regression model and the comparisons against it will show the gain of incorporating unlabeled instances.
- **Laplacian Manifold Regularization (Laplacian) [5], Hessian Energy Regularization (Hessian) [7], Parallel Field Regularization (PFR)**[8] are strong state-of-art baselines belonging to the popular field of manifold regularization. Comparing against them gives an insight into the state-of-art quality of our results.

– **Linear Latent Reconstructions (LLR)** [9] offers the possibility to understand the addiditive benefits of exploring nonlinear projections compared to plain linear ones.

5.2 Reproducibility

All our experiments were run in a three fold cross-validation mode and the hyper-parameters of our model were tuned using only train and validation data. The evaluation metric used in all experiments is the Mean Square Error (MSE).

SNF requires the tuning of seven hyper-parameters: the regularization weights $\lambda_Z, \lambda_V, \lambda_W$, the learning rates η_X, η_Y, the number of latent dimensions D and the degree of the polynomial kernel d. The search ranges of hyper-parameters are: $\lambda_Z, \lambda_V, \lambda_W \in \{10^{-6}, 10^{-5} \ldots 1, 10\}; \eta_X, \eta_Y \in \{10^{-5}, 10^{-4} \ldots 0.1\}; d \in \{1, 2, 3, 4\}$ while the latent dimensionality was set to one of 50%, 75% of the original dimensions. The maximum number of epochs was set to 1000. A grid search methodology was followed in finding the best combination of hyper-parameters. Please note that we followed exactly the same fair principle in computing the hyper-parameters of all baselines.

We selected eleven popular regression datasets in a random fashion from dataset repository websites. The selected datasets are AutoPrice, ForestFires, BostonHousing, MachineCPU, Mpg, Pyrimidines, Triazines, WisconsinBreast-Cancer from UCI[1] ; Baseball, BodyFat from StatLib[2]; Bears[3]. All the datasets were normalized between [-1,1] before usage.

5.3 Results

The experiments comparing the accuracy of our method SNF against the five strong baselines were conducted in scenarios with few labeled instances, as typically encountered in semi-supervised learning situations.

In the first experiment 5% labeled instances were selected randomly, while all other instances left unlabeled. Therefore, the competing methods had 5% target visibility and all methods, except LS-SVM, utilized the predictor variables of all the unlabeled instances. The results of the experiments are shown in Table 1. The metric of evaluation is the Mean Square Error (MSE), while both the mean MSE and the standard deviation are shown in each dataset-method cell. The winning method of each baseline is highlighted in bold. As it is distinguishable from the sum of wins, SNF outperforms the baselines in the majority of datasets (six in total), while the closest competing baseline wins in only two datasets. Furthermore in datasets such as AutoPrice, BodyFat and Mpg the improvement is significant. Even when SNF is not the winning method, the margin to the first method is not significant, as it occurs in the Baseball, ForestFires and WisconsinBreastCancer datasets.

[1] archive.ics.uci.edu
[2] lib.stat.cmu.edu
[3] people.sc.fsu.edu/~jburkardt/datasets/triola/

Table 1. Results - MSE - Real-life Datasets (5 % Labeled Instances)

Dataset	LS-SVM	Laplacian	Hessian	PFR	LLR	SNF
A.Price	0.075 ± 0.012	0.156 ± 0.059	0.102 ± 0.036	0.067 ± 0.024	0.090 ± 0.026	**0.049 ± 0.026**
B.ball	0.093 ± 0.019	0.131 ± 0.020	**0.072 ± 0.015**	0.082 ± 0.029	0.086 ± 0.022	0.073 ± 0.017
Bears	0.205 ± 0.046	0.407 ± 0.212	0.399 ± 0.218	0.380 ± 0.220	0.292 ± 0.146	**0.130 ± 0.032**
B.Fat	0.013 ± 0.007	0.081 ± 0.002	0.039 ± 0.009	0.019 ± 0.007	0.017 ± 0.006	**0.008 ± 0.002**
F.Fires	0.020 ± 0.004	**0.0137 ± 0.015**	0.0141 ± 0.015	**0.0137 ± 0.015**	0.018 ± 0.017	0.0139 ± 0.014
B.Hous.	0.087 ± 0.027	0.123 ± 0.027	0.104 ± 0.042	0.071 ± 0.027	0.110 ± 0.031	**0.069 ± 0.024**
M.Cpu	0.061 ± 0.041	0.053 ± 0.043	0.027 ± 0.016	0.018 ± 0.012	0.030 ± 0.021	**0.015 ± 0.007**
Mpg	0.044 ± 0.006	0.074 ± 0.009	0.055 ± 0.007	0.052 ± 0.008	0.062 ± 0.015	**0.041 ± 0.006**
Pyrim	0.131 ± 0.020	0.106 ± 0.043	**0.097 ± 0.050**	0.101 ± 0.055	0.152 ± 0.065	0.102 ± 0.053
Triaz.	0.191 ± 0.014	**0.160 ± 0.026**	0.165 ± 0.031	0.166 ± 0.035	0.196 ± 0.026	0.175 ± 0.039
WiscBC.	0.608 ± 0.074	0.356 ± 0.039	0.356 ± 0.042	**0.3499 ± 0.035**	0.429 ± 0.074	0.3504 ± 0.028
Wins	0	1.5	2	1.5	0	6

Table 2. Results - MSE - Real-life Datasets (10 % Labeled Instances)

Dataset	LS-SVM	Laplacian	Hessian	PFR	LLR	SNF
A.Price	0.051 ± 0.007	0.115 ± 0.054	0.084 ± 0.042	0.048 ± 0.018	0.062 ± 0.005	**0.037 ± 0.014**
B.ball	0.127 ± 0.035	0.092 ± 0.021	**0.068 ± 0.011**	0.080 ± 0.022	0.084 ± 0.014	0.072 ± 0.010
Bears	0.160 ± 0.053	0.182 ± 0.067	0.202 ± 0.018	0.156 ± 0.051	**0.067 ± 0.021**	0.071 ± 0.020
B.Fat	0.006 ± 0.001	0.058 ± 0.006	0.015 ± 0.007	0.008 ± 0.004	0.011 ± 0.004	**0.003 ± 0.002**
F.Fires	0.018 ± 0.001	0.0146 ± 0.014	0.0140 ± 0.015	0.0142 ± 0.014	0.016 ± 0.015	**0.0139 ± 0.015**
B.Hous.	0.067 ± 0.015	0.115 ± 0.026	0.095 ± 0.024	0.067 ± 0.004	0.071 ± 0.015	**0.061 ± 0.015**
M.Cpu	0.030 ± 0.007	0.039 ± 0.025	0.041 ± 0.030	0.020 ± 0.011	0.024 ± 0.015	**0.012 ± 0.003**
Mpg	0.071 ± 0.047	0.062 ± 0.009	0.048 ± 0.011	**0.038 ± 0.008**	0.066 ± 0.016	0.040 ± 0.004
Pyrim	0.076 ± 0.005	0.086 ± 0.042	0.076 ± 0.041	**0.069 ± 0.021**	0.107 ± 0.060	0.076 ± 0.030
Triaz.	0.265 ± 0.068	0.173 ± 0.041	**0.162 ± 0.033**	0.163 ± 0.038	0.169 ± 0.012	0.175 ± 0.009
WiscBC.	0.624 ± 0.044	**0.283 ± 0.027**	0.287 ± 0.017	0.284 ± 0.024	0.344 ± 0.079	0.307 ± 0.020
Wins	0	1	2	2	1	5

For the sake of completeness we repeated the experiments with another degree of randomly re-drawn labeled instances (10 % labeled instances). Table 2 presents the details of experiments over the selected eleven real-life datasets. The accuracy of SNF is prolonged even in this experiment. The sum of the winning methods (depicted in bold) shows that SNF wins in five of the datasets against the only two wins of the closest baseline. In particular the cases of BodyFat and MachineCpu demonstrate significant improvements in terms of MSE. As shown by the results, even in cases where our method is not the first, still it is close to the winner.

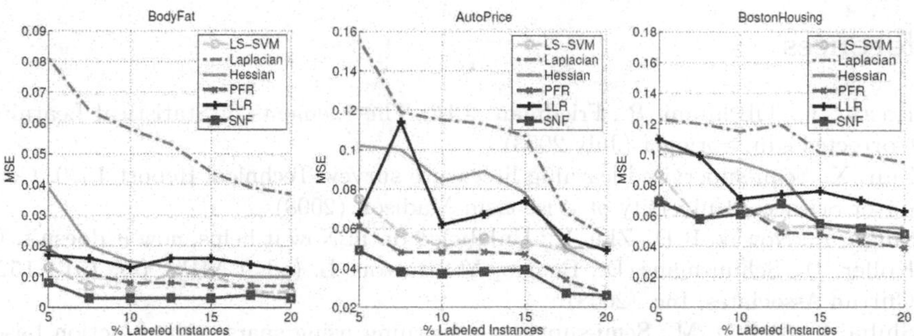

Fig. 1. Scale-up Experiments

In addition to the aforementioned results we extend our empirical analysis by conducting more fine-grained scale-up experiments with varying degree of labeled training instances. Figure 1 demonstrates the performance of all competing methods on a subset of datasets with a range of present labels varying from 5% up to 20%. SNF is seen to win in the earliest labeled percentages of the BostonHousing dataset (up to 10 %) while following in the later stages. On the contrary, we observe that our method dominates in all levels of label presence in the scaled-up experiments involving the BodyFat and AutoPrice datasets.

The accuracy of our method is grounded on a couple of reasons/observations. First of all, we would like to emphasize that each mentioned method is based on a different principle and *modus operandi.* Consequently, the dominance of a method compared to baselines depends on whether (or not) the datasets follow the principle of that particular method. Arguably the domination of SNF over manifold regularization baselines is due to the fact that our principle of mining hidden latent variables is likely (as results show) more present in general real-life datasets, therefore SNF is suited to the detection of those relations.

6 Conclusions

Throughout the present paper, a novel method that addresses the task of semi-supervised regression was proposed. The proposed method constructs a low-rank representation which jointly approximates the observed data via nonlinear functions which are learned in their dual formulation. A novel stochastic gradient descent technique is applied to learn the low-rank data using the obtained dual weights. Detailed experiments are conducted in order to compare the performance of the proposed method against five strong baselines over eleven real-life datasets. Empirical evidence over experiments in varying degrees of labeled instances demonstrate the efficiency of our method. The supervised nonlinear factorizations outperformed the manifold regularization state-of-art methods in the majority of experiments.

References

1. Hastie, T., Tibshirani, R., Friedman, J.H.: The Elements of Statistical Learning. Corrected edn. Springer (July 2003)
2. Zhu, X.: Semi-supervised learning literature survey. Technical Report 1530, Computer Sciences, University of Wisconsin-Madison (2008)
3. Singh, A., Nowak, R.D., Zhu, X.: Unlabeled data: Now it helps, now it doesn't. In: Koller, D., Schuurmans, D., Bengio, Y., Bottou, L. (eds.) NIPS, pp. 1513–1520. Curran Associates, Inc. (2008)
4. Sinha, K., Belkin, M.: Semi-supervised learning using sparse eigenfunction bases. In: NIPS, pp. 1687–1695 (2009)
5. Belkin, M., Niyogi, P., Sindhwani, V.: Manifold regularization: A geometric framework for learning from labeled and unlabeled examples. Journal of Machine Learning Research 7, 2399–2434 (2006)

6. Melacci, S., Belkin, M.: Laplacian support vector machines trained in the primal. Journal of Machine Learning Research 12, 1149–1184 (2011)
7. Kim, K.I., Steinke, F., Hein, M.: Semi-supervised regression using hessian energy with an application to semi-supervised dimensionality reduction. In: NIPS, pp. 979–987 (2009)
8. Lin, B., Zhang, C., He, X.: Semi-supervised regression via parallel field regularization. In: NIPS, pp. 433–441 (2011)
9. Menon, A.K., Elkan, C.: Predicting labels for dyadic data. Data Min. Knowl. Discov. 21(2), 327–343 (2010)
10. Mao, K., Liang, F., Mukherjee, S.: Supervised dimension reduction using bayesian mixture modeling. Journal of Machine Learning Research - Proceedings Track 9, 501–508 (2010)
11. Lawrence, N.: Probabilistic non-linear principal component analysis with gaussian process latent variable models. The Journal of Machine Learning Research 6, 1783–1816 (2005)
12. Suykens, J.A.K., Vandewalle, J.: Least squares support vector machine classifiers. Neural Processing Letters 9(3), 293–300 (1999)
13. Ye, J., Xiong, T.: Svm versus least squares svm. Journal of Machine Learning Research - Proceedings Track 2, 644–651 (2007)
14. Lin, T., Xue, H., Wang, L., Zha, H.: Total variation and euler's elastica for supervised learning. In: ICML (2012)
15. Ji, M., Yang, T., Lin, B., Jin, R., Han, J.: A simple algorithm for semi-supervised learning with improved generalization error bound. In: ICML (2012)
16. Nilsson, J., Sha, F., Jordan, M.I.: Regression on manifolds using kernel dimension reduction. In: ICML, pp. 697–704 (2007)
17. Ye, J.: Least squares linear discriminant analysis. In: ICML, pp. 1087–1093 (2007)
18. Pereira, F., Gordon, G.: The support vector decomposition machine. In: Proceedings of the 23rd International Conference on Machine Learning, ICML 2006, pp. 689–696. ACM, New York (2006)
19. Rish, I., Grabarnik, G., Cecchi, G., Pereira, F., Gordon, G.J.: Closed-form supervised dimensionality reduction with generalized linear models. In: Proceedings of the 25th International Conference on Machine Learning, ICML 2008, pp. 832–839. ACM, New York (2008)
20. Ciresan, D.C., Meier, U., Gambardella, L.M., Schmidhuber, J.: Convolutional neural network committees for handwritten character classification. In: ICDAR, pp. 1135–1139. IEEE (2011)
21. Urtasun, R., Darrell, T.: Discriminative gaussian process latent variable models for classification. In: International Conference in Machine Learning (2007)
22. Navaratnam, R., Fitzgibbon, A.W., Cipolla, R.: The joint manifold model for semi-supervised multi-valued regression. In: IEEE 11th International Conference on Computer Vision, ICCV 2007, pp. 1–8 (2007)
23. Hoffmann, H.: Kernel pca for novelty detection. Pattern Recogn. 40(3), 863–874 (2007)
24. Lee, H., Cichocki, A., Choi, S.: Kernel nonnegative matrix factorization for spectral eeg feature extraction. Neurocomputing 72(13-15), 3182–3190 (2009)

Learning from Crowds under Experts' Supervision

Qingyang Hu[1], Qinming He[1], Hao Huang[2], Kevin Chiew[3], and Zhenguang Liu[1]

[1] College of Computer Science and Technology,
Zhejiang University, Hangzhou, China
{huqingyang, hqm, zhenguangliu}@zju.edu.cn
[2] School of Computing, National University of Singapore, Singapore
huanghao@comp.nus.edu.sg
[3] Provident Technology Pte. Ltd., Singapore
kev.chiew@gmail.com

Abstract. Crowdsourcing services have been proven efficient in collecting large amount of labeled data for supervised learning, but low cost of crowd workers leads to unreliable labels. Various methods have been proposed to infer the ground truth or learn from crowd data directly though, there is no guarantee that these methods work well for highly biased or noisy crowd labels. Motivated by this limitation of crowd data, we propose to improve the performance of crowdsourcing learning tasks with some additional expert labels by treating each labeler as a personal classifier and combining all labelers' opinions from a model combination perspective. Experiments show that our method can significantly improve the learning quality as compared with those methods solely using crowd labels.

Keywords: Crowdsourcing, multiple annotators, model combination, classification.

1 Introduction

Crowdsourcing services such as Amazon Mechanical Turk have made it possible to collect large amount of labels at relatively low cost. Nonetheless, since the reward is small and the ability of workers is not certified, the labeling quality of crowd labelers is often much lower than that of an expert. In the worst case, some workers just submit random answers to get the fee deviously. One approach to dealing with low quality labels is repeated-labeling. Sheng *et al.* [16] empirically showed that under certain assumptions, repeated-labeling can improve the label quality. Thus in crowdsourcing, people may collect multiple labels $y_i^1, y_i^2, \ldots, y_i^L$ from L different labelers for one instance x_i, while in traditional supervised learning, one instance x_i corresponds to one label y_i.

The problem remains as how to learn a reliable predictive model with the unreliable crowd labels. Various methods have been proposed to infer the ground truth [4, 10] or learn from crowd labels directly [8, 15]. The basic idea is employing generative models for the labeling processes of crowd labelers. While these models are useful under certain conditions, their assumptions on labelers are not easy to verify for a certain task.

This situation motivates us to investigate making full use of opinions collected from crowds by incorporating some expert labels, which seems more sensible than trying to verify the behavior of each labeler. Intuitively, combining expert labels with crowd

V.S. Tseng et al. (Eds.): PAKDD 2014, Part I, LNAI 8443, pp. 200–211, 2014.

labels is expected to achieve higher learning quality than solely using crowd labels though, little work has been done under this configuration since most of the existing work has focused on crowd labels.

This paper proposes to improve the performance of crowdsourcing learning tasks with a minimum number of expert labels by maximizing the utilization of both the crowd and expert labels[1]. Our major contribution is a formalized framework for utilizing expert labels in crowdsourcing. Following a series of existing work [8, 15, 19], our work focuses on supervised classification problems.

Some existing models [8, 13, 15] are capable of combining expert labels by straightforward extensions. The major difference between our method and these models is that we use prior beliefs on experts much more explicitly.

1.1 An Illustrative Example

In what follows, we illustrate the limitation of crowd data with an example and explain the idea which forms the basis of our framework. Fig. 1(a) shows a synthetic dataset for binary classification. For each class, we sample 100 points respectively from two different Gaussian distributions, and get four underlying clusters. We simulate two labelers whose opinions differ in one cluster as shown in Figs. 1(b) & 1(c). Here no model that uses the crowd labels without extra information can weight one labeler over the other since there is simply not enough evidence. Nonetheless, these two labelers provide very informative labels. Labeler 1 actually gave all correct labels. If we can identify this fact by a few expert labels, we achieve an efficient method.

However, the problem is not trivial even for this toy data set. Supposing that we choose a controversial point and let an expert label it, we will find that Labeler 1 gave the correct answer. This is far from enough to conclude that Labeler 1 gave true labels for all controversial points given that in practice we only have crowd labels and are not aware of the underlying data distribution. Adding more expert labels may increase our confidence on Labeler 1, still a formalized mechanism is needed to combine the ground truth with crowd data.

We address the problem by a model combination process. We train a logistic regression classifier for each labeler separately with the labels provided by that labeler, thus get 2 classifiers. A data instance $\boldsymbol{x}_i$ will then get 2 predictions $\{f_1(\boldsymbol{x}_i), f_2(\boldsymbol{x}_i)\}$ from the 2 classifiers, where $f_\ell(\boldsymbol{x}_i)$ ($\ell \in \{1, 2\}$) is the posterior probability of the class colored in blue. We treat the values of $f_\ell(\boldsymbol{x}_i)$ as features in a new space, shown in Figure 1(d). This is referred to as the *intermediate feature space* [11]. The final prediction is made by another classifier in this intermediate feature space.

By summarizing the opinions of labelers using personal classifiers, the separation between classes becomes clearer and the controversial area is projected to the bottom right in the new space and becomes more compact. Incorporating expert label evidence in this space is much easier compared with the crowd labels in the original space. A few ground truth labels in the controversial area will enables most classifiers built in

[1] We assume that an expert always gives true labels and use the two terms 'expert labels' and 'ground truth' interchangeably. As experts can also make mistakes, this assumption is a simplification and may be relaxed in future work.

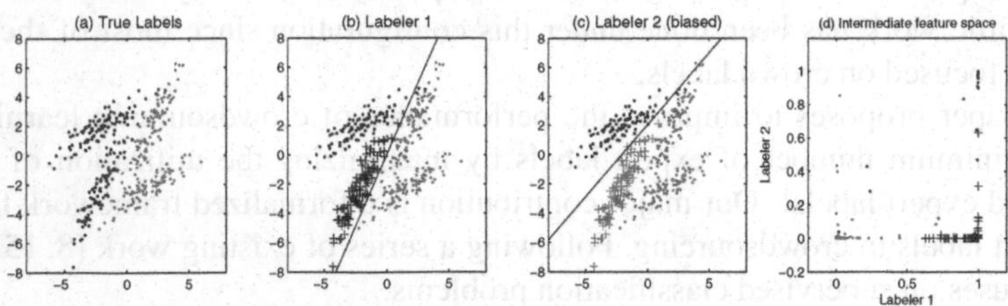

Fig. 1. An illustrative example. Instances labeled with cross(+) in (b)(c)(d) are controversial between the two labelers. These controversial data instances are gathered at the bottom right in the intermediate feature space as shown in (d).

this space to favor Labeler 1 over Labeler 2 naturally. We leave the the crucial step of combining expert evidence to the experiment section after we formalize our framework.

2 Related Work

With the arising of crowdsourcing services, crowd workers have shown their power in applications such as sentiment tracking [3], machine translation [1] and name entity annotating [5]. A key problem in crowdsourcing research is modeling data from multiple unreliable sources for inferring the ground truth. The problem has its origin in the early work [4] for combining multiple diagnostic test results. Recent work addressed problems with the same formulation by methods such as message transferring [10] and graphical models [13].

Our framework adopts the idea of learning a classifier from crowd data directly. Raykar *et al.* [15] and Yan *et al.* [19] treat true labels as hidden variables which are inferred by the EM algorithm. Kajino *et al.* [8] infer only the true classifier by personal classifiers without considering true labels explicitly. The nature of our method is similar to that of Kajino *et al.* [8], focusing on the final learning tasks and not being tangled with the correctness of a certain label.

To the best of our knowledge, very little work considered the case of learning from crowd and expert data simultaneously. Kajino *et al.* [9] addressed this problem by extending some existing models straightforwardly. Wauthier and Jordan [18] also used some expert labels. In their model crowd labels only make effects through the shared latent factors which express labelers. Our method differs from these work in both motivation and formulation.

We treat combining opinions of labelers as model combination. Getting the optimal combination of a group of pattern classifiers has been studied thoroughly for a long time and various methods have been proposed to employ the intermediate feature space. Merz [14] proposed to do feature extraction using singular decomposition in this space and Kuncheva *et al.* [12] proposed to combine classifiers giving soft labels using decision templates. In traditional model combination framework, multiple classifiers are obtained by different models trained on the *same* data set. Here the scenario is different, i.e., we have multiple unreliable label sets to train multiple classifiers, and we propose to use some reliable labels to combine them. Under the crowdsourcing setting the idea of absorbing the evidence of true labels in the intermediate feature space is also original.

3 Learning from Crowds and Experts

In this paper we focus on binary classification problems with crowdsourcing training data. The extension to multi-class cases is conceptually straightforward.

3.1 Problem Formulation

Formally, a crowdsourcing training set is denoted as $\mathcal{D} = \{(\boldsymbol{x}_i, \boldsymbol{y}_i)\}_{i=1}^N$, where instance $\boldsymbol{x}_i \in \mathbb{R}^D$ is a D-dimensional feature vector. We have L distinct labelers each of which gives labels to all N data instances.[2] The label given by the ℓth labeler for instance $\boldsymbol{x}_i$ is denoted as y_i^ℓ where $y_i^\ell \in \{-1, 1\}$. All labels corresponding to $\boldsymbol{x}_i$ are collected in the L-dimensional vector $\boldsymbol{y}_i$.

Different from most of the existing methods, we use some additional expert-labeled instances to improve the model quality. If there are N_0 expert labels, then the expert training set is $\mathcal{D}_0 = \{(\boldsymbol{x}_j, y_j^0)\}_{j=1}^{N_0}$ where $\boldsymbol{x}_j$ is again a D-dimensional feature vector and y_j^0 is the true label provided by the expert. Note that an expert-labeled instance $\boldsymbol{x}_j$ in $\mathcal{D}_0$ is not necessarily in $\mathcal{D}$. The task is to learn a reliable predictive function $f : \mathbb{R}^D \to [0, 1]$ for unseen data by taking both training sets $\mathcal{D}$ and $\mathcal{D}_0$ as inputs where $f(\boldsymbol{x}) = p(y = 1|\boldsymbol{x})$ is the posterior probability of the positive class. We denote the predictive function in this way for the convenience of the following steps.

3.2 Building Intermediate Feature Space

We extract the crowd opinions by treating labelers as personal classifiers. For the ℓth labeler, we use the personal training set $\mathcal{D}_\ell = \{(\boldsymbol{x}_i, y_i^\ell)\}_{i=1}^N$ to learn a classifier. Any classification model that expresses predictions as posterior probabilities of classes is compatible with our approach. Here we follow the work [8] and use a logistic regression model for each labeler, which is given by

$$\Pr[y = 1|\boldsymbol{x}, \boldsymbol{w}] = \boldsymbol{\sigma}(\boldsymbol{w}^{\mathrm{T}}\boldsymbol{x}) \tag{1}$$

where $\boldsymbol{w}$ is the model parameter and the logistic sigmoid function is defined as $\boldsymbol{\sigma}(a) = 1/(1 + e^{-a})$. We express all prediction functions of classifiers as an ensemble $\mathcal{F} = \{f_1, f_2, \cdots, f_L\}$ where $f_\ell(\boldsymbol{x})$ is the prediction of the classifier obtained from labeler ℓ on instance $\boldsymbol{x}$. The outputs of all L classifiers for a particular instance $\boldsymbol{x}_i$ is organized in an L-dimensional vector $[f_1(\boldsymbol{x}_i), f_2(\boldsymbol{x}_i), \cdots, f_L(\boldsymbol{x}_i)]^{\mathrm{T}}$, which is referred to as a *decision profile* [11]. In what follows, we denote this vector as $\boldsymbol{dp}_i$ with the ℓth element $dp_i^\ell = f_\ell(x_i)$. We treat values of dp_i^ℓ as features in a new feature space, namely the intermediate feature space, and use another classifier taking these values as inputs for making the final prediction.

---

[2] We assume at this point that all labelers give full labels to keep the notations simple. We will discuss the case of missing labels in Section 3.5.

3.3 Combination of Evidence from Crowds and Experts

The next step is to train a classifier in the intermediate feature space by utilizing expert labels. As expert labels are much more reliable than crowd labels, we should put more weights on them. However, if we discard crowd labels and use expert labels solely, building a stable model can be costly even in the more compact and representative intermediate feature space. Thus a balance has to be made between the crowd opinions and expert evidence.

We address the problem by imposing a Bayesian treatment on the model parameters of the classifier in the intermediate feature space. We use some straightforward combination of personal classifiers as the prior distribution of model parameters, and absorb expert label evidence by updating the posterior distribution sequentially. We believe that a fully Bayesian method is essential here for utilizing the prior distribution on parameters, which is informative in our framework as we will show later.

Specifically, we use the Bayesian logistic regression model [7] as our classifier in the intermediate feature space. The model achieved a tractable approximation of the posterior distribution over parameter $\boldsymbol{w}$ in Equation (1) by using accurate variational techniques. In our problem, the decision profile $\boldsymbol{dp}_i$ in the new space corresponding to the instance $\boldsymbol{x}_i$ in the original space is an $(L+1)$-dimensional vector consisting of all values of $dp_i^{\ell}, \ell = 1, \ldots, L$ and an additional constant 1 corresponding to the bias in parameter $\boldsymbol{w}$. The corresponding true label is y_i^0. The model assumes that the prior distribution over $\boldsymbol{w}$ is Gaussian with mean $\boldsymbol{\mu}$ and covariance matrix $\boldsymbol{\Sigma}$. Absorbing the evidence of expert-labeled instance $\boldsymbol{dp}$ and the true label y amounts to updating the mean and covariance matrix by

$$\boldsymbol{\Sigma}_{post}^{-1} = \boldsymbol{\Sigma}^{-1} + 2|\lambda(\xi)|\boldsymbol{dp} \cdot \boldsymbol{dp}^{\mathrm{T}} \tag{2}$$

$$\boldsymbol{\mu}_{post} = \boldsymbol{\Sigma}_{post}[\boldsymbol{\Sigma}^{-1}\boldsymbol{\mu} + (y/2)\boldsymbol{dp}] \tag{3}$$

where $\lambda(\xi) = [1/2 - \boldsymbol{\sigma}(\xi)]/2\xi$ and $\xi = [\boldsymbol{dp}^{\mathrm{T}}\boldsymbol{\Sigma}_{post}\boldsymbol{dp} + (\boldsymbol{dp}^{\mathrm{T}}\boldsymbol{\mu}_{post})^2]^{0.5}$. The update process is iterative and converges very fast (about two iterations) [7].

While one common criticism of the Bayesian approach is that the prior distribution is often selected on the basis of mathematical convenience rather than as a reflection of any prior beliefs [2], the prior distribution here is informative with a specific mean and an isotropic covariance matrix given by

$$\boldsymbol{\mu} = [-\frac{1}{2}, \frac{1}{L}, \ldots, \frac{1}{L}]^{\mathrm{T}} \tag{4}$$

$$\boldsymbol{\Sigma} = \alpha^{-1}\boldsymbol{I} \tag{5}$$

The mean is chosen such that all personal classifiers are combined by weighting them equally, and the bias is -0.5 to fit the shape of the logistic sigmoid function which is equal to 0.5 for $a = 0$.

There is a single precision parameter α governing the covariance matrix. We can interpret α as our confidence on the crowds. A large α will cause the prior distribution over $\boldsymbol{w}$ to peak steeply on the mean, thus the affect of absorbing one expert label will

Algorithm 1. Learning from crowd labelers and experts

1. **Input:** Crowd and expert training sets $\mathcal{D}$ and $\mathcal{D}_{\mathrm{O}}$;
2. Train the ensemble $\mathcal{F}$ of logistic regression classifiers defined by Equation (1) using $\mathcal{D}_\ell$ where $\ell = 1, \ldots, L$;
3. Use $\mathcal{F}$ to get predictions of data instances in $\mathcal{D}_{\mathrm{O}}$, collect results in $\mathcal{DP}$;
4. Initialize $\boldsymbol{\mu}$ and $\boldsymbol{\Sigma}$ by Equations (4) & (5);
5. **for** j=1 **to** N_{O} **do**
6. Calculate $\boldsymbol{\mu}_{post}$ and $\boldsymbol{\Sigma}_{post}$ by Equations (2) & (3) using the evidence from $\boldsymbol{dp}_j$ and y_j^0;
7. Set $\boldsymbol{\mu} = \boldsymbol{\mu}_{post}$, $\boldsymbol{\Sigma} = \boldsymbol{\Sigma}_{post}$;
8. **end for**
9. **Output:** Personal classifier ensemble $\mathcal{F}$, mean $\boldsymbol{\mu}$ and covariance matrix $\boldsymbol{\Sigma}$;

be relatively small, leading to a final classifier depending heavily on the mean of prior, which is the simple combination of personal classifiers. On the other hand, a small α means that the prior is close to uniform, causing the final classifier to make predictions mainly based on expert labels.

Intuitively, we should use a large α when personal classifiers are generally good, and use a small one when crowd labels are inaccurate. In a crowdsourcing scenario however, we usually do not have such knowledge. One alternative is to let α be related to the number of expert labels N_{O} given by $\alpha = 1/N_{\mathrm{O}}$. As this number increases, we decrease the confidence on crowds to let the final model put more weight on expert labels. Experiments show that with such selection of α, our model achieves relatively stable performance under various values of N_{O}.

Once the prior over $\boldsymbol{w}$ is chosen, we update its posterior distribution sequentially with Equations (2) & (3) by adding one expert label each time. If an instance $\boldsymbol{x}_j$ labeled by the expert is not in $\mathcal{D}$, we should first calculate its predictions $\boldsymbol{dp}_j$ by personal classifiers and use these values to update the model. We collect all $\boldsymbol{dp}_j$ in a set $\mathcal{DP} = \{\boldsymbol{dp}_j\}_{j=1}^{N_{\mathrm{O}}}$. The complete steps of learning our model are summarized in Algorithm 1.

3.4 Classification

To classify a new coming instance $\boldsymbol{x}_k$ using the above results, we firstly get the predictions $\boldsymbol{dp}_k$ of personal classifiers on $\boldsymbol{x}_k$, and calculate the predictive distribution of the true label y_k^0 in the intermediate feature space by marginalizing w.r.t. the final distribution $\mathcal{N}(\boldsymbol{w}|\boldsymbol{\mu}, \boldsymbol{\Sigma})$. The predictive likelihood is given by

$$\begin{aligned} \log P(y_k^0|\boldsymbol{x}_k, \mathcal{D}, \mathcal{D}_{\mathrm{O}}) &= \log \boldsymbol{\sigma}(\xi_k) - \frac{1}{2}\xi_k - \lambda(\xi_k)\xi_k^2 \\ &\quad - \frac{1}{2}\boldsymbol{\mu}^{\mathrm{T}}\boldsymbol{\Sigma}^{-1}\boldsymbol{\mu} + \frac{1}{2}\boldsymbol{\mu}_k^{\mathrm{T}}\boldsymbol{\Sigma}_k^{-1}\boldsymbol{\mu}_k + \frac{1}{2}\log\frac{|\boldsymbol{\Sigma}_k|}{|\boldsymbol{\Sigma}|} \end{aligned} \quad (6)$$

where subscript k assigned to $\boldsymbol{\mu}$ and $\boldsymbol{\Sigma}$ refers to the posterior distribution over $\boldsymbol{w}$ after absorbing the evidence of $\boldsymbol{dp}_k$ and y_k^0.

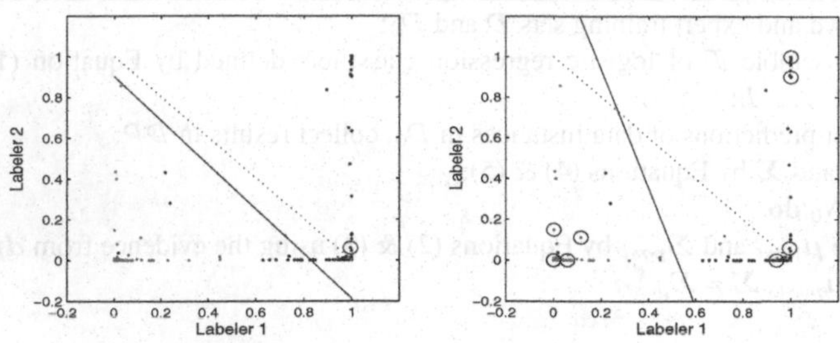

Fig. 2. Decision boundaries before and after absorbing expert label evidence. Dotted lines are means of prior distributions over $\boldsymbol{w}$, and solid lines are means of posterior distributions respectively after absorbing the label information of the circled instance(s) .

3.5 Missing Labels

In real crowdsourcing tasks, workers may label part of the instances instead of the whole set. Our model handles this problem naturally by training multiple personal classifiers independently. A worker only labels a few instances may lead to a pool personal classifier. But this is not fatal as he uses only a tiny proportion of the whole budget. Also in practice, we can avoid such cases simply by designing HITs with a moderate size.

4 Experiments

We use synthetic data to illustrate the process of absorbing expert evidence, and evaluate the performance of our method on both UCI benchmark data and real crowdsourcing data.

4.1 Synthetic Data

We complete our example in Figure 1 by illustrating the process of absorbing expert labels, shown in Figure 2. For clarity, we only show the decision boundaries given by means of the distributions over model parameter $\boldsymbol{w}$. Dotted lines are priors before adding expert labels. This prior is given by weighting each labeler equally following our framework.

In the left sub-figure, we add one expert label and get the posterior. Since the true label is blue, the decision boundary moves downward to suggest that data points near this labeled instance is more likely to be blue. In the right sub-figure, we add four expert labels for each class. The final decision boundary separates the actual class very well using merely eight expert labels. In this experiment we adjusted the model parameter α to get the best illustrative effect.

Table 1. Results on Waveform 1

Classifier	A_1	A_2	A_3
GT	0.853± 0.010		
MV	0.408± 0.123	0.638± 0.074	0.831± 0.007
AOC	0.490± 0.153	0.547± 0.101	0.831± 0.006
ML	0.437± 0.190	0.743± 0.063	0.842± 0.009
EL-10	0.718± 0.046	0.740± 0.045	0.740± 0.059
PCE-10	0.737± 0.051	0.742± 0.043	0.732± 0.051
CCE-10	0.725± 0.075	0.740± 0.068	0.816± 0.032
EL-20	0.756± 0.037	0.759± 0.034	0.783± 0.034
PCE-20	0.755± 0.033	0.768± 0.037	0.773± 0.048
CCE-20	0.801± 0.056	0.812± 0.057	0.822± 0.023
EL-50	0.792± 0.028	0.788± 0.033	0.798± 0.014
PCE-50	0.799± 0.025	0.796± 0.037	0.805± 0.016
CCE-50	0.816± 0.027	0.780± 0.039	0.833± 0.007
EL-100	0.797± 0.023	0.782± 0.023	0.767± 0.046
PCE-100	0.803± 0.008	0.811± 0.010	0.807± 0.015
CCE-100	0.831± 0.017	0.830± 0.024	0.829± 0.014

Table 2. Results on Spambase

Classifier	A_1	A_2	A_3
GT	0.924± 0.008		
MV	0.477± 0.327	0.641± 0.228	0.885± 0.013
AOC	0.535± 0.302	0.578± 0.208	0.879± 0.013
ML	0.510± 0.357	0.711± 0.302	0.925± 0.007
EL-10	0.672± 0.057	0.606± 0.113	0.665± 0.069
PCE-10	0.592± 0.095	0.641± 0.083	0.770± 0.049
CCE-10	0.857± 0.035	0.755± 0.165	0.890± 0.022
EL-20	0.860± 0.025	0.758± 0.033	0.755± 0.047
PCE-20	0.764± 0.080	0.708± 0.062	0.799± 0.046
CCE-20	0.891± 0.025	0.802± 0.087	0.894± 0.016
EL-50	0.830± 0.041	0.826± 0.032	0.831± 0.051
PCE-50	0.820± 0.053	0.803± 0.028	0.850± 0.017
CCE-50	0.900± 0.017	0.860± 0.053	0.895± 0.013
EL-100	0.860± 0.025	0.859± 0.025	0.858± 0.034
PCE-100	0.856± 0.025	0.861± 0.017	0.883± 0.010
CCE-100	0.891± 0.025	0.879± 0.031	0.903± 0.010

4.2 UCI Data

We test our method on three data sets from UCI Machine Learning Repository [6], Waveform 1(5000 points, 21 dimensions), Wine Quality(6497 points, 12 dimensions) and Spambase(4601 points, 57 dimensions). These data sets have moderate sizes which enable us to perform experiments when number of crowd labels varies.

Since multiple labelers for these UCI datasets are unavailable, we simulate L labelers for each dataset. We firstly cluster the data into L clusters using k-means and assign some labeling accuracy to each cluster for every labeler. Thus each labeler can have different labeling qualities for different clusters. We use an $L \times L$ matrix $\boldsymbol{A} = [a_{ij}]_{L\times L}$ to express the simulation process, in which a_{ij} is the probability that labeler i gives the true label for an instance in the jth cluster, thus a row corresponds to a labeler and a column to a cluster. We set $L = 5$ and use three different accuracy matrices $\boldsymbol{A}_1$, $\boldsymbol{A}_2$, and $\boldsymbol{A}_3$ to simulate different situations of labelers as follows.

$$\boldsymbol{A}_1 = \begin{bmatrix} 0\,1\,0\,1\,0 \\ 1\,1\,0\,0\,1 \\ 0\,0\,1\,1\,0 \\ 1\,0\,0\,1\,1 \\ 1\,0\,1\,1\,0 \end{bmatrix}, \boldsymbol{A}_2 = \begin{bmatrix} 0.3\;0.1\;0.8\;0.8\;0.8 \\ 0.3\;0.8\;0.1\;0.8\;0.8 \\ 0.3\;0.8\;0.8\;0.1\;0.8 \\ 0.3\;0.8\;0.8\;0.8\;0.1 \\ 0.8\;0.1\;0.1\;0.1\;0.1 \end{bmatrix}, \boldsymbol{A}_3 = \begin{bmatrix} 0.55\;0.55\;0.55\;0.55\;0.55 \\ 0.65\;0.65\;0.65\;0.65\;0.65 \\ 0.75\;0.75\;0.75\;0.75\;0.75 \\ 0.68\;0.68\;0.68\;0.68\;0.68 \\ 0.95\;0.95\;0.95\;0.95\;0.95 \end{bmatrix}.$$

$\boldsymbol{A}_1$ simulates severely biased labelers. $\boldsymbol{A}_2$ simulates labelers whose labels are both noisy and biased. $\boldsymbol{A}_3$ simulates simply noisy labels. Note that $\boldsymbol{A}_3$ satisfies the model assumption in the work by Raykar *et al.* [15].

We choose three baseline methods that learn with crowd data solely for comparison. To verify the ability of our method to utilize the crowd labels, we compare the results trained on expert labels solely. For comparison with existing methods we use the model proposed by Kajino *et al.* [9], which is a state-of-art model that addresses the same problem. We use the results trained on the original datasets which have all ground truth labels as the approximate upper bounds of the classification performance. Methods used in experiments are summarized as follows.

Table 3. Results on Wine Quality

Classifier	A_1	A_2	A_3
GT	0.743± 0.010		
MV	0.424± 0.119	0.571± 0.110	0.739± 0.007
AOC	0.582± 0.118	0.500± 0.109	0.740± 0.004
ML	0.417± 0.133	0.701± 0.020	0.739± 0.004
EL-10	0.550± 0.042	0.583± 0.047	0.582± 0.083
PCE-10	0.591± 0.035	0.578± 0.063	0.613± 0.033
CCE-10	0.634± 0.078	0.651± 0.092	0.715± 0.022
EL-20	0.623± 0.064	0.575± 0.075	0.623± 0.063
PCE-20	0.629± 0.019	0.604± 0.047	0.642± 0.041
CCE-20	0.679± 0.042	0.688± 0.047	0.720± 0.022
EL-50	0.666± 0.036	0.675± 0.040	0.682± 0.024
PCE-50	0.648± 0.019	0.644± 0.011	0.662± 0.022
CCE-50	0.687± 0.038	0.707± 0.025	0.722± 0.019
EL-100	0.707± 0.017	0.706± 0.017	0.711± 0.016
PCE-100	0.666± 0.011	0.665± 0.009	0.685± 0.020
CCE-100	0.718± 0.017	0.720± 0.010	0.733± 0.006

Table 4. Results under the variation of crowd label numbers on Spambase

Num.	50	100	200	500	1000
GT	0.826± 0.037	0.855± 0.023	0.878± 0.022	0.900± 0.009	0.913± 0.003
EL-50	0.835± 0.032				
MV	0.757± 0.057	0.734± 0.039	0.770± 0.021	0.849± 0.012	0.884± 0.012
AOC	0.587± 0.045	0.727± 0.025	0.798± 0.025	0.852± 0.010	0.880± 0.010
ML	0.807± 0.065	0.822± 0.025	0.876± 0.012	0.905± 0.005	0.921± 0.005
PCE-50	0.838± 0.025	0.839± 0.024	0.842± 0.016	0.837± 0.018	0.861± 0.026
CCE-50	0.792± 0.045	0.807± 0.045	0.820± 0.018	0.873± 0.008	0.903± 0.006

- **Majority Voting (MV)** method learns from the single-labeled training set estimated by majority voting.
- **All-in-One-Classifier (AOC)** treats all labels as in one training set.
- **Multiple Labelers (ML)** method [15] learns from crowd labels directly.
- Kajino *et al.* [9] extended their personal classifier model [8] to incorporate expert labels, which we refer to as **Personal Classifiers with Experts (PCE)**. **PCE-N_0** is the results trained with N_0 expert labels.
- We refer to our method as **Classifier Combination with Experts (CCE)**. **CCE-N_0** is the results after absorbing the evidence of N_0 expert labels.
- Training with expert labels solely is referred as **Expert Labels (EL)** classifiers. **EL-N_0** is the results trained with N_0 expert labels.
- **Ground Truth (GT)** classifier uses the original datasets for training.

For MV, AOC, GT, and EL, we use a logistic regression model respectively to train the classifiers. For PCE, CCE and EL, the set of expert labels are randomly chosen from the original datasets given the number of expert labels N_0 which is restricted to a small proportion of N. We divide each dataset into a 70% training set and a 30% test set and each result is averaged on 10 runs.

Tables 1–3 show the results for different datasets respectively. Results are in the form of classification accuracy and averaged on 10 trials. The GT classifier is independent of crowd labels thus it has only one result on each dataset. Our CCE outperforms EL, and also outperforms MV, ACL and ML in most cases. This validates the ability of CCE for combining crowd and expert labels. The only exception appears in ML under A_3 where labelers are not biased. CCE outperforms PCE with clear advantages. There are a number of cases that PCE performs worse than EL, which suggests that in the PCE model expert evidences are easily disturbed by inaccurate crowd labels.

Table 4 shows the results under the variation of numbers of crowd labels. We show the results on Spambase data under A_3 since under this situation all methods seem to work well. The top number of each column represents the number of labels provided by each labeler. This is also the number of expert labels used for GT. We use 50 expert labels for EL, PCE and CCE. EL has only one result as it is independent of crowd labels.

There is no surprise that ML performs very well in this experiment as the configuration here meets ML's model assumption. Yet we should not forget that ML fails in many cases as shown in Tables 1–3. We do not choose those cases because showing groups of failed results does not make any sense. Generally PCE and CCE outperform MV and AOC by using extra expert labels. CCE performs slightly worse than PCE when the number of crowd labels is small, while the performance raise of PCE is quite limited when using more crowd labels.

In summary, our method CCE achieved reasonable performance on different data sets with various labeler properties. The accuracy and stability of our CCE increase as we use more expert labels. On the other hand, learning solely from crowd labels is risky, especially when crowd labels are biased. PCE's performance is limited compared with CCE when we have enough crowd labels.

4.3 Affective Text Analysis Data

In this section we show results on the data for affective text analysis collected by Snow *et al.* [17]. The data is collected from Amazon Mechanical Turk. Annotators are asked to rate the emotions of a list of short headlines. The emotions include anger, disgust, fear, joy, sadness, surprise and the overall positive or negative valence. The former six are expressed with an interval $[0, 100]$ respectively while valence is in $[-100, 100]$. There is a total number of 100 headlines labeled by 38 workers. For each headline 10 workers rated for each of the seven emotions. Most workers labeled 20 or 40 instances thus more than one half labels are missing. All 100 instances are also labeled by the experts and have an average rating for each emotion, which we treat as ground truth.

We design the classification task which predicts the surprising level of a headline using other emotion ratings as features. We define that a headline of which the surprise rating is above 20 is a surprise, while others not, and use ratings of other six emotions provided by the experts to express a headline. Thus we get a binary classification task in a 6-dimensional feature space.

Figure 3 shows classification accuracy when continually adding expert labels. Results of MV, AOC and ML are not shown in this figure, which are three horizontal lines below GT and stay close to each other. PCE only performs similarly with EL, which collapses to GT when using all expert labels.

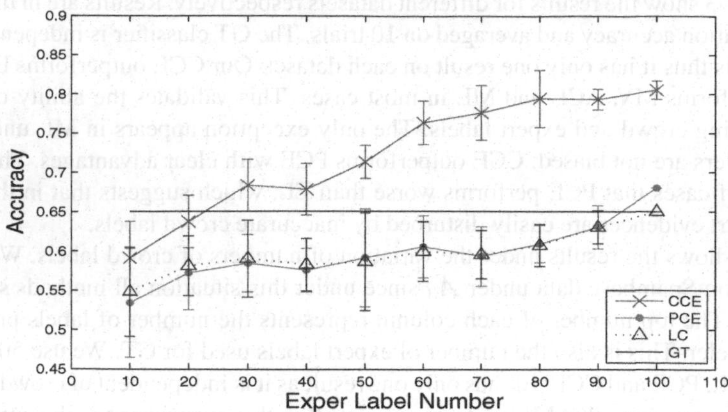

Fig. 3. Results on Affective Text Analysis data. The x-axis is the number of expert labels used while the y-axis is the classification accuracy.

The result of CCE is promising. The value of GT is 0.65, which suggests that according to the experts, there is no strong correlation between the surprising level and other emotions. However, CCE only uses about 20 expert labels to get a similar performance level with GT, and when adding more expert labels, CCE outperforms GT and achieves an accuracy up to 0.8. We attribute this fact to the power of our CCE model as a 'feature extractor'. Among the 38 workers, one or more of them did give ratings in manners that relate surprising levels to other emotions even if experts did not do so. Personal classifiers trained from these workers will then be able to predict the target and our model identifies these classifiers successfully using expert labels.

5 Conclusion and Future Work

In this paper, we have proposed a framework for improving the performance of crowdsourcing learning tasks by incorporating the evidence of expert labels with a Bayesian logistic regression classifier in the intermediate feature space. Experimental results have verified that by combining crowd and expert labels, our method has achieved better performance as compared with some existing methods, and has been stable under the variation of the number of expert labels and crowd labeler properties.

A promising direction of future work is to investigate actively querying for the expert labels, for which we can develop models by adopting basic ideas from active learning and considering the particular situation of crowdsourcing.

Acknowledgment. This work is supported by the National Key Technology R&D Program of the Chinese Ministry of Science and Technology under Grant No. 2012BAH94F03.

References

1. Ambati, V., Vogel, S., Carbonell, J.: Active learning and crowd-sourcing for machine translation. Language Resources and Evaluation (LREC) 7, 2169–2174 (2010)
2. Bishop, C.M., et al.: Pattern recognition and machine learning, vol. 4. Springer, New York (2006)
3. Brew, A., Greene, D., Cunningham, P.: Using crowdsourcing and active learning to track sentiment in online media. In: ECAI 2010, pp. 145–150 (2010)
4. Dawid, A.P., Skene, A.M.: Maximum likelihood estimation of observer error-rates using the EM algorithm. Applied Statistics, 20–28 (1979)
5. Finin, T., Murnane, W., Karandikar, A., Keller, N., Martineau, J., Dredze, M.: Annotating named entities in twitter data with crowdsourcing. In: Proceedings of the NAACL HLT 2010 Workshop on Creating Speech and Language Data with Amazon's Mechanical Turk, pp. 80–88. Association for Computational Linguistics (2010)
6. Frank, A., Asuncion, A.: UCI machine learning repository (2010)
7. Jaakkola, T., Jordan, M.: A variational approach to Bayesian logistic regression models and their extensions. In: Proceedings of the 6th International Workshop on Artificial Intelligence and Statistics (1997)
8. Kajino, H., Tsuboi, Y., Kashima, H.: A convex formulation for learning from crowds. In: Proceedings of the 26th AAAI Conference on Artificial Intelligence (2012) (to appear)
9. Kajino, H., Tsuboi, Y., Sato, I., Kashima, H.: Learning from crowds and experts. In: Proceedings of the 4th Human Computation Workshop, HCOMP 2012 (2012)
10. Karger, D.R., Oh, S., Shah, D.: Iterative learning for reliable crowdsourcing systems. In: Advances in Neural Information Processing Systems (NIPS 2011), pp. 1953–1961 (2011)
11. Kuncheva, L.I.: Combining pattern classifiers: Methods and algorithms (kuncheva, li; 2004)[book review]. IEEE Transactions on Neural Networks 18(3), 964 (2007)
12. Kuncheva, L.I., Bezdek, J.C., Duin, R.P.W.: Decision templates for multiple classifier fusion: an experimental comparison. Pattern Recognition 34(2), 299–314 (2001)
13. Liu, Q., Peng, J., Ihler, A.: Variational inference for crowdsourcing. In: Advances in Neural Information Processing Systems (NIPS 2012), pp. 701–709 (2012)
14. Merz, C.J.: Using correspondence analysis to combine classifiers. Machine Learning 36(1), 33–58 (1999)
15. Raykar, V.C., Yu, S., Zhao, L.H., Valadez, G.H., Florin, C., Bogoni, L., Moy, L.: Learning from crowds. The Journal of Machine Learning Research 11, 1297–1322 (2010)
16. Sheng, V.S., Provost, F., Ipeirotis, P.G.: Get another label? Improving data quality and data mining using multiple, noisy labelers. In: Proceeding of the 14th ACM SIGKDD International Conference on Knowledge Discovery and Data Mining, pp. 614–622 (2008)
17. Snow, R., O'Connor, B., Jurafsky, D., Ng, A.Y.: Cheap and fast—but is it good? evaluating non-expert annotations for natural language tasks. In: Proceedings of the Conference on Empirical Methods in Natural Language Processing, pp. 254–263. Association for Computational Linguistics (2008)
18. Wauthier, F.L., Jordan, M.I.: Bayesian bias mitigation for crowdsourcing. In: Advances in Neural Information Processing Systems (NIPS 2011), pp. 1800–1808 (2011)
19. Yan, Y., et al.: Modeling annotator expertise: Learning when everybody knows a bit of something. In: Proceedings of 13th International Conference on Artificial Intelligence and Statistics (AISTATS 2010), vol. 9, pp. 932–939 (2010)

A Robust Classifier for Imbalanced Datasets

Sori Kang and Kotagiri Ramamohanarao

Department of Computing and Information Systems, The University of Melbourne, Parkville Victoria 3052 Australia

Abstract. Imbalanced dataset classification is a challenging problem, since many classifiers are sensitive to class distribution so that the classifiers' prediction has bias towards majority class. Hellinger Distance has been proven that it is skew-insensitive and the decision trees that employ Hellinger Distance as a splitting criterion have shown better performance than other decision trees based on Information Gain. We propose a new decision tree induction classifier (HeDEx) based on Hellinger Distance that is randomized ensemble trees selecting both attribute and split-point at random. We also propose hyperplane as a decision surface for HeDEx to improve the performance. A new pattern-based oversampling method is also proposed in this paper to reduce the bias towards majority class. The patterns are detected from HeDEx and the new instances generated are applied after verification process using Hellinger Distance Decision Trees. Our experiments show that the proposed methods show performance improvements on imbalanced datasets over the state-of-the-art Hellinger Distance Decision Trees.

1 Introduction

In machine learning, the classification problem aims to predict class labels of new unseen examples on the basis of previously observed training datasets. Many methods have been proposed to generate high accuracy classifiers, but most classification methods show good performance on balanced class problems - the number of instances of classes is balanced - while yielding relatively poor performance on imbalanced class problems.

Imbalanced class distribution - one class (majority class, denoted as '-') vastly outnumbers the other class (minority class, denoted as '+') in training datasets - hinders the accuracy of classification of minority class, since typical classification algorithms, such as decision trees, intend to maximize the overall prediction accuracy and tend to have bias toward the majority class [1]. The class imbalance problem, however, is important, since imbalanced datasets are prevalent in real world (e.g. fraud/intrusion detection, medical diagnosis/monitoring) and the cost of misclassification for a minority class is usually much higher in many cases. For instance, since the examples of patient with a rare cancer are relatively very small, the classifier usually has a poor ability to predict the rare cancer. Therefore, the classifier can simply classify the patient with a rare cancer into the patient with a common cancer while the classifier keeps high accuracy. When

V.S. Tseng et al. (Eds.): PAKDD 2014, Part I, LNAI 8443, pp. 212–223, 2014.

such a misclassification happens, however, the misclassified patient may suffer from misdiagnosis.

There have been various approaches to tackle the class imbalance problems; kernel modification methods, sampling methods, and cost-sensitive methods. In this paper, we focus on methods that can apply to Decision Tree Induction Classification, as decision tree is one of the most effective methods for classification [2]. Based on the decision tree, there have been many studies that showed performance improvement on imbalanced datasets. However, there is still a room for improvement.

In this paper, we propose Hellinger Distance Extra Decision Tree (HeDEx) that employs Hellinger Distance as a split criterion and builds extremely randomized ensemble trees. Hellinger Distance Extra Decision Tree is named after Hellinger Distance Decision Tree [3] and Extra-Trees [4]. We also propose a novel oversampling method that helps not only our proposed decision tree but also other existing classifiers to get better performance for minority class. Our experiments show that HeDEx has generally better performance than other existing decision tree methods in terms of AUC and F-Measure for minority class. The proposed oversampling method improves the performance of F-Measure for minority class.

2 Related Work

Information gain and Gini index are used as the splitting criteria for the popular decision trees such as C4.5 and CART, respectively. However, several studies [5][6][7] have shown that these measures are skew-sensitive so that they have bias toward majority class. Equation 1 and 2 denote Entropy and Information Gain for binary class and binary split, respectively. In order to maximize the information we need to minimize the second term of Equation 2, since the first term $Entropy(S)$ is fixed for the dataset S. The second term of Equation 2 is denoted as $\frac{1}{|S|}(-|S_{1+}|log_2\frac{|S_{1+}|}{|S1|}-|S_{1-}|log_2\frac{|S_{1-}|}{|S1|})+\frac{1}{|S|}(-|S_{2+}|log_2\frac{|S_{2+}|}{|S2|}-|S_{2-}|log_2\frac{|S_{2-}|}{|S2|})$. Therefore, the minority class could not influence equally on Entropy like majority class, since $|S_{1+}|$ and $|S_{2+}|$ is relatively small in class imbalanced datasets. The classifier that uses Gini Index (Equation 4) also has the same problem. The studies also shows the skew-sensitivity of Information Gain using isometric form [8][3].

Cieslak et al. [3] proposed Hellinger Distance as a splitting criterion that is skew-insenstive. Equation 5 shows Hellinger Distance, and it shows that the class priors do not influence the distance calculation. Thus the minority class would not be ignored on distance calcuation. The experiments of Cieslak et al. showed that Hellinger Distance Decision Tree (HDDT) outperforms C4.4 - unpruned, uncollapsed C4.5 with Laplace smoothing - [3] in imbalanced class problems.

$$Entropy(S) = -\sum_{c=+,-}\frac{|S_c|}{|S|}log_2\frac{|S_c|}{|S|} \tag{1}$$

$$InfoGain(S) = Entropy(S) - \sum_{j=1,2}\frac{|S_j|}{|S|}Entropy(S_j) \tag{2}$$

$$InfoGainRatio(S) = \frac{InfoGain(S)}{\sum_{j=1,2} \frac{|S_j|}{|S|} log_2 \frac{|S_j|}{|S|}} \tag{3}$$

$$Gini(S) = 1 - \sum_{c=+,-} (\frac{|S_c|}{|S|})^2 \tag{4}$$

$$HellingerDisttance(S) = \sqrt{\sum_{j=1,2} (\sqrt{\frac{|S_{+j}|}{|S_+|}} - \sqrt{\frac{|S_{-j}|}{|S_-|}})^2} \tag{5}$$

Sampling methods are typically used to tackle imbalanced class problems. The experimental studies[9] have shown that using sampling methods generally improve the classifier performance. The sampling methods alter the class distribution to make the class distribution balanced.

Synthetic Minority Oversampling TEchnique (SMOTE) [10] takes the same approach with One-Sided Selection; mitigating bias toward majority class. The difference between two methods is that while One-Sided Selection removes the majority class data, SMOTE proposes creating synthesized minority class examples. Synthetic examples are generated by selecting a random point along the line between two minority class examples. Chawla et al. shows that this synthesized examples facilitate larger decision regions in feature space for minority class avoiding overfitting. Despite improved performance of SMOTE, the problem of SMOTE is overgeneralization, which means the region enlarged for minority class could be blindly generalized so that synthetic instances can lead to overlapping between classes. Many other synthetic sampling methods based on SMOTE have been proposed to address the overgeneralization problem of SMOTE; Border-line SMOTE [11], Safe-Level SMOTE [12], and LN-SMOTE [13].

Cluster-Based Oversampling [14] is the other method to solve the small disjunction problem by using clustering approach. It clusters the training data of each class separately and oversamples the minority instances per each cluster. In this idea, the new generated minority instances cannot be located in some majority class cluster. Thus, by clustering approach, it can handle both inter-class imbalance and between-class imbalance simultaneously.

The other approach for oversampling is pattern-based synthetic method. Alhammady et al. [15] proposed Emerging Patterns Decision Tree (EPDT) that is decision tree induction classifier using Emerging Patterns of minority class for generating new minority class instances. Emerging Pattern of minority class is a pattern whose support changes significantly from majority class to minority class in the training dataset.

Ensemble method is known to be very efficient to reduce variance by building multiple classifiers and averaging the classifiers output so that it allows us not to choose the classifier with a poor performance. There have been many ensemble methods based on Bagging [16] and Boosting [17]. In this review, however, we focus on Randomized ensemble trees that use a subset of attributes, since our method is based on Random Subspace [18].

Table 1. Comparison of Randomized Ensemble Methods

Ensemble method	**Training samples for tree**	**Candidate split-points at node**
Random Subspace	all training samples	all values of selected attributes
Random Forest	bootstrap replicas	all values of selected attributes
Extra-Trees	all training samples	$c = 1$, a random value of each selected attribute
HeDEx	all training samples	$c \geq 1$, random values of each selected attribute

Instead of using all attributes to find the optimal split-point at a tree node, Random Subspace and Random Forests randomly select a subset of attributes at a tree node and find the optimal split-point among only these selected attributes. The level of randomization is determined by the number K of attributes to be chosen at each node. The difference between two methods is the formation of the training examples of each tree. Random Subspace uses the entire training dataset for each tree so that only the randomly chosen attributes impact the variability of base classifiers. On the other hand, Random Forests is based on Bagging [16] - random sampling of training dataset with replacement - so that both attribute selection and training data impact the variability.

Extra-Trees [4] was proposed as extremely randomized trees. It randomly selects not only the number K of attributes but also a candidate split-point for each chosen attribute. The split-point at a tree node is determined among the number K of candidate split-points so that the split-point is sub-optimal for a corresponding node. Therefore, among three methods (Random Subspace, Random Forests, and Extra-Trees) Extra-Trees has the highest randomness. The advantage of Extra-Trees is computational efficiency while keeping comparative accuracy, reducing variance and slightly increasing bias.

3 Hellinger Distance Extra Decision Tree

In this paper, we propose Hellinger Distance Extra Decision Tree (HeDEx) that employs Hellinger Distance as a splitting criterion and build extremely randomized ensemble trees using entire training dataset for imbalanced dataset problems. HeDEx is basically based on Extra-Trees [4]. The main differences with other ensemble methods are that Extra-Trees and HeDEx randomly choose not only the attributes but also split-points for splitting the tree nodes and use the entire training examples (rather than a bootstrap replica) to build each base classifier. Due to the randomization on both attribute selection and split-point selection, Extra-Trees and HeDEx can achieve high level of varierity of trees even without sampling of training examples (bootstrap replicas).

The two main differences with Extra-Trees and HeDEx are splitting-criterion and the number of candidate split-points. Extra-Trees selects one candidate split-point for each attribute, while HeDEx selects more than one split-points, since we found from experiments that considering multiple candidate split-points to find a sub-optimal split-point showed better accuracy. The other difference is splitting

Table 2. HeDEx Parameters

Parameter	Description
M	the number of trees to be built
K	the number of attributes randomly selected at each node
C	the number of split-points randomly selected for each attribute
n_{min}	the minimum sample size at a leaf node

criterion. HeDEx uses Hellinger Distance for skew-insensitiveness, while Extra-Trees uses a score measure based on information gain. We provide the comparison table of existing random ensemble methods in Table 1. Notice that although the number of split-points c of HeDEx becomes equal to the number of values of the corresponding attribute in learning samples, HeDEx is different from Random Subspace. This is because HeDEx draws the split-points independently from the values in the learning samples.

We use weighted method as the aggregation rule for our HeDEx. Equation 6 shows how to predict the class label of a test example. C in Equation 6 is a set of class labels and $p_{t_i,c}$ is the probability of classifying as class label c in tree t_i such that $\sum_{c \in C} p_{t_i,c} = 1$ for each tree.

$$class = \arg\max_{c \in C} \sum_{i=1}^{M} p_{t_i,c} \tag{6}$$

3.1 Variant of Hellinger Distance Extra Decision Tree

We also propose a variant of HeDEx that employs different decision boundary from the original HeDEx. Hellinger Distance Extra Hyperplane Decision Tree (HeDExh) uses an arbitrary hyperplane as a decision surface, thus selects the optimal hyperplane at a node among K candidate hyperplane decision boundaries. To be brief, we could say that HeDEx uses single variable inequations when it tries to split the node according to the split-point, while HeDExh uses two variable inequations when it tries to split the node. The arbitrary hyperlane is determined by choosing two different points from feature space of the dataset; e.g. (v_{11}, v_{21}) and (v_{12}, v_{22}) where v_{1x} for attribute 1 and v_{2x} for attribute 2. The number of candidate hyperplane for a pair of chosen attributes is also an option parameter, C. The number of attributes to be used for hyperplane could be more than two. In this paper, we present 2-dimensional hyperplane only but the dimensionality can be extended.

4 Pattern-Based New Instance Creation

In this section, we propose the new method for generating minority instances based on patterns that are detected from HeDEx. This process consists of three parts; pattern detection, instance generation, and instance validation.

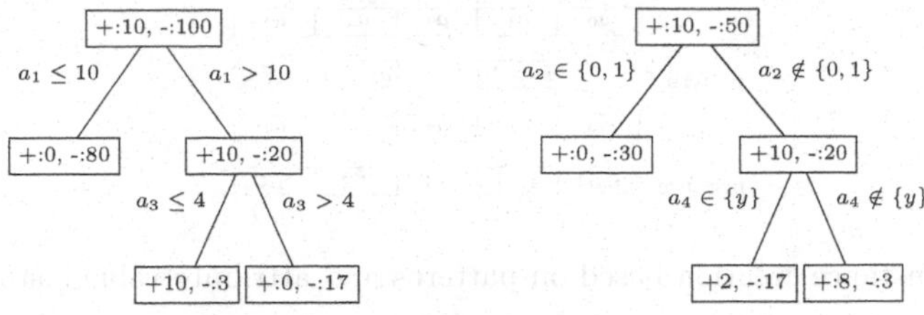

Fig. 1. An example of HeDEx for pattern detection

4.1 Pattern Detection

The main idea of pattern detection is that in decision tree the splitting criteria at each node of tree represent the patterns of the examples at the node. The patterns are detected from HeDEx that are built using the training samples. As HeDEx builds sub-optimal trees, it has a variety of patterns that are not highly coupled with training examples. After building HeDEx, we look at the leaf node of each tree, and if the number of minority instances is greater than that of majority, we build a pattern according to the split rules from root to the leaf node. For example, from the trees in Figure 1, we can build patterns; $\{a_1 > 10$ and $a_3 \leq 4\}$ and $\{a_2 \notin \{0,1\}$ and $a_4 \notin \{y\}\}$, respectively.

The patterns detected from multiple HeDEx trees get scores with respect to the strength of the pattern, and a pattern with higher score has higher probability of being selected for generating new instances. The equations 7 and 8 show how to get strength score of a pattern. For example, for the pattern detected from Figure 1 has $GrowthRate_{c_+} = {}^{10}/_{3}$ and $Strength_{c_+} = 10 * {}^{10/3}/({}^{10}/_{3}+1)$ for minority class.

$$Strength_{c_+} = Support_{c_+} * \frac{GrowthRate_{c_+}}{GrowthRate_{c_+} + 1} \tag{7}$$

$$GrowthRate_{c_+} = \frac{Support_{c_+}}{Support_{c_-}} \tag{8}$$

4.2 Instance Generation

New instances are generated based on the patterns. If the selected pattern does not cover all attributes, the values of missing attributes are generated at random but based on the distribution of the values of the training examples. Figure 2 shows an example of generating a new instance. As the pattern 1 has the rule $\{a_1 > 10$ and $a_3 \leq 4\}$, we generate random values of a_1 and a_3 according to the rule; $10 < v_1 \leq max(a_1)$ and $min(a_3) \leq v_3 \leq 4$. For the remaining attributes, we randomly draw the values such that $v_i \sim histogram(a_i)$ for ith attribute.

For the nominal attributes, we select the attribute values at random according to the rule. For example, the second pattern of Figure 1 is $\{a_2 \notin \{0,1\}$ and $a_4 \notin \{y\}\}$. The value of a_2 is drawn uniformly from $Set_{a_2} \backslash \{0,1\}$ and the value

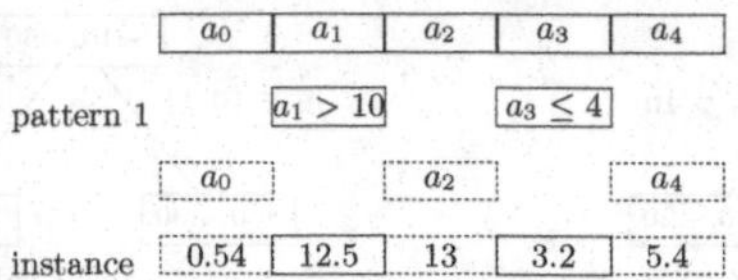

Fig. 2. Instance creation based on patterns and attribute value histogram

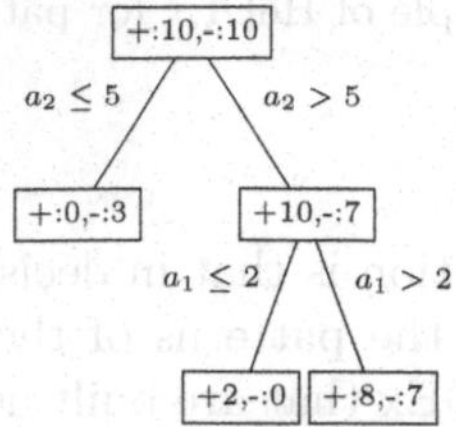

Fig. 3. An example of HDDT for instance validation

of a_4, is drawn uniformly from $Set_{a_4} \backslash \{y\}$ where Set_{a_i} is the set of attribute a_i's distinct values. For remaining attributes, we randomly draw each value such that $v_i \sim histogram(a_i)$ for ith attribute.

4.3 Instance Validation

We propose the instance validation phase to ensure that the newly generated instances are not noisy for minority class. The main idea of this validation phase is that if a classifier cannot distinguish instances of two different classes well, which can mean that the instances of two different classes are similar.

After generating new instances, we make a synthetic training dataset using the original minority instances and newly generated instances to validate new instances. We set the original minority instances as positive class and the new instances as negative class. We build a Hellinger Distance Decision Tree and visit each leaf node to check the instance distribution. Figure 3 shows the example. The right-most leaf node illustrates that the negative instances are not distinguishable from the positive instances, so we can add these negative instances (newly generated instances) as the instances for minority class to the training examples for building a classifier. On the other hand, the left most node contains only the negative instances, which implies that the three negative instances are well distinguishable from the positive class instances so that the three instances could become noisy if we add these instances to training examples. In that case, we do not include such instances for our training examples. Notice that we employ Hellinger Distance Decision Tree for validating instances, since HDDT builds a tree with optimal, discriminative splitting criteria.

Table 3. Datasets for imbalanced classes

Dataset	Classes	Type	Features	Instances	CV
compustat	2	numeric	20	10358	0.93
covtype	2	numeric	10	38500	0.86
estate	2	numeric	12	5322	0.76
german.numer	2	numeric	24	1000	0.4
ism	2	numeric	6	11180	0.95
letter	2	numeric	16	20000	0.92
oil	2	numeric	49	937	0.91
page	2	numeric	10	5473	0.8
pendigits	2	numeric	16	10992	0.79
phoneme	2	numeric	5	5400	0.41
PhosS	2	numeric	480	11411	0.89
satimage	2	numeric	36	6430	0.81
segment	2	numeric	19	2310	0.71
credit-g	2	both	20	1000	0.4
boundary	2	nominal	175	3661	0.93
breast-y	2	nominal	9	286	0.41
cam	2	nominal	132	18916	0.9
car	4	nominal	6	1728	1.25
dna	3	nominal	180	3186	0.48
glass	6	numeric	9	214	0.83
nursery	4	nominal	8	12961	0.60
page-5	5	numeric	10	5473	1.95
sat	6	numeric	36	6435	0.41

5 Experiements

The 23 datasets were chosen from the datasets that were used in HDDT experiments[1]. These datasets originate from UCI[2], LibSVM[3] and two other studies [8][10]. The 23 datasets have a variety of characteristics in terms of the level of imbalance and the number of attributes including both binary class and multi-class. Dataset 'page' and 'satimage' are originally the same dataset with 'page-5' and 'sat', respectively, but have different number of classes. In order to measure the level of imbalance in class distribution, we employ the coefficient of variance ($CV = \sigma/_{\mu}$) of class distribution like the study of HDDT [3]. Higher CV means higher skewness in class distribution. Table 3 provides details of the dataset used in this experiment.

We chose Bagging HDDT and Extra-Trees as the methods to be compared with our proposed methods. As Cieslak et al. showed in the study of HDDT [3] that Bagging HDDT outperforms other methods including C4.4, C4.4 combined with Bagging and Sampling methods, and HDDT combined with Sampling, we excluded other ensemble and sampling methods in this paper. We also use Bagging HDDT for the performance comparison of pattern-based oversampling while excluding other oversampling methods, since Bagging HDDT is proved that it has better performance than other oversampling methods[3]. We included Bagging C4.5 as a baseline and Logistic Regression for comparison purpose as it is insensitive to class distribution. Table 4 shows terminologies for algorithms that we use in this experiment. 5x2 cross-validation is used to evaluate each algorithm.

We used the default parameters that are mentioned on the corresponding papers for both Bagging HDDT and Extra-Trees; 100 unpruned trees (M = 100) with laplace smoothing and $n_{min} = 2$. For Extra-Trees and our methods the K number of candidate attributes is $K = \sqrt{n}$ as Extra-Trees recommended. HeDEx and HeDExh have another parameter for the number of candidate split-points, C. We set $C = 10$ as default, as we found that for most datasets above 10 candidate

[1] http://www3.nd.edu/~dial/hddt/

[2] http://archive.ics.uci.edu/ml/

[3] http://www.csie.ntu.edu.tw/~cjlin/libsvm/

Table 4. Algorithms

Abbr.	Algorithm
LR	Bagging Logistic Regression
C4.5	Bagging C4.5, pruned
HDDT	Bagging Hellinger Distance Decision Tree
ET	Extra-Trees
HeDEx	Hellinger Distance Extra Decision Trees
HeDExh	Hellinger Distance Extra Hyper Decision Trees
X^s	Algorithm X with pattern-based oversampling e.g. LR^s

Table 5. (Win/Draw/Loss) table of statistical significance test (corrected t-test) result at 95% for F-Measure$^+$

	LR	C4.5	ET	HDDT	HeDEx	HeDExh
LR		8/10/5	8/11/4	9/10/4	10/10/3	11/9/3
C4.5	5/10/8		3/18/2	3/19/1	7/15/1	7/15/1
ET	4/11/8	2/18/3		3/18/2	5/18/0	7/16/0
HDDT	4/10/9	1/19/3	2/18/3		3/19/1	5/18/0
HeDEx	3/10/10	1/15/7	0/18/5	1/19/3		1/22/0
HeDExh	3/9/11	1/15/7	0/16/7	0/18/5	0/22/1	
Total	19/50/46	13/77/25	13/81/21	16/84/15	25/84/6	**31/80/4**

	LR^s	C4.5^s	ET^s	HDDT^s	HeDEx^s	HeDExh^s
LR^s		9/12/2	11/12/0	9/12/2	12/10/1	11/12/0
C4.5^s	2/12/9		6/17/0	0/22/1	7/16/0	8/15/0
ET^s	0/12/11	0/17/6		0/19/4	2/21/0	3/20/0
HDDT^s	2/12/9	1/22/0	4/19/0		6/17/0	6/17/0
HeDEx^s	1/10/12	0/16/7	0/21/2	0/17/6		1/21/1
HeDExh^s	0/12/11	0/15/8	0/20/3	0/17/6	1/21/1	
Total	5/58/52	10/82/23	21/89/5	9/87/19	28/85/2	**29/85/1**

Table 6. (Win/Draw/Loss) table of statistical significance test (corrected t-test) result at 95% for AUC

	LR	C4.5	ET	HDDT	HeDEx	HeDExh
LR		9/12/2	13/9/1	9/12/2	14/8/1	12/10/1
C4.5	2/12/9		8/15/0	4/17/2	9/14/0	9/14/0
ET	1/9/13	0/15/8		0/18/5	2/21/0	4/19/0
HDDT	2/12/9	2/17/4	5/18/0		7/16/0	8/15/0
HeDEx	1/8/14	0/14/9	0/21/2	0/16/7		2/21/0
HeDExh	1/10/12	0/14/9	0/17/6	0/15/8	0/21/2	
Total	7/51/57	11/72/32	26/80/9	13/78/24	32/80/3	**35/79/1**

	LR^s	C4.5^s	ET^s	HDDT^s	HeDEx^s	HeDExh^s
LR^s		9/11/3	15/7/1	11/9/3	15/7/1	14/8/1
C4.5^s	3/11/9		7/16/0	2/19/2	7/16/0	8/15/0
ET^s	1/7/15	0/16/7		0/15/8	7/16/0	9/12/2
HDDT^s	3/9/11	2/19/2	8/15/0		8/15/0	9/14/0
HeDEx^s	1/7/15	0/16/7	0/16/7	0/15/8		3/20/0
HeDExh^s	1/8/14	0/15/8	2/12/9	0/14/9	0/20/3	
Total	9/42/64	11/77/27	32/66/17	13/72/30	37/74/4	**43/69/1**

split-points has no significant effects on the performance improvement. For the pattern-based oversampling method, the amount of newly generated minority instances makes the class distribution of final training dataset balanced, since we noticed that balanced distribution brought the best performance on HeDEx, although we cannot present the experiments results due to the space limitation.

The popular evaluation measure for imbalanced dataset problems is Area Under the Receiver Operating Characteristic curve (AUC), thus we employ AUC in our paper to compare different classifiers' performance. In addition, we present the comparison of F_1-Measure for the minority class to show the classifiers' performance toward minority class. In this paper, F-Measure$^+$ denotes F_1-Measure for the minority. We used corrected paired t-test to determine the statistical significance of performances of the compared algorithms.

Table 7. F-Measure$^+$ performance results for dataset in Table 3

	LR	LR^s	C4.5	C4.5^s	ET	ET^s	HDDT	HDDT^s	HeDEx	HeDEx^s	HeDExh	HeDExh^s
compustat	0.026	0.222v	0.101	0.292 v	0.064	0.322 v	0.136	0.298 v	0.153	0.340 v	0.222	0.355 v
covtype	0.369	0.476v	0.866	0.851	0.869	0.867	0.880	0.859	0.887	0.878	0.904	0.896
estate	0.047	0.216v	0.007	0.212 v	0.036	0.218 v	0.054	0.208 v	0.067	0.210 v	0.087	0.220 v
german.numer	0.542	0.593v	0.528	0.571	0.465	0.576 v	0.546	0.580	0.501	0.570	0.539	0.578
ism	0.543	0.507	0.634	0.603	0.612	0.628	0.647	0.613	0.645	0.631	0.647	0.629
letter	0.878	0.801*	0.948	0.925 *	0.943	0.945	0.957	0.940	0.965	0.959	0.965	0.961
oil	0.476	0.463	0.433	0.473	0.250	0.583	0.392	0.464	0.345	0.570	0.434	0.550
page	0.726	0.705	0.875	0.868	0.865	0.872	0.877	0.864	0.881	0.880	0.869	0.871
pendigits	0.910	0.860*	0.960	0.969	0.988	0.988	0.977	0.970	0.986	0.985	0.989	0.990
phoneme	0.522	0.626v	0.791	0.799	0.814	0.816	0.798	0.804	0.823	0.825	0.822	0.824
PhosS	0.178	0.239v	0.102	0.228 v	0.001	0.251 v	0.083	0.234 v	0.005	0.249 v	0.041	0.248 v
satimage	0.044	0.274v	0.634	0.626	0.614	0.640	0.647	0.641	0.654	0.655	0.647	0.655
segment	0.990	0.962	0.977	0.973	0.989	0.987	0.980	0.976	0.992	0.984	0.992	0.988
credit-g	0.528	0.571	0.491	0.553 v	0.393	0.557 v	0.534	0.558	0.502	0.567	0.476	0.563 v
boundary	0.162	0.174	0.009	0.146 v	0.000	0.147 v	0.000	0.142 v	0.000	0.154 v	0.006	0.158 v
breast-y	0.381	0.433	0.349	0.449 v	0.392	0.471	0.420	0.452	0.419	0.465	0.439	0.485
cam	0.185	0.317v	0.015	0.250 v	0.005	0.276 v	0.022	0.241 v	0.014	0.281 v	0.023	0.294 v
car	0.855	0.842	0.603	0.739	0.859	0.806	0.865	0.791	0.913	0.837	0.926	0.842
dna	0.894	0.912	0.913	0.910	0.930	0.929	0.895	0.895	0.930	0.923	0.924	0.914
glass	0.633	0.684	0.687	0.723	0.634	0.628	0.420	0.554	0.635	0.666	0.489	0.607
nursery	0.770	0.730	0.672	0.793 v	0.713	0.891 v	0.004	0.280 v	0.984	0.956	0.984	0.963
page-blocks-5	0.579	0.576	0.805	0.743	0.791	0.660	0.782	0.730	0.809	0.752	0.784	0.641
sat	0.456	0.549v	0.656	0.649	0.657	0.649	0.650	0.646	0.669	0.659	0.670	0.658
(win/draw/loss)[1]		(9/12/2)		(8/14/1)		(8/15/0)		(6/17/0)		(5/18/0)		(6/17/0)

[1] The result of statistical significance test (corrected paired t-test) at 95% against algorithm X^s. v means win and * means loss.

Table 8. AUC performance results for dataset in Table 3

	LR	LR^s	C4.5	C4.5^s	ET	ET^s	HDDT	HDDT^s	HeDEx	HeDEx^s	HeDExh	HeDExh^s
compustat	0.7843	0.7899	0.8811	0.8595	0.9211	0.8924 *	0.8988	0.8719 *	0.9212	0.8986 *	0.9281	0.9039 *
covtype	0.9026	0.9035	0.9918	0.9894	0.9945	0.9929 *	0.9940	0.9912 *	0.9952	0.9938 *	0.9966	0.9955 *
estate	0.6245	0.6183	0.6400	0.6301	0.6439	0.6433	0.6281	0.6292	0.6328	0.6363	0.6336	0.6381
german.numer	0.7792	0.7802	0.7750	0.7770	0.7786	0.7798	0.7794	0.7781	0.7787	0.7795	0.7819	0.7806
ism	0.9181	0.9264	0.9278	0.9342	0.9494	0.9481	0.9391	0.9435	0.9493	0.9472	0.9481	0.9464
letter	0.9861	0.9825*	0.9973	0.9987	0.9998	0.9994	0.9988	0.9981	0.9999	0.9997	0.9999	0.9997
oil	0.8870	0.9072	0.8964	0.8905	0.9206	0.9164	0.9058	0.8922	0.9251	0.9192	0.9268	0.9256
page	0.9474	0.9203*	0.9903	0.9898	0.9915	0.9908	0.9916	0.9903	0.9922	0.9914	0.9919	0.9914
pendigits	0.9907	0.9864*	0.9992	0.9982	1.0000	0.9999	0.9988	0.9972	0.9999	0.9999	1.0000	1.0000
phoneme	0.8125	0.8129	0.9398	0.9401	0.9531	0.9513	0.9451	0.9437	0.9554	0.9534	0.9567	0.9553
PhosS	0.7269	0.7420v	0.7431	0.7285	0.7715	0.7594	0.7459	0.7313	0.7697	0.7525	0.7673	0.7510
satimage	0.7657	0.7565	0.9480	0.9416	0.9563	0.9483 *	0.9537	0.9479 *	0.9591	0.9523 *	0.9607	0.9532 *
segment	0.9999	0.9988*	0.9977	0.9977	0.9999	0.9999	0.9971	0.9965	0.9999	0.9999	1.0000	0.9999
credit-g	0.7618	0.7670	0.7568	0.7591	0.7794	0.7773	0.7755	0.7709	0.7775	0.7766	0.7817	0.7818
boundary	0.7357	0.7270	0.5548	0.6691	0.6802	0.6817	0.6732	0.6527	0.6888	0.6811	0.7092	0.6999
breast-y	0.6284	0.6389	0.6754	0.6639	0.6648	0.6718	0.6552	0.6651	0.6659	0.6729	0.6624	0.6679
cam	0.8312	0.8289	0.7045	0.7315	0.7775	0.7805	0.7806	0.7512 *	0.7892	0.7847	0.8051	0.7987
car	0.9893	0.9892	0.9831	0.9821	0.9954	0.9940	0.9941	0.9940	0.9971	0.9967	0.9980	0.9974
dna	0.9855	0.9873	0.9888	0.9880	0.9925	0.9923	0.9797	0.9782	0.9925	0.9915	0.9917	0.9908
glass	0.8410	0.8402	0.8763	0.8790	0.9184	0.9176	0.8772	0.8699	0.9090	0.9095	0.8992	0.8983
nursery	0.9882	0.9871*	0.9958	0.9955	0.9984	0.9989 v	0.9479	0.9367	0.9998	0.9998	0.9999	0.9999
page-blocks-5	0.9889	0.9862*	0.9895	0.9885	0.9916	0.9913	0.9907	0.9895	0.9924	0.9920	0.9917	0.9916
sat	0.9802	0.9811	0.9889	0.9887	0.9897	0.9891 *	0.9887	0.9885	0.9901	0.9897 *	0.9905	0.9900 *
(win/draw/loss)[1]		(1/16/6)		(0/23/0)		(1/18/4)		(0/19/4)		(0/19/4)		(0/19/4)

[1] The result of statistical significance test (corrected paired t-test) at 95% against algorithm X^s. v means win and * means loss.

6 Experiment Results

6.1 Comparison of Methods

Table 5 shows win/draw/loss counts for F-Measure$^+$ of 23 datasets comparing the algorithm in the column versus the algorithm in the row. As can be seen, HeDEx has almost the same or better performance in terms of F-Measure$^+$ than HDDT and has bigger improvements from C4.5 than HDDT does. The only loss of HeDEx against HDDT and C4.5 is the dataset PhoSs. This is due to that Extra-Trees' performance becomes poorer as the dimensionality of dataset becomes higher. However, we noticed that HeDExh overcomes this disadvantage of extremely randomized trees. We also noticed that Logistic Regression has better performance on imbalanced dataset with higher dimensionality than other classifiers we compared. In terms of both AUC presented in Table 6 and F-Measure$^+$, HeDEx and HeDExh are the best option among others on average. Table 7 and 8 show the figures of F-Measure$^+$ and AUC for each method.

6.2 Effects of Pattern-Based Oversampling

Table 7 and Table 8 show the effects of pattern-based oversampling in terms of F-Measure for minority class and AUC. Pattern-based oversampling helps the performance improvement especially when the dataset has higher dimensionality. The F-Measure$^+$ is improved from 150% to more than 3000% compared to not using pattern-based oversampling. For F-Measure$^+$, we noticed that the degree of performance improvement due to oversampling varies among the classifiers. Decision Trees based on Hellinger Distance splitting criterion (HDDT, HeDEx, HeDExh) generally have lower improvements on performance than other classifiers such as LR, C4.5 and Extra-Trees.

In terms of AUC, however, pattern-based oversampling shows statistically significant losses on performance for 4 datasets among 23 datasets. That is why oversampling favors to minority class and AUC weights more on majority due to the class distribution. This result is similar to other studies [3][7] in which authors said that sampling is not helpful if the classifier employs a skew-insensitive split criterion.

7 Conclusion

In this paper, we have proposed a new decision tree induction classifier (HeDEx) that combines extremely randomized tree (Extra-Trees [4]) with multiple candidate split-points and Hellinger Distance as a splitting criterion for imbalanced dataset problems. We also have proposed a variant of HeDEx that employs a hyperplane decision surface (HeDExh). The main contribution of our proposed methods is that they build robust decision trees against imbalanced datasets with high computational efficiency. Moreover, because of choosing sub-optimal split-points at each node, the ensemble trees produced are all independent from each other. Due to the diversity of the shape of trees we can gather the variety of patterns from training examples, and the patterns are used to generate new minority class instances.

Overall, we ensure that Hellinger Distance is skew-insensitive as a splitting criterion, since applying Hellinger Distance to Extra-Trees shows improvement in the performance for imbalanced datasets. Moreover, we also verify that randomization at both attribute and split-point selection improves the performance of decision tree methods especially when combined with Hellinger Distance. Therefore, HeDEx and HeDExh for imbalanced dataset gives better prediction ability at lower or similar computational cost, respectively. In addition, as shown in study of HDDT[8], HeDEx can also be employed in the case of balanced datasets. Thus, we recommend HeDEx should be considered as one of classification methods to be chosen.

References

1. Hido, S., Kashima, H., Takahashi, Y.: Roughly balanced bagging for imbalanced data. Stat. Anal. Data Min. 2(5-6), 412–426 (2009)
2. Provost, F., Domingos, P.: Tree induction for probability-based ranking. Mach. Learn. 52(3), 199–215 (2003)
3. Cieslak, D.A., Hoens, T.R., Chawla, N.V., Kegelmeyer, W.P.: Hellinger distance decision trees are robust and skew-insensitive. Data Min. Knowl. Discov. 24(1), 136–158 (2012)
4. Geurts, P., Ernst, D., Wehenkel, L.: Extremely randomized trees. Machine Learning 63(1), 3–42 (2006)
5. Drummond, C., Holte, R.C.: Exploiting the cost (in)sensitivity of decision tree splitting criteria. In: Proceedings of the Seventeenth International Conference on Machine Learning, pp. 239–246. Morgan Kaufmann (2000)
6. Flach, P.A.: The geometry of roc space: understanding machine learning metrics through roc isometrics. In: Proceedings of the Twentieth International Conference on Machine Learning, pp. 194–201. AAAI Press (2003)
7. Liu, W., Chawla, S., Cieslak, D.A., Chawla, N.V.: A Robust Decision Tree Algorithm for Imbalanced Data Sets. In: SDM, pp. 766–777. SIAM (2010)
8. Cieslak, D.A., Chawla, N.V.: Learning decision trees for unbalanced data. In: Daelemans, W., Goethals, B., Morik, K. (eds.) ECML PKDD 2008, Part I. LNCS (LNAI), vol. 5211, pp. 241–256. Springer, Heidelberg (2008)
9. Van Hulse, J., Khoshgoftaar, T.M., Napolitano, A.: Experimental perspectives on learning from imbalanced data. In: Proceedings of the 24th International Conference on Machine Learning, ICML 2007, pp. 935–942. ACM, New York (2007)
10. Chawla, N.V., Bowyer, K.W., Hall, L.O., Kegelmeyer, W.P.: Smote: synthetic minority over-sampling technique. J. Artif. Int. Res. 16(1), 321–357 (2002)
11. Han, H., Wang, W.-Y., Mao, B.-H.: Borderline-SMOTE: A new over-sampling method in imbalanced data sets learning. In: Huang, D.-S., Zhang, X.-P., Huang, G.-B. (eds.) ICIC 2005, Part I. LNCS, vol. 3644, pp. 878–887. Springer, Heidelberg (2005)
12. Bunkhumpornpat, C., Sinapiromsaran, K., Lursinsap, C.: Safe-level-SMOTE: Safe-level-synthetic minority over-sampling tEchnique for handling the class imbalanced problem. In: Theeramunkong, T., Kijsirikul, B., Cercone, N., Ho, T.-B. (eds.) PAKDD 2009. LNCS, vol. 5476, pp. 475–482. Springer, Heidelberg (2009)
13. Maciejewski, T., Stefanowski, J.: Local neighbourhood extension of smote for mining imbalanced data. In: 2011 IEEE Symposium on Computational Intelligence and Data Mining (CIDM), pp. 104–111 (2011)
14. Jo, T., Japkowicz, N.: Class imbalances versus small disjuncts. SIGKDD Explor. Newsl. 6(1), 40–49 (2004)
15. Alhammady, H., Ramamohanarao, K.: Using emerging patterns and decision trees in rare-class classification. In: Fourth IEEE International Conference on Data Mining, ICDM 2004, pp. 315–318 (2004)
16. Breiman, L.: Bagging predictors. Machine Learning 24(2), 123–140 (1996)
17. Schapire, R.E.: The strength of weak learnability. Machine Learning 5(2), 197–227 (1990)
18. Ho, T.K.: The random subspace method for constructing decision forests. IEEE Transactions on Pattern Analysis and Machine Intelligence 20(8), 832–844 (1998)

Signed-Error Conformal Regression

Henrik Linusson, Ulf Johansson, and Tuve Löfström

School of Business and Informatics, University of Borås, Borås, Sweden*

{henrik.linusson,ulf.johansson,tuve.lofstrom}@hb.se

Abstract. This paper suggests a modification of the Conformal Prediction framework for regression that will strenghten the associated guarantee of validity. We motivate the need for this modification and argue that our conformal regressors are more closely tied to the actual error distribution of the underlying model, thus allowing for more natural interpretations of the prediction intervals. In the experimentation, we provide an empirical comparison of our conformal regressors to traditional conformal regressors and show that the proposed modification results in more robust two-tailed predictions, and more efficient one-tailed predictions.

Keywords: Conformal Prediction, prediction intervals, regression.

1 Introduction

Conformal Prediction (CP) [1] is a framework for producing reliable confidence measures associated with the predictions of an underlying classification or regression model. Given a confidence level $\delta \in (0, 1)$, a conformal predictor outputs prediction regions that, in the long run, contain the true target value with a probability of at least $1 - \delta$. Unlike Bayesian models, CP does not rely on any knowledge of the *a priori* distribution of the problem space; and, compared to the PAC learning framework, CP is much more resiliant to noise in the data.

Clearly, the motivation for using CP is the fact that the resulting prediciton regions are guaranteed to be valid. With this in mind, it is vital to fully understand what validity means in a CP context. Existing literature (e.g. [1]) provides a thorough explanation of how the validity concept relates to conformal classification, but leaves something to be desired regarding conformal regression.

In this paper, we identify an inherent but non-obvious weakness associated with the most common type of inductive conformal regressor — conformal regressors where the nonconformity score is based on the absolute error of a predictive regression model (Abs. Error CP Regression, or AECPR). Specifically we show that when the underlying model has a skewed error distribution, AECPR produces *unbalanced* prediction intervals — prediction intervals with no guarantee regarding the distribution of errors above and below the prediction interval —

* This work was supported by the Swedish Foundation for Strategic Research through the project High-Performance Data Mining for Drug Effect Detection (IIS11-0053) and the Knowledge Foundation through the project Big Data Analytics by Online Ensemble Learning (20120192).

V.S. Tseng et al. (Eds.): PAKDD 2014, Part I, LNAI 8443, pp. 224–236, 2014.

and argue that this limits the expressiveness of AECPR models. We suggest to instead produce two one-tailed conformal predictors, one for the low end of the prediction interval and one for the high end. The only modification needed is to use a nonconformity score based on the signed error of the underlying model. Once we have these two conformal predictors, they can either be used to output valid one-tailed predictions, or combined to create two-tailed prediction intervals that exhibit a stronger guarantee of validity than standard AECPR prediction intervals. In addition, we show that the suggested approach is more robust, i.e., less sensitive to outliers. In particular when δ is small, AECPR may be seriously affected by outliers, resulting in very conservative (large) prediction intervals.

2 Background

CP was originally introduced in [2], and further developed in [3], as a transductive approach for associating classification predictions from Support Vector Machine models with a measure of confidence. Vovk, Gammerman & Shafer provide a comprehensive guide on conformal classification in [1], and Shafer & Vovk provide an abridged tutorial in [4]. Since its introduction, CP has been frequently applied to predictive modeling and used in combination with several different classification and regression algorithms, including Ridge Regression [5] k-Nearest Neighbors [6], Artificial Neural Networks [7, 8] and Evolutionary Algorithms [9].

In [10] and [5], Papadopoulos proposes a modified version of CP based on inductive inference called Inductive Conformal Prediction (ICP). In ICP, only one predictive model is generated, thus avoiding the relative computational inefficiency of (transductive) conformal predictors.

Conformal predictors have been applied to a number of problems where confidence in the predictions is of concern, including prediction of space weather parameters [8], estimation of software project effort [11], early diagnostics of ovarian and breast cancers [12], diagnosis of acute abdominal pain [13] and assessment of stroke risk [14].

2.1 Conformal Prediction

Given a set of training examples $Z = ((x_1, y_1), ...(x_l, y_l))$, and a previously unseen input pattern x_j, the general idea behind CP is to consider each possible target value $\tilde{y}$ and determine the likelihood of observing $(x_j, \tilde{y})$ in Z.

To measure the likelihood of observing $(x_j, \tilde{y})$ in Z, a conformal predictor first assigns a *nonconformity score* $\alpha_i^{\tilde{y}}$ to each instance in the extended set $\hat{Z} = Z \cup \{(x_j, \tilde{y})\}$. This nonconformity score is a measure of the strangeness of each instance $(x_i, y_i) \in \hat{Z}$ compared to the rest of the set, and is, in a predictive modeling scenario, often based on the predictions from a model generated using a traditional machine learning algorithm, referred to as the *underlying model* of the conformal predictor. The underlying model is trained using $\hat{Z}$ as training data, and the nonconformity score for an instance $(x_i, y_i) \in \hat{Z}$ is defined as the level of disagreement (according to some error measure) between the prediction of the underlying model $\hat{y}_i$ and the true label y_i.

The nonconformity score $\alpha_j^{\tilde{y}}$ is compared to the nonconformity scores of all other instances in $\hat{Z}$ to determine how unusual $(x_j, \tilde{y})$ is according to the nonconformity measure used. Specifically, we calculate the p-value of $\tilde{y}$ using

$$p(\tilde{y}) = \frac{\#\left\{z_i \in \hat{Z} \mid a_i \geq a_j^{\tilde{y}}\right\}}{l+1} . \tag{1}$$

A key property of conformal prediction is that if $p(\tilde{y})$ is below some threshold δ, the likelihood of $\tilde{y}$ being the true label for x_j is at most δ if $\hat{Z}$ is i.i.d. If we select δ to be very low, e.g., 0.05, we can thus conclude that if $p(\tilde{y}) < 0.05$ it is at most 5% likely that $\tilde{y}$ is the true label for x_j. These p-values are calculated for each tentative label $\tilde{y}$, and the conformal predictor outputs a prediction region containing each label $\tilde{y}$ for which $p(\tilde{y}) > \delta$; i.e., a set of labels that contains the true label of x_j with probability $1-\delta$. Given that we already know the probability of any prediction region containing the true output for a test instance x_j, the goal in CP is not to maximize this probability, but rather to minimize the size of the prediction regions. In essence, CP performs a form of hypothesis testing. For each label $\tilde{y}$, we want to reject the null hypothesis that $(x_j, \tilde{y})$ is conforming with Z, and for every $\tilde{y}$ we are able to reject, we reduce the size of the prediction region, thus increasing the *efficiency* of the conformal predictor.

Since $(x_j, \tilde{y})$ is included in the training data for the underlying model, the model needs to be retrained for each tentative label $\tilde{y}$; as such, this form of CP suffers from a rather poor computational complexity. ICP, as described in the next subsection, solves this problem by dividing the data set into two disjunct subsets: a proper training set and a calibration set.

2.2 Inductive Conformal Prediction

An ICP needs to be trained only once, using the following scheme:

1. Divide the training set $Z = \{(x_1, y_1), ...(x_l, y_l)\}$ into two disjoint subsets:
   - a proper training set $Z' = \{(x_1, y_1), ..., (x_m, y_m)\}$ and
   - a calibration set $Z'' = \{(x_{m+1}, y_{m+1}), ..., (x_{m+q}, y_{m+q})\}$
2. Train the underlying model h_Z using Z' as training data.
3. For each calibration instance $(x_i, y_i) \in Z''$:
   - let h_Z predict the output value for x_i so that $\hat{y}_i = h_Z(x_i)$ and
   - calculate the nonconformity score a_i using the nonconformity function.

For a novel (test) instance we simply supply the input pattern x_j to the underlying model and calculate $\alpha_j^{\tilde{y}}$ using our nonconformity function. The p-value of each tentative label $\tilde{y}$ is then calculated by comparing $\alpha_j^{\tilde{y}}$ to the nonconformity scores of the calibration set:

$$p(\tilde{y}) = \frac{\#\left\{z_i \in Z'' \mid a_i \geq a_j^{\tilde{y}}\right\} + 1}{q+1} , \tag{2}$$

where q is the size of the calibration set. If $p(\tilde{y}) < \delta$, it is at most δ% likely that $\tilde{y}$ is the true output of x_j, and $\tilde{y}$ is thus excluded from the prediction region.

2.3 Inductive Conformal Regression

In regression it is not possible to consider every possible output value $\tilde{y}$, so we cannot explicity calculate the p-value for each and every $\tilde{y}$. Instead a conformal regressor must effectively work in reverse. First, the size of the $(1-\delta)$-percentile nonconformity score, $\alpha_{s(\delta)}$, is determined; second, the nonconformity function is used to calculate the magnitude of error that would result in x_j being given a nonconformity score at most $\alpha_{s(\delta)}$; i.e., the conformal regressor determines the largest error that would be commited by the underlying model when predicting y_j with probability $1-\delta$. To perform conformal regression, we first define a nonconformity function, typically using the absolute error, see e.g., [5–8]:

$$\alpha_i = |y_i - \hat{y}_i| \ . \tag{3}$$

Then, given a significance level δ and a set of calibration scores S, the goal is to find $\alpha_j^{\hat{y}}$ such that $P(\alpha_j^{\hat{y}} > \alpha_i \in S) \leq \delta$; i.e., the largest nonconformity score — and, due to the definition of (3), also the largest absolute error — with probability $1-\delta$. To do this, we simply sort the calibration scores in a descending order, and define the prediction interval as

$$\hat{Y}_j^\delta = (\hat{y}_j - \alpha_{s(\delta)}, \hat{y}_j + \alpha_{s(\delta)}), \tag{4}$$

where $s(\delta) = \lfloor \delta(q+1) \rfloor$, i.e., the index of the $(1-\delta)$-percentile in the sorted list of nonconformity scores. Since the underlying model's error is at most $\alpha_{s(\delta)}$ with probability $1-\delta$, the resulting interval covers the true target y_j with probability $1-\delta$. Note that when using (3) and (4) the conformal regressor will, for any specific significance level δ, always produce prediction intervals of the same size for every x_j; i.e., it does not consider the difficulty of a certain instance x_j. Papadopoulos et al. [5] suggest that prediction intervals can be *normalized* using some estimation of the difficulty of each instance, e.g., by using a separate model for estimating the error of the underlying predictor. In this paper, we will not consider normalized nonconformity scores, but leave them for future work.

3 Method

AECPR will, as described in the Background, always produce 'symmetrical' prediction intervals where the underlying model's prediction is the center of the interval, and the distance from the interval's center to either boundary is equal to $\alpha_{s(\delta)}$, i.e., the absolute error from the calibration set associated with the significance level $1-\delta$. If the errors of the underlying model are symmetrically distributed — i.e., the underlying model is equally likely to underestimate and overestimate the true output — AECPR will always yield optimal interval boundaries in the sense that neither boundary is overly optimistic nor overly pessimistic in relation to the error distribution of the model (as estimated on the calibration set). However, when the error distribution of the underlying model is skewed, it is possible for one of the boundaries to become overly optimistic, while the other becomes overly pessimistic, simply because the errors committed in one direction will influence the nonconformity scores, and consuequently the prediction intervals, in both directions. Figure 1 shows an example of a skewed error distribution, where the AECPR nonconformity scores of the underlying model (a neural network) fail to capture the model's tendency to produce smaller negative errors (overestimation) and larger positive errors (underestimation).

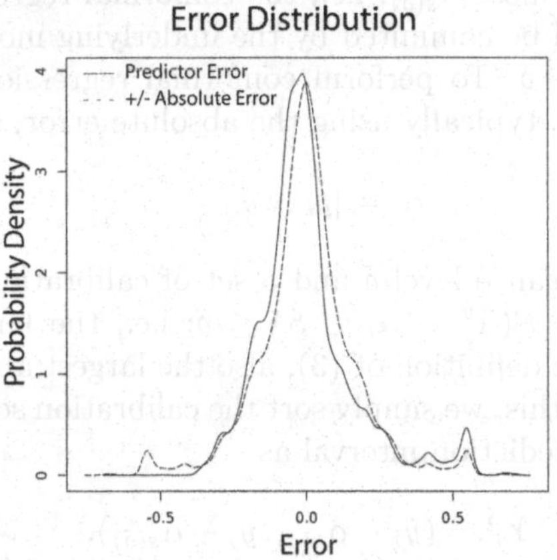

Fig. 1. Error distribution of an ANN on the Boston Housing data set. The signed error (solid line) approximates the skewed distribution of the ANN's error rate, while the absolute error (dashed line) assumes a symmetrical distribution when mirrored onto the negative range as per equations (3) and (4).

AECPR is proven valid [5], and thus there is no need to suspect that the mismatch between absolute error nonconformity scores and skewed error distributions would lead to invalid prediction intervals; however, it is necessary to be very specific about what validity means in the context of AECPR. Given that AECPR operates on the magnitude of the underlying model's error, the validity applies to the magnitude of errors and nothing else; i.e., AECPR guarantees that the absolute error of the underlying model is no larger than $\alpha_{s(\delta)}$ with probability $1-\delta$. However, without considering the underlying model's tendency to commit positive or negative errors, AECPR cannot provide information regarding how δ is distributed above and below the prediction interval.

Without this information, it is not possible for a user of AECPR to distinctly assess the validity of the prediction boundaries. To illustrate, consider a 95%-confidence prediction interval on the form $(-1, 1)$. If asked to assess the likelihood of the true value y_j being greater than 1, one is easily tempted to assume a probability of 2.5%, since intuitively, the probability of y_j being greater than the interval's upper boundary should be about the same as the probability of y_j being less than the interval's lower boundary. The true answer however, is that AECPR can only guarantee that there is at most a 5% probability of y_j being less than -1, and at most a 5% probability of y_j being greater than 1, since we have no information of the probabilities of the true value being higher or lower than the prediction interval's upper and lower boundaries respectively. In the following subsections, we expand on this argument, and propose a straightforward method for producing prediction intervals that possess a stronger guarantee of validity for the individual boundaries than for the full interval, while maintaining efficiency.

3.1 Validity of Interval Boundaries

If $\hat{Y} = (\hat{Y}_{low}, \hat{Y}_{high})$ is a valid prediction region at $\delta = d$, the one-tailed predictions $(-\infty, \hat{Y}_{high})$ and $(\hat{Y}_{low}, +\infty)$ must also be valid prediction regions at $\delta = d$, as they both cover at least the same error probability mass covered by $\hat{Y}$. However, without any knowledge of how the error probability d is distributed above and below $\hat{Y}$ it is not possible to assume that one specific one-tailed prediction is valid at any $\delta < d$. Hence, we can, in fact, only be confident in the one-tailed predictions with probability $1 - d$, i.e., if $\hat{Y} = (\hat{Y}_{low}, \hat{Y}_{high})$ is valid at $\delta = d$, then $(-\infty, \hat{Y}_{high})$ and $(\hat{Y}_{low}, +\infty)$ are valid at $\delta = d$, but may be invalid at $\delta < d$.

On the other hand, if $(-\infty, \hat{Y}_{high})$ and $(\hat{Y}_{low}, +\infty)$ are both known to be valid at $\delta = \frac{d}{2}$, we know by definition that either of these one-tailed predictions will be incorrect with probability at most $\frac{d}{2}$. Thus, if the two one-tailed predictions are combined into a prediction interval $\hat{Y}_c = (\hat{Y}_{low}, \hat{Y}_{high})$, we can guarantee that $\hat{Y}_c$ will be wrong in a specific direction with probability at most $\frac{d}{2}$, and in total with probability at most d. Now, we are able to express not only a confidence $1 - d$ for the interval, but also a greater confidence $1 - \frac{d}{2}$ for the individual boundaries. Hence, if $(-\infty, \hat{Y}_{high})$ and $(\hat{Y}_{low}, +\infty)$ are valid at $\delta = \frac{d}{2}$, then $\hat{Y}_c = (\hat{Y}_{low}, \hat{Y}_{high})$ must be valid at $\delta = d$. This follows from the fact that when two one-tailed predictions are combined into a two-tailed interval, the probability of the resulting interval being wrong is the sum of the probabilities of the boundaries being wrong.

Using AECPR, prediction intervals with boundaries guaranteed at $\frac{\delta}{2}$ can be constructed simply by creating a prediction region $\hat{Y}_j^{0.5\delta}$, and outputting it as a combined interval $\hat{Y}_{c,j}^{\delta}$. This is of course rather impractical — as $|\hat{Y}_j^{0.5\delta}| \geq |\hat{Y}_j^{\delta}|$, we're not only 'increasing' the guarantee of validity for the individual boundaries in $\hat{Y}_{c,j}^{\delta}$ compared to $\hat{Y}_j^{\delta}$, we're also effectively guaranteeing that our prediction interval is unnecessarily large! We would much rather output a combined interval such that $|\hat{Y}_{c,j}^{\delta}| \approx |\hat{Y}_j^{\delta}|$. To accomplish this, we are required to reduce the size of the predicted boundaries $\hat{Y}_{low}$ and $\hat{Y}_{high}$. We note the following: if the underlying model's predictions tend to underestimate the true output values, we are only required to adjust the upper boundary of the prediction intervals for them to remain valid; similarly, if the underlying model's predictions tend to overestimate, we are only required to adjust the lower boundary. Hence when predicting $\hat{Y}_{low}$ we only need to consider the negative errors made by the underlying predictor; and, when predicting $\hat{Y}_{high}$, we only need to consider the positive errors made by the underlying predictor. That is, we can construct the interval boundaries by considering only about half of the errors commited by the underlying model, thus reducing the expected size of each boundary while maintaining validity in the one-tailed predictions. This follows directly from the semantics of validity we are applying to the lower and upper boundary predictions — in both cases, we are only interested in guaranteeing validity in a single direction, i.e., for the upper boundary, we want to guarantee only that the probability of y_j being greater than $\hat{Y}_{high}$ is at most $1-\frac{\delta}{2}$, and for the lower boundary we want to guarantee that the probability of y_j being less than $\hat{Y}_{low}$ is at most $1-\frac{\delta}{2}$. Thus, if we are interested in predicting the upper boundary of y_j, we can define the nonconformity measure as $\alpha_i = y_i - \hat{y}_i$, i.e., a nonconformity function that returns larger nonconformity scores with larger positive errors, and define the prediction interval as $(-\infty, \hat{y}_j + \alpha_{s(\frac{\delta}{2})})$. In this case, we guarantee that the true value $y_j \leq \hat{y}_j + \alpha_{s(\frac{\delta}{2})}$ with probability $1-\frac{\delta}{2}$. Conversely, we can define the nonconformity score such that it increases with larger negative errors, i.e., $\alpha_i = \hat{y}_i - y_i$, output the prediction interval $(\hat{y}_j - \alpha_{s(\frac{\delta}{2})}, +\infty)$, and guarantee that $y_j \geq \hat{y}_j - \alpha_{s(\frac{\delta}{2})}$ with probability $1-\frac{\delta}{2}$.

From this point, we are able to do one of two things: we can output either $(-\infty, \hat{Y}_{high})$ or $(\hat{Y}_{low}, +\infty)$ as the prediction with confidence $1-\frac{\delta}{2}$; or, we can combine the two one-tailed predictions, and guarantee the boundaries at $1-\frac{\delta}{2}$, and the interval at $1-\delta$.

3.2 Signed Error CPR

To approximate the (potentially skewed) error distribution of the underlying model, we propose a nonconformity measure based on the signed error of the model; i.e., we define the nonconformity score as

$$\alpha_i = y_i - \hat{y}_i \ . \tag{5}$$

Just as in AECPR, we sort α in a descending order — note though, that while the sorted α in AECPR contains the absolute errors from largest to smallest, the sorted α in SECPR ranges from maximum positive error to maximum negative error. The prediction interval for a novel instance x_j is then formulated as

$$\hat{Y}_j^{\delta} = (\hat{y}_j + \alpha_{low(\frac{\delta}{2})}, \hat{y}_j + \alpha_{high(\frac{\delta}{2})}), \tag{6}$$

where $high(\frac{\delta}{2}) = \lfloor \frac{\delta}{2}(q+1) \rfloor$ and $low(\frac{\delta}{2}) = \lfloor (1-\frac{\delta}{2})(q+1) \rfloor + 1$. In effect, SECPR performs two simultaneous conformal predictions — one for each boundary of the interval. The boundaries are predicted with confidence $1-\frac{\delta}{2}$ and, when combined, form a prediction interval with confidence $1-\delta$.

3.3 Evaluation

The two methods (AECPR and SECPR) were evaluated on 33 publicly available data sets from the UCI [15] and Delve [16] repositories. Before experimentation all output values were scaled to $[0, 1]$, only to enhance interpretability in efficiency comparisons — with the outputs scaled to $[0, 1]$, the size of a prediction interval expresses the fraction of possible outputs covered by the interval. For each data set, 100 random sub-sampling tests were performed; in each iteration, a randomized subsample (20%) of the data set was used as the test set, and remaining instances were used for training and calibration. The calibration set size was defined as a function of the training set size: $|Z''| = 100 \left\lfloor \frac{|Z|}{100} \right\rfloor \times 0.1 + 199$.

Standard multilayer perceptron neural networks with $\left\lceil \sqrt{k} \right\rceil + 1$ hidden nodes were used as underlying models for the conformal predictors, where k is the number of input features of each data set. In each iteration, both AECPR and SECPR predictions were calculated from the same model and calibration instances.

4 Results

Given any $\delta \in (0, 1)$, an ICR should produce valid prediction intervals — in reality however, predictions with low confidence are rarely of interest. Thus, we choose to show the empirical validity and evaluate the efficiency of AECPR and SECPR at three commonly used confidence levels: 99%, 95% and 90%.

4.1 Validity of Intervals

As illustrated in Table 1, both methods produce prediction intervals that cover the true targets of the test set at or very close to the predefined significance levels; thus, in terms of interval coverage both methods are, as expected, valid.

Table 1. Mean coverage (portion of predictions that coincide with the true output of the test instances) for AECPR and SECPR at $\delta \in \{0.01, 0.05, 0.10\}$ on the 33 sets

	99%		95%		90%	
	Abs	Sign	Abs	Sign	Abs	Sign
abalone	.990	.990	.950	.949	.902	.902
anacalt	1.00	1.00	.943	.997	.920	.944
bank8fh	.989	.989	.950	.949	.900	.901
bank8fm	.990	.991	.950	.951	.902	.903
bank8nh	.989	.990	.951	.949	.901	.901
bank8nm	.990	.991	.950	.951	.900	.901
boston	.996	.992	.948	.958	.897	.897
comp	.992	.993	.952	.962	.904	.923
concrete	.989	.993	.947	.954	.898	.904
cooling	.990	.990	.949	.949	.898	.903
deltaA	.991	.991	.950	.951	.901	.902
deltaE	.990	.991	.951	.951	.901	.902
friedm	.990	.994	.948	.951	.902	.906
heating	.990	.991	.948	.950	.895	.897
istanbul	.991	.991	.952	.955	.903	.903
kin8fh	.990	.990	.952	.951	.902	.902
kin8fm	.990	.990	.951	.951	.901	.903

	99%		95%		90%	
	Abs	Sign	Abs	Sign	Abs	Sign
kin8nh	.990	.990	.950	.950	.901	.902
kin8nm	.990	.990	.950	.950	.900	.901
laser	.990	.991	.949	.948	.898	.901
mg	.991	.994	.951	.954	.905	.911
mortage	.990	.994	.950	.952	.906	.907
plastic	.999	1.00	.997	.997	.994	.993
puma8fh	.990	.990	.949	.950	.900	.902
puma8fm	.990	.990	.950	.950	.901	.902
puma8nh	.990	.990	.951	.951	.902	.903
puma8nm	.990	.990	.950	.950	.901	.902
quakes	.992	.995	.955	.970	.908	.934
stock	.990	.990	.948	.948	.899	.896
treasury	.991	.993	.952	.952	.903	.910
wineR	.991	.994	.956	.957	.910	.913
wineW	.993	.993	.954	.963	.911	.913
wizmir	.991	.994	.949	.954	.904	.908
mean	.991	.992	.952	.955	.905	.909
min	.989	.989	.943	.948	.895	.896

4.2 Validity of Boundaries

Here, we take a closer look at the coverage of the lower and upper boundaries of the intervals produced by AECPR and SECPR. Specifically, we expect AECPR and SECPR to have a clear difference in coverage of their one-tailed predictions $(\hat{Y}_{low}, +\infty)$ and $(-\infty, \hat{Y}_{high})$. We expect SECPR's boundaries to be valid at $1 - \frac{\delta}{2}$, and AECPR boundary coverage to vary between $1 - \frac{\delta}{2}$ and $1 - \delta$.

Table 2. Mean low/high coverage for AECPR and SECPR at $\delta \in \{0.01, 0.05, 0.10\}$

	99%				95%				90%			
	Abs		Sign		Abs		Sign		Abs		Sign	
	low	high	low	high	low	high	low	high	low	high	low	high
abalone	.999	.991	.995	.995	.989	.961	.974	.976	.968	.934	.949	.953
anacalt	1.00	1.00	1.00	1.00	.943	1.00	.997	1.00	.920	1.00	.944	1.00
bank8fh	.999	.990	.995	.995	.986	.964	.975	.974	.967	.933	.950	.951
bank8fm	.997	.993	.995	.995	.981	.970	.975	.976	.958	.944	.952	.951
bank8nh	1.00	.990	.995	.995	.995	.956	.974	.975	.981	.920	.950	.951
bank8nm	.998	.992	.995	.995	.983	.967	.976	.974	.959	.941	.951	.950
boston	.999	.997	.994	.998	.993	.955	.975	.983	.967	.930	.950	.947
comp	.995	.997	.997	.996	.960	.992	.979	.983	.919	.985	.953	.970
concrete	1.00	.989	.997	.996	.994	.953	.977	.976	.972	.927	.954	.950
cooling	1.00	.990	.996	.994	.997	.952	.976	.973	.984	.914	.954	.949
deltaA	.995	.996	.996	.996	.971	.978	.976	.975	.946	.956	.952	.949
deltaE	.994	.996	.996	.995	.975	.976	.976	.976	.953	.949	.951	.951
friedm	.995	.995	.996	.997	.978	.970	.974	.976	.956	.946	.951	.955
heating	1.00	.990	.996	.995	.998	.951	.976	.974	.982	.913	.949	.947
istanbul	.995	.996	.996	.994	.976	.976	.977	.978	.947	.955	.952	.951
kin8fh	.993	.997	.995	.995	.972	.980	.976	.975	.948	.954	.952	.951
kin8fm	.994	.996	.995	.995	.971	.979	.975	.976	.946	.955	.951	.952
kin8nh	.995	.995	.995	.994	.972	.979	.975	.975	.946	.954	.950	.951
kin8nm	.996	.994	.995	.995	.974	.975	.975	.975	.949	.952	.951	.950
laser	.995	.995	.994	.997	.973	.976	.973	.975	.947	.952	.950	.950
mg	.996	.995	.996	.998	.978	.972	.977	.977	.952	.953	.955	.956
mortage	1.00	.990	.998	.997	.998	.952	.976	.976	.995	.911	.955	.953
plastic	1.00	.999	1.00	1.00	1.00	.997	.998	.998	.998	.995	.996	.997
puma8fh	.995	.995	.995	.995	.972	.977	.974	.975	.946	.955	.950	.951
puma8fm	.994	.996	.995	.995	.973	.977	.975	.975	.950	.951	.952	.950
puma8nh	.993	.997	.995	.995	.972	.980	.976	.975	.947	.955	.951	.952
puma8nm	.994	.995	.995	.995	.974	.976	.976	.975	.949	.952	.952	.950
quakes	1.00	.992	.998	.997	1.00	.956	.991	.980	.998	.910	.979	.955
stock	.996	.994	.995	.994	.978	.970	.974	.974	.957	.942	.948	.948
treasury	1.00	.991	.996	.997	.999	.953	.973	.978	.993	.911	.956	.954
wineR	.994	.997	.997	.997	.976	.980	.977	.980	.952	.958	.955	.959
wineW	.996	.996	.996	.997	.984	.970	.978	.985	.961	.949	.955	.957
wizmir	.994	.997	.997	.997	.974	.976	.976	.978	.954	.950	.954	.954
mean	.997	.994	.996	.996	.981	.971	.977	.978	.960	.946	.954	.955
min	.989		.994		.943		.973		.910		.944	

Table 2 reveals that AECPR's interval boundaries show only a small deviance from $1 - \frac{\delta}{2}$-validity on average (across all data sets). However, more often than not, one of the boundaries is overly optimistic and the other overly pessimistic to compensate, and the reason the two interval boundaries appear valid on average is the simple fact that they are alternatingly optimistic and pessimistic. We also note that, in the worst cases (e.g. the treasury data set), one of the boundaries is valid only at $1 - \delta$. In contrast, SECPR interval boundaries show coverage at or very near $1 - \frac{\delta}{2}$-validity in both the average and worst cases.

To further illustrate, we let the D_p^δ be the average coverage of the boundary (low or high) that has the highest average coverage for data set D (the most pessimistic boundary), and we let D_o^δ be the average coverage of the boundary with the lowest average coverage for D (the most optimistic boundary). E.g., for abalone, AECPR has $D_o^{99} = 0.991$ and $D_p^{99} = 0.999$; and SECPR has $D_o^{99} = 0.995$ and $D_p^{99} = 0.995$. We then calculate the mean coverage $\bar{D}_o^\delta$ and $\bar{D}_p^\delta$ for each δ, and for both AECPR and SECPR (Table 3).

Table 3. Mean optimistic and pessimistic boundary coverage of AECPR and SECPR

	$\bar{D}_o^{99}$	$\bar{D}_p^{99}$	$\bar{D}_o^{95}$	$\bar{D}_p^{95}$	$\bar{D}_o^{90}$	$\bar{D}_p^{90}$
AECPR	.994	.997	.967	.985	.939	.966
SECPR	.996	.996	.977	.979	.952	.957

In Table 3 we can clearly see that, as we argued in the Method, AECPR intervals show a tendency towards having an overly pessimistic boundary and an overly optimistic boundary, in such a way that the error probability δ above and below the interval is unevenly distributed. Furthermore, we are not given any information regarding the distribution of δ, and thus as noted in the Method and supported by the empirical evidence, AECPR can only guarantee the validity of its interval boundaries at $1 - \delta$.

SECPR intervals can guarantee the interval boundaries at $1 - \frac{\delta}{2}$; so, this statement is supported by the empirical evidence. We also note that SECPR tends to produce balanced intervals — i.e., intervals where δ is evenly distributed above and below the prediction intervals. More importantly, even in the cases where the intervals are not perfectly balanced, we have already defined the boundaries to be valid at $1 - \frac{\delta}{2}$.

4.3 Efficiency

We can choose to compare the interval sizes of AECPR and SECPR (Table 4) based on two different criteria: either we compare intervals that share the same guarantee of validity for the full interval; or, we compare intervals that share the same guarantee of validity for the individual boundaries. That is, we either compare the 99% SECPR intervals to the 99% AECPR intervals, and so on, and remember that SECPR provides a stronger guarantee of validity for the boundaries than does AECPR; or, we compare the 90% SECPR intervals to the 95% AECPR intervals, and so on, and remember that the two methods in this case provide the same guarantee for the one-tailed prediction boundaries.

First, we consider predictions that share the same guarantee of validity for the full interval. Here, we note that on average SECPR produces tighter intervals than AECPR at the 99% confidence level, due to SECPR taking into account the sign of the underlying model's error. If the underlying model commits large errors in only one direction, only one of the interval boundaries predicted by SECPR is affected, while both boundaries are affected in AECPR. Thus, for data sets where the underlying model commits *outlier errors* — atypical errors that are of

Table 4. Mean-median interval sizes for AECPR and SECPR at $\delta \in \{0.01, 0.05, 0.10\}$

	99%		95%		90%	
	Abs	Sign	Abs	Sign	Abs	Sign
abalone	.528	.486	.324	.322	.238	.248
anacalt	1.77	1.00	1.18	.985	.728	.707
bank8fh	.543	.540	.379	.371	.295	.297
bank8fm	.266	.255	.191	.186	.152	.153
bank8nh	.793	.720	.452	.440	.327	.338
bank8nm	.390	.374	.224	.221	.158	.162
boston	1.11	.969	.712	.707	.488	.554
comp	.790	.685	.419	.405	.286	.303
concrete	1.00	.933	.753	.780	.646	.673
cooling	1.07	.910	.787	.779	.655	.665
deltaA	.269	.283	.162	.164	.126	.127
deltaE	.317	.333	.214	.215	.175	.175
friedm	.289	.317	.210	.213	.175	.179
heating	1.04	.952	.904	.840	.776	.752
istanbul	.503	.542	.338	.344	.274	.274
kin8fh	.387	.383	.286	.287	.238	.239
kin8fm	.163	.162	.119	.120	.099	.100
kin8nh	.628	.631	.482	.482	.405	.407
kin8nm	.549	.551	.395	.396	.321	.323
laser	.625	.605	.291	.281	.208	.215
mg	.787	.837	.513	.529	.401	.411
mortage	1.15	.915	.895	.791	.663	.721
plastic	.993	.976	.983	.973	.975	.968
puma8fh	.765	.768	.570	.571	.470	.472
puma8fm	.359	.361	.266	.266	.223	.224
puma8nh	.762	.756	.552	.554	.448	.451
puma8nm	.369	.365	.253	.255	.205	.207
quakes	1.08	.871	.709	.639	.492	.527
stock	.748	.733	.597	.604	.530	.530
treasury	1.13	.889	.807	.763	.556	.649
wineR	.841	1.10	.556	.567	.448	.449
wineW	.741	.756	.546	.552	.403	.394
wizmir	.507	.558	.418	.424	.383	.384
mean	.705	.652	.500	.486	.393	.402
std dev	.349	.260	.266	.242	.212	.216

a much larger magnitude than typical errors — in only one direction, SECPR will produce tighter intervals than AECPR. It must be noted that while this applies to all significance levels, most of the outlier errors will, for lower signficance levels, be excluded from the prediction intervals anyway. Consuequently, at the 95% confidence level we observe a similar pattern, but the effect is much less pronounced. At 90%, the effect is all but gone, and SECPR is instead slightly less effective than AECPR on almost all data sets. A Wilcoxon signed-ranks test at $\alpha = 0.05$ shows that, in terms of efficiency, SECPR is significantly better than AECPR for 99%-confidence predictions, while AECPR is significantly better than SECPR for 90%-confidence predictions. Thus, at 90% confidence or lower, we can expect that the strengthened guarantee of validity provided by SECPR is accompanied with a small but significant decrease in efficiency.

Second, we consider one-tailed predictions, specifically at the 95%-confidence level (i.e., AECPR at 95%, SECPR at 90%); here, we can clearly see that SECPR produces tighter intervals than AECPR for all data sets. Hence, simply by ensuring that the boundaries are affected only by the relevant errors of the underlying model, we can significantly increase the efficiency of one-tailed predictions.

5 Conclusions

In this paper, we have shown that conformal regressors based on the absolute error of the underlying model produce unbalanced prediction intervals — intervals with no guarantee for the distribution of error above and below the intervals — and that this unbalance leads to prediction intervals with weak guarantees of validity for the individual interval boundaries. To address this issue, we have proposed a straightforward approach for producing prediction intervals based on the signed error of the underlying model (SECPR) that can provide a stronger guarantee of validity for the interval boundaries. Also, we have shown that SECPR is less sensitive to outlier errors than AECPR, resulting in more efficient prediction intervals at the highest confidence levels. Finally, we show that, when expected to provide the same guarantees of boundary validity as AECPR, SECPR produces much more efficient prediction intervals than corresponding AECPR models.

References

1. Vovk, V., Gammerman, A., Shafer, G.: Algorithmic learning in a random world. Springer Verlag, DE (2006)
2. Gammerman, A., Vovk, V., Vapnik, V.: Learning by transduction. In: Proceedings of the Fourteenth Conference on Uncertainty in Artificial Intelligence, pp. 148–155. Morgan Kaufmann Publishers Inc. (1998)
3. Saunders, C., Gammerman, A., Vovk, V.: Transduction with confidence and credibility. In: Proceedings of the Sixteenth International Joint Conference on Artificial Intelligence (IJCAI 1999), vol. 2, pp. 722–726 (1999)
4. Shafer, G., Vovk, V.: A tutorial on conformal prediction. The Journal of Machine Learning Research 9, 371–421 (2008)
5. Papadopoulos, H., Proedrou, K., Vovk, V., Gammerman, A.: Inductive confidence machines for regression. In: Elomaa, T., Mannila, H., Toivonen, H. (eds.) ECML 2002. LNCS (LNAI), vol. 2430, pp. 345–356. Springer, Heidelberg (2002)
6. Papadopoulos, H., Vovk, V., Gammerman, A.: Regression conformal prediction with nearest neighbours. Journal of Artificial Intelligence Research 40(1), 815–840 (2011)
7. Papadopoulos, H.: Inductive conformal prediction: Theory and application to neural networks. Tools in Artificial Intelligence 18, 315–330, 2 (2008)
8. Papadopoulos, H., Haralambous, H.: Reliable prediction intervals with regression neural networks. Neural Networks 24(8), 842–851 (2011)
9. Lambrou, A., Papadopoulos, H., Gammerman, A.: Reliable confidence measures for medical diagnosis with evolutionary algorithms. IEEE Transactions on Information Technology in Biomedicine 15(1), 93–99 (2011)
10. Papadopoulos, H.: Inductive conformal prediction: Theory and application to neural networks. Tools in Artificial Intelligence 18, 315–330 (2008)
11. Papadopoulos, H., Papatheocharous, E., Andreou, A.S.: Reliable confidence intervals for software effort estimation. In: AIAI Workshops, pp. 211–220. Citeseer (2009)
12. Devetyarov, D., Nouretdinov, I., Burford, B., Camuzeaux, S., Gentry-Maharaj, A., Tiss, A., Smith, C., Luo, Z., Chervonenkis, A., Hallett, R., et al.: Conformal predictors in early diagnostics of ovarian and breast cancers. Progress in Artificial Intelligence 1(3), 245–257 (2012)
13. Papadopoulos, H., Gammerman, A., Vovk, V.: Reliable diagnosis of acute abdominal pain with conformal prediction. Engineering Intelligent Systems 17(2), 127 (2009)

14. Lambrou, A., Papadopoulos, H., Kyriacou, E., Pattichis, C.S., Pattichis, M.S., Gammerman, A., Nicolaides, A.: Assessment of stroke risk based on morphological ultrasound image analysis with conformal prediction. In: Papadopoulos, H., Andreou, A.S., Bramer, M. (eds.) AIAI 2010. IFIP AICT, vol. 339, pp. 146–153. Springer, Heidelberg (2010)
15. Bache, K., Lichman, M.: UCI machine learning repository (2013), `http://archive.ics.uci.edu/ml`
16. Rasmussen, C.E., Neal, R.M., Hinton, G., van Camp, D., Revow, M., Ghahramani, Z., Kustra, R., Tibshirani, R.: Delve data for evaluating learning in valid experiments (1996), `http://www.cs.toronto.edu/~delve`

Multi-Instance Learning from Positive and Unlabeled Bags

Jia Wu[1,3], Xingquan Zhu[2], Chengqi Zhang[1], and Zhihua Cai[3]

[1] Centre for Quantum Computation & Intelligent Systems, FEIT, University of Technology Sydney, NSW 2007, Australia
[2] Dept. of Computer & Electrical Engineering and Computer Science, Florida Atlantic University, Boca Raton, FL 33431, USA
[3] Dept. of Computer Science, China University of Geosciences, Wuhan, China
jia.wu@student.uts.edu.au, chengqi.zhang@uts.edu.au, xzhu3@fau.edu, zhcai@cug.edu.cn

Abstract. Many methods exist to solve multi-instance learning by using different mechanisms, but all these methods require that both positive and negative bags are provided for learning. In reality, applications may only have positive samples to describe users learning interests and remaining samples are unlabeled (which may be positive, negative, or irrelevant to the underlying learning task). In this paper, we formulate this problem as positive and unlabeled multi-instance learning (puMIL). The main challenge of puMIL is to accurately identify negative bags for training discriminative classification models. To solve the challenge, we assign a weight value to each bag, and use an Artificial Immune System based self-adaptive process to select most reliable negative bags in each iteration. For each bag, a most positive instance (for a positive bag) or a least negative instance (for an identified negative bag) is selected to form a positive margin pool (PMP). A weighted kernel function is used to calculate pairwise distances between instances in the PMP, with the distance matrix being used to learn a support vector machines classifier. A test bag is classified as positive if one or multiple instances inside the bag are classified as positive, and negative otherwise. Experiments on real-world data demonstrate the algorithm performance.

Keywords: Multi-instance learning, unlabeled bags, classification.

1 Introduction

Multi-instance learning (MIL) [1] is a special type of learning task where each observation contains a bag of instances. A bag is labeled positive if one or multiple instances inside the bag are positive, and negative otherwise. The uniqueness of not requiring labels for individual instances makes multi-instance learning very suitable for applications without label information for individual instances. Because the genuine positive instance(s) inside each positive bag is unknown, the main challenge of multi-instance learning is to leverage bag labels and constraints to derive accurate classification models. Roughly, existing MIL methods

[2] can be separated into the following three categories: (1) instance-based modeling, which finds most positive and most negative instances from bags to derive MIL models; (2) bag-based modeling, which directly builds classification models at the bag level; and (3) hybrid approaches, which use instances and bags to confine the learning space to build classification models.

For all existing MIL methods, one prerequisite is that training data must contain both positive and negative bags to derive discriminative models. In reality, many applications may only have positive bags to indicate users learning interests and the remaining bags are unlabeled (which may be positive, negative, or irrelevant to the learning task). For example, during an image retrieval process [3], users may click one or multiple images which are interesting to them (the clicked images can be regarded as positive bags), but majority images remain unchecked, so we do not know whether those images do not contain users retrieval concepts (*i.e.* negative bags) or users simply overlook the images. In this case, there is no negative bag but only positive and unlabeled bags are available.

When only positive and unlabeled bags are available, a straightforward solution for MIL is to propagate a bag's label to each instance inside the bag, so the problem can be solved by using standard Positive-Unlabeled learning [4]. Indeed, this simple solution is ineffective because not all instances inside a positive bag are positive, so some instances will be mislabeled and deteriorate the classification accuracy. A slightly more intelligent way is to use Positive and Unlabeled learning strategy [5] to first treat all unlabeled bags as negative bags and train an MIL classifier, and then iteratively refine identified negative bags by using trained MIL classifiers. This solution is still ineffective mainly because the identification of negative bags only relies on the MIL classifiers but does not take the unique MI bag constraints into consideration, so directly training MIL classifiers using positive bags and identified negative bags will severely deteriorate the classification accuracy.

The above observations motivate a very practical learning task where only positive and unlabeled bags are available for multi-instance learning. In this paper, we formulate this problem as positive and unlabeled multi-instance learning (puMIL), where the key challenge is twofold: (1) **MIL learning with unreliable bag labels:** Although it is always possible to identify some unlabeled bags as negative bags, the labels of identified negative bags are unreliable. Directly building MIL classifiers from positive and identified negative bags may result in low classification accuracy. This reality calls for new MIL learning frameworks capable of handling unreliable bag labels; and (2) **Tacking uncertainty inside positive bags:** For MIL, the genuine labels of instances inside a positive bag are unknown, although at least one instance has to be positive. Finding "most positive instances" plays a significant role for multi-instance learning. In a puMIL setting, this process is further complicated because no negative bags are available to help identify positive instance(s) in a positive bag.

In this paper, we propose a self-adaptive learning framework to tackle the above challenges for puMIL. More specifically, we assign a weight value to each bag, and use an Artificial Immune System based search process to update bag

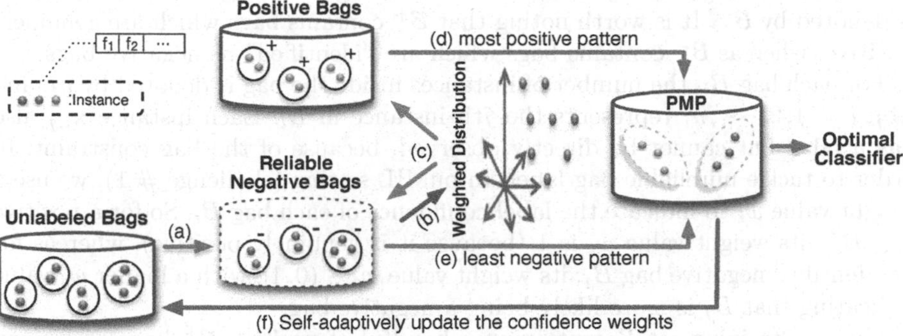

Fig. 1. A conceptual view of the proposed puMIL framework: Given a number of positive and unlabeled bags, puMIL starts from assigning a weight value for each bag. (a): the weight values help identify some unlabeled bags as reliable negative bags. (b): the obtained reliable negative bags help build a weighted distribution, which is used to measure instances in positive and reliable negative bags, respectively, (c). After that, the "most positive pattern" (d) for positive bags and the "least negative pattern" (e) for reliable negative bags are selected to form a Positive Margin Pool (PMP), which helps build a weighted SVM classifier for classification. Based on the constructed PMP, a self-adaptive strategy (discussed in Section 4.3) is used to update weight value of each bag (f). The iteration process continues, with the objective of improving the quality of reliable negative bags and the quality of PMP, and achieving optimal accuracy.

weight values and identify the most reliable negative bags in each iteration. For each bag, a "most positive instance" (for a positive bag) or a "least negative instance" (for an identified negative bag) is selected to form a positive margin pool (PMP). A weighted kernel function is used to calculate pairwise distances between instances in the PMP, with the distance matrix being used to learn a weighted support vector machines classifier. Experiments on real-world positive and unlabeled bags confirm the effectiveness of the proposed design.

The remainder of the paper is structured as follows. Preliminary and problem statement are addressed in Section 2, followed by the overall framework in Section 3. Section 4 introduces detailed algorithms, followed by experiments in Section 5. We conclude the paper in Section 6.

2 Preliminaries and Problem Statement

Denote $\mathcal{B} = \{B_1, \cdots, B_n\}$ a bag set with n bags, and B_i is the ith bag in the set. The bag set contains both positive and unlabeled bags, with B_i^+ and B_j^u indicating a positive and an unlabeled bag, respectively (for ease of representation, we also use B_i to denote a bag). Let $Y = [y_1, \cdots, y_n]$ where y_i is the label of B_i. In generic MIL settings, a positive and a negative bag's label can be denoted by $y_i = +1$ and $y_i = -1$, respectively. In a puMIL setting, an unlabeled bag B_j^u's label is denoted by $y_j = 0$. The collections of positive and unlabeled bag sets can be denoted by $\mathcal{B}^+$ and $\mathcal{B}^u$, respectively. During the learning process, the algorithm may identify a set of unlabeled bags as a negative bag set, which is denoted by $\mathcal{B}^-$. It is worth noting that $\mathcal{B}^+$ contains bags which are genuinely positive, whereas $\mathcal{B}^-$ contains bags which are "identified" as negative bags.

For each bag B_i, the number of instances inside the bag is denoted by n_i, and $\mathbf{x}_{i,j}, j = 1, 2, \cdots, n_i$ represents the jth instance in B_i. Each instance $\mathbf{x}_{i,j}$ also has a label but cannot be directly observed, because of the bag constraint. In order to tackle unreliable bag labels in puMIL setting (challenge # 1), we use a weight value w_i to indicate the label confidence of each bag B_i. So for a positive bag B_i^+, its weight value w_i is 1 (because it is genuinely positive), whereas for an identified negative bag B_j, its weight value $w_j \in (0, 1]$, with a higher w_j value indicating that B_j is more likely being a negative bag.

Given a training set $\mathcal{B}$ containing a handful of positive $\mathcal{B}^+$ bags and many unlabeled $\mathcal{B}^u$ bags, puMIL learning **aims** to build a prediction model from $\mathcal{B}$ to accurately predict previously unseen bags with maximum F-score.

3 Overall Framework

Figure 1 illustrates the conceptual view of the proposed puMIL framework, which includes two major steps to solve key challenges identified in Section 1.

Positive Margin Pool (PMP): In order to carry out multi-instance learning with positive and unlabeled bags only, we propose to use maximum margin idea [6] to build a positive margin pool (PMP) by modeling the distributions of the positive and unlabeled bags. More specifically, we utilize bag weight values to identify some unlabeled bags as most reliable negative bags. After that, we select "most positive patterns" from positive bags and "least negative patterns" from identified negative bags to form a positive margin pool. Because PMP contains signature patterns with respect to positive and identified negative bags, it will help differentiate decision boundaries to separate positive bags.

Self-Adaptive Bag Weight Updating: In order to properly identify most negative bags from unlabeled bags, we assign a weight value to each bag and will employ a self-adaptive iterative process to update bag weight values to find reliable negative bags. The proposed iterative bag weight updating process is based on clonal selection theory [7] in Artificial Immune Systems. It iteratively searches different weight values, including mutations and clones of weight values from the previous iteration, to find the one which optimizes the object function (*i.e.* the F-score in our case). As a result, the self-adaptive weight updating can ensure high quality negative bags are identified to form positive margin pool (PMP) for multi-instance learning without negative bags. The weight updating is detailed in Section 4.3.

4 Positive and Unlabeled Multi-Instance Learning

4.1 Optimization Framework with Unreliable Bag Labels

To handle unreliable bag labels from identified negative bags, we propose to build a Positive Margin Pool (PMP) which contains "most positive patterns" from positive bags and "least negative patterns" from identified negative bags. Our multi-instance learning is then achieved by building an optimization framework based on the PMP, under the condition that negative bags are unreliable and therefore need to be carefully handled. In Eq. (1), we propose our objective function in handling unreliable bag labels.

$$\begin{aligned} &\min_{\boldsymbol{\omega},\mathbf{x}^B} \frac{1}{2}||\boldsymbol{\omega}||^2 + C\sum\nolimits_{i=1}^{m} \ell(\mathbf{x}_i^B, w_i y_i; \boldsymbol{\omega}) \\ &s.t.\ \ \ell(\mathbf{x}_i^B, w_i y_i; \boldsymbol{\omega}) = max(0, 1 - w_i y_i \boldsymbol{\omega}^\top \mathbf{x}_i^B) \end{aligned} \tag{1}$$

where $\mathbf{x}_i^B$ denotes a "most positive pattern" or a "least negative pattern" in PMP with m denoting the number of instances in PMP. w_i denotes weight confidence of the ith instance in PMP. In our design, because only one instance is selected from each bag, the weight confidence of the instance is equal to the weight confidence value of the bag. More specifically, if the instance is from a positive bag, its weight confidence $w_i = 1$. For instances from identified negative bags, their weight values are $w_j \in (0, 1]$, with a higher w_j value indicating that B_j is more likely being a negative bag.

The cost function in Eq. (1) is the linear SVM in the primal formulation, where $\ell(\mathbf{x}_i^B, w_i y_i; \boldsymbol{\omega})$ is the hinge loss function used by the SVM classifier. The above primal form can be formulated into a dual version as

$$\begin{aligned} &\max_{\alpha,\mathbf{x}^B} \sum\nolimits_i \alpha_i - \frac{1}{2}\sum\nolimits_{i,j} \alpha_i w_i y_i (\mathbf{x}_i^B)^\top \mathbf{x}_j^B \alpha_j w_j y_j \\ &s.t.\ \ \sum\nolimits_i \alpha_i = 1\ and\ \alpha_i \geq 0 \end{aligned} \tag{2}$$

where α_i is the Lagrangian multiplier. Indeed, the above optimization problem defines a nonlinearly constrained nonconvex optimization task. It contains two sub-problems: (1) learning continuous variables α which is equivalent to finding hyper-plane $\boldsymbol{\omega}$, and (2) selection of instances $\mathbf{x}^B$ for PMP which is equivalent to finding optimal weight values $\mathbf{w}$. To the best of our knowledge, no direct solution exists to find its global optimal. In this paper, we derive an self-adaptive approach to find optimal bag weights for unlabeled bags, and then use generic SVM optimization to solve the above optimization problem.

In the following, we first explain the construction of the PMP, and then introduce detailed self-adaptive process for finding optimal bag weight values.

4.2 PMP: Positive Margin Pool

The main purpose of building a Positive Margin Pool is to identify some "most positive patterns" (for positive bags) and "least negative patterns" (for possible negative bags), so PMP can help build classifiers to differentiate positive *vs.* negative instances for multi-instance classification. This also provides solutions to tackle unreliable bag labels obtained from unlabeled bags.

The construction of the PMP is motivated by the margin principle, which states that samples close to the decision boundary play critical roles in improving the performance of the underlying classifier. In our puMIL setting, we assign a weight confidence value w_i to each bag, so it can help identify unlabeled bags which are most likely being negative as "reliable negative bags" and then extract most positive patterns from positive bags and least negative patterns from identified negative bags to form PMP.

According to MIL definition, a negative bag does not contain positive instances, and negative instances in negative bags can have very general distributions, we use a Weighted Kernel Density Estimator [8] (WKDE) to model the distributions of negative instances in reliable negative bags as follows.

$$p(\mathbf{x}|\mathcal{B}^-) = \frac{1}{\sum_i n_i^-} \sum_{i,j} K(w_x \mathbf{x}, w_j \mathbf{x}_{i,j}^-) \tag{3}$$

where w_x denotes weight of the bag to which the instance x belongs to. $\mathbf{x}_{i,j}^-$ denotes the jth instance of the ith reliable negative bag with its size to be n_i^-, and we use an isotropic Gaussian kernel K. Depending on whether the underlying bag is positive or is identified as negative, the "most positive pattern" or "least negative pattern" $\mathbf{x}_i^B$ in PMP can be obtained by Eq.(4), with $\mathcal{B}^-$ denoting the set of all identified reliable negative bags.

$$\mathbf{x}_i^B = \operatorname*{arg\,min}_{\mathbf{x}_{i,j} \in B_i, j=1,\cdots,n_i} p(\mathbf{x}_{i,j}|\mathcal{B}^-) \tag{4}$$

After that, the optimization problem in Eq. (2) can be solved through standard linear SVM. In other words, once weight values are fixed, a weighted SVM classifier can be trained from PMP for classification.

4.3 Self-Adaptive Bag Weight Optimization

In the proposed puMIL learning framework, a weight confidence value w_i is assigned to each bag to determine whether an unlabeled bag is likely being a reliable negative bag or not. In this section, we propose a self-adaptive process to search optimal bag weight values. Because the aim of puMIL is to propose a learning framework to maximize the performance measure (F-score), a good weight combination should corresponds to a high F-score, which trades off the precision P and the recall R: F-score$= 2 \times P \times R/(P+R)$. Accordingly, we drive a self-adaptive strategy to obtain optimal weights based on a clone selection theory in Artificial Immune Systems.

When pathogens invade the body, antibodies that are produced from B-cells will respond for the detection of a foreign antigen. This response process can be explained by clonal selection theory. More specifically, immune individuals with high affinity will gradually increase during the clone and mutation process. Meanwhile, some immune individuals will polarize into memory individuals, which will be evolved towards final optimal. In our solution, antigen is simulated as training bag set $\mathcal{B}$, while the antibody, presented by confidence weight vector $\hat{\mathbf{w}}$ will experience a form of clone and mutation. The evolving optimization process will help discover optimal bag weight values $\mathbf{w}^*$ which will lead to the best affinity (*i.e.* F-score). The details of the optimal weight search process are described as:

Algorithm 1. puMIL: Positive and Unlabeled Multi-Instance Learning

Input:
Antibody Population (weight vectors): $\hat{\mathcal{W}} = \{\hat{\mathbf{w}}_1, \cdots, \hat{\mathbf{w}}_L\}$;
Antigen Population (bags): $\mathcal{B} = \mathcal{B}^+ \cup \mathcal{B}^u$;
Clone Factor c; Threshold T; Maximum Evolution Iterations M;
Output:
The target class label y_s of a test bag B_s;
// **Training Phase:**
$\hat{\mathcal{W}} \leftarrow$ Initialize L weight vectors (Step (1) in Section 4.3).
$t \leftarrow 1$;
while $\{(t \leq M$ and $f[\hat{\mathbf{w}}_c^{t+1}] - f[\hat{\mathbf{w}}_c^t] \geq T)$ **OR** $(t=1)\}$ **do**
$(\mathcal{B}_i^-)^t \leftarrow$ Use $\hat{\mathbf{w}}_i^t$ to find unlabeled bags with the largest weight values to form reliable negative bag set (which has the same number of bags as $\mathcal{B}^+$).
$(\mathcal{D}_i^-)^t \leftarrow$ Generate a weighted distribution from the $(\mathcal{B}_i^-)^t$ via Eq. (3).
$PMP_i^t \leftarrow$ Apply $(\mathcal{D}_i^-)^t$ to $\mathcal{B}^+$ and $(\mathcal{B}_i^-)^t$ and obtain positive margin pool via Eq. (4).
$f[\hat{\mathbf{w}}_i^t] \leftarrow$ Apply PMP_i^t to $\mathcal{B}^u$ and calculate the affinity score for $\hat{\mathbf{w}}_i^t$ via Eq. (2).
$\hat{\mathbf{w}}_c^t \leftarrow$ Apply $f[\hat{\mathbf{w}}_i^t]$ to $\hat{\mathcal{W}}^t$ and find weight vector $\hat{\mathbf{w}}_c^t$ with the highest affinity score.
$\hat{\mathcal{W}}^{t+1} \leftarrow$ Apply $\hat{\mathbf{w}}_c^t$ to $\hat{\mathcal{W}}^t$ via Steps (2-4) in Section 4.3 and update weight population.
$\hat{\mathbf{w}}_c^{t+1} \leftarrow$ Apply $f[\hat{\mathbf{w}}_i^{t+1}]$ to $\hat{\mathcal{W}}^{t+1}$ and update the best weight vector.
$t \leftarrow t+1$.
end while
$\mathbf{w}_c^* \leftarrow \hat{\mathbf{w}}_c^t$ // The final optimal weight vector
// **Testing Phase:**
$\mathcal{H} \leftarrow$ Apply the global optimal $\mathbf{w}_c^*$ to all instances in the underlying PMP to build a weighted SVM classier $\mathcal{H}(\mathbf{x}^B) = \boldsymbol{\omega}^\top \mathbf{x}^B + b$ with $\boldsymbol{\omega} = \sum_i \alpha_i w_{c,i}^* y_i \mathbf{x}_i^B$.
$y_s \leftarrow$ Apply $\mathcal{H}$ to each instance in test bag B_s to predict its bag label.

1) **Initialization:** Given an MI set $\mathcal{B}$ with n bags, a set of weight vectors $\hat{\mathcal{W}} = \{\hat{\mathbf{w}}_1, \cdots, \hat{\mathbf{w}}_L\}$ are randomly generated, with each vector $\hat{\mathbf{w}}_i =< \hat{w}_{i,1}, \cdots, \hat{w}_{i,k}, \cdots, \hat{w}_{i,n} >$ where $\hat{w}_{i,k}$ represents a weight value of bag $B_k \in \mathcal{B}$. If B_k is a positive bag, $\hat{w}_{i,k}$ is set to 1. If B_k is an unlabeled bag, $\hat{w}_{i,k}$ is a random value within range $(0,1]$. It is worth noting that each weight vector $\hat{\mathbf{w}}_i$ contains initial weight value for all bags. Because there are L weight vectors, each bag will have L initial weight values.

2) **Antibody Clone:** For each weight vector $\hat{\mathbf{w}}_i^t$ in the tth generation, if we use their weight values as the bag weights as defined in Eq. (1), each $\hat{\mathbf{w}}_i^t$ will correspond to an SVM classifier with an affinity score (*i.e.* F-score) on the unlabeled bag set $\mathcal{B}^u$. This score allows us to assess the quality of weight vectors and use clone selection to find optimal weight vector. During antibody clone process, the individual $\hat{\mathbf{w}}_c^t \in \hat{\mathcal{W}}^t$ with the best affinity score will be selected as the memory antibody to be further cloned. To ensure a fixed population size in every generation, $\hat{\mathbf{w}}_c^t$ will be cloned under the clone factor c to replace weight vectors $\hat{\mathbf{w}}_j^t \in \hat{\mathcal{W}}^t$ with low affinity under the same rate c.

3) **Antibody Mutation:** In order to maintain the diversity of the weight vectors, the mutation operation should be applied. Specifically, for any $\hat{\mathbf{w}}_i^t$ from the tth generation, the new variation individual $\mathbf{v}_i^{t+1}$ can be generated as follows:

$$\mathbf{v}_i^{t+1} = \hat{\mathbf{w}}_i^t + F * N(0,1) * (\hat{\mathbf{w}}_c^t - \hat{\mathbf{w}}_i^t) \tag{5}$$

Among them, N(0,1) is a normally distributed random variable within [0,1]. $F = 1 - f[\hat{\mathbf{w}}_i^t]$, as the variation factor, can be adaptively obtained according to different clones [9] where $f[\hat{\mathbf{w}}_i^t]$ denotes the affinity of $\hat{\mathbf{w}}_i^t$.

4) **Antibody Update:** This process determines whether the variation of weight vectors (from above steps (2) and (3)) can be used to replace a target weight vector in the next generation. In our algorithm, we adopt a greedy search strategy. Only if the affinity of $\mathbf{v}_i^{t+1}$ is better than that of the target weight vector $\hat{\mathbf{w}}_i^t$, the new weight vector is then selected as the offspring.

4.4 puMIL Framework

Algorithm 1 reports the detailed process of the proposed puMIL framework, which combines the (1) PMP construction (Section 4.2); and (2) adaptive weight optimization (Section 4.3), to iteratively refine unlabeled bags for learning.

During the initial process, puMIL will initialize the bag weight vectors $\hat{\mathcal{W}}$ (line 1). During each *while* loop, puMIL will first select the "reliable negative bags" $(\mathcal{B}_i^-)^t$ (this set has the same size as the number of positive bags $\mathcal{B}^+$) for each weight vector $\hat{\mathbf{w}}_i^t$ in $\hat{\mathcal{W}}^t$. So $(\mathcal{B}_i^-)^t$ consists of unlabeled bags with the largest weight values in $\hat{\mathbf{w}}_i^t$ (line 4). After that, puMIL will form a positive margin pool for each weight vector (lines 5-6). The clone selection theory will be employed to update the weight vectors in order to find a better weight value in the next iteration (lines 7-10). The evolutionary process will repeat until (1) the algorithm surpasses the pre-set maximum number M, or (2) the same result is obtained for a number (*e.g.* T) of consecutive iterations.

After obtaining the best weight vector $\mathbf{w}_c^*$, puMIL uses the discovered optimal weight values to obtain PMP, and then builds a weighted SVM classifier $\mathcal{H}$. A test bag is classified as positive if one or multiple instances inside the bag are classified as positive, and negative otherwise.

5 Experiments

To evaluate the effectiveness of our puMIL framework, we use F-score as the evaluation metric (its definition is given in Section 4.3). For benchmark data sets used in the experiments, 30% of bags are randomly selected as testing set in each run, with the remaining bags being used as training set. All results are based on the average performance over 10 repetitions. Besides, the two parameters M and T in Algorithm 1 are set to 50 and 0.001, respectively.

Because there is no existing method for positive and unlabeled multi-instance learning, for comparison purposes we implement following baseline approaches from bag- and instance-level perspectives. The former directly employs Positive-Unlabeled learning [5] at the bag level, and the latter propagates bag label to instances and transfer MIL as a generic Positive-Unlabeled learning problem.

Bag-Level Approaches: **(a) MILR+MISVM-MI:** This method first labels all unlabeled bags as negative bags, and then uses MI learning approach MILR [10], which outputs probability estimation, to obtain a set of identified negative bags. After that, it runs MISVM [6] iteratively on the positive set and refined negative set until converges; and **(b) Spy+MISVM-MI:** The difference between Spy based PU learning [5] on bag-level and MILR+MISVM-MI is in its initialization.

Spy+MISVM-MI randomly samples a set of positive bags as "spies", and marks them as unlabeled bags. Because spies are genuinely positive and behave similarly to unknown positive samples the unlabeled set, adding spies allows algorithm to infer characteristics of unknown positive bags in the unlabeled set. Previous studies [5] on text documents have demonstrated good performance of this PU learning strategy.

Instance-Level Approaches: **(a) NB+SVM-MI**. The variation of traditional PU setting in [5,11] is used for comparison. Specifically, a Naive Bayes classifier [12] is used to obtain identified negative instance set from unlabeled instance set. After that, an iterative process is used to train SVM classifier from positive instances and identified negative instances; and **(b) Spy+SVM-MI** Similar to the approaches used in [5,11], during the initialization process, Spy+SVM-MI randomly selects a set of positive instances as "spies", and adds them into the unlabeled set. After that it follows the same procedure as NB+SVM-MI.

For ease of understanding, we also refer to MILR+MISVM-MI and NB+SVM-MI as "no-spy" based approaches, and Spy+MISVM-MI and Spy+SVM-MI are called "spy" based approaches.

5.1 Drug Activity Prediction

The objective of drug activity prediction is to predict whether a drug molecule can bind well to a target protein related to certain disease states, which are primarily determined by the shape of the molecule. Our drug activity prediction data set, MUSK1 [1], is a benchmark used for MI learning. It contains 476 instances grouped into 92 bags (45 inactive and 47 active), with each instance being described by a 166-dimensional feature vector. In our experiments, we use $r \times 100\%$ active bags as positive bags, and the remaining active bags and all inactive bags are treated as unlabeled bags. The results for the two types of baselines (bag-level and instance-level) and proposed puMIL with different r values, varying from 0.1 to 0.7, are reported in Figures 2(A) and 2(B), respectively.

Overall, "no-spy" based approaches achieve competitive F-score as "spy" based methods, which demonstrates that simply adding "spy" to unlabeled bags may not help differentiate positive *vs.* negative bags. Meanwhile, the proposed puMIL achieves better performances compared to other methods, especially for bag-level baselines. For small r values, such as $r = 0.2$ or less, puMIL demonstrates very significant performance gain, compared to other baselines. This further assets that when the number of positive bags is very limited, leveraging useful information from unlabeled bags, like puMIL does, can be very useful for positive and unlabeled multi-instance learning.

5.2 Scientific Publication Retrieval

The DBLP data set consists of bibliography data in computer science[1]. Each record in DBLP contains a number of attributes such as abstract, authors, title, and references [13]. To build a puMIL task, we select papers published in

[1] `http://dblp.uni-trier.de/xml/`

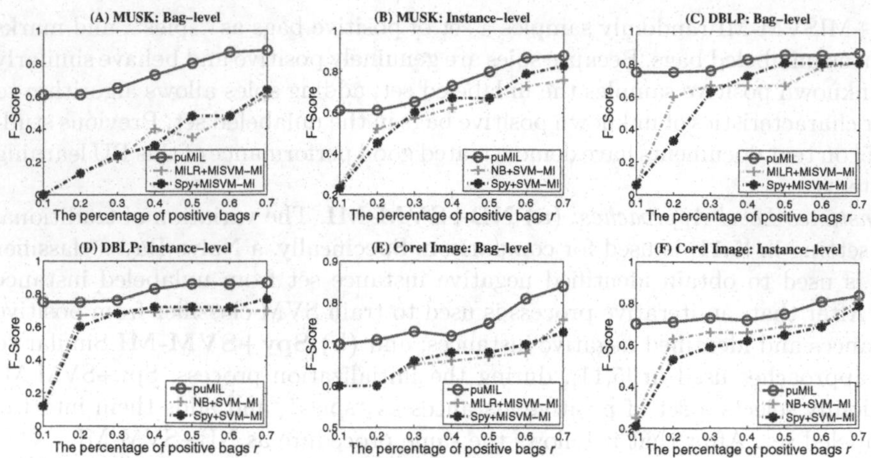

Fig. 2. F-score comparisons on **MUSK, DBLP, and Corel Image data sets** with respect to different r values. Where (A) and (B) represent MUSK data, (C) and (D) represent DBLP data, and (E) and (F) represent Corel Image data.

Artificial Intelligence field (AI: IJCAI, AAAI ,NIPS, UAI, COLT, ACL, KR, ICML, ECML and IJCNN) as positive bags and randomly select papers from other fields, such as computer vision, multimedia, pattern recognition as unlabeled bags. A "bag-of-words" representation based on TFIDF [14] is adopted to convert the abstract of a paper to an instance. So each paper is a bag and each instance inside the bag denotes either the paper's abstract or the abstract of a reference cited in the paper.

In our experiments, we choose 200 papers in total (which correspond to 200 bags), with each paper containing 1 to 10 references. For all 200 papers, the total number of references cited (*i.e.* instances) is 1136. Each instance is described by a 4497-dimensional feature vector. To vary the number of positive bags, we randomly select $r \times 100\%$ of AI bags (varying from 0.1 to 0.7) as positive bag set, and the remaining AI bags and all other bags are used as unlabeled bags. In Figures 2(C) and 2(D), we report the results with respect to different r values. The results show that "spy" based approaches can achieve a slightly better performance than 'no-spy" versions when r is greater than 0.4. This is possibly because that when a large number of positive bags are used, the "spies" selected from the positive bags will not reduce the role of positive bags. Meanwhile, puMIL clearly outperforms all baselines, especially when only a small portion of positive bags are labeled (*i.e.* r=0.1). This suggests that puMIL is effective over a wide range of percentage of labeled positive bags.

5.3 Region-Based Image Annotation

In the third experiment, we report puMIL's performance for automatic image annotation tasks. The original data are color images from the Corel data set [15] that have been preprocessed and segmented using Blobworld system [16].

Fig. 3. Example images from the Corel image database used in the experiment. The first and the second rows show positive and negative objects, respectively. Each image is considered as a bag with each region denoting an instance.

An image contains a set of segments (or blobs), each characterized by color, texture, and shape descriptors. In this case, each image is considered as a bag, and each region inside the image denotes an instance. Some example images from the database are shown in Figure 3. In our experiment, we use category "tiger" as positive bags (100 bags with 544 instances) and randomly draw 100 photos of other animals to form unlabeled bags with 676 instances. Each instance is described by a 230-dimensional feature vector which represent color, texture, and shape of the region. To validate the performance of puMIL with respect to different number of positive bags, we randomly select $r \times 100\%$ "tiger" images (varying from 0.1 to 0.7) as positive bags, and combine remaining "tiger" images and all other images as unlabeled bag set.

In Figures 2(E) and 2(F), we compare the performance of puMIL with two types of baselines by using different portions of positive bags. The results show that bag-level baselines can have a high F-score than instance-level methods when the percentage of positive bag is very small (*e.g.* $r < 0.2$). Because instance-level approaches directly assign bag labels to all instances in positive bags, for a small number of positive bags, the mislabeled instances in positive bags will have severe impact on the classifiers trained from labeled data. As a result, the classifier may not be able to accurately differentiate positive vs. negative instances. However, as the r values become sufficiently large, instance-level methods demonstrate better performance than bag-level approaches. By properly utilizing information in unlabeled bags, puMIL consistently outperforms all baselines for different percentages of positive bags.

6 Conclusions

In this paper, we formulated a unique multi-instance learning task, which only has positive and unlabeled bags available for multi-instance learning. This problem setting is more general but significantly more challenging than traditional multi-instance learning because no negative bags exist for deriving discriminative classification models. To address the challenge, we proposed a puMIL framework which self-adaptively selects some reliable negative bags (from unlabeled bags) and further selects some representative patterns from the positive bags and identified negative bags to help train SVM classifiers. The classifiers will further help refine the selection of negative bags and iteratively lead to updated classifiers for classification. Our main technical contribution, compared to existing research, is threefold: (1) a general framework to handle multi-instance learning with only positive and unlabeled bags; (2) an effective algorithm to identify reliable negative bags from unlabeled bags; and (3) an effective approach for utilizing unreliable labels derived for unlabeled bags.

Acknowledgments. The work was supported by the Key Project of the Natural Science Foundation of Hubei Province, China (Grant No. 2013CFA004), and the National Scholarship for Building High Level Universities, China Scholarship Council (No. 201206410056).

References

1. Dietterich, T., Lathrop, R., Lozano-Pérez, T.: Solving the multiple instance problem with axis-parallel rectangles. Artif. Intell. 89, 31–71 (1997)
2. Zhou, Z., Zhang, M., Huang, S., Li, Y.: Multi-instance multi-label learning. Artificial Intelligence 176, 2291–2320 (2012)
3. Qi, X., Han, Y.: Incorporating multiple SVMs for automatic image annotation. Pattern Recogn. 40, 728–741 (2007)
4. Zhang, B., Zuo, W.: Learning from positive and unlabeled examples: A survey. In: ISIP, pp. 650–654 (2008)
5. Liu, B., Dai, Y., Li, X., Lee, W.S., Yu, P.S.: Building text classifiers using positive and unlabeled examples. In: ICDM, Washington, DC, USA, p. 179 (2003)
6. Andrews, S., Tsochantaridis, I., Hofmann, T.: Support vector machines for multiple-instance learning. In: NIPS, pp. 561–568 (2003)
7. Shang, R., Jiao, L., Liu, F., Ma, W.: A novel immune clonal algorithm for mo problems. IEEE Trans. Evol. Comput. 16(1), 35–50 (2012)
8. Fu, Z., Robles-Kelly, A., Zhou, J.: Milis: Multiple instance learning with instance selection. IEEE Trans. Pattern Anal. Mach. Intell. 33(5), 958–977 (2011)
9. Zhong, Y., Zhang, L.: An adaptive artificial immune network for supervised classification of multi-/hyperspectral remote sensing imagery. IEEE Trans. Geosci. Remote Sens. 50(3), 894–909 (2012)
10. Ray, S., Craven, M.: Supervised versus multiple instance learning: an empirical comparison. In: ICML, New York, NY, USA, pp. 697–704 (2005)
11. Zhao, Y., Kong, X., Yu, P.S.: Positive and unlabeled learning for graph classification. In: ICDM, pp. 962–971 (2011)
12. Friedman, N., Geiger, D., Goldszmidt, M.: Bayesian network classifiers. Mach. Learn. 29(2-3), 131–163 (1997)
13. Tang, J., Zhang, J., Yao, L., Li, J., Zhang, L., Su, Z.: Arnetminer: extraction and mining of academic social networks. In: KDD, pp. 990–998 (2008)
14. He, J., Gu, H., Wang, Z.: Bayesian multi-instance multi-label learning using gaussian process prior. Mach. Learn. 88(1-2), 273–295 (2012)
15. Li, J., Wang, J.Z.: Real-time computerized annotation of pictures. IEEE Trans. Pattern Anal. Mach. Intell. 30(6), 985–1002 (2008)
16. Carson, C., Thomas, M., Belongie, S., Hellerstein, J.M., Malik, J.: Blobworld: A system for region-based image indexing and retrieval. In: Huijsmans, D.P., Smeulders, A.W.M. (eds.) VISUAL 1999. LNCS, vol. 1614, pp. 509–517. Springer, Heidelberg (1999)

Mining Contrast Subspaces*

Lei Duan[1,6], Guanting Tang[2], Jian Pei[2], James Bailey[3], Guozhu Dong[4], Akiko Campbell[5], and Changjie Tang[1]

[1] School of Computer Science, Sichuan University, China
[2] School of Computing Science, Simon Fraser University, Canada
[3] Dept. of Computing and Information Systems, University of Melbourne, Australia
[4] Dept. of Computer Sci & Engr, Wright State University, USA
[5] Pacific Blue Cross, Canada
[6] State Key Laboratory of Software Engineering, Wuhan University, China
{leiduan,cjtang}@scu.edu.cn, {gta9,jpei}@cs.sfu.ca,
baileyj@unimelb.edu.au, guozhu.dong@wright.edu,
acampbell@pac.bluecross.ca

Abstract. In this paper, we tackle a novel problem of mining contrast subspaces. Given a set of multidimensional objects in two classes C_+ and C_- and a query object o, we want to find top-k subspaces S that maximize the ratio of likelihood of o in C_+ against that in C_-. We demonstrate that this problem has important applications, and at the same time, is very challenging. It even does not allow polynomial time approximation. We present CSMiner, a mining method with various pruning techniques. CSMiner is substantially faster than the baseline method. Our experimental results on real data sets verify the effectiveness and efficiency of our method.

Keywords: contrast subspace, kernel density estimation, likelihood contrast.

1 Introduction

Imagine you are a medical doctor facing a patient having symptoms of being overweight, short of breath, and some others. You want to check the patient on two specific possible diseases: coronary artery disease and adiposity. Please note that clogged arteries are among the top-5 most commonly misdiagnosed diseases. You have a set of reference samples of both diseases. Then, you may naturally ask "In what aspect is this patient most similar to cases of coronary artery disease and, at the same time, dissimilar to adiposity?"

* This work was supported in part by an NSERC Discovery grant, a BCIC NRAS Team Project, NSFC 61103042, SRFDP 20100181120029, and SKLSE2012-09-32. Work by Lei Duan and Guozhu Dong at Simon Fraser University was supported by an Ebco/Eppich visiting professorship. All opinions, findings, conclusions and recommendations in this paper are those of the authors and do not necessarily reflect the views of the funding agencies.

V.S. Tseng et al. (Eds.): PAKDD 2014, Part I, LNAI 8443, pp. 249–260, 2014.

The above motivation scenario cannot be addressed well using existing data mining methods, and thus suggests a novel data mining problem. In a multidimensional data set of two classes, given a query object and a target class, we want to find the subspace where the query object is most likely to belong to the target class against the other class. We call such a subspace a *contrast subspace* since it contrasts the likelihood of the query object in the target class against the other class. Mining contrast subspaces is an interesting problem with many important applications. As another example, when an analyst in an insurance company is investigating a suspicious claim, she may want to compare the suspicious case against the samples of frauds and normal claims. A useful question to ask is in what aspects the suspicious case is most similar to fraudulent cases and different from normal claims. In other words, finding the contrast subspace for the suspicious claim is informative for the analyst.

While there are many existing studies on outlier detection and contrast mining, they focus on collective patterns that are shared by many cases of the target class. The contrast subspace mining problem addressed here is different. It focuses on one query object and finds the customized contrast subspace. This critical difference makes the problem formulation, the suitable applications, and the mining methods dramatically different. We will review the related work and explain the differences systematically in Section 2.

To tackle the problem of mining contrast subspaces, we need to address several technical issues. First, we need to have a simple yet informative contrast measure to quantify the similarity between the query object and the target class and the difference between the query object and the other class. In this paper, we use the ratio of the likelihood of the query object in the target class against that in the other class as the measure. This is essentially the Bayes factor on the query object, and comes with a well recognized explanation [1].

Second, the problem of mining contrast subspaces is computational challenging. We show that the problem is MAX SNP-hard, and thus does not allow polynomial time approximation methods unless P=NP. Therefore, the only hope is to develop heuristics that may work well in practice.

Third, one could use a brute-force method to tackle the contrast mining problem, which enumerates every non-empty subspace and computes the contrast measure. This method, however, is very costly on data sets with a non-trivial dimensionality. One major obstacle preventing effective pruning is that the contrast measure does not have any monotonicity with respect to the subspace-superspace relationship. To tackle the problem, we develop pruning techniques based on bounds of likelihood and contrast ratio. Our experimental results on real data sets clearly verify the effectiveness and efficiency of our method.

The rest of the paper is organized as follows. We review the related work in Section 2. In Section 3, we formalize the problem, and analyze it theoretically. We present a heuristic method in Section 4, and evaluate our method empirically using real data sets in Section 5. We conclude the paper in Section 6.

2 Related Work

Our study is related to the existing work on contrast mining, subspace outlier detection and typicality queries. We review the related work briefly here.

Contrast mining discovers patterns and models that manifest drastic differences between datasets. Dong and Bailey [2] presented a comprehensive review. The most renowned contrast patterns include emerging patterns [3], contrast sets [4] and subgroups [5]. Although their definitions vary, the mining methods share heavy similarity [6].

Contrast pattern mining identifies patterns by considering all objects of all classes in the complete pattern space. Orthogonally, contrast subspace mining focuses on one object, and identifies subspaces where a query object demonstrates the strongest overall similarity to one class against the other. These two mining problems are fundamentally different. To the best of our knowledge, the contrast subspace mining problem has not been systematically explored in the data mining literature.

Subspace outlier detection discovers objects that significantly deviate from the majority in some subspaces. It is very different from our study. In contrast subspace mining, the query object may or may not be an outlier. Some recent studies find subspaces that may contain substantial outliers. Böhm *et al.* [7] and Keller *et al.* [8] proposed statistical approaches *CMI* and *HiCS* to select subspaces for a multidimensional database, where there may exist outliers with high deviations. Both *CMI* and *HiCS* are fundamentally different from our method. Technically, they choose subspaces for all outliers in a given database, while our method chooses the most contrasting subspaces for a query object.

Our method uses probability density to estimate the likelihood of a query object belonging to different classes. There are a few density-based outlier detection methods, such as [9–12]. Our method is inherently different from those, since we do not target at outlier objects at all.

Hua *et al.* [13] introduced a novel top-k *typicality query*, which ranks objects according to their typicality in a data set or a class of objects. Although both [13] and our work use density estimation methods to calculate the typicality/likelihood of a query object with respect to a set of data objects, typicality queries [13] do not consider subspaces at all.

3 Problem Formulation and Analysis

In this section, we first formulate the problem. Then, we recall the basics of kernel density estimation, which can estimate the probability density of objects. Last, we investigate the complexity of the problem.

3.1 Problem Definition

Let $D = \{D_1, \ldots, D_d\}$ be a d-dimensional space, where the domain of D_i is $\mathbb{R}$, the set of real numbers. A *subspace* $S \subseteq D$ ($S \neq \emptyset$) is a subset of D. We also call D the *full space*.

Consider an object o in space D. We denote by $o.D_i$ the value of o in dimension D_i ($1 \leq i \leq d$). For a subspace $S = \{D_{i_1}, \ldots, D_{i_l}\} \subseteq D$, the *projection* of o in S is $o^S = (o.D_{i_1}, \ldots, o.D_{i_l})$. For a set of objects $O = \{o_j \mid 1 \leq j \leq n\}$, the *projection* of O in S is $O^S = \{o_j^S \mid o_j \in O, 1 \leq j \leq n\}$.

Given a set of objects O, we assume a latent distribution $\mathcal{Z}$ that generates the objects in O. For a query object q, denote by $L_D(q \mid \mathcal{Z})$ the likelihood of q being generated by $\mathcal{Z}$ in full space D. The posterior probability of q given O, denoted by $L_D(q \mid O)$, can be estimated by $L_D(q \mid \mathcal{Z})$. For a non-empty subspace S ($S \subseteq D$, $S \neq \emptyset$), denote by $\mathcal{Z}^S$ the projection of $\mathcal{Z}$ in S. The *subspace likelihood* of object q with respect to $\mathcal{Z}$ in S, denoted by $L_S(q \mid \mathcal{Z})$, can be estimated by the posterior probability of object q given O in S, denoted by $L_S(q \mid O)$.

In this paper, we assume that the objects in O belong to two classes, C_+ and C_-, exclusively. Thus, $O = O_+ \cup O_-$ and $O_+ \cap O_- = \emptyset$, where O_+ and O_- are the subsets of objects belonging to C_+ and C_-, respectively. Given a query object q, we are interested in how likely q belongs to C_+ and does not belong to C_-. To measure these two factors comprehensively, we define the *likelihood contrast* as $LC(q) = \frac{L(q|O_+)}{L(q|O_-)}$.

Likelihood contrast is essentially the Bayes factor[1] of object q as the observation. In other words, we can regard O_+ and O_- as representing two models, and we need to choose one of them based on query object q. Consequently, the ratio of likelihoods indicates the plausibility of model represented by O_+ against that by O_-. Jeffreys [1] gave a scale for interpretation of Bayes factor. When $LC(q)$ is in the ranges of < 1, 1 to 3, 3 to 10, 10 to 30, 30 to 100, and over 100, respectively, the strength of the evidence is negative, barely worth mentioning, substantial, strong, very strong, and decisive.

We can extend likelihood contrast to subspaces. For a non-empty subspace $S \subseteq D$, we define the likelihood contrast in the subspace as $LC_S(q) = \frac{L_S(q|O_+)}{L_S(q|O_-)}$. To avoid triviality in subspaces where $L_S(q \mid O_+)$ is very small, we introduce a minimum likelihood threshold $\delta > 0$, and consider only the subspaces S where $L_S(q \mid O_+) \geq \delta$.

Given a multidimensional data set O in full space D, a query object q, and a minimum likelihood threshold $\delta > 0$, and a parameter $k > 0$, the *problem of mining contrast subspaces* is to find the top-k subspaces S ordered by the subspace likelihood contrast $LC_S(q)$ subject to $L_S(q \mid O_+) \geq \delta$.

3.2 Kernel Density Estimation

We can use kernel density estimation [14] to estimate likelihood $L_S(q \mid O)$. In this paper, we adopt the Gaussian kernel, which is natural and widely used in density estimation. Given a set of objects O, the density of a query object q in subspace S, denoted by $\hat{f}_S(q, O)$, can be estimated as

$$\hat{f}_S(q, O) = \hat{f}_S(q^S, O) = \frac{1}{|O|\sqrt{2\pi}h_S} \sum_{o \in O} e^{\frac{-dist_S(q,o)^2}{2h_S^2}}$$

[1] Generally, given a set of observations Q, the plausibility of two models M_1 and M_2 can be assessed by the Bayes factor $K = \frac{Pr(Q|M_1)}{Pr(Q|M_2)}$.

where $dist_S(q,o)^2 = \sum_{D_i \in S} (q.D_i - o.D_i)^2$ and h_S is a bandwidth parameter. Silverman [15] suggested that the optimal bandwidth value for smoothing normally distributed data with unit variance is $h_{S\_opt} = A(K)|O|^{-1/(|S|+4)}$, where $A(K) = \{4/(|S|+2)\}^{1/(|S|+4)}$.

As the kernel is radially symmetric and the data is not normalized in subspaces, we can use a single scale parameter σ_S in subspace S and set $h_S = \sigma_S \cdot h_{S\_opt}$. As Silverman suggested [15], a possible choice of σ_S is the root of the average marginal variance in S.

Using kernel density estimation, we can estimate $L_S(q \mid O)$ as

$$L_S(q \mid O) = \hat{f}_S(q, O) = \frac{1}{|O|\sqrt{2\pi}h_S} \sum_{o \in O} e^{\frac{-dist_S(q,o)^2}{2h_S^2}} \tag{1}$$

Correspondingly, the likelihood contrast of object q in subspace S is given by

$$LC_S(q, O_+, O_-) = \frac{\hat{f}_S(q, O_+)}{\hat{f}_S(q, O_-)} = \frac{|O_-|h_{S_-}}{|O_+|h_{S_+}} \cdot \frac{\sum_{o \in O_+} e^{\frac{-dist_S(q,o)^2}{2h_{S_+}^2}}}{\sum_{o \in O_-} e^{\frac{-dist_S(q,o)^2}{2h_{S_-}^2}}} \tag{2}$$

We often omit O_+ and O_- and write $LC_S(q)$ if O_+ and O_- are clear from context.

3.3 Complexity Analysis

We have the following theoretical result. It can be proved by a reduction from the emerging pattern mining problem [3], which is MAX SNP-hard [16]. Limited by space, we omit the details here.

Theorem 1 (Complexity). *The problem of mining contrast subspaces is MAX SNP-hard.*

The above theoretical result indicates that the problem of mining contrast subspaces is even hard to approximate – it is impossible to design a good approximation algorithm. In the rest of the paper, we turn to practical heuristic methods.

4 Mining Methods

In this section, we first describe a baseline method that examines every possible non-empty subspace. Then, we present a bounding-pruning-refining method that expedites the search substantially.

4.1 A Baseline Method

A baseline method enumerates all possible non-empty spaces S and calculates the exact values of both $L_S(q \mid O_+)$ and $L_S(q \mid O_-)$. Then, it returns the top-k subspaces S with the largest $LC_S(q)$ values. To ensure the completeness and efficiency of subspace enumeration, the baseline method traverses the set enumeration tree [17] of subspaces in a depth-first manner.

$L_S(q \mid O_+)$ is not monotonic in subspaces. To prune subspaces using the minimum likelihood threshold δ, we develop an upper bound of $L_S(q \mid O_+)$. We sort all the dimensions in their standard deviation descending order. Let $\mathcal{S}$ be the set of children of S in the subspace set enumeration tree using the standard deviation descending order. Define $L^*_S(q \mid O_+) = \frac{1}{|O_+|\sqrt{2\pi}\sigma'_{min} h'_{opt\_min}} \sum_{o \in O_+} e^{\frac{-dist_S(q,o)^2}{2(\sigma_S h'_{opt\_max})^2}}$, where $\sigma'_{min} = min\{\sigma_{S'} \mid S' \in \mathcal{S}\}$, $h'_{opt\_min} = min\{h_{S'\_opt} \mid S' \in \mathcal{S}\}$, and $h'_{opt\_max} = max\{h_{S'\_opt} \mid S' \in \mathcal{S}\}$. We have the following result.

Theorem 2 (Monotonic density bound). *For a query object q, a set of objects O, and subspaces S_1, S_2 such that S_1 is an ancestor of S_2 in the subspace set enumeration tree using the standard deviation descending order in O_+, $L^*_{S_1}(q \mid O_+) \geq L_{S_2}(q \mid O_+)$.*

Using Theorem 2, in addition to $L_S(q \mid O_+)$ and $L_S(q \mid O_-)$, we also compute $L^*_S(q \mid O_+)$ for each subspace S. Once $L^*_S(q \mid O_+) < \delta$ in a subspace S, all super-spaces of S can be pruned.

Using Equations 1 and 2, the baseline algorithm computes the likelihood contrast for every subspace where $L_S(q \mid O_+) \geq \delta$, and returns the top-$k$ subspaces. The time complexity is $O(2^{|D|} \cdot (|O_+| + |O_-|))$.

4.2 A Bounding-Pruning-Refining Method

For a query object q and a set of objects O, the *ϵ-neighborhood* ($\epsilon > 0$) of q in subspace S is $N^\epsilon_S(q) = \{o \in O \mid dist_S(q,o) \leq \epsilon\}$. We can divide $L_S(q \mid O)$ into two parts, that is, $L_S(q \mid O) = L_{N^\epsilon_S}(q \mid O) + L^{rest}_S(q \mid O)$. The first part is contributed by the objects in the ϵ-neighborhood, that is, $L_{N^\epsilon_S}(q \mid O) = \frac{1}{|O|\sqrt{2\pi}h_S} \sum_{o \in N^\epsilon_S(q)} e^{\frac{-dist_S(q,o)^2}{2h_S^2}}$, and the second part is by the objects outside the ϵ-neighborhood, that is, $L^{rest}_S(q \mid O) = \frac{1}{|O|\sqrt{2\pi}h_S} \sum_{o \in O \setminus N^\epsilon_S(q)} e^{\frac{-dist_S(q,o)^2}{2h_S^2}}$.

Let $\overline{dist_S}(q \mid O)$ be the maximum distance between q and all objects in O in subspace S. We have,

$$\frac{|O| - |N^\epsilon_S(q)|}{|O|\sqrt{2\pi}h_S} \cdot e^{-\frac{\overline{dist_S}(q,O)^2}{2h_S^2}} \leq L^{rest}_S(q \mid O) \leq \frac{|O| - |N^\epsilon_S(q)|}{|O|\sqrt{2\pi}h_S} \cdot e^{-\frac{\epsilon^2}{2h_S^2}}$$

Using the above, we have the following upper and lower bounds of $L_S(q \mid O)$ using ϵ-neighborhood.

Theorem 3 (Bounds). *For a query object q, a set of objects O and $\epsilon \geq 0$,*

$$LL_S^\epsilon(q \mid O) \leq L_S(q \mid O) \leq UL_S^\epsilon(q \mid O)$$

where

$$LL_S^\epsilon(q \mid O) = \frac{1}{|O|\sqrt{2\pi}h_S}\left(\sum_{o \in N_S^\epsilon(q)} e^{\frac{-dist_S^\epsilon(q,o)^2}{2h_S^2}} + (|O| - |N_S^\epsilon(q)|)e^{-\frac{\overline{dist_S(q,O)}^2}{2h_S^2}}\right)$$

and

$$UL_S^\epsilon(q \mid O) = \frac{1}{|O|\sqrt{2\pi}h_S}\left(\sum_{o \in N_S^\epsilon(q)} e^{\frac{-dist_S^\epsilon(q,o)^2}{2h_S^2}} + (|O| - |N_S^\epsilon(q)|)e^{-\frac{\epsilon^2}{2h_S^2}}\right)$$

We obtain an upper bound of $LC_S(q)$ based on Theorem 3 and Equation 2.

Corollary 1 (Likelihood Contrast Upper Bound). *For a query object q, a set of objects O_+, a set of objects O_-, and $\epsilon \geq 0$, $LC_S(q) \leq \frac{UL_S^\epsilon(q|O_+)}{LL_S^\epsilon(q|O_-)}$.*

Using Corollary 1, for a subspace S, if there are at least k subspaces whose likelihood contrast are greater than $\frac{UL_\epsilon^S(q|O_+)}{LL_\epsilon^S(q|O_-)}$, then S cannot be a top-k subspaces of the largest likelihood contrast.

Using the ϵ-neighborhood, $L_S^*(q \mid O_+)$ is computed by

$$L_S^*(q \mid O_+) = \frac{\sum_{o \in N_S^\epsilon(q)} e^{\frac{-dist_S^\epsilon(q,o)^2}{2(\sigma_S h'_{opt\_max})^2}} + (|O_+| - |N_S^\epsilon(q)|)e^{-\frac{\epsilon^2}{2(\sigma_S h'_{opt\_max})^2}}}{|O_+|\sqrt{2\pi}\sigma'_{min}h'_{opt\_min}}$$

Our bounding-pruning-refining method, CSMiner (for Contrast Subspace Miner), conducts a depth-first search on the subspace set enumeration tree. For a candidate subspace S, CSMiner calculates $UL_S^\epsilon(q \mid O_+)$ and $LL_S(q \mid O_-)$ using the ϵ-neighborhood. If $UL_S^\epsilon(q \mid O_+)$ is less than the minimum likelihood threshold, S cannot be a contrast subspace. Otherwise, CSMiner checks whether the likelihood contrasts of the current top-k subspaces are larger than the upper bound of $LC_S(q)$. If not, CSMiner refines $L_S(q \mid O_+)$ and $L_S(q \mid O_-)$ by involving objects that are out of the ϵ-neighborhood. S will be added into the current top-k list if its likelihood contrast is larger than one of the current top-k ones. Algorithm 1 gives the pseudo-code of CSMiner. Due to the hardness of the problem shown in Theorem 1 and the heuristic nature of this method, the time complexity of CSMiner is $O(2^{|D|} \cdot (|O_+| + |O_-|))$, the same as the exhaustive baseline method. However, as shown by our empirical study, CSMiner is substantially faster than the baseline method.

Computing ϵ-neighborhood is critical in CSMiner. The distance between objects increases when dimensionality increases. Thus, the value of ϵ should not be

Algorithm 1. $CSMiner(q, O_+, O_-, \delta, k)$

Input: q: a query object, O_+: the set of objects belonging to C_+, O_-: the set of objects belonging to C_-, δ: a likelihood threshold, k: positive integer
Output: k subspaces with the highest likelihood contrast

let Ans be the current top-k list of subspaces, initialize Ans as k null subspaces associated with likelihood contrast 0
for each subspace S in the subspace set enumeration tree, searched in the depth-first manner **do**
    if $UL^{\epsilon}_S(q \mid O_+) \geq \delta$ and $\exists S' \in Ans$ s.t. $\frac{UL^{\epsilon}_S(q|O_+)}{LL^{\epsilon}_S(q|O_-)} > LC_{S'}(q)$ **then**
        calculate $L_S(q \mid O_+)$, $L_S(q \mid O_-)$ and $LC_S(q)$; // refining
        if $L_S(q \mid O_+) \geq \delta$ and $\exists S' \in Ans$ s.t. $LC_S(q) > LC_{S'}(q)$ **then**
            insert S into the top-k list
        end if
    end if
    if $L^*_S(q \mid O_+) < \delta$ **then**
        prune all super-spaces of S;
    end if
end for
return Ans;

Table 1. Data set characteristics

Data set	# objects	# attributes
Breast Cancer Wisconsin (BCW)	683	9
Climate Model Simulation Crashes (CMSC)	540	18
Glass Identification (Glass)	214	9
Pima Indians Diabetes (PID)	768	8
Waveform	5000	21
Wine	178	13

fixed. The standard deviation expresses the variability of a set of data. For subspace S, we set $\epsilon = \sqrt{r \cdot \sum_{D_i \in S} (\sigma^2_{D_i+} + \sigma^2_{D_i-})}$ $(r \geq 0)$, where $\sigma^2_{D_i+}$ and $\sigma^2_{D_i-}$ are the marginal variances of O_+ and O_-, respectively, on dimension D_i ($D_i \in S$), and r is a system defined parameter. Our experiments show that r can be set in the range of $0.3 \sim 0.6$, and is not sensitive.

5 Empirical Evaluation

In this section, we report a systematic empirical study using real data sets to verify the effectiveness and efficiency of our method. All experiments were conducted on a PC computer with an Intel Core i7-3770 3.40 GHz CPU, and 8 GB main memory, running Windows 7 operating system. All algorithms were implemented in Java and compiled by JDK 7.

5.1 Effectiveness

We use 6 real data sets from the UCI machine learning repository [18]. We remove non-numerical attributes and all instances containing missing values. Table 1 shows the data characteristics.

Table 2. Distribution of $LC_S(q)$ in BCW

		$LC_S^{out}(q)$					
		< 1	$[1,3)$	$[3,10)$	$[10,10^2)$	$\geq 10^2$	Total
$LC_S^{in}(q)$	$< 10^2$	0	0	0	2	21	23
	$[10^2,10^3)$	6	7	5	8	11	37
	$[10^3,10^4)$	176	37	18	15	18	264
	$[10^4,10^5)$	99	7	6	4	5	121
	$\geq 10^5$	38	25	87	82	6	238
Total		319	76	116	111	61	683

Table 3. Distribution of $LC_S(q)$ in Glass

		$LC_S^{out}(q)$					
		< 1	$[1,3)$	$[3,10)$	$[10,10^2)$	$\geq 10^2$	Total
$LC_S^{in}(q)$	< 10	0	4	0	2	4	10
	$[10,10^2)$	11	70	26	6	4	117
	$[10^2,10^3)$	2	24	5	3	2	36
	$[10^3,10^4)$	0	0	4	0	1	5
	$\geq 10^4$	0	23	14	6	3	46
Total		13	121	49	17	14	214

Table 4. Distribution of $LC_S(q)$ in PID

		$LC_S^{out}(q)$					
		< 1	$[1,3)$	$[3,10)$	$[10,10^2)$	$\geq 10^2$	Total
$LC_S^{in}(q)$	< 1	0	0	1	1	0	2
	$[1,3)$	0	124	99	19	2	244
	$[3,10)$	17	241	54	4	0	316
	$[10,10^2)$	28	146	19	4	0	197
	$\geq 10^2$	1	8	0	0	0	9
Total		46	519	173	28	2	768

Table 5. Distribution of $LC_S(q)$ in Wine

		$LC_S^{out}(q)$					
		< 1	$[1,3)$	$[3,10)$	$[10,10^2)$	$\geq 10^2$	Total
$LC_S^{in}(q)$	$< 10^3$	2	22	10	13	9	56
	$[10^3,10^4)$	0	17	11	6	2	36
	$[10^4,10^5)$	0	10	4	2	2	18
	$[10^5,10^6)$	0	5	5	2	0	12
	$\geq 10^6$	4	21	15	12	4	56
Total		6	75	45	35	17	178

Table 6. Distribution of $LC_S(q)$ in CMSC

		$LC_S^{out}(q)$					
		$[10,10^2)$	$[10^2,10^3)$	$[10^3,10^4)$	$[10^4,10^5)$	$\geq 10^5$	Total
$LC_S^{in}(q)$	$< 10^3$	2	6	41	15	0	64
	$[10^3,10^4)$	4	28	47	17	4	100
	$[10^4,10^5)$	7	38	44	17	7	113
	$[10^5,10^6)$	1	30	36	10	3	80
	$\geq 10^6$	4	82	75	16	6	183
Total		18	184	243	75	20	540

Table 7. Distribution of $LC_S(q)$ in Waveform

		$LC_S^{out}(q)$					
		$[1,3)$	$[3,10)$	$[10,10^2)$	$[10^2,10^3)$	$\geq 10^3$	Total
$LC_S^{in}(q)$	< 10	0	8	24	10	7	49
	$[10,10^2)$	88	462	695	222	98	1565
	$[10^2,10^3)$	235	686	956	299	104	2280
	$[10^3,10^4)$	151	346	383	71	23	974
	$\geq 10^4$	36	46	45	5	0	132
Total		510	1548	2103	607	232	5000

For each data set, we take each record as a query object q, and all records except q belonging to the same class as q forming the set O_1, and records belonging to the other classes forming the set O_2. Using CSMiner, we compute for each record (1) the *inlying contrast subspace* taking O_1 as O_+ and O_2 as O_-, and (2) the *outlying contrast subspace* taking O_2 as O_+ and O_1 as O_-. In this experiment, we only compute the top-1 subspace. For clarity, we denote the likelihood contrasts of inlying contrast subspace by $LC_S^{in}(q)$ and those of outlying contrast subspace by $LC_S^{out}(q)$. The minimum likelihood threshold is set to 0.001.

Tables 2 $\sim$ 7 list the joint distributions of $LC_S^{in}(q)$ and $LC_S^{out}(q)$ in each data set. As expected, for most objects $LC_S^{in}(q)$ are larger than $LC_S^{out}(q)$. However, interestingly a good portion of objects have strong outlying contrast subspaces. For example, in CMSC, more than 50% of the objects have outlying contrast subspaces satisfying $LC_S^{out}(q) \geq 10^3$. Moreover, we can see that, except PID,

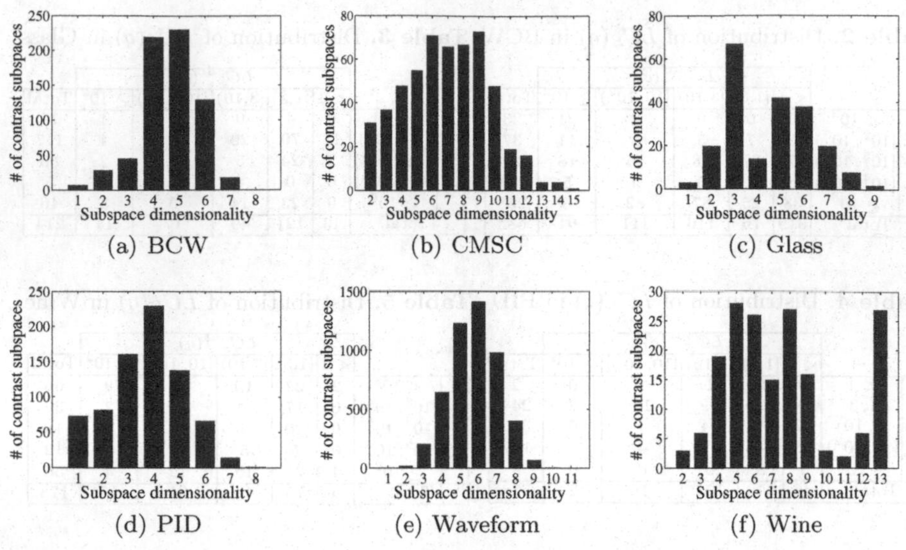

Fig. 1. Dimensionality distributions of inlying contrast subspaces

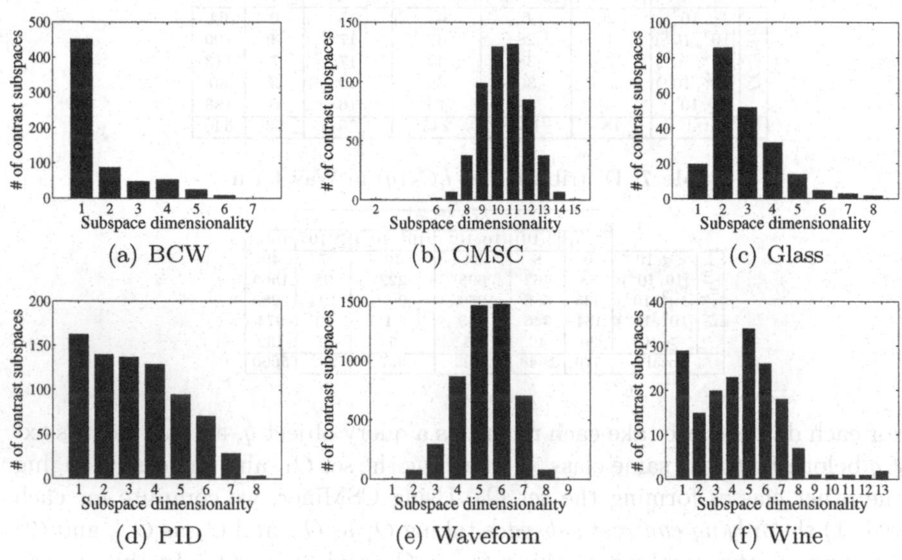

Fig. 2. Dimensionality distributions of outlying contrast subspaces

a non-trivial number of objects in each data set have both strong inlying and outlying contrast subspaces (e.g., $LC_S^{in}(q) \geq 10^4$ and $LC_S^{out}(q) \geq 10^2$).

Figures 1, 2 show the distributions of dimensionality of inlying and outlying contrast subspaces, respectively. The dimensionality distribution is an interesting feature characterizing a data set. For example, in most cases the dimensionality of contrast subspaces follows a two-side bell-shape distribution. However, in BCW and PID, the outlying contrast subspaces tend to have low dimensionality.

5.2 Efficiency

To the best of our knowledge, there is no previous method tackling the exact same mining problem. Therefore, we evaluate the efficiency of only CSMiner and the baseline method. Limited by space, we report the results on the Waveform data set only, since it is the largest one with the highest dimensionality. We randomly select 100 records from Waveform as query objects, and report the average runtime. The results on the other data sets follow similar trends.

Figure 3(a) shows the runtime (in logarithmic scale) with respect to the minimum likelihood threshold δ. As δ decreases, the runtime increases exponentially. However, the heuristic pruning techniques in CSMiner expedites the search substantially in practice. Figures 3(b) and 3(c) show the scalability on data set size and dimensionality. CSMiner is substantially faster than the baseline method.

CSMiner uses a user defined parameter r to define ϵ-neighborhood. Figure 4 shows the relative runtime with respect to r. The runtime of CSMiner is not very sensitive to r in general. Experimentally, the shortest runtime of CSMiner happens when r is in $[0.3, 0.6]$. Figure 5 illustrates the relative runtime of CSMiner with respect to k, showing that CSMiner is linearly scalable with respect to k.

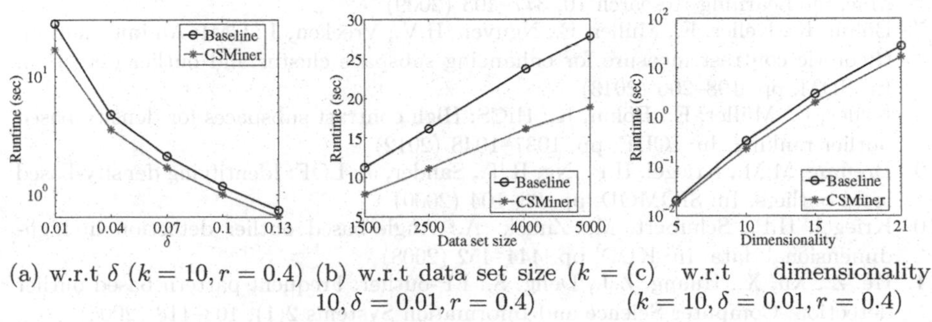

(a) w.r.t δ ($k = 10, r = 0.4$) (b) w.r.t data set size ($k = 10, \delta = 0.01, r = 0.4$) (c) w.r.t dimensionality ($k = 10, \delta = 0.01, r = 0.4$)

Fig. 3. Scalability test

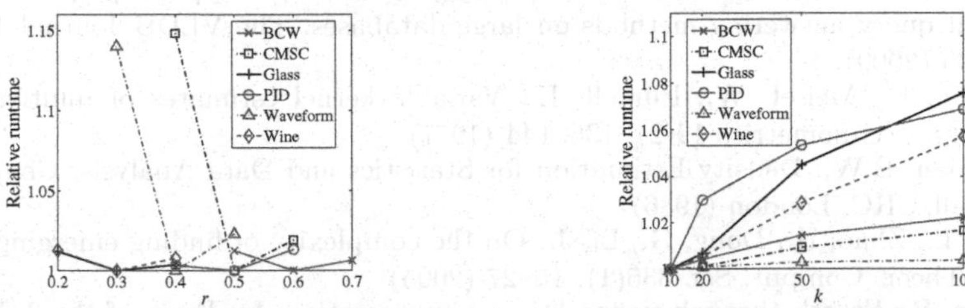

Fig. 4. Relative runtime w.r.t r ($k = 10, \delta = 0.01$)

Fig. 5. Relative runtime w.r.t k ($\delta = 0.01$)

6 Conclusions

In this paper, we studied a novel and interesting problem of mining contrast subspaces to discover the aspects that a query object most similar to a class and dissimilar to the other class. We showed theoretically that the problem is very challenging, and cannot even be approximated in polynomial time. We presented a heuristic method based on upper and lower bounds of likelihood and likelihood contrast. Our experiments on real data sets show that our method expedites contrast subspace mining substantially comparing to the baseline method.

References

1. Jeffreys, H.: The Theory of Probability, 3rd edn., Oxford (1961)
2. Dong, G., Bailey, J. (eds.): Contrast Data Mining: Concepts, Algorithms, and Applications. CRC Press (2013)
3. Dong, G., Li, J.: Efficient mining of emerging patterns: discovering trends and differences. In: KDD, pp. 43–52 (1999)
4. Bay, S.D., Pazzani, M.J.: Detecting group differences: Mining contrast sets. Data Mining and Knowledge Discovery 5(3), 213–246 (2001)
5. Wrobel, S.: An algorithm for multi-relational discovery of subgroups. In: Komorowski, J., Żytkow, J.M. (eds.) PKDD 1997. LNCS, vol. 1263, pp. 78–87. Springer, Heidelberg (1997)
6. Novak, P.K., Lavrac, N., Webb, G.I.: Supervised descriptive rule discovery: A unifying survey of contrast set, emerging pattern and subgroup mining. Journal of Machine Learning Research 10, 377–403 (2009)
7. Böhm, K., Keller, F., Müller, E., Nguyen, H.V., Vreeken, J.: CMI: An information-theoretic contrast measure for enhancing subspace cluster and outlier detection. In: SDM, pp. 198–206 (2013)
8. Keller, F., Müller, E., Böhm, K.: HiCS: High contrast subspaces for density-based outlier ranking. In: ICDE, pp. 1037–1048 (2012)
9. Breunig, M.M., Kriegel, H.P., Ng, R.T., Sander, J.: LOF: Identifying density-based local outliers. In: SIGMOD, pp. 93–104 (2000)
10. Kriegel, H.P., Schubert, M., Zimek, A.: Angle-based outlier detection in high-dimensional data. In: KDD, pp. 444–452 (2008)
11. He, Z., Xu, X., Huang, Z.J., Deng, S.: FP-outlier: Frequent pattern based outlier detection. Computer Science and Information Systems 2(1), 103–118 (2005)
12. Aggarwal, C.C., Yu, P.S.: Outlier detection for high dimensional data. ACM Sigmod Record 30, 37–46 (2001)
13. Hua, M., Pei, J., Fu, A.W., Lin, X., Leung, H.F.: Top-k typicality queries and efficient query answering methods on large databases. The VLDB Journal 18(3), 809–835 (2009)
14. Breiman, L., Meisel, W., Purcell, E.: Variable kernel estimates of multivariate densities. Technometrics 19(2), 135–144 (1977)
15. Silverman, B.W.: Density Estimation for Statistics and Data Analysis. Chapman and Hall/CRC, London (1986)
16. Wang, L., Zhao, H., Dong, G., Li, J.: On the complexity of finding emerging patterns. Theor. Comput. Sci. 335(1), 15–27 (2005)
17. Rymon, R.: Search through systematic set enumeration. In: Proc. of the 3rd Int'l Conf. on Principles of Knowledge Representation and Reasoning, pp. 539–550 (1992)
18. Bache, K., Lichman, M.: UCI machine learning repository (2013)

Forward Classification on Data Streams

Peng Wang[1,2,3], Peng Zhang[3], Yanan Cao[3], Li Guo[3], and Bingxing Fang[4]

[1] Institute of Computing Technology, Chinese Academy of Science, China
[2] University of Chinese Academy of Science, China
[3] Institute of Information Engineering, Chinese Academy of Science, China
[4] Chinese Academy of Engineering, China

peng860215@gmail.com, {zhangpeng,caoyanan,guoli}@iie.ac.cn, fangbx@cae.cn

Abstract. In this paper, we explore a new research problem of predicting an incoming classifier on dynamic data streams, named as *forward classification*. The state-of-the-art classification models on data streams, such as the incremental and ensemble models, fall into the *retrospective classification* category where models used for classification are built from past observed stream data and constantly lag behind the incoming unobserved test data. As a result, the classification model and test data are temporally inconsistent, leading to severe performance deterioration when the concept (joint probability distribution) evolves rapidly. To this end, we propose a new *forward classification* method which aims to build the classification model which fits the current data. Specifically, forward classification first predicts the incoming classifier based on a line of recent classifiers, and then uses the predicted classifier to classify current data chunk. A learning framework which can adaptively switch between forward classification and retrospective classification is also proposed. Empirical studies on both synthetic and real-world data streams demonstrate the utility of the proposed method.

Keywords: Data stream classification, linear dynamic system, concept drifting.

1 Introduction

Data stream classification is an important tool for real-time applications. For example, data stream classification is popularly used in real-time intrusion detection, spam filtering, and malicious website monitoring. Compared to data mining models, data stream classification models face extra challenges from the unbounded stream data and the continuously evolving concept (joint probability distribution)[21,18] underneath stream data.

In data stream scenarios, the classification ability of a stream classification model generally decreases with time because of the concept evolving reality[1]. For example, in data streams, a classification model c_N built at time stamp N may classify its synchronous data chunk D_N accurately, but its accuracy on incoming data chunk D_{N+1} may deteriorate significantly. This is because that data distributions in D_{N+1} may be significantly different from the training samples

V.S. Tseng et al. (Eds.): PAKDD 2014, Part I, LNAI 8443, pp. 261–272, 2014.

collected at time stamp N (in this paper, samples, records, and instances are interchangeable terms). As a result, in data stream classification, it is important to build the classifier which fits the concept of current data.

Unfortunately, existing data stream classification models, including the incremental models[5,8] and ensemble models[12], are based on the assumption that same data is first used for training then for testing. However, in real-world applications, we have to classify the incoming data first and the labeled samples of the incoming data for training tend to lag behind (for example, for fraud behavior classification in bank, typically it will take days or weeks to find whether the user was actually a fraud or not.). As a result, the classifier does not temporally consistent with test data, as illustrated in Fig.1. This type of approaches regard concepts of data stream as sequence of recurring events, so they can only model the recurring probability of old concepts/classifiers, but cannot forecast a new classifier not showing up before. In this paper, we refer to this type of classification models as ***retrospective prediction***, i.e., uses models directly trained from past stream data to classify incoming data. To synchronize the

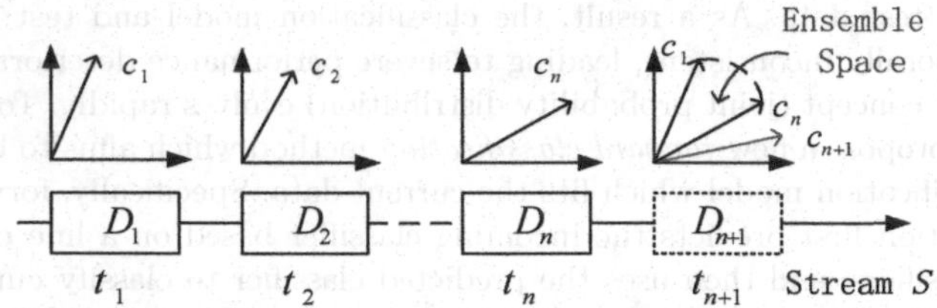

Fig. 1. The ensemble (retrospective) model can describe all the past concepts. However, it fails to describe the incoming concept(classifier) c_{N+1} that never appeared before.

classification model and test data on data streams, we present a novel ***forward classification*** method. Forward classification uses past classifiers to predict an incoming classifier, which is further used to classify incoming test data. Compared to the retrospective classification, the classifier used to classify incoming test data is not directly trained from historical stream records.

The main challenge of forward classification is to accurately predict the incoming classifier based on the past classifiers, which demands to model the evolution trend underneath the classifiers built from historical stream data. In this paper, we assume that concept evolution is a Markov process, i.e., the current concept of data stream is probabilistically determined by its previous state. This assumption is commonly used in data stream research [17]. Then, based on the observation that the classification boundaries of all the past classifiers can be represented as continuous vectors, we propose to use Linear Dynamic System (LDS)[3,11] as the solution. In this way, tracking the evolving concept is tantamount to learning the LDS model based on all the past observed classifiers (continuous vectors), and predicting the incoming classifier is equivalent to inferring the next state of the system.

Forward classification does not always outperform retrospective classification. As a model's performance is closely related to its version space [6,15], we design a flexible learning framework, which can adaptively switch between the forward classification and the retrospective classification, which is based on ensemble learning. In doing so, our learning framework is robust under different concept drifting patterns. We also demonstrate the effectiveness of the proposed method by experiments on both synthetic and real-world data streams.

The remainder of the paper is structured as follows: Section 2 introduces the modeling of the forward classification using Linear Dynamic System (LDS). Section 3 conducts the experiments. Section 4 surveys related work. We conclude the paper in Section 5.

2 Model for Forward Classification

In this section, we first describe concept evolution with a graphical model. Then, we discuss how to use Linear Dynamic System (LDS) as the solution. Finally, a forward classification framework is proposed. Consider a data stream S consisting of infinite number of records (x, y), where $x \in R^d$ is the feature vector and y is the class label. Assume the records arrive chunk-by-chunk. The records arrive at time stamp n are denoted as data chunk D_n. The classifier built on D_n is denoted as c_n. The concept at time stamp n is the joint probability distribution $p(x, y|n)$.

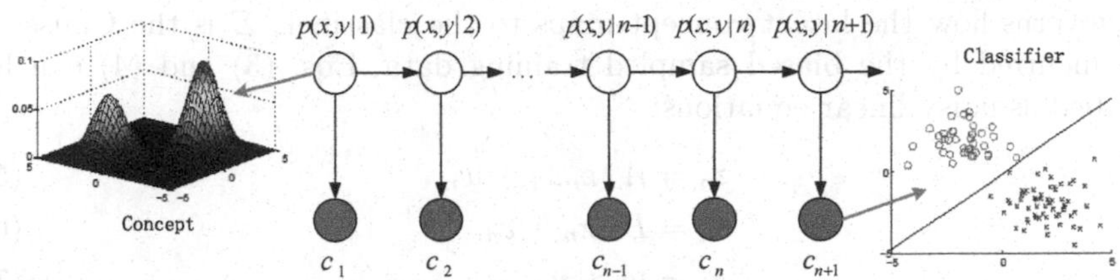

Fig. 2. Graphical model for concept description

Fig. 2 is the graphical model describing the concept evolution under the Markov assumption. The solid gray circles stand for the classifiers. The hollow circles represent the hidden concepts. The graph can be decomposed into two processes: a evolution process $p(x, y|n-1) \longrightarrow p(x, y|n)$ that describes the concept evolution between two neighboring concepts, and a modeling process $p(x, y|n) \longrightarrow c_n$ that describes the classifier training from the labeled training data. Noise is also involved in modeling process because the training set usually is a small biased data set sampled from the hidden concept. Based on graph model forward classification is formally defined as:

Forward classification: Given W historical classifiers $C = \{c_{N-W+1}, \cdots, c_N\}$ which are built consecutively from data stream S, the forward classification aims to predict the incoming classifier c_{N+1}:

$$f : (\{c_{N-W+1}, \cdots, c_N\}) \longrightarrow \widehat{c_{N+1}} \tag{1}$$

Here, c_{N+1} is the correct incoming classifier but cannot be known before hand while $\widehat{c_{N+1}}$ is our prediction. To solve the prediction problem, we use *Linear Dynamic System(LDS)* as the solution. In a deterministic LDS, a set of linear equations, governs the system evolution. Generally, the evolution of hidden concept is a stochastic process instead of a deterministic one. Thus, we add a random variable (denoted as w) to model the randomness as shown in Eq. (2).

$$z_{n+1} = A \cdot z_n + w_{n+1}. \tag{2}$$

To model the concept drifting in data stream with LDS, we assume the classifier model c_n can be converted to a vector of fixed length, such as linear classifier model $y = \omega x + b$ can be represented by the vector $[\omega, b]$. And we assume the concept $p(x, y)$ can be represented by a vector z. So the concept drifting process equates to the evolving of z. Moreover, to model the probabilistic dependence of $p(z_n|z_{n-1})$ and $p(c_n|z_n)$, we assume that probabilities $p(z_n|z_{n-1})$ and $p(c_n|z_n)$ follow Gaussian distributions:

$$p(z_n|z_{n-1}) = \mathcal{N}(z_n|A \cdot z_{n-1}, \Gamma) \tag{3}$$

$$p(c_n|z_n) = \mathcal{N}(c_n|B \cdot z_n, \Sigma) \tag{4}$$

where A represents the transform matrix that governs how the concept evolves, $A \cdot z_{n-1}$ is the mean value of z_n, Γ is the covariance of the Gaussian noise incurred by the irregular concept evolution. B represents the transform matrix that governs how the latent concept maps to the classifier, Σ is the Gaussian noise incurred by the biased sampled training data. Eqs. (3) and (4) can be described as noisy linear equations:

$$z_n = A \cdot z_{n-1} + w_n \tag{5}$$

$$c_n = B \cdot z_n + v_n \tag{6}$$

$$z_1 = \mu_0 + u \tag{7}$$

where Eq. (7) describe the initial state in LDS, $w_n \sim \mathcal{N}(w_n|0, \Gamma)$, $v_n \sim \mathcal{N}(v_n|0, \Sigma)$ and $u \sim \mathcal{N}(u|0, V_0)$ are the Gaussian noises.

We have described classifier prediction problem with LDS, then we will show how to learn the model and the resulting forward classification framework.

2.1 Model Learning

The learning problem [10,13] is to find the optimal parameter θ that maximizes the likelihood function of the observations $C = \{c_{N-W+1}, \cdots, c_N\}$, as in Eq. (8),

$$\widehat{\theta} = arg \max \log p(C|\theta). \tag{8}$$

where $p(C|\theta)$ is a marginal distribution of the joint distribution $p(C, Z|\theta)$ *w.r.t.* Z, as in Eqs. (9) and (10).

$$p(C|\theta) = \int_Z p(C, Z|\theta) dZ \tag{9}$$

$$p(C, Z|\theta) = p(z_1|\mu_0)\Big[\prod_{i=2}^{N} p(z_i|z_{i-1}, A)\Big]\prod_{j=1}^{N} p(c_j|z_j, B) \tag{10}$$

Eq. (10) is comprised of three parts. The first part is the probability of the initial state, the second part is probability of the concept evolution, and the last part is the probability of mapping the latent variables to classifiers. All the three parts are under the Gaussian distribution assumption.

Because $p(C|\theta) = \int_Z p(C, Z|\theta)dZ$ is very difficult to calculate, finding optimal solution is hardly achievable. Therefore, we use the Expectation Maximization (EM) algorithm to maximize $\log p(C|\theta)$. The EM algorithm starts with well-selected initial values for the parameters θ^{old}. Then, in the E-step, we use θ^{old} to find the posterior distribution of the latent variables $p(Z|C, \theta^{old})$. We then take the expectation of the log likelihood *w.r.t* the posterior distribution $p(Z|C, \theta^{old})$. In the M-step, we aim to find θ^{new} that maximizes $Q(\theta, \theta^{old}) = \mathbb{E}_{Z|\theta^{old}}[\ln p(C, Z|\theta)]$. The E-step and M-step are recursively executed until $Q(\theta, \theta^{old})$ converges. The details for the E-step and M-step of EM algorithm for learning LDS can be found in [3]. The basic process is summarized in Algorithm 1. The future classifier $\widehat{c_{N+1}}$ can be predicted based on the parameters θ and estimated current hidden state $\widehat{z_N}$ learned above:

$$\widehat{c_{N+1}} = B \cdot A \cdot \widehat{z_N} \tag{11}$$

Algorithm 1. Learning LDS

Require: A set of classification model in time sequential $C = \{c_{N-W+1}, \cdots, c_N\}$;
Initial value θ^{old};
The up bound of iterations M;
The convergence threshold ϵ.
Ensure: $\theta^{new} = \{\mu_0^{new}, V_0^{new}, A^{new}, \Gamma^{new}, B^{new}, \Sigma^{new}\}$;
The expected latent state $\widehat{Z_N}$ for N time block.

Complete-data likelihood Q; $Q_{pre} \longleftarrow +\infty$, $Q_{new} \longleftarrow 0$;
$i \longleftarrow 0$;
while $|Q_{pre} - Q_{new}| > \epsilon$ AND $i < M$ **do**
  $i \longleftarrow i + 1$
  $Q_{pre} \longleftarrow Q_{new}$
  $\{\mathbb{E}[z_n], \mathbb{E}[z_n z_{n-1}^T], \mathbb{E}[z_n z_n^T]\} \longleftarrow$ E-step(θ^{old}, C)
  $\theta^{new} \longleftarrow$ M-step$([\mathbb{E}[z_n], \mathbb{E}[z_n z_{n-1}^T], \mathbb{E}[z_n z_n^T]], C, \theta^{old})$
  Evaluate $Q_{new} \longleftarrow Q(\theta^{new}, \theta^{old})$ based on θ^{new}, C
end while
return θ^{new}, $\widehat{z_N} = \mathbb{E}[z_N]$

2.2 Method Comparison and Selection

In this part, we try to answer the following questions: does the forward classification always outperform retrospective classification (e.g., ensemble method)?

If the answer is NO, then how to select proper methods for real-world data streams? Here we denote the predicted classifier in forward classification as c_f, and we use ensemble method as a typical example of retrospective method. As a classifier can be mapped to a point in hyperspace, the ensemble classifier c_e will lie at the center of the most recently W classifiers, i.e. $c_e = (1/W)\sum_{i=1}^{W} c_i$.

Intuitively, if a data stream evolves continuously with stable patterns, i.e., the transform matrix A is time-invariant, the classifier is predictable; otherwise, the predicted classifier will overfit to the fake pattern. Except the irregular evolving pattern, the random noise will also make it difficult to learn the correct evolving pattern. So forward classification will not dominate retrospective method on all data streams. What's worse, there is no analytic solution for LDS, so c_f cannot be directly compared to c_e. It raise the problem how to adaptively select between these methods in real-world data stream?

We solve the problem from the view of *version space*. Version space, in concept learning or induction, refers to a subset of all hypotheses that are consistent with the observed training examples. For data stream, all possible classifiers form the version space. As concept drifts continuously, for an incoming data chunk at time $N+1$, the classifier c_{N+1} may have many possibilities, namely the version space of c_{N+1} can be large. According to Tong's theory in [16], the larger version space is, the less accurate the classifier tends to be. We illustrate different version space on different concept evolving scenarios in Fig. 3. We can notice that, for data stream with clear evolving patterns, the version space of c_f is smaller than c_e; while for data stream **without** clear evolving pattern, the version space of c_f is bigger than c_e.

Based on the Gaussian noise assumption, c_f obeys Gaussian distribution, i.e. $c_f \sim \mathcal{N}(BA\widehat{z_N}, \Psi_f)$. We take the volume within standard deviation from $BA\widehat{z_N}$ as the version space, which is determined by Ψ_f. According the analysis in [3], we have $c_f = \widehat{c_{N+1}} = \mathcal{N}(BAz_N, BP_NB^T + \Sigma)$, where $P_N = AV_NA^T + \Gamma$. So $\Psi_f = BP_NB^T + \Sigma$. On the other hand, the covariance of c_e, denoted as Ψ_e, is $\Psi_e = (1/W)\sum_{i=1}^{W}(c_i - c_e)(c_i - c_e)^T$. The volume of version space of the classifier can be calculated by $\zeta = \prod \lambda_i$, where λ_i is the eigenvector of Ψ. By comparing the version space of c_f and c_e, we can adaptively decide which method to adopt for the data stream on hand.

2.3 The Learning Framework

In this part, we introduce the classification framework which combines retrospective and forward classification. In data streams, it is often very hard to immediately obtain labeled records for model updating. In contrast, the proposed framework can avoid such a shortcoming, by tracing the trend of concept drifting and forecasting the model that reflects the current concept, then select the proper classifier based on criteria of version space. Learning from data streams contains both training and testing processes. Our framework mainly focuses on the training process. The framework is summarized in Algorithm 2.

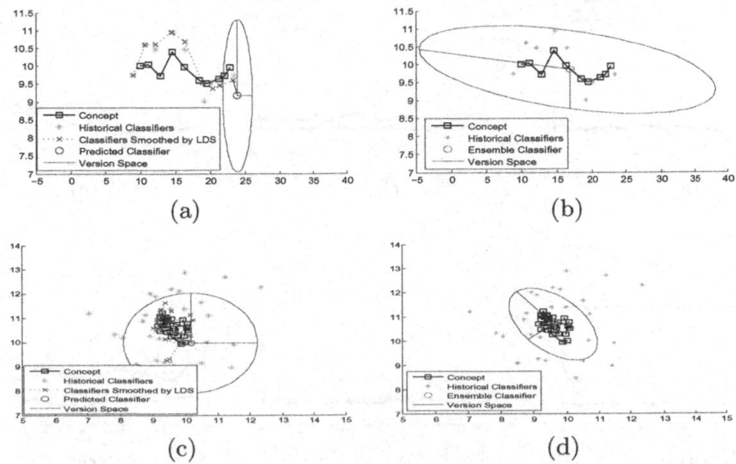

Fig. 3. Comparisons *w.r.t.* version space on different concept evolving scenarios. (a) c_f on stream with clear concept evolving pattern; (b) c_e on stream with clear concept evolving pattern; (c) c_f on stream **without** clear pattern; (d) c_e on stream **without** clear pattern

---

Algorithm 2. Learning Framework

Require: Data stream $\mathcal{D}$;
  Time window size β;
  maximum size of classifier set W.

Build initial classification model sets $C = \{c_1\}$;
N $\longleftarrow$ 1;
Calculate proper initial value θ^{init}.
while *true* **do**
  $[\theta^{new}, \mathbb{E}[z_N]] \longleftarrow$ learnLDS(C, θ^{init});
  Predict Model $\widehat{c_{N+1}} \longleftarrow B^{new} A^{new} \mathbb{E}[z_N]$;
  Calculate c_f and c_e and their corresponding version space ζ_f and ζ_e.
  Compared version space and send proper classifier to test process for classifying $N+1$ data chunk.
  Sleep during $N+1$ time window as labeled records cannot be get immediately.
  Build c_{N+1} based on the labeled records in $N+1$ time window;
  $C \longleftarrow C \cup c_{N+1}$;
  if $|C| > W$ **then**
    $C \longleftarrow C - \{c_{N-W}\}$;
  end if
  $\theta^{init} \longleftarrow \theta^{new}$, $N \longleftarrow N+1$;
end while

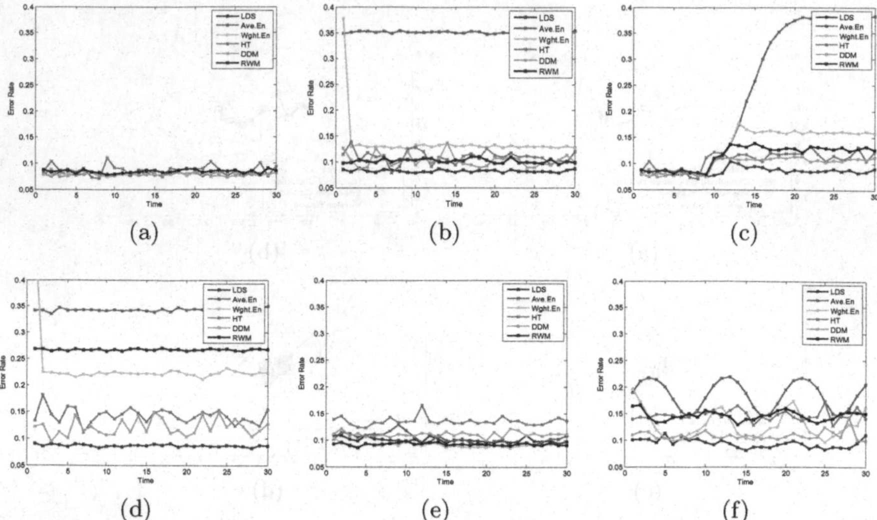

Fig. 4. Comparisons *w.r.t.* different concept evolving scenarios. (a) No concept drift; (b) Shift in one direction with speed $0.25/t$; (c) Hybrid drifting; (d) Spin with an angular speed $\frac{\pi}{3.6}/t$; (e) Random walk; (f) Period drift with a *sin* function.

Efficiency Analysis. The time cost of the proposed framework comes from two parts: training base classifiers, and predicting a future classifier. For simplicity, we take the cost of the first part as a constant value $O(1)$. The second part contains two sub-parts: an EM learning process and a prediction process. The E-step, which using the forward and backward recursions, has $O(Wd_zd_c)$ time cost, where W is the size of the historical classifier set, and d_z is the dimension of latent concept while d_c the classifier. The M-step directly updates the parameters, with time complexity of $O(W)$. Since we set the max number of iterations for EM as I, the time complexity of EM learning process is $O(WId_zd_c)$. In addition, from Eq.(11), the time complexity of predicting a future classifier is $O(1)$. To sum up, the total time complexity for a loop in the learning framework is $O(WId_zd_c)$.

3 Experimental Study

In this section, we first introduce the benchmark methods, followed by the test-bed. The test results on both synthetic and real-world data sets and the analysis will be given in the end.

We compare our method with four state-of-the-art classification method on data stream: ensemble learning[2,19], incremental learning[14], drift detection method(DDM)[7] and random walk model[9]. All of them fall into the category of retrospective learning.

3.1 Data Stream Test-Bed

In our experiment, we adopt both synthetic data stream generator and real world data streams as our test-bed.

Evolving Gaussian Generator. As we have described in the previous sections, the concept of data stream is a joint distribution $p(x, y)$. So we can generate a evolving data stream by generating records according to certain distribution and changes the distribution as time passes. For simplicity, we assume a binary classification problem, where the positive and negtive records are generated according to the Gaussian distribution. To simulate the concept evolving, we gradually change mean of the distribution as time passes. Particularly, we generate data stream with 6 types of concept drifting: *stay still*, *shift*, *hybrid*, *spin*, *random walk* and *periodic variation*.

Rotating Hyperplane is used as a test bed for CVFDT[8] models. A hyperplane in a d-dimensional space is a set of points x that satisfies $\sum_{i=1}^{d} w_i x_i = w_0$, where x_i is the ith dimension of the vector x. Records satisfying $\sum_{i=1}^{d} w_i x_i \geq w_0$ are labeled as positive. Otherwise, negative. To simulate concept evolution, we let each weight attribute $w_i = w_i + d\sigma$ change with time, where σ denotes the probability that the direction of change is reversed and d denotes that the change degree.

Sensor Stream. Sensor stream[20] contains information (temperature, humidity, light, and sensor voltage) collected from 54 sensors deployed in Intel Berkeley Research Lab. The learning task is to correctly identify the sensor ID based on the sensor data. This dataset can be downloaded from website[1].

Power Supply. Power Supply stream [20] contains hourly power supply of an Italian electricity company. The learning task is to classify the time the current power supply belongs to. This dataset can be downloaded from website[2].

3.2 Results

In Fig.4 we report the algorithm performance *w.r.t.* different types of concept drifting scenarios. From Fig.4(a), we can observe that our model is as good as the classic methods if there is no concept drifting in the stream data. Fig.4(b) indicates that for concept shifting data stream, our method outperforms others. This is because when concept drifting follows stable pattern, our method can track the pattern of the changes and more accurately predict future classifiers. In Fig.4(c), the drifting pattern is unstable. For example, before $t = 10$, the concept stays still, classifiers $\{c_1, \cdots, c_{10}\}$ are determined by the parameter θ_{still} while $\{c_{10}, \cdots, c_N\}$ are determined by θ_{drift}. When $\theta_{still} \longrightarrow \theta_{drift}$, the error rate of LDS arises, as the classifier predicted with θ_{still} for the future concept. As the drifting records increase, the LDS model can gradually tracking the concept evolution. So the error rate gradually decreases. In Fig.4(d), the concept's evolving rate is fast. We can observe that LDS significantly outperforms other methods for the fast evolving data stream. In Fig.4(e), the concept of data stream evolves in a random walk manner. We can see that LDS can handle the noise factor well for it can switch to retrospective method when the evolving

[1] `http://www.cse.fau.edu/~xqzhu/Stream/sensor.arff`

[2] `http://www.cse.fau.edu/~xqzhu/Stream/powersupply.arff`

Table 1. Error rate comparisons among different methods

DataSet	Average EN	Weighted EN	H.T.	DDM	RWM	LDS
Gaussian Static	$\mathbf{0.077_{\pm 0.001}}$	$0.079_{\pm 0.001}$	$0.087_{\pm 0.004}$	$0.079_{\pm 0.001}$	$0.0841_{\pm 0.001}$	$0.079_{\pm 0.002}$
Gaussian Shift	$0.351_{\pm 0.003}$	$0.138_{\pm 0.017}$	$0.109_{\pm 0.004}$	$0.103_{\pm 0.004}$	$0.102_{\pm 0.003}$	$\mathbf{0.082_{\pm 0.004}}$
Gaussian Hybrid	$0.085_{\pm 0.002}$	$0.221_{\pm 0.046}$	$0.087_{\pm 0.005}$	$\mathbf{0.085_{\pm 0.002}}$	$0.094_{\pm 0.003}$	$0.111_{\pm 0.010}$
Gaussian Spin	$0.343_{\pm 0.001}$	$0.230_{\pm 0.012}$	$0.143_{\pm 0.005}$	$0.121_{\pm 0.006}$	$0.267_{\pm 0.001}$	$\mathbf{0.086_{\pm 0.007}}$
Gaussian Random	$0.093_{\pm 0.003}$	$0.091_{\pm 0.002}$	$0.105_{\pm 0.008}$	$0.095_{\pm 0.018}$	$\mathbf{0.087_{\pm 0.001}}$	$0.090_{\pm 0.005}$
Gaussian Period	$0.218_{\pm 0.031}$	$0.113_{\pm 0.014}$	$0.117_{\pm 0.004}$	$0.094_{\pm 0.003}$	$0.096_{\pm 0.003}$	$\mathbf{0.089_{\pm 0.003}}$
Hyperplane	$0.173_{\pm 0.043}$	$0.110_{\pm 0.027}$	$0.176_{\pm 0.031}$	$0.107_{\pm 0.021}$	$0.103_{\pm 0.024}$	$\mathbf{0.100_{\pm 0.023}}$
Power Supply	$0.064_{\pm 0.027}$	$0.070_{\pm 0.028}$	$0.077_{\pm 0.035}$	$0.055_{\pm 0.019}$	$0.073_{\pm 0.033}$	$\mathbf{0.052_{\pm 0.026}}$
Sensor	$0.042_{\pm 0.006}$	$0.039_{\pm 0.004}$	$0.037_{\pm 0.009}$	$0.034_{\pm 0.003}$	$0.079_{\pm 0.022}$	$\mathbf{0.032_{\pm 0.020}}$

pattern is blurred. Random walk model also perform well and it can filter out the random noise of $\{c_1, \cdots, c_N\}$, and track the proper concept, as in this model, $A \equiv I$ so it will not learn false drifting patterns. In Fig.4(f), the evolving pattern of the data stream is changing periodically. the LDS model is robust to this kind of concept drifting. In summary, our learning framework outperforms other benchmark methods for data streams with regular evolving patterns.

In Table 1, we summarize and compare the performance of different methods on all data streams. Overall, for data streams having regular evolving patterns, the performance of the proposed LDS model outperforms other benchmark methods. For data streams whose evolving is not a stable Markov process, learning methods such as drift detection and random walk perform better.

Efficiency. From our experiments, we observe that the EM algorithm converges within 100 iterations, so M is manually set to 100. In most cases, when $W > 50$, the predicted classifier is stable, so W is manually set to 50. With these settings, on our PC with 2.8G Hz CPU, the time cost for predicting the classifier is less than 1 minute. In real-world applications, it usually takes hours for the concept having detectable change, so the framework is efficient.

4 Related Work

Existing data stream classification models can be categorized into three groups: online / incremental models[8,14], ensemble learning[12,18,19] and drift detection methods [7]. We briefly describe them based on the development trace. In the simplest situation, where the concept of data stream remains stable, for each time window, the classifier has prediction variance due to limited training samples. We can use majority voting method to ensemble these classifiers, because random variance tends to compensate each other. This is the fundamental framework for retrospective classification. For data stream with concept drifting, the majority voting for ensemble is inappropriate. An alternative solution is to use weighted ensemble, where each classifier is weighted according to their consistence with the most recent observed training data. For incremental or DDM method, they keep updating the classifier using newly arriving records, enabling the learning model to adapt to new concepts. For retrospective learning models [17,4], they regard concepts of data stream as a sequence of recurring events and use the most probable concept in the future to classify incoming stream data. Unfortunately, existing weighing and updating approaches, including the proactive learning framework, cannot forecast a completely new classifier. Thus, they cannot synchronize the classifier to the evolving stream data.

Forward classifier prediction method, on the other hand, uses probabilistic model to define time evolution of the concept. By using the probabilistic model, we can approximate the distribution that maximizes the posterior probability of the model [3]. Moreover, we can predict the optimal incoming classifier by adopting the inference method. That is to say, forward classifier prediction method is able to derive better classification results.

5 Conclusions

In this paper, we present a novel *forward classification* method for classifying evolving stream data. Due to the temporal evolving of data streams, simply learning classification models from historical data, as existing *retrospective classification* methods do, is inadequate and inaccurate. So a proper classification design is to capture the evolution trend underneath stream data and use it to predict a future classifier for classification. With this vision and the assumption that the concept evolving can be characterized by Markov process, we propose to model the trend of classifiers using the *Linear Dynamic System*, through which we can model the concept drifting and predict incoming classifier. We also notice that forward classification is not overwhelmingly better than retrospective classification. Then we design the learning framework, which adaptively switches between the forward classification based on LDS and basic ensemble learning method, so it is robust to different types of data streams. Experiments on synthetic and real-world streams demonstrate that our framework outperforms other methods in different types of concept drifting scenarios.

Acknowledgments. This work was supported by the NSFC (No. 61003167), IIE Chinese Academy of Sciences (No. Y3Z0062101), 863 projects (No. 2011AA010703 and 2012AA012502), 973 project (No. 2013CB329606), and the Strategic Leading Science and Technology Projects of Chinese Academy of Sciences (No. XDA06030200).

References

1. Aggarwal, C.C., Han, J., Wang, J., Yu, P.S.: On demand classification of data streams. In: Proc. of 10th ACM SIGKDD, New York, USA, pp. 503–508 (2004)
2. Bifet, A., Holmes, G., Pfahringer, B., Kirkby, R., Gavaldà, R.: New ensemble methods for evolving data streams. In: Proc. of the 15th ACM SIGKDD, New York, USA, pp. 139–148 (2009)
3. Bishop, C.M.: Pattern Recognition and Machine Learning. Springer Science+Business Media (2006)
4. Chen, S., Wang, H., Zhou, S., Yu, P.S.: Stop chasing trends: discovering high order models in evolving data. In: Proc. of the 24th IEEE International Conference on Data Engineering, ICDE 2008, pp. 923–932. IEEE (2008)

5. Domingos, P., Hulten, G.: Mining high-speed data streams. In: Proc. of the Sixth ACM SIGKDD, KDD 2000, pp. 71–80. ACM, New York (2000)
6. Dubois, V., Quafafou, M.: Concept learning with approximation: Rough version spaces. In: Alpigini, J.J., Peters, J.F., Skowron, A., Zhong, N. (eds.) RSCTC 2002. LNCS (LNAI), vol. 2475, pp. 239–246. Springer, Heidelberg (2002)
7. Gama, J., Medas, P., Castillo, G., Rodrigues, P.: Learning with drift detection. In: Bazzan, A.L.C., Labidi, S. (eds.) SBIA 2004. LNCS (LNAI), vol. 3171, pp. 286–295. Springer, Heidelberg (2004)
8. Hulten, G., Spencer, L., Domingos, P.: Mining time-changing data streams. In: Proc. of the Seventh ACM SIGKDD, KDD 2001, pp. 97–106. ACM, New York (2001)
9. Jaakkola, M.S.T., Szummer, M.: Partially labeled classification with markov random walks. In: Advances in Neural Information Processing Systems (NIPS), vol. 14, pp. 945–952 (2002)
10. Kalman, R.E.: A new approach to linear filtering and prediction problems. Journal of Basic Engineering 82(5311910), 35–45 (1960)
11. Katok, A., Hasselblatt, B.: Introduction to the modern theory of dynamical systems, Cambridge (1996)
12. Opitz, D., Maclin, R.: Popular ensemble methods: An empirical study. Journal of Artificial Intelligence Research 11, 169–198 (1999)
13. Zarchan, P., Musoff, H.: Fundamentals of Kalman filtering: a practical approach. American Institute of Aeronautics Astronautics (2005)
14. Pfahringer, B., Holmes, G., Kirkby, R.: New options for hoeffding trees. In: Orgun, M., Thornton, J. (eds.) AI 2007. LNCS (LNAI), vol. 4830, pp. 90–99. Springer, Heidelberg (2007)
15. Sverdlik, W., Reynolds, R.: Dynamic version spaces in machine learning. In: Proceedings of the Fourth International Conference on Tools with Artificial Intelligence, TAI 1992, pp. 308–315 (November 1992)
16. Tong, S., Koller, D.: Support vector machine active learning with applications to text classification. The Journal of Machine Learning Research 2, 45–66 (2002)
17. Yang, Y., Wu, X., Zhu, X.: Combining proactive and reactive predictions for data streams. In: Proc. of the Eleventh ACM SIGKDD, KDD 2005, pp. 710–715. ACM (2005)
18. Zhang, P., Zhu, X., Shi, Y.: Categorizing and mining concept drifting data streams. In: Proceedings of the 14th ACM SIGKDD, KDD 2008, pp. 812–820. ACM, New York (2008)
19. Zhang, P., Zhu, X., Shi, Y., Guo, L., Wu, X.: Robust ensemble learning for mining noisy data streams. Decision Support Systems 50(2), 469–479 (2011)
20. Zhu, X.: Stream data mining repository (2010), `http://www.cse.fau.edu/~xqzhu/stream.html`
21. Zliobaite, I.: Learning under concept drift: an overview. Technical Report (2009)

Shingled Graph Disassembly: Finding the Undecideable Path*

Richard Wartell[1], Yan Zhou[2], Kevin W. Hamlen[2], and Murat Kantarcioglu[2]

[1] Mandiant

[2] Computer Science Department, The University of Texas at Dallas
{rhw072000,yan.zhou2,hamlen,muratk}@utdallas.edu

Abstract. A probabilistic finite state machine approach to statically disassembling x86 machine language programs is presented and evaluated. Static disassembly is a crucial prerequisite for software reverse engineering, and has many applications in computer security and binary analysis. The general problem is provably undecidable because of the heavy use of unaligned instruction encodings and dynamically computed control flows in the x86 architecture. Limited work in machine learning and data mining has been undertaken on this subject. This paper shows that semantic meanings of opcode sequences can be leveraged to infer similarities between groups of opcode and operand sequences. This empowers a probabilistic finite state machine to learn statistically significant opcode and operand sequences in a training corpus of disassemblies. The similarities demonstrate the statistical significance of opcodes and operands in a surrounding context, facilitating more accurate disassembly of new binaries. Empirical results demonstrate that the algorithm is more efficient and effective than comparable approaches used by state-of-the-art disassembly tools.

Keywords: Binary analysis, disassembly, reverse-engineering, probabilistic finite state machines.

1 Introduction

Statistical data mining techniques have found wide application in domains where statistical information is valuable for solving problems. Examples include computer vision, web search, natural language processing, and more. A recent addition to this list is *static disassembly* [1,2]. Disassembly is the process of translating byte sequences to human-readable assembly code. Such translation is often deemed a crucial first step in software reverse engineering and analysis.

* The research reported herein was supported in part by AFOSR awards FA9550-12-1-0082 & FA9550-10-1-0088, NIH awards 1R0-1LM009989 & 1R01HG006844, NSF awards #1054629, Career-CNS-0845803, CNS-0964350, CNS-1016343, CNS-1111529, & CNS-1228198, ARO award W911NF-12-1-0558, and ONR award N00014-14-1-0030.

V.S. Tseng et al. (Eds.): PAKDD 2014, Part I, LNAI 8443, pp. 273–285, 2014.

Although all binary-level debuggers perform *dynamic disassembly* to display assembly code for individual runs of target programs, the much more challenging task of static disassembly attempts to provide assembly code for all possible runs (i.e., all reachable instructions). Static disassembly is therefore critical for analyzing code with non-trivial control-flows, such as branches and loops. Example applications include binary code optimization, reverse engineering legacy code, semantics-based security analysis, malware analysis, intrusion detection, and digital forensics. Incorrectly disassembled binaries often lead to incorrect analyses, and therefore bugs or security vulnerabilities in mission-critical systems.

Static disassembly of binaries that target Intel-based architectures is particularly challenging because of the architecture's heavy use of variable-length, unaligned instruction encodings, dynamically computed control-flows, and interleaved code and data. *Unalignment* refers to the fact that Intel chipsets consider all memory addresses to be legal instruction starting points. When some programs compute the destinations of jumps dynamically using runtime pointer arithmetic, statically deciding which bytes are part of reachable instructions and which are (non-executed) static data reduces from the halting problem. As a result, the static disassembly problem for Intel architectures is provably Turing-undecidable in general.

Production-level disassemblers and reverse engineering tools have therefore applied a long history of evolving heuristics to generate best-guess disassemblies. Such heuristics include fall-through disassembly, various control-flow and dataflow analyses, and compiler-specific pattern matching. Unfortunately, even after decades of tuning, these heuristics often fail even for non-obfuscated, non-malicious, compiler-generated software. As a result, human analysts are often forced to laboriously guide the disassembly process by hand using an interactive disassembler [3]. When binaries are tens or hundreds of megabytes in size, the task quickly becomes intractable.

Wartell et al. recently proposed to apply machine learning and data mining to address this problem [1]. Their approach uses statistical data compression techniques to reveal the semantics of a binary in its assembly form, yielding a segmentation of code bytes into assembly instructions and a differentiation of data bytes from code bytes. Although the technique is effective and exhibits improved accuracy over the best commercial disassembler currently available [4], the compression algorithm suffers high memory usage. Thus, training on large corpora can be very slow compared to other disassemblers.

In this paper, we present an improved disassembly technique that is both more effective and more efficient. Rather than relying on high-order context semantic information (which leads to long training times), we leverage a finite state machine with transitional probabilities to infer likely execution paths through a sea of bytes. Our main contributions include a graph-based static disassembly technique; a simple, efficient, but effective disassembler implementation; and an empirical demonstration of the effectiveness of the approach.

Our high-level strategy involves two linear passes: a preprocessing step which recovers a conservative superset of potential disassemblies, followed by a filtering step in which a state machine selects the best disassembly from the possible candidates. While the resulting disassembly is not guaranteed to be fully correct (due to the undecidability of the general problem), it is guaranteed to avoid certain common errors that plague mainstream disassemblers. Our empirical analysis shows our simple, linear approach is faster and more accurate than the observably quadratic-time approaches adopted by other disassemblers.

The rest of the paper proceeds as follows. Section 2 discusses related work in static disassembly. Section 3 presents our graph-based static disassembly technique. Section 4 presents experimental results, and Section 5 concludes and suggests future work.

2 Related Work

Existing disassemblers mainly fall into three categories: linear sweep disassemblers, recursive traversal disassemblers, and the hybrid approach. The GNU utility *objdump* [5] is a popular example of the linear sweep approach. It starts at the beginning of the text segment of the binary to be disassembled, decoding one instruction at a time until everything in executable sections is decoded. This type of disassembler is prone to errors when code and data bytes are interleaved within some segments. Such interleaving is typical of almost all production-level Windows binaries generated by non-GNU compilers.

IDA Pro [3, 4] follows the recursive traversal approach. Unlike linear sweep disassemblers, it decodes instructions by traversing the static control flow of the program, thereby skipping data bytes that may punctuate the code bytes. However, not all control flows can be predicted statically. When the control flow is constructed incorrectly, some reachable code bytes are missed, resulting in disassemblies that omit significant blocks of code.

The hybrid approach [6] combines linear sweep and recursive traversal to detect and locate disassembly errors. The basic idea is to disassemble using the linear sweep algorithm and verify the output using the recursive traversal algorithm. While this helps to eliminate some disassembly errors, in general it remains prone to the shortcomings of both techniques. That is, when the sweep and traversal phases disagree, there is no clear indication of which is correct; the ambiguous bytes therefore receive an error-prone classification.

Wartell et al. recently presented a machine learning- and data mining-based approach to the disassembly problem [1]. Their approach avoids error-prone control-flow analysis heuristics in favor of a three-phase approach: First, executables are segmented into subsequences of bytes that constitute valid instruction encodings as defined by the architecture [7]. Next, a language model is built from the training corpus with a statistical data model used in modern data compression. The language model is used to classify the segmented subsequence as code or data. Finally, a set of pre-defined heuristics refines the classification results. The experimental results demonstrate substantial improvements over IDA Pro's traversal-based approach. However, it has the disadvantage of high memory usage due to the large statistical compression model. This significantly slows the disassembly process relative to simple sweep and traversal disassemblers.

Our disassembly algorithm presented in this paper instead adopts a *probabilistic finite state machine* (FSM) [8, 9] approach. FSMs are widely used in areas such as computational linguistics, speech processing, and gene sequencing. Although the transitions of probabilistic FSMs are non-deterministic, they are labeled with probabilities given training data. For any given byte stream, there is more than one trace through the FSM. By querying the FSM, the likelihood of each trace can be computed, revealing the most probable path of reachable opcode and operand sequences in an executable.

3 Disassembler Design

Our machine learning approach to disassembly frames the disassembly problem as follows:

Problem Definition. Given an arbitrary string of bytes, which subset of the bytes is the most probable set of potentially reachable instruction starting points, where "probable" is defined in terms of a given corpus of correct binary disassemblies?

Figure 1 shows the architecture of our disassembly technique. It consists of a *shingled disassembler* that recovers the (overlapping) building blocks (*shingles*) of all possible valid execution paths, a finite state machine trained on binary executables, and a graph disassembler that traces and prunes the shingles to output the maximum-likelihood classification of bytes as instruction starting points, instruction non-starting points, and data.

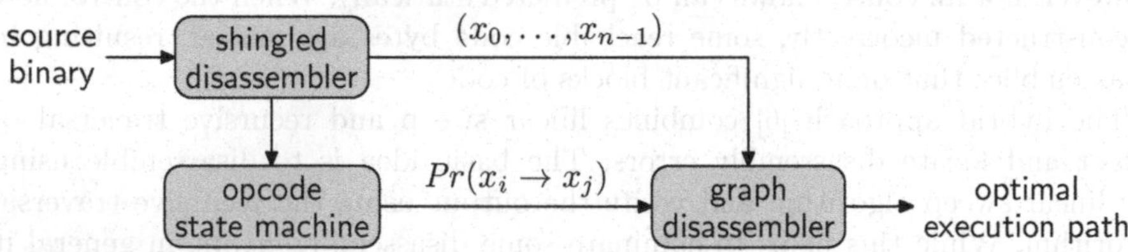

Fig. 1. Disassembler architecture

3.1 Shingled Disassembler

Since computed branch instructions in x86 have their targets established at runtime, every byte within the code section can be a target and thus must be considered as executable code. This aspect of the x86 architecture allows for *instruction aliasing*, the ability for two instructions to overlap each other. Therefore, we refer to a disassembler that retains all possible execution paths through a binary as a shingled disassembler.

Definition 1 ***Shingle***
A shingle *is a consecutive sequence of bytes that decodes to a single machine instruction. Shingles may overlap.*

The core functionality of the shingled disassembler is to eliminate bytes that are clearly data (because all flows that contain them lead to execution of bytes that do not encode any valid instruction), and to compose a byte sequence that retains information for generating every possible valid shingle of the source binary. This is a major benefit of this approach since the shingled disassembly encodes a superset of all the possible valid disassemblies of the binary. In later sections, we discuss how we apply our graph disassembler to prune this superset until we find the most probable byte classifications. In order to define what consists of a valid execution path, we must first discuss a few key concepts.

Definition 2 ***Fall through***[1]
Shingle x (conditionally) falls through *to shingle y, denoted $x \rightharpoondown y$, if shingle y is located adjacent to and after instruction x, and the semantics of instruction x do not (always) modify the program counter. In this case, execution of instruction x is (sometimes) followed by execution of instruction y at runtime.*

Definition 3 ***Unconditional Branch***
A shingle is an unconditional branch *if it only falls through when its operand explicitly targets the immediately following byte. Unconditional branch instructions for x86 include `jmp` and `ret` instructions.*

Unconditional branch instructions are important in defining valid disassemblies because the last instruction in any disassembly must be an unconditional branch. If this is not the case, the program could execute past the end of its virtual address space.

Definition 4 ***Static Successor***
A control-flow edge (x, y) is static *if $x \rightharpoondown y$ holds or if x is a conditional or unconditional branch with fixed (i.e., non-computed) destination y. An instruction's* static successors *are defined by $S(x) = \{y \mid (x, y) \text{ is static}\}$.*

Definition 5 ***Postdominating Set***
The (static) postdominating set *$P(x)$ of shingle x is the transitive closure of S on $\{x\}$. If there exists a static control-flow from x to an illegal address (e.g., an address outside the address space or whose bytes do not encode a legal instruction), then $P(x)$ is not well defined and we write $P(x) = \perp$.*

Definition 6 ***Valid Execution Path***
All paths in $P(x)$ are considered valid execution paths *from x.*

[1] At first glance, it would seem that we could strengthen our defintion of fall-throughs to any two instructions that do not have an unconditional branch instruction between them. However, there are cases where a compiler will place a `call` and `jcc` instruction followed by data bytes. A common example of this is `call [IAT:ExceptionHandler]` since the exception handler function will never return.

The x86 instruction set does not make use of every possible opcode sequence; therefore certain bytes cannot be the beginning of a code instruction. For example, the `0xFF` byte is used to distinguish the beginning of one of 7 different instructions, using the byte that follows to distinguish which instruction is intended. However, `0xFFFF` is an invalid opcode that is unused in the instruction set. This sequence of bytes is common because any negative offset in two's complement that branches less than `0xFFFF` bytes away starts with `0xFFFF`. The shingled disassembler can immediately mark any shingle whose opcode is not supported under the x86 instruction set as *data*. A shingle that is marked as *data* is either used as the operand of another instruction, or it is part of a data block within the code section. Execution of the instruction would cause the program to crash.

Lemma 1. ***Invalid Fall-through***
$\langle \forall x, y :: x \rightharpoondown y \wedge y := \emptyset \rightarrow x := \emptyset \rangle$*, in which* $\emptyset$ *stands for data bytes.*

Any time that we encounter an address that is marked data, all fall-throughs to that instruction can be marked as data as well. Direct branches also fall into this definition. All direct `call` and `jmp` instructions imply a direct executional relationship between the instruction and its target. Therefore, any shingle that targets a shingle previously marked as data is also marked as data.

Definition 7 ***Sheering***
A shingle x *is* sheered *from the shingled disassembly when* $\forall y :: x \rightharpoondown y$, x *and all* y *are marked as data in the shingled disassembly.*

Figure 2 illustrates how our shingled disassembler works. Given a binary of byte sequence `6A 01 51 56 8B C7 E8 B6 E6 FF FF`..., the shingled disassembler performs a single-pass, ordered scan over the byte sequence. Data bytes and invalid shingles are marked along the way. Figure 2(a) demonstrates the first series of valid shingles, beginning at the first byte of the binary. Figure 2(b) starts at the second byte, which falls through to a previously disassembled shingle. The shingle with byte `C7` is then marked as data (shaded in Figure 2(c)) since it is an invalid opcode. Figure 2(d) shows an invalid shingle since it falls through to an invalid opcode `FF FF`. Our shingled disassembler marks the two shingles `B6` and `FF` as invalid in the sequence. Figure 2(e) shows another valid shingle that begins at the ninth byte of the binary. After completing the scan, our shingled disassembler has stored information necessary to produce all valid paths in $P(x)$.

The secondary function of the shingled disassembler is to collect local statistics called code/data modifiers that are specific to the executable. These modifiers keep track of the likelihood that a shingle is code or data in this particular executable. The following heuristics are used to update modifiers:

1. If the shingle at address a is a long direct branch instruction with a' as its target, the address a' is more likely to be a code instruction. We apply this heuristic with short direct branches as well, but with less weight since two byte instructions are more likely to be seen within other instruction operands.

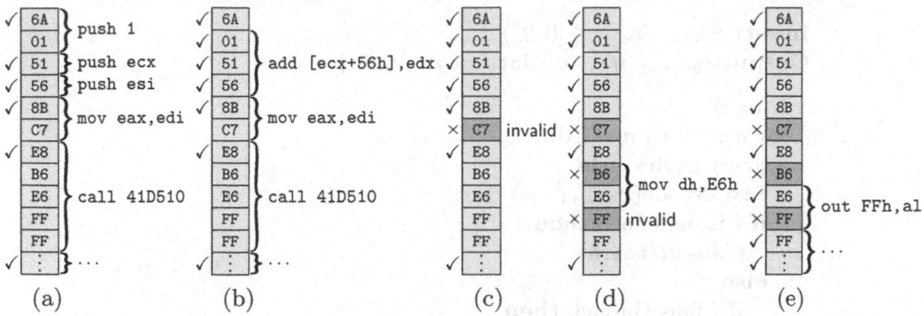

Fig. 2. Shingled disassembly of a sample byte sequence: (a) a shingle sequence beginning at the first byte; (b) a shingle sequence beginning at the second byte; (c) a non-shingle that starts with an invalid opcode; (d) a shingle that falls through to an invalid opcode; and (e) a shingle sequence beginning at the ninth byte

2. If three shingles sequentially fall-through to each other and match one of the most common instruction opcode sequences, each of these three addresses is more likely to be code. Common sequences include function prologues, epilogues, etc.
3. If bytes at address a and $a+4$ both encode addresses that reference shingles within the code section of the binary, the likelihood that addresses a through $a+7$ are data is very high. Shingles a through $a+7$ are marked as data, as well as any following four byte sequences that match this criteria. This is most likely a series of addresses referenced by a conditional branch elsewhere in the code section.

The pseudocode for generating a shingled disassembly for a binary is shown in Figure 3. For simplicity, the heuristics used to update modifiers are not described in the pseudocode. Lines 1–17 construct a static control-flow graph G in which all edges are reversed. A distinguished node bad is introduced with outgoing edges to all shingles that do not encode any valid instruction, or that branch to static, non-executable addresses. Lines 18–20 then mark all addresses reachable from bad as data. The rest are possible instruction starting points.

3.2 Opcode State Machine

The state machine is constructed from a large corpus of pre-tagged binaries, disassembled with IDA Pro v6.3. The byte sequences of the training executables are used to build an opcode graph, consisting of opcode states and transitions from one state to another. For each opcode state, we label its transition with the probability of seeing the next opcode in the training instruction streams. The opcode graph is a probabilistic finite state machine (FSM) that encodes all the correct disassemblies of the training byte sequences annotated with transition probabilities. The accepting state of the FSM is the last unconditional branch seen in the binary.

Figure 4 shows what this transition graph might look like if the x86 instruction set only contained four opcodes: 0x01 through 0x04. Each directed edge in the

```
Input: x_0, ..., x_{n-1} ∈ [0, 2^8)
Output: y_0, ..., y_{n-1} ∈ {data, maybe_code}

G := ∅
for a := 0 to n − 1 do
  y_a := maybe_code
  i := decode(x_a x_{a+1} ···)
  if i is undefined then
    G.insert(bad, a)
  else
    if i falls through then
      if a + |i| < n then G.insert(a + |i|, a)
      else G.insert(bad, a)
    endif
    if i is a static jump/branch then
      if is_exec_ok(dest(i)) then G.insert(dest(i), a)
      else G.insert(bad, a)
    endif
  endif
endfor
foreach a ∈ depth_first_search(G, bad) do
  y_a := data
endfor
```

Fig. 3. Shingled disassembly algorithm

graph between opcode x_i and x_j implies that a transition between x_i and x_j has been observed in the corpus, and the edge weight of $x_i \rightarrow x_j$ is the probability that given x_i, the next instruction is x_j. It is also important to note the node *db* in the graph which represents data bytes. Any transition from an instruction to data observed in the corpus will be represented by a directed edge to the *db* node. The graph for the full x86 instruction set includes more than 500 nodes, as each observed opcode must be included.

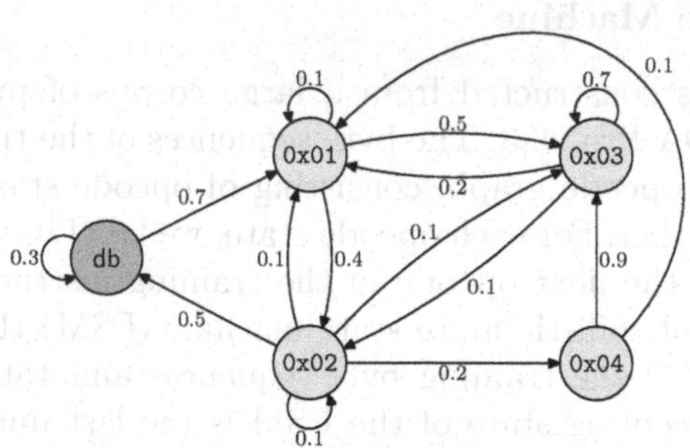

Fig. 4. Instruction transition graph: 4 opcodes

3.3 Maximum-Likelihood Execution Path

We name the output of the shingled disassembler a *shingled binary*. The shingled binary of the source executable encodes within it up to 2^n possible valid disassemblies. Our graph disassembler is designed to scan the shingled binary and prune shingles with lower probabilities. By using our graph disassembler, we can find the maximum-likelihood set of byte classifications by tracing the shingled binary through the opcode finite state machine. At every receiving state, we check which preceding path (predecessor) has the highest transition probability. For example in Figure 2, the 5th byte (8B) is the receiving state of two preceding addresses: byte 1 (see Figure 2(a)) and byte 2 (see Figure 2(b)). We compute the transition probability from each of the two addresses and sheer the one with a lower probability.

Theorem 1. *The graph disassembler always returns the maximum-likelihood byte classifications among the set S of all valid shingles.*

Proof. Each byte in the shingled binary is a potential receiving state of multiple predecessors. At each receiving state, we keep the best predecessor with the highest transition probability. Therefore, when we reach the last receiving state—the accepting state, which represents the last unconditional brach instruction—we find the shingle with the highest probability as the best execution path.

The transition probability of a predecessor consists of two parts: the global transition probability taken from the opcode state machine and the local modifiers, and local statistics of each byte being code or data based on several heuristics. This is important because runtime reference patterns specific to the binary being disassembled are included in distinguishing the most probable disassembly path.

Let r be a receiving state of a transition triggered at x_i in the shingled binary, let $Pr(pred(x_i))$ be the transition probability of the best predecessor of x_i, and let cm and dm be the code and data modifiers computed during shingled disassembly. The transition probability to r is as follows:

$$Pr(r) = Pr(pred(x_i)) * cm/dm$$

if x_i is a fall-through instruction, or

$$Pr(r) = Pr(pred(x_i)) * cm/dm * Pr(db_i) * Pr(db_r)$$

if x_i is a branch instruction, where $Pr(db_i)$ is the probability that x_i is followed by data and $Pr(db_r)$ is the probability that r is proceeded by data. Every branch instruction can possibly be followed by data. To account for this, when determining the best predecessor for each instruction, branch instructions are treated as fall-throughs to their following instruction and to data. Each branch instruction can be a predecessor to the following instruction or to any instruction that is on a 4-byte boundary and is reachable via data bytes.

Therefore, the transition probability of any valid shingle-path s resulting in a trace of $r_0, \ldots, r_i, \ldots, r_k$ is:

$$Pr(s) = Pr(r_0)Pr(r_1) \cdots Pr(r_i) \cdots Pr(r_k)$$

and the optimal execution path s^* is:

$$s^* = \arg\max_{s \in \mathcal{S}} Pr(s).$$

3.4 Algorithm Analysis

Our disassembly algorithm is much quicker than other approaches of comparable accuracy due to the small amount of information that needs to be analyzed. The time complexity of each of the three steps is as follows:

- Shingled disassembly: Lines 1–17 of Figure 3 complete in $O(n)$ time (where n is the number of bytes in executable sections) and construct a CFG G with at most $2n$ edges. The depth-first search in Lines 18–20 is linear in the size of G. We conclude that the algorithm in Figure 3 is $O(n)$.
- Sheering: Pruning invalid shingles also requires $O(n)$ time.
- Graph disassembly: The graph-based disassembler performs a single-pass scan over the shingled binary, and is therefore also $O(n)$.

Therefore, our disassembly algorithm runs in time $O(n)$, that is, linear in the size of the source binary executable.

4 Evaluation

A prototype of our shingled disassembler was developed in Windows using Microsoft .NET C#. Testing of our disassembly algorithm was performed on an Intel Xeon processor with six 2.4GHz cores and 24GB of physical RAM. We tested 24 difficult binaries with very positive results.

4.1 Broad Results

Table 1 shows the different programs on which we tested our disassembler, as well as file sizes and code section sizes. It also displays the number of instructions that the graph disassembler identified that IDA Pro didn't identify as code. Figure 5 shows the percentage of instructions that IDA Pro identified as code that our disassembler also identified as code.

Our disassembler runs in linear time in the size of the input binary. Figure 6 shows how many times longer IDA Pro took to disassemble each binary relative to our disassembler. Our disassembler is increasingly faster than IDA Pro as the size of the input grows.

Finally, for each binary we used Ollydbg to create and save the traces of executions. Tracing executions in this way does not reveal the ground truth of

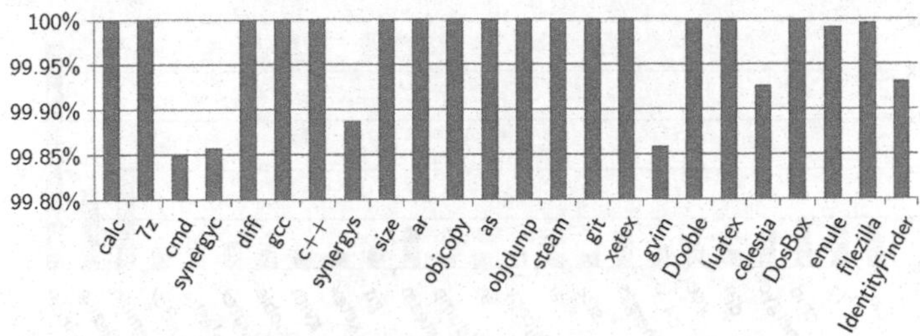

Fig. 5. Percent of instructions identified by IDA Pro that were also identified by our disassembler

Table 1. File Statistics

File Name	**File Size** (KB)	**Code Size** (KB)	**# Instr. Missed by IDA**
calc	114	75	1700
7z	163	126	680
cmd	389	129	5449
synergyc	609	218	12607
diff	1161	228	3002
gcc	1378	254	2760
c++	1380	256	2769
synergys	738	319	8061
size	1703	581	5540
ar	1726	593	8626
objcopy	1868	701	6293
as	2188	772	7463
objdump	2247	780	7159
steam	1353	860	16928
git	1159	947	9776
xetex	14424	1277	18579
gvim	1997	1666	19145
Dooble	2579	1884	57598
luatex	3514	2118	18381
celestia	2844	2136	24950
DosBox	3727	3013	24217
emule	5758	3264	52434
filezilla	7994	7085	79367
IdentityFinder	23874	12781	180176

non-executed bytes (which may be data or code), but the bytes that do execute are definitely code. We compared these results to the static disassembly yielded by our disassembler, by IDA Pro, and by the dynamic disassembly tool VDB/-Vivisect [10]. Both our disassembler and IDA Pro were 100% accurate against the execution paths that actually executed during the tests, but VDB/Vivisect exhibited much lower accuracies of around 15–35%. We also used VDB/Vivi-

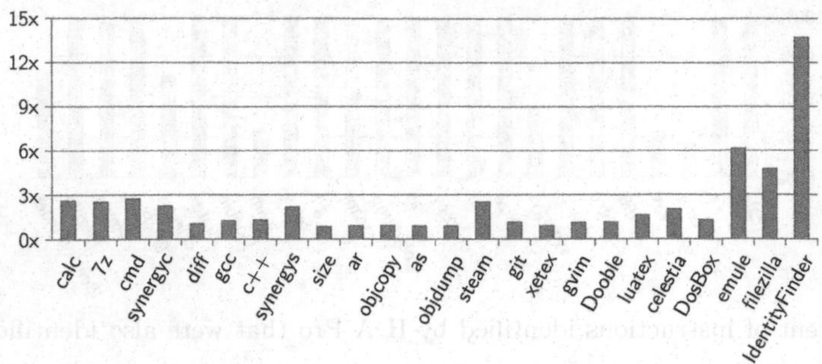

Fig. 6. Ratio of IDA Pro's disassembly time to our disassembly time

sect to dynamically trace command line tools, such as the Spec2000 benchmark suite and Cygwin, and obtained similar code coverages. This provides significant evidence that purely dynamic disassembly is not a viable solution to many disassembly problems where high code coverage is essential.

5 Conclusion

We presented an extremely simple yet highly effective static disassembly technique using probabilistic finite state machines. It finds the most probable set of byte classifications from all possible valid disassemblies. Compared to the current state-of-the-art IDA Pro, our disassembler runs in time linear in the size of the input binary. We achieve greater efficiency, and experiments indicate that our resulting disassemblies are more accurate than those yielded by IDA Pro.

We are currently working on extending our disassembler to instrument and record the actual execution traces of executables, for better estimation of ground truth and therefore more comprehensive evaluation of accuracy. One major challenge is to get high code coverage—the percentage of the code sections covered during each execution—especially for large applications. The instrumented execution traces would give us the advantage to verify all identified code sections in a controlled and automatic fashion.

References

1. Wartell, R., Zhou, Y., Hamlen, K.W., Kantarcioglu, M., Thuraisingham, B.: Differentiating code from data in x86 binaries. In: Gunopulos, D., Hofmann, T., Malerba, D., Vazirgiannis, M. (eds.) ECML PKDD 2011, Part III. LNCS, vol. 6913, pp. 522–536. Springer, Heidelberg (2011)
2. Krishnamoorthy, N., Debray, S., Fligg, K.: Static detection of disassembly errors. In: Proceedings of the 16th Working Conference on Reverse Engineering (WCRE), pp. 259–268 (2009)
3. Eagle, C.: The IDA Pro Book: The Unofficial Guide to the World's Most Popular Disassembler. No Starch Press, Inc., San Francisco (2008)

4. Hex-Rays: The IDA Pro disassembler and debugger, `http://www.hex-rays.com/idapro`
5. GNU Project.: Gnu binary utilities (2012), `http://sourceware.org/binutils/docs-2.22/binutils/index.html`
6. Schwarz, B., Debray, S., Andrews, G.: Disassembly of executable code revisited. In: Proceedings of the 9th Working Conference on Reverse Engineering (WCRE), pp. 45–54 (2002)
7. Intel: Intel® architecture software developer's manual (2011), `http://www.intel.com/design/intarch/manuals/243191.htm`
8. Vidal, E., Thollard, F., de la Higuera, C., Casacuberta, F., Carrasco, R.: Probabilistic finite-state machines – part I. IEEE Transactions on Pattern Analysis and Machine Intelligence 27(7), 1013–1025 (2005)
9. Vidal, E., Thollard, F., de la Higuera, C., Casacuberta, F., Carrasco, R.: Probabilistic finite-state machines – part II. IEEE Transactions on Pattern Analysis and Machine Intelligence 27(7), 1026–1039 (2005)
10. Invisigoth of KenShoto: Visipedia, `http://visi.kenshoto.com`

Collective Matrix Factorization of Predictors, Neighborhood and Targets for Semi-supervised Classification

Lucas Rego Drumond[1], Lars Schmidt-Thieme[1], Christoph Freudenthaler[1], and Artus Krohn-Grimberghe[2]

[1] ISMLL - Information Systems and Machine Learning Lab
University of Hildesheim, Germany
http://www.ismll.uni-hildesheim.de
[2] AIS-BI University of Paderborn, Germany
{ldrumond,schmidt-thieme,freudenthaler}@ismll.de, artus@aisbi.de

Abstract. Due to the small size of available labeled data for semi-supervised learning, approaches to this problem make strong assumptions about the data, performing well only when such assumptions hold true. However, a lot of effort may have to be spent in understanding the data so that the most suitable model can be applied. This process can be as critical as gathering labeled data. One way to overcome this hindrance is to control the contribution of different assumptions to the model, rendering it capable of performing reasonably in a wide range of applications. In this paper we propose a collective matrix factorization model that simultaneously decomposes the predictor, neighborhood and target matrices (PNT-CMF) to achieve semi-supervised classification. By controlling how strongly the model relies on different assumptions, PNT-CMF is able to perform well on a wider variety of datasets. Experiments on synthetic and real world datasets show that, while state-of-the-art models (TSVM and LapSVM) excel on datasets that match their characteristics and have a performance drop on the others, our approach outperforms them being consistently competitive in different situations.

Keywords: Semi-supervised classification; factorization models.

1 Introduction

In certain domains, the acquisition of labeled data might be a costly proccess making it difficult to exploit supervised learning models. In order to surmount this, the field of semi-supervised learning [2] studies how to learn from both labeled and unlabeled data. Given the small amount of available labeled data, semi-supervised learning methods need to make strong assumptions about the data distribution. The most prominent assumptions (briefly discussed in Section 2) are the cluster and the manifold assumption.

It is usually the case that, if such assumptions do not hold, unlabeled data may be seriously detrimental to the algorithms performance [3]. As a consequence, determining in advance good assumptions about the data and developing or choosing

V.S. Tseng et al. (Eds.): PAKDD 2014, Part I, LNAI 8443, pp. 286–297, 2014.

models accordingly may be as critical as gathering labeled data. In chapter 21 from [2] an extensive benchmark evaluation of semi-supervised learning approaches on a variety of datasets is presented resulting in no overall winner, i.e. no method is consistently competitive at all datasets, so that one has to rely on background knowledge about the data. General models that can work well on different kinds of data offer a means to circumvent this pitfall. One promising family of models that work in this direction are factorization models [10]. Such models are flexible enough to fit different kinds of data without overfitting (given that they are properly regularized). However, to the best of our knowledge, there is no systematic evaluation of the capabilities of factorization models as semi-supervised classifiers.

In this work, we show how semi-supervised learning can be approached as a factorization problem. By factorizing the predictor matrix, one can exploit unlabeled data to learn meaningful latent features together with the decision boundary. This approach however is suboptimal if the data is not linearly separable. Thus, we enforce neighboring points in the original space to still be neighbors in the learned latent space by factorizing also the adjacency matrix of the nearest neighbor graph. We call this model the Predictor/Neighborhood/Target Collective Matrix Factorization (PNT-CMF). While the state-of-the-art approaches may usually be very effective in some datasets, they perform poorly in others; we provide empirical evidence that PNT-CMF can profit from unlabeled data, making them competitive in settings where different model assumptions hold true. The main contributions of the paper are:

- We propose PNT-CMF, a novel model for semi-supervised learning that collectively factorizes the predictor, neighborhood and target relation.
- We devise a learning algorithm for PNT-CMF that is based on simultaneous stochastic gradient descent over all three relations.
- In experiments on both synthetic and real-world data sets we show that our approach PNT-CMF outperforms existing state-of-the-art methods for semi-supervised learning. Especially we show that while existing approaches work well for datasets with matching characteristics (cluster-like datasets for TSVM and manifold-like datasets for LapSVM), our approach PNT-CMF consistently performs competitive under varying characteristics.

2 Related Work

For a thorough survey of literature on semi-supervised learning in general, the reader is referred to [2] or [15]. In order to learn from just a few labeled data points, the models have to make strong assumptions about the data. One can categorize semi-supervised classification methods according to such assumptions. Historically, the first semi-supervised algorithms were based on the idea that, if two points belong to the same cluster, they have the same label. This is called the *cluster* assumption. If this assumption holds, it is reasonable to expect that the optimal decision boundary should stay in a *low density region.* Methods which fall into this category are the transductive SVMs [5] and the information regularization framework [11]. As pointed out by [15], this assumption does not hold true if, for instance, the data is generated by two highly overlapping gaussians. In this case, a *generative model* like the EM with mixture models [8] would be able to devise an appropriate classifier.

The second most relevant assumption is that data points lie on a low dimensional manifold [1]. One can think of the *manifold* assumption as the cluster assumption on the manifold. A successful approach implementing this assumption is the class of algorithms based on manifold regularization [1] [7], which regularizes the model by forcing points with short geodesic distances to have similar values for the decision function. Since the geodesic distances are computed based on the laplacian of a graph representation of the data, these methods can also be regarded as graph-based methods. This class of semi-supervised algorithms define a graph where the nodes are the data points and the edge weights are the similarity between them and are regularized to force neighboring nodes to have similar labels. Graph based methods like the one based on Gaussian Fields and Harmonic functions [16] and global consistency method [14] rely on the laplacian of the similarity graph to achieve this. These methods can be seen as special cases of the manifold regularization framework [1].

All of those methods have shown to be effective when their underlying assumptions hold true. However, when this is not the case, their performance might actually be worsened by unlabeled data. The factorization models proposed here are more flexible regarding the structure of the data since (i) they do not assme decision function lies in a low density region, but map the features to a space where they are easily separable instead and (ii) enforces neighboring points to have the same label by co-factorizing the nearest neighbor matrix, which contribution to the model can be adjusted so that the model is robust to datasets where this information is not relevant. Multi-matrix factorization as predictive models have been investigated by [10]. Previous work on the semi-supervised learning of factorization models has either focused on different tasks or had different goals from this work. While we are here focused on semi-supervised classification, previous work has focused on other tasks like clustering [12] and non-linear unsupervised dimensionality reduction [13]. A closer match is the work from Liu et al. [6] which approaches multi-label classification. Their method relies on ad-hoc similarity measures for instances and class labels that should be chosen for each kind of data. While their method only works for multi-label cases, the approach presented here deals with binary classification.

3 Problem Formulation

In a traditional supervised learning problem, data are represented by a predictors matrix X, where each row represents an instance predictor vector $\mathbf{x_i} \in \mathbb{R}^{|\mathcal{F}|}$, $\mathcal{F}$ being the set of predictors, and a target matrix Y, with each row $\mathbf{y_i}$ containing the values of the target variables for the instance i. Depending on the task, Y can take various forms. Since throughout the paper we will consider the binary classification setting, we assume Y to be a one dimensional matrix (i.e. a vector) $\mathbf{y} \in \{-1, +1\}^{|\mathcal{I}|}$, where $\mathcal{I}$ the set of instances.

Generally speaking, a learning model uses some training data $D^{\text{Train}} := (X^{\text{Train}}, \mathbf{y}^{\text{Train}})$ to learn a model that is able to predict the values in some test data $\mathbf{y}^{\text{Test}}$ given X^{Test}, all unseen when learning.

In the semi-supervised learning scenario, there are two distinct sets of training instances: the first comes with their respective labels, i.e. X_L^{Train} and $\mathbf{y}_L^{\text{Train}}$; the second is composed of training instances for which their respective labels are not known during training time, i.e. X_U^{Train}, thus the training data is composed by $D^{\text{Train}} := (X_L^{\text{Train}}, X_U^{\text{Train}}, \mathbf{y}_L^{\text{Train}})$.

At this point learning problems can again be separated in two different settings. In some situations, it is known during the learning time which instances should have their labels predicted. This is called *transductive* learning [4]. In other situations however, the focus is on learning a general model able to make predictions based on instances unknown during learning, which is called the *inductive* setting.

4 Factorization Models for Semi-supervised Classification

4.1 Classification as a Multi-matrix Factorization Task

Factorization models [10] decompose and represent a matrix as a product of two factor matrices $X \approx f(V, H)$ where $f : \mathbb{R}^{n\times k} \times \mathbb{R}^{m\times k} \to \mathbb{R}^{n\times m}$ is a function representing how two factors can be used to reconstruct the original matrix $X \in \mathbb{R}^{n\times m}$. One common choice for f is $f_X(V, H) := VH^\top$.

In some cases, two or more matrices need to be factorized at the same time. [10] propose a loss function for decomposing an arbitrary number of matrices. Be $\mathcal{M}$ the set of matrices to be factorized and Θ the set of factor matrices to be learned, each matrix $M \in \mathcal{M}$ is reconstructed by $M \approx f_M(\theta_{M_1}, \theta_{M_2})$, where $\theta_{M_1}, \theta_{M_2} \in \Theta$. Thus the overall loss is

$$J(\Theta) := \sum_{M\in\mathcal{M}} \alpha_M \, l_M(M, f_M(\theta_{M_1}, \theta_{M_2})) + Reg(\Theta) \tag{1}$$

where l_M is a loss function that measures the reconstruction error of matrix M, $0 \le \alpha_M \le 1$ is a hyperparameter defining the relative importance of each loss for the general objective function, and *Reg* is some regularization function.

In a classification problem the predictor matrix X and their respective targets $\mathbf{y}$ are given. A factorization model approximates X as a function of two latent factor matrices V, H, i.e. $X \approx f_X(V, H)$ and $\mathbf{y}$ as a function of $\mathbf{w} \in \mathbb{R}^{1\times k}$ and V, i.e. $\mathbf{y} \approx f_y(V, \mathbf{w})$. This way each row $\mathbf{v}_i$ of matrix V is a k-dimensional representation of the instance $\mathbf{x}_i$. The predicted targets are found in the approximating reconstruction of $\mathbf{y}$, i.e. $\mathbf{y} \approx V\mathbf{w}^\top$. This task can be defined as finding the factor matrices V, $\mathbf{w}$ and H that optimize a specific case of equation 1, namely:

$$J(V, \mathbf{w}, H) := \alpha\, l_X(X, f_X(V, H)) + (1 - \alpha) l_{\mathbf{y}}(\mathbf{y}, f_{\mathbf{y}}(V, \mathbf{w})) + Reg(V, \mathbf{w}, H) \quad (2)$$

For the purposes of this work, the approximation functions will be the product of the factor matrices, i.e.:

$$X \approx f_X(V, H) = VH^\top$$
$$\mathbf{y} \approx f_y(V, \mathbf{w}) = V\mathbf{w}^\top$$

This model allows for different possible choices for l_X and l_y. Since it is common to represent instances of a classification problem as real valued feature vectors (and this is the case for the datasets used in our experiments), we used the squared loss as l_X. For $l_{\mathbf{y}}$, a number of losses are suited for classification problems. We use the hinge loss here, since we are dealing with binary classification problems. In principle, any loss function can be used, so the one that best fits the task at hand should be selected.

One drawback of using the hinge loss is that it is not smooth, meaning that it is less easy to optimize. To circumvent this, we use the smooth hinge loss proposed by [9]:

$$h(y, \hat{y}) := \begin{cases} \frac{1}{2} - y\hat{y} & \text{if } y\hat{y} \leq 0, \\ \frac{1}{2}(1 - y\hat{y})^2 & \text{if } 0 < y\hat{y} < 1, \\ 0 & \text{if } y\hat{y} \geq 1 \end{cases} \quad (3)$$

4.2 Neighborhood Based Feature Extraction

The model presented so far is flexible enough for fitting a variety of datasets, but it still can not handle non-linear decision boundaries unless a very high number of latent dimensions is used, which are difficult to estimate from few labeled data. On the top of that, if the data follow the manifold assumption (i.e. data points lying next to each other on the manifold tend to have the same labels), the factorization model presented so far, will not be able to exploit this fact to learn better decision boundaries. Because we use a linear reconstruction of $\mathbf{y}$, if the data is not linearly separable in the learned latent space, the algorithm will fail to find a good decision boundary. This problem can be circumvented by forcing that the nearest-neighborhood relationship is maintained on the latent space. This works because the factorization of $\mathbf{y}$ forces the labeled points from different classes to be further apart from each other in the latent space. Forcing that the neighborhood in the original space is preserved in the latent space will make the unlabeled points to be "dragged" towards their nearest labeled neighbor, thus separating clusters or structures in the data, making it easier to find a good decision boundary that is linear in the latent space.

To accomplish this we construct a nearest neighbor graph and factorize its adjacency matrix $K \in \mathbb{R}^{n \times n}$, where each position k_{ij} is 1 if instance j is one of the N nearest neighbors of i and 0 otherwise. The nearest neighbor matrix is reconstructed as $K \approx f_K(V, V)$. This enforces that instances close to each other have similar latent features, thus being close in the latent space. We call this model the *Predictor/Neighborhood/Target Collective Matrix Factorization (PNT-CMF)*. Complexity control is achieved using Tikhonov regularization. The objective function we optimize PNT-CMF for in this work is shown in eq. 4, where $||\cdot||_F$ stands for the Frobenius norm.

$$\begin{aligned} J(V, H, \mathbf{w}) := \alpha_X \sum_{i \in \mathcal{I}} \sum_{f \in \mathcal{F}} (x_{i,f} - \mathbf{v}_i \mathbf{h}_f^\top)^2 + \alpha_K \sum_{i \in \mathcal{I}} \sum_{j \in \mathcal{I}} (k_{i,j} - \mathbf{v}_i \mathbf{v}_j^\top)^2 \\ + \alpha_Y \sum_{i \in \mathcal{L}} h(y_i, \mathbf{v}_i \mathbf{w}^\top) + \lambda_V ||V||_F^2 + \lambda_W ||\mathbf{w}||_F^2 + \lambda_H ||H||_F^2 \end{aligned} \tag{4}$$

The hyperparameter α_K controls the importance of the nearest neighbor relation to the model. This is a very important parameter since that, if the data does not have a cluster structure, or if it has a misleading cluster structure (i.e. points belonging to different classes in the same cluster), the factorization of K will harm the model more than help to improve its performance.

We point out that factorizing the nearest neighbor matrix is related to, but differs from, the concept of manifold regularization [1]. Manifold regularization forces instances close to each other to have similar values in the decision function. Forcing that neighbors have similar latent features causes the same effect in our multi-matrix factorization model. It has been observed however that, if the data does not follow the manifold or cluster assumption, manifold regularization based methods like Laplacian SVM and Laplacian Regularized Least Squares fail to find a good solution [2]. In this case the best solution is to set the laplacian regularization constant to zero, reducing the model to a fully supervised one, not taking advantage of the unlabeled data points. Here, by setting $\alpha_K = 0$ one still has a powerful semi-supervised model that simply does not rely on the neighborhood relation. In the experiments conducted in this paper we have some indication that it is possible to automatically estimate good values for α_K through model selection without any background knowledge on the dataset, although further investigation in this direction is needed.

4.3 Semi-supervised Learning of PNT-CMF

In a transductive setting, the test instances predictors are available at training time. A factorization model can naturally make use of this information by adding those predictors to the X matrix. If $\mathbf{y}$ is only partially observed, then the training data for the transductive factorization model is

$$X := \begin{bmatrix} X^{\text{Train}} \\ X^{\text{Test}} \end{bmatrix}, \mathbf{y} := \begin{bmatrix} \mathbf{y}^{\text{Train}} \\ ? \end{bmatrix}$$

Here non-observed values are denoted by a question mark. Learning a transductive model means optimizing a factorization model for equation 4 on the data above.

Algorithm 1. LearnPNT-CMF

procedure LEARNPNT-CMF
input: $X \in \mathbb{R}^{n \times m}, \mathbf{y} \in \mathbb{R}^{n \times l}, \lambda_V, \lambda_H, \lambda_{\mathbf{w}}, \alpha_X, \alpha_Y, \alpha_K, d, N$
$V, H, \mathbf{w} \sim \mathcal{N}(0, \sigma)$
$\mathcal{I} := \{1, ...n\}$
$\mathcal{F} := \{1, ...m\}$
$\mathcal{L} := \{1, ..., l\}$
K $\leftarrow$ computeNearestNeighborMatrix(X, d, N)
repeat
draw i from $\mathcal{I}$ and f from $\mathcal{F}$
$\mathbf{v_i} \leftarrow \mathbf{v_i} + \mu \left(\alpha_X \frac{\partial}{\partial \mathbf{v}_i} l_X(x_{i,f}, \mathbf{v}_i \mathbf{h}_f^\top) + \lambda_V \mathbf{v_i} \right)$
$\mathbf{h_f} \leftarrow \mathbf{h_f} + \mu \left(\alpha_X \frac{\partial}{\partial \mathbf{h}_f} l_X(x_{i,f}, \mathbf{v}_i \mathbf{h}_f^\top) + \lambda_H \mathbf{h_f} \right)$
draw i, j from $\mathcal{I}$, such that $i \neq j$
$\mathbf{v_i} \leftarrow \mathbf{v_i} + \mu \left(\alpha_K \frac{\partial}{\partial \mathbf{v_i}} l_K(k_{ij}, \mathbf{v}_i \mathbf{v}_j^\top) + \lambda_V \mathbf{v_i} \right)$
$\mathbf{v_j} \leftarrow \mathbf{v_j} + \mu \left(\alpha_K \frac{\partial}{\partial \mathbf{v_j}} l_K(k_{ij}, \mathbf{v}_i \mathbf{v}_j^\top) + \lambda_V \mathbf{v_j} \right)$
draw i from $\mathcal{L}$
$\mathbf{v_i} \leftarrow \mathbf{v_i} + \mu \left(\alpha_y \frac{\partial}{\partial \mathbf{v_i}} l_y(y_i, \mathbf{v}_i \mathbf{w}^\top) + \lambda_V \mathbf{v_i} \right)$
$\mathbf{w} \leftarrow \mathbf{w} + \mu \left(\alpha_y \frac{\partial}{\partial \mathbf{w}} l_y(y_i, \mathbf{v}_i \mathbf{w}^\top) + \lambda_W \mathbf{w} \right)$
until convergence
return $V, H, \mathbf{w}$
end procedure

To learn this model, a stochastic gradient descent algorithm is applied as shown in Algorithm 1. The algorithm starts by randomly initializing the parameters to be learned. The values for each one of them are drawn from a normal distribution with mean 0 and variance 0.001. Following this, the neighborhood is computed based on a distance measure d. In this paper we used the euclidean distance $d(\mathbf{x}_i, \mathbf{x}_j) := ||\mathbf{x}_i - \mathbf{x}_j||^2$. Next the parameters are updated in the direction of the negative gradient of each loss.

4.4 Learning Inductive Factorization Models for Classification

The same model could be used in an inductive setting, where the instances for which we want to make predictions are not known during training time. This can be achieved by simply setting X and $\mathcal{I}$ arguments in Algorithm 1 to X^{Train} and $\mathcal{L}$ respectively. In this process, the training data for the inductive model is

$$X := X^{\text{Train}}, \mathbf{y} := \mathbf{y}^{\text{Train}}$$

One drawback of this approach is that, in order to make out-of-sample predictions, the latent representations of the test instances need to be inferred for each test instance separately. In other words, for a previously unseen instance $\mathbf{x}_i$, its respective latent feature vector $\mathbf{v}_i$ is not computed during the training phase. We approach this problem by adding a fold-in step after learning the model.

The fold-in step takes a new instance x_i and maps it to the same latent feature space as the training instances. One straight-forward way to accomplish this is to minimize Equation 5.

$$\arg\min_{\mathbf{v}_i} J(\mathbf{x}_i, H, \mathbf{v}_i) := \alpha_X \, ||\mathbf{x}_i - \mathbf{v}_i H^\top||^2 + \alpha_K \, ||\mathbf{k}_i - \mathbf{v}_i V^\top||^2 + \lambda_V ||\mathbf{v}_i||^2 \quad (5)$$

5 Evaluation

The main goals of the experiments are: 1.) compare PNT-CMF against state-of-the-art semi-supervised classifiers; 2.) assess the robustness and competitiveness of our factorization model across datasets with different characteristics; 3.) observe how useful semi-supervised transductive factorization models are compared to their inductive supervised counterparts (i.e. we want to observe how much can factorization models benefit from unlabeled data).

5.1 Datasets

Chapelle et al. [2] have run an extensive benchmark analysis of semi-supervised learning models on 6 datasets, which are also used here. The *g241c*, *g241d*, *digit* and *usps* datasets have all 1500 instances and 241 predictors while the *bci* dataset, 400 instances and 117 predictors. Finally the *text* dataset has 1500 instances and 11960 predictors. The task for all data sets is binary classification. Here we note that we conduct experiments only on binary classification in order not to contaminate the results with influences from different tasks. The application and evaluation of the model on other tasks like multi-class and multi-label classification and regression is left for future work.

5.2 Setup

For the evaluation of the methods proposed here, we employed the same protocol used in the benchmark analysis presented by [2]. Each dataset, comes with two different sets of splits: the first one with 10 randomly chosen labeled training instances and the second with 100, each one with 12 splits. We used exactly the same splits as [2], which are available for download[1]. Each model was evaluated on a transductive semi-supervised setting. In order to be able to answer to question 3 posed in the beginning of this section, we also evaluated the models on a fully supervised setting.

The performance of the model was measured using the hinge loss since it is the measure the evaluated models are optimized for. We also evaluated the models on AUC but the results are suppressed here due to the lack of space. The findings from the experiments on both measures where however very similar.

[1] `http://olivier.chapelle.cc/ssl-book/benchmarks.html`

5.3 Baselines

We compare PNT-CMF against representative methods implementing the two most important assumptions of semi-supervised learning. As a representative of the manifold assumption we chose the Laplacian SVM (LapSVM) trained in the primal [7], since manifold regularization has been one of the most successful approaches for semi-supervised learning.

The second baseline is the transductive version of SVMs (TSVM) [5] which implements the low density (or cluster) assumption. On the inductive case, TSVM reduces to a standard SVM. Besides being representatives of their working assumptions, both LapSVM and TSVM optimize the same loss used for PNT-CMF in this paper. Other graph based methods [16][14] work under the same assumptions of LapSVM using a similar mathematical machinery (as discussed in section 2) but optimize the squared loss instead. By having all the competitor methods optimizing the same loss, the effects observed come from the models only and not from the usage of different losses.

At last, since PNT-CMF performs dimensionality reduction and LapSVM operates on a lower dimensional manifold, we add a fourth competitor method which incorporates dimensionality reduction to TSVM as well: we applied PCA dimensionality reduction to all datasets and ran TSVM using the transformed data. This method is called PCA+TSVM. For each dataset we used the first k PCA dimensions, k being the same number of latent dimensions used by PNT-CMF. As a TSVM implementation we used *SVMLight*[2]. For LapSVM we used the implementation from [7], which is also available for download[3].

5.4 Model Selection and Reproducibility of the Experiments

Model selection is a known problem for semi-supervised learning due to the low number of labeled instances available. We show that it is possible to estimate good hyperparameters for PNT-CMF even in the presence of only a few labeled data points. For **PNT-CMF**, each hyperparameter combination was evaluated through 5-fold cross-validation using **only the training data**. The code for PNT-CMF can be made available upon request to the first author. We gave the **baseline methods** a competitive advantage: use the same hyperparameter search approach but employing **both train and test data**. We also observe that the results for the competitor methods are consistent with the ones reported in the literature for the same datasets.

5.5 Results and Discussion

The hinge loss scores for the datasets with 10 and 100 labeled examples are shown in Figure 1. For each method and dataset, the average performance over the 12 splits is shown. The error bars represent the 99% confidence intervals. [2] divide

[2] `http://svmlight.joachims.org/`

[3] `http://www.dii.unisi.it/~melacci/lapsvmp/`

these datasets into two categories: the **manifold-like** and the **cluster-like**. The **manifold** group comprises the *digit*, *usps* and *bci* datasets in which the data lie near a low dimensional manifold. Algorithms like LapSVM are expected to excel in these datasets. *g241c*, *g241d* and *text* fall under the category of **cluster-like** datasets in which different classes do not share the same cluster thus making the optimal decision boundary to lie in a low density region, favoring algorithms like TSVM.

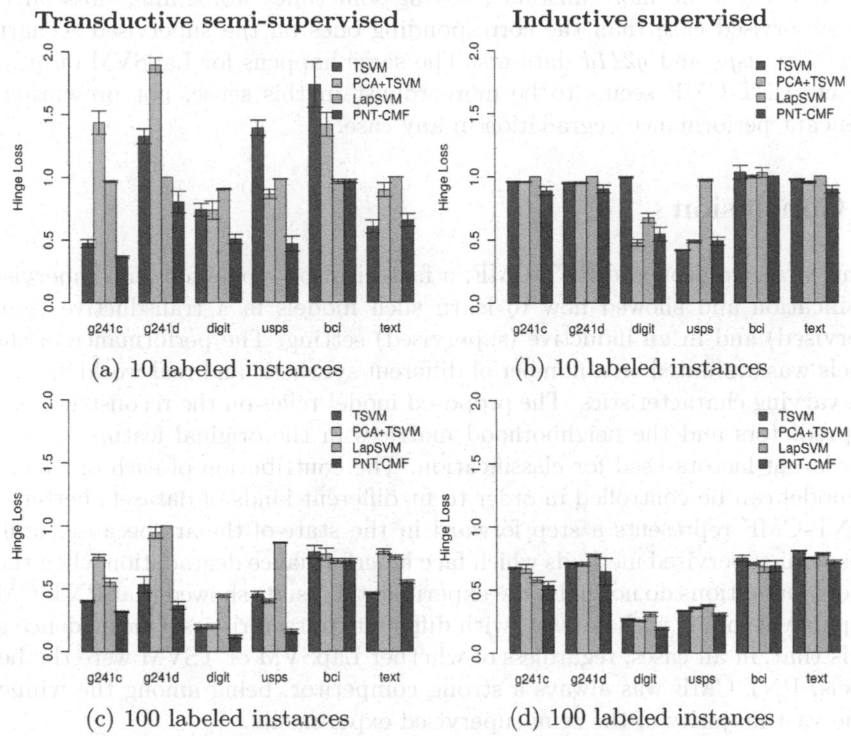

(a) 10 labeled instances (b) 10 labeled instances

(c) 100 labeled instances (d) 100 labeled instances

Fig. 1. Results for the Hinge Loss. The *lower* the better.

The left-hand side of Figure 1 shows how PNT-CMF performs in comparison to its competitors. By looking specially at this figure, one can see that TSVM is stronger in the *g241c* and *text* datasets while LapSVM is more competitive in the manifold-like data. One can also see that LapSVM is significantly weaker on the *g241c* and *text* sets where the data do not lie near a low dimensional manifold. PNT-CMF on the other hand is always either the statistically significant winner or is away from the winner by a non significant margin. Out of the 12 experiments, PNT-CMF is the sole winner in 8 of them and is one of the winning methods in all the 3 ties, with TSVM winning in one case, and only for 100 labeled instances. We show here the Hinge loss results because this was the measure all the models in the experiment are optimized for. As already said we also evaluated the AUC of these models in these experiments and the results were similar, with PNT CMF being the sole winner in 5 experiments and one of the winning methods in the other 7. This supports our claim that PNT-CMF can consistently work well under different assumptions. In accordance with our expectations, LapSVM is more competitive on the manifold-like datasets while TSVM on the cluster-like ones.

Finally, by comparing the right and left hand sides of Figure 1, we can have an idea of the effects of taking unlabeled data into account. One can observe that TSVM seems to be more unstable, having sometimes worse hinge loss on the semi-supervised case than the corresponding ones on the supervised scenarios for the *bci*, *usps* and *g241d* datasets. The same happens for LapSVM on *g241d* dataset. PNT-CMF seems to be more robust in this sense, not presenting a significant performance degradation in any case.

6 Conclusion

In this work we proposed PNT-CMF, a factorization model for semi-supervised classification and showed how to learn such models in a transductive (semi-supervised) and in an inductive (supervised) setting. The performance of such models was evaluated on a number of different synthetic and real-world datasets with varying characteristics. The proposed model relies on the reconstruction of the predictors and the neighborhood matrices in the original feature space to learn latent factors used for classification. The contribution of each of them to the model can be controlled in order to fit different kinds of datasets better.

PNT-CMF represents a step forward in the state-of-the-art because, unlike other semi-supervised methods which face a performance degradation when their model assumptions do not hold, the experimental results showed that PNT-CMF is capable of coping with datasets with different characteristics. One evidence for this is that, in all cases, regardless of whether LapSVM or TSVM were the best models, PNT-CMF was always a strong competitor, being among the winners in the vast majority of the semi-supervised experiments.

As future work we plan to investigate better factorization strategies for the matrix K, like different loss and reconstruction functions. Also the extension and evaluation of the model on regression, multi-class and multi-label classification tasks is an issue to be further investigated.

Acknowledgments. The authors gratefully acknowledge the co-funding of their work by the Multi-relational Factorization Models project granted by the Deutsche Forschungsgesellschaft[4]. Lucas Drumond is sponsored by a scholarship from CNPq, a Brazilian government institution for scientific development.

[4] `http://www.ismll.uni-hildesheim.de/projekte/dfg_multirel_en.html`

References

1. Belkin, M., Niyogi, P., Sindhwani, V.: Manifold regularization: A geometric framework for learning from labeled and unlabeled examples. The Journal of Machine Learning Research 7, 2399–2434 (2006)
2. Chapelle, O., Schölkopf, B., Zien, A. (eds.): Semi-Supervised Learning. MIT Press, Cambridge (2006)
3. Cozman, F., Cohen, I., Cirelo, M.: Semi-supervised learning of mixture models. In: 20th International Conference on Machine Learning, vol. 20, pp. 99–106 (2003)
4. Gammerman, A., Vovk, V., Vapnik, V.: Learning by Transduction. In: Proceedings of the Fourteenth Conference on Uncertainty in Artificial Intelligence, pp. 148–156. Morgan Kaufmann (1998)
5. Joachims, T.: Transductive Inference for Text Classification using Support Vector Machines. In: Proceedings of the 1999 International Conference on Machine Learning, ICML (1999)
6. Liu, Y., Jin, R., Yang, L.: Semi-supervised multi-label learning by constrained non-negative matrix factorization. In: Proceedings of the National Conference on Artificial Intelligence, vol. 21, p. 421. AAAI Press (2006)
7. Melacci, S., Belkin, M.: Laplacian Support Vector Machines Trained in the Primal. Journal of Machine Learning Research 12, 1149–1184 (2011)
8. Nigam, K., McCallum, A., Mitchell, T.: Semi-supervised text classification using EM. In: Chapelle, O., Schölkopf, B., Zien, A. (eds.) Semi-Supervised Learning, pp. 33–56. The MIT Press, Cambridge (2006)
9. Rennie, J.: Smooth Hinge Classification (February 2005), `http://people.csail.mit.edu/jrennie/writing/smoothHinge.pdf`
10. Singh, A.P., Gordon, G.J.: Relational learning via collective matrix factorization. In: Proceeding of the 14th ACM SIGKDD International Conference on Knowledge Discovery and Data Mining, pp. 650–658. ACM, New York (2008)
11. Szummer, M., Jaakkola, T.: Information regularization with partially labeled data. In: Advances in Neural Information Processing Systems, vol. 15, pp. 1025–1032 (2002)
12. Wang, F., Li, T., Zhang, C.: Semi-supervised clustering via matrix factorization. In: Proceedings of the 2008 SIAM International Conference on Data Mining, pp. 1–12. SIAM (2008)
13. Weinberger, K., Packer, B., Saul, L.: Nonlinear dimensionality reduction by semidefinite programming and kernel matrix factorization. In: Proceedings of the Tenth International Workshop on Artificial Intelligence and Statistics, pp. 381–388 (2005)
14. Zhou, D., Bousquet, O., Lal, T.N., Weston, J., Scholkopf, B.: Learning with local and global consistency. In: Advances in Neural Information Processing Systems, vol. 16, pp. 321–328 (2004)
15. Zhu, X.: Semi-supervised learning literature survey. Tech. Rep. 1530, University of Wisconsin, Madison (December 2006)
16. Zhu, X., Ghahramani, Z., Lafferty, J.: Semi-Supervised Learning Using Gaussian Fields and Harmonic Functions. In: Proceedings of the Twentieth International Conference on Machine Learning (ICML), pp. 912–919 (2003)

Ranking Tweets with Local and Global Consistency Using Rich Features

Zhankun Huang[1,2], Shenghua Liu[1,*], Pan Du[1], and Xueqi Cheng[1]

[1] Institute of Computing Technology, Chinese Academy of Sciences, Beijing, China
[2] University of Chinese Academy of Sciences, Beijing, China
{huangzhankun}@software.ict.ac.cn, {liushenghua,dupan,cxq}@ict.ac.cn

Abstract. Ranking tweets is more challenging in Microblog search because of content sparseness and lack of context. Traditional ranking methods essentially using Euclidean distance are limited to local structure. Manifold structure helps to rank with local and global consistency. However such structure is empirically built on content similarity in an unsupervised way, suffering from sparseness while being adopted in tweet ranking. In this paper, we explore rich features to alleviate content sparseness problem, where time locality feature is proposed to consider context dependency. We then propose a supervised learning model that aggregates rich features to construct manifold structure. A learning algorithm is then designed for solving the model by minimizing the loss of labeled queries. At last we conduct a series of experiments to demonstrate the performance on 109 labeled queries from TREC Microblogging. Compared with the well-known baselines and empirical manifold structure, our algorithm achieves consistent improvements on the metrics.

Keywords: time locality, rich features, manifold, supervised learning.

1 Introduction

Statistics from Wikipedia [17] indicate that, there are approximately 340 million tweets posted per day. Besides to reading tweets from their timelines or trending topics, people also need to search tweets on a topic. Different from traditional webpage search, tweet ranking in Microblog search is more challenging. Because of the limitation on length, tweets do not provide sufficient word occurrences. Tweets are context-dependent while lacking context makes context similarity cannot be measured directly. Nevertheless, tweets on a specific topic are usually temporally concentrative, especially when some big event breaks out, like terrorist attack. Hence tweets posted during a short time period can be regarded as having the same context. On the other hand, tweets posted in totally different time periods possibly refer to different topics though they contain some common words. As the examples from TREC Microblogging 2011 show in Table 1, both of them talked about "protest". However since they were posted in quite different

* Corresponding author.

V.S. Tseng et al. (Eds.): PAKDD 2014, Part I, LNAI 8443, pp. 298–309, 2014.

Table 1. Context-dependent tweets examples: both tweets were talking about "protest". But because of different contexts, they were related to different topics.

Relevant Tweet		Query
Post Time	**Content**	
Jan. 29 2011 18:14:21	Al Jazeera: **Protest**ors create a human shield around the Egyptian Museum to protect it.	Egyptian **protest**ers attack museum
Feb. 01 2011 12:46:40	Royal Palace: Jordan's king sacks government in wake of **protests**.	**protest**s in Jordan

contexts measured by post time, they were about different topics. The former one was related to "Egyptian protesters attack museum" and the latter one was about "protests in Jordan". Thus we use the time locality to approximate the context dependency in this paper.

Traditional ranking algorithms based on Euclidean distances are limited to the local structure [20]. Manifold structure has been employed for ranking problem [3,5,16,20], which maintains local and global consistency simultaneously. Generally [19,20], manifold structure is built on content similarity empirically. Unlike traditional webpages, tweets are extremely short and lack context. Content similarity alone cannot well describe the relevance between tweets. The manifold graph of all related tweets for a query may be disconnected due to sparseness, and mixed with much noise because of different contexts. Therefore, rich features, especially the time locality feature are explored to weight the network of manifold structure.

In this paper, we aggregate rich features to construct the manifold structure in a supervised way, with time locality as an approximation of context dependency. The similarities between query and tweets or tweets themselves are measured from different views in terms of each feature, so that several heterogeneous structures are obtained, which are then linearly combined to model the objective function with local and global consistency. A supervised algorithm is then designed to learn the combination coefficients. As seen later, the manifold structure leant from rich features better benefits the ranking than being determined empirically on a single content similarity. A series of experiments are conducted on the dataset of tweet corpus and labeled queries officially provided by TREC Microblogging. The evaluation results show that, in terms of P@10, P@20, P@30, and AUC (area under ROC curve), our approach can outperform the ranking models of Lucene and RankSVM.

The rest of the paper is organized as follows: Section 2 discusses the related work. Section 3 describes our approach in detail. Section 4 introduces the features we use for ranking. Section 5 demonstrates our evaluation results and we make conclusion in Section 6.

2 Related Works

Due to the features of short, sparseness and frequency, Microblog search is very different from traditional web search. As Teevan [14] explored, Twitter queries were shorter, but contained longer words, more specialized syntax, and more references to people. In addition, unlike the URLs in social bookmarking sites which have provided high values for search engine, the shorten URLs in tweets can either be high in quality or spam [9]. We will not expand tweets by extracting the contents that the URLs link to, and treat whether a tweet contains URL as a feature.

User oriented search has attracted more and more attentions. Instead of presenting a simple list of tweet messages, TweetMotif [12] provided a facet search interface to allow user to navigate tweets by topics, by which searched tweets are grouped into subtopics extracted by language model. Another personalized tweet ranking method [15] was proposed by ranking the incoming tweets based on the likelihood user may retweet them, and ranking the users given a tweet based on their willingness to retweet it. Using four groups of features: Author-based, Tweet-based, Content-based, User-based, a Coordinate Ascent learning to rank algorithm was trained to rank tweets according to users' need. To address the problem of sparseness Naveed [11] ignored length normalization for tweets. In his point of view, document length normalization for short documents like tweets may introduce an unmotivated bias. Abel [1] enriched the semantics of tweets by extracting facet values from tweets, where external resources were used in order to create facet and facet-value pairs. Massoudi [10] used time-dependent query expansion to overcome the disadvantage of redundancy-based IR methods over short documents retrieval. Duan [6] adopted learning to rank algorithm to select a feature set from both content and non-content features of tweets and proposed a ranking model based on RankSVM algorithm. In our approach, we treat the length of tweet as a feature, and use rich features to alleviate the problem of sparseness. Specifically, the weight of each feature is not given empirically, instead they are determined through a supervised way.

Jabeur [8] measured the conditional probability of relevance using Bayesian network model. In his study, time magnitude of tweet was estimated by query terms occurrence in the temporal neighborhood. His experimental results showed that time magnitude was a primordial feature for ranking tweets. In this paper, we introduce into time locality feature to simulate the context of tweet. For the tweets posted in a short time period, we assume that they are in the same context. Zhang [18] proposed a transductive framework that generated training data while no labeled data was available, and boosted the ranking by adding confident unlabeled data during iteratively training. In our approach, after the manifold structured is learnt, all the tweets to be ranked are treated as no labeled data and the query is treated as labeled data.

In some recent works, manifold structure has been employed for ranking problems and clustering problems[13] in consideration of local and global consistency. Zhou [20] proposed a universal ranking algorithm with respect to the intrinsic manifold structure collectively revealed by a great amount of data. Cheng and Du [3,5] introduced sink point into manifold structure to address diversity as well as relevance and importance. Wan and Yang [16] proposed a novel extractive approach based on manifold-ranking of sentences. Generally, manifold structure is built on content similarity empirically, which suffers from sparseness in tweet ranking. In our research, we leverage rich features to construct tweet manifold, including content, time locality and several intrinsic features, which are combined in a supervised way[4].

3 Our Approach

Ranking on data manifolds was proposed by Zhou [20]. Let f denote a score function, which assigns data x_i with ranking score f_i. And y_i is the initial ranking score for data x_i. For the labeled data $y_i = 1$ and for the unlabeled data $y_i = 0$. Let W denote an affinity matrix where W_{ij} is the similarity between data x_i and x_j. We name W as the similarity matrix. Note that W_{ii} is set to 0 to avoid self-reinforcement. Symmetrically normalize W by $S = D^{-1/2}WD^{-1/2}$ where D is a diagonal matrix with D_{ii} equal to the sum of the i-th row of W. As [20] illustrates, the final score function f^* can be achieved directly by:

$$f^* = (1-\alpha)(I-\alpha S)^{-1}y \,. \tag{1}$$

or in an iterative way using the following iterative function:

$$f_{t+1} = \alpha S f_t + (1-\alpha)y \,. \tag{2}$$

where α is a parameter for trade-off.

For tweet ranking, we then construct a weighted network over the manifold structure of tweets, where the nodes in the network are the unlabeled tweets and labeled query, and edges reflect the similarities between query and tweets or tweets themselves. Then we perform a graph-based semi-supervised method to rank tweets by their distances to the query on the intrinsic manifold structure. In the original thought of manifold regularization framework, the similarity matrix W is empirically built on content similarity. In our approach, we do not determine W in advance. Instead we achieve it through a supervised learning process leveraging rich features.

For each feature ϕ, we can define some similarity measurements. Then the similarity matrix W for our approach is defined as follow:

$$W = a^T \mathcal{F} = \sum_{i=1} a^i F^i \,. \tag{3}$$

where each F^i is a similarity matrix achieved by a kind of similarity measurement over some feature, and a is the linear coefficient vector needed to be learnt.

Given a corpus $CO = \{(q^i, T^i) | i = 1, \ldots, N\}$, where q^i denotes the i-th query and T^i denotes the set of labeled tweets for q^i, $P^i = \{p_1, \ldots, p_j\}$ and $O^i = \{o_1, \ldots, o_j\}$ denotes the set of relevant and irrelevant tweets in T^i. To find out the optimal solution for coefficient a, we construct the following optimization problem:

$$min_a L(a) = \frac{1}{2}\lambda \|a\|^2 + \sum_{i=1}^{N} \sum_{\substack{o \in O^i \\ p \in P^i}} l(f_o^i - f_p^i) \,. \tag{4}$$

where f^i is the final score function for q^i obtained from Eqn. (1) directly or Eqn. (2) iteratively, and λ is the regularization parameter that trades off between the model complexity and the fit of the model. $l(x)$ is the loss function which assigns a non-negative penalty to the violation of constraint $f_p^i > f_o^i$. Here we use Wilcoxon-Mann-Whitney (WMW) loss $l(x) = (1 + e^{-\frac{x}{b}})^{-1}$ since it shows good performance in our approach.

To minimize (4) with respect to parameter a, we use gradient descent method to minimize the loss and find the optimal solution for a. The gradient of $L(a)$ with respect to a is as follow:

$$\frac{\partial L(a)}{\partial a} = \lambda a + \sum_{i=1}^{N} \sum_{\substack{o \in O^i \\ p \in P^i}} \frac{\partial l(\delta_{op})}{\partial \delta_{op}} \left(\frac{\partial f_o^i}{\partial a} - \frac{\partial f_p^i}{\partial a} \right). \quad (5)$$

where $\delta_{op} = f_o^i - f_p^i$. Next we need to know $\frac{\partial f^i}{\partial a}$, $i.e. \frac{\partial f^*}{\partial a}$.

When f_t in Eqn. (2) converges to f^*, we have:

$$\begin{aligned} f^* &= \alpha S f^* + (1 - \alpha) y \,. \\ \frac{\partial f^*}{\partial a} &= \alpha \left(\frac{\partial S}{\partial a} f^* + S \frac{\partial f^*}{\partial a} \right). \\ \frac{\partial f^*}{\partial a} &= \alpha (I - \alpha S)^{-1} \left(\frac{\partial S}{\partial a} f^* \right). \end{aligned}$$

where $\frac{\partial S}{\partial a}$ is a supervector, elements in which are matrices, and can be expressed as: $\frac{\partial S}{\partial a} = \left[\frac{\partial S}{\partial a^1}, \frac{\partial S}{\partial a^2}, \ldots, \frac{\partial S}{\partial a^k} \right]$. Here we denote D^i as a diagonal matrix with (j, j)-element equal to the sum of the j-th row of the i-th feature matrix F^i for a specific query, then we have:

$$\begin{aligned} \frac{\partial S}{\partial a^i} &= \frac{\partial D^{-\frac{1}{2}}}{\partial a^i} W D^{-\frac{1}{2}} + D^{-\frac{1}{2}} \frac{\partial W}{\partial a^i} D^{-\frac{1}{2}} + D^{-\frac{1}{2}} W \frac{\partial D^{-\frac{1}{2}}}{\partial a^i} \\ &= -\frac{1}{2} D^{-\frac{3}{2}} D^i W D^{-\frac{1}{2}} + D^{-\frac{1}{2}} F^i D^{-\frac{1}{2}} - \frac{1}{2} D^{-\frac{1}{2}} W D^{-\frac{3}{2}} D^i \,. \end{aligned}$$

For the corpus we are going to introduce in the following, tweets are categorized into highly relevant, minimally relevant and irrelevant. To adapt to this multi-label situation, Eqn. (4) and Eqn. (5) are needed to be modified as follow:

$$min_a L(a) = \frac{1}{2} \lambda \|a\|^2 + \sum_{i=1}^{N} \sum_{\substack{o \in O^i \\ m \in M^i}} l(f_o^i - f_m^i) + \sum_{i=1}^{N} \sum_{\substack{o \in O^i \\ p \in P^i}} l(f_o^i - f_p^i) + \sum_{i=1}^{N} \sum_{\substack{m \in M^i \\ p \in P^i}} l(f_m^i - f_p^i) \,. \quad (6)$$

$$\frac{\partial L(a)}{\partial a} = \lambda a + \sum_{i=1}^{N} \sum_{\substack{o \in O^i \\ m \in M^i}} \frac{\partial l(\delta_{om})}{\partial \delta_{om}} \left(\frac{\partial f_o^i}{\partial a} - \frac{\partial f_m^i}{\partial a} \right) + \sum_{i=1}^{N} \sum_{\substack{o \in O^i \\ p \in P^i}} \frac{\partial l(\delta_{op})}{\partial \delta_{op}} \left(\frac{\partial f_o^i}{\partial a} - \frac{\partial f_p^i}{\partial a} \right)$$
$$+ \sum_{i=1}^{N} \sum_{\substack{m \in M^i \\ p \in P^i}} \frac{\partial l(\delta_{mp})}{\partial \delta_{mp}} \left(\frac{\partial f_m^i}{\partial a} - \frac{\partial f_p^i}{\partial a} \right) . \quad (7)$$

where O^i, M^i and P^i denote the set of irrelevant, minimally relevant and highly relevant tweets for q^i respectively.

Finally the optimization problem in (6) can be solved with a standard quasi-Newton method. Here Broyden-Fletcher-Goldfarb-Shanno (BFGS) [2] is employed for the learning process. Note that the optimization problem in Eqn. (6) is generally non-convex, we try several different start points to find a good solution which though may not be a global optimal solution.

Table 2 shows the overview of the learning algorithm for manifold structure, in which derivative of $L(a)$ is denoted as $dL(a)$.

Table 2. Learning algorithm of manifold structure for ranking tweets

Algorithm. Learning of Manifold Structure

Given an initial point a=a_0 and an initial Hessian matrix $B = B_0 = I$.
loop
  evaluate $dL(a)$ using Eqn.(7).
  if $norm(dL(a)) < \varepsilon$:
    break.
  else:
    obtain a descent direction p by solving $Bp = -dL(a)$.
    perform a linear search to find an acceptable step size s.
    update a using $a = a + sp$.
    update Hessian matrix B as BFGS algorithm says.
end loop
return fina a.

4 Features Description

To represent the similarity matrix W, we explore rich features to measure the similarities between query and tweets or tweets themselves from different views in terms of all these features, so that rich features can be exploited. The available features can be derived from, for example, content, context, length etc.

Number of co-occurrence word, cosine similarity, Dice coefficient and Jaccard similarity are employed to measure the content similarity from different views. Note that by weighting the words in tweets and query using different approaches, we compose different types of features and similarities, such as cosine similarity using tf weight and cosine similarity using tf-idf weight. All these similarities will be combined together to better capture the content similarity.

As addressed in Section 1, for ranking tweets, time locality is an important feature that can help us address the criterion of context awareness. Thus the context dependency can be approximated as the following equation (8) in terms of time locality.

$$1 - \frac{|pt_i - pt_j|}{max\{|pt_i - pt_j|| \forall i > j\}} \tag{8}$$

where pt_i is the post time of tweet t_i. For the query, the post time is represented as querying time.

Several intrinsic features are extracted as well. Here we only introduce some of them. "*oov_ratio*" is the out-of-vocabulary word count over the total word count, "*unique_word_ratio*" is similar to "*oov_ratio*", "*begin_with_at*" means whether the content of tweet starts with "@", "*has_rt*" means whether the tweet is posted by retweeting others. On the other hand, we combine these intrinsic features into vector as a new feature, which is denoted as "*intrinsic_vector*", and use cosine similarity on it. Because of lack of intrinsic features for the query, the similarities between query and tweets in terms of intrinsic features are set to 0.

The rich features for similarity measurement derived from content, time locality and intrinsic features are listed in Table 3. Note that we may not know in advance which feature is better than others, or whether a feature is useless and can be removed. Through a supervised learning approach, we learn the combination of these features and assign useful features higher weights.

Table 3. Rich features for similarity measurement derived from content, time locality and intrinsic features

Content Features	Time Locality Feature	Intrinsic Features
co_occurrence_bool		
co_occurrence_tf		oov_ratio
co_occurrence_idf		doc_len
co_occurrence_tfidf		unique_word_ratio
cos_sim_bool		entropy
cos_sim_tf	time_locality	has_url
cos_sim_tfidf		has_hashtag
dice_bool		begin_with_at
dice_idf		has_rt
jaccard_bool		intrinsic_vector
jaccard_idf		

5 Experiments

Our experiments are conducted based on the tweet corpus and labeled queries officially provided by TREC Microblogging. We compare our approach with the well-known baseline methods including different ranking models in Lucene,

RankSVM and conventional manifold approach. We then evaluate the effectiveness of the features we selected for our approach, such as time locality and intrinsic features.

5.1 Corpus

From the TREC Microblogging, we have got about 113,928 labeled tweets on 109 topic queries, out of 7,443,387 tweets totally. The first 49 topics are treated as training set and the remained 60 query topics are for testing. We preprocessed the labeled tweets simply by removing the "*@user*", "*#tag#*", URL, stop words, punctuations and transferring all the letters into lower case. The tweets are officially labeled as highly relevant, relevant, and irrelevant with integers 2, 1, and 0 respectively.

5.2 Baseline Methods

Apache Lucene is a well-known information retrieval software library, which has been widely recognized for its utility in the implementation of Internet search engines. We compare our approach with Lucene 4.2 using default, BM25 and LMDirichlet models, which are denoted as **Lu-Default**, **Lu-BM25**, and **Lu-LMD** respectively for comparisons. It is worth noticing that real-time constraint is added to make sure that only the tweets posted earlier than querying time are returned. The parameters of BM25 and LMDirichlet have been well tuned.

We use RankSVM as another baseline method. Support vector machines (SVMs) has been proved to be effective for classification and regression analysis. RankSVM is an instance of structured SVM used for ranking problem. For RankSVM, we select almost the same features employed in our approach to represent each tweet including content, time locality and intrinsic features.

Lastly, we also implement conventional manifold approach as baseline, by constructing manifold structure on single content similarity. And the edges with similarity less than a threshold $\varepsilon \in (0, 1)$ are dropt. We implement **Manifold-Cosine** using cosine similarity with tfidf weight "cos_sim_tfidf" in Table 3.

5.3 Evaluation Metrics

We use the Microblog track task metrics for evaluation. The main metrics for the task are the receiver operating characteristic (ROC) curve [7] and P@K. The ROC curve shows precision versus fallout for every possible score threshold while P@K gives a simple measure of ranking effectiveness.

To evaluate the result that the ROC curve explains, area under the curve (AUC) metric is employed. The formulation to calculate the AUC value for a ROC curve is as follow:

$$AUC = \frac{\sum_{i \in positive\ class} rank_i - \frac{M \times (M+1)}{2}}{M \times N}$$

where M is the number of positive samples and N is the number of negative samples.

As people usually prefer to browse the results in the top returned by search engines, we only consider the top K tweets in ranking list. The value P@K is calculated as follows:

$$P@K = \frac{|\{relevant\, tweets\, in\, top\, K\, results\}|}{K}$$

5.4 Evaluation Results

In the experiments, we list P@10, P@20 and P@30 for precision metrics and denote our approach as **Manifold-Rich**.

Table 4. Comparison with baselines

Methods	P@10	incr(%)	P@20	incr(%)	P@30	incr(%)	AUC	incr(%)
Lu-Default	0.1966	44.8	0.1695	49.0	0.1475	52.8	0.7970	11.6
Lu-BM25	0.2271	25.4	0.1932	30.7	0.1678	34.3	0.7985	11.4
Lu-LMD	0.1576	80.6	0.1254	101.4	0.1192	89.1	0.7838	13.4
RankSVM	0.2627	8.4	0.2102	20.1	0.1836	22.8	0.8909	-0.2
Manifold-Cosine	0.0695	309.6	0.0907	178.4	0.0819	175.2	0.8711	2.1
Manifold-Rich	0.2847	–	0.2525	–	0.2254	–	0.8892	–

The precision comparisons between Manifold-Rich and the baseline methods are listed in Table 4. The increasing percentages, "incr(%)", of Manifold-Rich are listed right next to the corresponding metric columns. It is seen that Manifold-Rich outperforms Lucene significantly no matter it uses default, LMDirichlet or BM25 models. Compared with Lu-LMD, Manifold-Rich achieves significant improvements of 80.6%, 101.4% and 89.1% respectively on P@10, P@20 and P@30. And compared with Lu-Default, Manifold-Rich gets improvements of 44.8%, 49.0% and 52.8% respectively. Lastly compared with Lu-BM25 which gets the best precision among the three Lucene models, the improvements of our approach on P@10, P@20 and P@30 are 25.4%, 30.7% and 34.3% respectively. It is seen that only using content similarity between query and tweet in Lucene is not sufficient for ranking tweets due to the content sparseness. By leveraging the local and global structure on rich features, we can achieve much better ranking results. Compared with RankSVM, Manifold-Rich achieves improvements of 8.4%, 20.1% and 22.8% with respect to P@10, P@20 and P@30. The manifold baseline Manifold-Cosine is constructed by dropping those edges with weight less than an empirical threshold 0.05. It is seen that Manifold-Rich outperforms Manifold-Cosine with improvements of 309.6%, 178.4%, 175.2% with respect to P@10, P@20, P@30, which indicates that the manifold structure leant from rich features better benefits the tweet ranking than being empirically determined on a single content similarity. It worth noticing that with the statistical significance testing of t-test, the p-value is less than 0.01, which indicates that our approach outperforms others significantly. It is seen that in terms of AUC value, our approach receives much better results than the ranking models in Luene. Especially compared with Lu-BM25, the AUC improvement of Manifold-Rich is 11.4%. The AUC metrics of manifold approaches are close to each other. Though the ROC curve of RankSVM gains a little more AUC value than that of Manifold-Rich, considering the impressive improvements on precision, our approach is still reliable.

5.5 Effectiveness of Rich Features

To show the effectiveness of the rich features, we conduct a set of experiments with some selections of features in our approach, such as time locality feature and intrinsic features. Traditional ranking approaches mainly rely on the content similarity between document and query for ranking. Thus we implement **Manifold-Content** to evaluate the performance of our approach only with content feature without an empirical threshold as Manifold-Cosine does. We use all the content features in Table 3 to measure the content similarity. To evaluate the effectiveness of time locality feature for context awareness, we then use both content features and time locality feature in Table 3 to construct the manifold structure, and compare it with Manifold-Content. This version is denoted as **Manifold-Content-Context**.

The comparisons between different feature selections of our approach are demonstrated in Table 5. The cells of "incr(%)" give the increasing percentages of Manifold-Rich comparing to others. Compared with Manifold-Content, Manifold-Rich achieves improvements of 33.3%, 41.9% and 40.0% with respect to P@10, P@20 and P@30. It shows that instead of only using content similarity, with rich features to construct the manifold structure, the ranking results are much better. Comparing Manifold-Content-Context with Manifold-Content, we find that with time locality feature taken into account for context awareness, Manifold-Content-Context achieves promising improvements of 22.2%, 24.7% and 18.6% on P@10, P@20 and P@30 respectively, which indicates that the time locality feature has somehow captured the missing context of tweets and is valuable for constructing the manifold. Therefore, by considering both content and time locality features, we can address the criteria of content relevance and context awareness simultaneously for ranking tweets.

Table 5. Comparison between our approaches with different features

Methods	P@10	incr(%)	P@20	incr(%)	P@30	incr(%)	AUC	incr(%)
Manifold-Content	0.2136	33.3	0.1780	41.9	0.1610	40.0	0.8811	0.9
Manifold-Content-Context	0.2610	9.1	0.2220	13.7	0.1910	18.0	0.8876	0.2
Manifold-Rich	0.2847	–	0.2525	–	0.2254	–	0.8892	–

6 Conclusions

In this paper, we employ rich features to construct manifold structure on tweets in a supervised way and then rank tweets on it. The similarities between query and tweets or tweets themselves are estimated from different views in terms of each feature, resulting in heterogeneous structures, which are then combined to model the objective function. Especially, by combining content similarity and context similarity together, we can address the criteria of content relevance and context awareness simultaneously. A learning algorithm is designed to solve the model to minimize the loss of the labeled ranked tweets. Experimental results demonstrate that our approach outperforms the ranking models in Lucene significantly and achieves higher precision than RankSVM trained on the common relevance features. In addition, we experimentally show that manifold structure learnt from rich features better benefits the ranking than being determined empirically only on content similarity.

Acknowledgments. Thanks a lot for the supply of corpuses from TREC Microblogging. This paper is partially supported by National Grand Fundamental Research 973 Program of China (No. 2012CB316303, 2013CB329602), National Natural Science Foundation of China (No. 61202213, 61202215, 61173064, 61173008, 61232010), Projects of Development Plan of the State High Technology Research (No. 2012AA011003), Beijing nova program (No. Z121101002512063).

References

1. Abel, F., Celik, I., Houben, G.-J., Siehndel, P.: Leveraging the semantics of tweets for adaptive faceted search on twitter. In: Aroyo, L., Welty, C., Alani, H., Taylor, J., Bernstein, A., Kagal, L., Noy, N., Blomqvist, E. (eds.) ISWC 2011, Part I. LNCS, vol. 7031, pp. 1–17. Springer, Heidelberg (2011), http://dx.doi.org/10.1007/978-3-642-25073-6_1
2. Broyden, C.G.: A Class of Methods for Solving Nonlinear Simultaneous Equations. Mathematics of Computation 19, 577 (1965)
3. Cheng, X.Q., Du, P., Guo, J., Zhu, X., Chen, Y.: Ranking on data manifold with sink points. IEEE Transactions on Knowledge and Data Engineering 25(1), 177–191 (2013)
4. Du, P., Guo, J., Cheng, X.: Supervised Lazy Random Walk for Topic-Focused Multi-document Summarization. In: IEEE International Conference on Data Mining, pp. 1026–1031 (2011)
5. Du, P., Guo, J., Zhang, J., Cheng, X.: Manifold ranking with sink points for update summarization. In: International Conference on Information and Knowledge Management, pp. 1757–1760 (2010)
6. Duan, Y., Jiang, L., Qin, T., Zhou, M., Shum, H.Y.: An Empirical Study on Learning to Rank of Tweets. In: International Conference on Computational Linguistics, pp. 295–303 (2010)
7. Fawcett, T.: An introduction to roc analysis. Pattern Recognition Letters 27(8), 861–874 (2006)

8. Jabeur, L.B., Tamine, L., Boughanem, M.: Uprising microblogs: a bayesian network retrieval model for tweet search, pp. 943–948 (2012)
9. Kandylas, V., Dasdan, A.: The utility of tweeted urls for web search. In: Proceedings of the 19th International Conference on World Wide Web, pp. 1127–1128. ACM (2010)
10. Massoudi, K., Tsagkias, M., de Rijke, M., Weerkamp, W.: Incorporating Query Expansion and Quality Indicators in Searching Microblog Posts (2011)
11. Naveed, N., Gottron, T., Kunegis, J., Alhadi, A.C.: Searching microblogs: coping with sparsity and document quality. In: Proceedings of the 20th ACM International Conference on Information and Knowledge Management, pp. 183–188. ACM (2011)
12. O'Connor, B., Krieger, M., Ahn, D.: TweetMotif: Exploratory Search and Topic Summarization for Twitter. In: International Conference on Weblogs and Social Media (2010)
13. Souvenir, R., Pless, R.: Manifold Clustering. In: International Conference on Computer Vision, vol. 1, pp. 648–653 (2005)
14. Teevan, J., Ramage, D., Morris, M.R.: TwitterSearch: a comparison of microblog search and web search. In: Web Search and Data Mining, pp. 35–44 (2011)
15. Uysal, I., Croft, W.B.: User oriented tweet ranking: a filtering approach to microblogs, pp. 2261–2264 (2011)
16. Wan, X., Yang, J., Xiao, J.: Manifold-Ranking Based Topic-Focused Multi-Document Summarization. In: International Joint Conference on Artificial Intelligence, pp. 2903–2908 (2007)
17. Wikipedia: Twitter. `en.wikipedia.org/wiki/Twitter`
18. Zhang, X., He, B., Luo, T.: Transductive learning for real-time twitter search. In: Sixth International AAAI Conference on Weblogs and Social Media (2012)
19. Zhou, D., Bousquet, O., Lal, T.N., Weston, J., Schölkopf, B.: Learning with local and global consistency. In: Advances in Neural Information Processing Systems, vol. 16(753760), p. 284 (2004)
20. Zhou, D., Weston, J., Gretton, A., Bousquet, O., Sch, B.: Ranking on Data Manifolds. In: Neural Information Processing Systems (2004)

Fast Triangle Core Decomposition for Mining Large Graphs

Ryan A. Rossi

Purdue University
rrossi@purdue.edu

Abstract. Large triangle cores represent dense subgraphs for which each edge has at least $k - 2$ triangles (same as cliques). This paper presents a fast algorithm for computing the triangle core decomposition on big graphs. The proposed triangle core algorithm adapts both the computations and representation based on the properties of the graph. In addition, we develop a fast edge-based parallel triangle counting algorithm, which lies at the heart of the triangle core decomposition. The proposed algorithm is orders of magnitude faster than the currently available approach. We also investigate and propose fast methods for two variants of the triangle core problem: computing only the top-k triangle cores fast and finding the maximum triangle core number of the graph. The experiments demonstrate the scalability and effectiveness of our approach on 150+ networks with up to 1.8 billion-edges. Further, we apply the proposed methods for graph mining tasks including finding dense subgraphs, temporal strong components, and maximum cliques.

Keywords: Triangle-core decomposition, parallel triangle counting, maximum clique, temporal strong components, triangle-core ordering.

1 Introduction

Consider a graph $G = (V, E)$. A k-core of G is a maximal induced subgraph of G where each vertex has degree at least k. The k-core number of a vertex v is the largest k such that v is in a k-core. There is a linear time $O(|E|+|V|)$ algorithm to compute the k-core decomposition [2]. Due to this efficient algorithm, and a simple, but often powerful, interpretation, k-cores are frequently used to study modern networks [6,1,10]. Important k-core related quantities include the size of the 2-core compared with the graph, the distribution of k-core sizes, the largest k-core, and many others. In particular, the maximum value of k such that there is a $(k-1)$-core in G is denoted as $K(G)$ and provides an upper-bound on the largest clique in G, hence $\omega(G) < K(G) + 1$.

An equivalent definition of a k-core is a maximal induced subgraph of G where each vertex is incident on at least k edges. This definition then generalizes to any motif, and in particular, a k triangle core is a maximal induced subgraph of G for which each edge $(u, v) \in E$ participates in at least $k-2$ triangles. There is also a polynomial time algorithm for triangle cores, which is $O(|T|) = O(|E|^{3/2})$ in the worst case.

V.S. Tseng et al. (Eds.): PAKDD 2014, Part I, LNAI 8443, pp. 310–322, 2014.

Triangle cores were first proposed by Cohen [4,3], and a faster algorithm was recently proposed, see [17,16]. In that work, however, they explicitly store an array of triangles in order to check whether or not a triangle has been processed. This storage limits scalability as even graphs of a few thousand vertices may have billions of triangles (see Table 3). In our algorithms, we accomplish the same type of check implicitly by carefully ordering and indexing. Further, we develop a data-driven optimizer to adapt the data structures and computations based on the input graph and its properties. We also propose *parallel* edge and vertex-based triangle counting algorithms and use them for the triangle core decomposition. These algorithms significantly speed up the triangle core algorithms. In addition, we find that counting triangles on edges rather than vertices enables better load balancing for graphs where the number of triangles are not uniformly distributed, as is the case for dense and sparse graphs. Finally, unlike the previous approach, we utilize an efficient compressed edge-based representation to perform fast computations over the edges. This reduces the runtime and memory requirements considerably (see comparison in Section 5).

This paper is the first to use triangle cores for graph mining tasks such as finding the largest cliques, temporal strong components, and discovering dense subgraphs (see Section 6). We find that in many cases, the maximum triangle core number gives the exact maximum clique in G, especially for large sparse graphs. We also investigate two useful variants of the triangle core problem: the top-k triangle core problem and the maximum triangle core problem. Both of which we leverage for the above applications. For these problems, our parallel approach uses a fast heuristic clique finder to obtain a lower-bound on the maximum triangle core number, which is surprisingly tight for sparse graphs.

2 Preliminaries

2.1 Wedges, Triangles, and Cliques

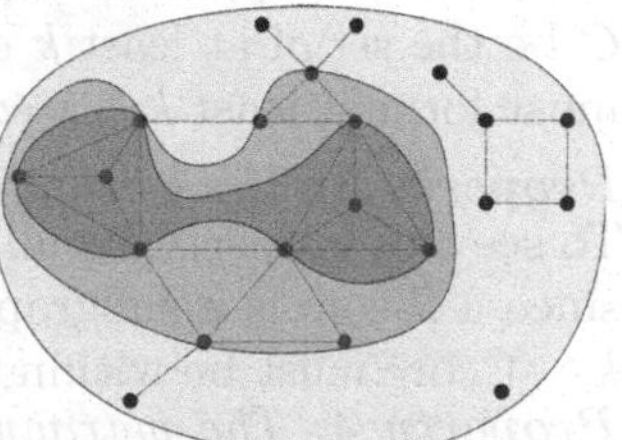

Fig. 1. Triangle cores 2, 3, and 4

A wedge is a 2-length path. The number of wedges W_u centered at u is given by $W_u = d_u(d_u - 1)/2$ where $d_u = |N(u)|$ is the degree. A wedge $\{(u,w),(w,v)\}$ forms a *triangle* if there exists an edge (u,v). Let $tr(u)$ and $tr(u,v)$ be the number of triangles centered at vertex u and edge (u,v), respectively. Further, $tr(G) = \sum_{u \in V} tr(u)$ and $W(G) = \sum_{u \in V} |W_u|$ are the number of triangles and wedges in G, respectively. Using these, we define the density of triangles in G as $\kappa(G) = tr(G)/W(G)$. Observe that if $\kappa(H)$ is close to 1, then H is dense, and a large fraction of vertices must form a clique.

2.2 Triangle Core Decomposition

Definition 1 (Triangle Core): *Consider a graph $G = (V,E)$. A k triangle-core is an edge-induced subgraph of G such that each edge participates in $k-2$ triangles.*

Hence, a triangle core must contain $k-2$ triangles, which is the same requirement for a clique of size k, since each edge in the k triangle core has two vertices

(u, v) with at least $k-2$ edges. By the above definition, a triangle is a 3-core, and a clique of size 4 is within the 4-core (See Figure 1).

Definition 2 (Maximal Triangle Core): *A subgraph $H_k = (V|E(F))$ induced by the edge-set F is a* maximal triangle core *of order k if $\forall (u, v) \in F : tr_H(u, v) - 2 > k$, and H_k is the maximum subgraph with this property.*

Definition 3 (Triangle Core Number): *The* triangle core number *denoted $T(u, v)$ of an edge $(u, v) \in E$ is the largest triangle core containing that edge.*

Definition 4 (Maximum Triangle Core): *The* maximum triangle core *of G denoted $T(G)$ is the largest number of $k-2$ triangles for a triangle core of order k to exist.*

2.3 Cliques, Triangle Cores, and K-cores

The relationship between cliques, triangle-cores and k-cores is defined more precisely in this section. Suppose H is a clique of size k, then each edge $(u, v) \in E(H)$ in that clique has two vertices (u, v) that have $k-2$ edges, which also form $k-2$ triangles.

***Property* 1:** *Each clique of size k is contained within a k triangle core of G.*

This is a direct consequence of Definition 1 and implies $\omega(G) \leq T(G)$. To understand the relationship between the k-core and the triangle core, we show the relationship between vertex degree and a triangle core number below.

***Property* 2:** *Each vertex in the k triangle core has degree $d_v \geq k-1$.*

To see this property, suppose a vertex v has an edge to u in the k triangle core, then v and u by definition, must also share k vertices that form triangles. Let C be the set of at least k common vertices such that $u, v \notin C$. Thus, $|C|$ and u must form at least $k+1$ edges with v.

***Property* 3:** *The k triangle core is contained within the $(k-1)$-core.*

To see this property, recall that a T_k triangle core must contain $k-2$ edges and since a k-core is a subgraph where all vertices have at least k degree, then the $k-1$ core must be within the k triangle core.

***Property* 4:** *The maximum triangle core number is bounded above by K and below by $\omega(G)$, giving rise to the bounds: $\omega(G) \leq T(G) \leq K(G) + 1 \leq d_{max}$*

This follows directly from the above properties and is used in Section 6.

3 Algorithms

This section describes our exact parallel triangle algorithms and their implementation for large graphs.

Table 1. Parallel triangle algorithms

				time (sec)	
Graph	$\lvert V\rvert$	$\lvert E\rvert$	$\lvert T\rvert$	VERTEX	EDGE
HOLLYWOOD	1.1M	56M	15B	36.2	**23.5**
ORKUT	3.0M	106M	1.6B	40.5	**31.9**
LIVEJOUR	4.0M	28M	251M	3.91	**2.09**
SINAWEIBO	58M	261M	638M	3148	**2194**
TWITTER10	21M	265M	51.8B	5092	**2462**
FRIENDSTER	65M	1.8B	12.5B	43591	***1947**

3.1 Parallel Triangle Counting

Since triangle counting is at the heart of the triangle core computation, we propose a parallel edge-based triangle counting algorithm using shared-memory. In particular, triangle counting is parallelized via the edges (instead of the vertices, see [8]), where jobs are broken up into independent edge computations and distributed dynamically to available workers (see Alg. 1). Our approach offers several benefits. First, the graph can be split up into many smaller *independent edge computations* that can be performed in parallel. We note that the problem with vertex triangle counters is load balancing, see Table 2 where a few vertices may have millions of triangles[1]. Our approach distributes work more evenly along the edges of the graph. For instance, the max number of triangles in sinaweibo (Chinese microblogging site) on any edge is only 17K compared to *8.3 million* for the vertices. As we see later, this approach alleviates many of the problems that arise with vertex-centric parallelization (see Table 1).

Algorithm 1. Parallel Edge Triangle Counting

1 Set p to be the number of workers (threads)
2 Init arrays $X_k, 0 \leq k \leq p$ to be $|V|$ each of all zeros.
3 Set tr to be an array of size $|E|$
4 Set max to be array of length p
5 **for each** $(u, v) \in E$ **in parallel do**
6   **for each** $w \in N(v)$ **do** $X_k(w) = (u, v)$
7   **for each** $w \in N(u)$ **do**
8   **if** $v = w$ **then continue**
9   **if** $X_k(w) = (u, v)$ **then** add 1 to $\texttt{tr}(u, v)$
10   **if** $\texttt{tr}(u, v) > \texttt{max(k)}$ **then** $\texttt{max(k)} = \texttt{tr}(u, v)$
11 **for each** thread $k = 2$ **to** p **do**
12   **if** max(1) > max(k) **then** max(1) = max(k)
13 **return** tr and max(1)

The parallel edge triangle counting algorithm is given in Alg. 1. The triangle counting computations are performed by dynamically allocating blocks of edges to workers. The workers then compute the number of triangles centered at each edge and store the counts into the edge-indexed array. Locks are avoided by assigning unique ids to the edges, which map to unique positions in the triangle counting array (See Section 3.2 for more details). Each worker also maintains a hash table X_k for $O(1)$ lookups. In particular, line 6 marks the neighbors of v with the unique edge id of (u, v), which is used later in line 9 to determine if a triangle exists. We avoid resetting X_k each time by using the unique edge id. After a worker finishes a job (i.e., block of edges) it requests more work and the process continues as above. Note that each worker also maintains (locally) the total number of triangles processed and the maximum number of triangles at any given edge. Upon completion of the parallel-for loop, the global max is realized in serial (lines 11-12). This takes only $O(p)$ where p is the number of worker nodes. The expected overall time is $O(|E|^{3/2}/p)$ since locks are avoided completely (See Figure 5).

3.2 Triangle Core Decomposition

Edge CSC Format. Edge-based compressed sparse column (ECSC) format adds two arrays to the traditional CSC format to allow for both vertex and

[1] We observe that vertices have highly skewed triangle counts resembling a triangle power-law [7].

edge-based computations. The ECSC graph representation is designed carefully to optimize performance and space while also providing $O(1)$ time access to edges, neighboring edges, and their vertices. In particular, ECSC helps maximize data locality (in disk, ram, cache, core) while minimizing interaction. Both are critical for parallel edge-centric graph algorithms (see [13] for examples). Indeed, the practical importance of ECSC was previously shown in Table 1.

To understand ECSC, we provide an intuitive illustration in Figure 2. The `eid` array is an non-unique edge-indexed array (from the neighbor array) where duplicate edges in the neighbor array are assigned a unique edge id stored in `eid` that acts as a *pointer into the emap array.* This provides two important features. First, observe that `eid` is a surjective function that maps duplicate edges from CSC to their unique position in `emap`. In addition, `eid` also points to the `emap` array where the *vertices* of a specific edge are *consecutively* stored and can be accessed in $O(1)$ time. This function provides the flexibility of using CSC while also giving random access to unique edges and their vertices in $O(1)$. Note that Edge-CSC is efficient to construct taking $O(|E|)$ time (when reading from disk or from CSC).

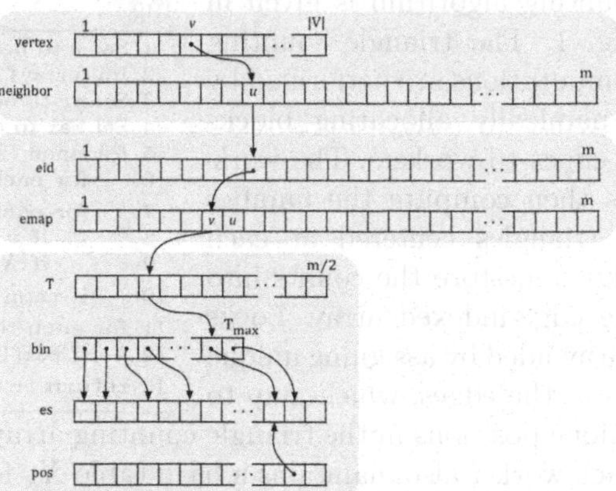

Fig. 2. Illustration of the arrays and pointers used in ECSC for computing the triangle core decomposition. Note $m = |E|$.

Triangle Core Arrays. The triangle core decomposition uses ESCS and four additional smaller arrays. This reduced storage cost is due to `emap` which provides $O(1)$ time access to a unique edge and its vertices (stored consecutively). The edge-indexed T array stores for each edge (u, v) the number of triangles that it participated. An index k of the `emap` array also directly indexes T using $k/2$ since T stores only the unique edges. The `bin` array is indexed by an integer representing triangles, and stores the starting position for each unique number of edge triangles. Hence `y = bin[x]` points directly to the starting position in the `es` array for which the edge given by `es[y]` must have exactly x triangles. The `es` array contains the edge ids sorted by the number of triangles whereas the `pos` array stores the location of a given edge in the sorted `es` array.

The proposed triangle core algorithm is given in Alg. 2. Edges are undirected and the neighbors of each vertex are sorted by degree (to improve caching). The algorithm begins by computing the number of triangles for each edge in parallel. We denote this here by T since it will contain the triangle core numbers upon completion of Alg 2. The lines 2–3 initialize various vars and arrays. Note that the `proc` array is of size m and is used for marking the processed edges. Afterwards, the edges are sorted by their triangle counts in T using bucket sort (lines 4–15). In particular, line 5 counts how many edges will be in each `bin` where a `bin` is a set of edges with the same triangle count. Thus, bins are numbered from 0 to T_{max}. Given the `bin` sizes, lines 6–9 setup the starting positions for each bin using the previously computed counts. For each edge, we index into bin with its triangle count and assign it to the position in `es` returned by bin, then increase that bins position by one (lines 10–13). Finally, lines 14–15 fix the starting positions of the bins.

Now that edges are sorted by their triangle counts, line 17 starts removing each edge in increasing order of triangles (using `es`). Line 18 retrieves the next edge for processing along with its vertices u and v in $O(1)$ time using `emap`. Next, line 19 selects v and instead of simply marking the neighbors with a 1 (i.e., $X[w] = 1$), we store the position+1 in which it appeared in the neighbor array. Let us note that the value in X maps directly to a unique edge using `eid` and `emap`. For each $w \in N(u)$, we index X and check in $O(1)$ time if w is a neighbor of v (lines 20–21). If so, then the vertices u, v, and w form a triangle, otherwise we continue searching for a match. Using the `eid` array, we retrieve the unique edge ids for the edges (u,w) and (v,w) in $O(1)$ time from ECSC (lines 22–23) and check if these edges have been processed (line 24). If either of the edges were processed before, then the triangle has been implicitly removed in a previous iteration, and we continue with the next neighboring edge. However, if both edges have not been removed, then we have found a valid triangle to remove. For each of the neighboring edges with a larger triangle core number, we decrease its number of triangles by 1 and move it one bin to the left. All these operations are $O(1)$ time using ECSC. Observe that each of the neighboring edges (u, w) and (v, w) are swapped with the first edge in its bin, respectively. We then swap their positions in the `pos` array and increase the previous bin and decrease the current bin of the edge by 1. Lines 20–33 are repeated for all neighboring edges of (u,v) that form unprocessed triangles with a triangle count that is currently larger than its own. Finally, the edge is marked as processed and X is reset (line 34–35).

Algorithm 2. Detailed Triangle Core Algorithm

```
1  Set T = ParallelEdgeTriangles(G)
2  Set pos and proc arrays to be length |E| of all 0.
3  Set the bin array to be of size T_max initialized to 0.
4  for e = 0 to |E| do bin[T(e)]++
5  Set start = 0
6  for x = 0 to T_max do
7      n = bin[x]
8      bin[x] = start
9      start = start + n
10 for e = 0 to |E| do
11     pos[e] = bin[T[e]]
12     tris[pos[e]] = e
13     bin[T[e]]++
14 for t = T_max to 0 do bin[t] = bin[t-1]
15 bin[0] = 0
16 Set X to be an array of length |V| containing zeros.
17 for i = 0 to |E| do
18     k = es[i]    v = emap[2k]    u = emap[2k+1]
19     for w ∈ N(v) do X[w] = w_pos + 1
20     for w ∈ N(u) do
21         if X[w] is not marked then continue
22         uw = eid[j]
23         uv = eid[X[w] - 1]
24         if proc[uw] or proc[uv] are marked then
25             continue
26         for adj ∈ {uw, uv} do
27             if T[adj] > T[k] then
28                 tw = T[adj]    and   pw = pos[adj]
29                 ps = bin[tw]   and   xy = es[ps]
30                 if xy ≠ wv then
31                     pos[adj] = ps   and   es[pw] = xy
32                     pos[xy] = pw    and   es[ps] = adj
33                 bin[tw]++  and  T[adj]--
34     proc[k] = 1
35     for w ∈ N(v) do X[w] = 0
```

4 Problem Variants

Two variants are investigated and space-efficient multi-threaded algorithms suitable for CPU and GPU parallelization are proposed. Both variants arise from real-world motivations, e.g., finding maximum cliques and dense subgraphs in big data (See Section 6).

4.1 Top K Triangle Cores

Using property 3 as a basis, we propose a parallel method that leverages a relationship between the k-core and triangle core for computing only the top-k triangle cores. Key to our approach is the heuristic clique finder which allows us to obtain a fast lower-bound.

Algorithm 3 Top-K Triangle Cores

```
K = CoreNumbers(G)
max = HeuMaxClique(G,K)
for each u ∈ V in parallel do
    if K(u) ≥ max then add u to W
Set H = G(W) - the induced graph from W.
tr = ParallelEdgeTriangles(H)
for each u,v ∈ E_H in parallel do
    if tr(u,v) < max then
        prune (u,v) from H
        for each neighboring edge ((w,v)or(w,u)) do
            set tr(w,u) to be one less
            prune (w,u) if tr(w,u) < max
T = TriangleCores(H)
```

Problem 1 (Top K Triangle Cores): *Given a graph G and an integer $k > 2$, find the set of edges that have triangle core numbers greater than k.*

The output is a subgraph H such that $u, v \in E(H)$ if $T(u, v) \geq k$ and an edge-indexed array indicating the triangle core numbers of each edge. Note that we use this variant later for the maximum clique problem and finding dense subgraphs (see Section 6.3).

The first step in Alg 3 computes the k-core numbers denoted K of G. This gives us an upper bound on the triangle core and thus if k is larger we set it to be the maximum k-core. Next, for each vertex $v \in V$ we add it to the vertex set W if it satisfies $K(u) + 1 \geq k$. Let H be an explicit vertex-induced subgraph from G using the vertex set W. Triangle counts for the edges are computed in parallel and if an edge $(u, v) \in E(H)$ has less triangles than the integer k given as input, $\mathrm{tr(v,u)} \leq \mathrm{k}$, then we can safely prune it, and update the graph, and the triangle count of any edge (w, v) or (w, u) that formed a triangle through (v, u). This smaller graph is then given to the triangle core routine. Instead of selecting k arbitrarily, we use a heuristic clique finder [12] to obtain a tight lower bound denoted $\tilde{\omega}(G)$. Observe that $\tilde{\omega}(G) \leq \omega(G) \leq T(G) \leq K(G) + 1$ where computing $\tilde{\omega}(G)$ is $O(|E| \cdot d_{\max})$ time. See Table 2 for heuristic runtimes.

4.2 Max Triangle Core

In many applications, the cost of computing the full triangle core decomposition is too expensive and/or not needed. For instance, the maximum triangle core may significantly speedup the termination of maximum clique algorithms when used for pruning. Thus, we solve the following problem instead:

Algorithm 4 Dense Triangle Subgraph

```
K = CoreNumbers(G)
k = HeuristicClique(G)
for each u ∈ V in parallel do
    if K(u) ≥ k then add u to W
Set H = G(W) to be the induced graph from W.
T = TriangleCores(H)
Set F = ∅
for each u ∈ V(H) in parallel do
    for each w ∈ N(u) do
        if T(u, w) ≥ k then
            add edge (u, w) to F
            add 1 to d_u
    if d_u ≤ k then Remove all edges of u from F
return the edge-induced graph from F
```

Problem 2 (Maximum Triangle Core Number): *Given a graph G, find the maximum triangle core number $T(G)$ directly.*

Our approach starts by computing the k-cores, then scans the vertices in order of their removal times from the k-core algorithm. A vertex v is added to W if $K(v) > K - \delta$ where for triangles we set $\delta = 3$. We then compute $H = (W, E[W])$ and give this smaller graph to the triangle core routine. Observe that $\tilde{T}$ is a fast but accurate approximation of the maximum triangle core.

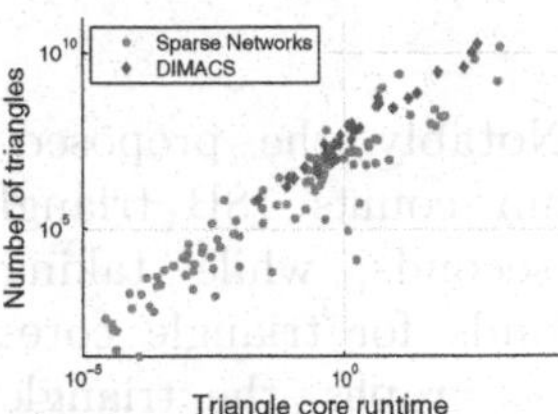

Fig. 3. Runtime vs. number of triangles

5 Experiments

In this section, we systematically investigate the performance and accuracy of our methods on over 150 graphs. The experiments are designed to answer the following questions:

- ▷ **Triangle core decomposition.** How fast is our algorithm for computing the triangle core decomposition? and how does it scale up for both dense and large sparse graphs of up to a billion edges.
- ▷ **Parallel edge triangle counting.** Does our parallelization scheme work? Is the edge triangle counting algorithm faster than vertex triangle counters? and do they have better load balancing?
- ▷ **Performance of variants.** How fast can we solve each variant? and how does the performance compare to the full triangle core decomposition?
- ▷ **Clique upper bound.** How tight are the triangle core upper bounds compared to the k-core? For which types of graphs are they better?

For the experiments, we used a 2 processor, Intel E5-2760 system with 16 cores and 256 GB of memory. Our algorithms never came close to using all the memory.

5.1 Performance of Proposed Algorithm

For the first question, we evaluate the performance of the proposed triangle core algorithm on over 150 graphs of all types. The results in Tables 2 and 3 demonstrate the effectiveness of the proposed methods for both large sparse graphs and dense graphs, respectively.

Table 2. Performance of heuristic, parallel triangle count algorithm, and triangle cores

Graph	$\|V\|$	$\|E\|$	$\|T\|$	$d_{\max}$	$\bar{\kappa}$	$tr_{\max}$	tr_{avg}	K	T	$\tilde{\omega}$	$\tilde{\omega}$ s.	tr s.	T s.
DBLP-2010	226k	716k	4.8M	238	0.64	5.9k	21	75	75	75	0.02	0.10	0.29
CITESEER	227k	814k	8.1M	1.4k	0.68	5.4k	35	87	87	87	0.05	0.09	0.43
MATHSCI	333k	821k	1.7M	496	0.41	1.6k	5.2	25	25	25	0.11	0.14	0.53
HOLLYWOOD	1.1M	56M	15B	11k	0.77	4.0M	13k	2209	2209	2209	1.57	23.5	727.2
YOUTUBE	496k	1.9M	7.3M	25k	0.11	151k	14	50	19	14	0.42	0.71	15.92
FLICKR	514k	3.2M	176M	4.4k	0.17	525k	343	310	153	49	0.21	1.26	18.67
ORKUT	3.0M	106M	1.6B	27k	0.17	1.3M	525	231	75	45	13.8	31.9	770.5
LIVEJOUR	4.0M	28M	251M	2.7k	0.26	80k	62	214	214	214	3.2	2.09	54.3
TWITTER10	21M	265M	51.8B	698k	0.04	44M	2.4k	1696	1153	174	145.7	2462	39535
FRIENDSTER	65M	1.8B	12.5B	5.2k	0.16	190	158k	305	129*	129	561	1947	45247
TEXAS84	36k	1.6M	34M	6.3k	0.19	141k	922	82	62	48	0.070	0.34	4.80
PENN94	42k	1.4M	22M	4.4k	0.21	68k	520	63	48	44	0.05	0.22	3.27
P2P-GNUT	63k	148k	6.1k	95	0.01	17	0.1	7	4*	4	0.01	0.01	0.02
RL-CAIDA	191k	608k	1.4M	1.1k	0.16	6.0k	7	33	19	15	0.04	0.06	0.33
AS-SKITTER	1.7M	11M	86M	35k	0.26	565k	50	112	68	64	0.37	4.33	70.6
IT-2004	509k	7.2M	1.0B	469	0.82	93k	19k	432	432	432	0.09	1.36	6.88
WIKIPEDIA	1.9M	4.5M	6.7M	2.6k	0.16	12k	3	67	31*	31	0.66	0.54	3.54

Table 3. Performance on dense DIMAC graphs.

graph	$\|E\|$	$\|T\|$	$tr_{\max}$	K	T	tr s.	T sec
C500-9	112k	45M	98k	433	373	0.08	0.58
C2000-5	1.0M	500M	288k	941	435	1.65	21.02
C1000-9	450k	365M	385k	875	764	0.58	5.23
C4000-5	4.0M	4.0B	1.1M	1910	899	12.09	177.75
C2000-9	1.8M	2.9B	1.5M	1759	1549	3.02	59.32
P-HAT15K3	569k	274M	372k	505	314	0.66	9.41
P-HAT15K	847k	741M	676k	930	597	1.31	23.83
MAN-81	5.5M	18B	5.5M	3281	3241	8.59	315.84
KELLER6	4.6M	10B	3.6M	2691	2084	7.92	252.37

Notably, the proposed algorithm counts 18B triangles in 8 seconds, while taking 315 seconds for triangle cores. For these graphs, the triangle core algorithm adapts the graph representation and computation to better exploit the structure. This includes using an adj structure for $O(1)$ time lookups, selecting specialized subroutines based on densities, among many other optimizations. Indeed, the triangle core algorithm is shown to be fast and scalable as it easily handles graphs with billions of edges while using a small memory footprint. Fig. 3 compares dense and sparse graphs directly. This indicates that many of the sparse graphs perform better than expected whereas dense graphs deviate much less from the diagonal.

Performance profile plots. We also compare with the implementation of Zhang et al. [17] using performance profiles [5]. Figure 4 evaluates the performance on 24 graphs with up to 1.8 billion edges. In all cases, the proposed algorithm outperforms the recent state-of-the-art algorithm [17]. We note that for a few large graphs, the baseline algorithm runs out of memory and eventually terminates.

5.2 Parallel Edge Triangle Counting

The proposed parallel edge triangle counting algorithm is fast and scalable as shown in Table 2. Moreover, we also find that it outperforms a recent MapReduce triangle counting algorithm [15]. In that work, it takes 319 seconds using a MapReduce cluster of 1636 nodes to compute triangles for LiveJournal whereas it takes us only 2.09 seconds (See Table 1).

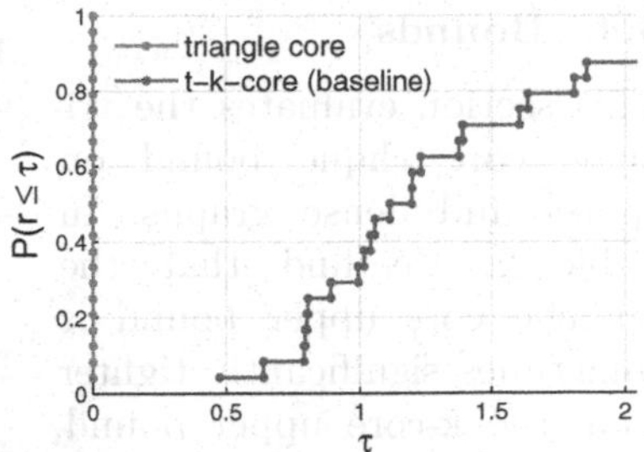

Fig. 4. In all cases, our algorithm is significantly faster than the recently proposed algorithm of Zhang et al. [17]. The vertical line formed from our algorithm indicates it outperformed the state-of-the-art on each of the 24 graphs tested. Further, the other algorithm fails on 3 of the largest graph as illustrated by the right-most point on the curve.

Speedup. Fig 5 shows that our approach scales well, especially for sparse graphs. Observe that for dense graphs, triangles are more uniformly distributed and thus the improved load balancing from our approach does not help as much. We also find that on average, less time is spent per triangle for dense graphs.

Edge vs. vertex triangle counting. We parallelize a vertex triangle counter to evaluate the performance of our parallel edge-based algorithm. The superiority of the parallel edge triangle counting algorithm is clearly shown in Table 1. In one example, the friendster social network of 1.8 billion edges takes only 1,947 seconds whereas the vertex triangle counter takes 43,591 seconds. The edge triangle counter is 22x faster.

5.3 Performance of Variants

Results are shown in Table 4. We compare the greedy maximum triangle core to the exact triangle core decomposition on the basis of speed and accuracy. Table 4 clearly demonstrates the effectiveness of the greedy maximum triangle core procedure. In some instances, a speedup of over 650x is observed while also returning the exact maximum triangle core number. For the top-k cores, we find that these are generally 1-20x faster than the full triangle core decomposition while guaranteed to be exact, as previously shown in Property 3. The large table of results are omitted due to brevity, but are later used for finding temporal strong components.

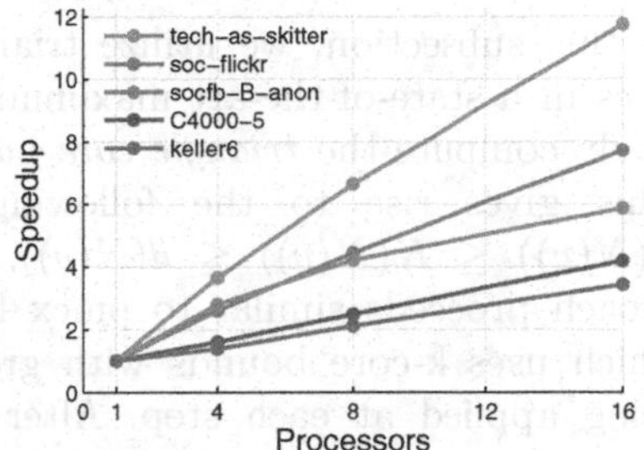

Fig. 5. Speedup of our parallel edge triangle counting algorithm

5.4 Bounds

Table 4. Performance and accuracy of greedy maximum triangle core

	density		triangle core		time (sec)	
graph	ρ	greedy	T	greedy	exact	greedy
AS-SKITTER	7.7e-06	0.50	68	68	65.39	0.10
FB-ANON-A	4.9e-06	0.06	30	29	83.56	0.31
STANFORD	0.008	0.04	58	58	2.0	0.87
LIVEJOUR.	3.4e-06	0.99	212	212	60.95	0.03
FLICKR	2.4e-05	0.44	151	151	22.55	3.16
HOLLYWOOD	9.8e-05	0.99	2207	2207	918.7	92.5
DBLP12	2.0e-05	0.99	112	112	0.85	0.007

This section evaluates the triangle core clique bound on sparse and dense graphs. In Table 2, we find that the triangle core upper bound is sometimes significantly tighter than the k-core upper bound, especially for sparse graphs that exhibit a weak power-law including social, facebook, and technological networks. Further, Table 3 shows that for dense graphs, the triangle core bound is almost always much tighter than the k-core bound. In addition, we use the triangle core upper bound to validate a fast heuristic clique finder. We find three such cases and mark them with a star in Table 2. Note this excludes the cases where the k-core upper bound also verifies the large clique as optimal. Besides the maximum triangle core number, we also found the full distribution of triangle core numbers interesting.

6 Applications

This section demonstrates the effectiveness of the proposed triangle core algorithms for a variety of graph mining applications.

6.1 Maximum Clique Algorithms

In this subsection, we utilize triangle core numbers to prune and order vertices in a state-of-the-art maximum clique algorithm. In particular, we aggressively compute the *triangle core numbers* of each vertex induced neighborhood. This gives rise to the following: $\omega \leq T(N(v)) \leq K(N(v)) \leq d(N(v))$. The approach proceeds similar to pmcx from [12] which uses k-core bounds with greedy coloring applied at each step. After pruning the vertex neighborhoods, we compute the density of the subgraph denoted $\rho(N(v))$ and use *neighborhood triangle cores* only if $\rho(N(v)) > 0.85$, otherwise, we proceed same as pmcx above. Intuitively, if density is large enough, then edges are pruned using the neighborhood cores. We compare trmc against two recent state-of-the-art finders: bbmc [14] and pmcx [12]. Let us note that bbmc uses a bitset encoding for set intersections and was shown to outperform all others tested [11]. In Fig. 6 we find that trmc outperforms bbmc on every graph, but outperforms pmcx on only a few instances.

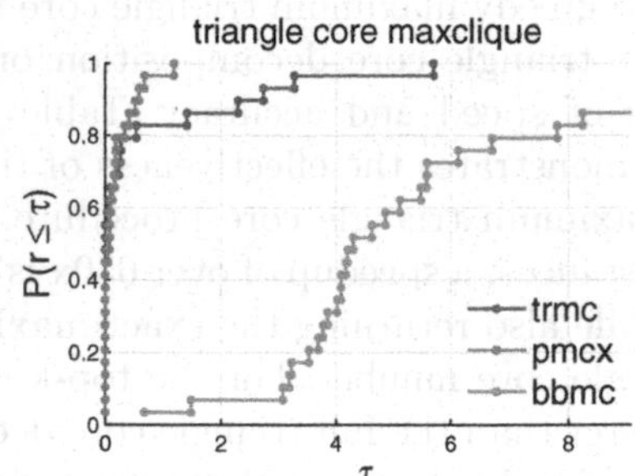

Fig. 6. Performance profile plot. trmc outperforms bbmc on all tested graphs and beats pmcx on a few.

6.2 Temporal SCC

Table 5. Max temporal-SCC via triangle cores

graph	bounds $\tilde{\omega}$	K	T	time (seconds) $\tilde{\omega}$ s	K s	T s	$\hat{T}$ s
FB-MESSAGES	707	707	707	0.15	0.01	6.72	1.54
REALITY	1236	1236	1236	0.28	0.03	323	11.21
TWITTER-COP	581	583	581	0.21	0.01	5.85	1.11

We use the algorithms from section 3 to explore the effectiveness of the top-k triangle cores for computing the largest temporal strongly connected component (TSCC), which is a known NP-hard problem [9]). The top-k triangle core algorithm in Alg 4 uses the k-cores and a large clique from a heuristic maximum clique finder [12] to prune the triangle cores without ever computing them. Table 5 shows the performance of our full triangle core algorithm compared with the top-k triangle core algorithm. Strikingly, in all cases, the triangle core upper bound gives the *exact size* of the largest TSCC. Moreover, the top-k triangle core algorithm is significantly faster than the full algorithm. For instance, the REALITY graph in Table 5 takes 323 seconds to compute the full decomposition, whereas the fast top-k algorithm takes only 11 seconds.

6.3 Dense Subgraph Mining

A particularly useful property of the *greedy maximum triangle core* is that it returns a relatively dense subgraph. In Table 4 we find the density of the as-skitter subgraph returned by the greedy maximum triangle core method is 0.50 with 345 vertices with an average degree of 173. Concerning performance, our greedy maximum triangle core procedure is clearly much faster while also accurate. In addition, Figure 7 summarizes the results from Alg 4. The top-k triangle cores are shown to be better at finding dense subgraphs.

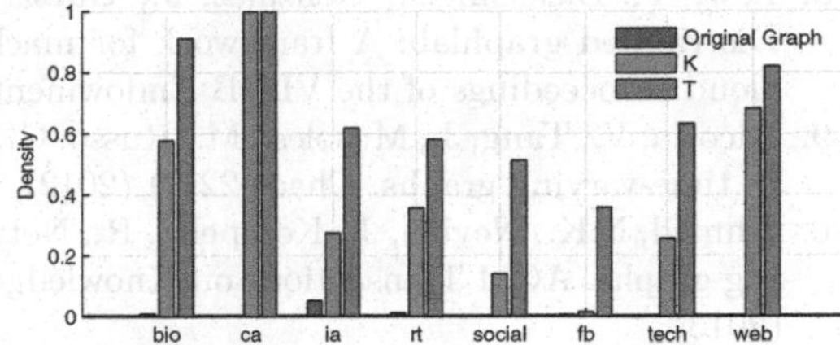

Fig. 7. Networks are grouped by type (bio, social, etc) and densities from each are averaged.

7 Conclusion

We have described a fast scalable algorithm for computing the *triangle core decomposition* and systematically investigated its scalability and effectiveness on a variety of large sparse networks. The proposed algorithm is shown to be orders of magnitude faster than the current state-of-the-art while also using less space. In addition, the algorithm was designed for graphs of arbitrary size and density, allowing it to be used for a variety of applications. We also proposed a *parallel edge triangle enumeration algorithm* and showed that it is significantly faster than vertex-based methods. Future work will investigate the proposed family of triangle core ordering techniques for use in greedy coloring methods and clique finding algorithms.

References

1. Alvarez-Hamelin, J.I., Dall'Asta, L., Barrat, A., Vespignani, A.: Large scale networks fingerprinting and visualization using the k-core decomposition. In: NIPS (2005)
2. Batagelj, V., Zaversnik, M.: An o(m) algorithm for cores decomposition of networks. arXiv preprint cs/0310049 (2003)
3. Cohen, J.D.: Graph twiddling in a mapreduce world. Computing in Science & Engineering 11(4), 29–41 (2009)
4. Cohen, J.D.: Trusses: Cohesive subgraphs for social network analysis. National Security Agency Technical Report (2008)
5. Dolan, E.D., Moré, J.J.: Benchmarking optimization software with performance profiles. Mathematical Programming 91(2), 201–213 (2002)
6. Dorogovtsev, S.N., Goltsev, A.V., Mendes, J.F.F.: K-core organization of complex networks. Physical Review Letters 96(4), 040601 (2006)
7. Kang, U., Meeder, B., Faloutsos, C.: Spectral analysis for billion-scale graphs: Discoveries and implementation. In: Huang, J.Z., Cao, L., Srivastava, J. (eds.) PAKDD 2011, Part II. LNCS, vol. 6635, pp. 13–25. Springer, Heidelberg (2011)
8. Low, Y., Bickson, D., Gonzalez, J., Guestrin, C., Kyrola, A., Hellerstein, J.M.: Distributed graphlab: A framework for machine learning and data mining in the cloud. Proceedings of the VLDB Endowment 5(8), 716–727 (2012)
9. Nicosia, V., Tang, J., Musolesi, M., Russo, G., Mascolo, C., Latora, V.: Components in time-varying graphs. Chaos 22(2) (2012)
10. Ahmed, N.K., Neville, J., Kompella, R.: Network sampling: From static to streaming graphs. ACM Transactions on Knowledge Discovery from Data (TKDD), 1–54 (2013)
11. Prosser, P.: Exact algorithms for maximum clique: A computational study. arXiv:1207.4616v1 (2012)
12. Rossi, R.A., Gleich, D.F., Gebremedhin, A.H., Patwary, M.A.: Fast maximum clique algorithms for large graphs. In: WWW Companion (2014)
13. Rossi, R.A., McDowell, L.K., Aha, D.W., Neville, J.: Transforming graph data for statistical relational learning. JAIR 45, 363–441 (2012)
14. San Segundo, P., Rodríguez-Losada, D., Jiménez, A.: An exact bit-parallel algorithm for the maximum clique problem. Comput. Oper. Res. 38, 571–581 (2011)
15. Suri, S., Vassilvitskii, S.: Counting triangles and the curse of the last reducer. In: WWW, pp. 607–614. ACM (2011)
16. Wang, J., Cheng, J.: Truss decomposition in massive networks. Proceedings of the VLDB Endowment 5(9), 812–823 (2012)
17. Zhang, Y., Parthasarathy, S.: Extracting analyzing and visualizing triangle k-core motifs within networks. In: ICDE, pp. 1049–1060 (2012)

Super-Graph Classification

Ting Guo[1] and Xingquan Zhu[2]

[1] Centre for Quantum Computation & Intelligent Systems, FEIT, University of Technology, Sydney, NSW 2007, Australia
[2] Dept. of Computer & Electrical Engineering and Computer Science, Florida Atlantic University, Boca Raton, FL 33431, USA
ting.guo-1@student.uts.edu.au, xzhu3@fau.edu

Abstract. Graphs are popularly used to represent objects with dependency structures, yet all existing graph classification algorithms can only handle simple graphs where each node is a single attribute (or a set of independent attributes). In this paper, we formulate a new super-graph classification task where each node of the super-graph may contain a graph (a single-attribute graph), so a super-graph contains a set of interconnected graphs. To support super-graph classification, we propose a *Weighted Random Walk Kernel* (WRWK) which generates a product graph between any two super-graphs, and uses the similarity (kernel value) of two single-attribute graph as the node weight. Then we calculate weighted random walks on the product graph to generate kernel value between two super-graphs as their similarity. Our method enjoys sound theoretical properties, including bounded similarity. Experiments confirm that our method significantly outperforms baseline approaches.

Keywords: Graph classification, kernel, super-graph.

1 Introduction

Many applications, such as social networks and citation networks, commonly use graph structure to represent data entries (*i.e.* nodes) and their structural relationships (*i.e.* edges). When using graphs to represent objects, all existing frameworks rely on two approaches to describe node content (1) **node as a single attribute:** each node has only one attribute (single-attribute node). A clear drawback of this representation is that a single attribute cannot precisely describe the node content [6]. This representation is commonly referred to as a *single-attribute graph* (Fig.1 (A)). (2) **node as a set of attributes:** use a set of independent attributes to describe the node content (Fig.1 (B)). This representation is commonly referred to as an *attributed graph* [2,3,8].

Indeed, in many applications, the attributes/properties used to describe the node content may be subject to dependency structures. For example, in a citation network each node represents one paper and edges denote citation relationships. It is insufficient to use one or multiple independent attributes to describe detailed information of a paper. Instead, we can represent the content of each paper as a graph with nodes denoting keywords and edges representing contextual correlations between keywords (*e.g.* co-occurrence of keywords in different sentences

V.S. Tseng et al. (Eds.): PAKDD 2014, Part I, LNAI 8443, pp. 323–336, 2014.

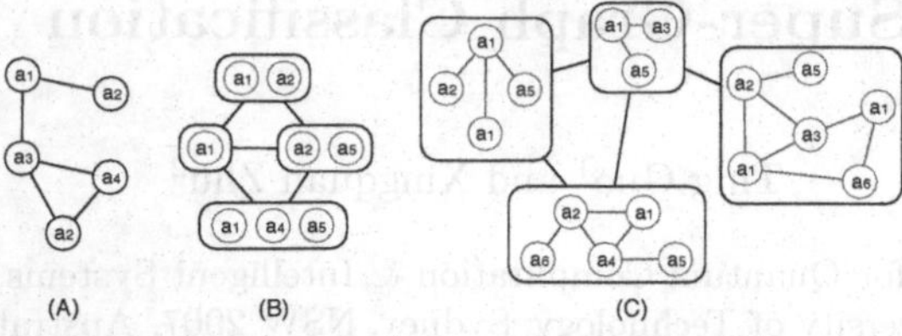

Fig. 1. (A): a single-attribute graph; (B): an attributed graph; and (C): a super-graph

or paragraphs). As a result, each paper and all references cited in this paper can form a super-graph with each edge between papers denoting their citation relationships. In this paper, we refer to this type of graph, where the content of the node can be represented as a graph, as a "*super-graph*". Likewise, we refer to the node whose content is represented as a graph, as a "*super-node*".

To build learning models for super-graphs, the mainly challenge is to properly calculate the distance between two super-graphs.

- **Similarity between two super-nodes:** Because each super-node is a graph, the overlapped/intersected graph structure between two super-nodes reveals the similarity between two super-nodes, as well as the relationship between two super-graphs. Traditional hard-node-matching mechanism is unsuitable for super-graphs which require soft-node-matching.
- **Similarity between two super-graphs:** The complex structure of super-graph requires that the similarity measure considers not only the structure similarity, but also the super-node similarity between two super-graphs. This cannot be achieved without combining node matching and graph matching as a whole to assess similarity between super-graphs.

The above challenges motivate the proposed Weighted Random Walk Kernel ($WRWK$) for super-graphs. In our paper, we generate a new product graph from two super-graphs and then use weighted random walks on the product graph to calculate similarity between super-graphs. A weighted random walk denotes a walk starting from a random weighted node and following succeeding weighted nodes and edges in a random manner. The weight of the node in the product graph denotes the similarity of two super-nodes. Given a set of labeled super-graphs, we can use an weighted product graph to establish walk-based relationship between two super-graphs and calculate their similarities. After that, we can obtain the kernel matrix for super-graph classification.

2 Problem Definition

Definition 1. ***(Single-attribute Graph)*** *A single-attribute graph is represented as* $g = (V, E, Att, f)$, *where* $V = \{v_1, v_2, \cdots, v_n\}$ *is a finite set of vertices,* $E \subseteq V \times V$ *denotes a finite set of edges, and* $f : V \rightarrow Att$ *is an injective function from the vertex set* V *to the attribute set* $Att = \{a_1, a_2, \cdots, a_m\}$.

Definition 2. ***(Super-graph and Super-node)*** *A super-graph is represented as* $G = (\mathcal{V}, \mathcal{E}, \mathcal{G}, \mathcal{F})$, *where* $\mathcal{V} = \{V_1, V_2, \cdots, V_N\}$ *is a finite set of graph-structured*

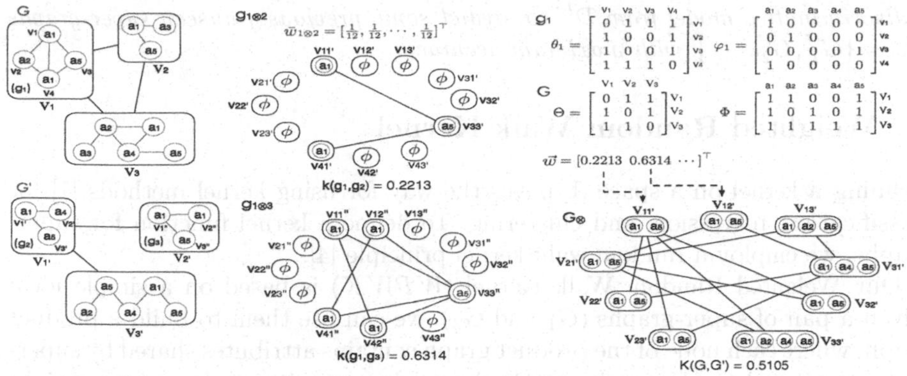

Fig. 2. $WRWK$ on the super-graphs (G, G') and the single-attribute graphs (g_1, g_2, g_3), where $g_{1\otimes 2}$ and $g_{1\otimes 3}$ are the single-attribute product graphs for (g_1, g_2) and (g_1, g_3), respectively, and $G_\otimes$ is the super product graph for (G, G'). (θ_1, φ_1) and (Θ, Φ) are the adjacency and attribute matrices for g_1 and G, respectively. $\bar{w}_{1\otimes 2}$ and $\boldsymbol{w}$ are the weight vectors for $g_{1\otimes 2}$ and $G_\otimes$, respectively. Each element of $\boldsymbol{w}$ is equal to the $WRWK$ of two super-nodes (as dotted arrows show). A random walk on single-attribute product graph, say $g_{1\otimes 3}$, is equivalent to performing simultaneous random walks on graphs g_1 and g_3, respectively. Assume that $v_1(a_1)$ means node v_1 with attribute a_1 in g_1. The walk $v_1(a_1) \to v_3(a_5) \to v_4(a_1)$ in g_1 can match the walk $v_{1''}(a_1) \to v_{3''}(a_5) \to v_{2''}(a_1)$ in g_3. The corresponding walk on the single-attribute product graph $g_{1\otimes 3}$ is: $v_{11''}(a_1) \to v_{33''}(a_5) \to v_{42''}(a_1)$. $k(g_1, g_2)$ and $k(g_1, g_3)$ are the kernels. $k(g_1, g_2) < k(g_1, g_3)$ means g_3 is more similar to g_1 than g_2. Likewise, $K(G, G')$ is the $WRWK$ of G and G'.

nodes. $\mathcal{E} \subseteq \mathcal{V} \times \mathcal{V}$ denotes a finite set of edges, and $\mathcal{F} : \mathcal{E} \to \mathcal{G}$ is an injective function from $\mathcal{E}$ to $\mathcal{G}$, where $\mathcal{G} = \{g_1, g_2, \cdots, g_M\}$ is the set of single-attribute graphs. A node in the super-graph, which is represented by a single-attribute graph, is called a Super-node.

Formally, a single-attribute graph $g = (V, E, Att, f)$ can be uniquely described by its attribute and adjacency matrices. The attribute matrix φ is defined by $\varphi_{ri} = 1 \Leftrightarrow a_i \in f(v_r)$, otherwise $\varphi_{ri} = 0$. The adjacency matrix θ is defined by $\theta_{ij} = 1 \Leftrightarrow (v_i, v_j) \in E$, otherwise $\theta_{ij} = 0$. Similarly, the adjacency matrix Θ of an super-graph $G = (\mathcal{V}, \mathcal{E}, \mathcal{G}, \mathcal{F})$ is defined by $\Theta_{ij} = 1 \Leftrightarrow (V_i, V_j) \in \mathcal{E}$, otherwise $\Theta_{ij} = 0$. However, because of the complicated structure of super-graph, we cannot use an unique matrix to describe its super-node information. To calculate conveniently as this article shows later, an super attribute matrix $\Phi \in \mathbb{R}^{N \times S}$ is defined for super-graph G, where $N = |\mathcal{V}|$ and $S = |Att_{g_1} \cup Att_{g_2} \cup \cdots \cup Att_{g_M}|$. $\Phi_{ri} = 1 \Leftrightarrow a_i \in Att_{g_r}$, otherwise $\Phi_{ri} = 0$. Examples of the attribute and adjacency matrices for a single-attribute graph and a super-graph are shown in the top right corner of Fig. 2.

A super-graph G_i is a labeled graph if a class label $L(G_i) \in \mathcal{Y} = \{y_1, y_2, \cdots, y_x\}$ is assigned to G_i. A super-graph G_i is either labeled (G_i^L) or unlabeled (G_i^U).

Definition 3. ***(Super-graph Classification)*** *Given a set of labeled super-graphs $\mathcal{D}^L = \{G_1^L, G_2^L, \cdots\}$, the goal of super-graph classification is to learn* *a discriminative model from $\mathcal{D}^L$ to predict some previously unseen super-graphs $\mathcal{D}^U = \{G_1^U, G_2^U, \cdots\}$ with maximum accuracy.*

3 Weighted Random Walk Kernel

Defining a kernel on a space $\mathcal{X}$ paves the way for using kernel methods [7] for classification, regression, and clustering. To define a kernel function for super-graphs, we employ a random walk kernel principle [4].

Our Weighted Random Walk Kernel ($WRWK$) is based on a simple idea: Given a pair of super-graphs (G_1 and G_2), we can use them to build a product graph, where each node of the product graph contains attributes shared by super-nodes in G_1 and G_2. For each node in the product graph, we can assign a weight value (which is based on the similarity of the two super-nodes generating the current node). Because a weighted product node means the common weight of the node appeared in both G_1 and G_2, we can perform random walks through these nodes to measure the similarity by counting the number of matching walks (the walks through nodes containing intersected attribute sets) and combining weights of the nodes. The larger the number, the more similar the two super-graphs are. Adding weight value to the random walk is meaningful. It provides a solution to take graph matching of super-nodes into consideration to calculate the similarity between super-graphs. We consider the similarity between any two super-nodes as the weight value instead of just using 1/0 hard-matching. The node similarities between different super-nodes represent the relationship between two graphs in a more precise way, which, in turn, helps improve the classification accuracy. Accordingly, we divide our kernel design into two parts based on the similarity of super-nodes and super-graphs.

3.1 Kernel on Single-Attribute Graphs

We firstly introduce the weighted random walk kernel on single-attribute graphs.

Definition 4. ***(Single-Attribute Product Graph)*** *Given two single-attribute graphs $g_1 = (V_1, E_1, Att_1, f_1)$ and $g_2 = (V_2, E_2, Att_2, f_2)$, their single-attribute product graph is denoted by $g_{1\otimes 2} = (V^*, E^*, Att^*, f^*)$ ($g_\otimes$ for short), where*

- $V^* = \{v | v = < v_1, v_2 >, v_1 \in V_1, v_2 \in V_2\}$;
- $E^* = \{e | e = (u', v'), u' \in V^*, v' \in V^*, f^*(u') \neq \phi, f^*(v') \neq \phi,$
  $u' = < u_1, u_2 >, v' = < v_1, v_2 >, (u_1, v_1) \in E_1, (u_2, v_2) \in E_2\}$;
- $Att^* = Att_1 \cup Att_2$;
- $f^* = \{f^*(v) | f^*(v) = f(v_1) \cap f(v_2), v = < v_1, v_2 >, v_1 \in V_1, v_2 \in V_2\}$.

In other words, $g_\otimes$ is a single-attribute graph where a vertex v is the intersection between a pair of nodes in g_1 and g_2. There is an edge between a pair of vertices in $g_\otimes$, if and only if an edge exists in corresponding vertices in g_1 and g_2, respectively. An example is shown in Fig. 2 ($g_{1\otimes 2}$). In the following, we show that an inherent property of the product graph is that performing a weighted random walk on the product graph is equivalent to performing simultaneous random walks on g_1 and g_2, respectively. So the single-attribute product graph provides an effective way to count the number of walks combining weight values on nodes between graphs without expensive graph matching.

To generate $g_\otimes$'s adjacency matrix $\theta_\otimes$ from g_1 and g_2 by using matrix operations, we define the *Attributed Product* as follow:

Definition 5. ***(Attributed Product)*** *Given matrices $B \in \mathbb{R}^{n\times n}$, $C \in \mathbb{R}^{m\times m}$ and $H \in \mathbb{R}^{n'\times m'}$, the attributed product $B \boxtimes C \in \mathbb{R}^{nm\times nm}$ and the column-stacking operator $vec(H) \in \mathbb{R}^{n'm'}$ are defined as*

$$B \boxtimes C = [vec(B_{*1}C_{1*})\ vec(B_{*1}C_{2*})\ \cdots\ vec(B_{*n}C_{m*})],$$

$$vec(H) = [H_{*1}^\top\ H_{*2}^\top\ \cdots\ H_{*n}^\top]^\top,$$

where H_{*i} and H_{j*} denote i^{th} column and j^{th} row of H, respectively.

Based on Def. 5, the adjacency matrix $\theta_\otimes$ of the single-attribute product graph $g_\otimes$ can be directly derived from $g_1(\theta_1, \varphi_1)$ and $g_2(\theta_2, \varphi_2)$ as follow:

$$\theta_\otimes = (\theta_1 \boxtimes \theta_2)\,\bar{\wedge}\,vec(\varphi_1\varphi_2^\top) \tag{1}$$

where

$$B\,\bar{\wedge}\begin{bmatrix} x_1 \\ \vdots \\ x_n \end{bmatrix} = \begin{bmatrix} B_{11}\wedge x_1 & \cdots & B_{1n}\wedge x_1 \\ \vdots & \ddots & \vdots \\ B_{n1}\wedge x_n & \cdots & B_{nn}\wedge x_n \end{bmatrix},$$

"$\wedge$" is a conjunction operation ($a \wedge b = 1$ *iff* $a = 1$ and $b = 1$).

To better assess the similarity between two single-attribute graphs, we assign each node a weight value, with weight indicating the importance of a node in the graph. So a weighted random walk can be calculated by multiply all the weight values of the nodes along the walk. Then we can calculate the weighted random walk counting by using matrix operation. More specifically, for the adjacency matrix θ_x of a graph g_x, each element $[\theta_x^z]_{ij}$ of this n^{th} power matrix gives the number of walks of length z from v_i to v_j in g_x. If we have the weight vector of g_x as $\bar{w}_x = [w_1, w_2, ...w_n]^\top$. Each element $[(\bar{w}_x\ \bar{w}_x^\top \odot \theta_x)^z]_{ij}$ corresponds to the total weight value of all random walks with length z from v_i to v_j in g_x, where "$\odot$" denotes element-wise multiplication (the same as ".*" operator in Matlab). For simplicity, we set an uniform distributions for all the statistical weights of nodes as $w_i = \frac{1}{n}$. So the weight vector $\bar{w}$ denote the weights over the vertices of $g_\otimes$ generated from g_1 and g_2 with the node sizes of n_1 and n_2, respectively, where

$$\bar{w} = [\frac{1}{n_1n_2}, \frac{1}{n_1n_2}, \cdots, \frac{1}{n_1n_2}]^\top \in \mathbb{R}^{n_1n_2}$$

According to Eq. (1), performing a weighted random walk on the single-attribute product graph $g_\otimes$ is equivalent to performing random walks on graphs g_1 and g_2 simultaneously. After $g_\otimes$ is generated, *Weighted Random Walk Kernel*

(WRWK), which computes the similarity between g_1 and g_2, can be defined with a sequence of weights $\delta = \delta_0, \delta_1, \cdots$ ($\delta_i \in \mathbb{R}$ and $\delta_i \geq 0$ for all $i \in \mathbb{N}$):

$$k(g_1, g_2) = \rho \sum_{i,j=1}^{n_1 n_2} \left[\sum_{z=0}^{\infty} \delta_z \left(\bar{w}\, \bar{w}^\top \odot \theta_\otimes \right)^z \right]_{ij} \tag{2}$$

where n_1 and n_2 are the node sizes of g_1 and g_2, respectively. ρ is the control parameter used to make the kernel function convergence. As a result, kernel values are upper bounded (the proof is given in Section 5). For simplicity, we set $\rho = \frac{1}{n_1 n_2}$. We call $\Gamma_\otimes = \bar{w}\ \bar{w}^\top \odot \theta_\otimes$ the weighted adjacency matrix of single-attribute product graph.

To compute the *WRWK* for single-attribute graph, as defined in Eq. (2), a diagonalization decomposition method [4] can be used. Because $\Gamma_\otimes$ is a symmetric matrix, the diagonalization decomposition of $\Gamma_\otimes$ exists: $\Gamma_\otimes = \mathcal{T}\mathcal{H}\mathcal{T}^{-1}$, where the columns of $\mathcal{T}$ are its eigenvectors, and $\mathcal{H}$ is a diagonal matrix of corresponding eigenvalues. The kernel defined in Eq. (2) can then be rewritten as:

$$k(g_1, g_2) = \rho \sum_{i,j=1}^{n_1 n_2} \left[\sum_{z=0}^{\infty} \delta_z \left(\mathcal{T}\mathcal{H}^z \mathcal{T}^{-1} \right) \right]_{ij} = \rho \sum_{i,j=1}^{n_1 n_2} \left[\mathcal{T} \left(\sum_{z=0}^{\infty} \delta_z\ \mathcal{H}^z \right) \mathcal{T}^{-1} \right]_{ij} \tag{3}$$

By setting $\delta_z = \lambda^z / z!$ in Eq. (3), and use $e^x = \sum_{z=0}^{\infty} x^z / z!$, we have,

$$k(g_1, g_2) = \rho \sum_{i,j=1}^{n_1 n_2} \left[\mathcal{T} e^{\lambda \mathcal{H}}\ \mathcal{T}^{-1} \right]_{ij} \tag{4}$$

The diagonalization decomposition can greatly expedite *WRWK* kernel computation. An example of *WRWK* kernel is shown in Fig. 2.

3.2 Kernel on Super-Graphs

The *WRWK* of single-attribute graph helps calculate the similarity between two graphs, so it can be used to calculate the similarity between two super-nodes. Given a pair of super-graphs G_1 and G_2, assume we can generate a new product graph $G_\otimes$ whose nodes are generated by super-nodes in G_1 and G_2, and weight value of each node is the similarity between the super-nodes which generate this node, then the same process shown in Section 3.1 can be used to calculate the *WRWK* for super-graphs G_1 and G_2 to denote their similarity.

Definition 6. ***(Super Product Graph)*** *Given two super-graphs $G_1 = (\mathcal{V}_1, \mathcal{E}_1, \mathcal{G}_1, \mathcal{F}_1)$ and $G_2 = (\mathcal{V}_2, \mathcal{E}_2, \mathcal{G}_2, \mathcal{F}_2)$, their Super Product Graph is denoted by $G_{1\otimes 2} = (\mathcal{V}^*, \mathcal{E}^*, \mathcal{G}^*, \mathcal{F}^*)$ ($G_\otimes$ for short), where*

- $\mathcal{V}^* = \{V | V =< V_1, V_2 >, V_1 \in \mathcal{V}_1, V_2 \in \mathcal{V}_2\}$;
- $\mathcal{E}^* = \{e | e = (V, V'), V \in \mathcal{V}^*, V' \in \mathcal{V}^*, \mathcal{F}^*(V) \neq \phi, \mathcal{F}^*(V') \neq \phi,$
  $V =< V_1, V_2 >, V' =< V_1', V_2' >, (V_1, V_1') \in \mathcal{E}_1, (V_2, V_2') \in \mathcal{E}_2\}$;
- $\mathcal{G}^* = \mathcal{G}_1 \cup \mathcal{G}_2$;
- $\mathcal{F}^* = \{\mathcal{F}^*(V) | \mathcal{F}^*(V) = \mathcal{F}(V_1) \cap \mathcal{F}(V_2), V =< V_1, V_2 >, V_1 \in \mathcal{V}_1, V_2 \in \mathcal{V}_2\}$.

An example of super product graph is shown in Fig. 2 ($G_{\otimes}$). Similar to Eq. (1), the adjacency matrix $\Theta_{\otimes}$ of the super product graph $G_{\otimes}$ can be directly derived from $G_1(\Theta_1, \Phi_1)$ and $G_2(\Theta_2, \Phi_2)$ as follows:

$$\Theta_{\otimes} = (\Theta_1 \boxtimes \Theta_2) \barwedge vec(\Phi_1 \Phi_2^{\top}) \tag{5}$$

Because we use the similarity between two super-nodes as the weight value of the node in the super product graph, the kernel value may increase infinitely with the increasing size of the super-graph. So for super-graph kernel, we add a control variable η to limit the range of the super-graph kernel value. Then the Weighted Random Walk Kernel, which computes the similarity between G_1 and G_2, can be defined with a sequence of weights $\sigma = \sigma_0, \sigma_1, \cdots$ ($\sigma_i \in \mathbb{R}$ and $\sigma_i \geq 0$ for all $i \in \mathbb{N}$):

$$K(G_1, G_2) = \eta \sum_{i,j=1}^{N_1 N_2} \left[\sum_{z=0}^{\infty} \sigma_z \, (\boldsymbol{w}\, \boldsymbol{w}^{\top} \odot \Theta_{\otimes})^z \right]_{ij} \tag{6}$$

where N_1 and N_2 are the node sizes of G_1 and G_2, respectively, and

$$\boldsymbol{w} = [k(g_1, g_1)\ k(g_1, g_2)\ \cdots\ k(g_{N_1}, g_{N_2})]^{\top}$$

Similar to WRWK of single-attribute graphs in Eq. (4), the WRWK on super-graphs can be calculated by setting $\sigma_z = \gamma^z / z!$, we have,

$$K(G_1, G_2) = \eta \sum_{i,j=1}^{N_1 N_2} \left[\widetilde{\mathcal{T}} e^{\gamma \widetilde{\mathcal{H}}}\, \widetilde{\mathcal{T}}^{-1} \right]_{ij} \tag{7}$$

where $\boldsymbol{w}\, \boldsymbol{w}^{\top} \odot \Theta_{\otimes} = \widetilde{\mathcal{T}} \widetilde{\mathcal{H}} \widetilde{\mathcal{T}}^{-1}$. To ensure kernel function convergence, we set

$$\eta = \frac{1}{N_1 N_2} e^{\frac{1 - N_1^2 N_2^2}{N_1 N_2} \gamma e^{2\lambda}} \tag{8}$$

where γ is the parameter of $WRWK$ on super-graphs, and λ is the parameter of $WRWK$ on single-attribute graphs which is given in previous sub-section. The $WRWK$ of super-graphs is also upper bounded with proof showing in Section 5.

4 Super-Graph Classification

The $WRWK$ provides an effective way to measure the similarity between super-graphs. Given a number of labeled super-graphs, we can use their pair-wise similarity to form an kernel matrix. Then generic classifiers, such as Support Vector Machine (SVM), Decision Tree (DT), Naive Bayes (NB) and Nearest Neighbour (NN), can be applied to the kernel matrix for super-graph classification.

Algorithm 1 shows the framework of using $WRWK$ to train a classifier.

5 Theoretical Study

Theorem 1. *The Weighted Random Walk Kernel function is positive definite.*

Algorithm 1. WRWK Classifier Generation

Input: Labeled super-graph set $\mathcal{D}^L$, Kernel parameters λ and γ
Output: Classifier ζ
Initialize: Kernel matrix $M \leftarrow [\,]$

for any two super-graphs G_a and G_b in $\mathcal{D}^L$ **do**
    $\boldsymbol{w} \leftarrow [\,]$
    for any g_x in G_a and g_y in G_b **do**
        $q \leftarrow (x-1)N_b + y$ **// q is the element index of $\boldsymbol{w}$**
        if $Att_x \cap Att_y = \phi$ **then**
            $\boldsymbol{w}_q \leftarrow 0$
        else
            $\rho \leftarrow \frac{1}{n_x n_y}$, $\bar{w} \leftarrow [\frac{1}{n_x n_y} \; \frac{1}{n_x n_y} \; \cdots \; \frac{1}{n_x n_y}]$, $\theta_{\otimes} \leftarrow (\theta_x \boxtimes \theta_y) \barwedge vec(\varphi_x \varphi_y^{\top})$
            $\boldsymbol{w}_q \leftarrow k(g_x, g_y)$ **// $k(g_x, g_y)$ is calculated by eq. (4)**
        end if
    end for
    $\eta \leftarrow \frac{1}{N_a N_b} e^{\frac{1-N_a^2 N_b^2}{N_a N_b} \gamma e^{2\lambda}}$, $\Theta_{\otimes} \leftarrow (\Theta_a \boxtimes \Theta_b) \barwedge vec(\Phi_a \Phi_b^{\top})$
    $M_{ab} \leftarrow K(G_a, G_b)$ **// $K(G_a, G_b)$ is calculated by eq. (7)**
end for
$\zeta \leftarrow TrainClassifier(\mathcal{C}, M, \bar{L})$
/* $\mathcal{C}$ is the learning algorithm. $\bar{L}$ is the class label vector of $\mathcal{D}^L$ */

Proof. As the random walk-based kernel is closed under products [4] and the $WRWK$ can be written as the limit of a polynomial series with positive coefficients (as Eqs. (4) and (7)), the $WRWK$ function is positive definite.

Theorem 2. *Given any two single-attribute graphs g_1 and g_2, the weighted random walk kernel of these two graphs is bounded by $0 < k(g_1, g_2) < e^{\lambda}$, where λ is the parameter of the weighted random walk kernel.*

Proof. Because $k(g_1, g_2)$ is a positive definite kernel by Theorem 1, so $k(g_1, g_2) > 0$. Then we only need to show that the upper bound of $k(g_1, g_2)$ is e^{λ}.

Based on the definition of WRWK, assume the node sizes of two single-attribute graphs g_1 and g_2 are n_1 and n_2, respectively, the number of random walks on $g_1 \otimes g_2$ must be not greater than that of the complete connect graph g^c which has $n_1 \times n_2$ nodes. We assume that $g^c = g_1' \otimes g_2'$, where g_1' and g_2' are with n_1 and n_2 nodes respectively. So we have $k(g_1, g_2) < k(g_1', g_2')$. More specifically,

$$k(g_1', g_2') = \rho \sum_{i,j=1}^{n_1 n_2} \left[\sum_{z=0}^{\infty} \delta_z \; (\bar{w}' \; \bar{w}'^{\top} \odot \theta_{\otimes}')^z \right]_{ij} = \rho \sum_{z=0}^{\infty} \{ \delta_z \sum_{i,j=1}^{n_1 n_2} \left[(\bar{w}' \; \bar{w}'^{\top} \odot \theta_{\otimes}')^z \right]_{ij} \}$$

Because $\rho = \frac{1}{n_1 n_2}$, $\bar{w}' = [\frac{1}{n_1 n_2}, \frac{1}{n_1 n_2}, \cdots, \frac{1}{n_1 n_2}]^{\top}$, and g^c is a complete connect graph, where the diagonal elements are equal to 0 and other element in the adjacency matrix are equal to 1.

So

$$\sum_{i,j=1}^{n_1 n_2} \left[(\bar{w}' \, \bar{w}'^{\top} \odot \theta'_{\otimes})^z\right]_{ij} = (\frac{n_1 n_2 - 1}{n_1^2 n_2^2})^{(z-1)}(1 - \frac{1}{n_1 n_2})$$

Then

$$k(g'_1, g'_2) = \frac{1}{n_1 n_2} \sum_{z=0}^{\infty} [\delta_z (\frac{n_1 n_2 - 1}{n_1^2 n_2^2})^{(z-1)}(1 - \frac{1}{n_1 n_2})]$$

$$= \frac{1}{n_1 n_2}(1 - \frac{1}{n_1 n_2}) \sum_{z=0}^{\infty} [\delta_z (\frac{n_1 n_2 - 1}{n_1^2 n_2^2})^{(z-1)}]$$

Because $\delta_z = \frac{\lambda^z}{z!}$, we have

$$k(g'_1, g'_2) = \frac{1}{n_1 n_2}(1 - \frac{1}{n_1 n_2}) \sum_{z=0}^{\infty} [\frac{\lambda^z}{z!} (\frac{n_1 n_2 - 1}{n_1^2 n_2^2})^{(z-1)}]$$

$$= \frac{1}{n_1 n_2}(1 - \frac{1}{n_1 n_2})(\frac{n_1^2 n_2^2}{n_1 n_2 - 1}) \sum_{z=0}^{\infty} [\frac{\lambda^z}{z!} (\frac{n_1 n_2 - 1}{n_1^2 n_2^2})^z]$$

Because $e^x = \sum_{z=0}^{\infty} x^z / z!$, we have

$$k(g_1, g_2) < k(g'_1, g'_2) = e^{\frac{\lambda(n_1 n_2 - 1)}{n_1^2 n_2^2}} < e^{\lambda}$$

Theorem 3. *The weighted random walk kernel between super-graphs* G_1 *and* G_2 *is bounded by* $0 < K(G_1, G_2) < e^{\gamma e^{2\lambda} - 2\lambda}$*, where* λ *and* γ *are weighted random walk kernel parameters for single-attribute graph and super-graph, respectively.*

Similar to Theorem 2, Theorem 3 can be derived.

6 Experiments and Analysis

6.1 Benchmark Data

DBLP Dataset: DBLP dataset consists of bibliography data in computer science (http://arnetminer.org/citation/). Each record in DBLP is a scientific publication with a number of attributes such as abstract, authors, year, venue, title, and references. To build super-graphs, we select papers published in Artificial Intelligence (AI: IJCAI, AAAI, NIPS, UAI, COLT, ACL, KR, ICML, ECML and IJCNN) and Computer Vision (CV: ICCV, CVPR, ECCV, ICPR, ICIP, ACM Multimedia and ICME) fields to form a classification task. The goal is to predict which field (AI or CV) a paper (*i.e.* a super-graph) belongs to by using the abstract of each paper (*i.e.* a super-node) and abstracts of references (*i.e.* other super-nodes), as shown in Fig. 3. An edge (undirected) between two super-nodes indicates a citation relationship between two papers. For each paper, we use fuzzy cognitive map (E-FCM) [5] to convert paper abstract into a graph (which represents relations between keywords with weights over a threshold (*i.e.* edge-cutting threshold)). This graph representation has shown better performance than simple bag-of-words representation [1]. We select 1000 papers

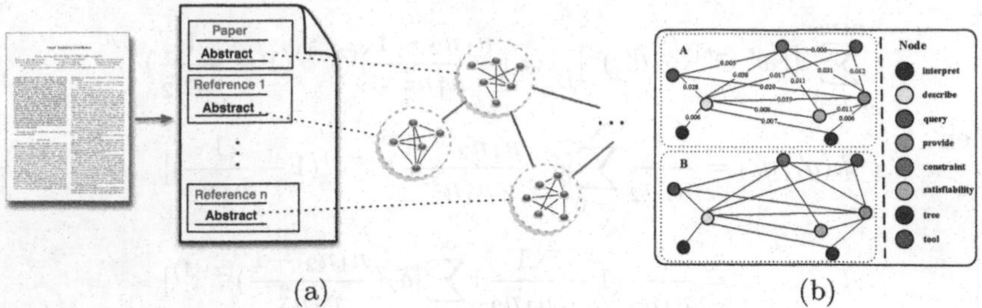

Fig. 3. An example of using super-graph representation for scientific publications. The left figure shows that each paper cites a number of references. For each paper (or each reference), its abstract can be converted as a graph. So each paper denotes a super-node, and the citation relationships between papers form a super-graph. The figure to the right shows a graph representation of paper abstract with each node denoting a keyword. A: The weight values between nodes indicate correlations between keywords. B: by using proper threshold, we can convert each abstract as an undirected unweighted graph.

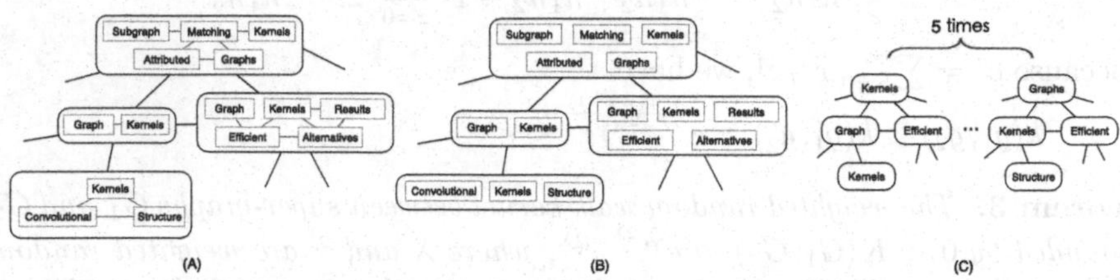

Fig. 4. Super-graph and comparison graph representations. A super-graph (A) can be represented as an attributed graph by discarding edges in each super-super as shown in (B). We also simplify a super-graph as a set of single-attribute graphs by retaining only one attribute in each super-node, as shown in (C).

(500 in each class), each of which contains 1 to 10 references, to form 1000 super-graphs.

Beer Review Dataset: The online beer review dataset consists of review data for beers (http://beeradvocate.com/). Each review in the dataset is associated with some attributes such as appearance score, aroma score, palate score, and taste score (rating of the product varies from 1 to 5), and detailed review texts. Our goal is to classify each beer to the right style (*Ale* vs. *Not Ale*) by using customer reviews. The graph representation for reviews is similar to the sub-graphs in DBLP dataset. Each review is represented as a super-node. The edge between super-nodes are built using following method: Because a review has four rating scores in appearance, aroma, palate, and taste, we use these four scores as a feature vector for each review. If two reviews's distance in the feature space (Euclidean distance) is less than 2, an edge is used to link two reviews (*i.e* super-nodes). We choose 1000 beer products, half from Ale and the rest is from Large and Hybrid Style, to form 1000 super-graphs for classification.

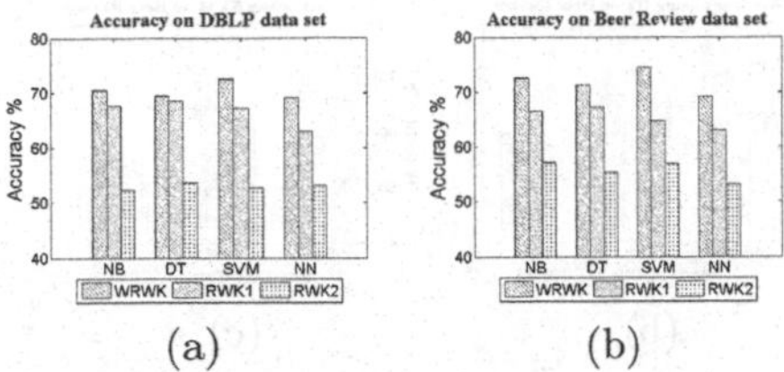

Fig. 5. Classification accuracy on DBLP and Beer review datasets *w.r.t.* different classification methods (NB, DT, SVM, and NN). For DBLP dataset, the number of super-nodes in each super-graph varies within 1-10 and the edge-cutting threshold for single-attribute graph is 0.001. For Beer Review dataset, the number of super-nodes in each super-graph varies within 30-60 and the edge-cutting threshold is 0.001. $WRWK$ uses super-graph representation (Fig.4(A)), $RWK1$ and $RWK2$ use traditional random walk kernel method with attributed graph representation (Fig. 4(B)) and single-attribute graph representation (Fig. 4(C)), respectively.

6.2 Experimental Settings

Baseline Methods: Because no existing method can handle super-graph classification, for comparison purposes, we use two approaches to generate traditional graph representations from each super-graph: (1) randomly selecting one attribute (*i.e* one word) in each super-node (and ignoring all other words) to form a single-attribute graph, as shown in Fig. 4 (C). We repeat random selection for five times with each time generating a set of single-attribute graphs from super-graphs. Each single-attribute graph set is used to train a classifier and their majority *voted accuracy* on test super-graphs is reported in the experiments. (2) We use the attribute set of the single-attribute graph in each super-node as multi-attributes of the super-node to generate an attributed graph (which is equal to removing all the edges from the super-graph as shown in Fig. 4 (B)). For all the two graph representations, we use the traditional random walk kernel (RWK) to measure the similarity between any two graphs [4].

We use 10 times 10-fold cross-validation classification accuracy to measure and compare the algorithm performance. To train classifiers from graph data, we use Naive Bayes (NB), Decision Tree (DT), Support Vector Machines (SVM) and Nearest Neighbor algorithm (NN). Majority examples are based on the kernel parameter settings: $\lambda = 100$ and $\gamma = 100$.

6.3 Results and Analysis

Performance on Standard Benchmarks: In Fig. 5, we report the classification accuracy on the benchmark datasets. The experimental results show that $WRWK$ constantly outperforms traditional RWK method, regardless of the type of graph representations used by RWK. This is mainly because in traditional graph representations each node only has one attribute or a set of independent attributes, whereas single-attribute nodes (or multiple independent attributes) cannot precisely describe the node content. For super-graphs, the

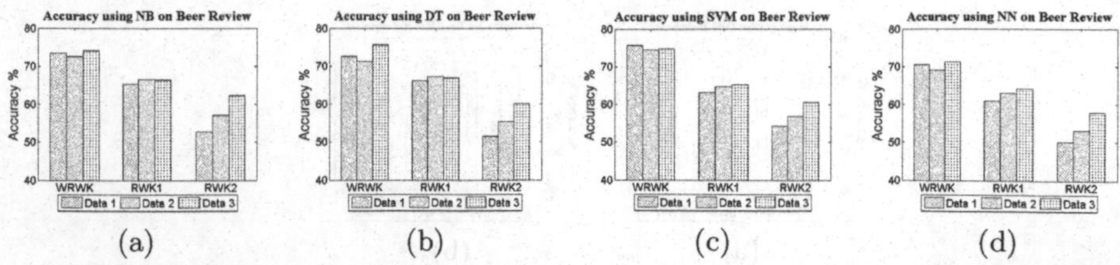

(a) (b) (c) (d)

Fig. 6. Classification accuracy on Beer review dataset *w.r.t.* different datasets and classification methods (NB, DT, SVM, and NN). Figures correspond to three types of super-graph structure on Beer review. *Data 1*: the number of super-nodes in each super-graph is 2-30 and the edge-cutting threshold is 0.00001. *Data 2*: the number of super-nodes in each super-graph is 30-60 and the edge-cutting threshold is 0.001. *Data 3*: the number of super-nodes in each super-graph is > 60 and the threshold is 0.1.

graph associated to each super-node provides an effective way to describe the node content. In $WRWK$ method, we consider the similarity between any two super-nodes as the weight value instead of just using 1/0 hard matching to represent whether there is an intersection of attribute sets between two nodes. The soft matching node similarities between super-nodes captures the relationship between two graphs in a more precise way. This, in turn, helps improve the classification accuracy.

Performance under different Super-Graph Structures: To demonstrate the performance of our $WRWK$ method on super-graphs with different characteristics, we construct super-graphs on Beer review dataset by using different super-node sizes and different structures of single-attribute graph (the structure of the single-attribute node is controlled by the edge-cutting threshold) as shown in Fig. 6 (a)-(d). The result shows that our $WRWK$ method is stable on super-graphs with different structures.

Performance *w.r.t.* Changes in Super-Nodes and Walks: The proposed weighted random walk kernel relies on the similarity between super-nodes and the common walks between two super-graphs to calculate the graph similarity. This raises a concern on whether super-node similarity or walk similarity (or both) plays a more important role in assessing the super-graph similarity.

In order to resolve this concern, we design the following experiments. In the first set of experiments (Fig. 7 (a) and (b)), we fix super-graph edges, and change edges inside super-nodes, which will impact on the super-node similarities. If this results in significant changes, it means that super-node similarity plays a more important role. In the second set of experiments (Fig. 7 (c)), we fix super-nodes but vary the edges in super-graphs (by randomly removing edges), which will impact on the common walks between super-graphs. If this results in significant changes, it means that walk plays a more important role than super-nodes.

Fig. 7 (a) and (b) report the algorithm performance with respect to the edge-cutting threshold. The accuracy decreases dramatically with the increase of the edge-cutting threshold for both datasets. When the edge-cutting threshold is set to 0.0001 on DBLP and 0.00001 on Beer Review, the single-attribute graph

Fig. 7. The performance *w.r.t.* different edge-cutting thresholds on DBLP and Beer Review datasets by using *WRWK* method. In (a) and (b), the node size of super-graph varies from 0-10 on DBLP dataset and 30-60 on Beer Review dataset. The x-axis shows the value of the edge-cutting threshold and the average node degree in its corresponding single-attribute graph is reported in the parentheses. In (c), the average number of edges cut from each super-graph varies from 0 to 100.

in each super-node is almost a complete graph and four methods achieve the highest classification accuracy. As the threshold is set to 0.01 on DBLP and 0.1 on Beer review, the single-attribute graph in each super-node is very small and contains very few edges. As a result, the accuracies are just around 65%. This demonstrates that *WRWK* heavily relies on the structure information of each super-node to assess the super-graph similarities.

Fig. 7 (c) reports the algorithm performance by fixing the super-nodes but gradually removing edges in the super-graph of Beer review dataset (as Fig. 5 (b)). Similar to the result in Fig. 7 (a) and (b), the classification accuracy decreases when edges are continuously removed from the super-graph (even if the super-nodes are fixed). From Figs. 7, we find that graph structure of super-graph is as important as that of super-nodes. This is mainly attributed to the fact that our weighted random walk kernel relies on both super-node similarity and walks in the super-graph to calculate graph similarities.

7 Conclusion

In this paper, we formulated a new super-graph classification problem. Due to the inherent complex structure representation, all existing graph classification methods cannot be applied for super-graph classification. In the paper, we proposed a weighted random walk kernel which calculates the similarity between two super-graphs by assessing (a) the similarity between super-nodes of the super-graphs, and (b) the common walks of the super-graphs. Our key contribution is twofold: (1) a weighted random walk kernel considering node and structure similarities between graphs; and (2) an effective kernel-based super-graph classification method with sound theoretical basis.

References

1. Angelova, R., Weikum, G.: Graph-based text classification: learn from your neighbors. In: ACM SIGIR, pp. 485–492 (2006)
2. Cai, Y., Cercone, N., Han, J.: An attribute-oriented approach for learning classification rules from relational databases. In: ICDE, pp. 281–288 (1990)
3. Cheng, H., Zhou, Y., Yu, J.X.: Clustering large attributed graphs: A balance between structural and attribute similarities. ACM TKDD 5(2) (2011)
4. Gärtner, T., Flach, P.A., Wrobel, S.: On graph kernels: Hardness results and efficient alternatives. In: COLT, pp. 129–143 (2003)
5. Luo, X., Xu, Z., Yu, J.: Building association link network for semantic link on web resources. IEEE Trans. Autom. Sci. Eng. 8(3), 482–494 (2011)
6. Riesen, K., Bunke, H.: Graph classification and clustering based on vector space embedding. World Scientific Publishing Co., Inc. (2010)
7. Schölkopf, B., Smola, A.J.: Learning with kernels. MIT Press (2002)
8. Xu, Z., Ke, Y., Wang, Y., Cheng, H., Cheng, J.: A model-based approach to attributed graph clustering. In: ACM SIGMOD, pp. 505–516 (2012)

Subtopic Mining via Modifier Graph Clustering

Hai-Tao Yu and Fuji Ren

The University of Tokushima, Japan
yu-haitao@iss.tokushima-u.ac.jp,
ren@is.tokushima-u.ac.jp

Abstract. Understanding the information need encoded in a user query has long been regarded as a crucial step of effective information retrieval. In this paper, we focus on *subtopic mining* that aims at generating a ranked list of subtopic strings for a given topic. We propose the *modifier graph* based approach, under which the problem of subtopic mining reduces to that of graph clustering over the modifier graph. Compared with the existing methods, the experimental results show that our modifier-graph based approaches are robust to the sparseness problem. In particular, our approaches that perform subtopic mining at a fine-grained term-level outperform the baseline methods that perform subtopic mining at a whole query-level in terms of *I-rec*, *D-nDCG* and *D#-nDCG*.

Keywords: Subtopic mining, Kernel-object, Modifier graph.

1 Introduction

Recent studies show that billions of daily searches are made by web users. Instead of formulating natural language queries, the vast majority of users are submitting short queries with little or no context, which are often ambiguous and/or underspecified. For example, the query *Harry Potter* could refer to a *book* or a *movie*. For the *movie*, a user may be interested in the *main character* or the *reviews*. In view of the above-mentioned facts, the technique of *search result diversification* (e.g., [1, 13]) has been proposed and attracted significant attention. It provides a diversified search result, which features a trade-off between *relevance* (ranking more relevant web pages in higher positions) and *diversity* (satisfying users with different information needs). For a better diversified search result, an accurate estimation of the information needs or subtopics underlying a given topic, which is referred to as *subtopic mining* [17, 19], becomes an important problem. Since the query log records the historical search behaviors performed by massive users (e.g., clicked web pages in response to a query, subsequently reformulated queries, etc), many researchers [10, 14, 16] performed subtopic mining via query log mining and have shown that the click information (e.g., page co-click and session co-occurrence) are valuable for determining the topical relevance of queries. However, as shown by Bonchi et al. [5], many methods have been proposed for and tested on head queries, which work poorly for unseen queries or queries with sparse click information. As query frequency is known to obey the power-law, these methods leave a large ratio of queries uncovered.

V.S. Tseng et al. (Eds.): PAKDD 2014, Part I, LNAI 8443, pp. 337–347, 2014.

In this paper, we propose an effective approach that performs subtopic mining at the term level, which outperforms baseline methods that work at the query-level. Subsequently, Section 2 discusses the related work. Section 3 formalizes the target problem. Section 4 details the proposed approach. Section 5 discusses the experimental results. We conclude our work in section 6.

2 Related Work

To capture the underlying subtopics encoded within user queries, considerable studies [6, 7, 15, 21, 22, 23] have been conducted from various aspects. For example, Beeferman and Berger [2] viewed the query log as a bipartite graph, and the obtained clusters were interpreted as subtopics covering diverse queries. The research by Jones and Klinkner [11] showed that users' search tasks are interleaved or hierarchically organized. They studied how to segment sequences of user queries into a hierarchical structure. Sadikov et al. [16] and Radlinski et al. [14] tried to infer the underlying subtopics of a query by clustering its refinements. The web page co-click and session co-occurrence information were viewed as indicators of topical relevance. Hu et al. [10] interpreted a subtopic as a set of keywords and URLs, their key intuition was that: many users add one or more additional keywords to expand the query to clarify their search intents.

Different from the above methods that have treated a whole query as the minimum analysis unit. A number of studies (e.g., [20, 21]) performed intent analysis at a term-level. For example, Wang et al. [21] studied how to extract broad aspects from query reformulations, each broad aspect is represented by a set of keywords. Rather than raw terms based subtopic analysis, we perform subtopic mining by encapsulating intent roles [24]. We believe that our approach achieves a robust subtopic mining, especially for tail or unseen queries.

3 Problem Formalization

The *Intent* tasks [17, 19] of *NTCIR-9*[1] and *NTCIR-10*[2] are the prototypes on which our formalization builds. The core notations are: (1) **Topic** ($\ddot{T}$): it refers to an input instance for testing; (2) **Subtopic** ($\ddot{t}$): it refers to a possible information need or an intent underlying a topic; (3) **Subtopic string** (*tStr*): it is viewed as an expression of a subtopic. For example, for the topic *Harry Potter*, *Harry Potter fiction* and *Harry Potter reading* are two subtopic strings about the subtopic *book*. A topic and a subtopic string can either be a real query or a query-like string derived from other resources, e.g., query suggestions.

Formally, for a given topic $\ddot{T}$, suppose there are k possible subtopics $X = \{\ddot{t}_1, ..., \ddot{t}_k\}$. Let $Y = \{tStr_1, ..., tStr_n\}$ be a set of n subtopic strings with respect to X. The problem of subtopic mining is formalized as: finding a ranked list of subtopic strings L that ranks subtopic strings representing more popular

[1] http://research.nii.ac.jp/ntcir/ntcir-9/index.html
[2] http://research.nii.ac.jp/ntcir/ntcir-10/index.html

subtopics at higher positions and includes as many subtopics as possible. The quality of L depends on its consistency with the ideal ranked list L^* (e.g., a ranked list created by human assessors).

4 Subtopic Mining through Modifier Graph Clustering

4.1 Intent Role Oriented Modifier Graph

Let Γ, Q, s, q, t, d denote the query log, query set, a session, a query, a term and a web page respectively, $D(q) = \{d\}$ denote the clicked web pages of q, $t \succ q$ denote that t is a composing term of q, $q \in s$ denote that q occurred in s. Fig. 1 shows some click behaviors, where q_1 and q_2 occur in session s_k, d_1, d_2 and d_3 are the clicked web pages. Some query log oriented notations are given as: (1) **Co-Session**: For $q_i \neq q_j$, iff $\exists s$ that meets $q_i \in s$ and $q_j \in s$, q_i and q_j are co-session queries, i.e., $CoSession(q_i, q_j)$, e.g., q_1 and q_2 in Fig. 1; (2) **Co-Click**: For $q_i \neq q_j$, iff $D(q_i) \cap D(q_j) \neq \emptyset$, q_i and q_j are co-click queries, i.e., $CoClick(q_i, q_j)$, e.g., q_1 and q_2 in Fig. 1; (3) **Co-Query**: For $t_i \neq t_j$, iff $\exists q$ that meets $t_i \succ q$ and $t_j \succ q$, t_i and t_j are co-query terms, i.e., $CoQuery(t_i, t_j)$, e.g., t_2 and t_3 in Fig. 1; (4) **Term-Level Co-Session**: For $t_i \neq t_j$, iff $\exists q_m \in Q$ and $\exists q_n \in Q$ that meet $CoSession(q_m, q_n) \wedge t_i \succ q_m \wedge t_j \succ q_n$, t_i and t_j are co-session terms, i.e., $TCoSession(t_i, t_j)$, e.g., t_1 and t_2 in Fig. 1; (5) **Term-Level Co-Click**: For $t_i \neq t_j$, iff $\exists q_m \in Q$ and $\exists q_n \in Q$ that meet $CoClick(q_m, q_n) \wedge t_i \succ q_m \wedge t_j \succ q_n$, t_i and t_j are co-click terms, i.e., $TCoClick(t_i, t_j)$, e.g., t_1 and t_3 in Fig. 1. In this paper, the statistics of co-session and co-click of a query log that are defined at a query level, are called *query-level knowledge in query log* (QKQL). In contrast, the statistics of co-query, term-level co-session and term-level co-click that are defined at a term level are called *term-level knowledge in query log* (TKQL).

Rather than raw terms based analysis, we encapsulate the terms with intent roles (*Kernel-object & modifier*) by Yu and Ren [24]. Kernel-object (*ko*) refers to the term that abstracts the core object of the underlying subtopic encoded in a query. Modifier (*mo*) refers to the co-appearing terms with kernel-object, which explicitly specify user's interested aspects. A query that can be represented with kernel-object and modifier is defined as a *role-explicit* query. Otherwise, a *role-implicit* query. Because topic and subtopic string are query-like strings, thus, they can be analogously classified as role-explicit ones and role-implicit ones. Table 1 shows 3 role-explicit subtopic strings, the 1st column (*In/Out*) implies whether they are included in query log SogouQ(Section 5.1). For convenience, we directly use kernel-object and modifier to refer to the terms annotated as kernel-object and modifier respectively. Moreover, when determining the kernel-object and modifier for a given role-explicit query, topic or subtopic string, the method by Yu and Ren [24] is used as a black box in our study, which selects the annotation with the maximum likelihood as the optimal annotation using a generative model.

Definition 1 (Co-kernel-object Elements). *Given a kernel-object ko, co-kernel-object elements refer to a set of role-explicit subtopic strings that share the same kernel-object, denoted as CoKO(ko)={tStr}.*

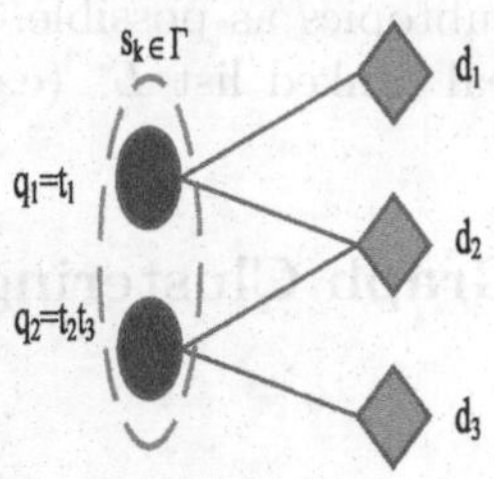

Fig. 1. Example click behaviors

Table 1. Role-explicit subtopic strings

In/Out	Annotated subtopic string
In	哈利波特游戏(*Harry Potter game*) — 哈利波特: *ko*, 游戏: *mo*; *Harry Potter*: *ko*, *game*: *mo*
In	哈利波特小说(*Harry Potter fiction*) — 哈利波特: *ko*, 小说: *mo*; *Harry Potter*: *ko*, *fiction*: *mo*
Out	哈利波特阅读(*Harry Potter reading*) — 哈利波特: *ko*, 阅读: *mo*; *Harry Potter*: *ko*, *reading*: *mo*

Definition 2 (Modifier Graph). *Modifier graph* $G_{ko} = \{V, E\}$ *is an undirected weighted graph derived from co-kernel-object elements* $CoKO(ko)$, *(i) The node set consists of modifiers, i.e.,* $V = \{mo|tStr \in CoKO(ko), mo \succ tStr\}$; *(ii) The edge set is given as:* $E = \{e = (mo_i, mo_j)|\delta_{TKQL}(mo_i, mo_j) > 0\}$, *where* $\delta_{TKQL}(mo_i, mo_j) = \lambda_1 \frac{CoQuery(mo_i, mo_j)}{max_V CoQuery+1} + \lambda_2 \frac{TCoSession(mo_i, mo_j)}{max_V TCoSession+1} + \lambda_3 \frac{TCoClick(mo_i, mo_j)}{max_V TCoClick+1}$, *namely, the TKQL of a pair of modifiers is normalized by the maximum ones (*$max_V CoQuery$, $max_V TCoSession$ *and* $max_V TCoClick$*) respectively, and the edge weight is a linear combination of the normalized values.*

Assuming that $CoKO(ko =$ 哈利波特$(HarryPotter))$ merely consists of the three subtopic strings in Table 1, Fig. 2 shows the corresponding modifier graph based on the TKQL of SogouQ in Table 2(given as *CoQuery:TCoSession:TCoClick*). Obviously, 小说(fiction) and 阅读(reading) are strongly interacted (a weight of 0.9991); 游戏(game) and 小说(fiction) (a weight of 0.0313), 游戏(game) and 阅读(reading) (a weight of 0.0049) are weakly interacted. If we perform graph clustering over this modifier graph, 游戏(game) is likely to be grouped into a cluster. 小说(fiction) and 阅读(reading) are likely to be grouped into another cluster. Many studies (e.g., [4, 5, 14, 22]) have shown that the QKQL helps to determine the topical relevance among queries. Analogously, the TKQL derived from QKQL manifests the topical relevance among terms. Put another way, the larger value of $\delta_{TKQL}(mo_i, mo_j)$, the more relevant subtopics indicated by mo_i and mo_j. Thus, 小说(fiction) and 阅读(reading) express a relevant subtopic, 游戏(game) expresses a different subtopic. Because the subtopics underlying the co-kernel-object elements depend on the composing modifiers, we can further deduce that 哈利波特阅读(Harry Potter reading) and 哈利波特小说(Harry Potter fiction) express the same or similar subtopic, while 哈利波特游戏(Harry Potter game) express a different subtopic. Not surprisingly, the partition of a modifier graph uncovers the prior unknown subtopics of a set of co-kernel-object elements.

4.2 Modifier Graph Construction and Clustering

Given the kernel-object ko with respect to a topic $\ddot{T}$, we construct the modifier graph G_{ko} through the following two steps: **Step-1**: Obtaining sufficient

Table 2. TKQL derived from SogouQ

	游戏(game)	小说(fiction)	阅读(reading)
游戏(game)	×	7:78:20	1:13:1
小说(fiction)	7:78:20	×	700:993:3637
阅读(reading)	1:13:1	700:993:3637	×

Fig. 2. A modifier graph

co-kernel-object elements $CoKO(ko)$. We first manage to obtain a set of candidate subtopic strings, say, $\{ctStr\}$, which is further used as the base for generating $CoKO(ko)$. Here *recall* is the key point, it ensures the coverage of different subtopics. The adopted resources are: (1) *Query log*; (2) *Query suggestions* for topic $\ddot{T}$; (3) *Noun phrase (NP) and verb phrase (VP) segments* extracted from the result snippets for topic $\ddot{T}$ by a search engine. Specifically, all the queries or segments that include *ko* as a substring are selected as candidate subtopic strings. The query suggestions are directly selected as candidate subtopic strings. According to Definition 1, the co-kernel-object elements are all role-explicit ones that can be represented with kernel-object and modifiers. When deriving the $CoKO(ko)$ based on a set of candidate subtopic strings, we relax the restriction as: regardless of whether a candidate subtopic string is role-explicit or role-implicit, once it includes *ko* as a substring, we intuitively assume that it is a co-kernel-object element. The substring *ko* is directly regarded as its kernel-object, the remaining parts are directly segmented into modifiers (stop-words are discarded). **Step-2**: Adding weighted edges. Given the co-kernel-object elements $CoKO(ko)$, the modifier graph G_{ko} can be obtained by adding an edge for each pair of distinct modifiers mo_i and mo_j iff $\delta_{TKQL}(mo_i, mo_j) > 0$.

Suppose π denotes a parameter-free graph clustering algorithm, given an modifier graph G_{ko}, a group of modifier clusters are generated, say, $MC = \{mc_k\}$, where $mc_k = \{mo_i\}$ denotes a cluster of modifiers. Corresponding to each modifier cluster mc_k, we generate a cluster of subtopic strings, say, $sc_k = \{tStr_m\}$, which is used to represent a subtopic, e.g., t_k. Thus, a set of subtopic string clusters $SC = \{sc\}$ can be obtained based on the set of modifier clusters MC. Specifically, for each modifier $mo_i \in mc_k$, we select the subtopic string $tStr_m$ that meets $mo_i \succ tStr_m$, and add $tStr_m$ into the corresponding subtopic string cluster sc_k. For example, given the two modifier clusters by clustering the modifier graph in Fig. 2 (Section 4.1), i.e., $mc_1 = \{$游戏$(game)\}$ and $mc_2 = \{$小说$(fiction)$, 阅读$(reading)\}$, for mc_1, $sc_1 = \{$哈利波特游戏$(HarryPottergame)\}$ will be generated due to 游戏$(game) \succ$ 哈利波特游戏$(HarryPottergame)$, and so does $sc_2 = \{$哈利波特阅读$(HarryPotterreading)$, 哈利波特小说$(HarryPotterfiction)\}$.

When allocating the subtopic string of which the composing modifiers belong to different modifier clusters, we obey the following priority rules: (1) Add this subtopic string into the subtopic string cluster, of which the corresponding modifier cluster encloses more composing modifiers; (2) If two modifier clusters enclose the same number of modifiers of a particular subtopic string, add this subtopic string into the subtopic string cluster, of which the corresponding modifier cluster encloses the most frequent modifier. The idea of the above rules are straightforward, i.e., a larger number of modifiers means a larger expressive power, and a more frequency modifier carries a larger expressive power than a rare modifier.

4.3 Generating the Ranked List

Definition 3 (Expression Power). *Expression power of a subtopic string is defined as the generating probability of its composing modifiers, which measures its effectiveness of describing the subtopic represented by the subtopic string cluster it belongs to. It is given as:* $EP(tStr) = \prod_{mo_i \succ tStr} p(mo_i)$.

By $p(mo_i) = \frac{|mo_i|+1}{\sum_{\{mo_j \in V\}} |mo_j| + |V|^+}$, where $|mo|$ denotes modifier frequency in V and $|V|^+$ denotes the number of distinct modifiers, we assume that there is a probability distribution for modifiers, which is sampled repeatedly by users to form mutually-independent modifiers to specify kernel-object oriented subtopics.

Definition 4 (Subtopic Popularity). *Subtopic popularity is defined as the likelihood of a particular subtopic. Formally, it is given as* $SP(\ddot{t}_k) = SP(sc_k) = \frac{|sc_k|}{\sum_{sc_s \in SC} |sc_s|}$*, where* $\ddot{t}_k$ *denotes the subtopic represented by the subtopic string cluster* sc_k*,* $|sc_k| = \sum_{tStr_m \in sc_k} |tStr_m|$ *and* $|tStr_m|$ *denotes the frequency of* $tStr_m$ *in* $CoKO(ko)$.

Based on Definition 4, the gain value of putting a subtopic string $tStr_r$ at the r-th slot is given as: $G(r) = \frac{SP(sc_k)}{log(r+1)}$, where $tStr_r \in sc_k$, i.e., the gain value builds upon the subtopic popularity of the subtopic string cluster that $tStr_k$ belongs to. The discounted cumulative gain of L with a cutoff r is given as: $DCG(r) = \beta * \frac{N_{1+}(r)}{|SC|} + (1-\beta) * \sum_{j=1}^{r} G(j)$, where $N_{1+}(r)$ denotes the number of distinct clusters that the top-r subtopic strings cover. If we view the cluster number $|SC|$ as the number of possible subtopics, $\frac{N_{1+}(r)}{|SC|}$ is used to represent the subtopic diversity. $\sum_{j=1}^{r} G(j)$ aims to rank subtopic strings that indicate popular subtopics at higher positions. Analogous to the metric of *D#-nDCG* [18], we let $\beta = 0.5$. Inspired by the Maximum Marginal Relevance (MMR) [8] criterion that combines relevance and novelty in the context of text retrieval, Algorithm 1 generates the target ranked list L by iteratively selecting the subtopic string that achieves the maximum margin.

In particular, by Steps 1-6, from the subtopic string cluster that achieves the maximum subtopic popularity, the subtopic string that has the maximum expression power is selected as the first element of L. By Steps 7-12, given the top-$(j-1)$ subtopic strings, the subtopic string that achieves the maximum margin (i.e., $DCG(j) - DCG(j-1)$) when being selected as the j-th element of L is put at the j-th slot of L. Because the discounted cumulative gain value (DCG) combines both subtopic diversity and relevance, the obtained L thus captures a trade-off between subtopic diversity and popularity.

Algorithm 1. The algorithm for generating the ranked list L

Input: η: required size of L; $SC = \{sc_k\}$: the set of subtopic string clusters.

```
for each sc_k ∈ SC do
    Rank the subtopic strings of sc_k in order of their decreasing expression power;
end for
Select the subtopic string cluster sc* that meets: sc* = arg max_{sc_k ∈ SC} SP(sc_k);
Select the subtopic string tStr* that meets: tStr* = arg max_{tStr_m ∈ sc*} EP(tStr_m);
Put tStr* at the first slot of the target ranked list L and remove it from sc*;
Set j = 2;
while j ≤ η and |SC| > 0 do
    Select tStr* that meets: tStr* = arg max_{sc_n ∈ SC} (DCG(j) − DCG(j − 1));
    Put tStr* at the j-th slot of L and remove tStr* from the cluster it belongs to;
    j++;
end while
```

5 Experiments

5.1 Experimental Setup

The publicly available topic set and resources (e.g., Chinese query log SogouQ, query suggestions and the ideal ranked list for each topic) from the Intent task of NTCIR-10 [17] are adopted, the top-100 result snippets of Google are used. Yu and Ren [24] have shown that the role-implicit queries are often question queries or verbose queries that merely require a specific answer. In this paper, we focus on role-explicit topics that generally have multiple subtopics, 6 role-implicit topics (i.e., 0244, 0245, 0249, 0256, 0270, 0283) are excluded, e.g., *0256: 什么是 RTF*(what is RTF). The remaining 92 topics are used as *Topic-Set-A*.

Fig. 3 (in Section 5.2) shows to what extent we can rely on the QKQL. The x-axes denote the count of instances that appeared in SogouQ. The y-axes denote the count of topics. The asterisk corresponds to co-session queries, the circle corresponds to candidate subtopic strings. We found that: 46 topics have co-session queries less than 10, 38 topics have candidate subtopic strings in SogouQ less than 10. For topics of this kind, the so called *sparseness problem* is a big trouble. By excluding the 38 topics from Topic-Set-A, of which the count of candidate subtopic strings in SogouQ is less than 10, we built the *Topic-Set-B*.

Two typical methods [9, 16] for query intent inference are compared. Deng et al. [9] proposed the *Click Frequency Inverse Query Frequency model* for query clustering (denoted as *CFIQF*). Under CFIQF, each query is represented by a vector of clicked documents. Sadikov et al. [16] performed intent inference by clustering the refinements of a given query (denoted as *RC*). A Markov graph is built based on the document click and session co-occurrence of the refinements. Finally, the problem of refinement clustering is reduced to the problem of Euclidean-vector clustering. As did by Sadikov et al. [16], the complete-linkage clustering algorithm is used for the two baselines.

The metric *D#-nDCG* [18] suggested by NTCIR-10 [17] is used as the test metric, which is a linear combination of *I-rec* and *D-nDCG*. I-rec indicates the proportion of subtopics covered by the top subtopic strings and measures diversity. D-nDCG measures overall relevance across subtopics. In our study, we evaluate I-rec, D-nDCG, D#-nDCG with cutoff values of 10 and 20.

When quantifying the edges of a modifier graph, we let $\lambda_1 = \lambda_2 = \lambda_3 = \frac{1}{3}$, i.e. co-query, term-level co-session and term-level co-click were treated equally. For modifier graph clustering, two parameter-free graph clustering algorithms *Louvain* [3] and *LinLog* [12] have been tested, the corresponding modifier graph based approaches are denoted as *MG-Lou* and *MG-Lin* respectively.

5.2 Experimental Results

Based on Topic-Set-B, Table 3 shows the results with cutoff values of l = 10&20.

Table 3. Topic-Set-B based comparison with cutoff values of l = 10&20

	I-rec@10	D-nDCG@10	D#-nDCG@10	I-rec@20	D-nDCG@20	D#-nDCG@20
CFIQF(5)	0.1782	0.1753	0.1768	0.1782	0.1168	0.1475
CFIQF(10)	0.3066	0.2688	0.2877	0.3066	0.1767	0.2417
CFIQF(15)	0.3093	0.2715	0.2904	0.3903	0.2328	0.3116
RC(5)	0.0353	0.0333	0.0343	0.0353	0.0216	0.0285
RC(10)	0.0595	0.0501	0.0548	0.0595	0.0327	0.0461
RC(15)	0.0806	0.0672	0.0739	0.1024	0.0504	0.0764
MG-Lou	0.4375	**0.4375**	0.4375	0.5579	**0.3975**	0.4777
MG-Lin	**0.4650**	0.4275	**0.4462**	**0.5878**	0.3921	**0.4899**

As the complete-linkage clustering algorithm requires a predefined cluster number, different values (e.g., 5 is marked as (5)) were tested for the baselines. As shown in Table 3, the baseline methods exhibit different performances due to different cluster numbers. For example, the D#-nDCG@10 value of CFIQF is 0.2904 (4th column) with a cluster number of 15, which is better than the values generated with cluster numbers of 5 and 10. However, the number of subtopics indeed varies from topic to topic. A predefined cluster number can't guarantee a natural subtopic mining. RC operates by clustering the query refinements (essentially co-session queries) of a given topic, its low performance implies that the clustered refinements greatly deviates from the ranked list by human assessors. By clustering the subtopic strings via co-click information, CFIQF shows a better performance than RC.

Different from the baseline methods that rely on QKQL, the modifier graph based approaches make use of TKQL. The results show that both MG-Lou and MG-Lin outperform the baselines in terms of I-rec, D-nDCG and D#-nDCG. The reasons are: The proposed approaches work at a term level instead of a whole query level. Based on the TKQL, they can determine the subtopics of subtopic strings from different resources rather than merely the queries of a query log. For example, though 哈利波特阅读(Harry Potter reading) (Table 1) is not included in SogouQ, we can determine its underlying subtopic based on its composing modifier 阅读(reading). On the contrary, the baseline methods will fail due to no click information. Leveraging on the two ad-hoc graph clustering algorithms, our modifier graph based approaches can determine the possible subtopics per topic instead of a uniform cluster number.

Based on Topic-Set-A, Table 4 shows the results with cutoffs of $l = 10\&20$.

Table 4. Topic-Set-A based comparison with cutoff values of $l = 10\&20$

	I-rec@10	D-nDCG@10	D#-nDCG@10	I-rec@20	D-nDCG@20	D#-nDCG@20
CFIQF(5)	0.1130	0.1042	0.1086	0.1130	0.0714	0.0922
CFIQF(10)	0.1449	0.1352	0.1401	0.1449	0.0890	0.1169
CFIQF(15)	0.1506	0.1430	0.1468	0.2142	0.1242	0.1692
RC(5)	0.0145	0.0160	0.0153	0.0145	0.0104	0.0125
RC(10)	0.0391	0.0355	0.0373	0.0391	0.0230	0.0311
RC(15)	0.0411	0.0373	0.0392	0.0543	0.0282	0.0413
MG-Lou	0.4111	**0.4253**	**0.4182**	**0.5334**	**0.4024**	**0.4679**
MG-Lin	**0.4121**	0.4117	0.4119	0.5309	0.3945	0.4627

Compared with Topic-Set-B, Topic-Set-A contains 38 topics that have sparse QKQL. This is the very reason why the performances of the baseline methods are greatly impacted. For example, the best performance of CFIQF (D#-nDCG@20 at the 7th column) is decreased from 0.3116 (Table 3) to 0.1692 (Table 4). In contrast, our modifier graph based approaches exhibit stable performances. For example, the best performances (results in bold) in Tables 3 and 4 do not differ a lot, which demonstrates that the modifier graph based approaches are robust to the sparseness problem.

To understand the impact that different graph clustering algorithms may have on modifier graph clustering, Fig. 4 illustrates the cluster number (i.e., the number of subtopic string clusters) distribution based on Topic-Set-A. The asterisk represents the official number of subtopics per topic. The circle and triangle represent the cluster number per topic under the Louvain algorithm and the LinLog algorithm respectively.

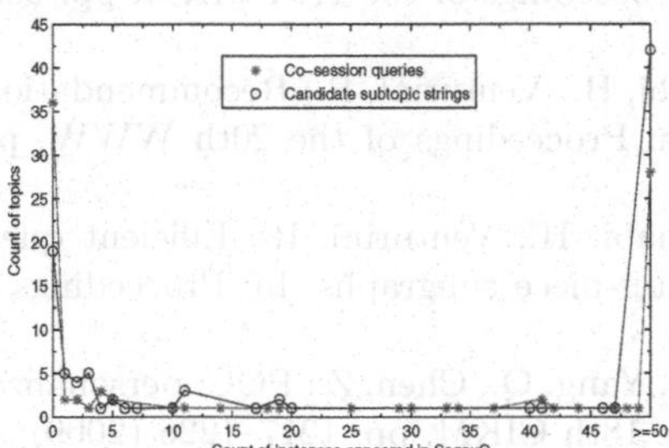

Fig. 3. The extent we can rely on QKQL

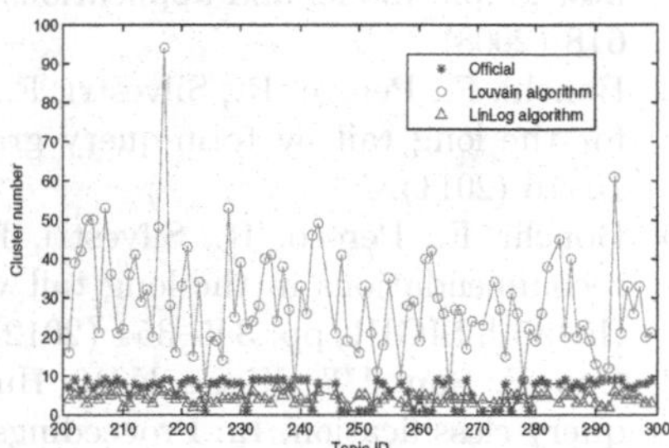

Fig. 4. Cluster number distribution

From Fig. 4, we found that: The two algorithms group the same modifier graph into different number of clusters due to different optimization criteria, and the LinLog algorithm generally outputs smaller numbers of clusters. Unfortunately, both of the two algorithms commonly generate different cluster numbers compared with the official subtopic number. If we directly regard the cluster number as the subtopic number (as we did in this paper), the approximated subtopic recall value would not be accurate enough. A particular algorithm should be devised for better clustering the modifier graph.

6 Conclusions and Future Work

In this paper, we performed subtopic mining via modifier graph clustering, which makes use of TKQL rather than QKQL. Compared with the baselines that treat a whole subtopic string or query as the minimum analysis unit, our modifier graph based approaches achieve a better performance in terms of I-rec, D-nDCG and D#-nDCG. A limitation of our current study is that the proposed approach is tested against Chinese topics. Whether it works effectively against English topics is planned as the future work. Moreover, devising a specific algorithm for modifier graph clustering would be an interesting future research direction.

Acknowledgements. This research has been partially supported by the Ministry of Education, Science, Sports and Culture, Grant-in-Aid for Scientific Research (A), 22240021.

References

[1] Agrawal, R., Gollapudi, S., Halverson, A., Ieong, S.: Diversifying search results. In: Proceedings of the 2nd WSDM, pp. 5–14 (2009)
[2] Beeferman, D., Berger, A.: Agglomerative clustering of a search engine query log. In: Proceedings of the 6th KDD, pp. 407–416 (2000)
[3] Blondel, V.D., Guillaume, J.L., Lambiotte, R., Lefebvre, E.: Fast unfolding of communities in large networks. Journal of Statistical Mechanics (2008)
[4] Boldi, P., Bonchi, F., Castillo, C., Donato, D., Gionis, A., Vigna, S.: The query-flow graph: model and applications. In: Proceedings of the 17th CIKM, pp. 609–618 (2008)
[5] Bonchi, F., Perego, R., Silvestri, F., Vahabi, H., Venturini, R.: Recommendations for the long tail by term-query graph. In: Proceedings of the 20th WWW, pp. 15–16 (2011)
[6] Bonchi, F., Perego, R., Silvestri, F., Vahabi, H., Venturini, R.: Efficient query recommendations in the long tail via center-piece subgraphs. In: Proceedings of the 35th SIGIR, pp. 345–354 (2012)
[7] Cao, B., Sun, J.T., Xiang, E.W., Hu, D.H., Yang, Q., Chen, Z.: PQC: personalized query classification. In: Proceedings of the 18th CIKM, pp. 1217–1226 (2009)
[8] Carbonell, J., Goldstein, J.: The use of mmr, diversity-based reranking for reordering documents and producing summaries. In: Proceedings of the 21st SIGIR, pp. 335–336 (1998)
[9] Deng, H., King, I., Lyu, M.R.: Entropy-biased models for query representation on the click graph. In: Proceedings of the 32nd SIGIR, pp. 339–346 (2009)
[10] Hu, Y., Qian, Y., Li, H., Jiang, D., Pei, J., Zheng, Q.: Mining query subtopics from search log data. In: Proceedings of the 35th SIGIR, pp. 305–314 (2012)

[11] Jones, R., Klinkner, K.L.: Beyond the session timeout: automatic hierarchical segmentation of search topics in query logs. In: Proceedings of the 17th CIKM, pp. 699–708 (2008)

[12] Noack, A.: Energy models for graph clustering. Journal of Graph Algorithms and Applications 11(2), 453–480 (2007)

[13] Radlinski, F., Dumais, S.: Improving personalized web search using result diversification. In: Proceedings of the 29th SIGIR, pp. 691–692 (2006)

[14] Radlinski, F., Szummer, M., Craswell, N.: Inferring query intent from reformulations and clicks. In: Proceedings of the 19th WWW, pp. 1171–1172 (2010)

[15] Ren, F., Sohrab, M.G.: Class-indexing-based term weighting for automatic text classification. Information Sciences 236, 109–125 (2013)

[16] Sadikov, E., Madhavan, J., Wang, L., Halevy, A.: Clustering query refinements by user intent. In: Proceedings of the 19th WWW, pp. 841–850 (2010)

[17] Sakai, T., Dou, Z., Yamamoto, T., Liu, Y., Zhang, M., Song, R.: Overview of the NTCIR-10 INTENT-2 task. In: Proceedings of NTCIR-10 Workshop, pp. 94–123 (2013)

[18] Sakai, T., Song, R.: Evaluating diversified search results using per-intent graded relevance. In: Proceedings of the 34th SIGIR, pp. 1043–1052 (2011)

[19] Song, R., Zhang, M., Sakai, T., Kato, M.P., Liu, Y., Sugimoto, M., Wang, Q., Orii, N.: Overview of the NTCIR-9 INTENT task. In: Proceedings of NTCIR-9 Workshop Meeting, pp. 82–105 (2011)

[20] Song, Y., Zhou, D., He, L.: Query suggestion by constructing term-transition graphs. In: Proceedings of the 5th WSDM, pp. 353–362 (2012)

[21] Wang, X., Chakrabarti, D., Punera, K.: Mining broad latent query aspects from search sessions. In: Proceedings of the 15th KDD, pp. 867–876 (2009)

[22] Wen, J.R., Nie, J.Y., Zhang, H.J.: Clustering user queries of a search engine. In: Proceedings of the 10th WWW, pp. 162–168 (2001)

[23] Yin, X., Shah, S.: Building taxonomy of web search intents for name entity queries. In: Proceedings of the 19th WWW, pp. 1001–1010 (2010)

[24] Yu, H., Ren, F.: Role-explicit query identification and intent role annotation. In: Proceedings of the 21st CIKM, pp. 1163–1172 (2012)

Net-Ray: Visualizing and Mining Billion-Scale Graphs

U. Kang[1], Jay-Yoon Lee[2], Danai Koutra[2], and Christos Faloutsos[2]

[1] KAIST, Daejeon 305-701, Korea
ukang@cs.kaist.ac.kr
[2] Carnegie Mellon University, Pittsburgh PA 15213, USA
{lee.jayyoon,danai,christos}@cs.cmu.edu

Abstract. How can we visualize billion-scale graphs? How to spot outliers in such graphs quickly? Visualizing graphs is the most direct way of understanding them; however, billion-scale graphs are very difficult to visualize since the amount of information overflows the resolution of a typical screen.

In this paper we propose NET-RAY, an open-source package for visualization-based mining on *billion-scale* graphs. NET-RAY visualizes graphs using the spy plot (adjacency matrix patterns), distribution plot, and correlation plot which involve careful node ordering and scaling. In addition, NET-RAY efficiently summarizes scatter clusters of graphs in a way that finds outliers automatically, and makes it easy to interpret them visually.

Extensive experiments show that NET-RAY handles very large graphs with billions of nodes and edges efficiently and effectively. Specifically, among the various datasets that we study, we visualize in multiple ways the YahooWeb graph which spans 1.4 billion webpages and 6.6 billion links, and the Twitter who-follows-whom graph, which consists of 62.5 million users and 1.8 billion edges. We report interesting clusters and outliers spotted and summarized by NET-RAY.

1 Introduction

Applying algorithms to find patterns in the data is one way of understanding it, but "a picture is worth a thousand words". How can we visualize big graphs with billions of nodes and edges? And, more importantly, how can we spot and plot outliers in such graphs quickly? Big graphs are everywhere: the World Wide Web, social network, biological network, phone call network, and many more. Visual mining on big graphs is a crucial tool for data miners to communicate with people outside: e.g., executives, government officials, domain experts, etc. In the case of outlier detection, visualization helps non-data miners understand the nature and seriousness of outliers.

In this paper, we propose NET-RAY, an open source package (available in http://kdd.kaist.ac.kr/netray) implemented on top of MAPREDUCE for visual mining on big graphs. NET-RAY provides the following plots:

1. *Spy plot* (=adjacency matrix showing the nonzero patterns) of graphs, as shown in Fig. 1(a) which visualizes the adjacency matrix of US Patent graph.
2. *Distribution plot* of graph features including in/out-degrees and triangles. For example, see Fig. 1(b) for the distribution of triangles in YahooWeb graph.

V.S. Tseng et al. (Eds.): PAKDD 2014, Part I, LNAI 8443, pp. 348–361, 2014.

3. *Correlation plot* of graph features including in-degree vs. out-degree, degree vs. triangle, and degree vs. PageRank. For instance, Fig. 1(c) shows the correlation plots between degree and Triangles of Twitter graph.

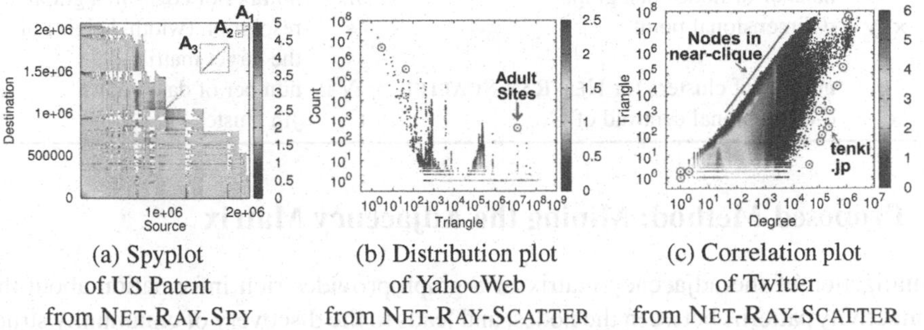

(a) Spyplot of US Patent from NET-RAY-SPY (b) Distribution plot of YahooWeb from NET-RAY-SCATTER (c) Correlation plot of Twitter from NET-RAY-SCATTER

Fig. 1. NET-RAY in action. **(a)** Spy plot (adjacency matrix pattern) of US Patent graph reveals communities which are labeled as A_1, A_2, and A_3. **(b)** Triangle distribution plot of YahooWeb graph reveals adult sites which are pointed by other adult sites. **(c)** Degree vs. Triangle correlation plot of Twitter graph reveals near-cliques (in the upper part of the plot), and anomalous nodes with few triangles like tenki.jp (details in Section 4).

NET-RAY uses those plots for various graph mining tasks including finding communities, discovering correlations, detecting anomalies, and visualization, as shown in Fig. 1. NET-RAY tackles two challenges. First, real world Web-scale graphs contain too much data, spanning Terabytes, for a standard single-machine plotting tools (e.g. gnuplot) to process. Moreover, the amount of information is too much to show on a standard screen with limited resolution, and thus careful reorganization and scaling of data are required. Second, even after presenting the data on the screen, finding representative outliers is difficult. NET-RAY solves the problem by distributed projection, careful ordering and scaling, and efficient summarization of representative outliers. The main contributions of this paper are the following:

1. **Method.** We propose NET-RAY, an open-source package for visualizing and mining big graphs. NET-RAY includes two algorithms: NET-RAY-SPY and NET-RAY-SCATTER. The former effectively visualizes adjacency matrices of graphs by careful ordering of nodes, and scaling values and axes. The latter efficiently finds representative outliers from big graphs.
2. **Scalability.** NET-RAY scales linearly with the number of machines and the edges in the graph.
3. **Discovery.** We employ NET-RAY to analyze large, real world data including the YahooWeb with 6.6 billion edges and total size of 0.11TB, as well as a Twitter graph with 1.8 billion edges and size of 24.2GB. We present interesting discoveries including communities and anomalous nodes. To the best of our knowledge, NET-RAY is the first work in visual mining for billion scale graphs.

The rest of the paper is organized typically: proposed methods in Sections 2 and 3, discovery results in Section 4, related works in Section 5, and conclusion in Section 6. Table 1 lists the symbols and their definitions used in this paper.

Table 1. Table of symbols

Symbol	Definition	Symbol	Definition
n	number of nodes in a graph	m	number of edges in a graph
$\mathbf{x}, \mathbf{y}$	d-dimensional point	s	resolution (width, height) of the target matrix
k	number of clusters for NET-RAY-SCATTER	N	number of data points
$\mathbf{c}_j$	d-dimensional centroid of C_j	C_j	jth cluster

2 Proposed Method: Mining the Adjacency Matrix

Visualization of the adjacency matrix of a graph provides rich information about the connectivity patterns between the nodes, and leads to the discovery of community structures. For small graphs, visualizing the adjacency matrix is tractable. However, visualizing the adjacency matrix of very large graphs poses several challenges. First, the size of the adjacency matrix can easily go beyond the resolution of a typical screen. For example, the adjacency matrix size of a 1 billion node graph becomes 1 billion by 1 billion; exactly visualizing the matrix requires 1 billion $\times$ 1 billion pixels which are too many to be shown on a typical screen. We address the challenge by projecting the original matrix into a small matrix which can be shown on a typical screen. For example, the 1 billion by 1 billion matrix can be projected into a 1000 by 1000 matrix, where an element of the small matrix is set to the number of nonzeros in the corresponding submatrix of the big matrix. However, this projection poses the second challenge: the small matrix will be almost full in most cases, as shown in Fig. 2 (a).

In this section, we describe our proposed NET-RAY-SPY method to address the two challenges of mining the *adjacency matrix*; the first challenge of resolution is handled in Section 2.1, and the second "full matrix" challenge is handled in Sections 2.2 and 2.3.

2.1 Projection

To handle the problem that the adjacency matrix is much larger than the screen, we project the original adjacency matrix into a small matrix which can be easily shown in a screen. In the following we assume a graph G with n nodes and m edges, and the target matrix of size s by s (e.g. $s = 1000$).

Let (x, y) be an edge in the graph, where $1 \leq x, y \leq n$. We project each edge by mapping (x, y) to the element $(\lceil x \cdot \frac{s}{n} \rceil, \lceil y \cdot \frac{s}{n} \rceil)$ in the target matrix. Note that both of the mapped values ($\lceil x \cdot \frac{s}{n} \rceil$ and $\lceil y \cdot \frac{s}{n} \rceil$) are in the range of $[1, s]$.

2.2 Node Reordering

Projection (Section 2.1) successfully decreases the data size, but it has a drawback: the target matrix becomes full in most cases, thereby giving a false impression that the graph is a clique or a near-clique, as in Fig. 2 (a). Identifying the communities in the graph becomes difficult in this case. To overcome this problem we propose to cluster the nonzero elements in the adjacency matrix of the original graph.

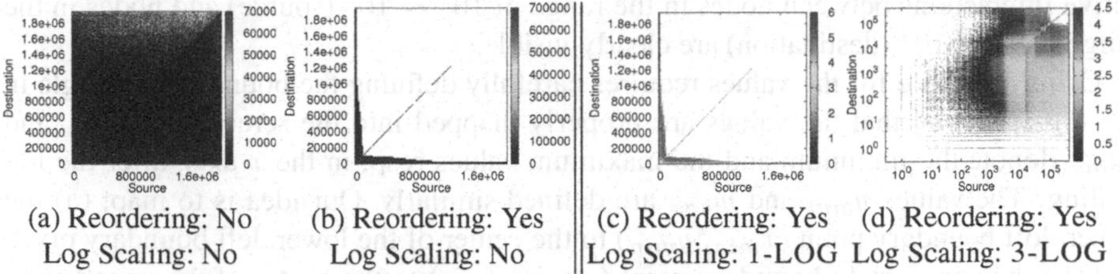

(a) Reordering: No Log Scaling: No (b) Reordering: Yes Log Scaling: No (c) Reordering: Yes Log Scaling: 1-LOG (d) Reordering: Yes Log Scaling: 3-LOG

Fig. 2. Effect of ordering and scaling in visualizing the adjacency matrix of Weibo-KDD graph. **(a,b)**: before applying NET-RAY-SPY. **(c,d)**: after applying NET-RAY-SPY. The color bar in (c,d) are in log scale of base 10. Note the node ordering and value/axis scaling provide rich information on the connectivity and activity patterns of the adjacency matrix.

Any edge clustering method (e.g. Metis [16], CrossAssociation [6], etc.) can be plugged into NET-RAY-SPY. By default, NET-RAY-SPY uses SlashBurn [13] for clustering the nonzeros in the adjacency matrix, since it provides the best performance in terms of compression. SlashBurn reorders the nodes and assigns small node ids to high degree nodes, and clusters the nonzero elements of the adjacency matrix into the left, bottom, and diagonal area of the spy plot, thereby making huge empty areas in the spy plot. For example, see Fig. 2 (b) where SlashBurn clusters the nonzero elements of the adjacency matrix to show the connectivity patterns between the nodes.

2.3 3-LOG Scaling

In addition to the node reordering, we handle the challenge of "full matrix" by scaling the x and y axes, as well as the numerical *value* of each element into log scale. We describe our proposed methods: "1-LOG" (value) and "3-LOG" (value, x, and y) scaling.

1-LOG: Value Scaling. In many real world adjacency matrices, the distribution of the nonzero elements is skewed: i.e., some areas of the adjacency matrix are very dense, while others are very sparse. For this reason, the simple linear scaling of the values loses the information on the subtle differences of the small values. For example, in Fig. 2 (a,b), most of the elements are colored blue, since few elements have the largest values (~70000) and most others values are smaller than 10% of the largest value. The red dots denoting the elements with the highest values are located near (0,0), but they are hardly visible. To resolve the problem, we use the log-scaling on the values, which we call the 1-LOG method. 1-LOG method shows the skewed distributions more effectively; e.g., in Fig. 2(c) we see that the area near (0,0) is very dense (red dots). Note the numbers along the color palette denote values raised to the power of 10 (e.g. 6 means 10^6).

3-LOG: Value and Axis Scaling. The nonzero pattern of the reordered adjacency matrix is also skewed after node reordering: most of the elements are clustered around the origin (0,0), two axes, and the diagonal line, thereby leaving many empty spaces as shown in Fig. 2(c). To better utilize the empty space, we additionally use log-scale for the two axes, and thus create the 3-LOG (value, x, and y) plot. For example, Fig. 2 (d) shows the 3-LOG plot of the Weibo-KDD graph (described in Section 4). Note the active interactions between nodes in the range of $10^4 \sim 10^5$ (source) and nodes in the range of $10 \sim 10^4$ (destination) are clearly visible.

Using log scale for the values requires carefully defining the bounding rectangle in the log-space, so that the values are properly mapped into the screen. Let x_{min} and x_{max} denote the minimum and the maximum values resp. in the x axis, after the log scaling. The values y_{min} and y_{max} are defined similarly. Our idea is to map: (a) the lower, left boundary point (x_{min}, y_{min}) to the center of the lower, left boundary pixel, and (b) the upper, right boundary point (x_{max}, y_{max}) to the center of the upper, right boundary pixel. Then remaining points (x, y) are mapped naturally to

$$(\lceil (s-1)\frac{x - x_{min}}{x_{max} - x_{min}} + \frac{1}{2}\rceil, \lceil (s-1)\frac{y - y_{min}}{y_{max} - y_{min}} + \frac{1}{2}\rceil). \tag{1}$$

In very large graphs with billions of nodes and edges, the number of points easily exceeds billions; thus, it takes long to compute the mapping. In Section 3.1 we describe a distributed algorithm to compute the mapping.

3 Proposed Method: Mining the Scatter Cluster

In this section we describe NET-RAY-SCATTER, our proposed method for visualizing and mining *scatter plots* including distribution and correlation plots. We first describe the distributed projection method for visualizing very large scatter plots, and then the summarization/outlier detection method.

3.1 Distributed Projection

Algorithm. For very large graphs with billions of nodes and edges, using a single machine for projection using Equation (1) takes very long. To speed up the task, our natural choice is to design a distributed algorithm for the task: specifically, we use MAPREDUCE, a popular distributed data processing platform. We designed and implemented a two-stage MAPREDUCE algorithm for the task. Given a set $\{(x, y)\}$ of data points, the first stage finds the minimum and the maximum of each dimension: x_{min}, x_{max}, y_{min}, and y_{max}. The second stage uses the values x_{min}, x_{max}, y_{min}, and y_{max}, computed from the first stage, to find the mapping given by Eq. (1). Note that the same distributed algorithm can be used for digitizing the spy plot described in Sec. 2; in that case, the first MAPREDUCE stage is omitted since the minimum and maximum values are known a priori (log 1 and log n, resp.).

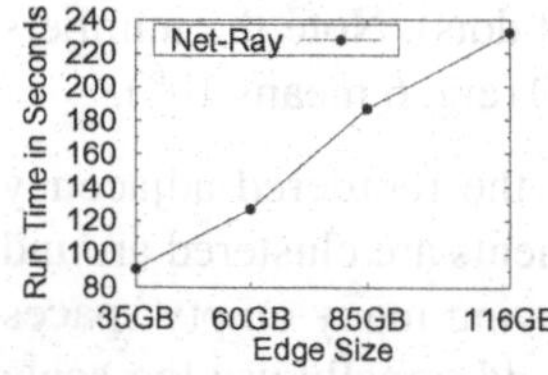

(a) Running time vs. edges

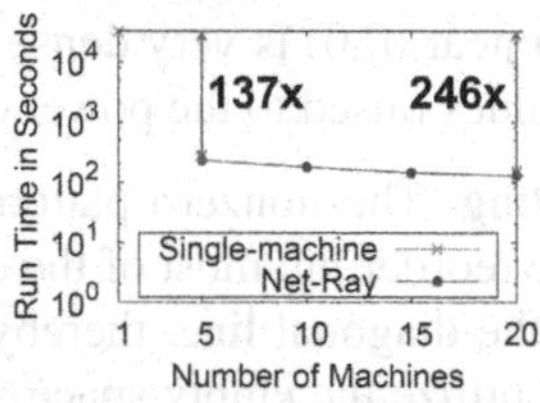

(b) Running time vs. machines

Fig. 3. Scalability of NET-RAY on YahooWeb graph. **(a)** NET-RAY scales linearly with the edges. **(b)** NET-RAY with 5 and 20 machines is 137× and 246× faster than the single-machine counterpart, resp.

Scalability. Fig. 3 shows the scalability of NET-RAY-SCATTER on the YahooWeb graph listed in Table 2. The Hadoop-based implementation was run on the OCC-Y Hadoop cluster described in Section 4, while the single-machine implementation on a machine with two dual-core Intel Xeon 3GHz CPUs and 4GB memory. Fig. 3(a) shows the running time vs. file size, where the number of reducer machines is fixed to 5. We see that the running time scales linearly on the edges size. Fig. 3(b) presents the running time comparison between a single-machine implementation and NET-RAY-SCATTER on Hadoop. Note that NET-RAY-SCATTER with 5 and 20 machines is 137× and 246× faster than the single-machine counterpart, resp. The running time with 20 machines is 1.8× (not 4×) faster than with 5 machines, due to overhead of running Hadoop jobs.

3.2 Summarization and Mining

The projected scatter plot still has many points: for example, the 1000 by 1000 scatter plot has at maximum 1 million points. Among these points, how can we automatically determine the representative points and outliers? We formally define the problem.

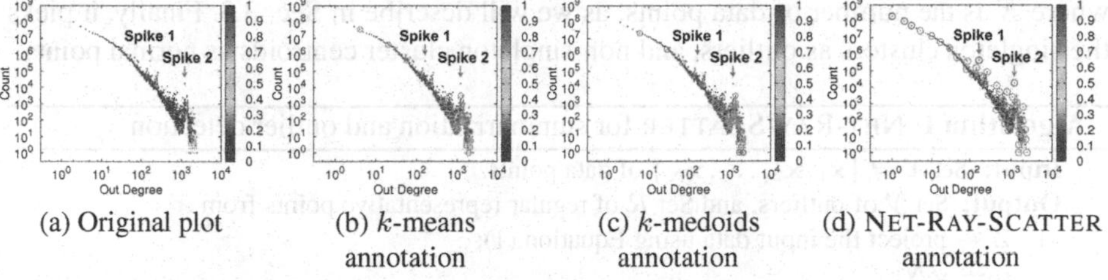

(a) Original plot (b) k-means annotation (c) k-medoids annotation (d) NET-RAY-SCATTER annotation

Fig. 4. Comparison of k-means, k-medoids and NET-RAY-SCATTER for spotting outliers in the out-degree distribution plot of YahooWeb graph. We use $k = 42$, following the choice of the parameter described in Section 3.3. Green circles denote the centroids of clusters containing more than 2 points. Red circles denote the centroids of singleton clusters. Notice that NET-RAY-SCATTER spots the two outstanding spikes of our interests, while k-means and k-medoids fail to detect them.

Problem 1. Given N points $\mathbf{x}_1, ..., \mathbf{x}_N$, find k representative points including top outliers (outlier score of a point x_j is the distance from x_j to its nearest neighbor). □

The choice of the representative points depends heavily on how we want to summarize the data. We list the desired properties of our summarization.

- **P1: Pick from Input.** The output of the summarization should be a subset of the input data points, since we want to give representative examples of the data.
- **P2: Outlier Detection.** The output of the summarization should contain outliers or extreme points so that we can use it for anomaly detection.
- **P3: Scalability.** The method should be fast and scalable.

Our proposed NET-RAY-SCATTER method uses the following main ideas: 1) use a clustering algorithm to compute k clusters where k is carefully chosen (details in

Section 3.3), 2) use the center points of the k clusters as the summaries, and 3) use singleton clusters (having 1 point in their clusters) as the outliers. The main question is, which clustering algorithm should we use to satisfy the three desired properties?

Our answer is to use the k-center [10] algorithm. Given a set $\mathcal{X} = \{\mathbf{x}_1, \mathbf{x}_2, \ldots, \mathbf{x}_N\}$ of N points, the k-center chooses a set of k points from $\mathcal{X}$ as cluster centers, $\mathbf{c}_1, \ldots, \mathbf{c}_k$ to minimize the objective function $\max_j \max_{\mathbf{x} \in C_j} ||\mathbf{x} - \mathbf{c}_j||$, where $|| \cdot ||$ denotes L_1 or L_2 norm. The effect of the $max()$ term is that an outlier, far from the rest of the points, is better to form a singleton cluster; otherwise the maximum distance between the points in the same cluster increases dramatically. Using the k-center for the summarization satisfies all the desired properties **P1**, **P2**, and **P3**. **P1** is satisfied since only the subset of the input data points are chosen. **P2** is satisfied since both singleton clusters (from outliers) and non-singleton clusters (from normal points) are spotted. Furthermore, **P3** is satisfied since the greedy version of the k-center algorithm [10] runs in $O(kN)$ time. Note that other algorithms like k-means and k-medoids do not satisfy all the properties: it can be shown that k-means violates **P1** and **P2**, and k-medoids violates **P2**.

NET-RAY-SCATTER is shown in Algorithm 1. NET-RAY-SCATTER first projects the input data using Equation (1). Then it applies the k-center algorithm with $k = \sqrt{N}$, where N is the number of data points, as we will describe in Sec. 3.3. Finally, it picks the singleton clusters as outliers, and non-singleton cluster centroids as normal points.

Algorithm 1: NET-RAY-SCATTER for summarization and outlier detection

---

Input: Set $\mathcal{X} = \{\mathbf{x}_1, \mathbf{x}_2, \ldots, \mathbf{x}_N\}$ of data points.
Output: Set $\mathcal{Y}$ of outliers, and Set $\mathcal{R}$ of regular representative points from $\mathcal{X}$.

```
1: Z ← project the input data using Equation (1);
2: k ← √N;
3: C ← k-center on Z;
4: 𝒴 ← singleton clusters from C;
5: ℛ ← non-singleton clusters' centroids from C;
6: return 𝒴, ℛ;
```

As an example of the outlier detection capability of NET-RAY-SCATTER, see Fig. 4 (d) whose red circles denote singleton clusters. Note that the red circles include the two outliers pointed by red arrows. Moreover, the 5 red circled points are exactly the points with top 5 outlier scores (distance from a point to its nearest neighbor), showing the effectiveness of NET-RAY-SCATTER. The effect of outliers in k-center was previously discussed in Charikar et al. [7]; however, the focus on the work is to perform 'robust' clustering by ignoring outliers when building normal clusters for small k. In contrast, NET-RAY-SCATTER uses sufficiently large k (details in Section 3.3) so that the outliers, which form their own clusters, are detected automatically.

3.3 Discussion

Parameter Choice. How to choose the parameter k for NET-RAY-SCATTER? Obviously, k should be greater or equal to the number of outliers that we want to find. The question is, how many outliers do we want to find? Our main target for using NET-RAY-SCATTER is to detect anomalous spikes in distribution or correlation plots. To detect the spike, we fix a value v in an axis (e.g., fix x coordinate), and investigate the set Y of points having the value v in the axis. If a point y in Y deviates significantly from the rest of the points in Y, then y is treated as an outlier. Based on this motivation, we choose k as the number of *distinct* coordinates in either of the axes. Assuming uniform distribution of the points, the parameter is given by $k = \sqrt{N}$.

Complexity. Our method is scalable. Specifically:

Lemma 1. NET-RAY-SCATTER *runs in* $O(N^{1.5})$ *time.* □

Proof. (Sketch) The summarization step takes $O(kN)$ time with $k = \sqrt{N}$.

4 Discoveries

We present discovery results to answer the following questions.

Q1 What connectivity patterns and communities does NET-RAY-SPY reveal on real world graphs?
Q2 What patterns and anomalies does NET-RAY-SCATTER detect in the distribution plots of real world graphs?
Q3 What patterns and outliers does NET-RAY-SCATTER find in the correlation plots of real world graphs?

We use the graph data listed in Table 2, and for each graph we extract and analyze the following information:

Table 2. Order and size of networks

Graph	Nodes	Edges	File Size
YahooWeb	1,413,511,394	6,636,600,779	116 GB
Twitter	62,539,895	1,837,645,377	24.2 GB
Weibo-KDD	1,944,589	50,655,143	594 MB
US Patent	6,009,555	10,565,431	169 MB
WWW-Barabasi	325,729	1,497,134	20 MB

- Spy plot (original, 1-LOG and 3-LOG)
- Distribution plot (in-degree, out-degree, and triangle)
- Correlation plot (in vs. out-degree, degree vs. triangle, and degree vs. PageRank)

The features (degree, PageRank, and triangles [14]) of the graphs are extracted using the Pegasus graph mining package [15]. NET-RAY is run on the OCC-Y Hadoop cluster, run by Open Cloud Consortium [1], with total 928 cores and 1 Petabyte disk.

4.1 Spy Plots

Spy plots generated from NET-RAY-SPY provide rich information on the connectivity patterns and communities in graphs. Figs. 2 and 5 show spy plots of real world graphs. We have the following observations.

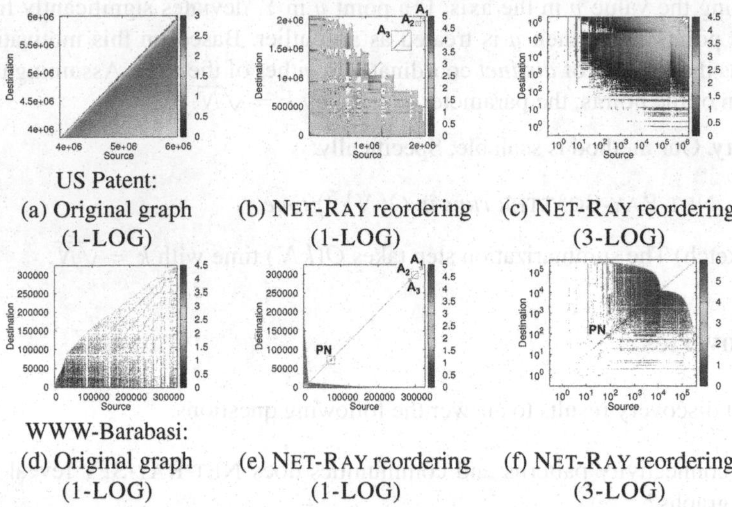

US Patent:
(a) Original graph (1-LOG) (b) NET-RAY reordering (1-LOG) (c) NET-RAY reordering (3-LOG)

WWW-Barabasi:
(d) Original graph (1-LOG) (e) NET-RAY reordering (1-LOG) (f) NET-RAY reordering (3-LOG)

Fig. 5. Spy plots of real world graphs, generated from NET-RAY-SPY. The spy plot of the US Patent graph in (a) shows that patents tend to cite neither too old nor too new patents. (b) and (e) show the 'near-spoke' structures (*sparse* sub-graph loosely connected to the rest of the graph) which are labeled as A_1, A_2, and A_3. (e) and (f) also show the 'peripheral near-clique' structures (*clique-like* subgraphs loosely connected to the rest of the graph) which are labeled as 'PN'.

Connectivity Patterns. The spy plot enables us to easily identify the connectivity patterns in the graph. The high activity regions (yellow and red colors) of Figs. 2 (c,d), and 5 (a,b,c) show the heavy interactions between the nodes. Especially, in Fig. 5 (a) we observe that patents usually cite other patents which are neither too new nor too old: for a fixed source id, the corresponding vertical line has a single mode distribution with the maximum around the center.

Community Identification. Spy plots enable the identification of communities; especially we present sparse or dense subgraphs loosely connected to the rest of the graph.

Observation 1. ***(Near-Spokes)*** *Reordering nodes by* NET-RAY-SPY *reveals "near-spokes"* (sparse *sub-graph loosely connected to the rest of the graph).* □

For example, see Fig. 5 (b,e) for the spy plots of US Patent and WWW-Barabasi. The three squares A_1, A_2, and A_3 show the adjacency matrices of the induced subgraphs of the three near spokes.

Observation 2. ***(Peripheral Near-Cliques)*** 1-LOG *and* 3-LOG *visualization of the WWW-Barabasi graph by* NET-RAY-SPY *reveal "peripheral near-cliques"* (clique-like *subgraphs loosely connected to the rest of the graph), marked as 'PN' in Fig. 5 (e,f).* □

4.2 Distribution Plots

Distribution plots generated from NET-RAY-SCATTER provide abundant information on the regularities that graphs nodes follow, as well as deviating patterns. Fig. 6 shows

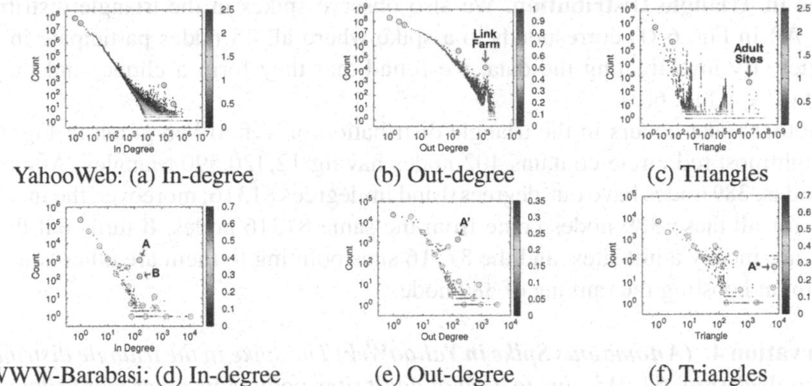

Fig. 6. Distribution plots of real world graphs, generated from NET-RAY-SCATTER. First column: in-degree distribution. Second column: out-degree distribution. Third column: triangle distribution. The red circles denote the singleton clusters. The green circles in (d-f) are the centroids of the non-singleton clusters; we omitted green circles from (a-c) for clarity. The spikes in these degree distributions often come from anomalous activities that need attention: e.g., (b) - a link farm in the YahooWeb graph, and (d-f) - a group of nodes belonging to cliques (marked A, A', A*, and B) in WWW-Barabasi.

the distributions of features in real world graphs. On the regularities, notice the power law-like slopes in the distributions of degree and triangles of real world graphs. It implies the formation of links and triangles are governed mostly by "rich-get-richer" process [17]. Furthermore, the distribution plots depict some "spikes" that deviate significantly from the fitting line of the majority of the nodes. We elaborate on the two types of spikes: one in the degree and the other in the triangle distributions.

Spikes In Degree Distribution. The spikes in the degree distribution often come from anomalous or special behaviors requiring attentions. The first observation is on the spike of the degree distributions in WWW-Barabasi (Fig. 6 (d,e)).

Observation 3. ***(Anomalous Spike in WWW-Barabasi)*** *Spikes in the in/out-degree distributions of WWW-Barabasi comes from cliques.* □

The spikes are observed at **A'** (in-degree 152, count 1192), **B** (in-degree 153, count 155), and **A** (out-degree 156, count 1353) of Fig. 6(d,e). It turns out that they form cliques, as we see in Fig. 7. Also, **A** is a subset of **A'**. Finally, we note that the out-degree distribution of the YahooWeb graph, shown in Fig. 6 (b), has a spike coming from a link farm [12].

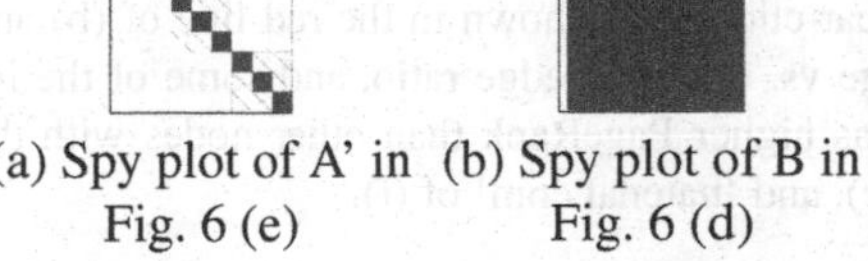

(a) Spy plot of A' in Fig. 6 (e) (b) Spy plot of B in Fig. 6 (d)

Fig. 7. Spy plots of the members of the spikes in distribution plots of Fig. 6(d,e). Many nodes belong to cliques; such nodes have same degree/triangle characteristics, and thus make spikes in the degree and triangle distributions.

Spikes in Triangle Distribution. We also observe spikes in the triangle distribution plots: **A*** in Fig. 6 (f) corresponds to a spike where all 45 nodes participate in 7239 triangles. By investigating the data, we found that they form a clique, and they are subset of **A** in Fig. 6(d).

Another spike occurs in the triangle distribution of YahooWeb graph in Fig. 6 (c). The rightmost red circle contains 402 nodes having 12,420,590 triangles. Among the 402 nodes, 389 nodes have out-degrees 0 and in-degrees 81316; moreover, the incoming edges for all these 389 nodes come from the same 81316 nodes. It turns out the 389 nodes are mostly adult sites, and the 81316 sites pointing to them are other adult sites that aim at boosting the ranking of 389 nodes.

Observation 4. ***(Anomalous Spike in YahooWeb)*** *The spike in the triangle distribution of YahooWeb (Fig. 6 (c)) is due to a set of adult sites pointed by other adult sites.* □

4.3 Correlation Plots

Correlation plots generated from NET-RAY-SCATTER provide opulent information on the communities, anomalous nodes, and correlation between features of nodes. Fig. 8 shows the correlation plots using the node degrees, PageRank scores, and participating triangles from real world graphs.

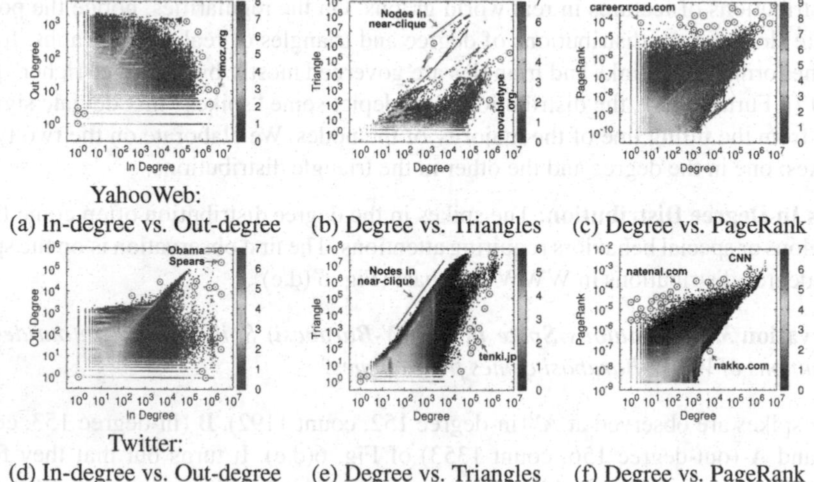

YahooWeb:
(a) In-degree vs. Out-degree (b) Degree vs. Triangles (c) Degree vs. PageRank

Twitter:
(d) In-degree vs. Out-degree (e) Degree vs. Triangles (f) Degree vs. PageRank

Fig. 8. Outliers in correlation plots give rich information on the structural patterns of graphs. **(Pattern 1)** Nodes with many incoming edges, but little friendship among the in-neighbors have few triangles, as shown in 'movabletype.org' of (b) and 'tenki.jp' of (e). **(Pattern 2)** On the other hand there are many nodes belonging to near-cliques, as shown in the red line of (b) and (e). **(Pattern 3)** If a node has high incoming edge vs. outgoing edge ratio, and some of the in-neighbors have high PageRank, then the node has higher PageRank than other nodes with the same degree, as shown in 'careerxroad.com' of (c), and 'natenal.com' of (f).

In and out-degrees. In Fig. 8 (a,d), popular nodes in graphs, like celebrities or portal sites, tend to have high in-degrees and small out-degrees.

Degree and Triangles. In Fig. 8(b,e) star-like nodes, which have very sparsely connected neighbors, are easily identified in the lower, right corner of the degree vs. triangle plots. Spammers are identified in this scheme since, by their nature, they often have random neighbors with very few triangles. Also near-clique communities are spotted in the upper left corner since a node with degree d can have $\binom{d}{2} \propto d^2$ triangles at most.

Degree and PageRank. In Fig. 8 (c,f), higher number of in-degree, rather than total degree, is correlated with higher PageRank in general. However, it is possible to boost PageRank with small number of in degrees by having an in-neighbor with a very high PageRank. E.g, in Fig. 8 (c) a page in 'careerxroad.com' has degree 3 with high PageRank since one of the in-neighbor has a very high PageRank.

5 Related Works

Although big graphs are ubiquitous, the existing visualization tools cannot handle efficiently billions of nodes. We review works on graph visualization and outlier detection.

Graph Visualization. Apolo [8] is a graph tool for attention routing, that interactively expands the vicinities of a few seed nodes. OPAvion [2], an anomaly detection system for large graphs consists of Pegasus (feature aggregation [15]), OddBall (outlier detection [3]) and Apolo. Here, in an attempt to understand the underlying patterns and detect outliers, we are interested in efficiently generating the spy and distribution/correlation plots of a graph, instead of plotting the graph itself. In [21], Shneiderman proposes simply scaled density plots to visualize scatter plots, and [4] presents sampling-based techniques for datasets with several thousands of points. [9] proposes an interactive graph visualization tool, but the focus is on only the adjacency matrix and the scalability is limited.

Clusters and Outliers in Spaces. For outliers, see LOF [5], LOCI [19], and angle-based methods [20]. For clustering, see methods like k-means in [11], k-harmonic means [22], k-medoids [18], and k-centers [10].

In general, there is not much work on visualization of the spy and scatter plots of features distributions and correlations for graphs with billions of nodes and edges.

6 Conclusion

In this paper, we tackle the problem of efficiently and effectively visualizing and mining billion-scale graphs. Our major contributions include:

1. **Method.** We propose NET-RAY, a carefully designed algorithm for visualizing and mining adjacency matrices and scatter plots from billion scale graphs.
2. **Scalability.** NET-RAY is linear on the number of machines and edges.
3. **Discovery.** We use NET-RAY to visualize large, real-world graphs and report interesting discoveries and anomalies, including near spokes, near cliques, and spikes.

Interesting future research directions include visual mining of dynamic graphs and complex high-dimensional data, like tensors.

Acknowledgments. Funding was provided by KAIST under project number G0413002. Funding was also provided by the U.S. ARO and DARPA under Contract Number W911NF-11-C-0088, by DTRA under contract No. HDTRA1-10-1-0120, by ARL under Cooperative Agreement Number W911NF-09-2-0053, and by the National Science Foundation under Grants No. IIS-1217559. The views and conclusions are those of the authors and should not be interpreted as representing the official policies, of the U.S. Government, or other funding parties, and no official endorsement should be inferred. The U.S. Government is authorized to reproduce and distribute reprints for Government purposes notwithstanding any copyright notation here on.

References

1. `http://opencloudconsortium.org`
2. Akoglu, L., Chau, D.H., Kang, U., Koutra, D., Faloutsos, C.: Opavion: mining and visualization in large graphs. In: SIGMOD (2012)
3. Akoglu, L., McGlohon, M., Faloutsos, C.: `Oddball`: Spotting anomalies in weighted graphs. In: Zaki, M.J., Yu, J.X., Ravindran, B., Pudi, V. (eds.) PAKDD 2010. LNCS, vol. 6119, pp. 410–421. Springer, Heidelberg (2010)
4. Bertini, E., Santucci, G.: By chance is not enough: Preserving relative density through non uniform sampling. In: Proceedings of the Information Visualisation (2004)
5. Breunig, M., Kriegel, H.-P., Ng, R.T., Sander, J.: Lof: Identifying density-based local outliers. In: SIGMOD (2000)
6. Chakrabarti, D., Papadimitriou, S., Modha, D.S., Faloutsos, C.: Fully automatic cross-associations. In: KDD (2004)
7. Charikar, M., Khuller, S., Mount, D.M., Narasimhan, G.: Algorithms for facility location problems with outliers. In: SODA (2001)
8. Chau, D.H., Kittur, A., Hong, J.I., Faloutsos, C.: Apolo: interactive large graph sensemaking by combining machine learning and visualization. In: KDD (2011)
9. Elmqvist, N., Do, T.-N., Goodell, H., Henry, N., Fekete, J.: Zame: Interactive large-scale graph visualization. In: IEEE Pacific Visualization Symposium, PacificVIS 2008 (2008)
10. Gonzalez, T.F.: Clustering to minimize the maximum intercluster distance. Theor. Comput. Sci. 38, 293–306 (1985)
11. Han, J., Kamber, M., Pei, J.: Data Mining: Concepts and Techniques, 3rd edn. Morgan Kaufmann Publishers Inc. (2011)
12. Kang, U., Chau, D.H., Faloutsos, C.: Mining large graphs: Algorithms, inference, and discoveries. In: ICDE (2011)
13. Kang, U., Faloutsos, C.: Beyond 'caveman communities': Hubs and spokes for graph compression and mining. In: ICDM (2011)
14. Kang, U., Meeder, B., Papalexakis, E., Faloutsos, C.: Heigen: Spectral analysis for billion-scale graphs. IEEE Transactions on Knowledge and Data Engineering 26(2), 350–362 (2014)
15. Kang, U., Tsourakakis, C., Faloutsos, C.: Pegasus: A peta-scale graph mining system - implementation and observations. In: ICDM (2009)
16. Karypis, G., Kumar, V.: MeTis: Unstructured Graph Partitioning and Sparse Matrix Ordering System, Version 4.0 (2009)
17. Newman, M.E.J.: Power laws, pareto distributions and zipf's law. Contemporary Physics (46), 323–351 (2005)

18. Ng, R.T., Han, J.: Efficient and effective clustering methods for spatial data mining. In: VLDB (1994)
19. Papadimitriou, S., Kitagawa, H., Gibbons, P.B., Faloutsos, C.: Loci: Fast outlier detection using the local correlation integral. In: ICDE (2003)
20. Pham, N., Pagh, R.: A near-linear time approximation algorithm for angle-based outlier detection in high-dimensional data. In: KDD (2012)
21. Shneiderman, B.: Extreme visualization: squeezing a billion records into a million pixels. In: SIGMOD (2008)
22. Zhang, B., Hsu, M., Dayal, U.: K-harmonic means - a spatial clustering algorithm with boosting. In: TSDM (2000)

Structure-Aware Distance Measures for Comparing Clusterings in Graphs

Jeffrey Chan*, Nguyen Xuan Vinh, Wei Liu, James Bailey, Christopher A. Leckie, Kotagiri Ramamohanarao, and Jian Pei

[1] Department of Computing and Information Systems, University of Melbourne, Australia
jeffrey.chan@unimelb.edu.au
[2] School of Computing Science, Simon Fraser University, BC Canada

Abstract. Clustering in graphs aims to group vertices with similar patterns of connections. Applications include discovering communities and latent structures in graphs. Many algorithms have been proposed to find graph clusterings, but an open problem is the need for suitable comparison measures to quantitatively validate these algorithms, performing consensus clustering and to track evolving (graph) clusters across time. To date, most comparison measures have focused on comparing the vertex groupings, and completely ignore the difference in the structural approximations in the clusterings, which can lead to counter-intuitive comparisons. In this paper, we propose new measures that account for differences in the approximations. We focus on comparison measures for two important graph clustering approaches, community detection and blockmodelling, and propose comparison measures that work for weighted (and unweighted) graphs.

Keywords: blockmodelling, community, clustering, comparison, structural, weighted graphs.

1 Introduction

Many data are relational, including friendship graphs, communication networks and protein-protein interaction networks. An important type of analysis that is graph clustering, which involves grouping the vertices based on the similarity of their connectivity. Graph clusterings are used to discover communities with similar interests (for marketing) and discovering the inherent structure of graphs. Despite the popularity of graph clustering, existing graph cluster comparison measures have focused on using membership based measures. To the best of the authors' knowledge, there are no work that evaluates if membership based comparison measures are appropriate for comparing graph clusterings. This analysis is important, as comparison measures are often used for comparing algorithms [1][2], form the basis of consensus clustering [3] and tracking algorithms [4], and knowing which properties measures possess allows us to evaluate if a measure is appropriate for a task, and if not, propose new ones that are. One of

* Corresponding author.

V.S. Tseng et al. (Eds.): PAKDD 2014, Part I, LNAI 8443, pp. 362–373, 2014.

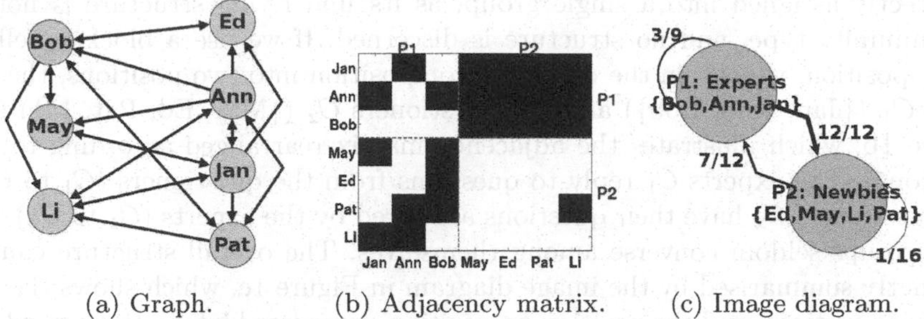

(a) Graph. (b) Adjacency matrix. (c) Image diagram.

Fig. 1. Example of an online Q&A forum. Each vertex corresponds to a forum user, and each edge represents replying between users (Figure 1a). For the corresponding adjacency matrix (Figure 1b), the positions induce a division of the adjacency matrix into blocks, delineated by the red dotted lines. Edge weights in Figure 1c are the expected edge probabilities between each pair of positions.

the important properties a measure should possess is the ability to be "spatially"/structurally aware. In traditional clustering of points, it was found that clusterings could be similar in memberships, but their points are distributed very differently. This could lead to counter-intuitive situations where such clusterings were considered similar [5][6]. We show the same scenario occurs when comparing graph clusterings. Therefore, in this paper, we address the open problems of analysing and showing why membership comparison measures can be inadequate for comparing graph clusterings and then proposing new measures that are aware of structural differences. We shortly illustrate why comparison measures should be structure aware, but we first introduce graph clustering.

Clustering in Graphs – Community Detection and Blockmodelling: Two popular approaches for graph clustering are community detection [7] and blockmodelling [8]. These approaches are used in many important clustering applications [9][1][2][4], and hence we focus on the comparison of communities and blockmodels in this paper. Community detection decomposes a graph into a community structure, where vertices from the same communities have many edges between themselves and vertices of different communities have few edges. Community structure has been found in many graphs, but it is only one of many alternatives for grouping vertices and possible graph structure (such as core-periphery). In contrast, an alternative approach is blockmodelling [8]. A blockmodel[1] partitions the set of vertices into groups (called *positions*), where for any pair of positions, there are either many edges, or few edges, between the positions.

As an example of the difference between the two approaches, consider a reply graph between a group of experts and questioners in a question and answer (Q&A) forum, illustrated in Figure 1. The vertices represent users, and the directed edges represent one user replying to another. If we use a popular community detection algorithm [7] to decompose the graph, all the vertices are

[1] We use the popular structural equivalence to assess similarity [8].

incorrectly assigned into a single group, as its underlying structure is not of a community type, and no structure is discerned. If we use a blockmodelling decomposition, we obtain the correct decomposition into two positions, the experts C_1 ({Jan, Ann, Bob}) and the questioners C_2 ({May, Ed, Pat, Li}) (see Figure 1b, which illustrates the adjacency matrix rearranged according to the positions). The experts C_1 reply to questions from the questioners (C_1 to C_2), the questioners C_2 have their questions answered by the experts (C_1 to C_2) and both groups seldom converse among themselves. The overall structure can be succinctly summarised by the image diagram in Figure 1c, which shows the positions as vertices, and the position-to-position aggregated interactions as edges. As can be seen, this graph has a two position bipartite structure, which the blockmodelling decomposition can find but a community decomposition fails to discover. The similarity definition of blockmodelling actually includes the community one, hence blockmodelling can be considered as a generalisation of community detection and comparison measures that work with blockmodels would work with comparing communities as well. Therefore we focus on measures that compare blockmodels, since they can also compare communities.

Comparing Communities and Blockmodels: As discussed earlier, almost all existing literature compares blockmodels (and communities) based on their positions and ignores any difference in the adjacency structure within positions and between positions. This can lead to unintuitive and undesirable behaviour. A blockmodel comparison measure should possess the three following properties, which we will introduce and motivate using two examples.

The first property is that the measure should be *sensitive to adjacency differences in the blocks.* Figures 2a (BM 11), 2b (BM 12) and 2c (BM 13) illustrate example 1. The vertices in each blockmodel are ordered according to their positions, and the boundary between the positions in the adjacency matrix is illustrated with red dotted lines. This effectively divides the adjacency matrix into *blocks*, which represent the position to position interactions. The positional differences from BM 12 and BM 13 to BM 11 are the same. However, the distribution of edges (frequency of edges and non-edges) are very different between BM 11 and BM 13 (there is one single dense block with all the edges in BM 11 (and BM 12), while the edges are distributed across the blocks more evenly in BM 13). Hence a measure should indicate BM 11 is more similar to BM 12 than to BM 13. Positional measures fail to achieve this.

The second property is that a measure should *account for differences in the edge distributions across all blocks.* In example 1, BM 11 and BM 12 have similar distributions across all blocks since their edges are concentrated in a block, while BM 13 is different because the edges are spread quite evenly across all blocks. We show that two existing comparison measures for blockmodels fail this property.

The third property is that a measure should be *sensitive to weighted edge distributions.* Many graphs have weights on the vertices and edges. It might be possible to binarise the weights, but this throws away important comparison information. For example, if the graph in the blockmodels of Figures 2d to 2f

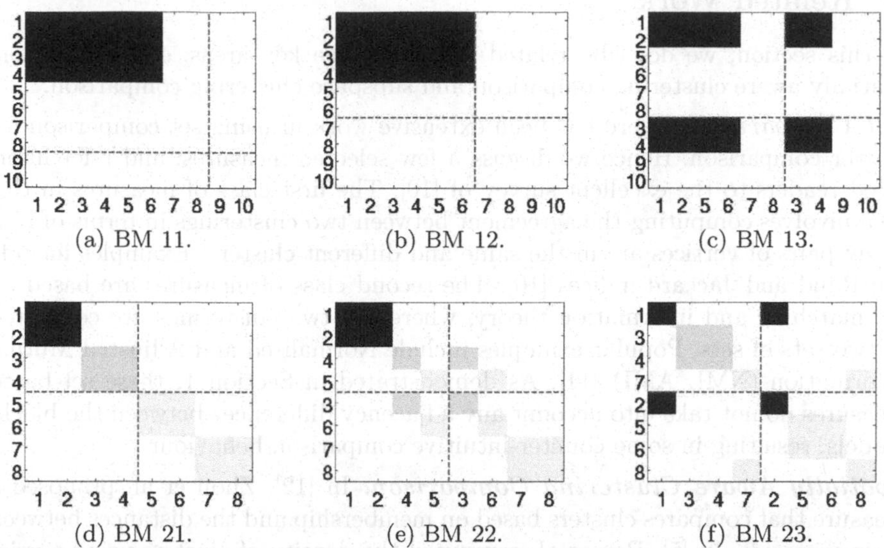

Fig. 2. Example to illustrate the strength and weaknesses of different blockmodel comparison measures. Each row corresponds to 3 blockmodels of the same graph.

(BM 21 to BM 23, dubbed example 2) were binarised, then BM 22 and 23 have exactly the same edge distribution differences from BM 21. But when the actual weights are taken into account (darker pixels in the figures represent larger valued weights), BM 22 will be correctly considered as more similar to BM 21, as most of the high-value edge weights in the top left block match, while in BM 23 these edges are evenly distributed among the four blocks.

These properties have important applications. For example in algorithm evaluation, the blockmodel output of the algorithms are compared to a gold standard. Imagine that the gold standard is BM 11 and two algorithms produced BM 12 and 13. Measures without these properties will incorrectly rank the two algorithms to be equally accurate since they have the same position differences to BM 11. In contrast, a measure that possess the first two properties will correctly rank the algorithm producing BM 12 as more accurate. Another application is in consensus blockmodelling, which finds a blockmodel that is the average of a set of blockmodels. If a measure does not possess the 3rd property and only considered position similarity, then the consensus blockmodel might have very different weight distribution to the other blockmodels and hence should not be considered as a consensus (e.g., BM 23 as a consensus of BM 21 and 22).

In summary, our contributions are: a) we propose three structural-based properties that a blockmodel comparison measure should possess; b) we analyse existing measures and show that they do not possess these properties; c) we propose new measures that satisfy these properties; and d) perform experiments on synthetic and real data to study the monotonicity of the new measures.

2 Related Work

In this section, we describe related work in three key areas: set comparison, spatially aware clustering comparison and subspace clustering comparison.

Set Comparison: There has been extensive work in using set comparison for cluster comparison. Hence we discuss a few selected measures, and refer interested readers to the excellent survey of [10]. The first class of measures in this area involves computing the agreement between two clusterings in terms of how many pairs of vertices are in the same and different clusters. Examples include the Rand and Jaccard indices [10]. The second class of measures are based on set matching and information theory, where the two clusterings are compared as two sets of sets. Popular examples include Normalised and Adjusted Mutual Information (NMI, AMI) [11]. As demonstrated in Section 1, these set-based measures do not take into account any adjacency differences between the blockmodels, resulting in some counter-intuitive comparison behaviour.

Spatially Aware Clustering Comparison: In [12], Zhou et al. proposed a measure that compares clusters based on membership and the distances between their centroids. In [5], Bae et al. computed the density of clusters using a grid, and then used the cosine similarity measure to compare the cluster distributions. Coen et al. [6] took a similar approach but used a transportation distance to compute the distances between clusters and between clusterings. All these measures depend on a notion of a distance between points (between clusters of the two clusterings). In blockmodel and community comparison, there are positions of vertices and the edges between them, but no notion of a distance between vertices across two blockmodels and hence existing spatial-aware measures cannot be applied for blockmodel comparison.

Subspace Clustering Comparison: In subspace clustering, the aim is to find a group of objects that are close in a subset of the feature space. In [13], Patrikainen and Meila proposed the first subspace validation measure that considered both the similarities in the object clusters and in the subspaces that the clusters occupied. They treated subspace clusters as sets of (object, feature) pairs, and then used set-based validation measures for comparing subspace clusters. When extended to compare the sets of (vertex,vertex) pairs between blockmodels, these measures are equivalent to comparing the positions (proof omitted due to lack of space), and hence have the same issues as the set comparison measures.

In summary, there has been much related work in cluster comparison, but none that address the unique problem of comparing blockmodels and communities. What is needed is a measure that is sensitive to differences in the adjacency structure as well as working in the relation spaces of graphs.

3 Blockmodelling (Graph Clustering) Background and Properties

In this section, we summarise the key ideas of blockmodelling (see [14][8] for more detail). A graph $G(V, E)$ consists of a set of vertices V and a set of edges E, where $E \in \{0,1\}^{|V|\times|V|}$ for unweighted graphs and $E \in \mathcal{N}^{|V|\times|V|}$ for weighted graphs[2]. The edge relation can be represented by an adjacency matrix $\mathbf{A}$ whose rows and columns are indexed by the vertices of G.

We use the Q&A example of Figure 1 to illustrate the notation. A blockmodel partitions a set of vertices into a set of positions $\mathcal{C} = \{C_1, C_2, \ldots, C_k\}$. We denote the number of positions in $\mathcal{C}$ by k. $\mathcal{C}$ can be alternatively specified by $\phi(.)$, a mapping function from vertices to the set of positions (i.e., $\phi : V \to \mathcal{C}$). A block $\mathbf{A}_{r,c}$ is a submatrix of $\mathbf{A}$, with the rows and columns drawn from C_r and C_c respectively, $C_r, C_c \in \mathcal{C}$, e.g., $\mathbf{A}_{1,1}$ defines the upper left submatrix in Figure 1b. Let $\Psi_{r,c}$ denote the random variable representing the probability mass function of the edge weights in block $\mathbf{A}_{r,c}$, e.g., $p(\Psi_{1,1} = 0) = \frac{7}{9}$, $p(\Psi_{1,1} = 1) = \frac{2}{9}$. This representation allow us to model both weighted and unweighted graphs. For simplicity, we introduce $\Upsilon_{r,c} = p(\Psi_{r,c} = 1)$ for unweighed graphs. We define $\mathbf{M}$ as the blockmodel image matrix that models inter-position densities, $\mathbf{M} : \mathcal{C} \times \mathcal{C} \to [0,1]$, $\mathbf{M}_{r,c} = \Upsilon_{r,c}$. For the rest of the paper, we use a superscript notation to distinguish two different instances of a variable (e.g., $G^{(1)}$ and $G^{(2)}$).

A blockmodel $\mathfrak{B}^{(l)}(\mathbf{C}^{(l)}, \mathbf{M}^{(l)})$ is defined by its set of positions $\mathcal{C}^{(l)}$ (the matrix version is $\mathbf{C}^{(l)}$) and its image matrix $\mathbf{M}^{(l)}$. The (unweighted) blockmodelling problem can be considered as finding a blockmodel approximation of $\mathbf{A}^{(l)}$ as $\mathbf{C}^{(l)}\mathbf{M}^{(l)}(\mathbf{C}^{(l)})^T$ that minimises a sum of squared errors[3][15], $||\mathbf{A}^{(l)} - \mathbf{C}^{(l)}\mathbf{M}^{(l)}(\mathbf{C}^{(l)})^T||^2$. We denote $\mathbf{C}^{(l)}\mathbf{M}^{(l)}(\mathbf{C}^{(l)})^T$ as $\hat{\mathbf{A}}^{(l)}$. Given two blockmodels , the distance between two blockmodels $\mathfrak{B}^{(1)}$ and $\mathfrak{B}^{(2)}$ is defined as $d(\mathfrak{B}^{(1)}, \mathfrak{B}^{(2)})$.

3.1 Desired Properties of Comparison Measures

In this section, we formalise the properties that a blockmodel comparison measure should possess (as discussed in Section 1). These properties allow us to formally evaluate the measures in the next section.

The following properties assume that there are three blockmodels with approximations $\hat{\mathbf{A}}^{(1)}$, $\hat{\mathbf{A}}^{(2)}$ and $\hat{\mathbf{A}}^{(3)}$ of the same graph, $\hat{\mathbf{A}}^{(1)} \neq \hat{\mathbf{A}}^{(2)} \neq \hat{\mathbf{A}}^{(3)}$ and they are not co-linear, i.e., $|\hat{\mathbf{A}}^{(1)} - \hat{\mathbf{A}}^{(3)}| \neq |\hat{\mathbf{A}}^{(1)} - \hat{\mathbf{A}}^{(2)}| + |\hat{\mathbf{A}}^{(2)} - \hat{\mathbf{A}}^{(3)}|$. They can be considered as measuring the sensitivity to differences in the structure (approximation) of the blockmodels.

P1: Approximation sensitivity: Given an **unweighted** graph and three blockmodels, a measure is approximation sensitive if $d(\hat{\mathbf{A}}^{(1)}, \hat{\mathbf{A}}^{(2)}) \neq d(\hat{\mathbf{A}}^{(2)}, \hat{\mathbf{A}}^{(3)})$.

P2: Block edge distribution sensitivity: Given $d_{mKL}(\hat{\mathbf{A}}^{(2)}, \hat{\mathbf{A}}^{(1)}) < d_{mKL}(\hat{\mathbf{A}}^{(3)}, \hat{\mathbf{A}}^{(1)})$, then a measure is block edge distribution sensitive if $d(\hat{\mathbf{A}}^{(2)}, \hat{\mathbf{A}}^{(1)}) < d(\hat{\mathbf{A}}^{(3)}, \hat{\mathbf{A}}^{(1)})$. $d_{mKL}(\hat{\mathbf{A}}^{(1)}, \hat{\mathbf{A}}^{(2)})$ is the KL divergence [16] for matrices, and is defined as $d_{mKL}(\hat{\mathbf{A}}^{(1)}, \hat{\mathbf{A}}^{(2)}) =$

[2] For simplicity, we assume a discrete set of weights. But this can easily be extended to continuous weight values.

[3] This is one of several popular blockmodelling objective formulations. See [1] for an objective for weighted graph.

$\sum_{i,j}^{|V^1|} \hat{\mathbf{A}}_{i,j}^{(1)} \log\left(\frac{\hat{\mathbf{A}}_{i,j}^{(1)}}{\hat{\mathbf{A}}_{i,j}^{(2)}}\right) - \hat{\mathbf{A}}_{i,j}^{(1)} + \hat{\mathbf{A}}_{i,j}^{(2)}$. It measures the difference in the edge (weight) distribution across all the blocks.

P3: Weight sensitivity: Given a **weighted** graph and three blockmodels[4], a measure is weight sensitive if $d(\hat{\mathbf{A}}^{(1)}, \hat{\mathbf{A}}^{(2)}) \neq d(\hat{\mathbf{A}}^{(2)}, \hat{\mathbf{A}}^{(3)})$.

In addition, we evaluate the important monotonicity property of the measures. Basically, we desire measures that increase in value as the compared blockmodels become more different. We measure "more different" by the minimum number of position changes to transform one blockmodel to another.

4 Blockmodel (Graph Clustering) Comparison Approaches

In this section, we describe existing measures, propose new blockmodel comparison approaches, and analyse their properties. Existing work for comparing blockmodels falls into two categories: *positional* and *reconstruction* measures. Positional measures compare the sets of positions associated with each blockmodel [10]. Reconstruction measures compare the blockmodel approximation of the graphs and can be expressed as $d(\mathfrak{B}^{(1)}, \mathfrak{B}^{(2)}) = d(\hat{\mathbf{A}}^{(1)}, \hat{\mathbf{A}}^{(2)})$. We next provide details on two existing reconstruction blockmodel measures.

4.1 Edge and Block Reconstruction Distances

The edge and block reconstruction distances were proposed in [4] and [8] respectively. The edge reconstruction distance [8] measures the difference in the expected edge probabilities across all edges (recall that $V^1 = V^2$):

$$d_{RE}(\mathfrak{B}^{(1)}, \mathfrak{B}^{(2)}) = \sum_{i}^{|V^1|} \sum_{j}^{|V^2|} |\Upsilon_{\phi(i),\phi(j)}^{(1)} - \Upsilon_{\phi(i),\phi(j)}^{(2)}| \quad (1)$$

The block reconstruction distance [4] measures the difference in block densities over all pairs of blocks, weighted by the overlap of the positions:

$$d_{RB}(\mathfrak{B}^{(1)}, \mathfrak{B}^{(2)}) = \sum_{r1,c1}^{k1} \sum_{r2,c2}^{k2} \frac{|C_{r1}^{(1)} \cap C_{r2}^{(2)}|}{n} \frac{|C_{c1}^{(1)} \cap C_{c2}^{(2)}|}{n} \cdot |\Upsilon_{r1,c1}^{(1)} - \Upsilon_{r2,c2}^{(2)}| \quad (2)$$

The two measures in fact differ only by a factor of $\frac{1}{n^2}$ (we prove this new result in Theorem 1). It is easier to understand and propose new measures based on the edge than the block reconstruction distance. But in terms of computational complexity, it takes $O(k^2)$ to compute the block reconstruction distance while $O(n^2)$ for the edge distance, hence it is more efficient to compute the block distance. Therefore Theorem 1 permits us to use whichever measure that is more convenient for the task at hand.

[4] $\hat{\mathbf{A}}$ consists of n^2 PMFs of a random variable with the event space of edge weights.

Theorem 1. $d_{RB}(\mathfrak{B}^{(1)}, \mathfrak{B}^{(2)}) = \frac{1}{n^2} d_{RE}(\mathfrak{B}^{(1)}, \mathfrak{B}^{(2)})$

Proof. The proof involves arithmetic manipulation. We detail the proof in a supplementary[5] paper due to space limitations.

Both these reconstruction distances possess the approximation sensitivity property (P1) but fail the block edge distribution property (P2). Reconsider the example of Figures 2a to 2c. $d_{RB}(BM11, BM12) = 0.21$ and $d_{RB}(BM11, BM13) = 0.18$. This means that the reconstruction distances incorrectly ranks BM 13 to be closer to BM 11 than BM 12. To explain why, we first state that the Earth Movers Distance (EMD) can be considered as a generalisation of the reconstruction measures (see Section 4.3). The EMD finds the minimum amount of mass to move from one PMF (of a block) to another. What this means is that the reconstruction distance considers the cost to move a unit of mass the same, whether the source PMF is a unimodal distribution or uniformly distributed one. Hence the reconstruction distances only consider the number of total units of mass moved and do not consider the differences in distribution of edge densities across all the blocks, leading them to fail property P2.

4.2 KL Reconstruction Measure

We propose to use the KL divergence [17] to compare the block densities (PMFs). Although it is similar in form to the P2 property definition, the proposed measure is in the form of the reconstruction distance and the KL divergence is a natural approach to measure distribution differences. It is defined as:

Definition 1. ***KL Reconstruction:***

$$d_{RKL}(\mathfrak{B}^{(1)}, \mathfrak{B}^{(2)}) = \sum_{r1,c1}^{\mathcal{C}^{(1)}} \sum_{r2,c2}^{\mathcal{C}^{(2)}} \frac{|C_{r1} \cap C_{r2}|}{n} \frac{|C_{c1} \cap C_{c2}|}{n} \cdot d_{KL}(\mathbf{A}^{(1)}_{r1,c1}, \mathbf{A}^{(2)}_{r2,c2}) \quad (3)$$

where $d_{KL}(\Psi^{(1)}_{r1,c1}, \Psi^{(2)}_{r2,c2}) = \sum_x p(\Psi^{(1)}_{r1,c1} = x) \log(\frac{p(\Psi^{(1)}_{r1,c1}=x)}{p(\Psi^{(2)}_{r2,c2}=x)}$.

The KL reconstruction measure can be considered as using the edge distributions (across the blocks) in $\mathfrak{B}^{(2)}$ to encode the edge distributions in $\mathfrak{B}^{(1)}$. This means it is sensitive to differences in the distribution of the block densities, and it gives the correct ranking for example 1 (see Section 5).

The KL divergence is asymmetric, hence $d_{RKL}(.)$ is also asymmetric, and can handle weighted graphs. The Jeffrey and Jenson-Shannon divergences [17] are symmetric versions of the KL divergence, and are included for comparison. Unfortunately they fail the block edge distribution property (see Section 5).

4.3 Weighted Block and Edge Reconstruction

In this section, we show how the edge/block reconstruction measures can be generalised to weighted graphs. We compare the blockmodels of weighted graphs

[5] Available at `http://people.eng.unimelb.edu.au/jeffreyc/`

by comparing the PMFs of the blocks. We desire a measure that can compare multi-valued PMFs and consider the difference in the actual weight values. One such measure is the Earth Mover's distance (EMD) [18], which also reduces to the reconstruction measures for unweighted graphs (see Theorem 2).

Definition 2. ***EMD Reconstruction Measure***

$$d_{REMD}(\mathfrak{B}^{(1)}, \mathfrak{B}^{(2)}) = \sum_{r1,c1}^{C^{(1)}} \sum_{r2,c2}^{C^{(2)}} \frac{|C_{r1} \cap C_{r2}|}{n} \frac{|C_{c1} \cap C_{c2}|}{n} \cdot d_{ED}(\Psi^{(1)}_{r1,c1}, \Psi^{(2)}_{r2,c2}) \quad (4)$$

where $d_{ED}(\Psi^{(1)}_{x,y}, \Psi^{(2)}_{a,b}) = \min_{\mathbf{M}} \sum_{u=0}^{L} \sum_{w=0}^{L} m_{u,w} \cdot d(u,w)$, subject to: **a)** $m_{u,e} \geq 0$; **b)** $\sum_w m_{u,w} = p(\Psi^{(1)}_{x,y} = u)$; **c)** $\sum_u m_{u,w} = p(\Psi^{(2)}_{a,b} = w)$ and **d)** $d(u,w) = |u - w|$.

We now show that the EMD reconstruction measure is in fact a generalisation of the reconstruction measures.

Theorem 2. *When we are comparing unweighted graphs,*

$$d_{ED}(\mathbf{A}^{(1)}_{x,y}, \mathbf{A}^{(2)}_{a,b}) = |\Upsilon^{(1)}_{x,y} - \Upsilon^{(2)}_{a,b}|$$

and $d_{REMD}(\mathfrak{B}^{(1)}, \mathfrak{B}^{(2)}) = d_{RB}(\mathfrak{B}^{(1)}, \mathfrak{B}^{(2)})$.

Proof. The proof involves arithmetic manipulation and reasoning on the constraints. Due to space limitations, please refer to supplementary for details.

Theorem 2 is a useful result, as it helps to explain why the block reconstruction distance fail property P2. In addition, it means EMD reconstruction distance can be used in place of block distance, since Theorem 2 tells us that the EMD distance is a generalisation of block one for unweighted graphs. At the same time, the EMD distance satisfies property P3 while the block one does not.

5 Evaluation of the Measures

In this section, we evaluate the measures using the proposed properties and empirically demonstrate their monotonicity (since it is difficult to prove this analytically). We also show how each of the measures perform in the examples from Section 1. In the experiments, we compare NMI, a popular and representative example of the positional measures, against the block (RB), EMD (REMD), KL (RKL), Jeffrey (RsKL) and JS (RJS) reconstruction distances.

5.1 Evaluation of the Examples

Table 1 shows the comparison results between the original blockmodel (BM *1) against the two other blockmodels (BM *2 and *3) of the examples in Figure 2. As can be seen, NMI cannot distinguish between the three blockmodels across all the datasets. RB fails to correctly rank the blockmodels of example 1, and fails to distinguish the weighted blockmodels of example 2. As expected, REMD has the same values as RB for example 1, but correctly classifies the relative ordering of example 2. RsKL and RJS fail example 1. RKL correctly distinguishes the ordered example 1, but like the other distributional measures (RsKL and RJS) cannot distinguish the blockmodels of example 2.

Table 1. Measure values when comparing the original blockmodel (BM*1) against the other blockmodels in the Karate club and other examples from Section 1. In each case, ideally $d(\text{BM }*2, \text{BM }*1)$ should be less than $d(\text{BM }*3, \text{BM }*1)$.

	Example 2		Example 3	
	BM 12	**BM 13**	**BM 22**	**BM 23**
NMI	0.3845	0.3845	0.1887	0.1887
RB	0.2100	0.1800	0.2188	0.2188
REMD	0.2100	0.1800	7.2266	10.5469
RKL	0.2803	4.4432	7.2407	7.2407
RsKL	5.5650	4.6501	9.0060	9.0060
RJS	0.1676	0.1259	0.2633	0.2633

5.2 Monotonicity Analysis

To evaluate monotonicity, we vary the membership of the positions for several real datasets. Each position change corresponds to a change in the membership of a vertex, and we ensure a vertex can only change membership once in each simulation. From a starting blockmodel, we generated 100 different runs, where up to 50% of the vertices change position. Table 2 shows the statistics of the three real datasets[6] on which we evaluate monotonicity.

Table 2. Statistics of the real datasets

Dataset	Vert #	Edge #	Weighted?	Pos. #
Karate	34	78	N	2
Football	115	613	N	12
C.Elegans	297	2359	Y	5

Figure 3 shows the results for REMD, RKL, RsKL and NMI (we plotted 1-NMI). As can be seen, all measures are monotonically increasing as the positions change. However, in the rare case where the adjacency approximation remains the same after a position splits into two, then the reconstruction distances can fail to distinguish the pre-split and post-split blockmodels.

5.3 Properties of the Measures

Table 3 shows the measures and the properties they have. For each measure, we prove (see supplementary) whether it possesses each of the properties.

[6] Available at `http://www.personal-umich.edu/~mejn/netdata`.

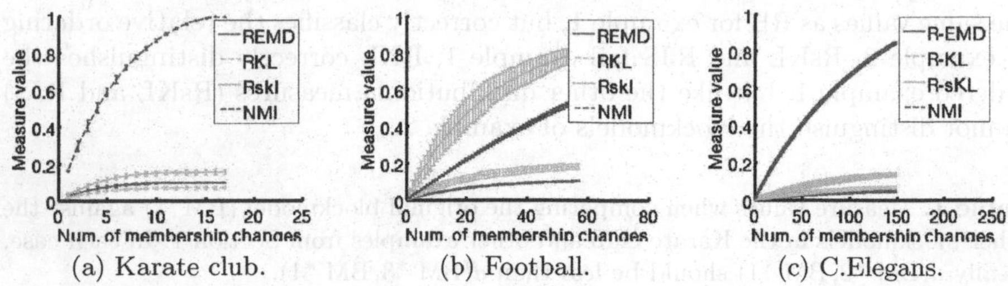

(a) Karate club. (b) Football. (c) C.Elegans.

Fig. 3. Evaluation of the monotonicity of the distance measures as the difference in the positions of the starting blockmodel and modified blockmodel increases

Table 3. List of blockmodel measures and their properties. *: The KL-based measures can fail when the distributions have the same shape but have different weight values. #: not strictly monotonic (can fail the co-incidence axiom).

Property	NMI	RB	REMD	RKL	RsKL	RJS
P1: Edge. Dist. Sensitivity	-	✓	✓	✓	✓	✓
P2: Block Dist. Sensitivity	-	-	-	✓	-	-
P3: Weight Sensitivity	-	-	✓	-*	-*	-*
Monotonicity	✓	✓#	✓#	✓#	✓#	✓#

Table 3 confirms the empirical evaluation of Section 5.1. The positional measures (e.g., NMI) do not consider blockmodel approximations and hence fail the structural sensitivity properties. The existing RB and RE distances cannot be applied to weighted graphs and are not block edge distribution sensitive. The proposed REMD generalises the reconstruction distances to weighted graphs, but still possesses the same assumptions as those distances, and hence fails the block edge distribution sensitivity property. The proposed KL-based distances are block edge distribution sensitive, but not weight sensitive as they measure distribution differences but ignore difference in weight values.

From these analyses, we recommend to use the KL reconstruction distance when comparing unweighted blockmodels, as it possess the first two properties. When comparing weighted graphs, the EMD distance might be preferred as it satisfies the weight sensitivity property. A further option is to combine several measures together via ideas from multi-objective optimisation, e.g., as a weighted linear sum, which we leave to future work.

6 Conclusion

Blockmodel comparison measures are used for validating blockmodelling and community detection algorithms, finding consensus blockmodels and other tasks that require the comparison of blockmodels or communities. In this paper, we have shown that popular positional measures cannot distinguish important differences in blockmodel structure and approximations because they do not reflect a number of key structural properties. We have also proposed two new measures, one based on the EMD and the other based on KL divergence. We formally proved these new measures possess a number of the desired structural properties and used empirical experiments to show they are monotonic.

Future work includes introducing new measures for evaluating mixed membership blockmodels [1] and evaluating multi-objective optimisation approaches for combining measures.

References

1. Airoldi, E.M., Blei, D.M., Fienberg, S.E., Xing, E.P.: Mixed membership stochastic blockmodels. J. of Machine Learning Research 9, 1981–2014 (2008)
2. Pinkert, S., Schultz, J., Reichardt, J.: Protein interaction networks–more than mere modules. PLoS Computational Biology 6(1), e1000659 (2010)
3. Lancichinetti, A., Fortunato, S.: Consensus clustering in complex networks. Nature 2(336) (2012)
4. Chan, J., Liu, W., Leckie, C., Bailey, J., Kotagiri, R.: SeqiBloc: Mining Multi-time Spanning Blockmodels in Dynamic Graphs. In: Proceedings of KDD, pp. 651–659 (2012)
5. Bae, E., Bailey, J., Dong, G.: A clustering comparison measure using density profiles and its application to the discovery of alternate clusterings. Data Mining and Knowledge Discovery 21(3), 427–471 (2010)
6. Coen, M.H., Ansari, H.M., Filllmore, N.: Comparing Clusterings in Space. In: Proceedings of ICML, pp. 231–238 (2010)
7. Rosvall, M., Bergstrom, C.T.: Maps of random walks on complex networks reveal community structure. Proceedings of PNAS 105, 1118–1123 (2008)
8. Wasserman, S., Faust, K.: Social Network Analysis: Methods and Applications. Cambridge Univ. Press (1994)
9. Chan, J., Lam, S., Hayes, C.: Increasing the Scalability of the Fitting of Generalised Block Models for Social Networks. In: Proceedings of IJCAI, pp. 1218–1224 (2011)
10. Halkidi, M., Batistakis, Y., Vazirgiannis, M.: On Clustering Validation Techniques. J. of Intelligent Information Systems 17(2/3), 107–145 (2001)
11. Vinh, N., Epps, J., Bailey, J.: Information theoretic measures for clusterings comparison: Variants, properties, normalization and correction for chance. The J. of Machine Learning Research 11, 2837–2854 (2010)
12. Zhou, D., Li, J., Zha, H.: A new Mallows distance based metric for comparing clusterings. In: Proceedings of the ICDM, pp. 1028–1035 (2005)
13. Patrikainen, A., Meila, M.: Comparing Subspace Clusterings. IEEE Trans. on Know. Eng. 18(7), 902–916 (2006)
14. Doreian, P., Batagelj, V., Ferligoj, A.: Generalized blockmodeling. Cambridge Univ. Press (2005)
15. Chan, J., Liu, W., Kan, A., Leckie, C., Bailey, J., Kotagiri, R.: Discovering latent blockmodels in sparse and noisy graphs using non-negative matrix factorisation. In: Proceedings of CIKM, pp. 811–816 (2013)
16. Lee, D., Seung, H.: Algorithms for non-negative matrix factorization. In: Proceedings of NIPS, pp. 556–562 (2000)
17. Cover, T., Thomas, J.: Elements of Information Theory. Wiley-Interscience (2006)
18. Rubner, Y., Tomasi, C., Guibas, L.: The earth mover's distance as a metric for image retrieval. International Journal of Computer Vision 40(2), 99–121 (2000)

Efficiently and Fast Learning a Fine-grained Stochastic Blockmodel from Large Networks

Xuehua Zhao[1,2], Bo Yang[1,2,*], and Hechang Chen[1,2]

[1] College of Computer Science and Technology, Jilin University, Changchun, China
[2] Key Laboratory of Symbolic Computation and Knowledge Engineer Ministry of Education, Jilin University, Changchun, China
ybo@jlu.edu.cn

Abstract. Stochastic blockmodel (SBM) has recently come into the spotlight in the domains of social network analysis and statistical machine learning, as it enables us to decompose and then analyze an exploratory network without knowing any priori information about its intrinsic structure. However, the prohibitive computational cost limits SBM learning algorithm with the capability of model selection to small network with hundreds of nodes. This paper presents a fine-gained SBM and its fast learning algorithm, named FSL, which ingeniously combines the component-wise EM (CEM) algorithm and minimum message length (MML) together to achieve the parallel learning of parameter estimation and model evaluation. The FSL significantly reduces the time complexity of the learning algorithm, and scales to network with thousands of nodes. The experimental results indicate that the FSL can achieve the best tradeoff between effectiveness and efficiency through greatly reducing learning time while preserving competitive learning accuracy. Moreover, it is noteworthy that our proposed method shows its excellent generalization ability through the application of link prediction.

Keywords: Network data mining, Social network analysis, Stochastic blockmodel, Model selection, Link prediction.

1 Introduction

As an important statistical network model, stochastic blockmodel (SBM) [1] enables us to reasonably decompose and then properly analyze an exploratory network with zero prior information about its intrinsic structure. That is, we believe, the most attractive merit of the model. Formally, a standard SBM is defined as a triple (K, Π, Ω), where K is the number of blocks, Π is an $K \times K$ matrix in which π_{ql} denotes the probability that a link from a node in block q connects to a node in block l, and Ω is an K-dimension vector in which ω_k denotes the probability that a randomly chosen node falls in block k.

SBM is able to approximate any mixture patterns of assortative and disassortative structures ubiquitously demonstrated by the real-world networks, once

* Corresponding author.

V.S. Tseng et al. (Eds.): PAKDD 2014, Part I, LNAI 8443, pp. 374–385, 2014.

it is properly parameterized through fitting observed networks. Besides, SBM is used either as a generative model to synthesize the artificial networks containing assortative communities, disassortative multi-partites and arbitrary mixtures of them, or as a prediction model to predict missing and spurious links. Therefore, SBM has attracted much attention of researchers from the domains of statistics and machine learning since it was firstly proposed by Fienberg and Wasserman in 1981[1]. So far, various extensions of SBM have been proposed to address the oriented tasks of network analysis, such as, multiple roles SBM [2], overlapping SBM [3], mixture SBM [4], scale-free SBM [5], hierarchical SBM [6], among others.

Although SBM has demonstrated superiority in structure analysis, the intractable time complexity of learning severely limits the model to those applications just involving very tiny networks. For the current available learning algorithms, given the number of blocks, K, that means we take no account of model selection, the time complexity is at least $O(K^2n^2)$ where n denotes the number of the nodes. Otherwise, it will get much more, say $O(n^5)$. In practice, if one uses conventional computers, given K, the algorithms just efficiently deal with the networks with at most thousands of nodes, otherwise at most hundreds of nodes, far from the scales of most real-world networks we are interested in. For most of real-world networks, usually, we have no priori knowledge about them, in this case, only the SBM learning algorithms with model selection provide us with the real powerful tools since it can automatically determine the "true" number of blocks. However, the prohibitive time complexity hinders us from effectively analyzing large networks with thousands of nodes.

To address this issue, on one hand, we present a fine-gained SBM named FSBM, to capture more details of networks, and on the other hand, propose a corresponding fast learning algorithm with the capability of model selection called FSL. The FSL ingeniously combines the Component-wise EM(CEM) with Minimum Message Length(MML) to achieve simultaneous rather than alternative execution of model selection and parameter estimation, being able to smartly select a "good" model from a huge problem space consisting of entire candidate models with a significantly reduced computational cost. To the best of our knowledge, this is the first effort in literature to propose a parallel learning process of SBM.

The rest of this paper is organized as follows: Section 2 reviews the related works. Section 3 presents the fine-gained SBM and its learning method. Section 4 validates proposed models and algorithms as well as demonstrates their applications. Finally, Section 5 concludes this work by highlighting our contributions.

2 Related Work

SBM learning has two subtasks: to learn the parameters of model (Θ, Ω) and to determine the number of the blocks (K), corresponding to parameter estimation and model selection, respectively. Model selection aims at selecting a model with a good tradeoff between description precision and model complexity. As we know, the complexity of a model is determined by the number of parameters in the model. In this sense, the model selection of SBM is actually the determination of a reasonable K, the number of blocks, since the quantity of parameters of SBM can be represented as a function of K.

Most existing algorithms adopt EM or variational EM to estimate the parameters of SBM [2–4, 7, 8] in which SBM utilizes a latent variable Z to indicate the group to which a node belongs. The time complexity of such methods is at least $O(n^2K^2)$. Currently, the available model selection methods applied to SBM fall into three categories including cross-validation(CV), bayesian-based criteria and minimum description length(MDL). The CV has been rarely used in the SBM learning algorithm since its extremely high computation cost [4]. According to the adopted approximation techniques, BIC, ICL and variation-based approximate evidence are three main bayesian based criteria. Airoldi et al. [4] used BIC to select optimal multiple role SBM. Daudin et al. proposed the SICL algorithm which adopted the ICL to select model [9]. Hofman et al. proposed the VBMOD algorithm which adopted variation based approximate evidence for model selection [10]. Latouche et al. proposed SILvb algorithm which adopted ILvb, a newly model selection criterion [11], which has the best performance among the current criteria by now. The MDL is the criterion derived from the information theory and formally coincides with the BIC. Very recently, Yang et al. proposed the GSMDL algorithm that adopted the MDL for model selection [6].

For the above criteria except CV, both of them naturally require that corresponding learning algorithms have to adopt a serial learning strategy. That is to say, the learning process will include the following three steps: firstly, to estimate the parameters of each model in model space, then to evaluate the learned model by a predefined certain criterion, finally to select the model with the best evaluation value as the optimal model. Hence, such a serial learning strategy requires estimating and evaluating each model in the model space even if the models in question are "bad". This leads to a extremely high time complexity since such strategy needs to estimate parameters not only for "good" models but for "bad" models. In general, the time complexity of the learning strategy, such as SICL [9], SILvb [11] and GSMDL [6], require at least $O(n^5)$.

Recently, Hofman and Wiggins proposed a VBMOD algorithm with time complexity $O(n^4)$ by reducing the quantity of parameters from K^2+K to $K+2$. So far, the VBMOD still remains the lowest time complexity among all available SBM learning algorithms with the capability of model selection. However, its low time complexity is obtained at the expense of flexibility. That is, the strategy of parameter reduction will restrict the VBMOD to deal with the networks with community structures.

Figueiredo et al. [12] ever applied the MML to gaussian mixture model, and their experiments showed that the accuracy of the MML outperforms that of the MDL/BIC, LEC or ICL criteria. Although their method was proposed towards to feature vector space and is not suitable for processing networks, the rationale behind it inspires us to propose a new SBM learning algorithm by designing an parallel integration of model selection and parameter estimation.

3 Model and Method

3.1 Fine-gained Stochastic Blockmodel

As the standard SBM only uses the block connection matrix to characterize the homogeneity and heterogeneity of links between nodes, it is difficult for SBM to capture more detailed structural information. To address this issue, we relax the parameter Π into the two parameters Θ and Δ, which respectively denote the connection probability from blocks to nodes and the connection probability from nodes to blocks. And based on the relaxation, we proposed the fine-gained stochastic blockmodel, FSBM.

Let $N = (V, E)$ be a directed and binary network where $V(N)$ denotes the set of nodes and $E(N)$ denotes the set of directed edges. Let $A_{n\times n}$ be the adjacency matrix of N where n denotes the number of nodes and $A_{ij} = 1$ if there exists an edge from the node i to the node j, otherwise $A_{ij} = 0$. In the case of undirected network, supposing there are two directed edges between nodes and $A_{ij}= A_{ji}$.

The FSBM is defined as $X = (K, Z, \Omega, \Theta, \Delta)$, where, K is the number of the blocks, Z is one $n \times K$ matrix in which z_{ik} denotes the block k to which the node i belongs, Ω is one K-dimension vector in which ω_k denotes the probability that a randomly chosen node falls in block k, Θ is one $K \times n$ matrix in which θ_{kj} denotes the probability that a link from a particular node in block k connects to vertex j, Δ is one $K \times n$ matrix in which δ_{kj} denotes the probability that a link from node j connects to a particular vertex in block k. If N is undirected network, then $\Theta = \Delta$. The block matrix of the FSBM can be obtained as follows:

$$\Pi_1 = \Theta Z D^{-1}, \qquad \Pi_2 = \Delta Z D^{-1} \tag{1}$$

where $D = diag(n\Omega)$. For undirected network, we have $\Pi_1 = \Pi_2 = \Pi$. Therefore, the standard SBM is a specific case of the FSBM.

The log-likelihood of the observed network N can be written as follows:

$$\log P(N|\Omega, \Theta, \Delta) = \sum_{i=1}^{n} \log \sum_{k=1}^{K} (\prod\nolimits_{j=1}^{n} f(\theta_{kj}, A_{ij}) f(\delta_{kj}, A_{ji}))\omega_k \tag{2}$$

where, $f(x, y) = x^y(1-x)^{(1-y)}$.

The log-likelihood for complete data can be written as follows:

$$\log P(N, Z|\Omega, \Theta, \Delta) = \sum_{i=1}^{n}\sum_{k=1}^{K} z_{ik}(\sum_{j=1}^{n} \log(f(\theta_{kj}, A_{ij}) f(\delta_{kj}, A_{ji})) + \log \omega_k) \tag{3}$$

3.2 Fast SBM Learning Method

The proposed fast SBM learning method, FSL, ingeniously combines the CEM algorithm [13] with the MML [12] together to achieve the parallel learning of parameter estimation and model selection. The FSL first obtains the estimated parameter value of one block by the CEM algorithm, then evaluates the block in terms of the MML. The bad block evaluated by the MML, that is the existence probability is zero, is directly annihilated and is not estimated in the next iteration. And so on, until the convergence. In the process, the sequential updating approach of the CEM algorithm and the MML directly evaluating one block provide the support of the parallel learning. Finally, parameters estimation and model selection are parallelly implemented in a time convergence process. In contrast to the serial learning algorithm, the FSL directly finds the "good" mode in the model space and effectively reduces the computational cost by preventing the algorithm estimating the parameters of the "bad" model.

Parameter Estimation. Although the CEM algorithm has almost the same time complexity as the EM algorithm, its faster convergence can reduce the time cost, but the major contribution is that the sequential updating approach of the CEM algorithm provides a smart framework for the parallel learning of parameter estimation and model selection. The CEM algorithm considers the decomposition of the parameter vector, and updates only one block at a time, letting the other parameters unchanged. The E-step and M-step are as follows:

E-step: Given the observed network N and h^{t-1} where h and t respectively denote the model parameters (Ω, Θ, Δ) and the current iteration step, compute the conditional expectation of complete log-likelihood, i.e. Q function.

$$Q(h, h^{t-1}) = \sum_{i=1}^{n}\sum_{k=1}^{K}\gamma_{ik}(\sum_{j=1}^{n}(\log f(\theta_{kj}, A_{ij}) + \log f(\delta_{kj}, A_{ji})) + \log \omega_k) \tag{4}$$

where $\gamma_{ik} = E[z_{ik}]$ denotes the posteriori probability of block k to which the node i belongs according to the model h^{t-1}

M-step: Update the parameters of the current block according to Eq. 5-7 which are obtained by maximizing Eq. 4:

$$\omega_k{}^{(t)} = \frac{\sum_{i=1}^{n}\gamma_{ik}}{n} \tag{5}$$

$$\theta_{kj}{}^{(t)} = \frac{\sum_{i=1}^{n}A_{ij}\gamma_{ik}}{\sum_{i=1}^{n}\gamma_{ik}} \tag{6}$$

$$\delta_{kj}{}^{(t)} = \frac{\sum_{i=1}^{n}A_{ji}\gamma_{ik}}{\sum_{i=1}^{n}\gamma_{ik}} \tag{7}$$

where $k = (t./K_{max}) + 1$, K_{max} is the number of the blocks and ./ denotes modulus operator. In the current step t, only the parameters of the k-th block are updated according to Eq. 5-7, respectively.

Model Selection. The MML is derived from information theory and the rational behind MML is that the shorter code the data has, the better the data generation model is. The particular MML criterion is as follows:

$$\widehat{h} = \arg\min_{h} \ell(-\log p(h) - \log p(N|h) + \frac{1}{2}\log|\mathbf{I}(h)| + \frac{c}{2}(1 + \log\frac{1}{12})) \tag{8}$$

where N denotes the observed network, h denotes the model parameters, c is dimension of h, $\mathbf{I}(h)$ is the information matrix and $|\mathbf{I}(h)|$ denotes its determinant.

Since FSBM contains the latent variable Z, the information matrix of FSBM can not be obtained analytically. We adopt the information matrix of the complete data log-likelihood, $\mathbf{I}_c(\Theta, \Delta)$, and assume the parameters of the blocks as a priori independent and also independent from the parameter ω. For each factor $p(\Theta_k, \Delta_k)$ and $p(\omega_1, ..., \omega_k)$, we adopt the noninformative Jeffrey' priori. Consequently, (8) becomes

$$\widehat{h} = \arg\min_h \ell(\frac{c}{2}\sum_{k=1}^{K}\log(\frac{n\omega_k}{12}) + \frac{K}{2}\log\frac{n}{12} + \frac{K(c+1)}{2} - \log p(N|h)) \quad (9)$$

where $\ell(*)$ is the cost function, K is the number of the blocks, c is the number of parameters specifying each block, h denotes the parameters (Ω, Θ, Δ).

When ω_k is zero, (9) will be nonsense, however, from the view of data coding, the parameters of zero-probability block do not contribute to coding-length [12]. Let k_{nz} is the number of non-zero probability blocks, the cost function becomes

$$\ell(h, N) = \frac{c}{2}\sum_{k:\omega_k>0}^{K}\log(\frac{n\omega_k}{12}) + \frac{k_{nz}}{2}\log\frac{n}{12} + \frac{k_{nz}(c+1)}{2} - \log p(N|h) \quad (10)$$

Minimizing the cost function (10) with respect to h constitutes the solutions of parameters estimated. As we can see, to minimize (10) with respect to θ, δ is the same as to minimize the -Q function since the terms except $-\log p(N|h)$ is dropped, and the difference is ω. The ω obtained by differentiating (10) with k_{nz} fixed is given as follows:

$$\widehat{\omega}_k^{(t)} = \frac{\max\left\{0, (\sum_{i=1}^{n}\gamma_{ik}) - \frac{c}{2}\right\}}{\sum_{j=1}^{K}\max\left\{0, (\sum_{i=1}^{n}\gamma_{ij}) - \frac{c}{2}\right\}} \quad (11)$$

where the γ_{ik} are given by the E-step, c is the dimension of parameters. Noting in (2) that any block for which $\widehat{\omega}_k = 0$ does not contribute to the log-likelihood.

We can see that (11) provides us an approach to evaluate the blocks, that is if $\widehat{\omega}_k^{(t)}$ is zero in the current step t, the block k is regarded as a "bad" block and may directly annihilate it. Based on the property of the MML and the sequentially updating of the CEM algorithm can commonly achieve the parallel learning of parameter estimation and model selection.

Algorithm Description and Complexity Analysis. A detailed pseudocode description of the FSL is listed in Table 1. After convergence of the CEM, there is no guarantee that a minimum of $\ell(h, N)$ has been found since the block annihilation in (11) does not consider the additional decrease in $\ell(h, N)$ caused by the decrease in k_{nz}. According to [12], we check if smaller values of $\ell(h, N)$ are achieved by setting to zero blocks that were not annihilated by (11). To this end, we simply annihilate the least probable block and rerun CEM until convergence.

Table 1. The FSL Algorithm

Algorithm 1. FSL

X=FSL(N, K_{min}, K_{max})

Input: N, K_{min}, K_{max}

Output: X_{best}

Initial: $\widehat{h}(0) \leftarrow \{\widehat{h}_1, ..., \widehat{h}_{k_{\max}}, \widehat{\omega}_1, ..., \widehat{\omega}_{k_{\max}}\}$; $t \leftarrow 0$; $k_{nz} \leftarrow k_{\max}$; $\ell_{\min} \leftarrow +\infty$; ε

$u_k^i \leftarrow p(N^{(i)}|\widehat{h}_k)$,for $k = 1, ..., k_{max}$ and $i = 1, ..., n$

while $K_{nz} \geq K_{\min}$ do

  repeat

  $t \leftarrow t + 1$

  for m=1 to $K_{\max}$ do

    $\gamma_k^i \leftarrow \widehat{\omega}_k u_k^{(i)} (\sum_{j=1}^{k_{\max}} \widehat{a}_j u_j^{(i)})^{-1}$, for i=1,,n

    $\widehat{\omega}_k \leftarrow \max\{0, (\sum_{i=1}^{n} \gamma_k^{(i)} - \frac{c}{2})\} \times (\sum_{j=1}^{k} \max\{0, (\sum_{i=1}^{n} \gamma_k^(i) - \frac{c}{2})\})^{-1}$

    $\{\widehat{\omega}_1, ..., \widehat{\omega}_m\} \leftarrow \{\widehat{\omega}_1, ..., \widehat{\omega}_m\}(\sum_{m=1}^{K_{\max}} \widehat{\omega}_m)^{-1}$

    if $\widehat{\omega}_k > 0$ then

      $\widehat{h_k} \leftarrow \arg\max_{h_k} \log p(N, \gamma|h)$

      $u_k^i \leftarrow p(N^{(i)}|\widehat{h}_k)$

    else

      $K_{nz} \leftarrow K_{nz} - 1$

    end if

  end for

  $\widehat{h}(t) \leftarrow \{\widehat{\theta}_1, ..., \widehat{\theta}_{K_{\max}}, \widehat{\delta}_1, ..., \widehat{\delta}_{K_{\max}}, \widehat{\omega}_1, ..., \widehat{\omega}_{K_{\max}}\}$

  $\ell[\widehat{h}(t), N] \leftarrow \frac{c}{2} \sum_{k:\widehat{\omega}_k > 0} \log \frac{n\widehat{\omega}_k}{12} + \frac{K_{nz}}{2} \log \frac{n}{2} + \frac{K_{nz}c + K_{nz}}{2} - \sum_{i=1}^{n} \log \sum_{k=1}^{K} \widehat{\omega}_k u_k^{(i)}$

  until $\ell[\widehat{h}(t-1), N] - \ell[\widehat{h}(t), N] < \varepsilon$

  if $\ell[\widehat{h}(t), N] < \ell_{\min}$ then

    $\ell_{\min} \leftarrow \ell[\widehat{h}(t), N]$

    $\widehat{h}_{best} \leftarrow \widehat{h}(t)$

    Z=compute(γ)

    $Xbest$=$(Z, \widehat{h}_{best})$;

  end if

  $k^* \leftarrow \arg\min_k \{\widehat{\omega}_k > 0\}$,$\widehat{\omega}_{k^*} \leftarrow 0$ $K_{nz} \leftarrow K_{nz} - 1$

end while

For the FSL, calculating the posterior of the latent variable Z and estimating parameters (Θ, Δ, Ω) are time-consuming, which dominate the time complexity of the whole computing process. We can analyse the time complexity by the detailed algorithm steps in Table 1. Given K, the time complexity is $O(In^2K)$, where I is the iterative steps. When K is unknown, in the worst case, the time complexity is $O(In^4)$. Among the available SBM learning algorithms with model selection, the FSL has the same low time complexity as the VBMOD, far less than the other algorithms, such as GSMDL [6] $O(n^5)$, SICL [9] $O(n^5)$, SILvb [11] $O(n^5)$, MBIC[2] $O(n^7)$, Shen [7] $O(n^6)$ and so on. Noting that the FSL

Table 2. Confusion matrices of blocks detected in network with community

	(a) $Q_{true}\backslash Q_{FSL}$							(b) $Q_{true}\backslash Q_{GSMDL}$							(c) $Q_{true}\backslash Q_{VBMOD}$						
	2	3	4	5	6	7	8	2	3	4	5	6	7	8	2	3	4	5	6	7	8
3	0	**100**	0	0	0	0	0	0	**98**	1	1	0	0	0	0	**100**	0	0	0	0	0
4	0	0	**100**	0	0	0	0	0	0	**100**	0	0	0	0	0	0	**100**	0	0	0	0
5	0	0	0	**97**	3	0	0	0	0	0	**100**	0	0	0	0	0	0	**100**	0	0	0
6	0	0	0	1	**94**	5	0	0	0	0	7	**89**	4	0	0	0	0	0	**100**	0	0
7	0	0	0	0	27	**68**	5	11	89	0	30	55	**15**	0	0	0	0	6	82	**12**	0

	(d) $Q_{true}\backslash Q_{SICL}$							(e) $Q_{true}\backslash Q_{SILvb}$						
	2	3	4	5	6	7	8	2	3	4	5	6	7	8
3	0	**100**	0	0	0	0	0	0	**100**	0	0	0	0	0
4	0	0	**100**	0	0	0	0	0	0	**100**	0	0	0	0
5	0	0	0	**100**	0	0	0	0	0	0	**100**	0	0	0
6	0	0	0	23	**77**	0	0	0	0	0	0	**100**	0	0
7	0	5	27	45	23	**0**	0	0	0	0	2	15	**83**	1

can analyse the networks with various structures, such as bipartite, multipartite and mixture structure, but the VBMOD just analysing the networks with the community.

4 Validation

In this section, we validate the proposed algorithm on the synthetic networks and real-world networks, and make comparisons with the other algorithms with model selection, which are GSMDL [6], SICL [9], VBMOD [10] and SILvb [11], respectively. We also validate the generalization ability by the application of link prediction. The experiments are run on the computer with dual-core 2GH CPU and 4GB RAM, and all the programs are implemented by Matlab 2010b.

4.1 Validation on the Synthetic Networks and Real-world Networks

Accuracy Validation. We use the same testing method that mentioned in [11] to validate the FSL. The SBM is used as generation model to produce three types of the synthetic networks, which respectively contain community, bipartite and multipartite structure. Each type of network is divides into 5 groups in which the number of the true blocks Q_{true} is 3, 4, 5, 6 and 7, respectively. And each group contains 100 networks with 50 nodes which are randomly produced.

Table 2-3 respectively show the confusion matrices of the results detected in networks with community and mixture structure. The results listed in Table 2 indicate that the FSL and SILvb are the best algorithm among the five algorithms, especially, when Q_{true} is 7, they can correctly find out 68 and 83 blocks, respectively. The results listed in Table 3 indicate that the VBMOD fails to correctly find any network structure, and the others can effectively find out the number of blocks in networks, especially, when Q_{true} is 7, the SILvb still exhibits

Table 3. Confusion matrices of blocks detected in networks with Multipartite

	(a) $Q_{true} \backslash Q_{FSL}$							(b) $Q_{true} \backslash Q_{GSMDL}$							(c) $Q_{true} \backslash Q_{VBMOD}$						
	2	3	4	5	6	7	8	2	3	4	5	6	7	8	2	3	4	5	6	7	8
3	0	**100**	0	0	0	0	0	0	**100**	0	0	0	0	0	100	**0**	0	0	0	0	0
4	0	0	**100**	0	0	0	0	0	0	**100**	0	0	0	0	0	100	**0**	1	0	0	0
5	0	0	0	**100**	3	0	0	0	0	0	**100**	0	0	0	0	100	0	**0**	0	0	0
6	0	0	0	0	**96**	4	0	0	0	0	9	**90**	1	0	0	0	100	0	**0**	0	0
7	0	0	0	0	9	**71**	20	0	0	0	19	50	**31**	0	0	0	0	96	4	**0**	0

	(d) $Q_{true} \backslash Q_{SICL}$							(e) $Q_{true} \backslash Q_{SILvb}$						
	2	3	4	5	6	7	8	2	3	4	5	6	7	8
3	0	**100**	0	0	0	0	0	0	**100**	0	0	0	0	0
4	0	0	**100**	0	0	0	0	0	0	**100**	0	0	0	0
5	0	0	0	**100**	0	0	0	0	0	0	**100**	0	0	0
6	0	0	2	8	**92**	0	0	0	0	0	0	**100**	0	0
7	1	0	8	36	49	**6**	0	0	0	0	0	21	**79**	1

the best performance, the FSL following behind. Finally, we can conclude that the SILvb outperforms other existing learning algorithms in the accuracy test. The FSL is slightly worse than SILvb, but much better than other algorithms. The VBMOD fails to analyse the networks with the other structures except the community.

Time Complexity Validation To validate the time complexity, we use the running time metric to evaluate the time complexity of the algorithms.

First, we test the running time in synthetic networks with different size. The Newman model is used to generate synthetic networks, and the parameters of model are set as follows: K=4, d=16 and z_{out}=2. Let s in turn take value 100, 200, 300, 400, 500, 600, 700 and 800. Finally, we randomly generate eight groups of networks and each group contains 50 networks with the same size. K_{min} and K_{max} are respectively set 1 and 10.

Fig. 1(a) shows the results of the average running time of five algorithms. As we can see, the running time of the FSL is obviously much less than other four algorithms. The VBMOD closely follows behind the FSL, however, the VBMOD achieves its low time complexity at the cost of accuracy since its too few parameters, $k+2$, enable it to only analyse the networks with the community. Noting that when the number of nodes is 3200, on average the FSL only takes 32s, the VBMOD, MSMDL, SICL and SILvb take 92s, 7513s, 153901s and 153244s, respectively.

Then, we demonstrate how the scale of model space affects computational cost using the football network with 115 nodes [14]. We fix K_{min}=1 and let K_{max} in turn take the value 10, 20, 30, 40, 50, 60, 70, 80, 90 and 100. Fig. 1(b) illustrates the relation of running time and the scale of model space. As we can see, the FSL significantly outperforms the other four algorithms.

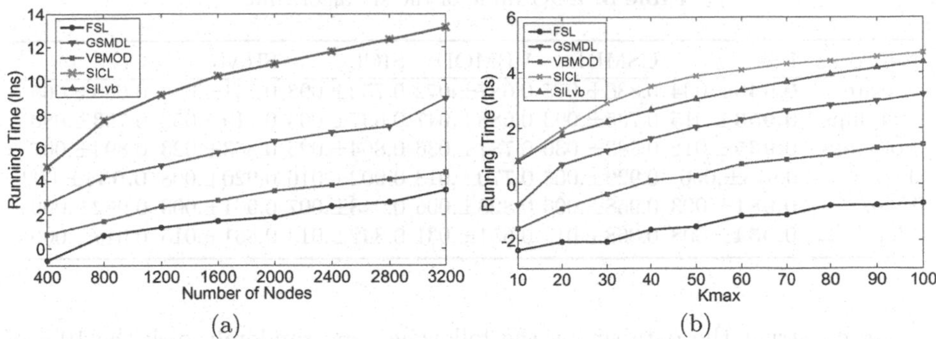

Fig. 1. The comparisons of running time of five algorithms. (a) shows the relation between time cost and network scale, (b) shows the relation between running time and K_{max}.

Table 4. The running time of the five algorithms(s)

Networks	FSL	GSMDL	VBMOD	SICL	SILvb
Karate	0.08	1.22	0.24	2.49	2.04
Dolphins	0.25	6.96	0.80	19.52	16.99
Polbooks	0.61	31.81	3.95	132.73	80.62
Jazz	3.59	605.57	36.77	1426.52	1006.53
Usair	57.83	7797.00	467.06	9429.49	8489.78
Metabolic	412.45	-	1567.97	-	-

4.2 Generalization Ability Validation

We first demonstrate the low computation cost of the FSL in the real-world networks, and use the learned parameters to test its generalization ability according to the performance of link prediction. In the experiment, we selected seven real-world networks, which respectively are Karate network [14], Dolphins network [15], Football network [14], Polbooks network (http://www.orgnet.com), Jazz network [17], Usair network [16] and Metabolic network [17].

Let K_{min}=1 and K_{max} take the number of nodes. The running time of the five algorithms is listed in Table 4. We can see that the running time of the FSL is significantly much lower than other four algorithms, the following is the VBMOD, and the SICL is the worst. Noting that the FSL only consumes 412.453s, but the VBMOD is 1567.97s in Metabolic network with 453 nodes, the running time of other algorithms are far more than that of the FSL, even fails to deal with the networks.

Then we evaluate the generalization ability of the models and algorithms in terms of the performance of link prediction. We also make comparisons with the CN algorithm [18] which is usually used as the baseline algorithm of link prediction. The AUC metric [18] is used to evaluate the performance of the algorithms. The SBM-based approaches to link prediction predict the missing links according to the learned parameters value related to connection probability.

Table 5. AUC value of the six algorithms

Networks	FSL	GSMDL	VBMOD	SICL	SILvb	CN
Karate	**0.946**±.034	0.836±.075	0.687±.072	0.752±.098	0.741±.084	0.680±.065
Dolphins	**0.953**±.015	0.735±.093	0.688±.047	0.697±.063	0.719±.052	0.793±.080
Polbooks	**0.935**±.012	0.883±.036	0.781±.036	0.864±.025	0.863±.023	0.894±.007
Jazz	0.953±.006	0.929±.006	0.770±.013	0.902±.016	0.920±.008	**0.954**±.002
Usair	**0.981**±.003	0.958±.006	0.828±.006	0.948±.007	0.961±.003	0.962±.006
Metabolic	**0.954**±.008	0.868±.013	0.651±.031	0.842±.013	0.861±.015	0.918±.006

We construct the data set by the following way: randomly pick the 10% of edges as test set from the network and the remaining 90% of edges as training set. For each network, we randomly sample 20 times and generate 20 groups of data.

The mean and standard deviation of AUC value of the prediction results are listed in Table 5. As we see, the results indicate that the FSL has the expected best performance among the six algorithms, and the performance of the VBMOD is the worst since the model only uses two parameters to describe the network structure so many information is missing. The main reasons are that the FSBM can capture the more detailed information of network structure than other SBMs, and the FSL can reasonably learn the FSBM.

5 Conclusions

In this paper, we proposed a fine-gained SBM and its fast learning algorithm with the capacity of model selection which adopts the parallel learning strategy to reduce the time complexity. To our best knowledge, this is the first time that the parallel learning strategy is proposed and applied to the SBM learning algorithm. We have validated the proposed learning algorithm on the synthetic networks and real-world networks. The results demonstrate that the proposed algorithm achieves the best tradeoff between effectiveness and efficiency through greatly reducing learning time while preserving competitive learning accuracy. In contrast to existing learning algorithms with model selection just dealing with networks with hundreds of nodes, the proposed algorithm can scale to the networks with thousands of nodes. Moreover, it is noteworthy that our proposed method demonstrates the excellent generalization ability with respect to link prediction.

Acknowledgments. This work was funded by the Program for New Century Excellent Talents in University under Grant NCET-11-0204, and the National Science Foundation of China under Grant Nos. 61133011, 61373053, 61300146, 61170092, and 61202308.

References

1. Holland, P., Laskey, K., Leinhardt, S.: Stochastic blockmodels: First steps. Social Networks 5(2), 109–137 (1983)
2. Airoldi, E., Blei, D., Fienberg, S., Xing, E.: Mixed membership stochastic blockmodels. The Journal of Machine Learning Research 9, 1981–2014 (2008)
3. Latouche, P., Birmel, E., Ambroise, C.: Overlapping stochastic block models with application to the French political blogosphere. The Annals of Applied Statistics 5(1), 309–336 (2011)
4. Newman, M., Leicht, E.: Mixture models and exploratory analysis in networks. Proceedings of the National Academy of Sciences of the United States of America 104(23), 9564–9569 (2007)
5. Karrer, B., Newman, M.: Stochastic blockmodels and community structure in networks. Physical Review E 83(1), 016107 (2011)
6. Yang, B., Liu, J., Liu, D.: Characterizing and Extracting Multiplex Patterns in Complex Networks. IEEE Transactions on Systems Man and Cybernetics, Part B-Cybernetics 42(2), 469–481 (2012)
7. Shen, H., Cheng, X., Guo, J.: Exploring the structural regularities in networks. Physical Review E 84(5), 056111 (2011)
8. Zhu, Y., Liu, D., Chen, G., Jia, H., Yu, H.: Mathematical modeling for active and dynamic diagnosis of crop diseases based on Bayesian networks and incremental learning. Mathematical and Computer Modelling 58(3), 514–523 (2013)
9. Daudin, J., Picard, F., Robin, S.: A mixture model for random graphs. Statistics and Computing 18(2), 173–183 (2008)
10. Hofman, J., Wiggins, C.: Bayesian approach to network modularity. Physical Review Letters 100(25), 258701 (2008)
11. Latouche, P., Birmele, E., Ambroise, C.: Variational Bayesian inference and complexity control for stochastic block models. Statistical Modelling 12(1), 93–115 (2012)
12. Figueiredo, M., Jain, A.: Unsupervised learning of finite mixture models. IEEE Transactions on Pattern Analysis and Machine Intelligence 24(3), 381–396 (2002)
13. Celeux, G., Chretien, S., Forbes, F., Mkhadri, A.: A component-wise EM algorithm for mixtures. Journal of Computational and Graphical Statistics 10(4), 697–712 (2001)
14. Girvan, M., Newman, M.: Community structure in social and biological networks. Proceedings of the National Academy of Sciences of the United States of America 99(12), 7821–7826 (2002)
15. Lusseau, D., Schneider, K., Boisseau, O., Haase, P., Slooten, E., Dawson, S.: The bottlenose dolphin community of Doubtful Sound features a large proportion of long-lasting associations - Can geographic isolation explain this unique trait? Behavioral Ecology and Sociobiology 54(4), 396–405 (2003)
16. Batageli, V., Mrvar, A.: Pajek datasets, `http://vlado.fmf.uni-lj.si/pub/networks/data/default.htm`
17. Duch, J., Arenas, A.: Community detection in complex networks using extremal optimization. Physical Review E 72(2), 027104 (2005)
18. Lu, L., Zhou, T.: Link prediction in complex networks: A survey. Physica A-Statistical Mechanics and Its Applications 390(6), 1150–1170 (2011)

Influence Propagation: Patterns, Model and a Case Study

Yibin Lin, Agha Ali Raza, Jay-Yoon Lee, Danai Koutra, Roni Rosenfeld, and Christos Faloutsos

School of Computer Science, Carnegie Mellon University, Pittsburgh, PA USA
{yibinl,araza,lee.jayyoon,danai,roni,christos}@cs.cmu.edu

Abstract. When a free, catchy application shows up, how quickly will people notify their friends about it? Will the enthusiasm drop exponentially with time, or oscillate? What other patterns emerge?

Here we answer these questions using data from the Polly telephone-based application, a large influence network of 72,000 people, with about 173,000 interactions, spanning *500MB* of log data and *200 GB* of audio data.

We report surprising patterns, the most striking of which are: (a) the FIZZLE pattern, i.e., excitement about Polly shows a power-law decay over time with exponent of -1.2; (b) the RENDEZVOUS pattern, that obeys a power law (we explain RENDEZVOUS in the text); (c) the DISPERSION pattern, we find that the more a person uses Polly, the fewer friends he will use it with, but in a reciprocal fashion.

Finally, we also propose a generator of influence networks, which generate networks that mimic our discovered patterns.

Keywords: social network mining, influence network, influence patterns.

1 Introduction

How will a catchy phone application propagate among people? Will the excitement about it spike, oscillate, or decay with time?

Information cascades, like the above one, appear in numerous settings, like blogs, trending topics in social networks, memes, to name a few. Social influence has been a topic of interest in the research community [38,30,16,23,21,9,4,8,11,27] because of the rise of various on-line social media and social networks. In this work, by *social influence* we refer to the fact that "individuals adopt a new action because of others".

Our current work tries to answer all these questions in a large dataset of hundreds of thousands of interactions. We obtained access to Polly data. Polly is a voice, telephone-based application which allows the sender to record a short message, choose among six funny manipulations of distorting his voice-message (faster, slower, high-pitch, etc.), and forward the modified recording to any of his friends[1]. Polly was devised as a platform for disseminating and popularizing voice-based information services for low-skilled, low-literate people in the developing world. We focus in two main problems, described informally as follows:

---

[1] For a brief video introduction to Polly, including demos of different voice effects, see http://www.cs.cmu.edu/~Polly/

V.S. Tseng et al. (Eds.): PAKDD 2014, Part I, LNAI 8443, pp. 386–397, 2014.

Informal Problem 1 (Pattern Discovery) *Consider a real-world influence network:* **Given** *who influences whom, and when,* **find** *general influence patterns this network obeys.*

Informal Problem 2 (Generator) *Create a realistic influence-network-generator:*
- **Given** *a friendship social network (who-likes-whom)*
- **Design** *a simple, local propagation mechanism*
- *so that we can generate realistic-looking influence networks.*

By "realistic" we mean that the resulting influence networks match our discovered patterns.

Figure 1 gives examples of a social network (who is friends with whom, in gray, directed edges), and a possible influence network (who sends messages to whom - in red; directed, time-stamped, multi-edges). For simplicity, only edges between 1 and 2 are shown with time-stamp and multi-edge structure.

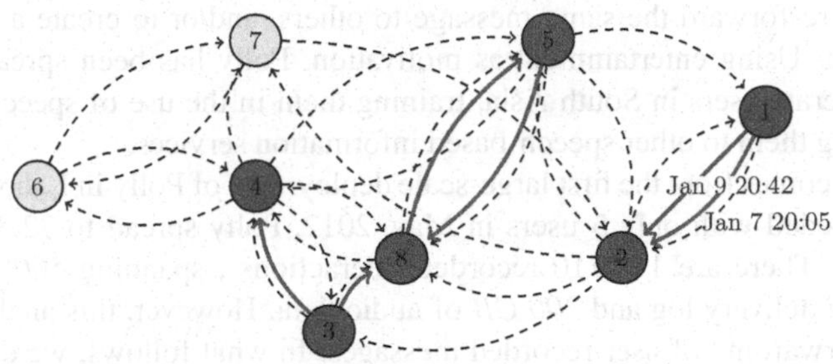

Fig. 1. Illustration of a 'base network' (in gray) and a possible 'influence network' (= cascades; in red). Here, the initial nodes ('seeds') are nodes 1 and 3.

The contributions of this work are the following:

- **Discovery** of three new patterns (laws): the FIZZLE, RENDEZVOUS, and DISPERSION pattern (Sections 3, 4, 5, respectively);
- **Generator and Analysis:** We propose a local, efficient propagation mechanism that simulates an influence graph on top of existing social network datasets (Enron, Facebook) or synthetic social network datasets [12]. Figure 1 illustrates the process of simulating influence network. We also did analysis on the DISPERSION pattern.

The importance of the former contribution is that patterns can help marketers and sociologists understand how influence propagates in a social network; they can also help spot anomalies, like spammers, or faulty equipment.

The importance of our second contribution is that a realistic generator is valuable for what-if scenarios, and reproducibility: publicly-available influence network datasets are notoriously difficult to obtain, due to privacy and corporate regulations; a good generator can serve as proxy.

Reproducibility

For privacy reasons, the Polly dataset that we used in this paper is not public. Thus, for reproducibility, we present experiments on public data (such the Enron Email network [22,1], and Facebook [2]) which exhibit similar behavior like our dataset. We also make our code open-source at: `https://github.com/yibinlin/inflood_generator/`.

Next, we describe the dataset (Sec. 2), our discoveries (Sec. 3, 4 and 5), our generator (Sec. 6), the related work (Sec. 7) and conclusions (Sec. 8).

2 Dataset Used

The dataset comes from Polly [34,33,32], a simple, telephone-based, voice message manipulation and forwarding system that allows a user to make a short recording of his voice, optionally modify it using a choice of funny sound effects, and have the modified recording delivered to one or more friends by their phone numbers. Each friend in turn can choose to re-forward the same message to others, and/or to create a new recording of his own. Using entertainment as motivation, Polly has been spreading virally among low-literate users in South Asia, training them in the use of speech interfaces, and introducing them to other speech-based information services.

The dataset comes from the first large-scale deployment of Polly in Lahore, Pakistan. After being seeded with only 5 users in May, 2012, Polly spread to 72,341 users by January, 2013. There are 173,710 recorded interactions , spanning *500 MB* of real-world message delivery log and *200 GB* of audio data. However, this analysis focuses only on the forwarding of user recorded messages. In what follows, we denote a user with a node and a forwarded message with a directed and dated edge. Hence we view our dataset as an influence network.

We have IRB[2]-approved access to the full, though anonymized, logs of interaction.

3 Discovered Pattern (P1): FIZZLE

Users may have been introduced to Polly by receiving a forwarded message from one of their friends, or simply by "word of mouth". Many such users may in turn call the system, experiment with it, and possibly send messages to their own friends. Most such users cease interacting with the system within a few days. Still, a significant number of users stay with the system for a long time. How does these users' activity change over time?

In the following analysis, we define a user's "system age" as the number of days elapsed after the user successfully sends out first message. Moreover, the *active senders* after n days are defined as the users who actively send out messages on the n-th day after they sent their first message. Figure 2 depicts the FIZZLE pattern: the number of active senders (that is, users that still send messages to their friends) vs. their system age. It also shows the count of messages they sent, as a function of their system age.

[2] Institutional Review Board.

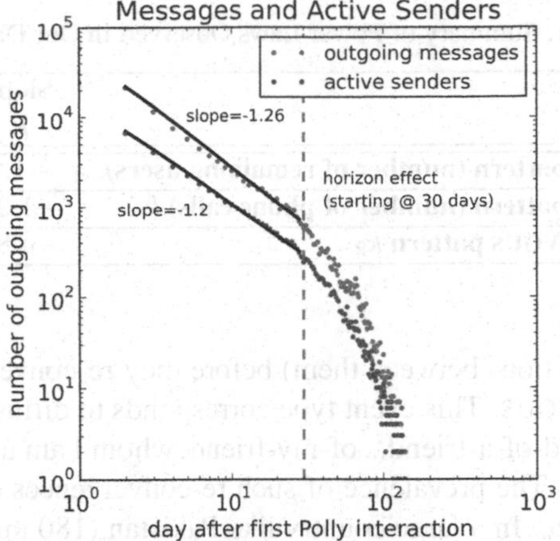

Activity (number of messages (red), remaining users (blue)) vs system age.

Fig. 2. The FIZZLE pattern (P1) - best viewed in color: the number of messages sent (in red) and count of active senders (in blue), versus system age. In both cases, the excitement follows a power law with exponent ≈ -1.2. The horizon effect is explained in Section 3.

Both follow power-law distribution with exponents of -1.2 and -1.26, respectively. This observation agrees with earlier results of the behavior of elapsed time in communication patterns (see [30]): there, Oliveira et al reported shows similar power-law patterns in mail and e-mail correspondences, but with slightly different exponents (1.5 and 1).

Observation 1 (P1). *The number of active senders $c(t)$ at system age t follows*

$$c(t) \propto t^{\alpha} \tag{1}$$

where $\alpha \approx -1.2$. Similarly for the count of messages $m(t)$ at system age t.

Horizon effect: In order to get accurate information about the FIZZLE pattern, new users who are introduced to the system later than 110th day after it was launched were excluded. In this paper, messages delivered within the first 140 days are analyzed. In other words, all the users shown in Figures 2 have passed "system age" of 30 (no matter whether they are still active or not) because they were introduced to the system at least 30 days before the end of our analysis scope. This is exactly the reason for the deviation from power-law "system ages" of 30 and above are unfairly handicapped.

The detailed power-law linear regression results for the FIZZLE pattern, as well as all our upcoming patterns, are listed in Table 1. Notice that they all have extremely high correlation coefficient (absolute value ≥ 0.95).

4 Discovered Pattern (P2): RENDEZVOUS

In a directed network, propagation from one source can take multiple paths to the same destination node. Of particular interest to us are two paths that diverge for a while (with

Table 1. Summary of Power Laws Observed in Our Dataset

Pattern	Slope k	Correlation Coefficient r
P1 The FIZZLE pattern (number of remaining users) k_1	−1.2	−0.994
P1 The FIZZLE pattern (number of phone calls) k_3	−1.26	−0.996
P2 The RENDEZVOUS pattern k_2	−4.88	−0.992

no intermediate connections between them) before they re-converge – an event which we here call RENDEZVOUS. This event type corresponds to diffusion into different social circles (e.g. a-friend-of-a-friend... of-my-friend, whom I am unlikely to know), followed by convergence. The prevalence of such re-convergences can shed light on the effective population size. In a large country like Pakistan (180 million people), the effective population size for our system may vary widely and is unknown a priori. Thus, we are interested in the prevalence of RENDEZVOUS as a function of the shortest path to the most recent common ancestor of two parents of a node, where path length and recency are both measured in terms of number of edges, rather than time. A node with k parents gives rise to $k \cdot (k-1)/2$ different RENDEZVOUS. Taking into account the edge from the common child to its parents, we have the following definition:

Definition 1 (n-RENDEZVOUS:). *An n-*RENDEZVOUS *is defined as a* RENDEZVOUS *where the* shorter *path from the two parents to their most recent ancestor is of length $n-1$.*

For example, Fig. 3(a) shows a RENDEZVOUS of length 2: the shortest leg from the final node Ω to the starting node A, is n=2 hops long.

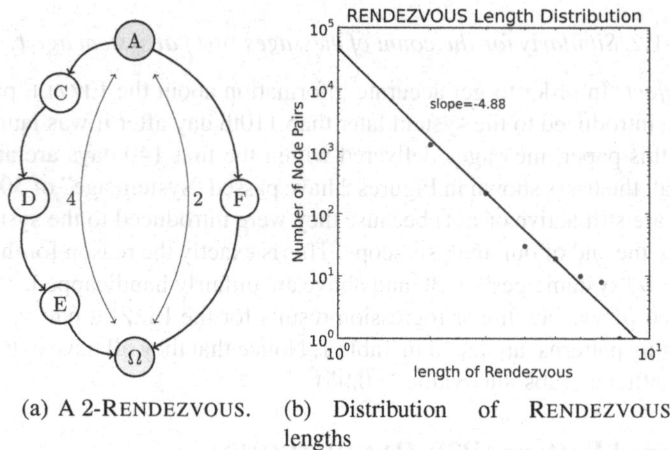

(a) A 2-RENDEZVOUS. (b) Distribution of RENDEZVOUS lengths

Fig. 3. (a) Example of a 2-RENDEZVOUS. (b) Distribution of RENDEZVOUS lengths follows a power law with exponent -4.88.

Figure 3(b) shows that the length distribution of RENDEZVOUS' in our dataset follows a power-law . Most of RENDEZVOUS have a length of 1, meaning that one of the parents *is* the most recent common ancestor, because it has a direct link to the other parent.

Observation 2 (P2). *The number of* RENDEZVOUS' $n(l)$ *at* RENDEZVOUS *length* l *follows*

$$n(l) \propto l^{\beta} \tag{2}$$

where $\beta \approx -4.88$.

5 Discovered Pattern (P3): DISPERSION

Let the "reciprocal activity" between two users be the smaller of the number of messages sent between them in either direction. Let the "activity profile" of a user be $\{m_1, m_2, m_3, ..., m_F\}$ $(m_1 > m_2 > m_3 > ... > m_F > 0)$, where m_i is the reciprocal activity between a user and one of his recipients.

Definition 2 (DISPERSION) *The* DISPERSION D *of a user with activity profile* $\{m_1, m_2, m_3, ..., m_F\}$ *is defined as the entropy* H *of the normalized count distribution:*

$$D(m_1, m_2, ..., m_F) = -\sum_{r=1}^{F} P_r * ln(P_r)$$

Where $P_r = {}^{m_r}/\sum_{k=1}^{F} m_k$.

Therefore, if a user has a high DISPERSION, she sends messages her friends more evenly than other users with the same number of friends, but lower DISPERSION.

Figure 4(a) shows that the real DISPERSION (entropy) is smaller than the "maximum dispersion" where a user sends messages each of her friends evenly. This means that long-term Polly users on average exhibit the DISPERSION pattern when they send messages to their friends.

We can explain the DISPERSION behavior using a closed-form formula, under the assumption that people send messages to their friends following a Zipf's distribution, which implies $P_r \propto 1/r$, to be specific, $P_r \approx 1/(r \times ln(1.78F))$[35, p. 33]. Based on this, we can derive that if we use integral as an approximation of the sum part of the entropy calculation:

Lemma 1. *The entropy* H *of a Zipf's distribution is given by:*

$$H \approx (C \times ln^2(F) + K \times ln(F)lnln(1.78F)), \quad \textit{where } F > 1. \tag{3}$$

The proof is omitted for brevity.

Observation 3 (P3). *Dispersion pattern can be modelled well by Zipf's law in our dataset.*

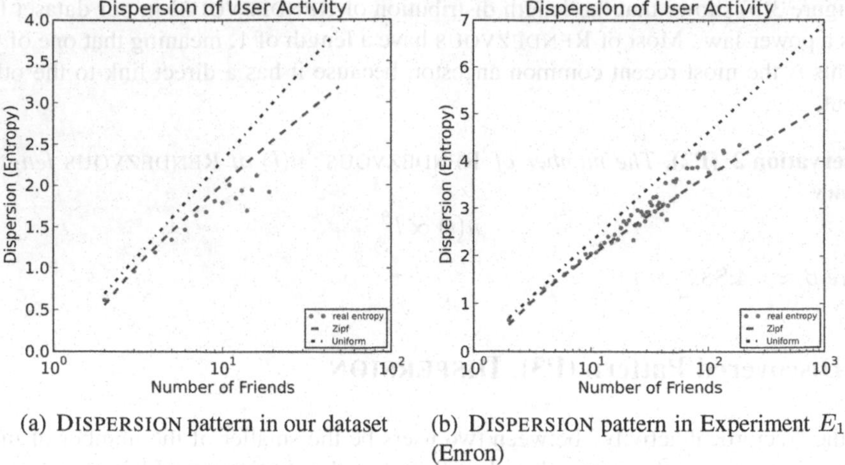

(a) DISPERSION pattern in our dataset (b) DISPERSION pattern in Experiment E_1 (Enron)

Fig. 4. (a) DISPERSION pattern found in the influence network. (b) DISPERSION pattern found in simulated influence network in Experiment E_1, see Section 6.

The mathematical analysis shows that the "friend contact" distribution of a user with F reciprocal friends will follow an expected entropy value proportional to the square of logarithm of F ($ln^2(F)$) other than ($ln(F)$) when we assume the distribution follows uniform distribution. The predicted entropy of Eq (3) matches reality much better than the uniformity assumption. As shown in Figure 4(a), the predicted entropy (the dashed blue curve) is a better match for the real data (red dots), while the uniformity assumption leads to the black-dotted line.

6 INFLOOD GENERATOR: Algorithm and Evaluation

First, we formally define the *base* and *influence* network:

Definition 3 (Base Network). *A Base Network (V_{base}, E_{base}) is the underlying social network of all people who are related to social information cascades. V_{base} is a set of individuals. E_{base} is a set of directed, weighted edges. The weights represent the strength of connections.*

Definition 4 (Influence Network). *An Influence Network (V_{infl}, E_{infl}) shows which node sent a system message to which node, and when. V_{infl} is a set of individuals. E_{infl} is a set of directed, timestamped edges of which the weight shows the number of times a node has been notified of the influence by another node.*

In our model, $V_{infl} \subseteq V_{base}$, $E_{infl} \subseteq E_{base}$, i.e., individuals can only be influenced by others they know.

We model all patterns by using INFLOOD GENERATOR. As mentioned above, Polly can be viewed as an influence network where people are notified of it from their base-network friends. After the notification, people may start forwarding messages.

Why we need a generator The best way to verify all our three patterns (FIZZLE, RENDEZVOUS and DISPERSION) is to study **other** influence network datasets. However, they are difficult to obtain, and some of them lack time stamp information (See Sec 1).

Details of INFLOOD GENERATOR The pseudo code of our influence-network generator, is given in Algorithm 1. In more details, on day 0, s_0 seed nodes ($s_0 = 5$) in the social network $G1$ (e.g., Facebook or Enron) are notified of Polly. Then, each following day t, every person u who has been notified of Polly has a probability, $P_{u,t}$, of calling some of her friends via Polly. The friends are sampled from the outgoing edges from $G1$ *independent of* no matter whether they have been already notified.

```
Input: G1 = A base network, T = simulated days, α = power-law decay factor, P_{u,0}:
       first-day infection probability of user u, day = 0.
Output: An influence network G2 on top of G1
begin
    notifiedUsers ←—(5 randomly chosen people in G1 with notified day t_u = 0);
    day ←— 0;
    while day < T do
        for each user u in notifiedUsers do
            coin ←— TAIL;
            if random() < P_{u,t} in Equation 4 then
                coin ←— HEAD;
            while coin == HEAD do
                f = a friend sampled from G1;
                if f is not in notifiedUsers then
                    add f into notifiedUsers with t_u = day
                else
                    //User u has sent a message to user f, Record this interaction.
                    //Details are omitted for simplicity.
                if random() >= P_{u,t} then
                    coin ←— TAIL;
        day ←— day + 1;
```

Algorithm 1. Pseudo Code of INFLOOD GENERATOR

The probability of a user u telling other people about Polly is given by

$$P_{u,t} = P_{u,0} \times (day - t_u)^{-1.0\times\alpha}, day > t_u \tag{4}$$

, where α parameter affects the exponent of the power-law that governs the decay of infection activities, and $P_{u,0}$ is defined the *first-day infection probability*.

Estimating $P_{u,0}$ $P_{u,0}$ depends on the total weight (number of communication messages) a user (node) has in base network $G1$. In fact, $P_{u,0}$ is set so that, in expectation, the number of edges user u makes on the first day $msg_{u,0}$ is proportional to the user's total weight of out-going connection strength in base network $G1$, i.e.

$$msg_{u,0} = c \times \sum_{v \in V_{G1}} w_{u,v} \tag{5}$$

Note that in Algorithm 1, the process of sending messages is a geometric distribution. Hence we can get $msg_{u,0} = 1/(1-P_{u,0})$ in expectation.

Hence, we set the $P_{u,0}$ to be:

$$P_{u,0} = 1 - \frac{1}{c \times \sum_{v \in V_{G1}} w_{u,v}} \tag{6}$$

This ensures that a high weight node has more simulated edges. It is also more realistic, because more social people may spread messages more easily. In our setting, $c = \frac{1}{4}$, i.e. a user will contact a quarter of her $G1$ out-degrees in expectation on the first day. This formula is based on experimental observations.

Evaluation of INFLOOD GENERATOR We tested INFLOOD GENERATOR in a number of networks. Here we use communication networks, such as Facebook, to be approximations of "real" base network. The results are presented in Table 2.

Table 2. Results of INFLOOD Simulations

Experiment	Base Network $G1$	$G2\ \lvert V\rvert$	$G2\ \lvert E\rvert$	FIZZLE slope k_1	RENDEZVOUS slope k_2
Polly	N/A	72,341	173,710	−1.2	−4.88
E_1	Enron [22,1]	19,829	227,659	−1.16	−8.39
E_2	Slashdot [3,17]	6,880	19,781	−1.18	−6.11
E_3	Facebook [2]	22,029	222,686	−1.16	−7.65

In all experiments, $\alpha = 1.17$, and the number of simulated days is $T = 140$.

In all cases, the correlation coefficients $|r|$ were high ($|r| > 0.93$). The FIZZLE slopes k_1 are calculated based only on the first 30 days of interactions of each user, exactly as we did for the real, Polly dataset. Recall that k_1 is the slope of the FIZZLE pattern, that is, the slope of the number of remaining active users, over time, in log-log scales. For the RENDEZVOUS pattern, the k_2 slope varies between experiments. This may be due to the small count of data points, see Fig 3(b). We also tested the INFLOOD GENERATOR on synthetic datasets, such as Erdös-Rényi graphs of various parameter settings. Notice that the RENDEZVOUS pattern is violated: the Erdös-Rényi graphs do *not* follow a power-law in their RENDEZVOUS plots.

Because the INFLOOD GENERATOR graph is big, we observe the DISPERSION pattern in Experiment E_1. Figure 4(b) shows that the entropy footprint grows well with the Zipf's distribution curve for users who have less than 50 friends. When the number of friends goes beyond 50, the entropy footprints seem less regular as the number of samples decreases.

Again, the INFLOOD GENERATOR code is open source, see Sec 1.

7 Related Work

Static graph patterns. These include the legendary 'six-degrees of separation' [29]; the skewed degree distribution [13], specially for telephone graphs [5]; the power law tails in connected components distributions; the power law PageRank distributions and bimodal radius plots [20]; the super-linearity rules [28], triangle patterns [39,19]. This list is by no means exhaustive; see [10] for more patterns. Algorithms for detecting these patterns have been proposed by multiple research teams, such as [18].

Temporal and influence patterns. Work on this topic encompasses the shrinking diameter and densification [25]; the power law for the mail response times of Einstein and Darwin, [30]; analysis of blog dynamics [16,26], and discovery of core-periphery patterns in blogs and news articles [15]; viral marketing [23,21]; meme tracking [24]; reciprocity analysis [14,6]; analysis of the role of weak and strong ties in information diffusion in mobile networks [31]; identification of important influencers [36]; prediction of service adoption in mobile communication networks [37]; information or cascade diffusion in social networks [9,4,8,38]; linguistic change in online forums, and predicting the user's lifespan based on her linguistic patterns [11]; peer and authority pressure in information propagation [7].

However, none of the above works reports anything similar to our discoveries, the RENDEZVOUS and the DISPERSION patterns.

8 Conclusions

We study a large, real influence network induced by the Polly system, with over 70,000 users (nodes), 170,000 interactions (edges), distilled from 500MB of log data and 200GB of audio data. Polly is a free, telephone-based, voice message application that has been deployed and used in the real world. Our contributions are as follows:

1. **Discovery** of new patterns in Polly:
   - P1: the 'enthusiasm' drops as a power law with time.
   - P2: The RENDEZVOUS pattern shows a power-law distribution.
   - P3: The DISPERSION pattern of users behaves like a Zipf distribution;
2. **Generator and Analysis**:
   - We propose the INFLOOD GENERATOR algorithm, which matches the observed patterns (P1, P2 and P3) in various communication networks. The code is open-sourced at `https://github.com/yibinlin/inflood_generator/`.
   - We give the derivation for the observed DISPERSION pattern

With respect to future work, a fascinating research direction is to estimate the underlying population size of our dataset, from the statistics of the RENDEZVOUS pattern.

Acknowledgements. The authors would like to thank Naoki Orii in CMU for suggestions and proof-reading.

This material is based upon work supported by the National Science Foundation under Grant No. IIS-1217559. Funding was also provided by the U.S. Army Research Office (ARO) and Defense Advanced Research Projects Agency (DARPA) under Contract Numbers W911NF-11-C-0088 and W911NF-09-2-0053. The content of the information in this document does not necessarily reflect the position or the policy of the

Government, and no official endorsement should be inferred. The U.S. Government is authorized to reproduce and distribute reprints for Government purposes notwithstanding any copyright notation here on.

References

1. Berkeley enron email analysis (2013)
2. Facebook wall posts network dataset - konect (August 2013)
3. Slashdot threads network dataset - konect (August 2013)
4. Agrawal, D., Budak, C., El Abbadi, A.: Information diffusion in social networks: Observing and influencing societal interests. PVLDB 4(12), 1512–1513 (2011)
5. Aiello, W., Chung, F., Lu, L.: A random graph model for massive graphs. In: STOC, pp. 171–180. ACM, New York (2000)
6. Akoglu, L., Vaz de Melo, P.O.S., Faloutsos, C.: Quantifying reciprocity in large weighted communication networks. In: Tan, P.-N., Chawla, S., Ho, C.K., Bailey, J. (eds.) PAKDD 2012, Part II. LNCS, vol. 7302, pp. 85–96. Springer, Heidelberg (2012)
7. Anagnostopoulos, A., Brova, G., Terzi, E.: Peer and authority pressure in information-propagation models. In: Gunopulos, D., Hofmann, T., Malerba, D., Vazirgiannis, M. (eds.) ECML PKDD 2011, Part I. LNCS, vol. 6911, pp. 76–91. Springer, Heidelberg (2011)
8. Barbieri, N., Bonchi, F., Manco, G.: Cascade-based community detection. In: WSDM, pp. 33–42 (2013)
9. Budak, C., Agrawal, D., El Abbadi, A.: Diffusion of information in social networks: Is it all local? In: ICDM, pp. 121–130 (2012)
10. Chakrabarti, D., Faloutsos, C.: Graph Mining: Laws, Tools, and Case Studies. Morgan Claypool (2012)
11. Danescu-Niculescu-Mizil, C., West, R., Jurafsky, D., Leskovec, J., Potts, C.: No country for old members: User lifecycle and linguistic change in online communities. In: WWW. ACM, New York (2013)
12. Erdös, P., Rényi, A.: On the evolution of random graphs. Publication 5, pp. 17–61, Institute of Mathematics, Hungarian Academy of Sciences, Hungary (1960)
13. Faloutsos, M., Faloutsos, P., Faloutsos, C.: On power-law relationships of the internet topology. In: SIGCOMM, pp. 251–262 (August-September 1999)
14. Garlaschelli, D., Loffredo, M.I.: Patterns of Link Reciprocity in Directed Networks. Phys. Rev. Lett. 93, 268701 (2004)
15. Rodriguez, M.G., Leskovec, J., Krause, A.: Inferring networks of diffusion and influence. In: KDD, pp. 1019–1028. ACM, New York (2010)
16. Gruhl, D., Guha, R.V., Liben-Nowell, D., Tomkins, A.: Information diffusion through blogspace. In: WWW Conference, New York, NY, pp. 491–501 (May 2004)
17. Gómez, V., Kaltenbrunner, A., López, V.: Statistical analysis of the social network and discussion threads in Slashdot. In: Proc. Int. World Wide Web Conf., pp. 645–654 (2008)
18. Jiang, D., Pei, J.: Mining frequent cross-graph quasi-cliques. ACM TKDD 2(4), 16:1–16:42 (2009)
19. Kang, U., Meeder, B., Faloutsos, C.: Spectral analysis for billion-scale graphs: Discoveries and implementation. In: Huang, J.Z., Cao, L., Srivastava, J. (eds.) PAKDD 2011, Part II. LNCS, vol. 6635, pp. 13–25. Springer, Heidelberg (2011)
20. Kang, U., Tsourakakis, C.E., Faloutsos, C.: Pegasus: mining peta-scale graphs. Knowl. Inf. Sys. 27(2), 303–325 (2011)
21. Kempe, D., Kleinberg, J., Tardos, É.: Maximizing the spread of influence through a social network. In: KDD, pp. 137–146. ACM, New York (2003)

22. Klimt, B., Yang, Y.: The enron corpus: A new dataset for email classification research. In: Boulicaut, J.-F., Esposito, F., Giannotti, F., Pedreschi, D. (eds.) ECML 2004. LNCS (LNAI), vol. 3201, pp. 217–226. Springer, Heidelberg (2004)
23. Leskovec, J., Adamic, L.A., Huberman, B.A.: The dynamics of viral marketing. TWEB 1(1) (2007)
24. Leskovec, J., Backstrom, L., Kleinberg, J.M.: Meme-tracking and the dynamics of the news cycle. In: KDD, pp. 497–506 (2009)
25. Leskovec, J., Kleinberg, J.M., Faloutsos, C.: Graphs over time: densification laws, shrinking diameters and possible explanations. In: KDD, pp. 177–187 (2005)
26. Leskovec, J., McGlohon, M., Faloutsos, C., Glance, N.S., Hurst, M.: Patterns of cascading behavior in large blog graphs. In: SDM (2007)
27. Leskovec, J., Singh, A., Kleinberg, J.: Patterns of influence in a recommendation network. In: Ng, W.-K., Kitsuregawa, M., Li, J., Chang, K. (eds.) PAKDD 2006. LNCS (LNAI), vol. 3918, pp. 380–389. Springer, Heidelberg (2006)
28. McGlohon, M., Akoglu, L., Faloutsos, C.: Weighted graphs and disconnected components: patterns and a generator. In: KDD, pp. 524–532 (2008)
29. Milgram, S.: The small world problem. Psychology Today 2, 60–67 (1967)
30. Oliveira, J.G., Barabási, A.-L.: Human dynamics: Darwin and Einstein correspondence patterns. Nature 437(7063), 1251 (2005)
31. Onnela, J.-P., Saramäki, J., Hyvönen, J., Szabó, G., Lazer, D., Kaski, K., Kertész, J., Barabási, A.-L.: Structure and tie strengths in mobile communication networks. Proc. Natl. Acad. Sci. USA 104(18), 7332–7336 (2007)
32. Raza, A.A., Haq, F.U., Tariq, Z., Razaq, S., Saif, U., Rosenfeld, R.: Job opportunities through entertainment: Virally spread speech-based services for low-literate users. In: SIGCHI, Paris, France, pp. 2803–2812. ACM (2013)
33. Raza, A.A., Haq, F.U., Tariq, Z., Saif, U., Rosenfeld, R.: Spread and sustainability: The geography and economics of speech-based services. In: DEV (2013)
34. Raza, A.A., Milo, C., Alster, G., Sherwani, J., Pervaiz, M., Razaq, S., Saif, U., Rosenfeld, R.: Viral entertainment as a vehicle for disseminating speech-based services to low-literate users. In: ICTD, vol. 2 (2012)
35. Schroeder, M.: Fractals, Chaos, Power Laws: Minutes from an Infinite Paradise. Henry Holt and Company (1992)
36. Subbian, K., Sharma, D., Wen, Z., Srivastava, J.: Social capital: the power of influencers in networks. In: AAMAS, pp. 1243–1244 (2013)
37. Szabo, G., Barabasi, A.: Network effects in service usage. ArXiv Physics e-prints (November 2006)
38. Tang, J., Sun, J., Wang, C., Yang, Z.: Social influence analysis in large-scale networks. In: KDD, pp. 807–816. ACM (2009)
39. Tsourakakis, C.E.: Fast counting of triangles in large real networks without counting: Algorithms and laws. In: ICDM (2008)

Overlapping Communities for Identifying Misbehavior in Network Communications

Farnaz Moradi, Tomas Olovsson, and Philippas Tsigas

Chalmers University of Technology, Göteborg, Sweden
{moradi,tomasol,tsigas}@chalmers.se

Abstract. In this paper, we study the problem of identifying misbehaving network communications using community detection algorithms. Recently, it was shown that identifying the communications that do not respect community boundaries is a promising approach for network intrusion detection. However, it was also shown that traditional community detection algorithms are not suitable for this purpose.

In this paper, we propose a novel method for enhancing community detection algorithms, and show that contrary to previous work, they provide a good basis for network misbehavior detection. This enhancement extends disjoint communities identified by these algorithms with a layer of auxiliary communities, so that the boundary nodes can belong to several communities. Although non-misbehaving nodes can naturally be in more than one community, we show that the majority of misbehaving nodes belong to multiple overlapping communities, therefore overlapping community detection algorithms can also be deployed for intrusion detection.

Finally, we present a framework for anomaly detection which uses community detection as its basis. The framework allows incorporation of application-specific filters to reduce the false positives induced by community detection algorithms. Our framework is validated using large *email networks* and *flow graphs* created from real network traffic.

1 Introduction

Network intrusion detection systems are widely used for identifying anomalies in network traffic. Anomalies are patterns in network traffic that do not conform to normal behavior. Any change in the network usage behavior, for example caused by malicious activities such as DoS attacks, port scanning, unsolicited traffic, and worm outbreaks, can be seen as anomalies in the traffic.

Recently, it was shown that network intrusions can successfully be detected by examining the network communications that do not respect the community boundaries [9]. In such an approach, normality is defined with respect to social behavior of nodes concerning the communities to which they belong and intrusion is defined as "*entering* communities to which one does not belong".

A community is typically referred to as a group of nodes that are densely interconnected and have fewer connections with the rest of the network. However,

V.S. Tseng et al. (Eds.): PAKDD 2014, Part I, LNAI 8443, pp. 398–409, 2014.

there is no consensus on a single definition for a community and a variety of definitions have been used in the literature [13,19,28]. For network intrusion detection, Ding et al. [9] defined a community as a group of source nodes that communicate with at least one common destination. They also showed that a traditional community detection algorithm which is based on a widely used definition, i.e., modularity, is not useful for identifying intruding nodes.

In this paper, we extend and complement the work of Ding et al. [9] by looking into other definitions for communities, and investigate whether the communities identified by different types of algorithms can be used as the basis for anomaly detection. Our hypothesis is that misbehaving nodes tend to *belong to multiple communities*. However, a vast variety of community detection algorithms partition network nodes into disjoint communities where each node only belongs to a single community, therefore they cannot be directly used for verifying our hypothesis. Therefore, we propose a simple novel method which enhances these disjoint communities with a layer of *auxiliary communities*. An auxiliary community is formed over the boundary nodes of neighboring communities, allowing nodes to be members of several communities. This enhancement enables us to show that, in contrary to [9], it is possible to use traditional community detection algorithms for identifying anomalies in network traffic.

In addition to traditional community detection algorithms, another class of algorithms exist which allow a node to belong to several overlapping communities [26]. In this study, we compare a number of such *overlapping algorithms* with our proposed enhancement method for non-overlapping community detection algorithms for network anomaly detection.

Finally, we propose a framework for network misbehavior detection. The framework allows us to incorporate different community detection algorithms for identifying anomalous nodes that belong to multiple communities. However, since legitimate nodes can also belong to several communities [28], application-specific filters can be used for discriminating the legitimate nodes from the anti-social nodes in the community overlaps, thus reducing the induced false positives.

We have evaluated the framework by using it for network intrusion detection and unsolicited email detection in large-scale datasets collected from a high-speed Internet backbone link. These types of misbehavior have traditionally been very hard to detect without inspecting the content of the traffic. To conclude, we show that by using our methodology, it is possible to effectively detect misbehaving traffic by only looking at the network communication patterns.

The remainder of the paper is organized as follows. Section 2 presents related work. Section 3 presents our proposed method for uncovering community overlaps. The framework is presented in Section 4. Section 5 summarizes our findings and experimental results. Finally, Section 6 concludes our work.

2 Related Work

Anomaly detection has been extensively studied in the context of different application domains [6]. In this study, we propose a new graph-based anomaly detection method for identifying network intrusion and unsolicited email in real network traffic. Although there has been considerable amount of research on detecting these types of misbehavior, it is still a challenge to identify anomalies by merely investigating communication patterns without inspecting their content.

A taxonomy of graph-based anomaly detection methods can be found in [2]. A number of previous studies have proposed methods for finding unusual subgraphs, anomalous substructure patterns, and outlier nodes inside communities in labeled graphs [11,21,14]. In this study, we merely use the graph structure and therefore we consider only plain graphs without any labels.

Akoglu et al. [3] proposed a method to assign anomaly scores to nodes based on *egonet* properties in weighted networks. Our framework allows us to incorporate such properties as application-specific filters. Sun et al. [25] proposed a method for identifying anomalous nodes that are connected to irrelevant neighborhoods in bipartite graphs. Ding et al. [9] showed that although finding the cut-vertices can be used for intrusion detection, more robust results can be achieved by using clustering coefficient in a one-mode projection of a bipartite network. Moreover, they showed that using a modularity maximization community detection algorithm [7] is not suitable for spotting network intruders.

In this paper, we revisit the problem of finding anomalous nodes in bipartite/unipartite plain graphs by using community detection algorithms. We deploy an alternative definition for an anomaly as suggested in [9] and confirm their finding that maximizing modularity is not suitable for identifying intruders on its own. However, we show that there are several types of algorithms which are useful for misbehavior detection if enhanced with auxiliary communities.

3 Community Detection

In this section, we introduce a novel approach which enables us to deploy existing community detection algorithms for identifying anomalies in network traffic.

3.1 Auxiliary Communities

In this paper, we introduce the concept of auxiliary communities. An auxiliary community is added over the boundary nodes of disjoint communities, forcing nodes to become members of more than one community.

The most basic approach is to introduce one auxiliary community for each *boundary edge* between two different communities. However, a *boundary node* can have multiple boundary edges. Therefore, an improvement over the above approach is to add only one auxiliary community over a boundary node and all its boundary edges, covering all its neighbors that are members of other external communities (Algorithm 1). Our approach can be further refined to consider the whole one-step neighborhood, i.e., *egonet*, of a boundary node as an auxiliary community instead of just its boundary neighbors.

Ding et al. [9] defined a community in a directed bipartite network as a group of source nodes that have communicated with at least one common destination.

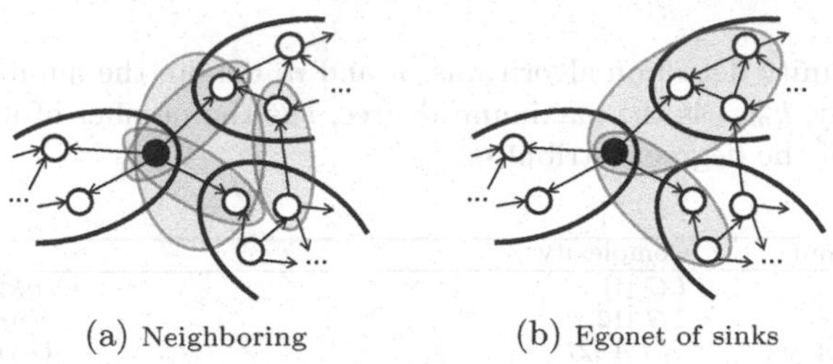

(a) Neighboring (b) Egonet of sinks

Fig. 1. Auxiliary communities

In a bipartite network, there are two distinct sets of source nodes and destination nodes. Based on this definition, the source nodes that belong to the egonet of a destination node form a community. In a unipartite network, a distinct set of source and destination nodes does not exist. Therefore, we apply the above definition of communities only to the sink nodes which have only incoming edges (Algorithm 2).

Algorithm 1. Neighboring Auxiliary Communities (NA)

Input: a graph $G(V, E)$; a non-overlapping community set $\mathcal{C}$;
Output: auxiliary community set $\mathcal{A}$;

```
for all v ∈ V do
    Com(v) = {C ∈ 𝒞 : v ∈ V(C)};
    for all u ∈ Neighbors(v) do
        if Com(v) ≠ Com(u) then
            A ← A ∪ {u, v};
        end if
    end for
    𝒜 ← 𝒜 ∪ A;
end for
return 𝒜
```

Algorithm 2. Egonet Auxiliary Communities of Sinks (EA)

Input: a graph $G(V, E)$; a non-overlapping community set $\mathcal{C}$;
Output: auxiliary community set $\mathcal{A}$;

```
for all v ∈ V do
    Com(v) = {C ∈ 𝒞 : v ∈ V(C)};
    for all u ∈ Neighbors(v) do
        if Com(v) ≠ Com(u) and Sink(u) then
            A ← Egonet(u);
            𝒜 ← 𝒜 ∪ A;
        end if
    end for
end for
return 𝒜
```

Figure 1 shows a comparison of the proposed methods for adding auxiliary communities. It can be seen that each approach places the intruding node (black node) in different auxiliary communities (grey communities). The main difference of our methods is that Algorithm 1 only adds neighboring auxiliary (NA) communities over the boundary nodes, whereas Algorithm 2 also allows the neighbors of the boundary sink nodes to be covered by egonet auxiliary (EA) communities. Therefore, a misbehaving node which is not in the boundary of its community can still belong to multiple communities by using Algorithm 2.

The complexity of adding auxiliary communities for a network with a degree distribution $p_k = k^{-\alpha}$, is $O(nk_{max}^{3-\alpha})$, where n is the number of nodes, k_{max} is the highest degree, and α is the exponent of the degree distribution.

Table 1. Community detection algorithms. n and m denote the number of nodes and edges, respectively, k_{max} is the maximum degree, t is the number of iterations, and α is the exponent of the degree distribution.

Algorithm	Complexity	
Overlapping	*LC* [1]	$O(nk_{max}^2)$
	LG [12]	$O(nm^2)$
	SLPA [27]	$O(tm)$
	OSLOM [18]	$O(n^2)$
	DEMON [8]	$O(nk_{max}^{3-\alpha})$
Non-Overlapping	*Blondel* (also known as Louvain method) [5]	$O(m)$
	Infomap [22]	$O(m)$
Auxiliary	*NA* (Neighboring Auxiliary Communities)	$O(nk_{max}^{3-\alpha})$
	EA (Egonet Auxiliary Communities)	$O(nk_{max}^{3-\alpha})$

3.2 Community Detection Algorithms

In this paper, we use a number of well-known and computationally efficient (overlapping) community detection algorithms, which are listed in Table 1. Our goal is to investigate which definition of a community and which types of algorithms are most suitable for network misbehavior detection.

LC and LG find overlapping communities in a graph based on the edges. LG, induces a *line graph* from the original network to which any non-overlapping algorithm can be applied. In this paper, we uses a weighted line graph with self-loops, E, and refer to LG using this graph as *LG(E)*. SLPA and OSLOM are both node-based methods and have very good performance [26]. Finally, DEMON is an state-of-the-art node-based, local, overlapping community detection algorithm.

The non-overlapping algorithms used in this study also have very good performance [17]. Blondel greedily maximizes modularity and unfolds a hierarchical community structure with increasing coarseness. In this study, we consider the communities identified at both the last and the first level of the hierarchy and refer to them as *Blondel* and *Blondel L1*, respectively. We also use the communities formed by Blondel as input to OSLOM, which modifies these communities in order to improve their statistical significance. Finally, Blondel L1 is also used to partition the nodes in the induced line graphs by LG(E).

4 Framework

This section presents our framework for community-based anomaly detection. Algorithm 3 shows the first component of our framework, where overlapping algorithms can be directly used, but non-overlapping algorithms only after being enhanced with auxiliary communities.

The second component of our framework consists of a set of graph properties which are used as *filters*. Our hypothesis is that intruding nodes are likely to be placed in community overlaps. However, non-misbehaving nodes can also belong to more than one community, and basing detection merely on community overlaps, can lead to false positives. Therefore, these filters are used to reduce the induced false positives by the community detection algorithms.

Algorithm 3. Community-based anomaly detection

Input: a graph $G(V, E)$; a community detection algorithm CD;
Output: a set AS of $\langle v, score(v)\rangle$;
Set $AS = \emptyset$; Set $\mathcal{C} = \emptyset$; Set $\mathcal{A} = \emptyset$;
$\mathcal{C} = CD(G)$;
if CD is non-overlapping **then**
  $\mathcal{A} \leftarrow Auxiliary(G, \mathcal{C})$;
  $\mathcal{C} \leftarrow \mathcal{C} \cup \mathcal{A}$;
end if
for all $v \in V$ **do**
  $score(v) \leftarrow Filters(v, G, \mathcal{C})$;
  $AS \leftarrow \langle v, score(v)\rangle$;
end for
return AS

Algorithm 4. Application-specific filters

Input: a node v; a graph $G(V, E)$; a set of communities $\mathcal{C}$; weights $w_i \in [0, 1]$ s.t. $\sum w_i = 1$; user-defined threshold values t_i, where i is the index of the property;
Output: an anomaly score $score(v)$;
$Coms(v) = \{C \in \mathcal{C} : v \in V(C)\}$;
$\phi_1(v) = |Coms(v)|$;
$\phi_2(v) = |Coms(v)|/|Neighbors(v)|$;
$\phi_3(v) = 1 - ClusteringCoeff(v)$;
$\phi_4(v) = OutDeg(v)/Deg(v)$;
$\phi_5(v) = Deg(v)/EdgeWeights(v)$;
$score(v) = \sum w_i \mathcal{I}(\phi_i(v), t_i)$;
return $score(v)$

The framework uses a simple method for combining the extracted properties. For each node v in the graph, the anomaly score is calculated as $score(v) = \sum_i w_i \mathcal{I}(\phi_i(v), t_i)$, where i is the index of the property which is being aggregated, w_i is a weight for property ϕ_i where $\sum w_i = 1$, and $\mathcal{I}(\phi_i(v), t_i)$ is an indicator function which compares the value of a graph property $\phi_i(v)$ to a corresponding threshold value t_i such that $\mathcal{I}(\phi_i(v), t_i) = \begin{cases} 1, & \phi_i(v) > t_i \\ 0, & \text{otherwise.} \end{cases}$ The threshold values and weights are dependent on the type of data and prior knowledge of normal behavior, which is necessary for anomaly detection and can be achieved from studies of anomaly-free data. Finally, the anomaly score $score(v)$ can be used to quantify to what extent a node v is anomalous.

The properties presented in Algorithm 4 are examples of community and neighborhood properties that we have used as filters in our experiments for intrusion and unsolicited email detection. The selection of appropriate filters depends on the application of anomaly detection.

Network intruders are normally not aware of the community structure of the network, and therefore communicate to random nodes in the network [23]. It is expected to be very expensive for attackers to identify the network communities, and even if they do, limiting their communication with the members in the same community can inversely affect their gain. Therefore, the number of communities per node, as well as the ratio of the number of communities per node over the number of its neighbors, which correspond to ϕ_1 and ϕ_2 in Algorithm 4, respectively, are expected to be promising properties for finding intruders.

The rest of the properties, are graph metrics that correspond to the social behavior of nodes and can be extracted from the direct neighborhood of the nodes. We have used these properties for detecting unsolicited email (Section 5.3) and therefore in the following we explain them in the context of spam detection.

The *clustering coefficient* of a node is known to have a lower value for spammers than legitimate nodes [15,20]. Property ϕ_3 calculates one minus the clustering coefficient so that spammers are assigned higher values. It has also been shown that spammers are mostly using randomized fake source addresses and therefore it is not expected that they receive many emails [16]. Property ϕ_4 calculates the ratio of the out-degree over the degree of the nodes, which is expected to be high for the spammers. Finally, it has been shown that spammers tend to use the fake source email address to send only a few spam, and target each receiving email address only once [16]. Therefore, the degree of a node over its edge weights, property ϕ_5, is expected to be higher for spammers than legitimate nodes, where the edge weights correspond to the number of exchanged emails.

5 Experimental Results

We have evaluated the usability of different algorithms in our framework using two different datasets which were generated from network traffic collected on a 10 Gbps Internet backbone link of a large national university network.

Flow Dataset. The flow level data was collected from the incoming network traffic once a week during 24 hours for seven weeks in 2010 [4]. The flows were used to generate bipartite networks where source and destination IP addresses form the two node sets. The malicious source addresses in the dataset were taken from the lists reported by DShield and SRI during the data collection period [10,24]. This dataset is used to compare our approach with the method proposed by Ding et al. [9] for network intrusion detection. The datasets are similar with respect to the ground truth and only differ with respect to the collection location and the sampling method used.

Email Dataset. This dataset is generated from captured SMTP packets in both directions of the backbone link. The collection was performed twice (2010 and 2011), where the duration of each collection was 14 consecutive days. This dataset was used for generating *email networks*, in which email addresses represent the nodes, and the exchanged emails represent the edges. The ground truth was obtained from a well-trained content-based filtering tool[1] which classified each email as legitimate (*ham*) or unsolicited (*spam*).

5.1 Comparison of Algorithms

In this section, we present a comparison of the algorithms using the email dataset. Figure 2a shows the percentage of ham and spam nodes (averaged over the 14 days in 2010), which are placed in multiple communities by different algorithms. It can be seen that many ham nodes belong to more than one community,

[1] SpamAssassin (`http://spamassassin.apache.org`) which provided us with an estimated false positive rate of less than 0.1% and a detection rate of 91.4%.

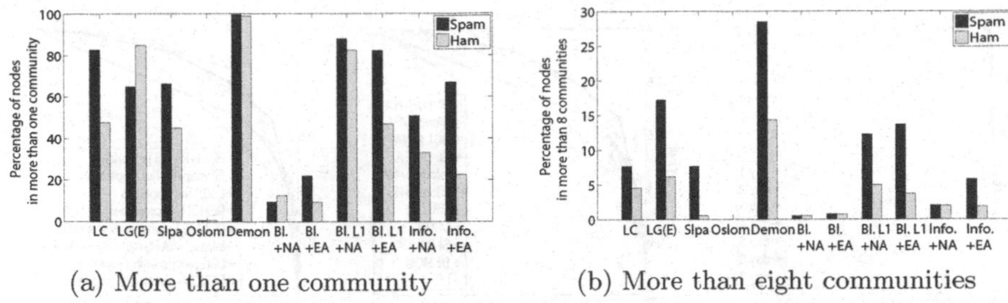

(a) More than one community (b) More than eight communities

Fig. 2. Percentage of nodes in multiple communities in email dataset (2010)

which is an expected social behavior. It can also be seen that, most algorithms place the majority of spammers into more than one community, except OSLOM and Blondel which tend to form very coarse-grained communities.

The figure also shows that, regardless of which non-overlapping algorithm being used, adding egonet auxiliary communities (Algorithm 2) places more spam than ham nodes into several communities compared to adding neighboring auxiliary communities (Algorithm 1). The reason is that NA communities are only added over the boundary nodes, however, EA communities also allow the neighbors of the boundary sink nodes to be covered by auxiliary communities.

Finally, Figure 2b shows that a higher percentage of spammers belong to more than eight communities compared to legitimate nodes. The same observation holds for the data collected in 2011. Therefore, we can confirm that both fine-grained algorithms enhanced with EA communities, and overlapping algorithms can be used to spot misbehaving nodes based on the number communities to which they belong.

5.2 Network Intrusion Detection

It has been shown that a non-overlapping community detection algorithm (which maximizes modularity) is not suitable for identifying intruders in network flow data [9]. In this study, we have further investigated the possibility of using different community detection algorithms, including a modularity-based one, by using auxiliary communities for network intrusion detection.

One example of network intrusion is port scanning, where a scanner searches for open/vulnerable services on selected hosts. Current intrusion detection systems are quite successful in identifying scanners. In this paper, we just verify the possibility of detecting scanners using a community-based technique.

We generated one bipartite graph from the flows collected for each day. As an example, the flow graph generated from the first day of data contained 51,720 source nodes sending 93,113 flows to 32,855 destination nodes. This includes 607 malicious nodes (based on DShield/SRI reports) that have sent 7,861 flows. We made the assumption that the malicious source nodes that have tried to communicate with more than 50 distinct destinations are suspected of scanning. Figure 3a shows the ROC curves for seven different days. These curves show the

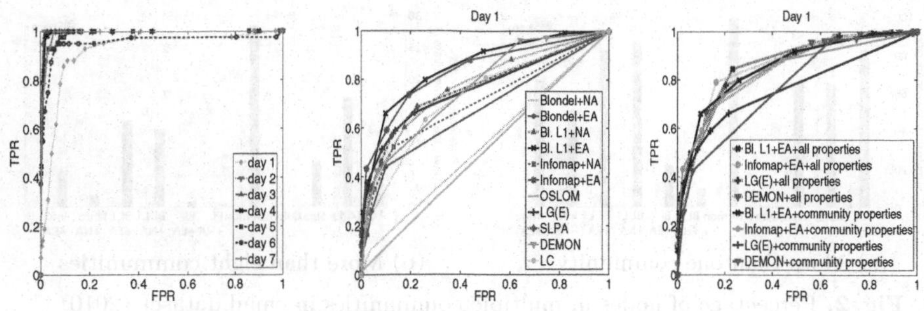

(a) Scanner detection using community properties (2010) (b) Spam detection using community properties (2010) (c) Spam detection using neighbor and community properties (2010)

Fig. 3. Performance of different algorithms for network misbehavior detection

trade-off between the true positive rate (TPR) and the false positive rate (FPR). We have used Blondel L1 enhanced with egonet auxiliary communities (EA), and have only used property ϕ_1, i.e., the number of communities to which a node belongs, as the filter. It can be seen that this approach yields high performance with mean area under curve (AUC) of 0.98, where around 90% to 100% of malicious scanners are detected with a FPR of less than 0.05. This observation confirms that our framework is successful in identifying scanners.

Network intrusion attacks are not limited to scanning attacks, therefore we have also tried to identify other malicious (DShield/SRI) sources and have compared our approach with the method proposed by Ding et al. [9]. Our experiments show that the performance of both methods are quite consistent with mean AUC 0.60 (standard error 0.009) for the method by Ding et al. and 0.62 (standard error 0.015) for our approach using LG(E) as the overlapping community detection and properties ϕ_1 and ϕ_2 as filters. Overall, these results confirm that the community structure of a network provides a good basis for network intrusion detection and both non-overlapping communities enhanced with EA communities and overlapping communities can indeed be used for this purpose.

5.3 Unsolicited Email Detection

Our experimental comparison of community detection algorithms in Section 5.1 showed that most of the studied algorithms place spammers into multiple communities. In this section, we investigate how these algorithms can be used in our framework to detect these spammers only by observing communication patterns.

For this study, we have generated one email network from the emails collected for each day. The community detection algorithms were applied to the undirected and unweighted giant connected component of each email network. The edge directions and weights were later taken into account for adding auxiliary communities and calculating different graph properties. We consider an email address to be a spammer if it has sent more than one spam to more than one recipient. As an example, the email network generated from the first day

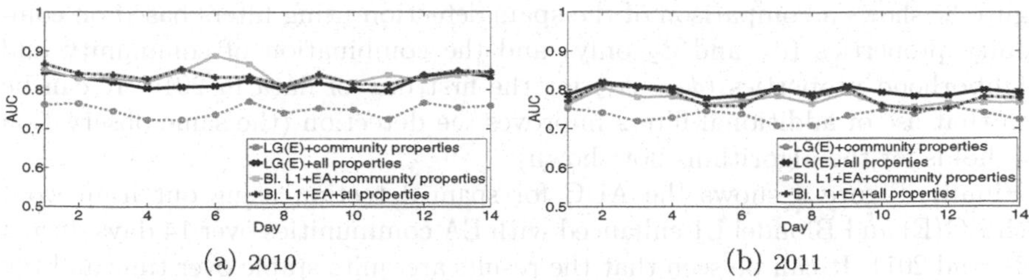

(a) 2010 (b) 2011

Fig. 4. Area under the ROC curve for spam detection over time

of data in 2010, contains 167,329 nodes and 236,673 edges, where 23,628 nodes were spammers sending 126,145 spam emails. It is important to note that the vast majority of the spammers have not sent large volumes of email and therefore a simple volume-based detection method would not be suitable for spammer detection.

Figure 3b shows the ROC curves for our spam detection method using different algorithms and the community-based properties ϕ_1 and ϕ_2. It can be seen that OSLOM, which aims at forming statistically significant communities fails to identify spamming nodes. It can also be seen that a node-based overlapping algorithm, SLPA, and an edge-based algorithm, LG(E), perform similarly, and the AUC (not shown in the figure) is identical for both algorithms (0.76).

Figure 3b also shows the ROC curves for non-overlapping algorithms which are enhanced with our auxiliary communities. It can be seen that Blondel, which aims at optimizing modularity, performs very poor. This observation is in accord with the observation in [9] that a modularity maximization algorithm is not suitable for anomaly detection due to its resolution limit. However, Blondel L1 (first level in the community hierarchy of Blondel), which forms finer granularity communities, performs dramatically better than its last level using either type of the auxiliary communities. Moreover, it can be seen that adding EA communities leads to better results compared to NA communities.

Overall, our experiments for different days in both email datasets showed that Blondel L1 and Infomap enhanced with EA, SLPA, LG(E), and DEMON all perform well with respect to placing spamming nodes into multiple communities. In practice, low false positive rates are essential for spam detection, therefore both Blondel L1 with EA communities and LG(E) that allow us to, on average, detect more than 25% and 20% of spamming nodes, respectively, for different days with very low FPR (less than 0.01) are the most suitable algorithms.

These results confirm that our method for adding EA communities to enhance non-overlapping algorithms yields not only comparable, but even better, results than an overlapping algorithm. Although both Blondel L1 with EA communities and LG(E) use the same modularity-based algorithm as their basis (we have applied Blondel L1 on the induced line graph of LG(E)), adding EA communities has also a lower complexity than inducing weighted line graphs (Table 1).

As mentioned earlier, our framework allows us to incorporate a number of application-specific filters to reduce the induced false positives (Algorithm 4).

Figure 3c shows a comparison of the spam detection using filters based on community properties (ϕ_1 and ϕ_2 only) and the combination of community and neighborhood properties (ϕ_1 - ϕ_5) for the first day of data in 2010. It can be seen that use of additional filters improves the detection (the same observation also holds for the algorithms not shown).

Finally, Figure 4 shows the AUC for spam detection using our framework with LG(E) and Blondel L1 enhanced with EA communities over 14 days during 2010 and 2011. It can be seen that the results are quite stable over time and the AUC of our method for adding EA communities compared to a more complex overlapping algorithm is much better when only community properties are used.

6 Conclusions

In this paper, we have evaluated the performance of community detection algorithms for identifying misbehavior in network communications. This paper extends and complements the previous work on community-based intrusion detection, by investigating a variety of definitions for a community, introducing auxiliary communities for enhancing traditional community detection algorithms, and showing that, in contrary to previous work, these algorithms can indeed be deployed as the basis for network anomaly detection.

We have also provided a framework for community-based anomaly detection which allows us to find the nodes that belong to multiple communities by either using auxiliary communities or overlapping algorithms. It also enables us to deploy neighborhood properties, which are indicative of social behavior, for discriminating the nodes that naturally belong to more than one community from the anti-social ones. The applicability of our framework for identifying network intrusions and unsolicited emails was evaluated using two different datasets coming from traffic captured on an Internet backbone link. Our experiments show that our framework is quite effective and provides a consistent performance over time. These results suggest that detecting community overlaps is a promising approach for identifying misbehaving network communications.

Acknowledgments. This work was supported by .SE – The Internet Infrastructure Foundation and SUNET. The research leading to these results has also received funding from the European Union Seventh Framework Programme (FP7/2007-2013) under grant agreement no. 257007.

References

1. Ahn, Y.-Y., Bagrow, J.P., Lehmann, S.: Link communities reveal multiscale complexity in networks. Nature 466(7307), 761–764 (2010)
2. Akoglu, L., Faloutsos, C.: Anomaly, event, and fraud detection in large network datasets. In: WSDM, p. 773. ACM Press (2013)
3. Akoglu, L., McGlohon, M., Faloutsos, C.: `oddball`: Spotting Anomalies in Weighted Graphs. In: Zaki, M.J., Yu, J.X., Ravindran, B., Pudi, V. (eds.) PAKDD 2010. LNCS, vol. 6119, pp. 410–421. Springer, Heidelberg (2010)

4. Almgren, M., John, W.: Tracking Malicious Hosts on a 10Gbps Backbone Link. In: Aura, T., Järvinen, K., Nyberg, K. (eds.) NordSec 2010. LNCS, vol. 7127, pp. 104–120. Springer, Heidelberg (2012)
5. Blondel, V.D., Guillaume, J.-L., Lambiotte, R., Lefebvre, E.: Fast Unfolding of Communities in Large Networks. Journal of Statistical Mechanics: Theory and Experiment 2008(10), P10008 (2008)
6. Chandola, V., Banerjee, A., Kumar, V.: Anomaly Detection: A Survey. ACM Computing Surveys 41, 1–72 (2009)
7. Clauset, A., Newman, M.E.J., Moore, C.: Finding community structure in very large networks. Physical Review. E 70(6 pt. 2), 066111 (2004)
8. Coscia, M., Rossetti, G., Giannotti, F., Pedreschi, D.: DEMON: a local-first discovery method for overlapping communities. In: ACM SIGKDD, p. 615 (2012)
9. Ding, Q., Katenka, N., Barford, P., Kolaczyk, E., Crovella, M.: Intrusion as (anti)social communication. In: ACM SIGKDD, p. 886 (2012)
10. DShield. Recommended block list (2010), `http://www.dshield.org/block.txt`
11. Eberle, W., Holder, L.: Anomaly detection in data represented as graphs. Intelligent Data Analysis 11(6), 663–689 (2007)
12. Evans, T., Lambiotte, R.: Line graphs, link partitions, and overlapping communities. Physical Review E 80(1), 1–8 (2009)
13. Fortunato, S.: Community detection in graphs. Physics Reports 486(3-5), 75–174 (2010)
14. Gao, J., Liang, F., Fan, W., Wang, C., Sun, Y., Han, J.: On community outliers and their efficient detection in information networks. In: ACM SIGKDD (2010)
15. Gomes, L., Almeida, R., Bettencourt, L.: Comparative Graph Theoretical Characterization of Networks of Spam and Legitimate Email. In: CEAS (2005)
16. Kreibich, C., Kanich, C., Levchenko, K., Enright, B., Voelker, G.M., Paxson, V., Savage, S.: On the Spam Campaign Trail. In: LEET, pp. 697–698 (2008)
17. Lancichinetti, A., Fortunato, S.: Community Detection Algorithms: A Comparative Analysis. Physical Review E 80(5), 1–11 (2009)
18. Lancichinetti, A., Radicchi, F., Ramasco, J.J., Fortunato, S.: Finding statistically significant communities in networks. PloS One 6(4), e18961 (2011)
19. Leskovec, J., Lang, K.J., Mahoney, M.: Empirical Comparison of Algorithms for Network Community Detection. In: WWW, p. 631 (2010)
20. Moradi, F., Olovsson, T., Tsigas, P.: Towards modeling legitimate and unsolicited email traffic using social network properties. In: SNS (2012)
21. Noble, C.C., Cook, D.J.: Graph-based anomaly detection. In: ACM SIGKDD, pp. 631–636 (2003)
22. Rosvall, M., Bergstrom, C.T.: Maps of random walks on complex networks reveal community structure. National Academy of Sci. 105(4), 1118–1123 (2008)
23. Shrivastava, N., Majumder, A., Rastogi, R.: Mining (Social) Network Graphs to Detect Random Link Attacks. In: ICDE, pp. 486–495. IEEE (2008)
24. SRI. International Malware Threat Center, most aggressive malware attack source and filters (2010), `http://mtc.sri.com/live_data/attackers/`
25. Sun, J., Qu, D., Chakrabarti, H., Faloutsos, C.: Neighborhood Formation and Anomaly Detection in Bipartite Graphs. In: ICDM, pp. 418–425 (2005)
26. Xie, J., Kelley, S., Szymanski, B.: Overlapping community detection in networks: the state of the art and comparative study. ACM Computing Surveys 45(4) (2013)
27. Xie, J., Szymanski, B.K.: Towards Linear Time Overlapping Community Detection in Social Networks. In: Tan, P.-N., Chawla, S., Ho, C.K., Bailey, J. (eds.) PAKDD 2012, Part II. LNCS, vol. 7302, pp. 25–36. Springer, Heidelberg (2012)
28. Yang, J., Leskovec, J.: Defining and evaluating network communities based on ground-truth. In: ICDM, pp. 745–754. IEEE (2012)

Fault-Tolerant Concept Detection in Information Networks

Tobias Kötter[1], Stephan Günnemann[1], Michael R. Berthold[2], and Christos Faloutsos[1]

[1] Carnegie Mellon University, USA
{koettert,sguennem,christos}@cs.cmu.edu
[2] University of Konstanz, Germany
berthold@ieee.org

Abstract. Given information about medical drugs and their properties, how can we automatically discover that Aspirin has blood-thinning properties, and thus prevents heart attacks? Expressed in more general terms, if we have a large information network that integrates data from heterogeneous data sources, how can we extract semantic information that provides a better understanding of the integrated data and also helps us to identify missing links? We propose to extract concepts that describe groups of objects and their common properties from the integrated data. The discovered concepts provide semantic information as well as an abstract view on the integrated data and thus improve the understanding of complex systems. Our proposed method has the following desirable properties: (a) it is *parameter-free* and therefore requires no user-defined parameters (b) it is *fault-tolerant*, allowing for the detection of missing links and (c) it is *scalable*, being linear on the input size. We demonstrate the effectiveness and scalability of the proposed method on real, publicly available graphs.

1 Introduction

If we have two sources about medical drugs, one source listing their medical properties, and the other describing their chemical behavior, how can we discover whether, for example, Aspirin has blood thinning properties - and can therefore also help prevent heart attacks and rheumatism, in addition to being a miracle painkiller? This is the precise focus of this work; Fig. 1 shows a selection of medical uses of Aspirin (Acetylsalicylic acid) that have been detected by our proposed algorithm.

In a nutshell, we want to integrate multiple sources (e.g. Wikipedia articles describing drugs), find hidden concepts (say, 'heart disease drugs') and discover missing connections (e.g. Aspirin is related to heart disease). Information networks [9] allow the integration of such heterogeneous data by modeling the relations between objects e.g. drugs and their properties. In comparison to heterogeneous information networks [17], for example, which *a-priori* assign each node to a certain type, information networks as defined in [9] do not a-priori categorize the integrated objects into different classes: any object can be a member described by its neighbors or a property describing the members to which it is related.

Given an information network (Fig. 2), we want to identify underlying concepts (Fig. 3) and analyze them. In order to do so we extract concept graphs [10] that allow the discovery and description of concepts in such networks.

V.S. Tseng et al. (Eds.): PAKDD 2014, Part I, LNAI 8443, pp. 410–421, 2014.

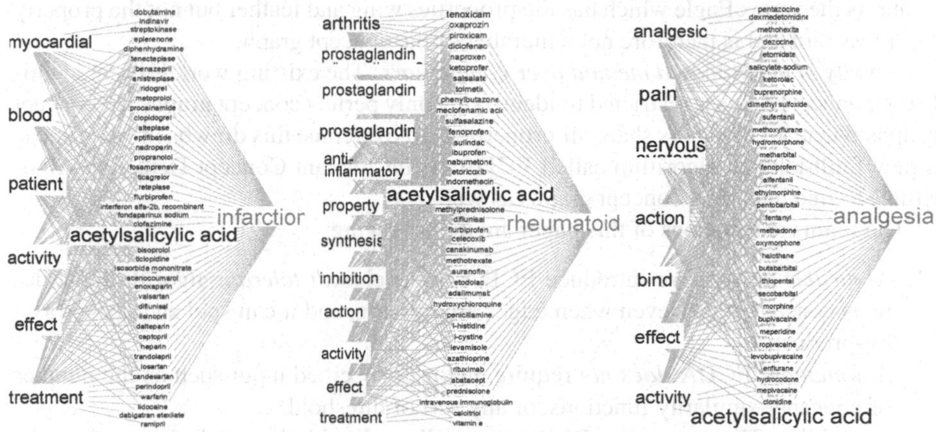

Fig. 1. Our method discovers Aspirin (Acetylsalicylic acid) as a member of the infarction, rheumatoid and analgesia (painkiller) concept graphs. See Sect. 4.1 for details.

Concept graphs allow the organization of information by grouping together objects (the *members*) that show common properties (the *aspects*), improving understanding of the concept graph and the actual concept it represents.

Concept graphs further support the abstraction of complex systems by identifying vertices that can be used as representatives for the discovered concepts. E.g., when considering the concept of animals, a number of members such as lion, bird, etc. and their characteristic aspects e.g. they are alive, can move, etc. come into our mind. In addition, they also support *context dependent vertex types* since the type of a vertex is defined by a given concept graph. Thus, a member of one concept graph could be an aspect or a concept in another concept graph. E.g., a bird might be a member of the general concept of animals but it could simultaneously be the concept representing all different kinds of birds.

An example of a concept graph representing the concept of flightless birds is depicted in Fig. 3. The concept graph consists of the members Penguin, Ostrich and Weka and their shared aspects wing and feather as well as its symbolic representation. It further

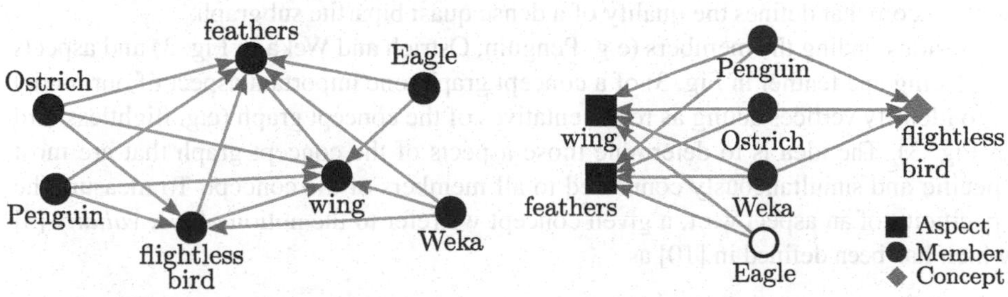

Fig. 2. Input network

Fig. 3. Detected concept graph

contains the vertex Eagle which has the properties wing and feather but not the property flightless bird and is therefore not a member of the concept graph.

Novelty of the proposed method over competitors: The existing work [10] on identifying concept graphs is restricted to identifying only perfect concept graphs i.e. concept graphs where all members share all properties. To overcome this drawback we propose a new fault-tolerant algorithm called FCDA (Fault-tolerant Concept Detection Algorithm) to find imperfect concept graphs as well.

The main contributions of this paper are the following:

1. *Algorithm design:* We introduce FCDA, a novel *fault-tolerant* algorithm that detects concept graphs even when edges are missing, and it can spot such edges as a by-product.
2. *Automation:* FCDA *does not* require any user specified input such as the number of concepts, similarity functions, or any sort of threshold.
3. *Scalability:* The run time of FCDA grows *linearly* with the number of vertices.
4. *Effectiveness:* We evaluate our method on two real world data sets. Our results show that FCDA detects meaningful concepts that help to understand the integrated data.

For easy usage, FCDA has been implemented in KNIME [4]. The source code and experiments are available at `http://cs.cmu.edu/~koettert/pakdd2014`.

2 Proposed Method

In this section, we describe our model for fault-tolerant concept graph detection. We assume that we have a network $G = (V, E)$ with vertices V and directed edges $E \subseteq V \times V$, where $(u, v) \in E$ states that the vertex u possesses the property v. We denote the predecessors of v with $N^-(v) := (V \times \{v\}) \cap E$, and its successors with $N^+(v) := (\{v\} \times V) \cap E$. Given such a network we want to find all underlying concept graphs.

The general idea of concept graphs is to find disjoint sets of vertices $V_M, V_A \subseteq V$ such that each vertex of the member set V_M is connected to many vertices in the aspect set V_A, and vice versa. The existing method [10], which is based on frequent itemsets, was defined to identify *perfect concept graphs*. A perfect concept graph is one that forms a quasi biclique where each member is connected to all aspects of the concept. Real word data, however, is often noisy and incomplete and thus might not contain all of the connections between the members and aspects of a concept. Our solution to finding these imperfect concept graphs is to replace the strict fully connectedness requirement with a score that defines the quality of a dense quasi bipartite subgraph.

Besides finding the members (e.g. Penguin, Ostrich and Weka in Fig. 3) and aspects (e.g. wing and feather in Fig. 3) of a concept graph, one important aspect of our model is to identify vertices acting as representatives of the concept graph (e.g. flightless bird in Fig. 3). The idea is to determine those aspects of the concept graph that are most specific and simultaneously connected to all members of the concept. To measure the specificity of an aspect w.r.t. a given concept we refer to the notion of *cue validity* [3] which has been defined in [10] as

$$cv(v, V_M) = \frac{|N^-(v) \cap V_M|}{|N^-(v)|}. \tag{1}$$

For example, the cue validity of the aspect, feather, for the concept graph, flightless bird, in Fig. 3 is $\frac{3}{4}$ whereas the cue validity of the concept representative, flightless bird, is 1. Thus, flightless bird is the most specific aspect that is connected to all members of the graph and should therefore be selected as the representative. Note that the cue validity of a vertex depends on the currently selected set of members V_M.

In general, our model allows multiple vertices as representatives providing they share the same specificity. We can therefore handle concept graphs where two or more terms are used interchangeably as representatives of the same concept (e.g. synonyms) and thus refer to the same members and aspects. Overall, we define a concept graph as:

Definition 1. ***Concept graph.*** *Given an information network $G = (V, E)$, a concept graph $C = (V_M, V_A, V_C)$ is a triplet of concept graph members $V_M \subseteq V$, concept graph aspects $V_A \subseteq (V \backslash V_M)$ and concept graph representatives $V_C = \{v \in V'_C \mid cv(v, V_M) = \max_{v' \in V'_C} cv(v', V_M)\}$ such that the subgraph $S = (V_M \cup V_A, (V_M \times V_A) \cap E)$ is connected and $V_C \neq \emptyset$ with $V'_C = \{v \in V_A \mid V_M \subseteq N^-(v)\}$ defines the set of aspects, which are connected to all members of the concept.*

2.1 Concept Graph Score

The previous definition of concept graphs is very loose in the sense that many sets of vertices fulfill the definition. Thus, our goal is to focus on the 'most interesting' concepts graphs. We measure the interestingness of concept graphs via a score that incorporates three properties desired for detecting interesting concept graphs. The size of the concept graph, i.e. the number of members $|V_M|$ and aspects $|V_A|$. The connectivity of its members and aspects, i.e. the number of (missing) links connecting its members and aspects. The specificity of the aspects for the given concept (see Eq. 1). In general, the larger the concept graph, the more densely connected the two sets and the more specific the aspects, the higher the score.

Definition 2. *The concept graph score cs is defined as follows*

$$cs(V_M, V_A) = \sum_{m \in V_M} \sum_{a \in V_A} \epsilon(a, m) \frac{|N^-(a) \cap V_M|}{|N^-(a)|} = \sum_{m \in V_M} \sum_{a \in V_A} \epsilon(a, m) cv(a, V_M) \tag{2}$$

with $\epsilon(a, m) = 1$ if $m \in N^-(a)$, otherwise -1.

Given this definition, we are now interested in finding those concept graphs that maximize the score.

2.2 The FCDA Algorithm

The fault-tolerant detection of concept graphs is similar to the detection of maximum quasi-bicliques, which is NP-complete [13]. Thus, we cannot expect to find an efficient algorithm computing an exact solution. To improve performance we propose a greedy algorithm that maximizes the quality score *cs* of a given concept graph.

Intuition behind our algorithm: (Part 1) For each vertex $c' \in V'_C$ with incoming edges, extract its predecessors (potential members) and their successors (potential aspects). (Part 2) Optimize the two discovered sets using Eq. 2. The details are as follows:

Part 1: Find Initial Concept Graph

Step 1. Extract the set of potential members V'_M for the potential concept $c' \in V'_C$ which is the set of its predecessors $V'_M = N^-(c')$.

Step 2. Extract the set of potential aspects V'_A which are the successors of the potential members V'_M that are equal or less specific than the potential concept vertex c' thus $V'_A = \left\{a \in \bigcup_{m \in V'_M} N^+(m) : |N^+(a)| \geq |N^-(c')|\right\}$.

Once we have extracted the set of potential aspects V'_A and potential members V'_M we have to identify the optimal subsets given the concept graph score cs (see Eq. 2). *Computational Speed considerations:* Obviously, enumerating all possible combinations to find the optimal solution is intractable that is why we follow an iterative approach.

Part 2: Refine Concept Graph

Step 3. Set $V^*_A = V'_A$, i.e. V^*_A contains all potential aspects.

Step 4. Find the subset of members V^*_M that maximizes the concept graph score cs for the current aspect set V^*_A by adding each member that improves the score, i.e. $V^*_M = \arg\max_{V^*_M \subseteq V'_M} cs(V^*_A, V^*_M)$.

Step 5. If the current concept graph score $cs(V^*_A, V^*_M) > cs(V_A, V_M)$ is the highest for the current potential concept c' set $V_M = V^*_M$ and $V_A = V^*_A$.

Step 6. Remove the aspect a' from V^*_A that has the least members in V'_M, i.e. the one with the lowest value for $|N^-(a') \cap V'_M|$, and start over from step 4 until V^*_A is empty.

Step 7. Once the aspect loop has been terminated, iterate over the remaining potential aspects $V'_A \setminus V_A$ and add all vertices to V_A that improve the score. This is necessary because an aspect might have been removed early that would improve the score due to vertices in V^*_A with a negative score. Accordingly, we iterate over the set $V'_M \setminus V_M$ to add vertices to V_M that improve the score.

Step 8. Finally check for vertices that are members of both sets V_A and V_M. This situation can arise as a vertex can have incoming and outgoing edges and thus end up in both sets. Remove a vertex $v \in V_M \cap V_A$ from the set that affects the score the least. E.g., if $v \in V_A \wedge v \in V_M \wedge cs(V_M/v, V_A) < cs(V_M, V_A/v)$ holds, remove v from V_M otherwise remove v from V_A.

By following these steps we extract the concept graphs and their scores for each potential concept $c' \in V'_C$. Since the computation of each potential concept $c' \in V'_C$ is independent of other concepts, computation is easily parallelized, which subsequently improves the run time significantly (Fig. 5c).

Efficient Incremental Score Computation. The above algorithm requires computing the concept graph score cs at multiple places. To speed up the computation we can make use of an incremental computation of the concept graph score. The following equations hold when adding or removing a single member or aspect from the concept graph:

• Adding a member m^+ to V_M:

$$cs(V_M \cup m^+, V_A) = cs(V_M, V_A) + \sum_{a \in V_A} \epsilon(a, m^+) \frac{|N^-(a) \cap (V_M \cup m^+)|}{|N^-(a)|} + \sum_{a \in V_A} \sum_{m \in V_M} \epsilon(a, m) \frac{|N^-(a) \cap m^+|}{|N^-(a)|}. \quad (3)$$

• Adding an aspect a^+ to V_A:

$$cs(V_M, V_A \cup a^+) = cs(V_M, V_A) + \sum_{m \in V_M} \epsilon(a^+, m) \frac{|N^-(a^+) \cap V_M|}{|N^-(a^+)|}. \quad (4)$$

• Removing a member m^- from V_M:

$$cs(V_M / m^-, V_A) = cs(V_M, V_A) - \sum_{a \in V_A} \epsilon(a, m^-) \frac{|N^-(a) \cap V_M|}{|N^-(a)|} - \sum_{a \in V_A} \sum_{m \in V_M / m^-} \epsilon(a, m) \frac{|N^-(a) \cap m^-|}{|N^-(a)|}. \quad (5)$$

• Removing an aspect a^- from V_A:

$$cs(V_M, V_A / a^-) = cs(V_M, V_A) - \sum_{m \in V_M} \epsilon(a^-, m) \frac{|N^-(a^-) \cap V_M|}{|N^-(a^-)|}. \quad (6)$$

2.3 Missing Link Recovery

The missing link recovery approach is based on the information from the discovered concept graphs. All of the connections that need to be added to a concept graph in order to make it fully connected are potentially missing links (see Fig. 4). However, not all of them are interesting or missing. That is why we provide two scoring functions that allow the global sorting of the recovered missing links based on their confidence and interestingness. Since the (missing) edge (m, a) between a member m and an aspect a can be missing in several concept graphs, we use $\mathbb{C}_{a,m}$ to denote all concept graphs that have the member $m \in V_M$ and the aspect $a \in V_A$.

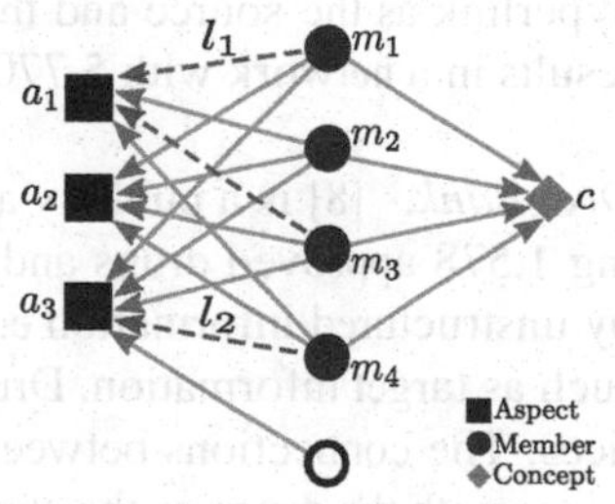

Fig. 4. Missing links example

Confidence. To measure the confidence of a missing link we can use the structure of the concept graphs based on the consideration that a single missing link in a large concept graph is much more likely to be an artifact than a link in a smaller concept graph that misses many links. E.g. the missing link l_1 in Fig. 4 has a lower confidence than the missing link l_2 since the aspect a_1 is only connected to two whereas the aspect a_3 is connected to three of the four concept members. We take the minimum to ensure that the possibility of the missing link is high for all concept graphs in $\mathbb{C}$.

Definition 3. *The confidence score of a missing link between a member m and an aspect a is defined as* $conf(m,a) = \min_{C(V_A,V_M,V_C)\in\mathbb{C}_{a,m}} \frac{|N^+(m)\cap V_A|}{|V_A|} \cdot \frac{|N^-(a)\cap V_M|}{|V_M|}$.

Interestingness. The interestingness of a recovered link is related to the cue validity (see Eq. 1) of the aspect. For example, a missing link to a very general aspect with a low cue validity is not as interesting as a missing link to a more specific concept with a high cue validity. E.g. The missing link l_1 in Fig. 4 is potentially more interesting than the missing link l_2 since the aspect a_1 has a cue validity of 1 whereas the aspect a_3 has a cue validity of $\frac{3}{4}$. Since the cue validity of an aspect is context dependent we take the minimum to ensure that the aspect is interesting from a global point of view.

Definition 4. *The global interestingness score of a missing link between a member m and an aspect a is defined as* $int(m,a) = \min_{C(V_A,V_M,V_C)\in\mathbb{C}_{a,m}} cv(a, V_M)$.

3 Experiments

This section demonstrates the quality of the proposed method based on networks that were extracted from two publicly available real world data sets from different domains.

3.1 Datasets

Wikipedia Selection for Schools. (Schools Wikipedia)[1] is a selection of the English Wikipedia for children. It has about 5500 articles organized into 154 subjects such as countries, religion, and science. Each article and each related subject is represented by a vertex. Hyperlinks are represented by directed edges with the article that contains the hyperlink as the source and the referenced article or subject as the target vertex. This results in a network with 5,770 vertices and 231,985 edges.

DrugBank. [8] is a publicly available data base with more than 6,000 entries describing 1,578 approved drugs and 5,000 experimental substances. Each drug is described by unstructured information e.g. textual descriptions as well as structured information such as target information. Drugs and their extracted properties are represented by vertices. The connections between drugs and their properties are represented by directed edges with the drugs as the source and the properties as target vertices. This results in a network with 18,574 vertices and 109,721 edges.

3.2 Quality

Since the proposed method is unique in its characteristics and most of the more similar methods require some kind of parameter to define the number of clusters, it is difficult to undertake a comparison. Therefore, we decided to use information that exists in each data source, defining specific groups of objects, to evaluate our results. In Schools Wikipedia we used the subject pages and the related subjects of each article. This leads

[1] `http://schools-wikipedia.org`

to a total of 154 groups including topics like dinosaurs, chemical elements, computer programming, and artists. For DrugBank we used the drug classes as well as the pharmaceutical and pathway based classification available from the DrugBank homepage. This leads to a total of 651 groups of which 584 corresponding vertices can be found in the network. Given the predefined groups of objects and the concept graphs detected by our method, we can compute the F_1 score [16] to measure the quality of our algorithm. The F_1 score for Schools Wikipedia is 0.803 with precision 0.769 and recall 0.924. For DrugBank the F_1 score is 0.859 with precision 0.863 and recall 0.862.

3.3 Scalability

In this section, we demonstrate the time complexity and parallelizability of FCDA experimentally. In order to measure the running time on different sized graphs we extracted several subgraphs of different sizes from the two data sets using snowball sampling [18]. Figure 5 shows the running time w.r.t. increasing number of vertices as well as the average speedup over all experiments.

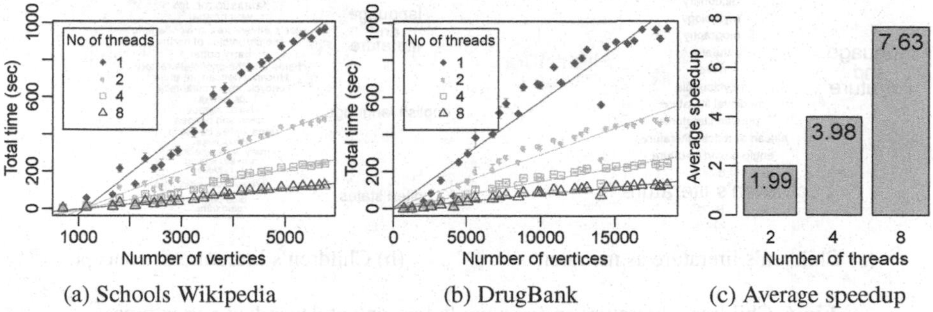

(a) Schools Wikipedia (b) DrugBank (c) Average speedup

Fig. 5. Run time of FCDA versus the number of vertices (average over 5 runs) and average speedup over all experiments

Usually the two vertex sets V_A and V_M form a Galois connection [7], which describes the correspondence between two partially ordered sets. It states that if one of the two sets increases in size, the size of the other set will decrease. Therefore, if we are looking at a very general concept with a lot of members the set of aspects is very small and vice versa. This behavior contributes to the efficiency of the algorithm.

4 Discoveries

This section describes the discoveries we were able to make by applying our proposed method to the two data sets mentioned above. The concept graph images always depict the **aspects** of the concept graph in the left column, the members in the middle column, and the concept representative in the right column. In addition, aspects are ordered based on their cue validity from highly specific at the top to more general at the bottom.

4.1 Concept Detection

Observation 1 (Aspirin as blood thinner). *The concept graphs in Fig. 1 demonstrate the ability to discover members e.g. Aspirin (Acetylsalicylic acid) that are part of different concepts e.g. painkillers, rheumatoid and infarction. It further shows the value of the extracted semantic information which helps to better understand the extracted concepts e.g that rheumatism is caused by inflammation.*

Observation 2 (Context dependent vertex types). *The ability to model context dependent vertex types (e.g. concept, aspect or member), of concept graphs and* FCDA *is necessary to model real world concepts. See Fig. 6 for an example of two concept graphs from Schools Wikipedia where the type of the children's literature vertex changes from member (Fig. 6a) to concept (Fig. 6b).*

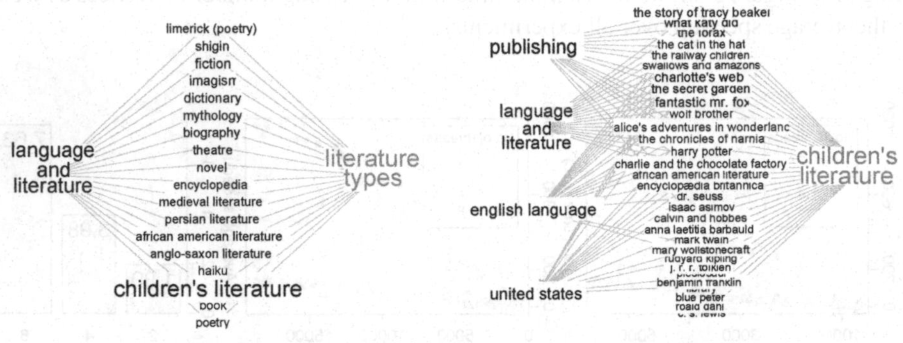

(a) Children's literature as member. (b) Children's literature as concept.

Fig. 6. Children's literature as an example for context dependent vertex types

4.2 Missing Link Recovery

Thanks to the fault-tolerance of FCDA we can identify missing links between an aspect and a member of a detected concept graph. Using the semantic information provided by the discovered concept graphs we can further assign a global interestingness and a confidence score to all potentially missing links. Table 1 shows the top 5 missing links for the DrugBank data set based on the combined scores.

Observation 3 (Haloperidol can impair the effect of antiparkinson agents). *A very interesting connection is the one between Haloperidol and 'antiparkinson agents'. This connection is not directly mentioned on the DrugCard. However, by searching for the two terms we learned that Haloperidol can impair the effect of antiparkinson agents [1]. Thus, our method successfully reveals novel information which is not encoded in the current data set but predicted based on the discovered concept graphs.*

Table 1. Top 5 missing links based on confidence and interestingness

Aspect (Property)	Member (Drug)	$conf$	int	$conf \cdot int$
anhydrases	trichlormethiazide	0.82	0.86	0.70
shigella	netilmicin	0.73	0.89	0.65
cell-cell	enflurane	0.72	0.8	0.58
aminoglycoside antibiotic	neomycin	0.72	0.8	0.57
antiparkinson agents	**haloperidol**	0.70	0.81	0.57

Observation 4 (Detected missing links are often true omissions). *The top 5 discovered missing links in Table 1 are true omissions. Most of them do not exist in the network due to the limitations of the used text mining algorithm. For example, the connection between 'aminoglycoside antibiotic' and Neomycin is mentioned on its DrugCard but does not exist in the network. However, thanks to the missing link recovery we are still able to recover these links.*

5 Related Work

Network-Based Approaches: *(Quasi)-clique detection* methods [12] aim at finding dense subgraphs in a given graph. These methods do not distinguish between the members and aspects of a subgraph but treat all vertices equally. Thus, these methods cannot be used to find concept graphs since, e.g., the members of a concept graph need not be connected at all. Methods [11] have been proposed for *quasi-biclique detection*, which explicitly distinguish between members and aspects. The existing methods, however, are not aware of the concept of cue validity and are not able to determine representative vertices for each concept. *Block-Modeling* techniques [2,5] try to find homogeneous blocks in the adjacency matrix of a graph. However, none of the existing approaches takes the cue validity into account and are not able to identify representative vertices. Moreover, most of the techniques require important parameters to be specified manually, e.g., the maximal fraction of missing edges allowed in a pattern.

Global Pattern Discovery: Finding vertices sharing similar properties might be regarded as an instance of *shared nearest neighbor* clustering. That is, using the distance function d that assigns low values to pairs of vertices sharing many neighbors and high values to pairs of vertices sharing no neighbors, any distance based clustering method as, e.g., k-medoid, might be used to find concept graphs. This principle, however, does not consider the current context of the subgraph as our method does. Additionally, most of the methods either do not identify representative vertices or do not allow for overlapping clusters or require certain parameters e.g. the number of clusters to detect.

Local Pattern Discovery: *Co-Clustering/Biclustering* [14] is the task of simultaneously clustering the rows and columns of a data matrix. In our scenario, the data matrix would correspond to the binary adjacency matrix. Co-clustering can be roughly divided into four categories [14]. According, our method is mostly related to the category for finding patterns with constant values, as we are interested in detecting dense (bipartite) subgraphs of the network. In this regard, co-clustering would reflect a certain kind of block-modeling sharing the same drawbacks as discussed above. *Frequent itemset mining* aims at finding groups of items that frequently occur together in a set of transactions. By considering each row of the adjacency matrix as a transaction, and each column as an item, frequent itemset mining can be used to detect bicliques in a network. This idea has been exploited by existing methods for concept graph detection [10]. However, when based on frequent itemsets, these methods are highly sensitive to errors and missing values in the data. While fault-tolerant extensions of frequent itemset mining [15] and, even more general, subspace clustering [6] have been proposed to handle missing data, such methods are often not scalable and require hard to set parameters. In contrast, our technique requires no user defined parameters and the computational costs increase linear in the number of vertices.

In conclusion, none of the described methods supports all of our goals, namely, scalability, fault-tolerance, parameter free, dynamic vertex types, and the detection of concept representatives.

6 Conclusions

We have proposed FCDA, the first fault-tolerant and parameter-free method for finding concept graphs in information networks. The detected concept graphs provide valuable information about the members of a concept and their characteristic properties, improving understanding of the concept graph and the represented concept itself. They further support the abstraction of complex systems by identifying vertices that can be used as representatives for the discovered concepts. In addition, they also support overlapping concept graphs and context dependent vertex types. The main contributions of our work include:

- *Algorithm design:* We introduces FCDA, a novel algorithm that detects concept graphs in information networks even when edges are missing, and it can spot such edges as a by-product.
- *Automation:* FCDA *is fully automatic*. It does not require any user specified input such as the number of concepts, similarity functions, or any sort of threshold.
- *Scalability:* The run time of FCDA grows *linearly* with the number of vertices.
- *Effectiveness:* We demonstrate that FCDA detects meaningful concepts that help to understand the integrated data in diverse real-world data sets. E.g. the discovery of Aspirin's benefit in the treatment of infarction and rheumatism in addition to it being a painkiller. We further show the potential of the algorithm for discovering missing relations such as the detection that Haloperidol can impair the effect of antiparkinson agents.

Acknowledgments. T. Kötter was supported by stipend KO 4661/1-1 of the "Deutsche Forschungsgemeinschaft" (DFG). S. Günnemann was supported by a fellowship within the postdoc-program of the German Academic Exchange Service (DAAD). This material is based upon work supported by the National Science Foundation under Grant No. IIS-1247489. Research was also sponsored by the Army Research Laboratory and was accomplished under Cooperative Agreement Number W911NF-09-2-0053.

Any opinions, findings, and conclusions or recommendations expressed in this material are those of the author(s) and do not necessarily reflect the views of the National Science Foundation, DARPA, or other funding parties. The U.S. Government is authorized to reproduce and distribute reprints for Government purposes notwithstanding any copyright notation here on.

References

1. Ahmed, S.P., Siddiq, A., Baig, S.G., Khan, R.A.: Comparative efficacy of haloperidol and risperidone: A review. Pakistan Journal of Pharmacology 24, 55–64 (2007)
2. Airoldi, E.M., Blei, D.M., Fienberg, S.E., Xing, E.P.: Mixed membership stochastic blockmodels. Journal of Machine Learning Research 9, 1981–2014 (2008)
3. Beach, L.R.: Cue probabilism and inference behavior. Psychological Monographs: General and Applied 78, 1–20 (1964)
4. Berthold, M.R., Cebron, N., Dill, F., Gabriel, T.R., Kötter, T., Meinl, T., Ohl, P., Sieb, C., Thiel, K., Wiswedel, B.: KNIME: The Konstanz Information Miner. In: Studies in Classification, Data Analysis, and Knowledge Organization (GfKL 2007). Springer (2007)
5. Chakrabarti, D., Papadimitriou, S., Modha, D.S., Faloutsos, C.: Fully automatic cross-associations. In: KDD, pp. 79–88 (2004)
6. Günnemann, S., Müller, E., Raubach, S., Seidl, T.: Flexible fault tolerant subspace clustering for data with missing values. In: ICDM, pp. 231–240 (2011)
7. Herrlich, H., Husek, M.: Galois connections. In: Mathematical Foundations of Programming Semantics, pp. 122–134 (1985)
8. Knox, C., Law, V., Jewison, T., Liu, P., Ly, S., Frolkis, A., Pon, A., Banco, K., Mak, C., Neveu, V., Djoumbou, Y., Eisner, R., Guo, A.C., Wishart, D.S.: Drugbank 3.0: a comprehensive resource for 'omics' research on drugs. Nucleic Acids Research 38, 1–7 (2010)
9. Kötter, T., Berthold, M.R.: From information networks to bisociative information networks. In: Berthold, M.R. (ed.) Bisociative Knowledge Discovery. LNCS (LNAI), vol. 7250, pp. 33–50. Springer, Heidelberg (2012)
10. Kötter, T., Berthold, M.R.: (Missing) concept discovery in heterogeneous information networks. In: Berthold, M.R. (ed.) Bisociative Knowledge Discovery. LNCS (LNAI), vol. 7250, pp. 230–245. Springer, Heidelberg (2012)
11. Li, J., Sim, K., Liu, G., Wong, L.: Maximal quasi-bicliques with balanced noise tolerance: Concepts and co-clustering applications. In: SDM, pp. 72–83 (2008)
12. Liu, G., Wong, L.: Effective pruning techniques for mining quasi-cliques. In: Daelemans, W., Goethals, B., Morik, K. (eds.) ECML PKDD 2008, Part II. LNCS (LNAI), vol. 5212, pp. 33–49. Springer, Heidelberg (2008)
13. Liu, X., Li, J., Wang, L.: Quasi-bicliques: Complexity and binding pairs. In: Hu, X., Wang, J. (eds.) COCOON 2008. LNCS, vol. 5092, pp. 255–264. Springer, Heidelberg (2008)
14. Madeira, S.C., Oliveira, A.L.: Biclustering algorithms for biological data analysis: A survey. IEEE/ACM Transactions on Computational Biology and Bioinformatics 1, 24–45 (2004)
15. Poernomo, A.K., Gopalkrishnan, V.: Towards efficient mining of proportional fault-tolerant frequent itemsets. In: KDD, pp. 697–706 (2009)
16. Rijsbergen, C.J.V.: Information Retrieval, 2nd edn. Butterworth-Heinemann, Newton (1979)
17. Sun, Y., Yu, Y., Han, J.: Ranking-based clustering of heterogeneous information networks with star network schema. In: KDD, pp. 797–806 (2009)
18. Thompson, S.: Sampling. Wiley Series in Probability and Statistics. John Wiley & Sons, Inc., New York (2002)

Characterizing Temporal Anomalies in Evolving Networks

N.N.R. Ranga Suri[1], M. Narasimha Murty[2], and G. Athithan[1,3]

[1] Centre for AI and Robotics (CAIR), Bangalore, India
{rangasuri,athithan.g}@gmail.com
[2] Dept of CSA, Indian Institute of Science (IISc), Bangalore, India
mnm@csa.iisc.ernet.in
[3] Presently working at Scientific Analysis Group (SAG), Delhi, India

Abstract. Many real world networks evolve over time indicating their dynamic nature to cope up with the changing real life scenarios. Detection of various categories of anomalies, also known as outliers, in graph representation of such network data is essential for discovering different irregular connectivity patterns with potential adverse effects such as intrusions into a computer network. Characterizing the behavior of such anomalies (outliers) during the evolution of the network over time is critical for their mitigation. In this context, a novel method for an effective characterization of network anomalies is proposed here by defining various categories of graph outliers depending on their temporal behavior noticeable across multiple instances of a network during its evolution. The efficacy of the proposed method is demonstrated through an experimental evaluation using various benchmark graph data sets.

Keywords: Mining network data, Evolving networks, Anomaly detection, Graph anomalies, Temporal outliers.

1 Introduction

Graph based representation of network structure gave rise to the rapid proliferation of research work on social network analysis. Similarly, graph based representation is preferred for modeling the dynamism inherent to many real life applications. Thus, graph mining plays an important role in any meaningful analysis of the network data. An emerging research problem related to graph mining is to discover anomalies [4,1], also known as outliers, in graph representation of the network data. Due to the networked nature of the data pertaining to many real life applications such as computer communication networks, social networks of friends, the citation networks of documents, hyper-linked networks of web pages, etc [7], anomaly/outlier detection in graphs [1,3,12] turns out to be an important pattern discovery activity. It provides the basis for realizing certain application specific tasks such as fraud detection in on-line financial transactions, intrusion detection in computer networks, etc [4].

V.S. Tseng et al. (Eds.): PAKDD 2014, Part I, LNAI 8443, pp. 422–433, 2014.

In case of IP networks comprising of several individual entities such as routers and switches, the behavior of the individual entities determines the ensemble behavior of the network [15]. In this context, network anomalies typically refer to circumstances when network operations deviate from normal behavior, resulting in many unusual traffic patterns noticeable in such network data. These traffic patterns arise due to different types of network abuse such as denial of service attacks, port scans, and worms; as well as from legitimate activity such as transient changes in customer demand, flash crowds, or occasional high-volume flows [10]. These traffic anomalies happen to be the major security concern to the network administrators and detecting such anomalies in evolving networks [14,11] is essential for ensuring a healthy operational status of these networks.

Connected with any dynamic network, the underlying graph representation undergoes time dependent changes in its connectivity structure. This will have a consequence on the local connectivity and thus influences the temporal behavior of various graph entities such as nodes, edges and sub-graphs. As a result, a graph entity that qualifies to be an outlier (irregular connectivity pattern) at the current instance may no more continue to be the same at a later point of time due to the changes in connectivity structure of the graph. Additionally, some new irregularities may arise in a different portion of the graph hinting at the presence of some new outliers. If the graph connectivity structure is altered further, some intermediate outliers may cease to exist and further new types of outliers may start appearing in the graph. Thus, it is important to track such evolving scenarios for a subjective analysis of the graph representation towards understanding and characterizing the outlier dynamics. Accordingly, study of evolving graphs/networks has been an active research problem as evident from some recent efforts in this direction [2,14].

Motivated by the above discussion, a novel framework for characterizing temporal anomalies in evolving networks is envisaged here. This involves identifying various graph outliers and characterizing their dynamics across multiple instances of the graph representation of a dynamic network at various points of time during its evolution. Accordingly, a novel algorithm is proposed in this paper for detecting various temporal outliers in a dynamic graph and for producing a semantic categorization of the detected outliers by defining different categories of temporal outliers.

Highlights of the work reported in this paper are as follows:

- The problem of detecting anomalies in dynamic network data is an active research area [11,14].
- Characterizing temporal anomalies (outliers) is essential in *many application contexts* such as ensuring the security of information networks.
- A thorough exploration of the *temporal behavior of graph outliers* by defining various categories of temporal outliers based on their occurrence patterns.
- A novel algorithm for effective detection of various temporal outliers and for analyzing their dynamics.

The rest of the paper is organized in four sections. Section 2 gives a brief discussion on the relevant literature. Subsequent section describes the proposed method for the study of outlier dynamics in evolving networks/graphs. An experimental evaluation of the proposed method is furnished in Section 4 bringing out its empirical findings. Finally, Section 5 concludes this paper with a discussion and a few directions for future work on this problem.

2 Related Literature

A method for spotting significant anomalous regions in dynamic networks has been proposed [11] recently. According to this method, the anomalous score of a weighted edge in an undirected connected graph is determined based on a statistical measure. The aim is to find contiguous regions having adjacent anomalous edges forming large regions of higher score, named as Significant Anomalous Regions (SAR) in a dynamic network. The anomaly score of such a region is quantified by aggregating the scores of the participating edges. Given an edge-weighted graph $G = (V, E, W)$, the anomaly score of a temporal network region $R = (G', [i, j])$, where $G' = (V', E')$ is a connected sub-graph of G, in the time interval $[i, j]$ is given by

$$score_G(R) = \sum_{e \in E'} \sum_{t=i}^{j} w^t(e) \tag{1}$$

where $w^t(e)$ is anomaly score associated with an edge e at time point t.

This method is basically an edge-centric approach where the anomalous scores of edges determine the anomalous regions in a graph. Also, computing aggregate score over time may not highlight the temporal characteristics of the primitive outliers such as node/edge outliers in a graph.

In another recent effort, a multi-graph clustering method [13] was proposed exploiting the interactions among different dimensions of a given network. Though the multi-graph scenario is conceptually different from multiple instances of an evolving graph, establishing a relationship between these two problem settings may enable leveraging the methods developed for one setting in effectively solving the other. Similarly, other recent efforts [2,14] on dynamic graphs are aimed at modeling some specific time varying properties of such graphs in a meaningful manner. A substructure-based network behavior anomaly detection approach, named as Weighted Frequent Sub-graphs method, was proposed [6] to detect the anomalies present in large-scale IP networks. According to this method, patterns of abnormal traffic behavior are identified using multivariate time series motif association rules mining procedure.

A method to capture the evolving nature of abnormal moving trajectories was proposed in [5] considering both current and past outlierness of each trajectory. Such a method is intended to explore the behaviors of moving objects as well as the patterns of transportation networks. It identifies the top-k evolving outlying trajectories in a real time fashion by defining an appropriate outlier score function. Unlike the method described in [11] computing a simple summation of the anomalous scores over time, this particular method utilizes a decay function in determining the evolving outlier score so as to mitigate the influence of the past trajectories.

2.1 Outlier Detection in Graphs

As mentioned in the previous section, the essential graph mining task here is to detect various anomalies/outliers present in the graph representation of a dynamic network at a specific point of time, i.e. detecting outliers in a single graph instance. Among the methods available in the literature addressing this requirement, the method proposed in [12] is an early one using the minimum description length (MDL) principle. The main idea of this method is that sub-graphs containing many common sub-structures are generally less anomalous than sub-graphs with few common sub-structures. Thus, it is suitable mainly for applications involving many common sub-structures such as the graphs describing the atomic structure of various chemical compounds. Similarly, the method proposed in [3] is meant for anomalous link (edge) discovery defining a novel edge significance measure.

A recent work on anomaly detection in graph data [1], known as 'OddBall', makes use of the notation of *egonet* of a node defined as the induced sub-graph of its 1-step neighbors. According to this method, the number of nodes N_i and the number of edges E_i of the egonet G_i of node i in a graph $G = (V, E)$ follow a power law relationship, named the Egonet Density Power Law (EDPL) defined as

$$E_i \propto N_i^{\alpha}, \quad 1 \leq \alpha \leq 2. \tag{2}$$

Consequently, two types of anomalous egonets, named as *near cliques* and *near stars*, are determined by measuring the amount of deviation from the power law relationship. Thus, the outlierness score of an egonet sub-graph G_i is computed as the distance to the least squares fitting line as defined in [1].

$$out\_score(G_i) = \frac{max(E_i, CN_i^{\alpha})}{min(E_i, CN_i^{\alpha})} log(|E_i - CN_i^{\alpha}| + 1) \tag{3}$$

Finally, the anomalous sub-graphs are indicated in a scatter plot, referred to as EDLP plot, showing their deviation from the fitting line.

The spectral approach for detecting subtle anomalies in graphs [16] is a more recent method, which is of relevance to our work. This approach detects the anomalous sub-graphs in the input graph by exploring the minor eigenvectors of the adjacency matrix of the graph. The underlying assumption is that there exist some minor eigenvectors with extreme values on some entries corresponding to the anomalies in the graph.

3 Proposed Method

Given the graph representation of network data, there may exist various types of graph outliers in the form of node outliers, edge outliers and sub-graph outliers as discussed in the previous section. The process of graph evolution over time results in some interesting observations in terms of these graph outliers. A framework for capturing this phenomenon is proposed here as shown in Figure 1. The rationale behind this setting is to explore various types of outliers present in a graph at various points of time independently and then to characterize them based on their appearance at different points of time during the network evolution.

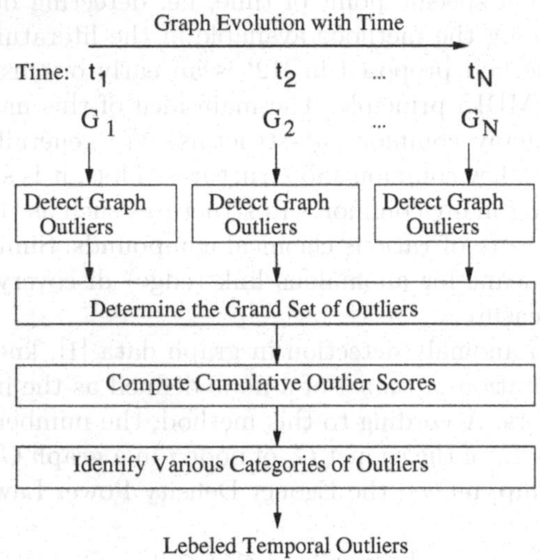

Fig. 1. Proposed framework for detecting temporal outliers

3.1 Preliminaries

The mathematical notation along with some definitions required for developing a novel algorithm for temporal outlier detection is introduced here. We use the two terms 'outliers' and 'anomalies' in an interchanging manner as they both indicate the data objects deviating from the normal data. Likewise, the two terms 'network' and 'graph' are used to refer to the same data object.

Let $G = \{G_1, G_2, \ldots, G_N\}$ be a dynamic graph data set comprising of N instances of an evolving graph corresponding to the analysis time period $[t_1, t_N]$ consisting of N discrete points of time. Similarly, let $\Psi_i = \{\psi_{i,1}, \psi_{i,2}, \ldots, \psi_{i,k}\}$ denote the top-k outliers present in the graph instance G_i at time point t_i and $S_i = \{s_{i,1}, s_{i,2}, \ldots, s_{i,k}\}$ be their outlier scores respectively.

Definition 1 (Temporal outlier). *A graph-based outlier that is present in one or more instances of an evolving network/graph.*

Definition 2 (Cumulative outlier score). *The cumulative outlier score of a temporal outlier is computed using its outlier scores at different points of time during the analysis.*

The exact computation of the cumulative outlier score depends upon the specific algorithmic logic considered based on the application context. The set of temporal outliers detected by the proposed algorithm is indicated by Ω and $\Phi = \{\phi_1, \phi_2, \dots\}$ represents the cumulative outlier scores of the detected temporal outliers.

In our characterization of the dynamics of the temporal outliers, we resort to a categorization of them based on their time varying characteristics. To this end, every category of temporal outlier that is noticed during the graph evolution is assigned an explicit label to mark it distinctly depending on its behavior. In order to capture various behavioral patterns of the temporal outliers and to convey the underlying semantics associated with each such pattern, we define the following five categories of temporal outliers noticeable in dynamic graphs.

Definition 3 (Halting outlier). *A temporal outlier O_H that may cease to exist over time during the network evolution.*
$\exists\, t_j \in [t_3, t_N]$ *s.t.* $(O_H \in \Psi_p,\ \forall t_p \in [t_1, t_j))\ \wedge\ (O_H \notin \Psi_j)$.

Definition 4 (Emerging outlier). *A temporal outlier O_E that starts appearing at some point of time.*
$\exists\, t_i, t_j \in [t_2, t_N]$ *s.t.* $(O_E \notin \Psi_p,\ \forall t_p \in [t_1, t_i])\ \wedge\ (O_E \in \Psi_q,\ \forall t_q \in [t_{i+1}, t_j])\ \wedge$ $(t_j > t_{i+1})$.

Definition 5 (Alternating outlier). *A temporal outlier O_A that is noticed at different points of time intermittently.*
$\exists\, t_i, t_j, t_k \in [t_1, t_N]$ *s.t.* $(O_A \in \Psi_i)\ \wedge\ (O_A \notin \Psi_j)\ \wedge\ (O_A \in \Psi_k)\ \wedge$
$(O_A \in \Psi_p,\ \forall t_p \in (t_i, t_j))\ \wedge\ (O_A \notin \Psi_q,\ \forall t_q \in (t_j, t_k))\ \wedge\ (t_k > t_j > t_i)$.

Definition 6 (Repeating outlier). *A temporal outlier O_R that keeps appearing at all points of time during the network evolution.*
$O_R \in \Psi_i, \forall\, t_i \in [t_1, t_N]$.

Definition 7 (Transient outlier). *A temporal outlier O_T that exists only at one point of time.*
$\exists\, t_i \in [t_1, t_N]$ *s.t.* $(O_T \in \Psi_i)\ \wedge\ (O_T \notin \Psi_j,\ \forall t_j \in [t_1, t_N])\ \wedge\ (t_j \neq t_i)$.

It is important to note that a halting outlier is different from a transient outlier, where the former one disappears after being present for more than one time point, while the later one exists precisely at one point of time. Also note that the set of top-k outliers Ψ_i present in a graph instance G_i may consist of one or more categories of temporal outliers defined above.

In view of the five categories of temporal outliers defined above, the proposed framework is named as *NetHEART* indicating that the detection of various network anomalies, by way of discovering Halting, Emerging, Alternating, Repeating and Transient (HEART) categories of outliers in dynamic networks, is essential for an effective management of these networks such as ensuring the security of the IP networks.

Algorithm 1. Detecting temporal outliers in evolving graphs

Input: A dynamic graph G with its N instances at different points of time.
Output: A ranked list Ω of temporal outliers and their scores Φ.

1: For each graph instance G_i perform the following steps.

(a) Construct the corresponding edge list L_i .
(b) Apply the graph outlier detection procedure described in [1] on L_i.
(c) Determine the set of top-k outliers Ψ_i along with their scores S_i.

2: Determine the grand set of outliers across all graph instances as

$$\Omega = \Psi_1 \cup \Psi_2 \cup \cdots \cup \Psi_N \tag{4}$$

3: For each temporal outlier $O_j \in \Omega$, determine its cumulative outlier score as

$$\phi_j = \frac{1}{N} \sum_{i=1}^{N} \xi_{i,j} \tag{5}$$

where

$$\xi_{i,j} = \begin{cases} s_{i,p} & \text{if } \exists p \ s.t. \ O_j = \psi_{i,p} \\ 0 & \text{otherwise.} \end{cases}$$

4: Determine the set of repeating outliers that exist across all graph instances as

$$\Omega^* = \Psi_1 \cap \Psi_2 \cap \cdots \cap \Psi_N \tag{6}$$

5: Similarly, determine all other categories of outliers based on the Ψ_i's.
6: For each outlier $O_j \in \Omega$, label it as per the category it belongs to.
7: Arrange the set Ω of temporal outliers in descending order of their scores.

3.2 Proposed Algorithm for Temporal Outlier Detection

According to the framework shown in Fig. 1, the initial task is to detect outliers present in a single graph instance. Though this task can be accomplished using any one of the established methods like [1,16], we have considered the method described in [1]. The rest of the framework deals with collating the instance specific outliers and characterizing their dynamics across multiple instances of a dynamic graph. Accordingly, a novel method for detecting various categories of temporal outliers in evolving graphs is proposed here as per the steps presented in Algorithm 1.

As per Equation 4, the grand set of outliers (Ω) consists of at most $N * k$ elements and at least k elements. Similarly, Equation 5 determines the cumulative outlier score of a temporal outlier by computing the average of its scores corresponding to various graph instances. Unlike the other recent methods [11,5], the detection of a temporal outlier is not affected by its cumulative outlier score. However, this score indicates the significance of a temporal outlier over all the graph instances.

The execution time of the proposed algorithm is mainly contributed by the time required to complete the first step. For a small k value with a reasonably small analysis time period ($N \leq 10$), the rest of the steps can be completed in a constant amount of time. Therefore, the computational complexity of the proposed algorithm is determined by that of the specific method employed for performing the first step, such as the method described in [1].

4 Experimental Evaluation

To demonstrate the efficacy of the proposed algorithm, an experimental evaluation has been carried out considering two real life dynamic graph data sets. For ease of illustration and simplicity in describing the results, only three graph instances ($N = 3$) have been considered corresponding to each one of these graph data sets as described below.

4.1 Data Sets

The DBLP Computer Science Bibliography from the University of Trier contains millions of bibliographic records bundled in a huge XML file [9]. Various sub-sets of these records concerned with AAAI publications have been considered here, denoted as DBLP-AAAI data set, as per the details furnished in Table 1.

Table 1. Details of the dynamic graph data sets

Data Set Name	Instance	# Nodes	# Edges	Graph Instance Details
DBLP-AAAI	G_1	5,581	25,276	Publications upto 2008
	G_2	6,219	28,622	Publications upto 2010
	G_3	7,551	35,628	Publications upto 2012
AS-2000	G_1	2,107	8,979	as19991231.txt
	G_2	3,570	14,783	as20000101.txt
	G_3	6,474	26,467	as20000102.txt

The graph of routers comprising the Internet can be organized into sub-graphs called Autonomous Systems (AS). Each AS exchanges traffic flows with some neighbors (peers). It is possible to construct a communication network of who-talks-to-whom from the BGP (Border Gateway Protocol) logs. The experiments in this paper have been carried out on three most recent graph files taken from the Autonomous Systems collection [8] as per the details furnished in Table 1, denoted as AS-2000 data set hereafter.

4.2 Outlier Detection and Characterization

As per the first step of the proposed algorithm, various node-based outliers present in each graph instance were detected using 'oddball' method [1]. The resulting top-k (=10) outliers on the DBLP-AAAI graph instances are shown in Fig. 2 indicated using node IDs along with their outlier scores. Subsequently, the grand set consisting of 13 temporal outliers ($|\Omega| = 13$) was determined (as per Equation 4) and the cumulative outlier score of each temporal outlier has also been computed accordingly (as per Equation 5). Then, various categories of temporal outliers (as defined in Section 3.1) present in this graph data set were identified as listed in Table 2, and the same have been depicted in Fig. 2 using a different line style for representing each category. Finally, a ranked list of labeled temporal outliers is produced as the output.

Similarly, the AS-2000 graph data set was also subjected to the computational steps listed in Algorithm 1, and various categories of temporal outliers thus detected have been furnished in Table 2 for completeness.

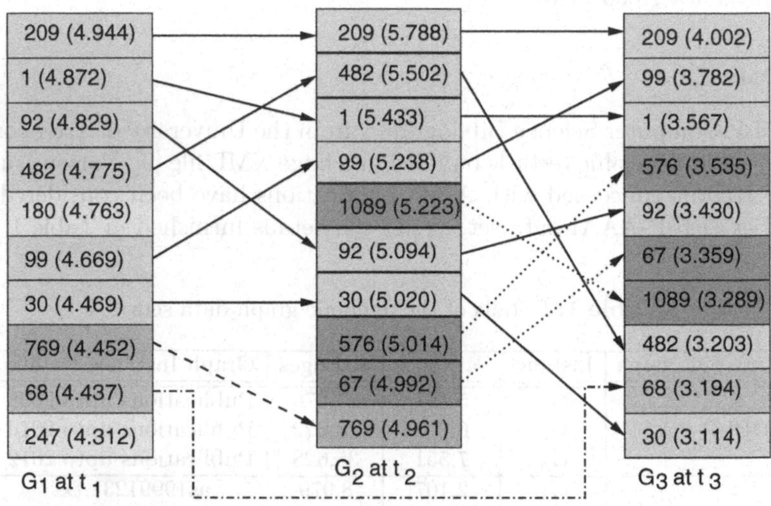

Fig. 2. Time varying characteristics of graph outliers in DBLP-AAAI data set

Table 2. Various categories of temporal outliers identified in dynamic graph data sets

Graph File	Repeating Outliers	Transient Outliers	Halting Outliers	Emerging Outliers	Alternating Outliers
DBLP-AAAI	209, 1, 92, 482, 99, 30	G_1:180, 247 G_2: — G_3: —	769	1089, 576, 67	68
AS-2000	4006, 4544 1746, 3563	G_1: 2551, 2041 145, 6467 G_2: 6261, 701 G_3: 5779, 5000 11329, 2828	721, 3847	6113, 668	—

It is important to note that the temporal behavior observed in this experimentation may vary with the number of top ranked outliers (k) considered corresponding to each graph instance and the number of instances (N) considered in each graph data set, as the outlier ranking is basically a relative measure.

4.3 Exploring the Outlier Dynamics

As brought out in Section 2.1, an EDPL plot [1] indicates two types of node-based anomalies, namely the near-star anomalies appearing below the fitting line and the near-clique anomalies above the fitting line. Accordingly, the EDPL plots obtained corresponding to different instances of the DBLP-AAAI data set are shown in Fig. 3. A keen observation at these plots indicates that as more and more new edges get added during the graph evolution from G_1 to G_3, one can notice the gradual appearance of more near-clique type of anomalies (triangle points marked in the upper portion of the plot) as shown in Fig. 3(b).

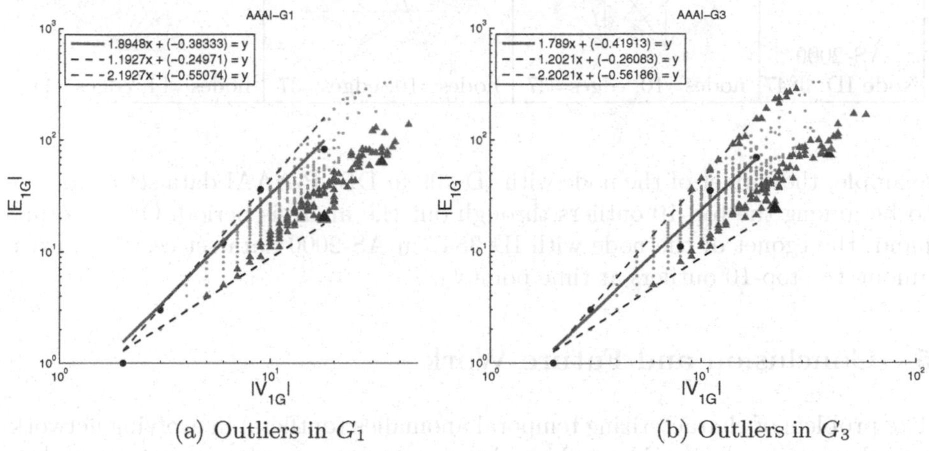

(a) Outliers in G_1 (b) Outliers in G_3

Fig. 3. Top 100 outliers detected in DBLP-AAAI data set

Temporal outliers present in an evolving graph are expected to highlight the inherent semantics of the graph connectivity patterns causing these outliers. To illustrate these semantic aspects, a few of the detected node-based outliers, in the form of egonet sub-graphs, are depicted in Table 3. The number of nodes and the number of edges present in each such sub-graph are also furnished along side in the same table. It is important to note that with the evolution of the graph over time, the underlying egonets have also undergone corresponding change in their connectivity structure. Though there is not any significant change expected in the node-based anomalous sub-graphs individually, their relative position in the top-k outlier rank list may change across various graph instances (refer to Fig. 2) due to the modified connectivity structure of the graph during evolution. For

Table 3. Sub-graphs of various node outliers in different graph instances

Graph File	Egonet in G_1 at t_1	Egonet in G_2 at t_2	Egonet in G_3 at t_3
DBLP-AAAI Node ID: 209	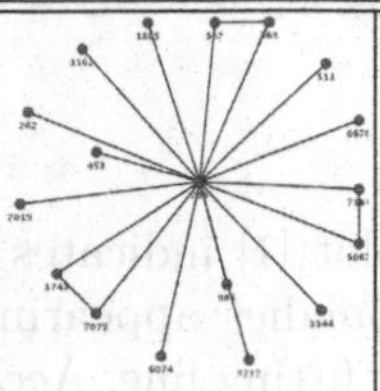 nodes=18, edges=20	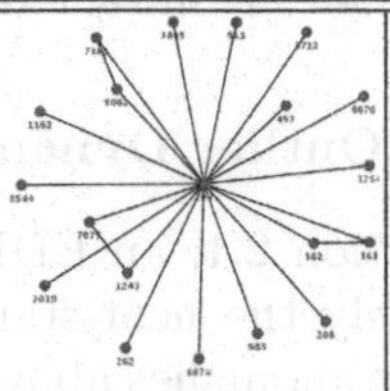 nodes=20, edges=22	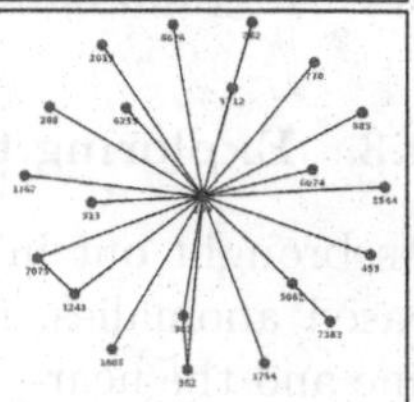 nodes=22, edges=24
AS-2000 Node ID: 3847	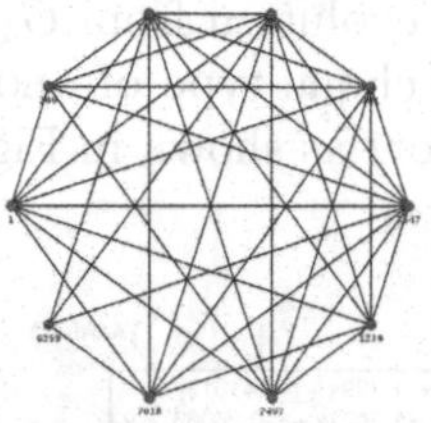 nodes=10, edges=37	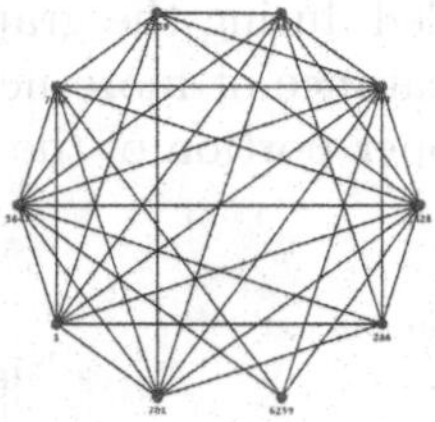 nodes=10, edges=37	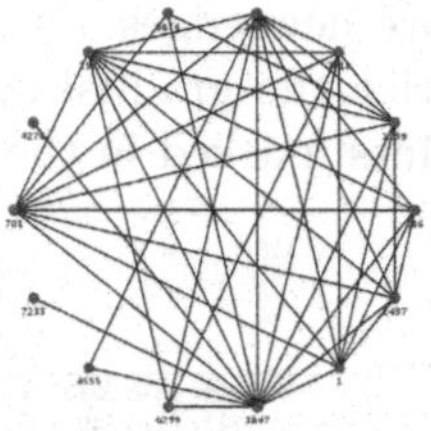 nodes=14, edges=44

example, the egonet of the node with ID 209 in DBLP-AAAI data set continues to be among the top-10 outliers through out the analysis period. On the other hand, the egonet of the node with ID 3847 in AS-2000 data set ceases to exist among the top-10 outliers at time point t_3.

5 Conclusion and Future Work

The problem of characterizing temporal anomalies (outliers) in evolving network (graph) data has been addressed here by proposing a novel framework for exploring their time varying behavior. The proposed method has its merit in defining various categories of temporal outliers taking into account their dynamics. An experimental evaluation of the proposed algorithm for temporal outlier detection has been carried out using two benchmark dynamic graph data sets. The experimental observations confirm the presence of various categories of temporal outliers as defined in this paper, demonstrating the effectiveness of the proposed method.

Employing alternative methods, such as the one described in [16], for discovering the anomalies present in the individual instances of an evolving graph and exploring their temporal behavior. Enhancement to the proposed categorization of temporal outliers by involving some graph in-variants may be other interesting direction to consider.

Acknowledgments. The authors would like to thank Director, CAIR for supporting this work.

References

1. Akoglu, L., McGlohon, M., Faloutsos, C.: `Oddball`: Spotting anomalies in weighted graphs. In: Zaki, M.J., Yu, J.X., Ravindran, B., Pudi, V. (eds.) PAKDD 2010. LNCS, vol. 6119, pp. 410–421. Springer, Heidelberg (2010)
2. Anagnostopoulos, A., Kumar, R., Mahdian, M., Upfal, E., Vandin, F.: Algorithms on evolving graphs. In: ACM ITCS, Cambridge, Massachussets, USA, pp. 149–160 (2012)
3. Chakrabarti, D.: AutoPart: Parameter-free graph partitioning and outlier detection. In: Boulicaut, J.-F., Esposito, F., Giannotti, F., Pedreschi, D. (eds.) PKDD 2004. LNCS (LNAI), vol. 3202, pp. 112–124. Springer, Heidelberg (2004)
4. Chandola, V., Banerjee, A., Kumar, V.: Anomaly detection: A survey. ACM Computing Surveys 41(3) (2009)
5. Ge, Y., Xiong, H., Zhou, Z.H., Ozdemir, H., Yu, J., Lee, K.C.: TOP-EYE: Top-k evolving trajectory outlier detection. In: ACM CIKM, Toronto, Canada, pp. 1733–1736 (2010)
6. He, W., Hu, G., Zhou, Y.: Large-scale ip network behavior anomaly detection and identification using substructure-based approach and multivariate time series mining. Telecommunication Systems 50(1), 1–13 (2012)
7. Kim, M., Leskovec, J.: Latent multi-group memebership graph model. In: ICML, Edinburgh, Scotland, UK (2012)
8. Leskovec, J.: Stanford large network dataset collection (2013), `http://snap.stanford.edu/data/index.html`
9. Ley, M.: Dblp - some lessons learned. PVLDB 2(2), 1493–1500 (2009)
10. Li, X., Bian, F., Crovella, M., Diot, C., Govindan, R., Iannaccone, G., Lakhina, A.: Detection and identification of network anomalies using sketch subspaces. In: ACM IMC, Rio de Janeiro, Brazil (2006)
11. Mongiovi, M., Bogdanov, P., Ranca, R., Singh, A.K., Papalexakis, E.E., Faloutsos, C.: Netspot: Spotting significant anomalous regions on dynamic networks. In: SDM, Austin, Texas, pp. 28–36 (2013)
12. Noble, C.C., Cook, D.J.: Graph-based anomaly detection. In: Proc. SIGKDD, Washington, DC, USA, pp. 631–636 (2003)
13. Papalexakis, E.E., Akoglu, L., Ienco, D.: Do more views of a graph help? community detection and clustering in multi-graphs. In: Fusion, Istanbul, Turkey, pp. 899–905 (2013)
14. Rossi, R.A., Neville, J., Gallagher, B., Henderson, K.: Modeling dynamic behavior in large evolving graphs. In: WSDM, Rome, Italy, pp. 667–676 (2013)
15. Thottan, M., Ji, C.: Anomaly detection in ip networks. IEEE Trans. on Signal Processing 51(8), 2191–2204 (2003)
16. Wu, L., Wu, X., Lu, A., Zhou, Z.: A spectral approach to detecting subtle anomalies in graphs. Journal of Intelligent Information Systems 41, 313–337 (2013)

An Integrated Model for User Attribute Discovery: A Case Study on Political Affiliation Identification

Swapna Gottipati[1], Minghui Qiu[1], Liu Yang[1,2], Feida Zhu[1], and Jing Jiang[1]

[1] School of Information Systems, Singapore Management University
[2] School of Software and Microelectronics, Peking University
{swapnag.2010,minghui.qiu.2010,liuyang,fdzhu,jingjiang}@smu.edu.sg

Abstract. Discovering user demographic attributes from social media is a problem of considerable interest. The problem setting can be generalized to include three components — users, topics and behaviors. In recent studies on this problem, however, the behavior between users and topics are not effectively incorporated. In our work, we proposed an integrated unsupervised model which takes into consideration all the three components integral to the task. Furthermore, our model incorporates collaborative filtering with probabilistic matrix factorization to solve the data sparsity problem, a computational challenge common to all such tasks. We evaluated our method on a case study of user political affiliation identification, and compared against state-of-the-art baselines. Our model achieved an accuracy of 70.1% for user party detection task.

Keywords: Unsupervised Integrated Model, Social/feedback networks, Probabilistic Matrix Factorization, Collaborative filtering.

1 Introduction

User demographic attributes such as gender, age, financial status, region are critically important for many business intelligence applications such as targeted marketing [1] as well as social science research [2]. Unfortunately, for reasons including privacy concerns, these pieces of user information are not always available from online social media platforms. Automatic discovery of such attributes from other observable user behavior online has therefore become an important research topic, which we call the user attribute discovery problem for short.

Existing work on detecting the user demographics on datasets such as blogs, micro-blogs and web documents [3,4,5,6,7,8] have mainly adopted the supervised approach and relied on either the connections among users such as user social network, or the language aspects in the data, or both. However, in many cases the interaction between the users and the topics is not effectively incorporated.

The first contribution of our work is that we proposed an integrated unsupervised model which takes into consideration all the three components integral to the user attribute discovery problem, namely the users, the topics and the feedback behavior between the user and the topics. In particular, besides social links

V.S. Tseng et al. (Eds.): PAKDD 2014, Part I, LNAI 8443, pp. 434–446, 2014.

between users, we exploit users' feedback on topics, which gives great insight into user affiliation that cannot be modeled by current approaches. Although illustrated with the case study on political affiliation, our model by design can be generalized for most other user attributes that are associated with users' behavior including but not limited to religion affiliation, technology affiliation, political affiliation etc., We present elaborated motivation for our model in Section 2.

The second contribution of our work is that we proposed a solution to a computational challenge common to the user attribute discovery problem: data sparsity — users might not participate in all the huge number of topics and the user social network could be sparse as sites such as forums and debates are not meant for maintaining social relations but to voice out public opinions. As such standard clustering [9] and community detection algorithms [10,11] would not give satisfactory results. We adopt collaborative filtering with probabilistic matrix factorization (PMF) [12], a technique that has been successfully applied for collaborative filtering-based recommendation tasks such as social recommendation [13]. In general, the intuition behind PMF is the assumption that, if two users have same rating/stance/opinion on item/topic/user, they tend to behave the same on other items/topics/users. PMF automatically discovers a low-rank representation for both users and items based on observed rating data. We then apply clustering algorithms on users to detect the communities of users

Lastly, we evaluated our method on data set collected from the CreateDebate site[1], and compared against state-of-the-art baselines as well as degenerative versions of our model. Among the various demographic attributes, the rapidly growing attention for user political affiliation is probably the most noticeable [3,14,15]. Therefore in this paper we study our model's performance on party affiliation detection problem. Our model improves the accuracy and gives promising results for political affiliation detection task.

2 Problem Setting

To motivate our integrated model, we present an analysis of the typical problem setting of user attribute discovery from social media data. We first present an overview of the data components, followed by a motivating example to demonstrate the importance of each element as well as the insufficiency of each if used alone. We finally point out the computational challenge of data sparsity which is common for a large class of user attributes including political affiliation as well as challenge of integrating model components in a principled approach.

2.1 Data Components

We define a typical problem setting of discovering user attributes from social media data to include three components — (I)users, (II)topics and (III)behaviors — which is illustrated in Figure 1. A topic here is defined as any item like a

[1] www.createdebate.com

movie or abortion that users can feedback/rate upon. Two kinds of *behaviors* are usually available from the social media data, (1) social behavior between users and (2) feedback behavior of users on topics, which we detail as follows.

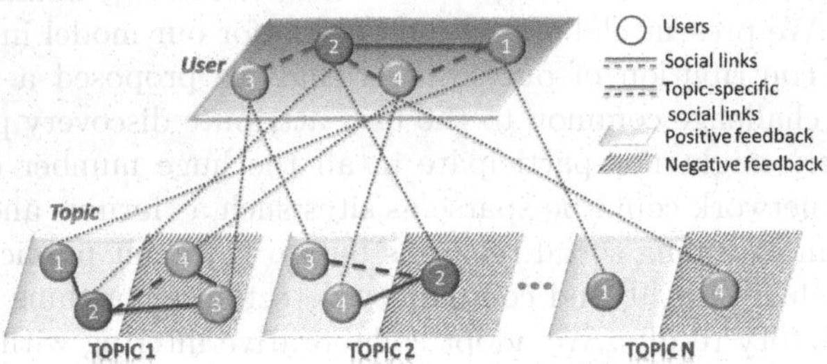

Fig. 1. An illustration of users, topics and behaviors. Social links are in user layer and topic-specific social links are in topic layers, where solid and dotted links represent positive and negative links respectively. Feedback behaviors are exhibited through user feedbacks on topics.

(1) Social behavior between users
The social behavior can be further categorized into two types. One type is the *topic-independent* one which is usually more long-term and stable, e.g., friendship/enmity, which we represent by *User Social Matrix.* Topic-independent social behavior is an important component used in several studies for prediction, recommendation and community detection tasks. The social friendship/enmity networks can be built from the friendship information or friendship/enmity information or sender/receiver or follower/followee information depending on the type of the network structure [14,16]. In Figure 1, under "User" layer, the social links represent the social matrix.

The other type is the *topic-specific* one reflecting users relationship on a particular topic, e.g., agreement/disagreement or thumbsup/thumbsdown on other user's feedback for a specific topic, which we represent by *User Interaction Matrix.* An important observation is that in forums or debate sites, users tend to dispute or agree with others on the debate issues by replying directly to the commenter. [17] observed that users not only interact with others who share same views, but also actively engage with whom they disagree. In Figure 1, under "Topic" layer, the topic-specific social links represent the interaction matrix. A pair of users exhibit different interactions across topics.

(2) Feedback behavior of users on topics
We focus on explicit user feedbacks such as ratings or stances on topics that can be observed as user opinions towards different opinion targets, represented by a *User Feedback Matrix.* In Figure 1, feedback behaviors are exhibited in user feedbacks on topics. This model, with slightly different variations, has been adopted by many previous work in social network and media analysis [13] [18]. The difference is that, while their problem is usually social recommendation, our task here is to discover users' implicit attributes.

2.2 Correlation Analysis

In Figure 2, we show the networks from our debate dataset using Gephi[2]. We use this data as an example to illustrate two observations.

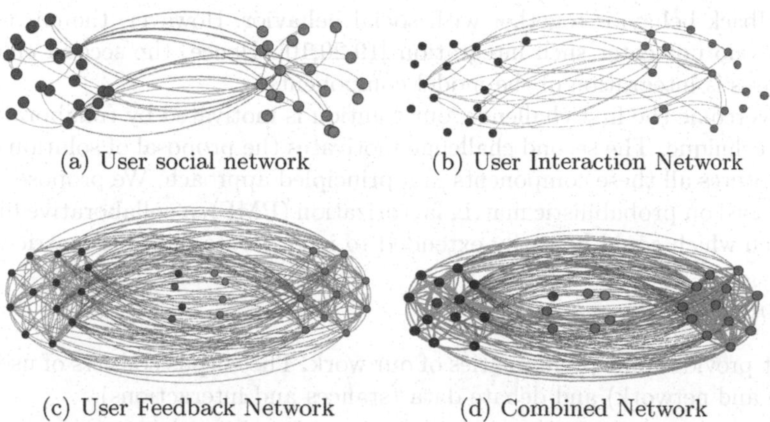

(a) User social network (b) User Interaction Network

(c) User Feedback Network (d) Combined Network

Fig. 2. Visualization of networks from a subset of our corpus. We used ground truth to represent the nodes and edges. Blue nodes are republicans, purple nodes are democrats, green edges represent positive(friendship/agreement) links and red edges represent negative(enmity/disagreement) links. For feedback and combined networks, a link between users in the network indicates whether these two users hold the same feedback most of the time(green edge means yes, red edge means no). Modular communities and nodes in the middle represents small individual communities that are misaligned.

(I) Each type of behavior provides an important insight into users' political affiliation; For example, Figure 2a), a social network shows more friendships (green edges) within parties than inter-parties. Figure 2b), an interaction network shows large disagreements (red edges) among users but very few agreements among them. In Figure 2c), a feedback network, we observe that within parties, the support on topics (green edges) is high compared to inter-parties. These observations indicate that both behaviors provide unique and important insights to users' attribute affiliation.
(II) Each type of behavior alone is not sufficient to accurately identify users' political affiliation. For example, Figures 2a) and 2b) consists of singletons which make the task of identifying the political affiliation harder. For combined network, Figure 2d), we have fewer misaligned users compared to feedback network indicating the benefits of combining social and feedback behavior.

Combining (I) and (II), we drive home the importance of an integrated model to consider all the networks.

[2] `http://wiki.gephi.org`

2.3 Computational Challenge

A common challenge in discovering user attributes from social media is data sparsity, which is actually shared by all data settings where the number of topics is huge and user participation is sparse. The debate data is a good case in point from the above analysis. Figure 2d) indicates the benefits of leveraging the feedback behavior together with social behavior. However, the current approaches do not cater such integration [19,20,10]. Hence, the second challenge that arises is integration of the model components.

To overcome the first challenge, our solution is motivated by collaborative filtering technique. The second challenge motivates the proposal of solution model that captures all these components in a principled approach. We propose a technique based on probabilistic matrix factorization (PMF), a collaborative filtering approach which could be easily extended to incorporate multiple matrices.

3 Solution

We first provide some preliminaries of our work. The corpus consists of user data (profile and network) and debate data (stances and interactions).

User Social Matrix: We use $\mathcal{S}$ to denote the user social network where each entry $s_{i,j}$ indicates the relationship between i-th user and j-th user (0 means enmity and 1 means friendship).

User Interaction Matrix: Each reply argument in our data set has an interaction information - *Disputed*, *Support* or *Clarified* to its recipient argument. We use $\mathcal{O}$ to represent agreement (*Support* and *Clarified*) and disagreement (*Disputed*) positions between users[3]. An entry $o_{i,k}$ in $\mathcal{O}$ equals to 1 means user i and user k mostly agree with each other and 0 otherwise. In case of such information is not available, we can use methods such as [21] to derive interaction network.

User Feedback Matrix: Feedback refers to user stances captured from *Side* status of a user. $\mathcal{R}$ represents the stances held by different users on the various opinion targets[4], where an entry $r_{i,m}$ is a stance score (0 for "Oppose" and 1 for "Support") indicating the i-th user's stance towards the m-th opinion target.

Given the matrices $\mathcal{R}$, $\mathcal{S}$ and $\mathcal{O}$, we perform probabilistic matrix factorization to derive a low-rank vector representation and based on which we apply clustering algorithms to detect political communities.

3.1 Probabilistic Matrix Factorization

The original matrix factorization model is proposed in [12]. The model is extended in [13] to incorporate user social network for recommendation. Our model

[3] Users with the same stance tend to dispute with each other indicating that stance matrix and interaction matrix do not overlap.

[4] An opinion target is defined as the topic that users can express their opinions on. It can be either a controversial topic like "Abortion" or "Gun Control" with "support" or "oppose" stances, or "Does God Exist?" with "Yes" or "No" votes.

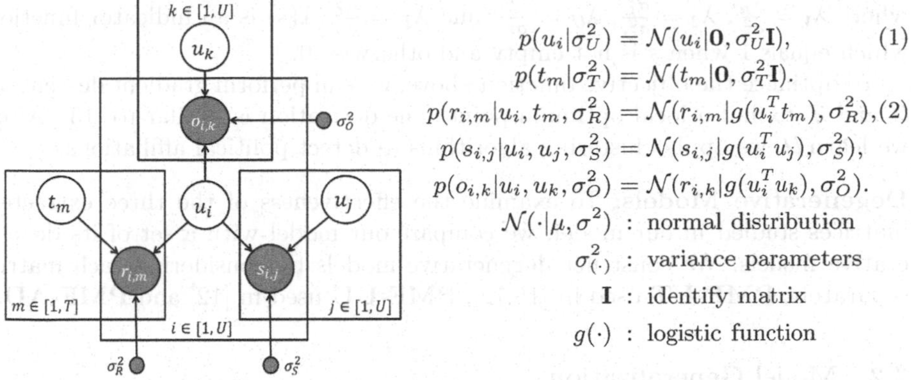

$$p(u_i|\sigma_U^2) = \mathcal{N}(u_i|\mathbf{0}, \sigma_U^2\mathbf{I}), \qquad (1)$$
$$p(t_m|\sigma_T^2) = \mathcal{N}(t_m|\mathbf{0}, \sigma_T^2\mathbf{I}).$$
$$p(r_{i,m}|u_i, t_m, \sigma_R^2) = \mathcal{N}(r_{i,m}|g(u_i^T t_m), \sigma_R^2), (2)$$
$$p(s_{i,j}|u_i, u_j, \sigma_S^2) = \mathcal{N}(s_{i,j}|g(u_i^T u_j), \sigma_S^2),$$
$$p(o_{i,k}|u_i, u_k, \sigma_O^2) = \mathcal{N}(r_{i,k}|g(u_i^T u_k), \sigma_O^2).$$

$\mathcal{N}(\cdot|\mu, \sigma^2)$: normal distribution
$\sigma_{(\cdot)}^2$: variance parameters
$\mathbf{I}$: identify matrix
$g(\cdot)$: logistic function

Fig. 3. Our probabilistic matrix factorization model on user stance and social behaviors. Priors over users and opinion targets are omitted for clarity.

is a direct extension on the model from [13], where we add one more user interaction matrix into the model. Another difference is that the user social network and interaction work are symmetric in our model.

Figure 3 shows the plate notation for the generative model. We assume that both users and opinion targets are profiled by K latent factors. Let $u_i \in \mathbb{R}^K$ denote the vector in the latent factor space for the i-th user, and $t_m \in \mathbb{R}^K$ denote the vector for the m-th opinion target. We assume u_i and t_m are generated by Gaussian distributions as in Eqn. 1. We assume the extracted matrices $\mathcal{R}$, $\mathcal{S}$ and $\mathcal{O}$ are generated by taking the product of the related entities. Specifically, the generation processes of the stance scores $r_{i,m}$ between i-th user and m-th opinion target, the polarity scores $s_{i,j}$ between i-th and j-th user and $o_{i,k}$ between i-th user and k-th user in the matrices $\mathcal{R}$, $\mathcal{S}$ and $\mathcal{O}$ are in Eqn. 2.

With this generative assumption, if two users are similar in terms of their dot product in the latent factor space, then they are more likely to have positive interactions or relations. Similarly, if two users share the same stance on an opinion target, then they are similar in the latent space. The latent factors can therefore encode user preferences and similarity between two users in the latent factor space reflects whether they share similar viewpoints.

Let $\mathcal{U}(K \times U)$ and $\mathcal{T}(K \times T)$ be user and opinion target matrices. To learn $\mathcal{U}$ and $\mathcal{T}$, we need to maximize the posterior of generating all the opinion matrices $\mathcal{R}$, $\mathcal{S}$ and $\mathcal{O}$ which is equivalent to minimize the following objective function:

$$\begin{aligned}
&\mathcal{L}(\mathcal{U}, \mathcal{T}, \mathcal{R}, \mathcal{S}, \mathcal{O}) \\
&= \frac{1}{2}\sum_{i=1}^{U}\sum_{m=1}^{T} \mathbb{I}(r_{i,m})(r_{i,m} - g(u_i^T t_m))^2 + \frac{\lambda_1}{2}\sum_{i=1}^{U}\sum_{j=1}^{U} \mathbb{I}(s_{i,j})(s_{i,j} - g(u_i^T u_j))^2 \\
&+ \frac{\lambda_2}{2}\sum_{i=1}^{U}\sum_{k=1}^{U} \mathbb{I}(o_{i,k})(o_{i,k} - g(u_i^T u_k))^2 + \frac{\lambda_U}{2}||\mathcal{U}||_F^2 + \frac{\lambda_T}{2}||\mathcal{T}||_F^2,
\end{aligned}$$

where $\lambda_1 = \frac{\sigma_R^2}{\sigma_S^2}$, $\lambda_2 = \frac{\sigma_R^2}{\sigma_O^2}$, $\lambda_U = \frac{\sigma_R^2}{\sigma_U^2}$, and $\lambda_T = \frac{\sigma_R^2}{\sigma_T^2}$, $\mathbb{I}(s)$ is an indicator function which equals 1 when s is not empty and otherwise 0.

To optimize the objective function above, we can perform gradient descent on $\mathcal{U}$ and $\mathcal{T}$ to find a local optimum point. The derivation is similar to [13]. After we learn $\mathcal{U}$, we apply clustering algorithms to detect political affiliations.

Degenerative Models: To examine the effectiveness of the three extracted matrices studied in our model, we compare our model with a set of its degenerative models. We construct degenerative models by considering each matrix separately: **PMF-UT** used in [15,12], **PMF-UU** used in [12] and **PMF-AD**.

3.2 Model Generalization

The attributes supported by our model are associated with user's behavior, where the users exhibit a debatable (support/oppose) behavior on topics. For example, attributes such as religion orientation, political leaning, technology affiliation etc, can be discovered through the users behavior in several debatable topics specific to the attribute. For model generality, the feedback behavior should be domain specific and aligned to the user attribute. Similarly, the corresponding social behavior should be captured in the same settings. For example, for discovering users' technology affiliation e.g., *Apple* vs *Microsoft*, the feedback behavior should be captured for the topics related to technology, e.g., operating system, software, usability, etc., but not on lifestyle or politics. In our experiments under Section , we show our motivation for defining feedback topics for political affiliation task. To support multi-valued attributes such as multiple parties, the users can be grouped into the multiple clusters.

4 Experiments

4.1 Dataset

We collected user profile information for registered users and the corresponding sociopolitical debates' information in which these users participated in Createdebate.com. In our experiments, we focus on political affiliation discovery task.

Table 1. Some statistics of the data. Interaction links are based on controversial debates.

Users	1773 (145 for evaluation - 63 Democrats and 82 Republicans))
Controversial debates	88
Controversial arguments	Abortion(1,423), gun control(1,148), same-sex marriage(1,000), death penalty(588), universal health care(438), higher taxes(418), total (5,012)
Social network links	1540 (68% friendship and 32% enmity links)
Interaction links	2830 (31% agreements and 69% disagreements)

Testbed. The statistics of the data are also shown in the Table 1. Recall that the feedback behavior should be domain specific. For our study, we use only the two-sided debates on 6 controversial issues which are specific to political domain and motivated by party's beliefs[5] listed in Table 1. Since, the sides of debates associated with the same topic can be flipped, they should be aligned manually. We engaged two judges to manually label the sides for 88 debates as suport/oppose to the topic and both of them had perfect agreement (Cohen's kappa coefficient is 1).

Matrix Generation. Recall that our solution model compromises of three matrices, **UU**, **AD** and **UT**. *User social matrix* **UU** (represents user-user matrix $\mathcal{S}$ in the solution) is generated from user friendship/enmity links. *User interaction matrix* **AD** (represents agreement/disagreement matrix $\mathcal{O}$ in the solution) is generated from agreement/disagreement (interaction) links among users. *User feedback matrix* **UT** (represents user-opinion target matrix $\mathcal{R}$ in the solution) is from the user stances ("Support/Oppose") on the debates (topics).

4.2 Political Affiliation Discovery Experiments

The main goal of our study is to discover the political affiliation of the user. Through this experiment, we would like to study not only the model performance but also the performance of feedback and social behaviors independently.

Experimental Settings. The ground truth on the users' political leaning is available from the users' profiles. We use all three matrices, **UU**, **AD** and **UT** described in Section 4.1. Apart from 3 degenerative baseline models described in Section 3, we consider 3 additional baselines, described as below:

Discussant Attribute Profile (DAP): A recent work [9] proposes to profile discussants by their attribute towards other targets and use standard clustering (K-Means) to cluster discussants, and achieves promising results on a similar task - subgroup detection. We thus incorporate the method on our task by profiling each user by his/her opinions towards other opinion targets and users.

Correlation Clustering (CC): Correlation clustering [20], aids to partition the corresponding signed network such that positive intra-group links and negative inter-group links are dense. It is also used in subgroup detection [19]. We use large scale correlation clustering tool [22] for our baseline.

Louvain Method (LM): Louvain method [23] is an efficient algorithm to find high modularity partitions of large networks, and is widely used in community detection. When the method produces more than two communities, we align small communities to one of the two largest communities by maximizing intra-community positive links and inter-community negative links. Since the method usually discovers too many communities when applied on sparse disconnected network (on UU and AD, more than 60 communities detected), we only apply it on UT and combined(UU+UT+AD) matrices.

[5] `http://www.diffen.com/difference/Democrat_vs_Republican`

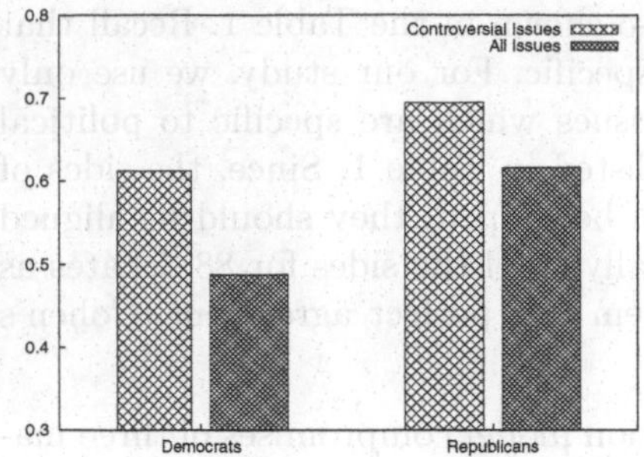

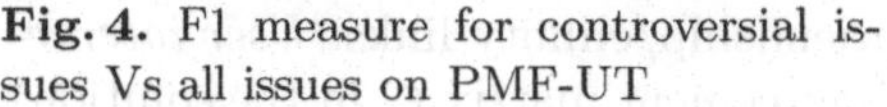
Fig. 4. F1 measure for controversial issues Vs all issues on PMF-UT

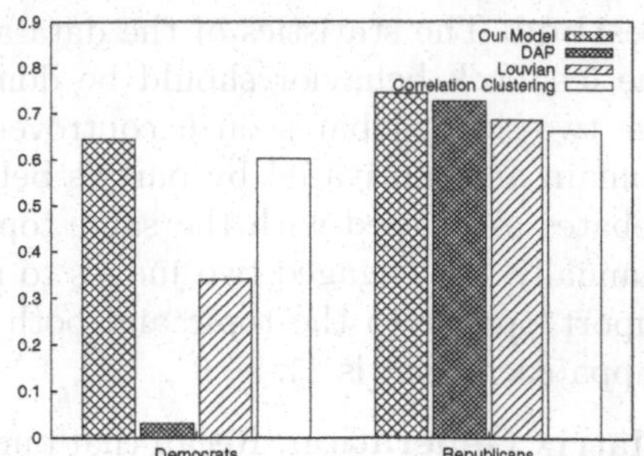

Fig. 5. F1 measure on party detection evaluation on all models using combined matrix (UU+UT+AD)

In all our experiments, for our model, we set the number of latent factors to 10 as we do not observe big difference when vary the latent factor size from 10 to 50. For the other parameters in probabilistic matrix factorization methods, we select the optimal setting for each method based on the average of 10 runs. λ_1, λ_2, λ_U and λ_T are chosen from $\{0.1, 0.01\}$. We use *Purity*, *Entropy* [24], *Accuracy* and *F1 score* to evaluate the models.

Results. We first present an overall performance of all the methods on combined (UU, AD, UT) matrix and show F1 measure on the party affiliation detection task in Figure 5. We observe that our model outperforms all the baselines on the task. It achieves F1 of 74.6% and 64.5% for republicans and democrats respectively. The best baseline, correlation clustering, achieves F1 of 66.2% and 60.2% for republicans and democrats respectively. In comparison, our model has 8.4% higher performance for republicans and 4.3% higher for democrats.

We present the detailed clustering results in Table 2. We observe that the combined matrix, (UU+UT+AD) has the highest performance for most of the baselines and our model. Our model outperforms all the baseline models with balanced clusters. From these results, it is evident that combining all the matrices is important for the political affiliation detection. Also, it is evident that the feedback behavior plays an important role in this task. In particular, from these results we observe that for data sets such as debates where social relations are

Table 2. Clustering results for political affiliation detection. Combined represents UU+UT+AD. P, E and A refer to Purity, Entropy and Accuracy, respectively.

Method	UU			UT			AD			Combined		
	P	E	A	P	E	A	P	E	A	P	E	A
CC	0.57	0.99	0.56	0.57	0.98	0.52	**0.57**	0.99	0.55	0.64	0.91	0.64
DAP	0.57	0.99	0.56	0.57	0.98	0.51	**0.57**	**0.98**	**0.57**	0.57	0.98	0.57
LM	N/A	N/A	N/A	0.61	0.95	0.59	N/A	N/A	N/A	0.64	0.93	0.63
PMF	**0.58**	**0.96**	**0.58**	**0.65**	**0.93**	**0.65**	0.56	0.98	0.54	**0.70**	**0.88**	**0.70**

sparse, the feedback behavior of participants aids to bridge the gaps and performs efficiently in political affiliation discovery.

Summary. Our model performs with promising results compared to baselines and original PMF model with an accuracy of 70.1%. We further experimented on three more degenerated baseline versions of our model, PMF-UUUT, PMF-ADUT and PMF-UUAD. In each of the baseline version, we remove one matrix from the original model to learn the latent factor representation. For PMF-UUUT, we choose UU and UT to learn the latent factor representation. This model is similar to the one used in [13]. For PMF-ADUT, matrices AD and UT are used and for PMF-UUAD we use UU and AD matrices. Our model on combined matrix still outperforms all these baseline degenerated models. Due to space constraints, we skip the details from Table 2.

4.3 Threats to Validity

Similar to other empirical studies, there are several threats to validity in interpreting the results. One such threat corresponds to the ability to link the stance(feedback) behavior to political affiliation. Our experimental results supports that leveraging stance behavior aids in political affiliation discovery and we used standard metrics for evaluation.

Another threat corresponds to the topics chosen for the feedback. "Do all issues that users participate can aid in detecting affiliation?" We crawled 1,332 debates and corresponding arguments, 10,833 on all issues. We study the performance of our model using users' leaning on controversial issues versus all issues and Figure 4 shows F1 measure on the affiliation detection. We observe that **UT** (controversial topics) outperforms $\mathbf{UT}_{all}$ (all topics). For republicans, F1 is 68% which is 6.9% higher than $\mathbf{UT}_{all}$ and for democrats, F1 is 60% which is 11.5% higher than $\mathbf{UT}_{all}$. The results indicate that a user's stances towards controversial topics have strong correlations with their political affiliation.

Another threat corresponds to the ability to generalize our results. For data generality, we evaluated with 145 users on political debates which is no way near the number of users available in all online debates. In future, we plan to extend our case study to include more users, debates and demographics. For attributes such as age, gender etc., the controversial topics should be carefully engineered and the current model cannot be applied directly. For model generality, an interesting future work is to study the cases where users can fall into more than one group and multi-party situations.

5 Related Work

User profiling studies examine users' interests, gender, sex, age, geo-localization, and other characteristics of the user profile. [25] aggregated social activity data from multiple social networks to study the users' online behavior. For user profiling, many studies took a supervised approach on various datasets; gender classification on blog data [6], age prediction on social networks [7] and location of origin prediction in twitter [8].

Similar research to user profiling studies is community or subgroup detection. [9] proposed a system that uses linguistic analysis to generate attitude vectors on ideological datasets. [23] used Louvain method which is based on modularity optimization to find high modularity partitions of large networks. Subgroup detection is studied in [19] using clustering based techniques. [26] studied both textual content and social interactions to find opposing network from online forums. In our work, besides user-user social links, we use feedback behavior which cannot be modeled by current community detection approaches.

Our proposed technique which is based on probabilistic matrix factorization (PMF) [12], a collaborative filtering based method originally used for recommendation tasks. The PMF method has been applied on social recommendation [13], news article [18] recommendation, relation prediction [27] [28] and modeling friendship-interest propagations [16]. In particular, [12] proposed a PMF model that combines social network information with rating data to perform social recommendation and [28] extended PMF for relation prediction task. Our model is a direct extension on [13] where we model three components: social, interaction and feedback. Besides this, our model assumes symmetric user social behavior.

Similar to our political affiliation task, a line of research was devoted to discover the political affiliations of informal web-based contents like news articles [29], weblogs [4], political speeches [30] and web documents [3]. Political datasets such as debates and tweets are explored for classifying user stances[31]. These applications are similar to our task as they are focussed on political content and relies on left-right beliefs.

For users' political affiliation identification on Twitter, using supervised approaches, [32,14,5] achieved high accuracy and [33] using semi-supervised label propagation method, achieved high accuracy. These studies report high performance just based on textual content or hashtags with strong nuances on political affiliations which are unique Twitter properties. Where as, we proposed an unsupervised approach and studied on data without special text characteristics. In our previous work [15], we exploited feedback behavior for the same task. However, the model performance degrades with high sparsity rate. In this work, we proposed a principled way to integrate social links and user's feedback.

6 Conclusion

In this paper, we proposed an unsupervised integrated approach based on probabilistic matrix factorization that combines social and feedback behavior features in a principled way to cater two major challenges - combining integral data components and data sparsity. Interesting future work is to study multiple-party cases and user demographics discovery such as technology or religion.

Acknowledgements. This research/project is supported by the Singapore National Research Foundation under its International Research Centre@Singapore Funding Initiative and administered by the IDM Programme Office.

References

1. Burke, R.: Hybrid recommender systems: Survey and experiments. User Modeling and User-Adapted Interaction 12(4), 331–370 (2002)
2. Behrman, J.R., Behrman, J., Perez, N.M.: Out of Sync? Demographic and other social science research on health conditions in developing countries. Demographic Research 24(2), 45–78 (2011)
3. Efron, M.: Using cocitation information to estimate political orientation in web documents. Knowl. Inf. Syst. 9(4) (2006)
4. Durant, K.T., Smith, M.D.: Mining sentiment classification from political web logs. In: WebKDD 2006 (2006)
5. Pennacchiotti, M., Popescu, A.M.: Democrats, republicans and starbucks afficionados: user classification in twitter. In: KDD 2011, pp. 430–438 (2011)
6. Yan, X., Yan, L.: Gender classification of weblog authors. In: AAAI 2006, pp. 228–230 (2006)
7. Peersman, C., Daelemans, W., Vaerenbergh, L.V.: Predicting age and gender in online social networks. In: SMUC, pp. 37–44 (2011)
8. Rao, D., Yarowsky, D., Shreevats, A., Gupta, M.: Classifying latent user attributes in twitter. In: SMUC 2010, pp. 37–44 (2010)
9. Abu-Jbara, A., Diab, M., Dasigi, P., Radev, D.: Subgroup detection in ideological discussions. In: ACL 2012, pp. 399–409 (2012)
10. Blondel, V.D., Loup Guillaume, J., Lambiotte, R., Lefebvre, E.: Fast unfolding of communities in large networks. Statistical Mechanics (2008)
11. Traag, V., Bruggeman, J.: Community detection in networks with positive and negative links. Physical Review E 80(3), 036115 (2009)
12. Salakhutdinov, R., Mnih, A.: Probabilistic matrix factorization. In: Advances in Neural Information Processing Systems (NIPS), p. 20 (2008)
13. Ma, H., Yang, H., Lyu, M.R., King, I.: Sorec: Social recommendation using probabilistic matrix factorization. In: Proc. of CIKM (2008)
14. Pennacchiotti, M., Popescu, A.M.: A machine learning approach to twitter user classification. In: ICWSM (2011)
15. Gottipati, S., Qiu, M., Yang, L., Zhu, F., Jiang, J.: Predicting user's political party using ideological stances. In: Jatowt, A., et al. (eds.) SocInfo 2013. LNCS, vol. 8238, pp. 177–191. Springer, Heidelberg (2013)
16. Yang, S.H., Long, B., Smola, A., Sadagopan, N., Zheng, Z., Zha, H.: Like like alike: joint friendship and interest propagation in social networks. In: WWW 2011 (2011)
17. Yardi, S., Boyd, D.: Dynamic Debates: An Analysis of Group Polarization Over Time on Twitter. Bulletin of Science, Technology & Society 30(5), 316–327 (2010)
18. Pan, R., Zhou, Y., Cao, B., Liu, N.N., Lukose, R., Scholz, M., Yang, Q.: One-class collaborative filtering. In: ICDM 2008 (2008)
19. Abu-Jbara, A., Radev, D.: Subgroup detector: a system for detecting subgroups in online discussions. In: ACL 2012, pp. 133–138 (2012)
20. Bansal, N., Blum, A., Chawla, S.: Correlation clustering. In: Machine Learning, pp. 238–247 (2002)
21. Galley, M., McKeown, K., Hirschberg, J., Shriberg, E.: Identifying agreement and disagreement in conversational speech: use of bayesian networks to model pragmatic dependencies. In: ACL 2004 (2004)
22. Bagon, S., Galun, M.: Large scale correlation clustering optimization. CoRR (2011)

23. Traag, V., Bruggeman, J.: Community detection in networks with positive and negative links. Physical Review E 80(3), 036115 (2009)
24. Manning, C.D., Raghavan, P., Schütze, H.: Introduction to Information Retrieval. Cambridge University Press (2008)
25. Benevenuto, F., Rodrigues, T., Cha, M., Almeida, V.: Characterizing user behavior in online social networks. In: ACM SIGCOMM 2009, pp. 49–62 (2009)
26. Lu, Y., Wang, H., Zhai, C., Roth, D.: Unsupervised discovery of opposing opinion networks from forum discussions. In: CIKM 2012, pp. 1642–1646 (2012)
27. Singh, A.P., Gordon, G.J.: Relational learning via collective matrix factorization. In: KDD 2008, pp. 650–658 (2008)
28. Qiu, M., Yang, L., Jiang, J.: Mining user relations from online discussions using sentiment analysis and probabilistic matrix factorization. In: NAACL (2013)
29. Zhou, D.X., Resnick, P., Mei, Q.: Classifying the political leaning of news articles and users from user votes. In: ICWSM (2011)
30. Dahllöf, M.: Automatic prediction of gender, political affiliation, and age in Swedish politicians from the wording of their speeches - a comparative study of classifiability. LLC 27(2), 139–153 (2012)
31. Somasundaran, S., Wiebe, J.: Recognizing stances in ideological on-line debates. In: NAACL HLT 2010, pp. 116–124 (2010)
32. Conover, M., Gonçalves, B., Ratkiewicz, J., Flammini, A., Menczer, F.: Predicting the political alignment of twitter users. In: SocialCom 2011 (2011)
33. Boutet, A., Kim, H.: What's in Twitter? I Know What Parties are Popular and Who You are Supporting Now! In: ASONAM 2012, vol. 2 (2012)

Programmatic Buying Bidding Strategies with Win Rate and Winning Price Estimation in Real Time Mobile Advertising

Xiang Li and Devin Guan

Drawbridge Inc, 2121 S. El Camino Real, San Mateo, CA, 94403, USA
{xiang,devin}@drawbrid.ge

Abstract. A major trend in mobile advertising is the emergence of real time bidding (RTB) based marketplaces on the supply side and the corresponding programmatic impression buying on the demand side. In order to acquire the most relevant audience impression at the lowest cost, a demand side player has to accurately estimate the win rate and winning price in the auction, and incorporate that knowledge in its bid. In this paper, we describe our battle-proven techniques of predicting win rate and winning price in RTB, and the corresponding bidding strategies built on top of those predictions. We also reveal the close relationship between the win rate and winning price estimation, and demonstrate how to solve the two problems together. All of our estimation methods are developed with distributed framework and have been applied to billion order numbers of data in real business operation.

Keywords: Mobile advertising, programmatic buy, real time bidding (RTB), win rate estimation, winning price estimation, bidding strategy.

1 Introduction

A recent trend in mobile advertising is the emergence of programmatic buying in real time bidding (RTB) based marketplace, where each advertiser bid on individual impression in real time. Unlike the conventional mediation type marketplace with pre-negotiated fixed clearing cost, the clearing price in the RTB marketplaces depends upon the bid that each advertiser submits. In second price auction [14], which is commonly used in the mobile advertising RTB marketplace, the clearing price is actually the second highest bid. While at one hand this unique behavior in RTB marketplaces gives advertiser greater flexibility in implementing their own biding strategy based on their own need and best interest, it also makes the market more competitive and demand solutions to some new problems that don't exist in mediation type marketplaces. Some of the specific problems are predicting the win rate given a bid, estimating the most likely clearing price in the second price auction setup, and optimizing the bid based on various estimates. In this paper, we describe our approaches of estimating the win rate and winning price given the bid and correspondingly how do we carry out bidding strategies in different RTB auction setups.

V.S. Tseng et al. (Eds.): PAKDD 2014, Part I, LNAI 8443, pp. 447–460, 2014.

The general workflow in a RTB marketplace is the following: As an end-user launches a mobile app, which we call *"property"*, a request will be sent to the ad-exchange or marketplace, and it will be further passed to all the bidders who trade in that ad-exchange. Those bidders get the information regarding the property, the available ad space to display on the property, and some basic information regarding the device, e.g. device type, os version etc., and they need to decide if they want to place a bid for this particular request, and if so, how much to bid. Once ad-exchange receives all the response from those bidders, it will pick the winner with the highest bid, determine the cost and notify the winner. As an analogy to financial market, an ad-exchange is on the "sell" or "supply" side, and a bidder is on the "buy" or "demand" side.

Win rate estimation refers to the problem of estimating the likelihood of winning an incoming auction request given a specific bid price. It imposes different level of difficulties to the ad-exchange on the sell side and the bidder on the buy side. For the ad-exchange, it has the complete observations regarding all the bids that it receive from all the participants in the auction and it can hereby construct a win rate estimate fair easily by taking a histogram on the winning bid. The task of win rate estimation is more challenging to a bidder in the RTB, in the sense that he only knows his own bid and the outcome of the auction, and he has no idea about other people's bid. In other words, there is a missing attribute situation to the bidder in RTB. Being able to utilize a huge quantity of data with partial missing attribute is the key to a successful demand side win rate estimation.

Winning price estimation is another unique problem in the second price auction type RTB, and is indeed an even more challenging problem to the bidder. In most of the RTB exchanges, a bidder will only know the winning price information if his bid win the auction. Most RTB exchanges, especially those with big volume, are highly competitive in terms of number of bidders participating in the auction, and that consequently push down the win rate for each individual bidder. For each bidder in the auction, his average win rate in general decreases as more competitors join the auction, and such decrease of the win rate further reduces the amount of positive data available to him for the winning price prediction and that makes his estimation more difficult, which could further reduces his win rate. Such vicious circle of win rate and winning price makes the winning price prediction a critical component in the RTB auction.

Despite all the above challenges on the win rate and winning price prediction tasks, there is one good news regarding those two problems, in the sense that those two are closely related to each other. With some statistical transformation, the solution to the one problem can be automatically applied to the other. Specifically, we will illustrate in section 4 that win rate is essentially the cumulative distribution function (cdf) of the corresponding winning price distribution. Solutions of one problem bring the solutions to the other. In this paper we choose to approach the problem from the win rate side, first model the win rate using a logistic regression model, and then take the derivative of win rate estimation to generate the distribution of the winning price, and use the expected value of the distribution under the bid price as the winning price estimate.

While win rate and winning price model enables a RTB bidder to predict the likelihood of success and the associated cost, the actual programmatic bidding has to be done through a bidding strategy. A bidding strategy is actually an optimization function, takes the input of expected revenue if winning the auction (i.e. ecpm estimate), win rate and winning price estimate, and generate the final bid price according to some pre-defined objective functions. Some specific strategies could be to maximize the actual revenue or the profit from the operation, or the combination of both revenue and profit.

The paper will be organized as the following. Section 2 reviews some of the existing works in the field. In section 3 and 4 we describe our effort of estimating the win rate and the winning price given a bid, respectively; and in section 5 we explain the various bidding strategies that we have tested. The experimental setup and results are illustrated in section 6, and we conclude in section 7 with our contributions.

2 Related Work

Our win rate estimation techniques are based on logistic regression methods, and there are some existing works of estimating various probabilistic events in advertising using logistic regression models. For example, there are those of using logistic regression models to estimate click through rate [9][10][12], television audience retention [11], contextual effect on click rates [12] and bounce rate in sponsored search [13] etc. To estimate the winning price in auction, there are in general two high level approaches: machine learning based approaches as in [3][5][6] and historical observation or simulation based approaches [5-8]. R. Schapire et al. [3] used boosting based conditional density estimation in Tac-2001 competition [4]. They treated the price estimation as a classification problem, discretized the price into different buckets, and used boosting approach to estimate the selling price. For bidding strategies, many existing work [3-8] are closely related to the corresponding winning price estimation tasks, mostly due to the fact that they are motivated by series of TAC competitions [4].

Despite all those existing work, we see very few publications regarding the estimation and bidding strategies in the setup that we are facing in our everyday business operations. For example, many of the existing winning price estimation work are based on the assumption that a buyer has complete observation regarding the past auction outcome [3], while this assumption is apparently not valid in our daily operation.

3 Win Rate Estimation

3.1 Logistic Regression Based Win Rate Estimation and Corresponding Features

The likelihood of winning an auction given a bid price depends on two high level factors. One is the characteristics of the incoming request, and the other is the bid that a bidder willing to pay. Not all the incoming requests are with the same value, and there are many different attributes that affect the quality of each incoming request. For two requests with the same bid price, the win rate could be drastically different depending on the attributes like the nature of the mobile app (property), the time of the request, the size of the available ad-space, the geo location of the user, and many other attributes. The second factor that affects our win rate is apparently the bid price itself: for the same request, the higher our bid is, the more likely we will win the auction. Those two factors have to be included in the win rate estimation as features.

As a probability term bounded between 0 and 1, win rate makes itself a perfect candidate of using logistic regression model [10], as in Eq. 1:

$$winRate = \frac{1}{1+e^{-z}}, \ where \ z = \beta + \sum_{i=0}^{n} \theta_i * x_i \quad (1)$$

, where *winRate* is the estimated win rate, β is the intercept, x_i is individual feature extracted from the request, and θ_i is the corresponding model weight.

With no surprise, the list of the features that we have constructed resonates with the two high level factors we mentioned above. We extracted various features regarding the nature of the request, our bid, and we combine them to make the win rate prediction. Some of the features are *"stand-alone"* features that describe single attribute of the request that we received, e.g. the name of the app; and others are what we called *"cross"* features that describe the inter-action between individual *"stand-alone"* features. For example, we have a *"cross"* feature to describe mobile app name and day of the week that we receive the request. With *"stand-alone"* and *"cross"* features all together, there are about 1 Million total features in our win rate model.

3.2 Scaling Up Win Rate Prediction with Distributed Machine Learning

One of the challenges we are facing with win rate prediction, as with many other machine learning tasks, is the scale of the data that is available to us. The number of daily requests that we receive from the ad-exchanges is at the order of billions, and literally for every request that we submit our bid, it can be used as training data for our win rate prediction model, either as negative or positive data, depending on if our bid win the auction or not. With such huge amount of available data, it will be a crime to down-sample and use only a small percentage of the data in order to fit into existing non-distributed machine learning toolkit. Instead, we choose to utilize as much of the data as possible to build our model, and as we will illustrate in section 6, more data does help on the prediction accuracy.

Our approach of utilizing such huge amount of data is to use distributed machine learning algorithm and toolkit, for example Mahout[1] and Vowpal Wabbit(VW) packages[2]. We have tried both packages, and we end up using the VW. VW is more specialized in the classification tasks using general linear learner, while Mahout focuses more on the recommendation, clustering and general machine learning tasks. From our own observations, VW is faster than Mahout, especially for the large-scale sparse training data that we have used. On average our win rate model utilizes about 1 or 2 billion of records as training data for each model update, and the training process can be finished in couple of hours using VW. VW is faster because it uses true parallel processing with message passing interface, while Mahout is built on top of the MapReduce framework.

3.3 Feature Selection, Regularization and Missing Feature in Win Rate Prediction

While the distributed machine learning algorithm and tools give us the capability of utilizing the huge amount of available data, we still need to answer some of the traditional machine learning questions before we can build an accurate prediction model. The first question is how to deal with large number of features that can be extracted from the available data. As mentioned earlier, the original number of unique features is in the order of million, and we need to avoid directly feeding all those features into model building process. Typically there are two high level approaches to address this high feature dimension problem: Feature selection and regularization. Feature selection techniques can effectively reduce the number of unique features before model building actually take places, but it has to be done off-line first and it's quite expensive. On the other hand, regularization techniques mix the feature selection process with model building process and are more efficient than the separate feature selection approaches. Within regularization, L1 regularization automatically decides the feature to get non-zero value during the model building process, and L2 regularization can put more emphasis into more discriminant features. In our case, we decided to do regularization directly during model training without a separate feature selection process.

The second question we need to answer is regarding the new attributes. Regardless of how frequently do we update the model, and even if we use online model updating process, it's still quite often that we will see new attribute values pop up from the request, and this is especially true in large ad-exchanges since they keeps on adding new mobile apps to their inventory. Almost every week, we see new app become available for bidding, and all the features associated with that new app will become undefined. This significantly affects the accuracy of our prediction model.

Our solution to address this new attribute problem is to add *"filler"* feature into the model building process. During the model training, we will remove certain features and replace those removed features with corresponding *"filler"* features to build model. When we make predictions for new attribute values, e.g. a new mobile app (property) name, we just use *"filler"* feature to represent the value and generate win rate estimate.

4 Winning Price Estimation

The cost in an auction depends on the format of the auction. In the first price auction, winning price is exactly the same as the bid. In the second price auction [14], winning price is the second highest bid that the ad-exchange receives. Due to business constrains, ad-exchange only notify the bidder what the winning price is if the bidder actually win the auction. If a bidder loses the auction, all the information he would know is that the winning price is at least the same as his bid, since otherwise he would win the auction. Also, as we mentioned earlier, because there are many bidders in some of the largest mobile ad-exchanges, the typical win rate for a bidder in such large mobile ad-exchange is in the order of single digit. All the above factors translate into two problems for machine learning based winning price estimation: 1) unbalanced distribution among positive and negative training data, and 2) missing value issue in the negative training data.

4.1 Linear Regression Based Winning Price Prediction

Maybe the simplest way of estimating the winning price is to use linear regression. Specifically, we could extract features from all the auctions that we have won, and try to fit a linear regression function as in Eq. (2). This approach has immediate problems in the followings three ways: 1) it assumes the relationship between winning price and all the features, including price, to be linear, which is hardly satisfied. 2) It only uses the positive data, the observed winning price to make prediction, and throw away all the negative data for the cases that we didn't win. While those negative data point don't contain as valuable information as the positive data point, they still provide us some information, e.g. the winning price is at least the same as our bid. 3) It's a deterministic process, for the request with same attribute values and same bid, it will always return one fixed winning price. In reality, winning price is determined based on the behavior of our competitors. Behavior change of our competitor causes the winning price to fluctuate, and we want a way to model this fluctuation if possible. Due to those limitations of linear regression, it was no surprise to see that it didn't perform well in winning price estimation task.

$$winningPrice = \theta_0 + \theta_1 x_1 + \theta_2 x_2 + \cdots + \theta_n x_n \quad (2)$$

4.2 Solving Win Rate and Winning Price Estimation Problems All Together

Fortunately, there is one good thing about winning price estimation, in the sense that it's not a separate problem and it actually relates to the win rate estimation problem. Imaging that we have a probability density function of the winning price distribution, as in the Fig. 1, then for every bid *b* as the vertical line in the Fig. 1, the probability of winning corresponds to the situation where the winning price is less than or equal to the bid *b*, which is the marked area on the left side of the bid line.

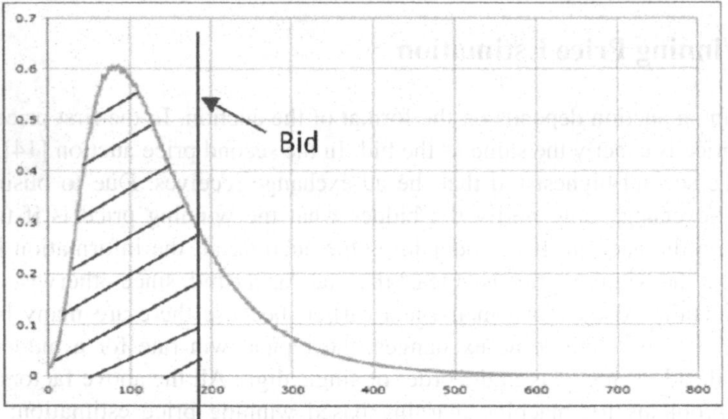

Fig. 1. Win rate given a bid can be calculated from the winning price distribution

In other words, for each bid *b*, the win rate is the same as this probability

$$P(winningPrice \leq b) \quad (3)$$

, which is the cumulative distribution function (cdf) of the probability density function (pdf) of the winning price. In other words, if we have a winning price distribution, and we take the integral up to the bid price, we will have the win rate; if we have the win rate distribution and we take the derivative with respective to the price, we have the winning price distribution. They are dual problems that can be solved with one uniform solution. Assuming the win rate estimation is based on logistic regression as in Eq. (1), we can re-format the Eq. (1) as Eq. (4)

$$winRate = \frac{1}{1+C*e^{-\theta_b*bid}} \quad (4)$$

C is a constant factor that covers the exponential term of all the features that are unrelated to the bid price, and $\theta_b$represents the model weights associated with bid price.

If we take the derivative of Eq. (4) with respect to the bid price, we will get:

$$\frac{d(winRate)}{d(bid)} = \frac{1}{(1+C*e^{-\theta_b*bid})^2} \cdot C * e^{-\theta_b*bid} * \theta_b \quad (5)$$

, and that is exactly the probability density function (pdf) of the winning price distribution given all the attributes of an incoming request.

$$f(x) = \frac{1}{(1+C*e^{-\theta_b*x})^2} \cdot C * e^{-\theta_b*x} * \theta_b \quad (6)$$

Having the closed form solution on the winning price distribution, we compute the actual winning price for each given bid by taking numerical approximation of the mean value in the region left to the bid *b* as winning price, as in Eq. (7). *f(x)* is the pdf of winning price distribution as in Eq. (6), and *E{}* is the expected value of a distribution.

$$winningPrice = E\{x * f(x)\}, where\ x \leq b \quad (7)$$

5 Bidding Strategy

Having the estimate of win rate and the winning price, we can derive our bidding strategy to automatically calculate the bid for each incoming request. Bidding strategy is indeed an optimization function, it takes as input the monetization capability on the individual request, the win rate estimation function and the corresponding winning price estimation for each bid, and produces the final bid based on specific business objectives as output. In the remaining of this section, we will illustrate some of the different bidding strategies that we use to drive different business objectives. For the annotation, we use *pRev* and *eRev* to represent the expected revenue given that we win the auction and expected revenue when we places the bid (those two revenue terms are different), *winRate(bid)* as win rate estimate for the specific bid, and *cost(bid)* as the cost for winning the auction. *pRev* is the "effective cost per-thousand impression" (CPM), i.e., what we charge our client for serving thousand impressions for them; *cost(bid)* is either the winning price in the second price auction, as discussed in section 4, or the bid price itself in the first price auction. Among those terms, *pRev* is a constant term independent of how do we construct the bid, *eRev, cost(bid)* and *winRate(bid)* are all monotonically non-decreasing functions of the bid price.

5.1 Strategy That Maximizes the Revenue

The first specific strategy is to maximize the revenue. If we assume that the term *pRev* is accurate, then whether or not we can realize this revenue *pRev* solely depends on if we could win the auction and serve the ad. In other words, the expected revenue for each incoming request *eRev* can be formulated as:

$$eRev = winRate(bid) * pRev \tag{8}$$

The corresponding bidding strategy that maximizes the revenue can then be formulated as:

$$bid^* = argmax\{winRate(bid) * pRev\} \tag{9}$$

Since *winRate(bid)* is a monotonically non-decreasing function with respect to the bid, the *bid** that maximize the Eq. (9) is simply the one that maximizes the win rate, which is *pRev*.

Bid with *pRev* is the optimal revenue generating strategy in the first price auction scenario, as we pay what we bid and we can maximize our bid to the extent that we don't lose money. On the other hand, most of the RTB ad-exchanges that we participate operate on the second price auction mechanism [14], bid with *pRev* may indeed lose opportunities since the cost of winning most of time is less than our bid, the *pRev*. In this case, we would bid at a higher bid price whose corresponding winning price is the same as the *pRev*. This can be described as in Eq. (10):

$$bid^* = argmax\{winRate(bid) * pRev\}, where\ argWinning(bid) \le pRev \tag{10}$$

argWinning(x) is a function that returns the bid whose corresponding winning price is the input price point *x*. We can use a linear search algorithm to approximate this function.

5.2 Strategy That Maximizes the Profit

While the strategy in section 5.1 maximizes the revenue, the daily operation of the business is to make profit, and it quite often that we want to have a bidding strategy that maximizes the profit produced from the business operation. Unlike the revenue, which is a monotonically increasing function with respect to our bid, profit depends on the difference between the revenue and the cost. *pRev* is a constant term, but the cost increases with our bid. At one hand, if we win the auction, the higher our bid the less profit we would realize; on the other hand, the higher our bid the higher likelihood of us winning the auction, and if we don't win the auction, the cost and revenue will all be 0. The optimal bid is the one that maximizes the joint effect of the two above factors, which can de described as in Eq. (11):

$$bid^* = argmax\{winRate(bid) * (pRev - cost(bid)\} \tag{11}$$

Depends on how the cost is constructed, *cost(bid)* would be either the *bid* as in first price auction or the winning price estimate of the bid as described in section 4.

One thing worth noting is that the strategy of maximizing the profit is different from the one that maximize the profit margin. Regardless whether profit margin is defined as the ratio of revenue minus cost over revenue, or directly the revenue over cost, it all has the fixed revenue term as either the denominator or the numerator, and the bid that maximizes the profit margin will simply be the one that minimize the cost, and that translates into 0 as the bid. While this yields the maximum value of profit margin in theory, we won't be able to realize this margin, since we won't win any auction with 0 bid price. This is another reason why we don't use profit margin as the objective for our bidding.

5.3 Strategy That Maximize the Combined Profit and Revenue Goal

Unlike the previous two bidding strategies that solely focus on either the profit or revenue, we could also combine those two factors together and maximize a combined objective during bidding. Specifically, we can mix the two objective functions together and use a weight alpha to control the blend of two strategies. If the alpha term is applied to the profit based objective function, then we can make the combined strategy as the following:

$$bid^* = argmax\left\{\begin{matrix} \alpha * winRate(bid) * (pRev - cost(bid)) \\ +(1-\alpha) * winRate(bid) * pRev \end{matrix}\right\} \tag{12}$$

The *winrate(bid)*pRev* term in the profit and revenue strategy component will indeed cancel the effect of alpha, and we will have the mixed bidding strategy as in Eq. (13):

$$bid^* = argmax\{winRate(bid) * (pRev - \alpha * cost(bid))\} \tag{13}$$

This mixed strategy covers both the revenue and profit objective during business operation. We can adjust the alpha value that controls the relative importance of profit v.s. revenue when we bid. When alpha equals to 0, this combined strategy becomes the one that maximize the revenue, as in section 5.1; when alpha becomes 1, this strategy falls back into the profit optimization strategy. In addition to this flexibility, this bidding strategy gives us another advantage in the sense that we can dynamically adjust the value of alpha in real time based on the performances of the bidding system so far. For example, we can set a goal on either the revenue or profit metric, check the progress of the bidding system toward the goal at fixed time intervals, and adjust the alpha value accordingly to hit the pre-defined revenue or profit goal.

6 Experimental Setup and Results

6.1 Evaluation Metrics

We tested the performance of our various estimation methods and bidding strategies using one of the leading mobile ad exchange platforms that we participate. On everyday we bid on more than a few billion requests in that ad exchange.

The performance metrics we used to evaluate our methods are the followings:

- For win rate estimation: We used the log-loss of predicted results collected from bidding data.
- For winning price estimation: We used two metrics, RMSE and ratio-RMSE. RMSE is computed based on predicted winning price and actual observed winning price; and ratio-RMSE is computed by first taking the ratio of predicted winning price over observed price, then compare it against value 1 and compute the corresponding RMSE. ratio-RMSE was introduced to offset the scale difference between winning prices. For the same 1 cent difference between the predicted and actual winning prices, the level of accuracy is completely different between the case with base price of 10 dollar and the case with base price 10 cents, ratio-RMSE normalizes the level differences between different data points.
- For bidding strategy: We looked at the revenue and profit margin per unit percentage of traffic. We took the baseline revenue and profit margin figures from maximizing revenue strategy as 1, and computed relative metrics on the revenue and profit margin from other bidding strategies.

6.2 Experiment Setup and Results for the Win Rate Estimation

Since we conducted the experiments using domain specific real business operation data, all the numbers reported here were relative performance metrics. Nevertheless, the findings from the experiments are still meaningful, especially for the purpose of comparing different approaches and identifying better modeling techniques.

In the win rate and winning price estimation, we conducted the experiments by splitting the data into training and testing set. Testing set has the data collected from 1 week time period during January 2013. We built models on the training set and tested the model on the testing data, all done in off-line fashion.

Our win rate estimation baseline approach was to simply use the historical observed win rate. We sliced all the auction winning and auction not winning instances observed in the past time period according to combination of their attributes. For each slice of data, we first segmented the bid price into discrete chunk, then under each chunk, we collected all the auction winning instances, divided by the total auction instances to produce the win rate for that specific price chunk given the combination of attributes.

We tested the baseline method with two different historical time windows. One was *baseline_7* and the other was *baseline_14*. *baseline_7* uses the past 7 days training data and *baseline_14* covers 2 week period instead.

One issue with our baseline approach is that we won't know the win rate for the new apps (new property) or new ad dimension. We used the fall back approach to handle that. Every time we can't find the historical win rate based on the feature combination look up mentioned above, we fall back to a combination with one less factor and check if we could find the win rate, and continue to fall back if the historical win rate is still missing.

The first experiment with our logistic regression model was to use the *"standalone"* features only. We built the win rate model using 7 days and 14 days sliding window historical data, and the performance is labeled as *logistic_standalone_7* and *logistic_standalone_14* respectively.

We then tested the performance of adding *"cross"* features into win rate estimation. The performances were summarized as *"logistic_all_7"* and *"logistic_all_14"* respectively.

Table 1 lists the performance metrics of all the win rate estimation methods.

From table 1, it's clear that our win rate model performs better than baseline. The best performing model is the all feature model using 14 days of historical data. Indeed, it almost reduced the log-loss of *baseline_7* by half. All the log-loss results were statistically significant.

Table 1. Performance comparison between various win rate estimation methods

Win rate estimation method	Log-loss
baseline_7	0.156
baseline_14	0.142
logistic_standalone_7	0.118
logistic_standalone_14	0.109
logistic_all_7	0.091
logistic_all_14	0.087

6.3 Experiment Setup and Results for the Winning Price Estimation

Experimental results for winning price estimation are listed in table 2.

The baseline results was achieved by slicing the historical data of winning price based on the same way that we sliced the data to get the baseline results for the win rate. We also applied the same fall back logic if there was any attribute value missing. Results are labeled as *price_baseline_7* and *price_baseline_14* in table 2.

The second result sets were based on linear regression approaches. The features used in linear regression winning price model were the same as those used in *logistic_all_7* (or *logistic_all_14*), and the linear regression coefficients were computed using 7 and 14 days of data as well. *linear_all_7* and *linear_all_14* in table 2 are the corresponding results.

The third set of results, *logistic_price,* came from logistic regression based methods. We calculated the expected value of winning price based on the distribution from the price point of 0 all the way up to the bid price, and used it as the winning price. *logistic_price_7* was generated using the 7 days of training data, and *logistic_prce_14* was generated using 14 days of training data.

Two baseline approaches using different time window yield very similar results, probably due to the fact the mean value of winning price for each combination didn't change that much from 1 week to 2 weeks time period. *"logistic_price"* based winning price out-performed the other two approaches, but the lift wasn't substantial. We believe that was due to the intrinsic high variance among the winning prices. Naturally, the winning price depends on our competitor's bidding behaviors, and their bidding behaviors can be heavily influenced by their business needs. We see quite often that the winning price for the same segment of traffic changes drastically from one individual request to another, in a very short time period of minute. Indeed, we computed an "oracle" experiment using the mean value of winning price observed per traffic segment in the testing data and measured its performance on the same testing set (in other words, train and test on the same testing set). The RMSE and ratio-RMSE was 22.8 and 0.34 respectively. This "oracle" experiment performance can be treated as the upper bound of all winning price estimation methods.

Table 2. Performance comparison between various winning price estimation methods

Winning price estimation method	RMSE	ratio-RMSE
price_baseline_7	33.76	0.59
price_baseline_14	33.74	0.59
linear_all_7	38.71	0.62
linear_all_14	37.76	0.59
logistic_price_7	31.72	0.51
logistic_price_14	31.57	0.52

6.4 Bidding Strategy Experiment Setup and Results

The evaluation of bidding strategy is slight different from the win rate and winning price. While all win rate and winning price prediction approaches can be evaluated offline using previously collected historical data, for bidding strategy it has to be evaluated in online environment against live traffic data. In our experiment, we created multiple testing buckets with the same percentage of traffic, and let each single bidding strategy drive the bid inside one bucket. We ran multiple of those testing buckets in parallel for one week to offset "day of the week" effect, and compared different testing buckets based on their relative revenue and profit measurements with respect to the strategy that maxsimize the revenue, and the results are summarized in table 3.

Table 3. Performance comparison between various bidding strategies

Bidding strategy	Normalized revenue	Normalized profit margin
Maximizing revenue	1	1
Maximizing profit	0.9	1.35
Combined strategy, alpha=0.3	0.97	1.1
Combined strategy, alpha=0.8	0.95	1.23

It's clear from table 3 that each strategy does what it supposed to do. We obtained the maximum revenue, with the sacrifice on profit margin using the revenue maximizing strategy, and vice verse for profit maximizing strategy.

It's also interesting to compare the relative gain on the revenue and profit margin between different strategies. In general, we observed that it was easier to gain on the profit margin side than to grow the revenue. In our specific example below, in order to grow the revenue by 10% relative from the profit maximizing strategy to the revenue maximizing strategy, we need to sacrifice nearly 35% of the profit margin. This is not a surprise to us. We need to bid more to get higher revenue scale, and a higher bid will result in higher cost per unit traffic. In other words, revenue grows linearly with the traffic, while cost grows faster than linear with the traffic scale.

7 Conclusions

We described our effort of estimating win rate, winning price and corresponding bidding strategies in real time bidding (RTB) based ad-exchange in mobile advertising. We explained our effort of building large scale logistic regression based win rate model, which is capable of handing order of billions real data and provide accurate win rate estimation. We have also demonstrated the dual relationship between win rate and winning price in the second price auction scenario, revealed that the two problems can be solved with one solution, and proposed a corresponidng winning price estimation method. Based upon the win rate and winning price estiamtion, we outlined various bidding strategies that we used in our daily operation. Comparison data from real business operation confimed the superiority of our proposed methods against various baseline approaches.

References

1. Apache software foundation, Scalable machine learning and data mining (2013), http://mahout.apache.org
2. Langford, J.: Wowpal wabbit (2013), https://github.com/JohnLangford/vowpal_wabbit/wiki
3. Schapire, R., Stone, P., McAllester, D., Littman, M., Csirik, J.: Modeling auction price uncertainty using boosting-based conditional density estimation. In: ICML 2002 (2002)
4. Wellman, M., Greenwald, A., Stone, P., Wurman, P.: The 2001 trading agent competition. Electronic Markets 13, 4–12 (2003)
5. Wellman, M., Reeves, D., Lochner, K., Vorobeychik, Y.: Price prediction in a trading agent competition. Journal of Artificial Intelligence Research 21, 19–36 (2004)
6. Putchala, R., Morris, V., Kazhanchi, R., Raman, L., Shekhar, S.: kavayaH: A trading agent developed for TAC-02. Tech. Rep., Oracle India (2002)
7. He, M., Jennings, N.: SouthamptonTAC: An adaptive autonomous trading agent. ACM Transactions on Internet Technology 3, 218–235 (2003)
8. Greenwald, A., Lee, S., Naroditskiy, V.: RoxyBot-06: Stochastic prediction and optimization in TAC travel. Journal of Artificial Intelligence Research 36, 513–546 (2009)

9. Hosmer, D., Lemeshow, S.: Applied logistic Regression, 2nd edn. John Wiley and Sons (2000)
10. Richardson, M., Dominowska, E., Ragno, R.: Predicting clicks: estimating the click-through rate for new ads. In: Proceedings of WWW 2007, pp. 521–530 (2007)
11. Interian, Y., Dorai-Raj, S., Navernioük, I., Opalinski, P., Kaustuv, Zigmond, D.: Ad quality on tv: Predicting television audience retention. In: ADKDD (2009)
12. Becker, H.: Modeling contextual factors of click rates. In: Proceedings of AAAI Conference on Artificial Intelligence, pp. 1310–1315 (2007)
13. Sculley, D., Malkin, R., Basu, S., Bayardo, R.: Predicting bounce rates in sponsored search advertisements. In: SIGKDD Conference on Knowledge Discovery and Data Mining (KDD), pp. 1325–1334 (2009)
14. Edelman, B., Ostrovsky, M., Schwarz, M., Fudenberg, T., Kaplow, L., Lee, R., Milgrom, P., Niederle, M., Pakes, A.: Internet advertising and the generalized second price auction: selling billions of dollars worth of keywords. American Economic Review 97 (2005)

Self-training Temporal Dynamic Collaborative Filtering

Cheng Luo[1], Xiongcai Cai[1,2], and Nipa Chowdhury[1]

[1] School of Computer Science and Engineering
[2] Centre of Health Informatics
University of New South Wales, Sydney NSW 2052, Australia
{luoc,xcai,nipac}@cse.unsw.edu.au

Abstract. Recommender systems (RS) based on collaborative filtering (CF) is traditionally incapable of modeling the often non-linear and non Gaussian tendency of user taste and product attractiveness leading to unsatisfied performance. Particle filtering, as a dynamic modeling method, enables tracking of such tendency. However, data are often extremely sparse in real-world RS under temporal context, resulting in less reliable tracking. Approaches to such problem seek additional information or impute all or most missing data to reduce sparsity, which then causes scalability problems for particle filtering. In this paper, we develop a novel semi-supervised method to simultaneously solve the problems of data sparsity and scalability in a particle filtering based dynamic recommender system. Specifically, it exploits the self-training principle to dynamically construct observations based on current prediction distributions. The proposed method is evaluated on two public benchmark datasets, showing significant improvement over a variety of existing methods for top-k recommendation in both accuracy and scalability.

Keywords: Temporal Recommender, Collaborative Filtering, Particle Filtering.

1 Introduction

Collaborative filtering [10] generates personalized recommendation to match users' interests and its performance can be improved by exploiting temporal information, as the tendency of user preferences and item attractiveness is not static [10,11]. There are four general approaches in temporal CF, i.e, heuristic, binning-based, online updating and dynamic-based approaches. Heuristic approach penalizes the importance of data before a pivot point[6,10], which tends to undervalue the past data. In binning-based approach, training and testing data could be from the same interval [11]. The prediction for users' interests is actually *post hoc* about what interests *would have been* in the past, rather than what interests would be in the future. Although online updating approach only uses past information to make prediction [8,12], it usually focuses on scalability and ignores the dynamics of user taste and item attractiveness. Dynamic-based approach explicitly models temporal dynamics by a stochastic state space model,

V.S. Tseng et al. (Eds.): PAKDD 2014, Part I, LNAI 8443, pp. 461–472, 2014.

which shows advantages over the other methods [13]. However, in such an approach, item attractiveness is assumed to be static with Gaussian distributions, which may be oversimplified. To overcome such problems, we first utilize particle filtering [16] as a dynamic technique to model non-Gaussian behaviors and track latent factors representing user preferences and item attractiveness.

Furthermore, under the temporal context, the data sparsity problem [19] becomes challenging as many users would be inactive for some consecutive time slots. Although matrix factorization [15] could realize sparsity reduction, the tendency tracked by particle filtering may be unreliable due to insufficient observations at every time step. Exploiting additional information, such as contextual information or common patterns [10], or imputing all or most missing data are two common approaches in CF to reduce sparsity. However, this extra information collection is usually not only infeasible in practice but also computational complexity. Therefore, we do not consider utilizing side information in this paper. Missing data are usually imputed as negative [18,14] or other heuristic values [10]. Deterministic optimization [17] and current model estimation [9] are also used to yield some reasonable imputed values. However, these methods are only developed under static context and based on some heuristic rules or point estimators. Meanwhile, all these methods impute all or most missing data, leading to an unaffordable computational complexity for succeeding recommendation algorithms, since it not only heavily reduces the scalability of algorithms but also influences the recommendation accuracy.

In this paper, we aim to solve the problems of data sparsity and scalability in particle filtering-based temporal dynamic recommender systems. We utilize self-training principle [22] to dynamically construct feedback data to enhance our particle filtering-based recommendation method. In particular, we use latent factors to compactly represent user preferences and item attractiveness respectively at each time step, whose initial settings are learned by probabilistic matrix factorization (PMF) [15]. Based on such a representation, we develop a novel self-training mechanism based on the distributions of the current personalized prediction to complement recent observations with negative items sampled from missing data. The mechanism is then cooperated with particle filtering techniques to simultaneously track the tendency of user preferences and item attractiveness. For top-k recommendation [21,4], a personalized list for each user is generated based on current user and item latent factors.

We discuss related work in Section 2. Particle filtering for PMF is described in Section 3. Self-training for the particle filtering based method is developed in Section 4. Experimental results are in Section 5. Conclusion is presented in Section 6.

2 Related Work

There have been few studies [13,11,20] on exploiting temporal dynamics to improve the performance of RS. Among these studies, [13] is the most related one to our work, which uses Kalman filter as temporal priors to track user latent factors. However, item latent factors are only updated but not tracked. Meanwhile, the usage of Kalman filter restricts the dynamic and observation functions to be linear and Gaussian. Particle filtering has been used to dynamically update a log-normal distribution that models user preferences [3] in music recommendation, assuming the staticness of item popularity. Moreover, the method is not based on latent factors, and very application-specific (otherwise, no proper features).

Conventional CFs with imputation [19] all suffer from the domination of imputed ratings. Sampling missing feedback is only used in non-temporal context and one class CF (OCCF) problem [5]. An OCCF problem can be deducted from our problem by setting, for example, relevant ratings as positive examples. User-oriented and item-oriented mechanisms, which only based on the times that items or users present, are proposed in [14] as sampling methods. In these methods, recommendation accuracy is compromised in order to boost the scalability. To obtain a more accurate recommendation, samples are selected based on pairwise estimation for OCCF to iteratively train the model parameters [21]. All of these sampling methods are developed under static context for OCCF. Unlike our proposed method, these algorithms do not aim to solve problems of scalability and sparsity at the same time.

3 Particle Filtering for Matrix Factorization

Probabilistic matrix factorization method, as a model-based approach in CF, has been widely used due to its simplicity and efficiency. In particular, assuming N users and M items, let $R \in \mathcal{R}^{N\times M}$ be a user-item preference matrix with an entry $r_{u,i}$ representing the rating given by user u to item i. Rating $r_{u,i}$ is generated with a Gaussian distribution $P(r_{u,i}|U_u, V_i)$ conditioned on K dimensional user and item latent factors $U \in \mathcal{R}^{N\times K}$ and $V \in \mathcal{R}^{M\times K}$. Prior distributions $P(U)$ and $P(V)$ are formulated to contain regularization terms [15]. These latent variables are further assumed to be marginally independent while any rating $r_{u,i}$ is assumed to be conditionally independent given latent vectors U_u and V_i for user u and item i [15]. The likelihood distribution over preference matrix R is,

$$P(R|U, V, \alpha) = \prod_{u=1}^{N}\prod_{i=1}^{M} Y_{u,i} \cdot \mathcal{N}(r_{u,i}|U_u V_i^T, \alpha^{-1}), \tag{1}$$

where $\mathcal{N}(x|\mu, \alpha^{-1})$ is a Gaussian distribution with mean μ and precision α, and $Y_{u,i}$ is an indicator variable with value 1 when rating $r_{u,i}$ is not missing and value 0 when the rating is not observed. Priors $P(U)$ and $P(V)$ are given as,

$$P(U|\alpha_U) = \textstyle\prod_{u=1}^{N} \mathcal{N}(U_u|0, \alpha_U^{-1} I) \quad p(V|\alpha_V) = \prod_{i=1}^{M} \mathcal{N}(V_i|0, \alpha_V^{-1} I).$$

Maximizing the log-posteriors over U and V is equivalent to minimizing the sum-of-square error function with quadratic regularization terms for PMF [15], leading to the following objective function,

$$E = \frac{1}{2}\sum_{u=1}^{N}\sum_{i=1}^{M} Y_{u,i}(r_{u,i} - U_u V_i^T)^2 + \frac{\lambda_U}{2}\sum_{u=1}^{N} ||U_u||_{Fro}^2 + \frac{\lambda_V}{2}\sum_{i=1}^{M} ||V_i||_{Fro}^2, \tag{2}$$

where $\lambda_U = \alpha_U/\alpha$, $\lambda_V = \alpha_V/\alpha$, and $||\cdot||_{\text{Fro}}$ denotes the Frobenius norm.

We use a state space approach [16] to track the tendency of user preferences and item attractiveness. With linear and Gaussian assumption, it is straightforward to define the state to be a joint vector of user and item latent factors, due to the existence of analytical and tractable solution. However, it is shown [15] that empirical distributions of the posterior Hence, we use particle filtering to simultaneously track these latent factors. Particle filtering iteratively approximates regions of high density as discrete sample points. As the number of particles goes to infinity, the approximation converges to the true distribution [16]. In practice, given d-dimensional state space, the number of required particles should be $O(2^d)$ to achieve a satisfiable result [16]. To make a compromise between the accuracy of user/item representation and tractability of particle filtering, we separately track latent factors for each user and item.

We assume that the tendency of user u's preference and item i's properties follows a first-order random walk driven by multivariate normal noise, due to the lack of prior knowledge. The transition functions at time t are as follows,

$$U_t^u = U_{t-1}^u + c_t^u, V_t^i = V_{t-1}^i + d_t^i, \tag{3}$$

where $c_t^u \sim \mathcal{N}(0, \sigma_U I)$ and $d_t^i \sim \mathcal{N}(0, \sigma_V I)$ are defined as unrelated Gaussian process noises. The posterior distribution of U_t^u is approximated by particle filtering with S particles as $P(U_t^u|R_{1:t}) = \sum_{s=1}^S w_{U,t}^{u,(s)} \delta(U_t^u - U_t^{u,(s)})$ where $w_{U,t}^{u,(s)}$ is a weight of s-th particle at time t, and $U_t^{u,(s)}$ is obtained by propagating $U_{t-1}^{u,(s)}$ using dynamics in Eq (3). The estimation of item v's latent factors at time t is obtained in a similar way. Using the transition prior in Eq (3) as the proposal distribution, the weight at time t for all the particles is evaluated recursively as $w_t = w_{t-1} \cdot P(R_t|U_t, V_t)$ where $P(R_t|U_t, V_t)$ is the observation function.

The observation function should reflect the ability of a particle to reconstruct given ratings. The objective function in Eq (2) is an immediate candidate in which $P(R|U, V, \theta) \propto e^{-E}$. However, this candidate function is sub-optimality for a top-k recommendation task, because an algorithm attempting to minimize the root-mean-squared-error in prediction does not have a satisfiable performance for a top-k recommendation task [4]. Moreover, the objective function in Eq (2) assumes that unobserved data in both training and testing cases are missing at random. That is, the probability that a rating to be missing is independent of its value. Nevertheless, it is shown [18] that feedback in RS is generally not missing at random (NMAR). Low ratings are much more likely to be missing than high ratings [18] because users are free to choose items to give feedback.

To design a suitable observation function, the key idea is to consider the ranking of *all* the items, no matter whether they are observed or not. By treating all missing data as negative with weights (wAMAN), the observation function over imputed ratings $\bar{R}_t^u$ for s-th particle of user u is as follows,

$$P(\bar{R}_t^u|U_t^{u,(s)}, \{V_t^{i,(s')}\}) = \sum_{s'=1}^{S'} exp\{-\sum_{i=1}^{M} w_{V,t}^{i,(s')} W_{u,i}(\bar{r}_t^{u,i} - U_t^{u,(s)}(V_t^{i,(s')})^T)^2\}, \tag{4}$$

where $\bar{r}_t^{u,i} = r_t^{u,i}$ if $Y^{u,i} = 1$ and $\bar{r}_t^{u,i} = r_m$ if $Y^{u,i} = 0$. The r_m is an imputed value for all the missing data, which is regarded as the average value of ratings in the complete but unknown data. Weight $W_{u,i}$ is defined to reflect the confidence over imputation and set as a global constant w_m for simplicity [18]. Latent factors $V_t^{i,(s')}$ and their weights $w_{V,t}^{i,(s')}$ represent s'-th particle for item i. The observation function over s'-th particle of item i is defined similarly. The distributions are no longer Gaussian after the introduction of w_m and an imputed value for all the missing data. Meanwhile, no regularization terms exist in Eq (4) because we obtain point mass approximation of posterior distributions via particle filtering attempting to avoid overfitting. This method is named as PFUV hereafter.

4 Personalized Self-training Method

In practice, feedback is usually unavailable before recommendation is made, which implies observation R_t at the current period is not available to estimate the tendency of user preferences and item attractiveness before recommendation. It is straightforward to use all the historic observations $R_{1:t-1}$ to approximate the estimation. However, the ratings would be dominated by the past information and cannot represent the recent tendency. An alternative approximation uses the most recent observation R_{t-1} instead. However, under temporal context, the ratings are too sparse for each user or item to track the current tendency. The data sparsity can be reduced by imputing all the missing data as shown in Eq (4). The observation function can be approximated by $P(\bar{R}_{t-1}|U_t^{u,(s)}, \{V_t^{i,(s')}\})$. However, the dynamics in this approximation will drift away from the true tendency due to the domination of imputed ratings in $\bar{R}_{t-1}$. Meanwhile, this approximation does not have a satisfiable scalability due to the usage of all the missing data.

Therefore, we exploit self-training principle [22] to solve the above mentioned problems. Instead of treating wAMAN, for each user at every time step, we will dynamically select a subset of missing items as negative samples to complement the user's most recent observation. This personalized and self-training procedure not only distinguishes the past and recent information but also avoids to dominate recent observation with imputed data. These samples are the most confident negative samples w.r.t the current prediction distribution for each user.

4.1 Self-training Sampling Method

Given user u and its current unobserved items $\mathcal{I}_t^{m,u}$, a set of $N_t^{n,u}$ items $\mathcal{I}_t^{n,u} \subseteq \mathcal{I}_t^{m,u}$ is selected by a multi-nominal distribution. The distribution is

$$P(x_1, \cdots, x_{N_t^{m,u}}|N_t^{n,u}, \theta_1, \ldots, \theta_{N_t^{m,u}}) \propto \theta_1^{x_1} \cdots \theta_{N_t^{m,u}}^{x_{N_t^{m,u}}}, \quad (5)$$

where $N_t^{m,u}$ is the number of unobserved items for user u until time t, $\{x_i|i \in \{1, \ldots, N_t^{m,u}\}\}$ represents the times that unobserved item i would be selected as negative, and $\{\theta_i|i \in \{1, \ldots, N_t^{m,u}\}\}$ is the probability that unobserved item i is disliked by user u. Without restricting x_i's to binary variables, this personalized selection is an adaptive mechanism. An unseen item with a high probability will be selected more frequently than those with lower probability. As the accumulation of w_m for the same negative sample in Eq (4), more emphasis will be placed on the sample. By imposing such restriction, $N_t^{n,u}$ different items will be chosen. We will adopt this restriction for simplicity.

Confidence Estimation. A candidate negative sample should have a small prediction error and a small estimation variance if the sample was negative. This criterion resembles the bias and variance decomposition of generalized errors [10]. Given the historic data $R_{1:t-1}$, we define θ_i as $P(\hat{r}_{u,i} = r_m \wedge var(\hat{r}_{u,i})|R_{1:t-1})$ where $\hat{r}_{u,i} = r_m$ represents the event that the predicted rating equal to the imputed value and $var(\hat{r}_{u,i})$ represents the variance of the prediction. Assuming prediction error and variance are conditionally independent given U_t and V_t,

$$\theta_i = \int P(\hat{r}_{u,i} = r_m|U_t, V_t)P(var(\hat{r}_{u,i})|U_t, V_t)P(U_t, V_t|R_{1:t-1})dU_t dV_t$$

$$\sim \sum_{s=1}^{S}\sum_{s'=1}^{S'} w_{U,t}^{u,(s)} w_{V,t}^{i,(s')} P(\hat{r}_{u,i}|U_t^{u,(s)}, V_t^{i,(s')})P(var(\hat{r}_{u,i})|U_t^{u,(s)}, V_t^{i,(s')}), \quad (6)$$

where the predicted joint distribution of latent factors is estimated using particle filtering described below, S and S' are the number of particles used to track user u's latent factors and item i's latent factors, respectively.

Predication with Canonical Particles. To estimate a particle's weight for U_t in Eq (6), we need a weight of a particle for V_t. Likewise, we use a weight of a particle for U_t to reweight a particle for item latent factors. As the computation over all possible pairs of user and item particles is too expensive, we resort to canonical particles [7] $\hat{U}_t$ and $\hat{V}_t$ to respectively represent the total effect of particles on the estimation for each user and item latent factors at time t.

In general, $\hat{U}_t$ and $\hat{V}_t$ can be any proper function taking as input $\{(w_{U,t}^s, U_t^s)|s \in 1 \ldots S\}$ and $\{(w_{V,t}^{s'}, V_t^{s'})|s' \in 1 \ldots S'\}$. To avoid the degeneracy problem [16] in particle filtering, we will resample particles proportional to their weights After resampling, we use the expectation of posterior distributions of U_t and V_t as canonical particles, which are estimated as $\hat{U}_t^u = \sum_{n=1}^{S} w_{U,t}^{u,(n)} U_t^{u,(n)}$ and $\hat{V}_t^i = \sum_{m=1}^{S} w_{V,t}^{i,(m)} V_t^{i,(m)}$ for user and item latent factors, respectively.

Combining with canonical particles $\hat{V}_t$ and Eq 6, the prediction distribution of user u's preference over item i is estimated as follows,

$$\theta_i \sim \sum_{s=1}^{S} w_{U,t}^{u,(s)} P(\hat{r}_{u,i} = r_m|U_t^{u,(s)}, \hat{V}_t^i)P(var(\hat{r}_{u,i})|U_t^{u,(s)}, \hat{V}_t^i), \quad (7)$$

where $U_t^{u,(s)}$ are obtained by propagated $U_{t-1}^{u,(s)}$ as in Eq (3). A small distance between the imputed value and the predicated rating usually means a high confidence that the item should be negative. Thus, the probability of prediction error is defined as $P(\hat{r}_{u,i} = r_m|U_t^{u,(s)}, \hat{V}_t^i) = exp\{-|U_t^{u,(s)}(\hat{V}_t^i)^T - r_m|\}$. In terms of variance estimation, the probability of prediction variance can be estimated by $P(var(\hat{r}_t^{u,i}|U_t^{u,(1)}, \ldots, U_t^{u,(S)}, \hat{V}_t^i))$ as $exp\{\frac{1}{S-1}\sum_{s=1}^{S}(U_t^{u,(s)}\hat{V}_t^i - \hat{U}_t^u\hat{V}_t^i)\}$.

4.2 Two-Phase Self-training Method

Considering the large size and high sparsity of user-item preference matrix, the previous sampling mechanism considering all the unobserved items is infeasible in practice. We use a two-phase approach to reduce the computational complexity.

In phase I, for each user u, we sample a subset $\mathcal{I}'^{n,u}_t$ from the unobserved items $\mathcal{I}^{m,u}_t$. Generally, these sampling schemes can be implemented in terms of any distribution that properly represents NMAR. It is shown in [18] that arbitrary data missing mechanism is NMAR as long as they are missing with a higher probability than relevant ratings do. We use a uniform distribution of $\mathcal{I}^{m,u}_t$ to avoid interfering prediction distribution, which has been extensively used to handle some large datasets [21,14]. This simple and efficient distribution is a rational choice as long as we set $N^{n,u}_t$ to be a reasonably large value, which will be discussed in Section 5. For simplicity, we set $|\mathcal{I}'^{n,u}_t| = 2 * N^{n,u}_t$.

In phase II, personalized probability θ_i will be computed only for candidates $\mathcal{I}'^{n,u}_t$. Based on Eq (5), negative samples will be selected and then combined with observed data R_{t-1} to construct a sparsity reduced data $\tilde{R}_t$ at time t. User and item latent factors are thus tracked by using this dynamically built data. We name this two-phase self-training method as ST-PFUV hereafter. Let $N^{n,u}_t = \hat{N}$ for simplicity. For computational complexity, at each time step, PFUV takes $O(KSMN)$. As sampling size $\hat{N}$ is usually comparable with $K \ll min(N, M)$, ST-PFUV takes $O(KS(M+N)\hat{N}) \approx O(K^2 S(M+N))$, which scales efficiently as a linear function of user and item size.

5 Experiments

The performance of the proposed algorithm is tested on the popular Movielens 100K [1] and HetRec MovieLens datasets [2]. Both are public benchmark datasets. MovieLens 100K contains 530 users and 1493 items with sparsity 93.3% (at the 16-th week in 1998). HetRec contains 1775 users and 9228 items with sparsity 95.1% (in December 2008). MovieLens spans 8 months with integer scale 1 - 5 while HetRec spans 12 years with half mark scales from 1 to 5.

Protocol. In our experiments, ratings are grouped into time frames based on their timestamps. All the ratings before a predefined time instance t_{test} are training data, and ratings after it are testing data. This setting is preferred over a random split over all the data as it is infeasible to make prediction utilizing information in the future in a real-world deployment. The training periods for MovieLens 100K and HetRec are Sept. ∼ Dec. 1997 and Sept. 1997 ∼ Dec. 2007, respectively. Their testing periods are the 1st ∼ 16th weeks in 1998 and Jan. ∼ Dec. 2008, respectively. Different units of time frame are selected to ensure that ratings for each user in a time slot are not too sparse. Therefore, the task in our experiments is to predict individual's ranking over all the items in next time frame t based on all the information upto $t-1$. All the algorithms are repeated 10 times and the average results are reported. They are all implemented in Matlab with 3.3G Hz CPU and 8G memory.

We use precision@k, recall@k [4] to measure recommendation relevance, which directly assess the quality of top-k recommendations. To measure user satisfaction in recommendation, we use top-k hitrate [4,18]. As top-k hitrate recall is proportional to top-k hitrate precision [18], we only test top-k hitrate recall and name it top@k. In order to test temporal performance, the temporal extensions of those metrics are defined. Conventional accuracy metrics adopted to RS can be found in [4,10], which are omitted here due to space limitation.

For user u, precision@k over month t is denoted as $prec(k, u, t)$. During the prediction at time $t-1$, instead of using all the items in testing data as conventional precision@k does, *only* items in month t are scanned to determined their relevance to a user. The temporal precision is defined as, $prec_{temp}(k) = \frac{1}{T*N}\sum_{t=1}^{T}\sum_{u=1}^{N} prec(k, u, t)$, which is the average on all users over all the time frames T. The temporal recall $recall_{temp}(k)$ and temporal hitrate $top_{temp}(k)$ are defined in a similar fashion. As a common practice, we treat items with rating 5 as relevant items, and measure $k = 10$ for precision. For hitrate, we set $k = 10$ and each relevant item is mixed with 500 randomly sampled unobserved items to avoid spending too much computational power on evaluation. We set $k = 100$ for recall as we are ranking all the items in temporal context. We tune all the model parameters under temporal recall and use the identical setting to test the performance of algorithms under temporal recall and hitrate.

Baseline Methods. In order to test the performance of the proposed algorithms (ST-PFUV) that balances imputation and the recent observations, we compare it with the following five algorithms as part of baseline methods: PFUV, ST-PFUV-User, TopPopular, PureSVD [4] and AllRank [18]. PFUV is used to verify the efficiency and scalability of our sampling methods on incorporating temporal dynamics. It is empirically shown [14] that user-oriented sampling based on user preference has better performance than uniform sampling, we therefore hybrid the PFUV method with user-oriented sampling, and name it ST-PFUV-User. As TopPopular is a non-personalized algorithm that ranks item higher when the item is more often rated as relevant, which is included to verify the benefits of considering personalized recommendations in temporal context. PureSVD and AllRank are state-of-art algorithms developed to pursuit the top-k recommendation task. They are selected to illustrate the ability of our sampling methods to cope with non-Gaussian behaviors. These baseline algorithms are designed without exploiting any temporal information. In our experiment, to make these algorithms dynamic, we retrain all the learned models at each time step. This simple extension, which is a common practice in real-world deployment, is important to make the comparison fair. To the best of our knowledge, most of developed algorithms in temporal RS are compared with static versions of some baseline algorithms. We name these dynamic extensions as DynTopPopular, DynPureSVD and DynAllRank. To confirm the necessity of exploiting temporal information, we also adopt PMF as the only static baseline method, which always predicts the ranking without updating model parameters. To balance the accuracy, variance and scalability, we set $S = S' = 1000$ and $K = 8$ for all the PFUV-based methods. The imputed value is set to $r_m = 2$ which is

Table 1. Results on MovieLens 100K under temporal accuracy metrics. The best performance is in italic font.

Method	$prec_{temp}(k)$	$recall_{temp}(k)$	$top_{temp}(k)$
PMF	0.0310	0.3301	0.1425
DynTopPop	0.0465	0.2910	0.1782
DynPureSVD	0.0233	0.3410	0.2806
DynAllRank	0.0546	0.4140	0.3162
PFUV	0.0401	0.3626	0.2909
ST-PFUV-User	0.0408	0.3630	0.2339
ST-PFU	0.0498	0.3794	0.3115
ST-PFUV	*0.0613*	*0.4403*	*0.3543*

(a) Temporal accuracy.

Method	$prec_{temp}(k)$	$recall_{temp}(k)$	$top_{temp}(k)$
PFUV-Rect-wAMAN	0.0423	0.3310	0.2400
PFUV-Hist-User	*0.0447*	*0.3648*	*0.2439*
DynAllRank-User	0.0164	0.2358	0.1151

(b) Effects of recent and historic data.

a lower value than the average observed ratings in MovieLens 100K and Hetrec datasets. All the PFUV-based methods are initialized by AllRank.

MovieLens 100K. Table 1a shows results of above methods under temporal accuracy metrics. By cross validation method [10], we set weight $w_m = 0.05$, regularization constant λ to 0.05 and $K = 12$ for DynAllRank, and $K = 10$ for DynPureSVD and PMF. To reduce variance in particle filtering, we set $\sigma_U = \sigma_V = 0.05$. By cross validation, we set $\hat{N} = 30$ for any method involving sampling. The low values in the table are due to the fact that few relevant items exist for each user in a time frame. Compared with these baseline methods, ST-PFUV has the best performance. All the improvement is statistically significant under paired t-tests [10] with $p < 0.01$. This result verifies that the efficiency of proposed self-training sampling mechanism on modeling temporal dynamics in data and systematically selecting informative negative samples for each user. To verify the benefit of tracking item latent factors, we set $V_t = V_0$ in ST-PFUV, and name it as ST-PFU. Table 1a also shows the benefit of tracking the tendency for both users and items.

To further illustrate the power of self-training and the importance of distinguishing recent and historic ratings, we test the performance of PFUV under two extra settings, where ratings consist of 1) the recent observation and wAMAN, 2) all the historic data and missing data sampled by user-oriented distribution. We name these methods as PFUV-Rect-wAMAN and PFUV-Hist-User. We also extend the best baseline method DynAllRank by replacing wAMAN with missing data sampled by user-oriented distribution. We name it DynAllRank-User. For easy comparison, DynAllRank-User has the same setting as DynAllRank. Table 1b shows

Table 2. Results on HetRec dataset under temporal recall metric. The best performance is in italic font.

Method	PMF	DynAllRank	ST-PFUV-User	ST-PFUV
$recall_{temp}(k)$	0.1718	0.2263	0.2020	*0.3036*
time (second)	432.9	947.9	*179.3*	198.7

(a) Temporal recall and scalability. $\hat{N} = 50$ for ST-PFUV and ST-PFUV-User.

$\hat{N}$	10	20	30	40	50	60	70
ST-PFUV-User	0.1754	0.1882	0.1977	0.2026	0.2020	0.2071	0.2041
ST-PFUV	0.2536	0.2826	0.2980	0.3016	*0.3036*	0.3013	0.3032

(b) Effect of different number of samples

the performance of these methods. Combining with the results in Table 1a, results in Table 1b further confirm the ability of our methods to balance information intrinsic in recent observations and the sparsity reduction introduced by imputation to better incorporate temporal and dynamic information.

HetRec MovieLens. In following experiments, we focus on the study of accuracy and scalability of ST-PFUV over a longer period, and the influence of the number of selected samples. We compare ST-PFUV with the best baseline method DynAllRank and the static PMF method. Similar to previous experiments, ST-PFUV has the best performance in accuracy. Due to space limitation, only temporal recall is shown here. By cross validation, we set $w_m = 0.08$, $K = 40$ and $\lambda = 0.12$ for DynAllRank, and $\lambda_U = \lambda_V = 0.1$ for ST-PFUV. For PMF, we set $K = 40$, step size for gradient descent as 0.005 and $\lambda = 0.05$.

Table 2a shows the results of these methods under temporal recall, where $\hat{N} = 50$. The results show that ST-PFUV significantly outperforms other methods in terms of recommendation accuracy. All the improvement is statistically significant under paired t-tests with $p < 0.01$. Compared with results on MovieLens 100K, ST-PFUV has much greater accuracy improvement over baseline methods, verifying its much better exploiting temporal and dynamic information, especially over a longer period. As a sequential approach, our proposed algorithms do not require retraining stages. Thus, we average the total running time (both retraining time and testing time) to compare the scalability as shown in Table 2a. These empirical results confirm that the proposed sampling methods are much faster than baseline methods, and the two-phase method ST-PFUV is comparable with its one-phase counterpart. Note that PMF has been among the fastest state-of-art CF methods.

The results with different number of samples are shown in Table 2b. Upto $\hat{N} = 50$, the performance of ST-PFUV is constantly being improved as the effect of sparsity reduction. The performance is not improved significantly when $\hat{N}$ is larger, as the built observations are cluttered by negative samples.

To further evaluate temporal behaviors of ST-PFUV, we define the average of accumulated improvement (AAI) over time. Let the performance of any two

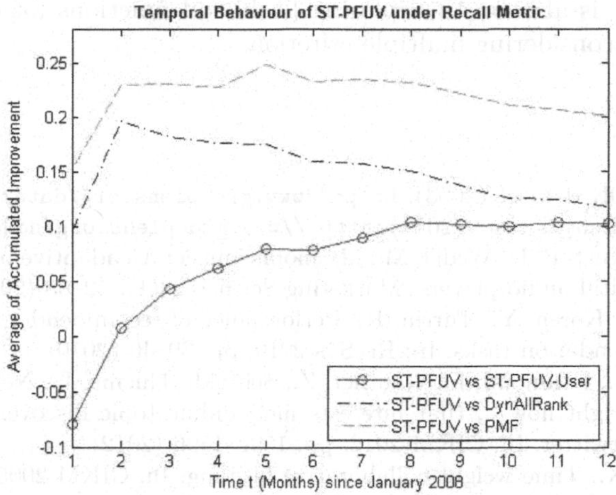

Fig. 1. Comparison of temporal behavior on HetRec dataset by AAI over time

methods under temporal recall in month t be $Rec_1(t)$ and $Rec_2(t)$, respectively. The AAI in month t_1 is $\frac{1}{t_1}\sum_{t=1}^{t_1}(Rec_1(t) - Rec_2(t))$. Figure 1 plots the AAI among ST-PFUV, ST-PFUV-User, DynAllRank and PMF methods. Except in the first month of red curve (ST-PFUV vs ST-PFUV-User), all the curves are above zero, showing that our method constantly outperforms baseline methods over time by selecting negative items via personalized self-training sampling scheme. Meanwhile, compared with DynPMF and DynAllRank, the tendency of dash and dot dash curves demonstrates that ST-PFUV is more efficient at exploiting the underlying temporal patterns. While baseline methods require longer training period, ST-PFUV performs well even if the training period is short (within 5 months) and accumulated amount of ratings is few.

6 Conclusion

In order to simultaneously solve the problems of data sparsity and scalability for temporal dynamic recommender systems, we have developed a novel two-phase self-training mechanism to dynamically construct a small but delicate set of observations from missing data. Cooperating with a particle filtering-based dynamic model, this work facilitates to track temporal dynamic user preference and item attractiveness in recommender systems.

The proposed algorithms are evaluated on two public benchmark datasets under the temporal accuracy metrics. The empirical results show that the proposed methods significantly improve recommendation performance over a variety of state-of-art algorithms. The experiments also illustrate the efficiency and scalability of the developed self-training temporal recommendation algorithms.

In future, we would like to investigate more sophisticated techniques to even better represent and learn the underlying dynamics of user preferences and item characteristics. It is also worth exploring likelihood functions for other recommendation tasks considering multiple criterion.

References

1. MovieLens 100K dataset (2003), http://www.grouplens.org/data/
2. HetRec MovieLens dataset (2011), http://www.grouplens.org/node/462
3. Chung, T.S., Rust, R.T., Wedel, M.: My mobile music: An adaptive personalization system for digital audio players. Marketing Science 28(1), 52–68 (2009)
4. Cremonesi, P., Koren, Y., Turrin, R.: Performance of recommender algorithms on top-n recommendation tasks. In: RecSys 2010, pp. 39–46 (2010)
5. Diaz-Aviles, E., Drumond, L., Gantner, Z., Schmidt-Thieme, L., Nejdl, W.: What is happening right now ... that interests me?: online topic discovery and recommendation in twitter. In: CIKM 2012, pp. 1592–1596 (2012)
6. Ding, Y., Li, X.: Time weight collaborative filtering. In: CIKM 2005, pp. 485–492 (2005)
7. Grimes, D.B., Shon, A.P., Rao, R.P.N.: Probabilistic bilinear models for appearance-based vision. In: ICCV 2003, pp. 1478–1485 (2003)
8. Hong, W., Li, L., Li, T.: Product recommendation with temporal dynamics. Expert Systems with Applications 39(16), 12398–12406 (2012)
9. Jeong, B., Lee, J., Cho, H.: An iterative semi-explicit rating method for building collaborative recommender systems. Expert Systems Applications 36(3), 6181–6186
10. Kantor, P.B.: Recommender systems handbook (2009)
11. Koren, Y.: Collaborative filtering with temporal dynamics. In: SIGKDD 2009, pp. 447–456 (2009)
12. Liu, N.N., Zhao, M., Xiang, E., Yang, Q.: Online evolutionary collaborative filtering. In: RecSys 2010, pp. 95–102 (2010)
13. Lu, Z., Agarwal, D., Dhillon, I.S.: A spatio-temporal approach to collaborative filtering. In: RecSys 2009, pp. 13–20 (2009)
14. Rong, P., Yunhong, Z., Bin, C., Liu, N.N., Lukose, R., Scholz, M., Qiang, Y.: One-class collaborative filtering. In: ICDM 2008, pp. 502–511 (2008)
15. Salakhutdinov, R., Mnih, A.: Bayesian probabilistic matrix factorization using markov chain monte carlo. In: ICML 2008, vol. 25, pp. 880–887 (2008)
16. Sanjeev Arulampalam, M., Maskell, S., Gordon, N., Clapp, T.: A tutorial on particle filters for online nonlinear/non-gaussian bayesian tracking. IEEE Transactions on Signal Processing 50(2), 174–188 (2002)
17. Sindhwani, V., Bucak, S.S., Hu, J., Mojsilovic, A.: One-class matrix completion with low-density factorizations. In: ICDM 2010, pp. 1055–1060 (2010)
18. Steck, H.: Training and testing of recommender systems on data missing not at random. In: SIGKDD 2010, pp. 713–722 (2010)
19. Su, X., Khoshgoftaar, T.M.: A survey of collaborative filtering techniques. Advances in Artificial Intelligent 2009, 4:2 (2009)
20. Xiong, L., Chen, X., Huang, T.-K., Schneider, J., Carbonell, J.G.: Temporal collaborative filtering with bayesian probabilistic tensor factorization. In: Proceedings of SIAM Data Mining (2010)
21. Zhang, W., Chen, T., Wang, J., Yu, Y.: Optimizing top-n collaborative filtering via dynamic negative item sampling. In: SIGIR 2013, pp. 785–788 (2013)
22. Zhu, X.: Semi-supervised learning literature survey (2006)

Data Augmented Maximum Margin Matrix Factorization for Flickr Group Recommendation

Liang Chen*, Yilun Wang*, Tingting Liang, Lichuan Ji, and Jian Wu

College of Computer Science,
Zhejiang University, China
{cliang,yilunwang,liangtt,jilichuan,wujian2000}@zju.edu.cn

Abstract. User groups on photo sharing websites, such as Flickr, are self-organized communities to share photos and conversations with similar interest and have gained massive popularity. However, the huge volume of groups brings troubles for users to decide which group to choose. Further, directly applying collaborative filtering techniques to group recommendation will suffer from *cold start* problem since many users do not affiliate to any group. In this paper, we propose a hybrid recommendation approach named Data Augmented Maximum Margin Matrix Factorization (DAM^{3}F), by integrating collaborative user-group information and user similarity graph. Specifically, Maximum Margin Matrix Factorization (MMMF) is employed for the collaborative recommendation, while the user similarity graph obtained from the user uploaded images and annotated tags is used as an complementary part to handle the *cold start* problem and to improve the performance of MMMF. The experiments conducted on our crawled dataset with 2196 users, 985 groups and 334467 images from *Flickr* demonstrate the effectiveness of the proposed approach.

1 Introduction

With the dramatic development of Web 2.0 and social network technologies, social media become more and more important as a way for users to obtain valuable information, express individual opinions, share experiences as well as keep in touch with friends. Online photo sharing sites, such as *Flickr*[1] and *Picasa Web Album*[2], become popular with numerous images uploaded every day (over 6 billion images in *Flickr*) [18]. User groups on such sites are self-organized communities to share photos and conversations with similar interest and have gained massive popularity. Joining groups facilitates flexibility in indexing and managing self photos, making them more accessible to the public and searching photos and users with similar interests. As the support of above view, Negoescu *et al.* provide an in-depth analysis of the structure of *Flickr* groups and the motivation of group activities [10][11].

---

* These two authors contributed equally to this work.

[1] http://www.flickr.com

[2] http://picasaweb.google.com

V.S. Tseng et al. (Eds.): PAKDD 2014, Part I, LNAI 8443, pp. 473–484, 2014.

Although the information contributed by user groups could greatly improve the user's browsing experience and enrich the social connections, the real situation is that many users rarely join in any group. By studying 3 million images and the respective users and groups crawled from *Flickr*, we discover an interesting fact that only 6.7% users have ever joined one image group and only 1.9% users have joined more than 5 groups. Thus, it is necessary to design a recommendation approach that could automatically recommend appropriate groups for users.

Group recommendation for photo sharing sites is a relatively novel scenario that have not been systematically studied. Current works commonly employs traditional recommendation techniques such as *collaborative filtering* [9] and *matrix factorization* [7,15] to recommend groups in a collaborative way. However, the recommendation performances of current work are not satisfied enough, due to two following points which are not yet considered carefully:

- **Sparse User-Group Matrix.** As discussed above, many users rarely join in any group. According to the statistic of our crawled dataset, the density of the user-group relationship matrix is only 0.46%. Simply implementing state-of-the-art recommendation techniques on such sparse dataset can't achieve satisfied result.
- **Cold Start Problem.** It should be noted that the *cold-start* problem in this paper means recommending groups to the users haven't joined in any group. The sparse user-group matrix makes it more difficultly to handle the *cold-start* problem.

As an early explorer in group recommendation for online photo sharing sites, we propose a hybrid recommendation approach named Data Augmented Maximum Margin Matrix Factorization (DAM^{3}F) to handle the above two problems, by integrating collaborative user-group relationship and user similarity graph. On one hand, Maximum Margin Matrix Factorization (MMMF) [17] is adopted for collaborative recommendation, by jointly learning the latent factors of users and groups from the original user-group relationship. As an improvement of tradition matrix factorization approaches, Maximum Margin Matrix Factorization uses hinge loss instead of sum-square loss and has been proven to be an effective approach for collaborative recommendation on sparse dataset. On the other hand, the user similarity graph obtained from the user uploaded images and annotated tags is used as an complementary part to handle the *cold start* problem and to improve the performance of MMMF. Specifically, graph regularization is introduced to preserve the user similarity, which provides a more interpretable way to characterize the users and groups. Further, a novel objective function is proposed which jointly consider the above issues, and an efficient optimization algorithm is provided to solve the objective function.

In particular, the main contributions of this paper can be summarized as follows:

1. This paper proposes a hybrid approach named DAM^{3}F to handle the *sparse user-group matrix* and *cold-start* problems in *Flickr* group recommendation.

2. A novel objective function is proposed by jointly considering the collaborative user-group information and the user similarity graph.
3. To evaluate the performance of the proposed approach, a real-world dataset consists of 2196 users, 985 groups and 334467 images is crawled from *Flickr*. The experimental results demonstrate that the proposed approach outperforms the state-of-the-art techniques in terms of six well-known evaluation metrics.

The rest of this paper is organized as follows: Section 2 gives a survey of related work in group recommendation. Section 3 shows the details of the proposed approach for *Flickr* group recommendation, while Section 4 reports the performance of DAM^{3}F based on real-world dataset. Finally Section 5 concludes this paper.

2 Related Work

User groups on social Websites, are self-organized communities to share photos and conversations with similar interest and have gained massive popularity. Group recommendation is an important paradigm that discovering the interesting groups for users, and attracts a lot of attention in recent years. In this section, we briefly introduce the related work in this area, by classifying them into three categories according to the employed approaches.

- ***Content-based recommendation*:** This category of methods recommends a group to a user based on the content of user or group, e.g., description of the group, the profile of the users interests, etc. Sihem *et al.* utilize the user profiles and propose a formal semantics that accounts for both item relevance to a group and disagreements among group members [1]. Liu *et al.* propose a tag-based group recommendation method on Flicker dataset by building a tag ranking system [8]. Kim *et al.* represent items with keyword features by a content-based filtering algorithm, and propose a community recommendation procedure for online readers [6].
- ***Collaborative filtering based recommendation*:** This category of methods was successfully applied in traditional recommender systems, and is based on the assumption that similar users are likely to attend similar groups. Chen *et al.* propose an improved collaborative filtering method named combinational collaborative filtering(CCF), which considers multiple types of cooccurrences in social data and recommends personal communities [4]. Zheng *et al.* implement several collaborative filtering methods, and provide a systematic experimental evaluation on Flicker group recommendation [20]. Yu *et al.* propose a collaborative filtering recommendation algorithm for Web communities, in which the latent links between communities and members are utilized to handle the sparsity problem [12].
- ***Hybrid recommendation*:** This category of methods combines several algorithms to recommend groups. Chen *et al.* compare association rule mining

(ARM) and latent dirichlet allocation (LDA) for the community recommendation, and find that LDA performs consistently better than ARM when recommending a list of more than 4 communities [3]. Chen *et al.* design a group recommendation system based on collaborative filtering, and employ genetic algorithm to predict the possible interactions among group members [5]. This strategy makes the estimated rating that a group of members might give to a group more correct. Zheng *et al.* propose a tensor decomposition model for Flickr group recommendation, which measures the latent relations between users and groups by considering both tags and users social relations [19]. Zheng *et al.* also propose an approach which combines the topic model and collaborative filtering, and this method is demonstrated to have better performance than traditional CF and negative matrix factorization [18].

The proposed approach DAM^{3}F in the paper is a hybrid one, in which we take the advantage of the above related works (e.g., user-annotated tags are utilized), and introduce some novel data (i.e., visual features extracted from the uploaded images) to further improve the performance of recommendation. In the technology aspect, we extend the traditional MMMF, and propose a novel objective function in which both user-group relationship and user similarity graph are considered. Further, an efficient optimization approach is proposed to solve the objective function.

3 DAM^{3}F Based Flickr Group Recommendation

In this section, we show the details of Data Augmented Maximum Margin Matrix Factorization (DAM^{3}F), which is the extension of the classical Maximum Margin Matrix Factorization approach by taking the uploaded images and user-annotated tags into consideration. To begin with, we give the main framework of DAM^{3}F in Figure 1.

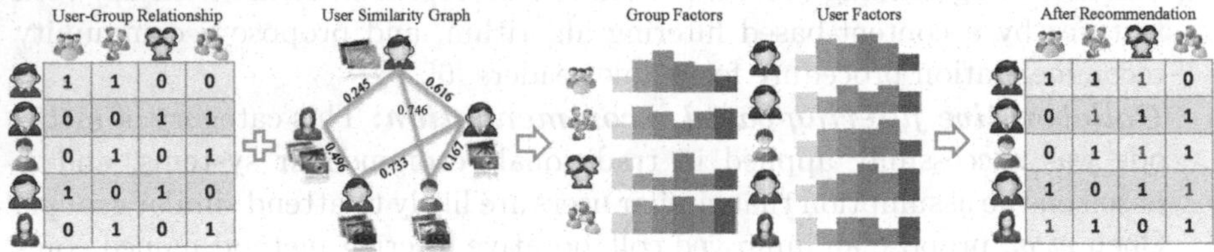

Fig. 1. The main framework of the proposed DAM^{3}F. The user-group relationship and user similarity graph are integrated in Data Augmented Maximum Margin Matrix Factorization framework to obtain the user and group latent factor matrices. Specifically, the similarity graph is computed based on the features extracted from images and annotated tags. Then the recommendation results can be calculated from the latent factors of users and groups.

3.1 Maximum Margin Matrix Factorization

Given the sets of M users, N images and P groups respectively, $R \in \mathbb{R}^{M \times P}$ is the affiliation matrix between users and groups, where $R_{ij} = 1$ means that the ith user is the member of the jth group and 0 otherwise. Furthermore, we use matrix $S \in \mathbb{R}^{M \times N}$ to denote the ownership between users and images where $S_{ik} = 1$ indicates that the kth image is uploaded by the ith user. By extracting the D dimension visual feature of the N images, we obtain the image feature matrix $X = [x_1, x_2, \ldots, x_N]^T \in \mathbb{R}^{N \times D}$. Given the information above, the aims of group recommendation is to recover a new affiliation matrix R_{rec} denote the relationship between users and groups, and more importantly to recommend new groups to users based on R_{rec}.

The traditional matrix factorization approaches for recommendation tries to factorize the affiliation matrix R into two $M \times K$ and $N \times K$ dimensional low-rank matrices U and G by:

$$\underset{U,G}{\operatorname{argmin}} \parallel R - UG^T \parallel_F + \lambda(\parallel U \parallel_F + \parallel G \parallel_F), \tag{1}$$

where $\parallel \cdot \parallel_F$ denotes the Frobenius norm and K is dimensionality of the latent factors of both users and groups. Besides, the regularization penalties $\lambda(\parallel U \parallel_F + \parallel G \parallel_F)$ is utilized to avoid over-fitting. Afterwards, the recommendation results can be obtained by calculating similarity between the latent factors of users and groups as $R_{rec} = UG^T$.

As for our group recommendation problem, $R_{rec_{ij}}$ only has two entries, i.e. 0 and 1, which indicates whether the ith user affiliates to the jth group. Therefore, comparing with traditional recommendation techniques which obtain the rating matrix, group recommendation is more appropriate to be formulated as a binary classification problem. Besides, since there are much more 0's than 1's in matrix R, the resulting recommendation results will be heavily biased towards 0 by using traditional matrix factorization approaches.

In order to overcome such limitations, Maximum Margin Matrix Factorization (MMMF) is proposed by replacing the sum-squared loss with hinge loss which has been widely used in classification application, such as Support Vector Machines. According to [13][17], the objective function of MMMF can be written as:

$$\underset{U,G}{\operatorname{argmin}} \, h(R - UG^T) + \lambda(\parallel U \parallel_F + \parallel G \parallel_F), \tag{2}$$

where $h(z) = (1 - z)_+ = max(0, 1 - z)$ corresponds to the hinge loss.

Above formulation can be also interpreted as simultaneous learning of feature vectors and linear classifiers. By viewing matrix U as the feature vectors of the users, matrix G can be regarded as linear classifiers that map the user feature vectors into binary labels that indicate whether the user is interested in that group. In addition, hinge loss is adopted for learning maximum margin classifiers with respect to each group.

3.2 Data Augmented Maximum Margin Matrix Factorization

Similar to other collaborative filtering algorithms, MMMF based recommendation still suffers *cold-start* problem, i.e., recommendation results for new users who have not joined groups tend to be very inaccurate. This problem could be solved, to some extent, by exploiting content information of the users, i.e., the features extracted from their uploaded images and the corresponding annotated tags. The basic assumption is: *if two users have joined in the same group, then their uploaded images to this group will probably be visually similar or semantically (tag-based) similar.* Based on this assumption, we can incorporate such user similarity graph into the MMMF based recommendation framework. Firstly, the feature vector f_i w.r.t. the i_{th} user can be calculated by averaging the feature vectors of all his images or tags as:

$$f_i = \frac{\sum_{j=1}^{N} S_{ij} X_j}{\sum_{j=1}^{n} S_{ij}}, \tag{3}$$

where S_{ij} denotes the j_{th} image uploaded by u_i, X_j means the visual or semantic (tag-based) feature of S_{ij}. It should be noted that the process of feature extraction is introduced in Section 4.1. Then we can construct the adjacency matrix W of the user similarity graph as follows:

$$W_{ij} = \begin{cases} exp(\frac{\|f_i - f_j\|}{t})^2, & if \ x_j \in \mathcal{N}(x_i) \ or \ \ x_i \in \mathcal{N}(x_j) \\ 0, & otherwise, \end{cases} \tag{4}$$

where $\mathcal{N}(x_i)$ denotes the k-nearest neighbor of x_i and the Heat Kernel is exploited to measure the similarity of two feature vectors. To guarantee that users who upload visually or semantically similar images will also obtain similar latent factors, we introduce the following graph regularization term:

$$\begin{aligned} &\frac{1}{2}\sum_{i,j} \| \ u_i - u_j \ \|_2 \ W_{ij} \\ =&\sum_{ij} u_i W_{ij} u_i^T - \sum_{ij} u_j W_{ij} u_j^T \\ =&\sum_{i} u_i D_{ii} u_i^T - \sum_{ij} u_j W_{ij} u_j^T \\ =&tr(U^T (D - W)U) \\ =&tr(U^T LU), \end{aligned} \tag{5}$$

where $tr(\cdot)$ denotes the matrix trace, $D_{ii} = \sum_i W_{ij}$ is a diagonal matrix and $L = D - W$ is the Laplacian matrix of the user similarity graph.

By leveraging the collaborating information and user similarity graph, we propose the Data Augmented Maximum Margin Matrix Factorization (DAM^{3}F) framework, which unifies maximum margin matrix factorization and graph regularization as:

$$\underset{U,G}{\operatorname{argmin}} \ h(R - UG^T) + \mu tr(U^T(\gamma L_1 + (1-\gamma)L_2)U) + \lambda(\| \ U \ \|_F + \| \ G \ \|_F) \tag{6}$$

where μ is the trade-off parameter between collaborating information and content information, L_1 is the Laplacian matrix of the image based user similarity graph (visually), L_2 is the Laplacian matrix of the tag based user similarity graph (semantically), and γ is the trade-off parameter between visual information and tag information.

Although the proposed objective function is not a convex function of U and G, but it is convex to one variable when the other one is fixed. Therefore, we could obtain the local optimal solution by alternatively updating the two variables using gradient descent methods.

Denoting the objective function as $J(U, G)$, we can calculate the gradient of $J(U, G)$. The partial derivative with respect to U is:

$$\frac{\partial J}{\partial U} = -h'(R - UG^T)G + 2\mu(\gamma L_1 + (1-\gamma)L_2)U + 2\lambda U \tag{7}$$

The partial derivative with respect to G is:

$$\frac{\partial J}{\partial G} = -h'(R - UG^T)^T U + 2\lambda G \tag{8}$$

Since the hinge loss function $h(z)$ is non-smooth at $z = 1$, following [13], we adopt smooth hinge instead of hinge loss for the ease of optimization. The further details of optimization process is omitted due to the space limitation.

4 Experiments

In this section, we evaluate the performance of the proposed approach on the real-world dataset crawled from *Flickr* by using kinds of metrics, and compare it with state-of-the-art approaches. Sepcifically, all experiments are conducted on a windows workstation with Intel 2.67GHz Xeon CPU and 32GB RAM by using Matlab 8.0.

4.1 Experiment Setup

To evaluate the performance of the proposed approach, we collect an image dataset from *Flickr* by using its API[3]. The details of this dataset could be found in Table 1. To obtain this dataset, we first select popular groups in *Flickr* by keyword searching. Then, active users of these groups and their uploaded images as well as the annotated tags are crawled, respectively. As for the process of feature extraction, we extract 81-dimensional color histogram/moments feature and 37-dimensional edge histogram feature to generate the visual feature to represent these images by using FELib[4], and employ Latent Dirichlet Allocation (LDA) to generate 50-dimensional semantic feature to represent the annotated

[3] `http://www.flickr.com/services/api/`

[4] `http://www.vision.ee.ethz.ch/~zhuji/felib.html`

Table 1. Overview of Dataset Crawled from Flickr

#Image	#Group	#User	#Tag	#Tag Token
334467	985	2196	3603353	239557

Table 2. Group Recommendation Performance Comparison ($\alpha = 60\%$)

Method	F_1 score	RMSE	P@5	P@10	MAP	MAE
CB	0.1702	0.0796	0.1315	0.4691	0.2149	0.3936
CF	0.1962	0.0811	0.1533	0.5909	0.2208	0.3923
SVD	0.2847	0.0863	0.2303	0.5719	0.3778	0.3963
NMF	0.3024	0.0799	0.2472	0.6589	0.4322	0.3912
MMMF	0.2992	0.0779	0.2478	0.7193	0.4316	0.3859
DAM^{3}F$_{tag}$	0.3184	0.0757	0.2612	0.7337	0.4504	0.3743
DAM^{3}F$_{visual}$	**0.3219**	**0.0755**	**0.2635**	**0.7353**	**0.4527**	**0.3722**

tags of the uploaded images. Due to the space limitation, we don't give the details of the tag-based feature extraction.

To evaluate the performance of group recommendation, we randomly sample $\alpha \times 100\%$ of the user-group assignments from the user-group affiliation matrix to generate the matrix R for training and use the full user-group affiliation matrix as the ground-truth for evaluation.

Evaluation Metrics. To comprehensively evaluate the performance of the proposed approach, we consider the following evaluation metrics: Precision@k (P@k), Mean Average Precision (MAP), Mean Absolute Error (MAE) [2], Root Mean Squared Error (RMSE), and F_1 score. In particular, k is chosen to 5 and 10 for P@k metric.

4.2 Recommendation Performance Comparison

In order to demonstrate the effectiveness of the proposed approach, we implement the following approaches and compare the performances:

1. **CB:** Content based recommendation by using the user similarity graph.
2. **CF:** Collaborative Filtering recommendation by using user-group relationship [14].
3. **SVD:** Singular Value Decomposition based recommendation by using user-group relationship[9].
4. **NMF:** Nonnegative Matrix Factorization based recommendation by using user-group relationship[15].
5. **MMMF:** Maximum Margin Matrix Factorization based recommendation by using user-group relationship[17].
6. **DAM^{3}F$_{tag}$:** One version of the proposed DAM^{3}F in this paper, while only sematic (tag-based) user similarity graph is combined to MMMF.

7. **DAM^{3}F$_{visual}$**: One version of the proposed DAM^{3}F in this paper, while only visual user similarity graph is combined to MMMF.

In this experiment, parameter μ is set empirically to 1 and λ is set to 0.1. The dimensionality of latent factors K is set to 200. It should be noted that the parameters of all the competitive methods have been fairly tuned using cross-validate, and the average evaluation results after 10-fold cross-validation are selected.

Table 2 shows the performance comparison of above group recommendation approaches when $\alpha = 60\%$ in terms of multiple evaluation metrics. From Table 2, it can be observed that the two versions of DAM^{3}F largely outperform the other state-of-the-art approaches in terms of multiple metrics. The superior performance of the proposed approach comes from two aspects, one is the selection of hinge loss for matrix factorization, while the other one is the integration of user similarity graph. Further, it can be discovered that the DAM^{3}F$_{visual}$ outperforms DAM^{3}F$_{tag}$. The reason is two-fold: 1) extracted visual feature is more explicit than the semantic (tag-based) feature; 2) User-annotated tags are inherent uncontrolled, ambiguous, and overly personalized. Thus, a pre-processing should be implemented before adopting user-annotated tags. In the following experiments, we implement a tag recommendation process to smooth the distribution of tagging data, and the performance comparison could be found in Section 4.4.

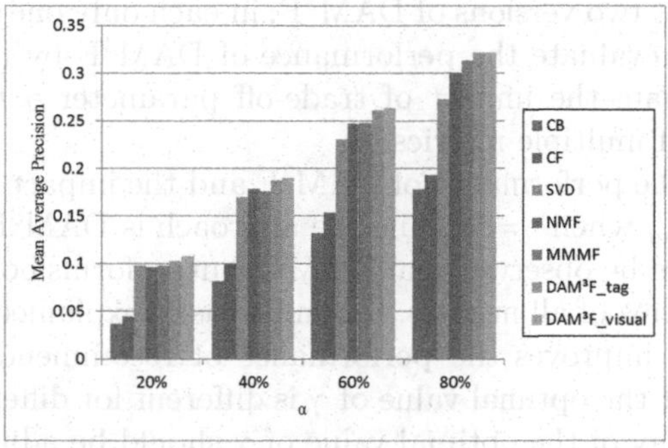

Fig. 2. Performance comparison with different α in terms of Mean Average Precision (MAP)

Figure 2 shows the group recommendation performance comparison of different approaches while the proportion of training data α varies from 20% to 80% in terms of MAP. The superior performance at different α further verified the effectiveness of the proposed approach. In addition, it can be observed that the two versions of DAM^{3}F are of greater advantage than other approaches when α gets smaller. The reason is that the user similarity regularization plays a more important role when the initial affiliation matrix is sparse, which is consistent with our motivation.

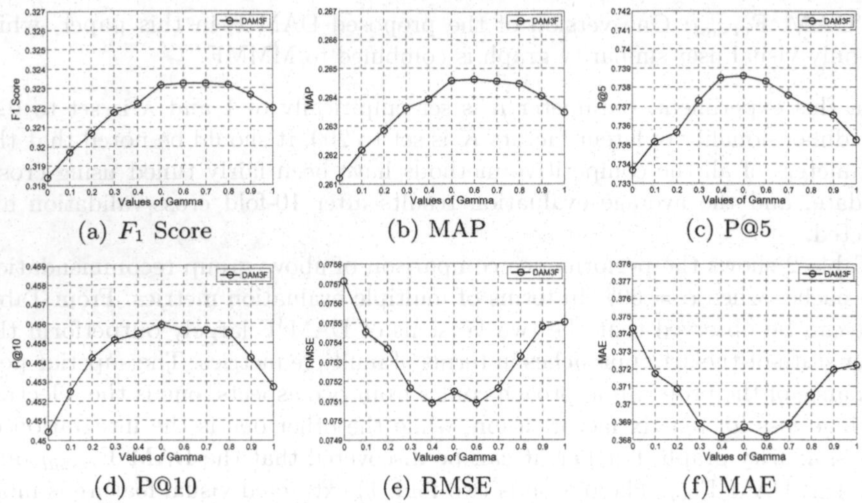

(a) F_1 Score (b) MAP (c) P@5

(d) P@10 (e) RMSE (f) MAE

Fig. 3. Impact of γ to The Performance of DAM^{3}F

4.3 Performance Evaluation of DAM^{3}F

In the proposed DAM^{3}F, we utilize both visual feature extracted from uploaded images and semantic feature extracted from annotated tags. In the above experiment, we implement two versions of DAM^{3}F, in each only one feature is utilized. In this section, we evaluate the performance of DAM^{3}F by utilizing these two features, and evaluate the impact of trade-off parameter γ to performance of DAM^{3}F in terms of multiple metrics.

Figure 3 shows the performance of DAM^{3}F and the impact of γ to it. The approach is DAM^{3}F$_{tag}$ when γ=0, while the approach is DAM^{3}F$_{visual}$ when γ=1. From Fig. 3, it could be observed that DAM^{3}F outperforms both DAM^{3}F$_{tag}$ and DAM^{3}F$_{visual}$ in terms of all metrics. It can be easily explained that the increase of relevant feature improves the performance of recommendation. Further, it could be found that the optimal value of γ is different for different metric. Thus, the selection strategy of the optimal value of γ should be adjusted according to the application scenario.

4.4 Impact of Tag Recommendation to DAM^{3}F

As discussed above, user-annotated tags are inherent uncontrolled, ambiguous, and overly personalized. Thus, a pre-processing should be implemented before adopting user-annotated tags. In this section, we implement two commonly accepted tag recommendation approach, i.e., *Sum* and *Vote* [16], to recommend relevant tags to the images with few tags and to delete irrelevant tags for the purpose of smoothing the tagging data distribution.

Table 3 show the performance comparison of DAM^{3}F with different tag recommendation approaches in terms of multiple metrics. From Table 3, it can

Table 3. Performance Comparison of DAM^{3}F with Different Tag Recommendation Approaches

Method	F_1 score	RMSE	P@5	P@10	MAP	MAE
DAM^{3}F+ Original Tag	0.3184	0.0757	0.2612	0.7337	0.4504	0.3743
DAM^{3}F+ Vote	0.3197	0.0754	0.2625	0.7358	0.4532	0.3709
DAM^{3}F+ Sum	**0.3202**	**0.0752**	**0.2628**	**0.7367**	**0.4534**	**0.3706**

be observed that the introduction of tag recommendation process improves the performance of DAM^{3}F in terms of all metrics. It can be easily understood as the irrelevant tags don't contribute to the representation of the uploaded images or even have negative effect, while the addition of relevant tags improves the representation quality of the tag-based feature.

5 Conclusion and Future Work

In this paper, we propose a hybrid approach for *Flickr* group recommendation by leveraging traditional collaborative recommendation with user similarity regularization. More specifically, the proposed Data Augmented Maximum Margin Matrix Factorization (DAM^{3}F) approach integrates the maximum margin matrix factorization with the user similarity graph calculated from their uploaded images and the annotated tags. Experiments implemented on the real-world dataset crawled from *Flickr* demonstrates the effectiveness of the proposed approach, by comparing it with state-of-the-art approaches in terms of multiple metrics. As a general framework, DAM^{3}F can be also applied to other recommendation tasks.

In our future work, we will try to employ more personal and contextual information of *Flickr* users in the framework of DAM^{3}F for the purpose of improving group recommendation performance.

Acknowledgements. This research was partially supported by the National Technology Support Program under grant of 2011BAH16B04, the National Natural Science Foundation of China under grant of 61173176, National Key Science and Technology Research Program of China 2013AA01A604.

References

1. Amer-Yahia, S., Roy, S.B., Chawlat, A., Das, G., Yu, C.: Group recommendation: Semantics and efficiency. Proceedings of the VLDB Endowment 2, 754–765 (2009)
2. Manning, C.D., Raghavan, P., Schuetze, H.: Introduction to Information Retrieval. Cambridge University Press (2008)
3. Chen, W.Y., Chu, J.C., Luan, J., Bai, H., Wang, Y., Chang, E.Y.: Collaborative filtering for orkut communities: Discovery of user latent behavior. In: Proceedings of the 18th International Conference on World Wide Web, pp. 681–690 (2009)

4. Chen, W.Y., Zhang, D., Chang, E.Y.: Combinational collaborative filtering for personalized community recommendation. In: Proceedings of the 14th ACM SIGKDD International Conference on Knowledge Discovery and Data Mining, pp. 115–123 (2008)
5. Chen, Y.L., Cheng, L.C., Chuang, C.N.: A group recommendation system with consideration of interactions among group members. In: Proceedings of the 18th International Conference on World Wide Web, vol. 34(3), pp. 2082–2090 (2008)
6. Kim, H.K., Oh, H.Y., Gu, J.C., Kim, J.K.: Commenders: A recommendation procedure for online book communities. Electronic Commerce Research and Applications 10(5), 501–509 (2011)
7. Koren, Y., Bell, R.M., Volinsky, C.: Matrix factorization techniques for recommender systems. IEEE Computer 42(8), 30–37 (2009)
8. Liu, D., Hua, X.S., Yang, L., Wang, M., Zhang, H.J.: Tag ranking. In: Proceedings of the 18th International Conference on World Wide Web, pp. 351–360 (2009)
9. Lu, L., Medo, M., Yeung, C.H., Zhang, Y.C., Zhang, Z.K., Zhou, T.: Recommender systems. Physics Reports 519, 1–49 (2012)
10. Negoescu, R.A., Gatica-Perez, D.: Analyzing flickr groups. In: CIVR, pp. 417–426 (2008)
11. Negoescu, R.A., Gatica-Perez, D.: Modeling flickr communities through probabilistic topic-based analysis. IEEE Transactions on Multimedia 12(5), 399–416 (2010)
12. Qian, Y., Zhiyong, P., Liang, H., Ming, Y., Dawen, J.: A latent topic based collaborative filtering recommendation algorithm for web communities. In: Proceedings of the Ninth Web Information Systems and Applications Conference, pp. 241–246 (2012)
13. Rennie, J.D.M., Srebro, N.: Fast maximum margin matrix factorization for collaborative prediction. In: ICML, pp. 713–719 (2005)
14. Resnick, P., Iacovou, N., Suchak, M., Bergstrom, P., Riedl, J.: Grouplens: an open architecture for collaborative filtering of netnews. In: Proceedings of the 1994 ACM Conference on Computer Supported Cooperative Work (1994)
15. Seung, D., Lee, L.: Algorithms for non-negative matrix factorization. In: Advances in Neural Information Processing Systems, vol. 13, pp. 556–562 (2001)
16. Sigurbjörnsson, B., van Zwol, R.: Flickr tag recommendation based on collective knowledge. In: Proceedings of the 17th International Conference on World Wide Web, pp. 327–336 (2008)
17. Srebro, N., Rennie, J.D., Jaakkola, T.: Maximum-margin matrix factorization. In: Advances in Neural Information Processing Systems, vol. 17(5), pp. 1329–1336 (2005)
18. Zheng, N., Bao, H.: Flickr group recommendation based on user-generated tags and social relations via topic model. In: Guo, C., Hou, Z.-G., Zeng, Z. (eds.) ISNN 2013, Part II. LNCS, vol. 7952, pp. 514–523. Springer, Heidelberg (2013)
19. Zheng, N., Li, Q., Liao, S., Zhang, L.: Flickr group recommendation based on tensor decomposition. In: Proceedings of the 33rd International ACM SIGIR Conference on Research and Development in Information Retrieval, pp. 737–738 (2010)
20. Zheng, N., Li, Q., Liao, S., Zhang, L.: Which photo groups should i choose? a comparative study of recommendation algorithms in flickr. Journal of Information Science 36(6), 733–750 (2010)

A Graph Matching Method for Historical Census Household Linkage

Zhichun Fu[1], Peter Christen[1], and Jun Zhou[2]

[1] Research School of Computer Science,
The Australian National University,
Canberra ACT 0200, Australia
{sally.fu,peter.christen}@anu.edu.au

[2] School of Information and Communication Technology,
Griffith University,
Nathan, QLD 4111, Australia
jun.zhou@griffith.edu.au

Abstract. Linking historical census data across time is a challenging task due to various reasons, including data quality, limited individual information, and changes to households over time. Although most census data linking methods link records that correspond to individual household members, recent advances show that linking households as a whole provide more accurate results and less multiple household links. In this paper, we introduce a graph-based method to link households, which takes the structural relationship between household members into consideration. Based on individual record linking results, our method builds a graph for each household, so that the matches are determined by both attribute-level and record-relationship similarity. Our experimental results on both synthetic and real historical census data have validated the effectiveness of this method. The proposed method achieves an F-measure of 0.937 on data extracted from real UK census datasets, outperforming all alternative methods being compared.

Keywords: graph matching, record linkage, household linkage, historical census data.

1 Introduction and Related Work

Historical census data capture valuable information of individuals and households in a region or a country. They play an important role in analysing the social, economic, and demographic aspects of a population. [2,17,19] Census data are normally collected on a regular basis, e.g. every 10 years. When linked over time, they provide insightful knowledge on how individuals, families and households have changed over time. Such information can be used to support a number of research topics in the social sciences.

Due to the benefit of historical census data linkage, and the fact that there are large amount of data available, automatic or semi-automatic linking methods have been explored by data mining researchers and social scientists [2,11,17,19].

V.S. Tseng et al. (Eds.): PAKDD 2014, Part I, LNAI 8443, pp. 485–496, 2014.

These methods treat historical census data linkage as a special case of record linkage, and apply string comparison methods to match individuals. Some researchers have linked historical census data with other types of data, and used Bayesian inference or discriminative learning methods to distinguish matched from non-matched records [20]. Although progress has been made in this area, the current solutions are far from practical in dealing with the ambiguity of data.

Difficulties of historical census data linkage come from several aspects. These include poor data quality caused by the census data collection and digitisation process, and large amount of similar values in names, ages and addresses. More importantly, the condition of individuals in a household may change significantly between two censuses. For example, people are born and die, get married, change occupation, or moved home. These problems are made more challenging in early historical census, i.e. those collected in the 19th or early 20th century, where only limited information about individuals were available. As a result, linking individuals is not reliable, and many false or duplicate matches are often generated. This is also a common problem in other record linkage applications, such as author disambiguation [8].

To tackle this problem, some methods have used the household information in the linkage process to help reducing erroneous matches. For example, a group linking method [16] has been applied to generate a household match score by combining similarity scores from each matched individual in a household [9]. This allows the detection of possible truth matches of both households and individuals by selecting candidates with the highest group linking score. When labeled data are available, Fu et al. developed a household classification method based on multiple instance learning [6,10]. In this method, individual links are considered as instances and household links as bags. Then a binary bag level classifier can be learned to distinguish matched and non-matched households.

Nonetheless, these household linking methods treated a household as a set of collected entities that correspond to individuals. They have not taken the structural information of households into consideration. While personal information, such as marital status, address and occupation, may change over time, surnames of females may change after marriage, and even ages may change due to different time of the year for census collection or input errors, the relationships between household members normally remain unchanged. This is the most stable structural information of a household. Such relationships include but are not limit to age difference, generation difference, and role-pairs of two individuals in a household. If the structural information can be incorporated into the linking model, the linking accuracy can be improved. Figure 1 shows an example on how household structure helps improve the household linking performance.

A graph-based approach is a natural solution to model the structural relationship between groups of records. During the past years, several graph matching methods have been proposed to match records. Domingos proposed a multi-relational record linage method to de-duplicate records [7]. This method defines conditional random fields, which are undirected graphical models, on all candidate record pairs. Then a chain of inference is developed to propagation matching

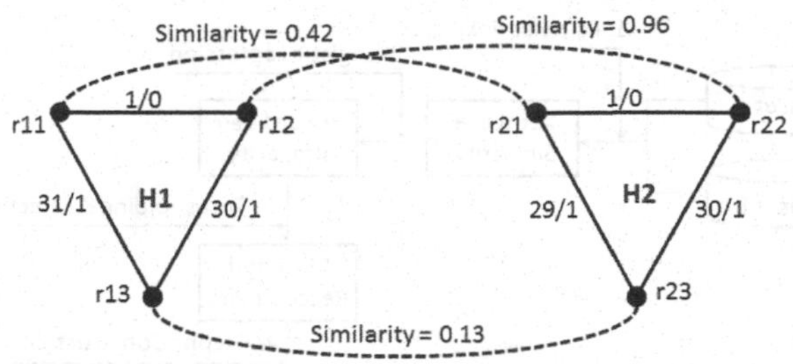

	ADDRESS	SN*	FN*	REL*	SEX	AGE
r11	goodshaw	trickettt	richard	head	m	32
r12	goodshaw	trickettt	elizabeth	wife	f	31
r13	goodshaw	trickettt	mary	daughter	f	1

H1

	ADDRESS	SN*	FN*	REL*	SEX	AGE
r21	goodshaw	trickett	richard	head	m	40
r22	goodshaw	trickett	elizabeth	wife	f	41
r23	goodshaw	trickett	m.	daughter	f	11

H2

*SN: Surname FN: First name REL: Relationship to head of household

Fig. 1. An example of structural information of household extracted from the historical census dataset. The edge attributes are age difference and generation difference of neighbouring vertices. The similarities between two pairs of records from two households are low. When the relationships between household members are considered, e.g. roles in a household, it is clear that these two households shall be matched.

information among linked records. Hall and Fienberg reported a method to build bipartite graphs and evaluate the confidence of different hypothetical record link assignments [12]. This method can be used to link datasets of moderate size. Nuray-Turan et al. built a graph model and labelled dataset to compute the strength of connections among linked candidate records [15]. A self-tuning approach is developed to update the model in a linear programming fashion. Furthermore, hierarchical graphical model have been proposed to cope with the potential structure in large amount of unlabeled data [18].

Because the goal of graph methods is to match or de-duplicate multiple records, all of them treat records that are linked to each other as vertices and links between them as edges. Therefore, the edges show the similarity between individual records. In our research, on the contrary, we build graphs on households. Specifically, the vertices in graph correspond to members in a household, while the edges shows the relationship between members in that household. Then we transform the household linking problem to a graph matching problem [3], i.e., household matching is determined not only by individuals, but also by the structure of their households.

The contribution of this paper is two-fold. First, we develop a graph matching method to match households in historical census datasets. Our method demonstrates excellent performance in finding potential household matches and removing multiple matches. Second, to generate more accurate record matching

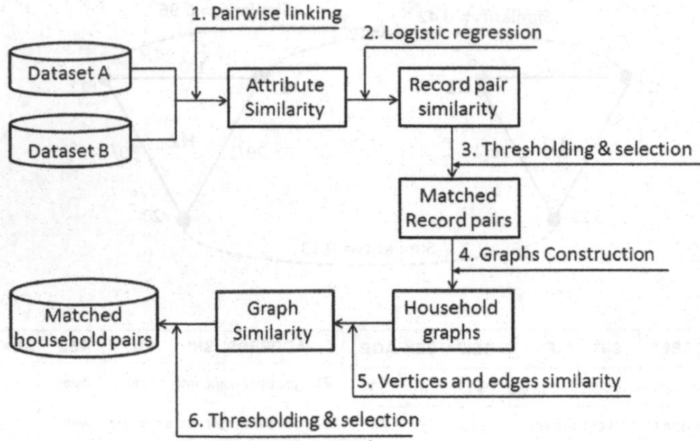

Fig. 2. Key steps of the proposed graph matching method

results, we adopted a logistic regression method to estimate the probability that two vertices across two household graphs are matched.

2 Graph-Based Household Matching

Given a query household, the goal of our work is to find the best matching household among a list of target households, and then to determine whether this match is a true match. In the proposed method, we show that this goal can be reached by a graph matching method, whose structure is summarised in Figure 2. The first step is record similarity calculation, whose results are used to find candidate matched record pairs. These records are then used to construct graph for each household. Graph matching is then performed based on vertex matching and graph similarity calculation.

2.1 Definition

Let H be a query household and $r_i \in H$ be the record of the ith member in this household, with $M = |H|$ be the total number of records in household H, and $1 \leq i \leq M$. Similarly, let $H' \in \mathcal{H}$ be a household amongst a list of target households in $\mathcal{H}$ to be linked with H, and $r'_j \in H'$ be the record of the jth member in H', with $M' = |H'|$ the number of records in household H', and $1 \leq j \leq M'$. If H and H' refers to the same household, they are matched. Otherwise, H and H' are not matched.

An undirected attributed graph $G = (V, E, \alpha, \beta)$ can be defined on H, where V is a set of vertices correspond to the household members. $E \in V \times V$ is a set of edges connecting vertex pairs, which show the relationship between household members. $\alpha = \{r_1, ...r_M\}$ and $\beta = \{r_{12}, ...r_{(M-1)M}\}$ are the attributes associated with vertices and edges respectively. In a similar manner, we can define a graph $G' = (V', E', \alpha', \beta')$ on household H'.

Once these household graphs are built, the household linking problem becomes a graph matching problem, such that the matched household can be identified based on graph similarity [21]. During this process, a key step is to generate a matching matrix for the graph pair such that vertices in V can be matched to vertices in V'. When labeled data is available, this problem can be solved by the quadratic or linear assignment method [3].

In the census household linking problem, domain knowledge tells that each individual in one household can only be matched to one individual in another household. In the following, we show how this domain knowledge is used to develop an efficient vertex matching method. Furthermore, we introduce a vertex matching method to match household members before graph construction, such that the sizes of graphs can be reduced.

2.2 Record Similarity

The historical census datasets used in this research contain attributes for each individual in a specific district as detailed in Section 4. Approximate string matching methods can be applied to these attributes to generate similarity values. During this process, a blocking technique [5] is used to remove those record pairs with low similarities, so that the cost of computation can be reduced.

Each attribute may have a different contribution in matching two records. In order to estimate the contribution from each attribute, we model the vertex matching problem as a binary classification problem, and solve it by logistic regression method. Assume we have T record pairs $\mathbf{x}_i$, $i = 1, 2, \cdots, T$, with label $y_i = +1$ for matched and $y_i = -1$ for non-matched classes. Let features of record pairs be x_{ij}, where $0 \le x_{ij} \le 1$, $j = 1, 2, \ldots, Q$, and Q is the number of similarities generated from different approximate string matching methods on the record attributes. A logistic regression model is given by

$$\log \left\{ \frac{p(y_i \mid \mathbf{x}_i)}{1 - p(y_i \mid \mathbf{x}_i)} \right\} = \mathbf{x}_i \mathbf{w} \tag{1}$$

where $\mathbf{w}$ is a vector of coefficients corresponding to the input variables. Then the maximum likelihood estimation of $\mathbf{w}$ is

$$\mathbf{w}^* = \arg\max_{\mathbf{w}} \left\{ \sum_{i=1}^{T} - \log(1 + \exp(-y_i \mathbf{x}_i \mathbf{w})) \right\} \tag{2}$$

which can be solved by iterative optimisation methods [13].

Once the optimal solution $\mathbf{w}^*$ is available, the posterior probability that a record pair is matched can be calculated as

$$P(y = 1 | \mathbf{x}_i) = \frac{1}{1 + \exp(-\mathbf{x}_i \mathbf{w}^*)} \tag{3}$$

Note that this posterior probability can be considered as the vertex similarity in the following graph model. It should also be pointed out that the logistic regression based record similarity is independent of graph matching, hence, can be used on any pairwise record comparison as long as a training set is available.

2.3 Record Linking

The outputs of the above step are record pair similarities. Here, we need to determine which record pairs may be a true match. Decisions can be made by comparing the vertex similarity with a threshold ρ, such that

$$P(y = 1|\mathbf{x}_i) > \rho \tag{4}$$

In our method, following the classic decision rule of logistic regression classification [1], we set $\rho = 0.5$.

After thresholding, low similarity record pairs are removed from consideration. In the remaining record pairs, the query record may still be linked to multiple target records. For this case, the record pair with the highest similarity shall be selected. In some cases, more than one record pairs may have the same highest similarity value, then all of the matched records are selected.

2.4 Graph Generation and Vertex Matching

After the record pair selection step, a graph can be generated for each household. Note that the record matching step can remove a large number of low probability links, such that individual links in a household without high probability do not need to be included in the graph generation. This allows small household graphs to be generated, which leads to high computation efficiency.

As mentioned previously, several target records may be selected for a query record at the record matching step. Therefore, one-to-many and many-to-one vertex mappings may be generated between two graphs. Then the optimal vertex to vertex correspondence has to be determined. Although such vertex matching can be done by supervised learning [3], in our method, we adopted the Hungarian algorithm [14], which is a more straightforward method with an $O(n^3)$ computational complexity, where n is the number of vertices. This algorithm generates the vertex matching that maximizes the sum of matched probabilities. The output of this step is graph pairs with one-to-one vertex mapping. Note that the average number of members in a household is less than 5, therefore, the complexity of this step is not a significant factor that affects the efficiency of our method.

2.5 Graph Similarity and Matching

In the previous record matching step, a record may be linked to multiple records in different households. Therefore, a graph containing the record may be linked to several other graphs. Similar to the record matching step, decisions also have to be made on which graph pair is a possibly a true match, and if there are multiple matches, which pair is the correct one. This requires the calculation of graph similarity. Here, we define the similarity between graph G and G' as

$$f(G,G') = \lambda f_v(V,V') + (1-\lambda) f_e(E,E') \tag{5}$$

where $f_v(V,V')$ and $f_e(E,E')$ are the total vertex similarity and total edge similarity, respectively, and λ is a parameter that controls the contribution from $f_v(V,V')$ and $f_e(E,E')$.

Note that vertex similarity has been generated in the record matching step from the output of the Hungarian algorithm. Let $sim_v(r_i, r_i')$ be the vertex similarity of the ith record pair $\mathbf{x}_i$ in the graph, and the total number of vertices in G be N, then

$$f_v(V,V') = \frac{\sum_{i=1}^{N} sim_v(r_i, r_i')}{N} \tag{6}$$

The calculation of total edge similarity is based on differences of edge attributes (details to be described later) between each pair of edges in the graph pair. Let r_{ijk} be the $k^{th}(k \in [1, ..., K])$ attribute of the edge r_{ij} which connects record r_i and r_j in graph G, and r'_{ijk} be the corresponding edge in graph G', then

$$f_e(E,E') = \frac{\sum_{i=1}^{L} sim_e(r_{ij}, r'_{ij})}{L} \tag{7}$$

where L is the number of edges in the graph. $sim_e(r_{ij}, r'_{ij})$ is the edge similarity, which is defined as follows

$$sim(r_{ij}, r'_{ij}) = \frac{\sum_{k=1}^{K} \tau_k sim_a(r_{ijk}, r'_{ijk})}{K} \tag{8}$$

where $sim_a(r_{ijk}, r'_{ijk})$ is the edge attribute similarity.

The graph similarity calculation allows selecting the optimal match from several target graph candidates. Then whether the selected graph G'^* is a true match of the query or not can be judged by the following condition:

$$f(G, G'^*) > \eta \tag{9}$$

If the graph similarity is larger than threshold η, then it is considered as true match. Note that parameters λ, τ and η can be learned from the training set by grid search.

3 Implementation Details

In this section, we give implementation details of several key steps in our method. Starting from the record similarity calculation, we adopted 10 combinations of attributes and approximate string matching methods to generate features of record pairs for the logistic regression model. The implementation of the string

Table 1. Record similarity using five attributes and various approximate string matching methods [5]

Attribute	Method
Surname	Q-gram / Jaccard / String exact match
First name	Q-gram / Jaccard / String exact match
Sex	String exact match
Age	Gaussian probability
Address	Q-gram / Longest common subsequence

matching methods follows the work done by Christen [4] and Fu et al. [9], except that the age similarity is based on probabilities generated by a Gaussian distribution on the age differences. A summary of these attributes and string matching methods are provided in Table 1. In calculating the total vertex similarity $f_v(V, V')$, an alternative method is the group linking approach proposed in [16]. We implemented this model and combined it with the total edge similarity for graph similarity calculation. Different from [9], we used the probability generated by the logistic regression step to calculate record similarity, instead of using an empirical record similarity calculation by adding the attribute-wise similarities. Then the group linking based graph vertex similarity is calculated using the following equation

$$f_v(V, V') = \frac{\sum_{i=1}^{L} sim_v(r_i, r'_i)}{M + M' - N}. \tag{10}$$

where M and M' are the numbers of household members in H and H' respectively. N is the set of record pairs matched between H and H' as defined in Equation (6). Note that different from the vertex similarity calculated in Equation (6), group linking takes the number of distinct household members into consideration rather than merely the matched members.

Equation (8) requires calculation of several edge attribute similarities. In the proposed method, such edge attributes are generated to reflect the structural property of households. In more detail, three attributes have been considered. They are age and generation differences between two household members connected by an edge, and the role pair between two household members. The calculation of age difference is straightforward. When comparing edges in two graphs, the edge similarity on this attribute is the probability generated by the Gaussian distribution of the difference of the age differences in two edges. The generation difference is based the relative generation with respect to that of the household head. A lookup table is built for this purpose. For example, as shown in Figure 1, a record with role value "wife" is in the same generation as the a record with "head", therefore their generation difference is 0. The generation difference between "head" and "son" or "daughter" is 1.

The role pairs are even more complex. We listed most of the possible role pairs between two household members and generated a lookup table to show how such role pair can change. For example, "wife-son" may change to "head-son" if the husband of a household died in between census and the wife became the head. When comparing two edges, binary values are generated for both generation difference and role pair attributes. If the corresponding generation difference value of two edges is different, the similarity is 0, otherwise, it is set to 1. For the role pair attribute, if a role pair change is has been recorded in the training data, we set the similarity to 1, otherwise, it is set to 0.

4 Experimental Results

In the experiments, we used six census datasets collected from the district of Rawtenstall in North-East Lancashire in the United Kingdom, for the period from 1851 to 1901 in ten-year intervals. These datasets are in digital form, and each census is a table that contains information of record for each individual. There are 12 attributes for each record, including address of the household, full name, age, sex, relationship to the household head, occupation and place of birth et al. These data were standardised and cleaned before applying the record/household linkage step as done in [9]. In total, there are 155,888 records which correspond to 32,686 households in the six datasets.

4.1 Results on Synthetic Data

We built a synthetic dataset from the real census dataset in order to evaluate the performance of our method. We manually labeled 1,250 matched household pairs from the 1871 and 1881 historical census datasets. The labels also include matched records in the matched households. These became the positive samples in the dataset. Then we built negative samples by randomly selecting households and records in the 1871 and 1881 datasets, we built links to labelled positive data. Because both household and individual follow one-to-one match, we are sure that these negative samples are true un-matched samples. In this way, we have generated a dataset with ground truth at both household and record levels.

In order to train the logistic regression model in Equation (1), λ in Equation (5), τ in Equation (8), and η in Equation (9), we split the synthetic dataset into a training and a testing test with equal number of households. After these parameters had been learned on the training set, we applied them to the graph matching model and evaluate the model on the testing set.

We compared our method (Graph Matching) with several baseline methods. The first baseline method (Highest Similarity) matches household based on the highest record similarity. If one query household is linked to several target households, the target household with the highest record similarity is selected. The second baseline (Vertex Similarity) builds household graphs using linked records. Then the household matching is determined only by the vertex similarity calculated by Equation (6). This is equivalent to calculating the mean record similarity on those records used to build graphs. The third method (Group Linking) is the group linking method [9] as defined by Equation (10). We replaced the vertex similarity with the group linking score in the graph matching step, so that the

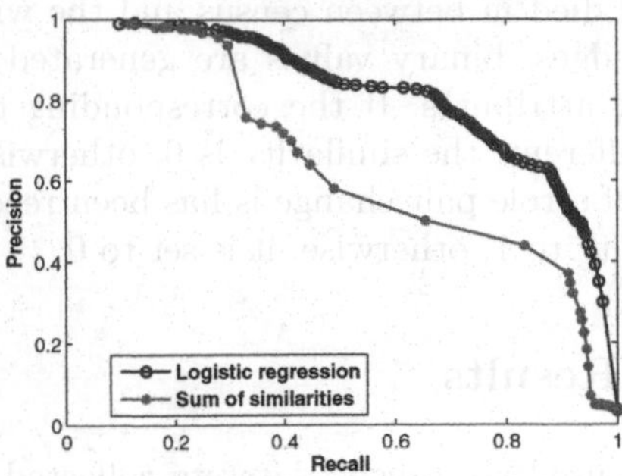

Fig. 3. Precision-recall curve for record linking

Table 2. Comparison of performance of the proposed method and baseline methods on the testing set. Highest values per measure are shown in bold.

	Precision	Recall	F-measure
Highest Similarity	0.6767	0.8608	0.7577
Vertex Similarity	0.6725	0.8544	0.7526
Group Linking	0.9522	0.8928	0.9216
Graph Matching	**0.9766**	0.8672	0.9186
Group Graph	0.9757	**0.9008**	**0.9368**

final decision of household matching is determined by the sum of group linking and edge similarity. We mark this method as "Group Graph".

Figure 3 shows the precision-recall curve of record matching, when the proposed logistic regression method is used to generate the similarity between pairs of records, or when the sum of the attribute-wise similarities generated by the approximate string matching methods are taken directly as similarity between pairs of records, which was the method adopted in [9]. The precision and recall values change with the thresholds used to determine whether two records are matched or not. This figure shows that the performance of logistic regression model has significantly outperformed the sum of similarity method. This is due to the training process that allows better modelling of the data distribution.

To show how effective the training step is, we used the trained logistic regression model, λ, τ, and ρ_G as the default values for the proposed graph matching method, and evaluated the method on the testing set using precision and recall values. We also calculate the F-measure, which allows balanced contribution from both precision and recall. The results from the methods being compared are summarised in Table 2. It can be seen that the graph matching method has generated the best F-score when combined with group linking for graph similarity calculation. Its performance is very close to the graph matching method as proposed in this paper, which has significantly outperformed record similarity based method. This shows that by considering the structure information of households, we can greatly improve the linking performance.

Table 3. Total household pairs found in historical census datasets

	1851–1861	1861–1871	1871–1881	1881–1891	1891–1901
Highest Similarity	2,509	3,136	3,708	3,938	4,109
Vertex Similarity	2,478	3,090	3,677	3,922	4,091
Group Linking	1,586	2,275	2,830	2,942	3,155
Graph Matching	1,409	1,995	2,462	2,523	2,784
Group Graph	1,493	2,117	2,688	2,756	2,982

Table 4. Unique household pairs found from the first datasets

	1851–1861	1861–1871	1871–1881	1881–1891	1891–1901
Highest Similarity	2,289	3,032	3,592	3,845	3,998
Vertex Similarity	2,289	3,032	3,592	3,845	3,998
Group Linking	1,584	2,272	2,827	2,942	3,136
Graph Matching	1,398	1,988	2,452	2,516	2,772
Group Graph	1,492	2,115	2,685	2,756	2,978

4.2 Results on Historical Census Datasets

Finally, we trained the graph model on the whole labelled data set, and applied it to all six historical census datasets. Similar to the experiment setting in [10], we classified all household and record links from any pair of consecutive census datasets, e.g. 1851 with 1861, 1861 with 1871, and so on. The matching results are displayed in Table 3 for the number of total household matches found on different datasets that include multiple matches of a household in another dataset, and in Table 4 for the number of unique household matches for which a household in one dataset is only matched to one household in another dataset. From the tables, it can be observed that both graph-based methods and the group linking method have generated much less total matches and unique matches than the record similarity based methods. Note that the difference between total matches and unique matches are duplicate matches. The results indicate that the proposed graph matching methods are very effective reduce number of duplicate matches.

5 Conclusion

In this paper, we have introduced a graph matching method to match households across time on the historical census data. The proposed graph model considers not only record similarity, but also incorporates the household structure into the matching step. Experimental results have shown that such structure information is very useful in household matching practise, and when combined with a group linking method, can generate very reliable linking outcome. This method can easily be applied to other group record linking applications, in which records in the same group are related to each other. In the future, we will develop graph learning methods on larger datasets, and incorporate more features for graph similarity calculation.

References

1. Bishop, C.: Pattern Recognition and Machine Learning. Springer (2006)
2. Bloothooft, G.: Multi-source family reconstruction. History and Computing 7(2), 90–103 (1995)
3. Caetano, T., McAuley, J., Cheng, L., Le, Q.V., Smola, A.: Learning graph matching. IEEE TPAMI 31(6), 1048–1058 (2009)
4. Christen, P.: Febrl: an open source data cleaning, deduplication and record linkage system with a graphical user interface. In: ACM KDD, Las Vegas, pp. 1065–1068 (2008)
5. Christen, P.: Data Matching - Concepts and Techniques for Record Linkage, Entity Resolution, and Duplicate Detection. Springer (2012)
6. Dietterich, T.G., Lathrop, R.H., Lozano-Perez, T.: Solving the multiple-instance problem with axis-parallel rectangles. Artificial Intelligence 89, 31–71 (1997)
7. Domingos, P.: Multi-relational record linkage. In: KDD Workshop, pp. 31–48 (2004)
8. Elmagarmid, A.K., Ipeirotis, P.G., Verykios, V.S.: Duplicate record detection: A survey. IEEE TKDE 19(1), 1–16 (2007)
9. Fu, Z., Christen, P., Boot, M.: Automatic cleaning and linking of historical census data using household information. In: IEEE ICDM Workshop, pp. 413–420 (2011)
10. Fu, Z., Zhou, J., Christen, P., Boot, M.: Multiple instance learning for group record linkage. In: Tan, P.-N., Chawla, S., Ho, C.K., Bailey, J. (eds.) PAKDD 2012, Part I. LNCS, vol. 7301, pp. 171–182. Springer, Heidelberg (2012)
11. Fure, E.: Interactive record linkage: The cumulative construction of life courses. Demographic Research 3, 11 (2000)
12. Hall, R., Fienberg, S.: Valid statistical inference on automatically matched files. In: Domingo-Ferrer, J., Tinnirello, I. (eds.) PSD 2012. LNCS, vol. 7556, pp. 131–142. Springer, Heidelberg (2012)
13. Hosmer, D.W., Lemeshow, S., Sturdivant, R.X.: Applied Logistic Regression, 3rd edn. Wiley (2013)
14. Munkres, J.: Algorithms for the assignment and transportation problems. Journal of the Society for Industrial and Applied Mathematics 5(1), 32–38 (1957)
15. Nuray-Turan, R., Kalashnikov, D.V., Mehrotra, S.: Self-tuning in graph-based reference disambiguation. In: Kotagiri, R., Radha Krishna, P., Mohania, M., Nantajeewarawat, E. (eds.) DASFAA 2007. LNCS, vol. 4443, pp. 325–336. Springer, Heidelberg (2007)
16. On, B.W., Koudas, N., Lee, D., Srivastava, D.: Group linkage. In: IEEE ICDE, Istanbul, Turkey, pp. 496–505 (2007)
17. Quass, D., Starkey, P.: Record linkage for genealogical databases. In: ACM KDD Workshop, Washington, DC, pp. 40–42 (2003)
18. Ravikumar, P., Cohen, W.W.: A hierarchical graphical model for record linkage. In: UAI, pp. 454–461 (2004)
19. Ruggles, S.: Linking historical censuses: a new approach. History and Computing 14(1+2), 213–224 (2006)
20. Sadinle, M., Fienberg, S.: A generalized Fellegi-Sunter framework for multiple record linkage with application to homicide record systems. Journal of the American Statistical Association 108(502), 385–397 (2013)
21. Zager, L., Verghese, G.: Graph similarity scoring and matching. Applied Mathematics Letters 21(1), 86–94 (2008)

Intervention-Driven Predictive Framework for Modeling Healthcare Data

Santu Rana*, Sunil Kumar Gupta*, Dinh Phung, and Svetha Venkatesh

Center for Pattern Recognition and Data Analytics,
Deakin University, Australia 3216
{santu.rana,sunil.gupta,dinh.phung,svetha.venkatesh}@deakin.edu.au

Abstract. Assessing prognostic risk is crucial to clinical care, and critically dependent on both diagnosis and medical interventions. Current methods use this augmented information to build a single prediction rule. But this may not be expressive enough to capture differential effects of interventions on prognosis. To this end, we propose a supervised, Bayesian nonparametric framework that simultaneously discovers the latent intervention groups and builds a separate prediction rule for each intervention group. The prediction rule is learnt using diagnosis data through a Bayesian logistic regression. For inference, we develop an efficient collapsed Gibbs sampler. We demonstrate that our method outperforms baselines in predicting 30-day hospital readmission using two patient cohorts - Acute Myocardial Infarction and Pneumonia. The significance of this model is that it can be applied widely across a broad range of medical prognosis tasks.

Keywords: Bayesian nonparametric, Healthcare data modelling.

1 Introduction

Medical interventions cure us, and keep us alive. They form the cornerstone of modern medical practice. Doctors carefully study the clinical observations related to our illness, and perform interventions. To formulate the most effective post-discharge care plan, they assess the prognosis. For example, what is the risk of readmission? How long will this person live? Answering these questions requires risk prediction models.

A patient's condition captures usual risk factors that can then be used in prognostic models. But medical interventions performed on patients are confounding, changing the outcome and thus prediction rules. For example, different cancer treatments (such as radiotherapy, chemotherapy or their combinations) have different prognosis profiles for the same tumor type [1]. Similarly, prognosis of cardiac patients for different procedures are different [2]. Thus interventions should be taken into account when developing prediction models.

Traditionally, in the healthcare community both the patient conditions and interventions are augmented together and a single prediction rule is learnt [3].

* Contributed equally.

V.S. Tseng et al. (Eds.): PAKDD 2014, Part I, LNAI 8443, pp. 497–508, 2014.

A single rule, however, may not be expressive enough to capture differential rules due to different interventions. Current predictive methods, such as logistic regression (LR), Support vector machine (SVM), Naïve Bayes (NB), and Random Forest (RF) require amalgamation of interventions with the patient condition variables, and suffer from the same limitation. At the other extreme, learning prediction rules for each intervention separately is not useful either - out of hundreds of unique interventions, all are not equally important and many of them are performed together (as groups of interventions) - for a variety of reasons including current treatment policies, hospital capacity and cost. This opens up the need to learn a set of intervention groups and group-specific prediction models.

Following this, we propose a nonparametric, supervised framework that uses a mixture distribution over interventions, learning a prediction model for each mixture component. A Dirichlet Process (DP) prior over interventions mixture is used allowing extraction of latent intervention groups, for which the number of groups is not known *a priori*. The outcome is then modeled as conditional on this latent grouping and patient condition data through a Bayesian logistic regression (B-LR). The use of DP also allows formation of new intervention groups when necessary, thus coping with changes in medical practice. In addition, the intervention based clustering inferred by the model is made predictive. This encourages formation of intervention groups that lead to a low prediction error. We refer to this model as DPM-LR. Efficient inference is derived for this model.

To evaluate our model, prediction of 30-day readmission on two retrospective cohorts of patients from an Australian hospital is considered: 2652 admissions related to Acute Myocardial Infarction (AMI) between 2007-2011 and 1497 admissions related to Pneumonia between 2009-2011. On both the cohorts, DPM-LR outperforms several baselines - dpMNL [4], Bayesian Logistic Regression, SVM, Naïve Bayes and Random Forest. We show that the intervention groups discovered using DPM-LR are clinically meaningful. We also illustrate that the highest risk factors identified by DPM-LR for different intervention groups are different, validating the necessity of intervention-driven predictive modeling.

In summary, our main contributions are:

- A nonparametric Bayesian, supervised prediction framework (DPM-LR) that explicitly models interventions and extracts latent groups by imposing a Dirichlet Process Mixture over interventions. The prognosis is modeled as conditional on this latent grouping and patient condition data through a Bayesian logistic regression (B-LR).
- Efficient inference for DPM-LR is derived and implemented.
- Validation on both synthetic and two real-world patient cohorts, demonstrating better performance by model over state-of-the-art baselines.

2 Background

Hospital readmissions are common and costly. The 30-day readmission rate among the Medicare beneficiaries in the USA is estimated at 18%, costing $17 billion [5]. Some hospital readmissions are considered avoidable and thus 30-day readmission rates are used for benchmarking across hospitals, with financial penalties for hospitals with high risk-adjusted rates [5]. Avoidable readmissions can be avoided by appropriately planning post-discharge care [6]. This requires accurate risk prediction.

Few models exist in the healthcare community to predict 30-day readmission risk in general medical patients [7,8,3]. All these methods employ Logistic Regression to derive a score based system for risk stratification using retrospective clinical and administrative data collected mainly from Electronic Health Records. Readmission prediction using other machine learning techniques such as SVM, Naïve Bayes and Random Forest have been studied respectively for heart-failure patients in [9] and for ten different diseases [10]. In all the methods, both the patients condition and interventions are augmented together to learn a single prediction rule.

A single rule, however, may not be sufficient to model the effect of different interventions. On the contrary, learning prediction rules for each intervention is not necessary - out of all the unique interventions, many of them are performed together and only a few latent groups exist. This gives rise to the need to learn the set of intervention groups and group-specific prediction models. The intervention grouping can be learnt using a mixture distribution with a Dirichlet Process prior to account for the unknown number of groups.

The use of Dirichlet process (DP) has been previously studied for modeling a set of classifiers under mixture model settings. In an attempt to develop a nonlinear classifier, Shahbaba and Neal [4] use DP as a prior for dividing data in clusters learning a separate linear classifier for each cluster. This model (dpMNL) learns nonlinear boundaries through a piecewise linear approximation. The idea from this model can be adapted for dividing patients for different intervention groups. Instead of using a single feature for both clustering and classification, we can use interventions to cluster the patients, and learn separate classifiers using patient condition features for each of the intervention groups.

3 Framework

We describe a prediction framework that learns a set of latent, predictive intervention groups and builds a prediction rule for each intervention group. In developing such a framework, our intention is to develop a predictive model that is flexible in modeling the effect of medical interventions on patient condition variables and outcome.

Typically, healthcare data has the following form: for each patient, we have a list of patient conditions (denoted by $\mathbf{x}$), a list of medical interventions (denoted by $\mathbf{i}$) and an outcome variable (denoted by y). We denote the data as $D = \{(\mathbf{x}_n, \mathbf{i}_n, y_n) \mid n = 1, \ldots, N\}$ where $\mathbf{x}_n \in \mathbb{R}^{M_x \times 1}$, $\mathbf{i}_n \in \mathbb{R}^{M_i \times 1}$.

To model the effect of interventions, we cluster the interventions into a set of predictive groups. A Dirichlet process mixture (DPM) over interventions is used to extract a set of latent intervention groups. The use of DPM allows us to form new intervention groups when necessary and thus copes with changes in hospital practices and policies. Further, the intervention-based clustering is made predictive so that it encourages formation of intervention groups that lead to a low predictive error. Given such clustering, we learn a separate classifier for each intervention group. We refer to this model as DPM-LR.

The generative process of DPM-LR can be described as follows: A random probability measure G is drawn from a Dirichlet process $\text{DP}(\alpha, H)$ where α is a positive concentration parameter and H is a fixed base measure. Since we are using a DP prior, the random measure G is discrete with probability one [11]. In stochastic process notation, we can write:

$$G \sim \text{DP}(\alpha, H), \ \psi_n \sim G, \ \{\mathbf{x}_n, \mathbf{i}_n, y_n\} \sim \psi_n \tag{1}$$

Stick-breaking construction of Dirichlet process [12] often provides more intuitive and clearer understanding of DP-based models. Using stick-breaking notation, the above generative process can be written as:

$$G = \sum_{k=1}^{\infty} \pi_k \delta_{\theta_k} \tag{2}$$

where θ_k are independent random variables (also called "atoms") distributed according to H. Further, δ_{θ_k} denotes an atomic measure at θ_k and π_k are the "stick-breaking weights" such that $\sum_k \pi_k = 1$. For our model, the variable θ_k takes values in a product space of two independent variables ϕ_k and $\mathbf{w}_k$. Thus, we can explicitly write $\theta_k \equiv \{\phi_k, \mathbf{w}_k\}$. For DPM-LR model, the ϕ_k can be interpreted as k-th "intervention topic" while the $\mathbf{w}_k$ is the classifier weight vector for k-th intervention topic. We model $\phi_k \sim \text{Dir}(\lambda)$, i.e. a Dirichlet distribution with parameter λ and $\mathbf{w}_k \sim \mathcal{N}(\mathbf{0}, \sigma_w^2 \mathbf{I})$, i.e. a multivariate normal distribution with zero mean and single standard deviation parameter σ_w. The two representations (the stochastic and the stick-breaking) can be tied by introducing an indicator variable z_n such that $\psi_n \equiv \theta_{z_n}$. We summarize the generative process as:

$$\pi \sim \text{GEM}(\alpha), \ (\phi_k, \mathbf{w}_k) \overset{\text{iid}}{\sim} H(\lambda, \sigma_w), \ H(\lambda, \sigma_w) = \text{Dir}(\lambda) \times \mathcal{N}(\mathbf{0}, \sigma_w^2 \mathbf{I}) \tag{3}$$

For $n = 1, \ldots, N$

$$z_n \sim \text{Discrete}(\pi), \ \mathbf{i}_n \mid z_n, \phi \sim \Pi_{m=1}^{M_i} \text{Discrete}(\phi_{z_n}) \tag{4}$$

$$y_n \mid \mathbf{x}_n, z_n, \mathbf{w} \sim \text{Ber}\left(f\left(\mathbf{w}_{z_n}^{\mathsf{T}} \mathbf{x}_n\right)\right) \tag{5}$$

where GEM distribution is named after the first letters of Griffiths, Engen and McCloskey [13]. Ber (.) and Dir (.) denote the Bernoulli and Dirichlet distributions, respectively and $f(.)$ denotes the logistic function. Graphical representations of DPM-LR is shown in Figure 1.

4 Inference

The inference of parameters in a fully Bayesian model is performed by sampling them from their joint posterior distribution, conditioned on the observations. For

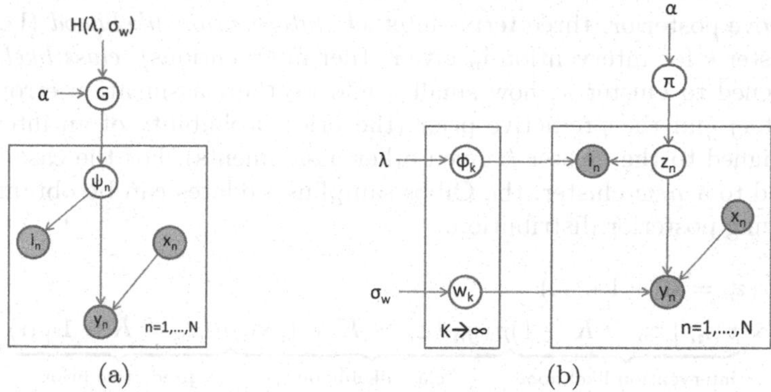

Fig. 1. Graphical representation of the DPM-LR (a) the stochastic process view (b) the stick-breaking view

DPM-LR model, this distribution does not take a closed form. A popular way to circumvent this problem is to approximate this distribution using Markov chain Monte Carlo (MCMC) sampling. Asymptotically, the samples obtained using MCMC are guaranteed to come from the true posterior distribution [14]. We use Gibbs sampling (a MCMC variant) - an algorithm that iteratively samples a set of variables conditioned upon the remaining set of variables and the observations. The MCMC parameter state space consists of the variables $\{\pi, z, \phi, \mathbf{w}\}$ and the hyperparameters λ, α and σ_w. To improve the sampler mixing, we integrate out π, ϕ and only sample variables $\{\mathbf{z}, \mathbf{w}\}$ and the hyperparameter α. The hyperparameters λ and σ_w are fixed to one. After the sampler convergence, we finally estimate ϕ as it provides useful insights into different intervention groups.

Since our model uses a Dirichlet process (DP) prior, Gibbs sampling of variable ϕ conditioned on other variables remains identical to the standard DP mixture model. However, due to the changes in the generative process caused by altering the model into a supervised setting, the Gibbs sampling updates for the variables z and $\mathbf{w}$ need to be derived.

4.1 Sampling z

We sample the variable z_n from Gibbs conditional posterior integrating out π and ϕ from the model. For the assignment of z_n, there are *two* possibilities: (1) the intervention $\mathbf{i}_n$ is assigned to an existing intervention cluster, i.e. given K clusters, z_n takes a value between 1 and K (2) the intervention $\mathbf{i}_n$ is assigned to a *new* intervention cluster, i.e. z_n is set to $K+1$. For the *former* case, the Gibbs sampling updates can be obtained from the following posterior distribution:

For $k = 1, \ldots, K$

$$p(z_n = k \mid \ldots) = p\left(z_n = k \mid \mathbf{z}^{-n}, \mathbf{i}^{-n}, \mathbf{i}_n, y_n, \mathbf{x}_n, \mathbf{w}\right) \tag{6}$$

$$\propto \underbrace{p\left(\mathbf{i}_n \mid z_n = k, \mathbf{z}^{-n}, \mathbf{i}^{-n}\right)}_{\text{intervention likelihood}} \underbrace{p\left(y_n \mid z_n = k, \mathbf{x}_n, \mathbf{w}\right)}_{\text{class likelihood}} \underbrace{p\left(z_n = k \mid \mathbf{z}^{-n}\right)}_{\text{predictive prior}}$$

In the above posterior, three terms interact: *intervention likelihood* (how likely is the cluster k for intervention $\mathbf{i}_n$ given other interventions), *class likelihood* (if $\mathbf{i}_n$ is assigned to cluster k, how small would be the classification error for the k-th cluster) and the predictive prior (the prior probability of an intervention being assigned to the cluster k given other assignments). For the case when z_n is assigned to a *new* cluster, the Gibbs sampling updates can be obtained from the following posterior distribution:

$$p(z_n = K+1 \mid \ldots) \tag{7}$$
$$\propto \underbrace{p(\mathbf{i}_n \mid z_n = K+1)}_{\text{intervention likelihood}} \underbrace{p(y_n \mid z_n = K+1, \mathbf{x}_n)}_{\text{class likelihood}} \underbrace{p(z_n = K+1 \mid \alpha)}_{\text{predictive prior}}$$

The class likelihood term in the above expression requires integrating out a Bernoulli likelihood with respect to $\mathbf{w}_{K+1}$. We approximate this integral numerically using Monte Carlo samples of $\mathbf{w}_{K+1}$.

4.2 Sampling $\mathbf{w}_k$

Using the generative process of (3-5), the Gibbs conditional posterior of $\mathbf{w}_k$ can be written as:

$$p(\mathbf{w}_k \mid ...) = p\left(\mathbf{y}^k \mid \mathbf{w}_k, \mathbf{X}^k\right) p\left(\mathbf{w}_k \mid \sigma_w\right)$$
$$\propto \left[\Pi_{i=1}^{n_k} \left(s_i^k\right)^{y_i^k} \left(1 - s_i^k\right)^{1-y_i^k}\right] e^{-\mathbf{w}_k^{\intercal}\mathbf{w}_k/2\sigma_w^2} \tag{8}$$

where we define $\mathbf{X}^k \triangleq \{\mathbf{x}_n \mid z_n = k\}$, which contains the patient condition data from the k-th intervention group and $\mathbf{x}_i^k$ is the i-th data column of $\mathbf{X}^k$. Further, we have $N^k \triangleq \#\{n \mid z_n = k\}$ and $s_i^k \triangleq f\left(\mathbf{w}_k^{\intercal}\mathbf{x}_i^k\right)$. The direct sampling from the above posterior is not possible as this does not reduce to any standard distribution. However, we can approximate the density using Laplace approximation [15,16]. The idea is to find the mode of the posterior distribution through an optimization procedure and then fitting a Gaussian with its mean at the computed mode. Instead of optimizing the posterior directly, we optimize the *logarithm* of the posterior (results are unaltered due to monotonicity of logarithm), for which it is possible to compute the first and the second derivatives in closed form. The first and the second derivatives of the log posterior are given as:

$$\nabla_{w_k} \ln p\left(\mathbf{w}_k \mid ...\right) = \sum_{i=1}^{n_k} \left(y_i^k - s_i^k\right) \mathbf{x}_i^k - \frac{1}{\sigma_w^2}\mathbf{w}_k \tag{9}$$

$$\nabla_{w_k}^2 \ln p\left(\mathbf{w}_k \mid ...\right) = -\mathbf{X}^k \mathbf{D}_s\left(\mathbf{w}_k\right) \left(\mathbf{X}^k\right)^{\intercal} - \frac{\mathbf{I}}{\sigma_w^2} \tag{10}$$

where $\mathbf{D}_s\left(\mathbf{w}_k\right) \triangleq \operatorname{diag}\left(\left[s_1^k\left(1 - s_1^k\right), \ldots, s_{N^k}^k\left(1 - s_{N^k}^k\right)\right]\right)$ is a diagonal matrix with entries between 0 and 1. For the above optimization, we use quasi-Newton

(L-BFGS) method as it converges faster compared to steepest-descent given good initializations. The optimization solution (denoted as $\mathbf{w}_k^*$) is used as mean of the approximating Gaussian. The covariance matrix of the Gaussian is computed (in closed form) by taking the negative of the inverse of the Hessian of the log posterior, i.e. $\Sigma_{\mathbf{w}_k}^* = -\left[\nabla_{w_k}^2 \ln p\left(\mathbf{w}_k \mid ...\right)\right]^{-1}$. Given $\mathbf{w}_k^*$ and $\Sigma_{\mathbf{w}_k}^*$, the posterior samples of $\mathbf{w}_k$ are drawn from $\mathcal{N}\left(\mathbf{w}_k^*, \Sigma_{\mathbf{w}_k}^*\right)$.

4.3 Sampling ϕ_k, α

Sampling ϕ_k is not necessary for the prediction. However, since it provides useful insights into different intervention groups, we finally estimate (after the sampler convergence) it as $\hat{\phi}_{m,k} = \frac{n_{m,k}+\lambda}{\sum_{m=1}^{M_i}(n_{m,k}+\lambda)}$ where $n_{m,k}$ is the number of occurrences of the m-th intervention in the k-th group. Sampling of the hyperparameter α remains same as in standard DPM model. Further details can be found in [17].

4.4 Prediction for New Observations

After training the model with data $D = \{(\mathbf{x}_n, \mathbf{i}_n, y_n) \mid n = 1, \ldots, N\}$, we have samples $\left\{\mathbf{w}^{(l)}, \mathbf{z}^{(l)}\right\}_{l=1}^{L}$. Given a new observation $\left\{\tilde{\mathbf{x}}, \tilde{\mathbf{i}}\right\}$, its outcome $\tilde{y}$ can be sampled from the following distribution:

$$p\left(\tilde{y} \mid \tilde{\mathbf{x}}, \tilde{\mathbf{i}}\right) \approx \frac{1}{L} \sum_{l=1}^{L} \sum_{\tilde{z}=1}^{K} p\left(\tilde{y} \mid \tilde{z}, \tilde{\mathbf{x}}, \mathbf{w}^{(l)}\right) p\left(\tilde{z} \mid \tilde{\mathbf{i}}, \mathbf{z}^{(l)}, \alpha\right) \tag{11}$$

The posterior $p\left(\tilde{z} \mid \tilde{\mathbf{i}}, \mathbf{z}^{(l)}, \alpha\right)$ can be computed similar to the corresponding terms in (6) as the model is not updated during the test phase.

5 Experiments

We perform experiments with a synthetic dataset and two hospital datasets. Baseline methods used for comparison are first presented followed by results on synthetic data. Finally, evaluation is performed on two patient cohorts.

5.1 Baselines

We compare the predictive performance of DPM-LR with the following methods: (a) Standard DP-Multinomial Logit model, with Gaussian observation model (dpMNL)[4]. The method learns a nonlinear classifier with data constructed by augmenting patients condition and intervention features (b) An adaptation of dpMNL with Multinomial observation model (referred to as dpMNL(MM)) (c) Bayesian Logistic Regression (B-LR) (d) SVM with linear kernel (Linear-SVM) (e) SVM with 3rd order polynomial kernel (Poly3-SVM) (f) Naïve Bayes (g) Random Forest. Weka implementation [18] is used for the SVM, Naive Bayes and the Random Forest. For all the baselines, the feature vector is created by merging the patient condition and intervention features.

5.2 Experiments with Synthetic Data

The synthetic dataset spans 5 years with 100 unique patients per year. Nine different interventions are considered. Six intervention topics are created from horizontal and vertical bar patterns of a 3x3 matrix (Fig 2a). Per patient "intervention" feature is synthesized by sampling an intervention topic from a uniform mixture distribution and then sampling 4 interventions from the selected intervention topic. Each intervention topic is considered as an intervention group. The classification weight vector of each group is sampled from a 50-variate Normal distribution. The "patient condition" feature is randomly sampled from a set of 10 distinct random binary vectors. The label (or outcome) is computed by combining the group-specific classifier with patient data following (5). The prediction task is to predict labels for the patients in the 5th year. Default settings from Weka is used for SVM, Naïve Bayes and the Random Forest.

DPM-LR outperforms all the baselines (Table 1) in terms of AUC (Area under the ROC curve). DPM-LR outperforms (AUC 0.942) the closest contender, Random Forest (AUC 0.873). The performance of standard dpMNL (AUC 0.630) with Gaussian observation model was poor, however, the adapted version with multinomial observation model did reasonably well (AUC 0.836). All the other methods performed poorly (AUC<0.750). Figure 2b shows the number of intervention topics sampled over 1000 Gibbs iterations (including 500 burnins). It can be seen that the convergence to the true number of topics is achieved quickly (*i.e.* the mode of the number of groups (K_m) remains unchanged after about 50 iterations), implying stable estimate of the posterior. Intervention topics inferred by the DPM-LR closely match true intervention topics (Figures 2a).

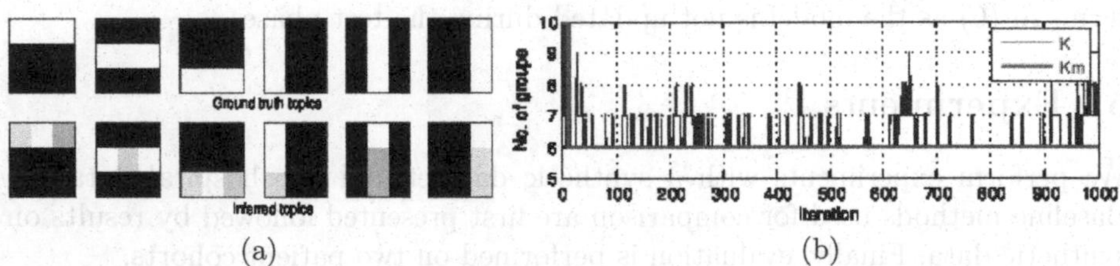

Fig. 2. Experiments on synthetic data (a) 6 intervention topics - True (top) and inferred (below) (b) Number of intervention topics (K) over Gibbs iterations and its running mode (K_m)

5.3 Experiments with Hospital Data

The data is collected from a large public hospital[1] in Australia. The hospital patient database provides a single point of access for information on patient hospitalizations, emergency department visits, in-hospital medications and treatments. Detailed records of these patient interactions with the hospital system are available through the EMR. This includes International Classification of Disease

[1] Ethics approval obtained through University and the hospital – Number 12/83.

Table 1. AUC for prediction on the synthetic dataset. Training is performed with 400 patients and testing with the remainder 100 patients.

Methods	DPM-LR	dpMNL	dpMNL (MM)	B-LR	Linear-SVM	Poly3-SVM	Naive Bayes	Random Forest
AUC	**0.942**	0.630	0.836	0.719	0.568	0.735	0.686	0.873

10 (ICD-10) codes[2], Diagnosis-related Group (DRG) codes of each admission, ICD-10 codes for each emergency visit, details of procedures, and departments that have been involved in the patient's care. Other information includes demographic data (age, gender, and occupation) and details of the patient's access to primary care facilities.

Cohort 1: Acute Myocardial Infarction (AMI). The patient cohort consists of 2652 consecutive admissions with confirmed diagnosis of Acute Myocardial Infarction (AMI) admitted between 1st January 2007 and 31st December 2011. For each patient, we have a sequence of interactions with the hospital system. Of these, the discharge corresponding to an admission with primary reason for admission as AMI is treated as assessment points (APs) from which prediction is made. Patient records prior to an AP are used to construct features. The "patient condition" feature contains demographic (age, gender and occupation) and disease information (ICD-10 codes) for each admission, accumulated at four different time scales - past 1 month, past 3 months, past 6 months and past 1 year. The "intervention" feature consists of procedure codes associated with only the current admission. The label is set to one if there are any readmissions in 30-day period following an AP with a cardiac related diagnosis. Readmission rate in this cohort varied from 11.7% (2007) to 4.8% (2011).

Experimental Results. Patient data from 2007-2010 are used for training and patient data from 2011 for testing. The comparative results with the baselines (Table 2) shows that DPM-LR outperforms all other methods.

DPM-LR is better (AUC 0.677) than the the closest contender dpMNL(MM) (AUC 0.641) by a significant margin. This is followed by dpMNL (AUC 0.635) and B-LR (AUC 0.607). All other methods have AUC less than 0.6. Surprisingly, more complex models such as SVM with polynomial kernel and the Random Forest perform the worst.

Table 3 lists the 5 strongest risk factors for the three intervention groups. These risk factors are the patient condition features that correspond to the largest positive weights in the linear regression model. We can see from the table that the strongest risk factors for different intervention groups are different. This vindicates the need of modeling intervention-specific prediction rules.

Cohort 2 - Pneumonia. This cohort consists of 1497 admissions with confirmed diagnosis of Pneumonia, admitted between 1st January 2009 and 31st

[2] `http://www.who.int/classifications/icd10/`

Table 2. AUC for 30-day readmission prediction for the AMI cohort. Patient data from 2007-2010 is used for training. Test year is 2011.

Methods	DPM-LR	dpMNL	dpMNL (MM)	B-LR	Linear-SVM	Poly3-SVM	Naive Bayes	Random Forest
AUC	**0.677**	0.635	0.641	0.607	0.576	0.516	0.577	0.566

Table 3. Strongest risk factors associated with a 30-day readmission risk in the AMI cohort for three main intervention groups - *Coronary angioplasty with stenting, Coronary artery bypass, and No intervention*

Intervention group	Top 5 strongest risk factors for readmission
Coronary angioplasty with stenting, Coronary angiography, Examination procedures on ventricle, Generalised allied health interventions	Hypertension in the past 1 month Retired and pensioner Congestive heart failure in the past 1 month Obesity in the past 1 month Fluid and electrolyte disorders in the past 1 month
Coronary artery bypass, Coronary angiography, Examination procedures on ventricle, Generalised allied health interventions	Metastatic cancer in the past 1 month Depression in the past 1 month Diabetes, complicated in the past 1 month Congestive heart failure in the past 3 month Peripheral vascular disease in the past 1 month
No intervention	Obesity in the past 1 month Metastatic cancer in the past 1 year Solid tumor without metastasis in the past 3 month Fluid and electrolyte disorders in the past 3 month Age above 90 years

December 2011. Similar to AMI, the discharges corresponding to an admission with primary reason for admission as Pneumonia is treated as the assessment points (APs) from which prediction is made. Patient records prior to an AP are used to construct the features, in a similar fashion as in the AMI cohort described in the previous section. The label is set to one if there are any readmissions in 30-day period following an AP with respiratory related diagnosis. Readmission rate in this cohort varied between 5-6% over the study years (2009-2011).

Experimental Results. The model is trained using patient data from 2009-2010 and then tested on patient data from 2011. The comparative results with the baselines are presented in Table 4. Once again, DPM-LR outperforms (AUC 0.667) the closest contender dpMNL(MM) (AUC 0.664).

DPM-LR learns two intervention groups. The risk factors corresponding to these two intervention groups are different (Table 5) - a point that was also observed for AMI cohort.

Table 4. AUC for 30-day readmission prediction (pneumonia). Training is with patient data from 2007-2010. Test is with patient data from 2011.

Methods	DPM-LR	dpMNL	dpMNL (MM)	B-LR	Linear-SVM	Poly3-SVM	Naive Bayes	Random Forest
AUC	**0.667**	0.590	0.664	0.640	0.523	0.511	0.635	0.561

Table 5. Strongest risk factors associated with a 30-day readmission risk in the pneumonia cohort for two main intervention groups

Intervention group	Top 5 strongest risk factors for readmission
Generalized allied health intervention, Administration of blood and blood products, Administration of pharmacotherapy.	Iron deficiency anaemia in the past 1 month Lower respiratory infection in the past 1 month Angina pectoris in the past 1 month Acute kidney failure in the past 1 month Fluid and electrolyte disorders in the past 1 month
No intervention	Intestinal disorders in the past 1 month Congestive heart failure in the past 1 month Age between 70-80 years Acute myocardial infarction in the past 1 month Acute kidney failure in the past 1 month

6 Conclusion

We present a novel predictive framework for modeling healthcare data in the presence of medical interventions. This framework automatically discovers the latent intervention groups and builds group-specific prediction rules. A Dirichlet process mixture used over the intervention groups ensures that new groups are created when a new intervention is introduced. The prediction rule is learnt using patients condition data through a Bayesian logistic regression. Efficient inference is derived for this model. Experiments demonstrate that this method outperforms state-of-the-art baselines in predicting 30-day hospital readmission on two cohorts - Acute Myocardial Infarction and Pneumonia. As a future work, it would be interesting to explore the performance improvement through sharing across various intervention groups using Bayesian shared subspace learning [19].

References

1. Al-Sarraf, M., LeBlanc, M., Giri, P., Fu, K.K., Cooper, J., Vuong, T., Forastiere, A.A., Adams, G., Sakr, W.A., Schuller, D.E., Ensley, J.F.: Chemoradiotherapy versus radiotherapy in patients with advanced nasopharyngeal cancer: phase iii randomized intergroup study. Journal of Clinical Oncology 16(4), 1310–1317 (1998)
2. Hannan, E.L., Racz, M.J., Walford, G., Jones, R.H., Ryan, T.J., Bennett, E., Culliford, A.T., Isom, O.W., Gold, J.P., Rose, E.A.: Long-term outcomes of coronary-artery bypass grafting versus stent implantation. New England Journal of Medicine 352(21), 2174–2183 (2005)

3. Donzé, J., Aujesky, D., Williams, D., Schnipper, J.L.: Potentially avoidable 30-day hospital readmissions in medical patientsderivation and validation of a prediction modelpotentially avoidable 30-day hospital readmissions. JAMA Internal Medicine 173(8), 632–638 (2013)
4. Shahbaba, B., Neal, R.: Nonlinear models using dirichlet process mixtures. The Journal of Machine Learning Research 10, 1829–1850 (2009)
5. Jencks, S.F., Williams, M.V., Coleman, E.A.: Rehospitalizations among patients in the medicare fee-for-service program. New England Journal of Medicine 360(14), 1418–1428 (2009)
6. Bradley, E.H., Curry, L., Horwitz, L.I., Sipsma, H., Wang, Y., Walsh, M.N., Goldmann, D., White, N., Piña, I.L., Krumholz, H.M.: Hospital strategies associated with 30-day readmission rates for patients with heart failure. Circulation: Cardiovascular Quality and Outcomes 6(4), 444–450 (2013)
7. Omar Hasan, M., Meltzer, D.O., Shaykevich, S.A., Bell, C.M., Kaboli, P.J., Auerbach, A.D., Wetterneck, T.B., Arora, V.M., Schnipper, J.L.: Hospital readmission in general medicine patients: a prediction model. Journal of General Internal Medicine 25(3), 211–219 (2010)
8. van Walraven, C., Dhalla, I.A., Bell, C., Etchells, E., Stiell, I.G., Zarnke, K., Austin, P.C., Forster, A.J.: Derivation and validation of an index to predict early death or unplanned readmission after discharge from hospital to the community. Canadian Medical Association Journal 182(6), 551–557 (2010)
9. Meadem, N., Verbiest, N., Zolfaghar, K., Agarwal, J., Chin, S.-C., Roy, S.B.: Exploring preprocessing techniques for prediction of risk of readmission for congestive heart failure patients (2013)
10. Cholleti, S., Post, A., Gao, J., Lin, X., Bornstein, W., Cantrell, D., Saltz, J.: Leveraging derived data elements in data analytic models for understanding and predicting hospital readmissions, vol. 2012, 103 (2012)
11. Ferguson, T.S.: A bayesian analysis of some nonparametric problems. The Annals of Statistics, 209–230 (1973)
12. Sethuraman, J.: A constructive definition of dirichlet priors. DTIC Document, Tech. Rep. (1991)
13. Pitman, J.: Combinatorial stochastic processes, vol. 1875. Springer (1875)
14. Gilks, W.R.: Full conditional distributions. In: Markov Chain Monte Carlo in Practice, pp. 75–88 (1996)
15. Breslow, N.E., Clayton, D.G.: Approximate inference in generalized linear mixed models. Journal of the American Statistical Association 88(421), 9–25 (1993)
16. Bishop, C.M., Nasrabadi, N.M.: Pattern recognition and machine learning, vol. 1. Springer, New York (2006)
17. Escobar, M., West, M.: Bayesian density estimation and inference using mixtures. Journal of the American Statistical Association 90(430), 577–588 (1995)
18. Hall, M., Frank, E., Holmes, G., Pfahringer, B., Reutemann, P., Witten, I.H.: The weka data mining software: An update. SIGKDD Explorations 11 (2009)
19. Gupta, S., Phung, D., Venkatesh, S.: A Bayesian nonparametric joint factor model for learning shared and individual subspaces from multiple data sources. In: Proc. of SIAM Int. Conference on Data Mining (SDM), pp. 200–211 (2012)

Visual Analysis of Uncertainty in Trajectories

Lu Lu[1], Nan Cao[2], Siyuan Liu[3], Lionel Ni[1], Xiaoru Yuan[4], and Huamin Qu[1]

[1] Hong Kong University of Science and Technology
{llu,huamin}@ust.hk
[2] IBM Thomas J Watson Research Center
[3] Heinz College, Carnegie Mellon University
[4] Peking University

Abstract. Mining trajectory datasets has many important applications. Real trajectory data often involve uncertainty due to inadequate sampling rates and measurement errors. For some trajectories, their precise positions cannot be recovered and the exact routes that vehicles traveled cannot be accurately reconstructed. In this paper, we investigate the uncertainty problem in trajectory data and present a visual analytics system to reveal, analyze, and solve the uncertainties associated with trajectory samples. We first propose two novel visual encoding schemes called the *road map analyzer* and the *uncertainty lens* for discovering road map errors and visually analyzing the uncertainty in trajectory data respectively. Then, we conduct three case studies to discover the map errors, to address the ambiguity problem in map-matching, and to reconstruct the trajectories with historical data. These case studies demonstrate the capability and effectiveness of our system.

Keywords: Uncertainty, trajectory, visual analysis.

1 Introduction

In recent years, there has been a dramatic increase in GPS-embedded devices used for navigation and tracking, which enables the collection of large volumes of GPS trajectories [8]. Trajectory data play important roles in urban planning, route recommendation, traffic analysis, and transportation management. Usually, trajectories are presented in two styles: curves (parameterized by time) in a 2D plane or trajectory samples that are discrete spatial-temporal points. The latter style is widely adopted in trajectory datasets since the cost of capturing and maintaining data is relatively low. However, the trajectories represented by samples often involve *uncertainty* which might appear as data imprecision due to sampling/measurement errors or fuzziness caused by pre-processing for preserving anonymity. Uncertainty in trajectories poses challenges for enhancing, reconstructing, and mining trajectories.

Uncertainty that appears as measurement errors has been studied for enhancing historical trajectories [9]. Limited by the technology used, the trajectory data are not precise due to measurement and sampling errors. Therefore, the recorded GPS positions often need to be matched with the given road network (referred to as road map). This process is called *map-matching*. Map-matching is integrated in many trajectory-based applications as a pre-processing module, which aligns trajectory samples with the given road networks. Quddus *et al.* [14] summarized the existing map-matching approaches

V.S. Tseng et al. (Eds.): PAKDD 2014, Part I, LNAI 8443, pp. 509–520, 2014.

and compared their performances. They addressed the measurement error problem and focused on the sampling errors. While map-matching methods are effective in solving the problem of sampling errors, few of them address the issues of inaccurate road maps and ambiguous selections of roads for trajectory samples.

Road maps, considered as the vital input to map-matching, are not always reliable due to two reasons. First, the update of road maps is not as frequent as the collection of trajectory data. Usually, a city digital map is only updated monthly or even less frequently, but the actual road changes happen everyday or even every hour in large cities. Thus, some road changes might not be incorporated in the road maps used for map-matching. Second, although the semi-automatic or fully-automatic methods used for road-map extraction are often effective, they might still fail to obtain correct road positions for various reasons, such as the low resolution of images, and the overlapping of roads. Therefore, road maps should be checked for errors before they are used for map-matching. Since road maps are complex and large, an effective visual analysis tool for revealing and fixing road map errors is needed.

Another kind of uncertainty is the ambiguity in the selection of roads to match samples. An essential step of map-matching methods is selecting an appropriate road for an off-road sample to align with. For that, the map-matching algorithms first select several road candidates with loose conditions, and then score the road candidates with specific cost functions, and finally choose the road with the highest score as the target. However, when the scores of the road candidates are similar, *e.g.*, the sample positions fall in the middle of multiple roads, all map-matchin algorithms will encounter the ambiguity problem as it is no longer clear which road should be chosen.

Uncertainty involving low sampling rates is also a serious issue for reconstructing trajectories. Some applications may need to reconstruct a vehicle's continuous route from its discrete trajectory samples. Due to low sampling rates, the collected trajectory samples can be very sparse. Between two consecutive sparse samples, there may exist several routes. Thus, additional information is needed to help choose routes to complete the reconstruction. Historical data could be very helpful. For example, we can investigate the historical data and check whether there are other relevant trajectories with denser sampling rates in the region of interest. With the help of relevant trajectories, we may have a better chance to find out the correct routes. In addition, by investigating the uncertainty patterns in the data, we may find ways to reduce the uncertainty and improve the data quality. For example, for areas dense with poor trajectory samples, we can add some road-side-units to improve the position accuracy and increase the sampling rates.

To solve the uncertainty problem, it is vital to keep humans in the loop and present all the relevant information to the users in an intuitive manner, especially for some fuzzy patterns and tricky cases. In this paper, we present our visual analytics solutions to the uncertainty problem in the trajectory data. Specifically, we propose two novel visual designs, *i.e.*, a *road map analyzer* for discovering potential errors in road maps, and an *uncertainty lens* for resolving the uncertainty in trajectory data. We demonstrate how to visually reveal the road map errors, resolve the ambiguities in map-matching, and reconstruct trajectories from sparse consecutive samples. We further test the effectiveness and usefulness of our approach with three case studies on real trajectory data.

The major contributions of our work are as follows:

- A visual design called road map analyzer to reveal errors in road maps based on observed trajectory samples. Our visual design is able to discover map errors such as road shifting and road missing.
- A visual design called uncertainty lens to reveal and resolve the uncertainty in the trajectory data. Our method integrates multiple factors (*e.g.*, speed, time, sparseness, direction) related to uncertainty into a coherent analytical framework.
- Case studies with real trajectories and digital maps to demonstrate the effectiveness and usefulness of the approach.

2 Related Work

Uncertainty Modeling. Kraak [7] proposed an interactive system to explore and visualize space-time data under a space time cube. Pfoser and Jensen[12] described the notion of the uncertainty in sampling error and the error across time. Later, a cylindrical model was presented by Trajcevski [16] to represent and capture the uncertainty for efficient querying. Most of the works solve the uncertainty problem in trajectory datasets, but few of them address the uncertainty in the context of specific trajectory-based applications. In our work, we deal with the uncertainty issue in the context of map-matching approaches, and address not only the uncertainty problem in trajectory datasets but also the uncertainty issue in map-matching approaches.

Uncertainty Visualization. Visualization techniques have also been developed for uncertainty, *e.g.*, glyphs, error bars, scale modulation, and ambulation. Pang *et al.* [11] proposed an uncertainty classification, studied its representation, and presented various approaches for its visualization. Fisher [3] surveyed the literature on uncertainty visualizations for bounded errors. Color and texture were considered to be the best choices for visualizing uncertainty [2][5]. The traditional representations of uncertainty usually consider uncertainty as a one-dimensional variable, but the uncertainty we want to address is more complicated. Traditional methods are inadequate for solving our problem.

Map-Matching. To deal with the errors in trajectory datasets, various map-matching approaches have been proposed and they can be categorized into three groups: geometric-based methods[15], topological-based methods [4][10], and statistical methods [6]. Geometric-based methods are effective in finding local matches, but sensitive to map errors. Extended from geometric-based methods, topological methods aim at matching the entire trajectory to road maps by using the topologies of road networks. Although topological methods are more robust than geometric-based methods, they still suffer from various errors associated with navigation sensors and road maps. Statistical models are employed in map-matching include Kalman Filter [6], and Bayesian classifier [13]. All map-matching approaches will encounter the ambiguity problem when the trajectory samples fall into the middle of multiple roads. Besides, few of the existing works take the map errors into consideration, which has a big impact on the map-matching results. In this paper, we present a novel visualization method to identify and fix the map errors and propose a visual-guided approach to resolve the ambiguities.

3 Visualization Design

3.1 Design Principles

We identify a few key principles to follow during the development of our designs: (1) To address the uncertainty in trajectories, maps should be used to facilitate the analysis but their errors should be identified and fixed first. The visualization schemes should help reveal and differentiate different map errors.

(2) To resolve the uncertainty in samples, multiple factors (the maps, other trajectories, and the traffic) should be taken into account and these factors should be put into a coherent analytical framework.

(3) The visualization mantra, "overview first, zoom and filter, then details-on-demand", will be followed.

3.2 Road Map Analyzer

As road maps are the critical reference for map-matching, road maps should be as accurate as possible and the errors in road maps need to be identified and corrected. After studying the map error problem and checking with domain experts, we focus on two specific cases are commonly seen in applications: (a) *Road Shifting*. In this case, the distribution of the samples forms a road shape, but the samples are not symmetric with respect to the road. (b) *Road Missing*. In this case, samples are roughly separated into multiple sets (usually two). The distribution of the samples forms two road shapes, and one of them is different from the observed road. It is likely that the road near the observed road is missing. Since the trajectory samples are sparse, we try to establish the relations among the samples. Given the set of the samples, we first establish the relations by building a tree considering the samples as the nodes and the distances between samples as the weights of the edges. The goal is to obtain a set of edges that minimize the sum of the weights. After establishing the relations, we get a set of edges that link the samples, called virtual edge. The virtual edges assign each pixel (on the display) they covered a density value by weights. Then, we generate a density map with the samples and their virtual edges. Finally we bound the density map by the bubble sets algorithm [1]. Fig. 1 is an example of the visualization of the boundary, called bubble view. Using the bubble view, we can easily detect the road error cases. For road shifting case, the bubble doesn't cover the observed road. For road missing case, the bubble not only covers the observed road, but also forms a trace linking with other roads. We also provide a statistical view that review statistical information of trajectory samples on road. In this view, we encode the distribution of the samples by plotting the samples into a tranform space. Each line represents a sample. It starts from a dot where indicates the average distance of the samples to the road, passes through its relative position to the road (left side or right side, by clockwise on the map), and ends on its projection on the road. Fig. 6 gives an example of the statistical view.

3.3 Uncertainty Lens

Identifying and resolving the uncertainty in trajectory data are our major goals. To achieve these goals, we design a novel visualization scheme, uncertainty lens, which

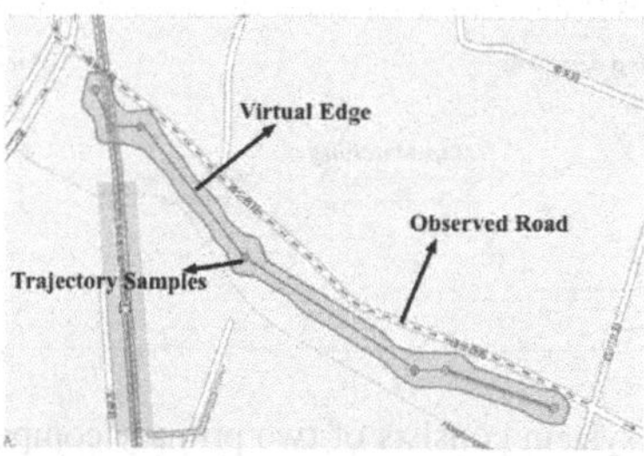

Fig. 1. Bubble view. The bubble view helps the users to identify the map errors, and it is generated from the trajectory samples around the observed road and the virtual edges linking the trajectory samples.

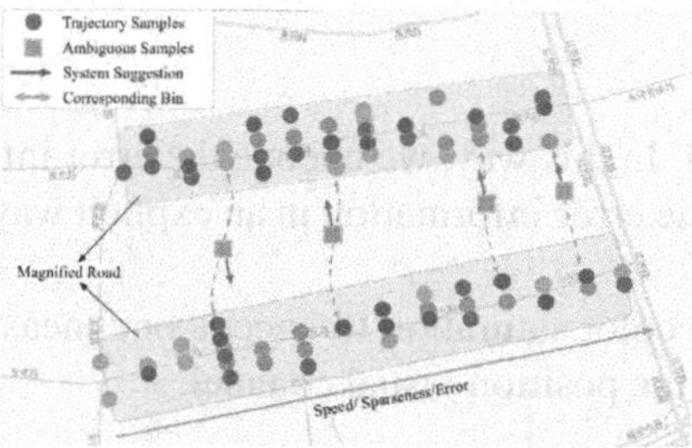

Fig. 2. Uncertainty lens. The uncertainty lens magnifies relative roads and sorts relative trajectory samples by their information (speed, error or sparseness). Based on the ordered information, the uncertainty lens provides suggestions for resolving ambiguities.

integrates multiple visualizations into one display to present all the important information that may help users resolve the uncertainty.

Fig.2 shows the design of the uncertainty lens which consists of relevant magnified roads with the trajectory samples, and a set of system suggestions. The uncertainty lens is overlaid on the geographical map. One straightforward way to observe the trajectory samples is plotting the samples on the geographical map. In this way, users can quickly find out the geographical information for the trajectory data and whether the samples fall on roads. However, as the number of trajectory samples is usually very large, the visualization becomes very cluttered. Therefore, we magnify the roads and re-order the trajectory samples in a certain way to get a better view. Additionally, users often need to consider other information besides maps to perform the analytical tasks, so this information should also be presented. After discussing with domain experts, we identify three features that are most important for analysis, including speed, error and sparseness.

The uncertainty lens magnifies relevant roads and displays trajectory samples on the relevant roads. For better analysis, the uncertainty lens encodes the speed, error, and sparseness by users' preferences. The encoding scheme is as follows.

- **Position:** We sort the trajectory sample by speed, error, or sparseness along the relevant road direction. At the same time, we keep the distances of the samples to

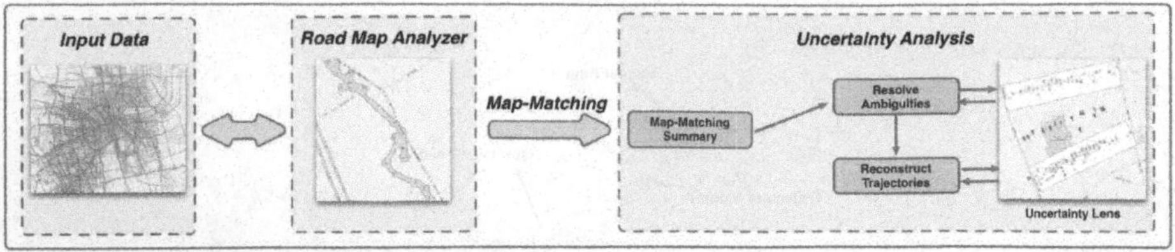

Fig. 3. System overview. Our system consists of two primary components. The input of the system is a set of trajectory samples and a digital map. The digital map is firstly processed by the module *road map analyzer*. The road map analyzer refines the digital map with the bubble views. Then a map-matching process is applied on the trajectory data with the refined map. The *uncertainty analysis* module analyzes the uncertainty of the trajectory data, resolves ambiguity, and reconstructs continuous trajectories.

the center of the road. In this way, we encode the error information implicitly. But users still can show the error information in an explicit way. The system uses speed to sort by default.

- **Color:** Users can set color saturation to encode one measurement that is different from the one used in the position visual channel.

4 System Overview and Workflow

Fig. 3 shows the overview of our system which consists of two primary components for three major tasks (*i.e.*, analyzing map errors, resolving ambiguities, and reconstructing trajectories). The input to the system is a set of trajectory samples and a digital map. The digital map is first processed by the module *road map analyzer*. The road map analyzer refines the digital map with the help of the bubble views. With the refined map, the map-matching process adjusts the observed positions of the trajectory samples, which partially solves the uncertainty problem in the data. After summarizing the results of the map-matching, the *uncertainty analysis* module identifies the ambiguous samples, analyzes the uncertainty of the ambiguous samples by using the uncertainty lens, and resolves the ambiguities. Then, to reconstruct trajectories, the system extracts the sparse consecutive samples with multiple route candidates. For selecting the best candidate route as a partial trajectory, the uncertainty lens provides historical trajectory information as hints for users to make decision.

Analyze Map Errors. Since the road network is large and complex, it is infeasible to review every road segment or display the visualizations of the roads in a size-limited screen. Therefore, our system computes the uncertainties of the roads and provides a list of the uncertainties. The list is shown in Fig. 4(b). The texts and the colors of the list cells refer to the identifications of the roads and their uncertainties respectively. The light color indicates the high uncertainty while the dark color indicates the low uncertainty (usually dark color suggests the correct road). The users can double click the list to select the roads. Once a road is selected to observe, the main viewer (Fig. 4(c)) automatically zooms to fit the display to the road and shows the bubble view of the observed road. If the users find any road errors, they can refine the roads by drawing the estimated roads manually.

Resolve Ambiguities. In our system, we leverage a map-matching approach [15] to adjust the observed position of the trajectory samples with the refined map. The core step in map-matching is to select an appropriate road for an off-road sample to align with by utilizing a specific cost function. In some cases, the selections are quite certain since the samples are close to a road. When the samples are reported in the middle of multiple roads, the selections become ambiguous (the samples in this case are called *ambiguous samples*). To resolve the ambiguities, we extract the ambiguous samples from the map-matching results. To locate the ambiguous samples, our system provides different levels of the map views. In low-level views, the system generates heatmaps that describe the distribution of the ambiguous samples. In high-level views, the system shows the ambiguous samples directly (using rectangles by default).

Once a set of ambiguous samples is focused, the uncertainty lens first auto-detects the relevant roads according to the directions of the ambiguous samples. Then, the uncertainty lens magnifies the relevant roads and displays the trajectory samples on the relevant roads. The users can sort the samples by their preferences. For example, they can sort the samples by speed and set the color of the samples by sparseness. After that, the system sorts the ambiguous samples accordingly and shows the suggestions for resolving the ambiguities. The users can resolve the ambiguities by clicking the ambiguous samples to follow the system suggestions or dragging the ambiguous samples to the roads depending on their own decisions.

Reconstruct Trajectories. Once the users load a trajectory, the system reconstructs the trajectory from the most sparse sample pair. To find out the routes between a sample pair, the system employs the time-space prism model [12]. By using this model, the possible movement region (of the trajectory) between two neighboring samples can be found. The moving object (such as vehicles) can only move within this region under a given speed. Therefore, the possible routes are confined within this region. For any sample pair, the system detects the possible routes. If there exists only one route, it continues to the next pair. If there exist multiple routes, the system shows the uncertainty lens for choosing a route as a partial trajectory. In the uncertainty lens, every possible route refers to a relevant route. Similar to resolving ambiguities, the users can sort the trajectory samples on the relevant routes. The system treats the trajectory between the neighboring samples as an ambiguous sample with information from this pair of samples. For example, the trajectory speed can be considered as the average speed of its adjacent samples. In this way, the system can make the suggestion for the users to choose a route.

5 Experiment and Discussion

The entire system was developed using Java, DaVinci and JOSM. We tested our system on a 1.8 GHz Intel Core i7 laptop with 4 GB DDR3. Raw map data were collected from OpenStreetMap.org. Trajectory data were collected from 4,000 taxis in a big city over an eight-month period. Each GPS record contains longitude, latitude, timestamp, and logs of other activities. The preprocessing of the data includes removing erroneous trajectories with impossible speeds and storing the valid trajectories in binary files to reduce the storage space and processing time.

Fig. 4 shows the system user interface. The system user interface consists of three components, including a tool bar that enables users to perform operations such as selecting and zooming, a road list that uses color to show the uncertainties of the roads, and a main viewer that provides the views of the geographical map, the road map analyzer, and the uncertainty lens.

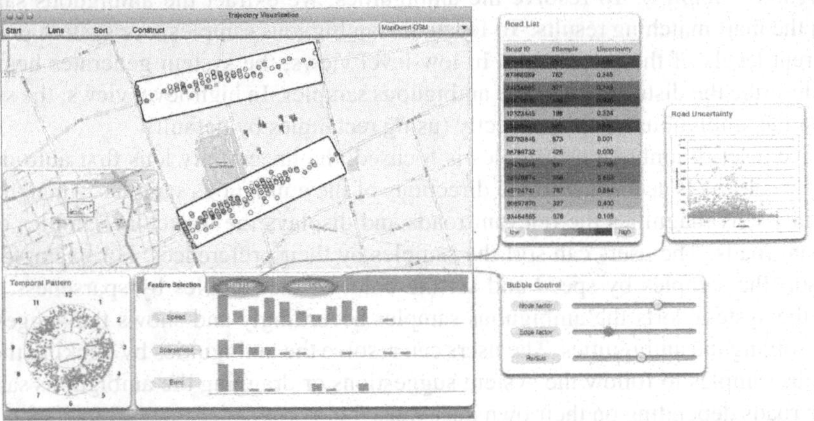

Fig. 4. System user interface. The system user interface consists of three components, including (a) a tool bar that enables users to perform operations such as selecting and zooming, (b) a road list that uses color to show the uncertainties of the roads, and (c) a main viewer that provides the views of the geographical map, the road map analyzer, and the uncertainty lens. In the road list, the light color indicates the roads with high uncertainty while the dark color suggests the roads with low uncertainty.

5.1 Experiment

To show the system usability, two transportation system researchers were invited to use our system to investigate the uncertainty in trajectory data. We have consulted the researchers when we design our system, so they are familiar with our design. We first gave a tutorial for about 15 minutes, then the researchers played with our system and did a test run to get familiar with the system user interface. After that, they were asked to perform three tasks: (1) analyzing map errors, (2) resolving ambiguities, and (3) reconstructing trajectories. The process for each researcher was about 40 minutes.

Analyzing Map Errors. In this task, the participants were asked to identify errors on the map. They first investigated some roads by clicking the road list. Fig. 5 shows three of the roads that the participant checked. The roads with different uncertainty showed different patterns in the bubble views. For example, the uncertainty of the road in Fig. 5 (b) was very low. The bubble aligned this road very well. Therefore, the participants considered this road was correct. For the case a participant found in Fig. 5 (c), it was with high uncertainty. He observed the bubble view (a zoom view shown in Fig. 6

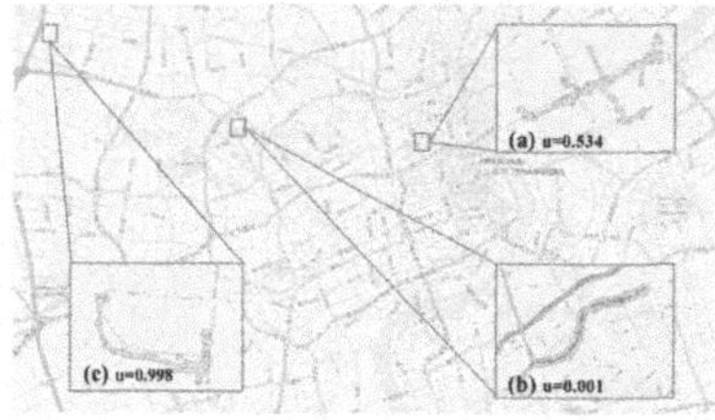

Fig. 5. Roads with different uncertainties. (a) Road uncertainty=0.534, the road is probably with errors. (b) Road uncertainty=0.001, the road is correct. (c) Road uncertainty=0.998, the road is with errors.

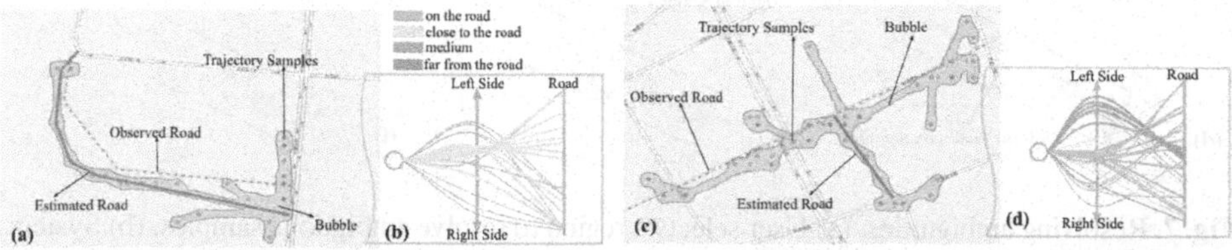

Fig. 6. Map error cases. (a) (b) Road shifting case. Most of the trajectory samples are reported slightly away from the observed road (red dashed lines). The actual road position (blue lines) is estimated using the bubble generated from the trajectory samples. (c) (d) Road missing case. The bubble forms a trace from the observed road to another road, which means that a road is absent from the map.

(a)) and recognized the road as a road shifting case. Because most of the trajectory samples were not falling on the road and the bubble didn't cover the road. He estimated the road position according to the bubble (Fig. 6 (a)). For an interesting error another participant found on Fig. 5 (a), the uncertainty of this road was medium. The uncertainty of this road was medium. The bubble covered the observed road, which confirmed the observed road is correct. However, the bubble formed a trace (without covering any road) connecting to another road. He considered that a road was absent from the map. Then he manually added the road on the map (Fig. 6(c)). The bubble view is effective at revealing road missing and estimating the actual positions of road with errors.

Resolving Ambiguities. To resolve the ambiguities, a participant located the ambiguous samples by using multiple level map views. He first selected a set of ambiguous samples in the region in Fig. 7(a). Then he opened the uncertainty lens. The uncertainty lens suggested two relevant roads (shown in Fig. 7(b)) and magnified them with the trajectory samples (see Fig. 7(c)). After that, he chose to sort the samples by speed and use the color to encode error. Following the setting by the participant, the system provided the suggestions for the ambiguous samples. Fig. 7(d) shows the system suggestions on the ambiguous samples by arrows. The arrows pointed to the road that the system recommended for the ambiguous samples. The participant resolved the ambiguities by clicking on the ambiguous samples (shown in Fig. 7(e)). Finally the ambiguous samples moved to the road (see Fig. 7(f)). From the task, the system's capability of resolving the ambiguities in map-matching was confirmed.

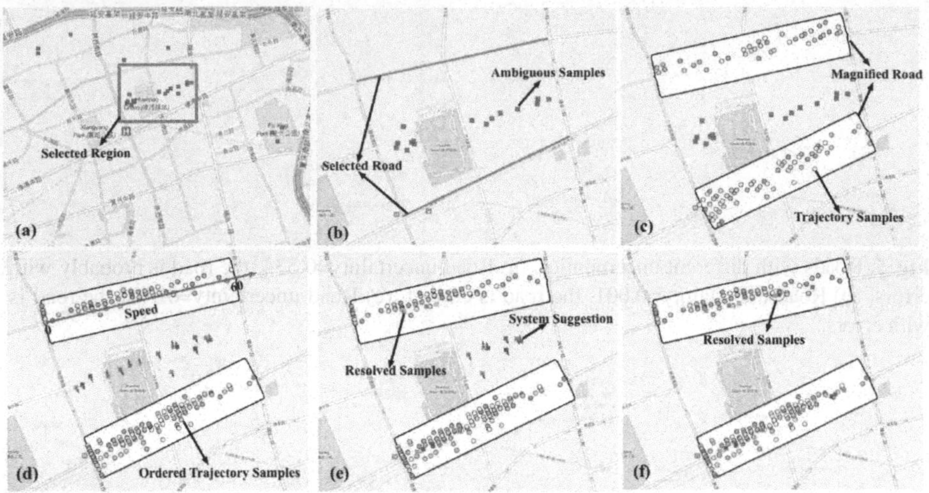

Fig. 7. Resolving ambiguities. (a) User selects a region to resolve ambiguous samples. (b) System suggests two candidate roads according to the directions of the ambiguous samples. (c) The system magnifies the candidate roads and shows the trajectory samples positioning on the roads. (d) Trajectory samples and ambiguous samples are sorted according to their speeds. And the system provides suggestions for the ambiguous samples. (e) User makes the decision according to the system suggestions. (f) Ambiguous samples are resolved and placed on the roads.

Reconstructing Trajectories. A participant started the task with loading a sparse trajectory. The system focused on the most sparse sample pair at the beginning. He reconstructed the trajectory pair by pair. Fig.8 shows a process of the participant reconstructing one trajectory between a pair of samples. The system focused on a partial trajectory that needed to be reconstructed. In Fig. 8 (a), this trajectory didn't pass any roads. The system computed the possible movement region for the sample pair and found out two possible routes (Route 1 and Route 2). Then, the participant opened the uncertainty lens. The uncertainty lens combined the roads in the routes and magnified the routes (Fig.8 (b)). After that, the participant sorted the trajectory samples on the routes, and obtained system suggestion (Fig.8 (c)). The system suggested the trajectory should pass Route 1. Finally, he reconstructed the trajectory by accepting system suggestion. Fig. 8 (d) depicts reconstruction result. From the task, the usefulness of our system for reconstructing the continuous trajectories was confirmed.

5.2 User Feedback

We consulted two experts who have expertise in intelligent transportation system research for comments after internally testing our system. They both have worked with trajectory data and are familiar with map-matching.

The well-designed visual encoding schemes to resolve uncertainty in trajectory data and the highly supported interactions with users are greatly appreciated by the experts. Especially the powerful encoding scheme to reveal errors in road maps as well as the method to reveal the uncertainty in the trajectory data received very positive comments.

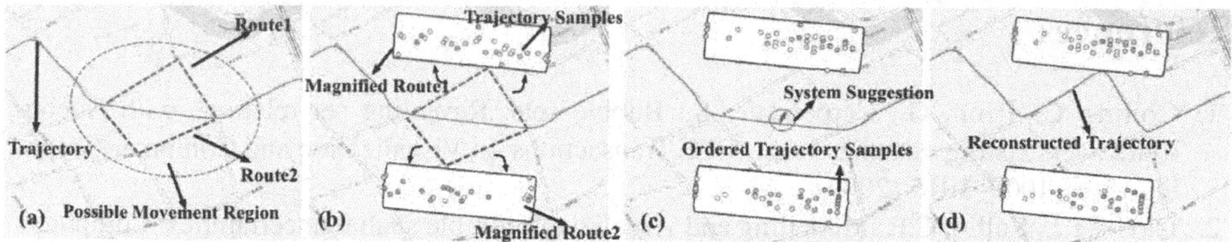

Fig. 8. Reconstructing a trajectory. (a) A sparse trajectory doesn't pass any roads. System computes the possible movement region for its adjacent samples. Within the possible movement region, the system finds out two possible routes. (b) Uncertainty lens combines the roads on the routes and magnifies the routes. It shows the trajectory samples on the routes at the same time. (c) Uncertainty lens re-orders the trajectory samples and the system provides a suggestion accordingly. (d) Reconstruction result.

These features largely improved the user experiences in traffic data exploration. Compared with their current map-matching analysis tools, which only plots the trajectory samples on the map to reconstruct the trajectory, our system makes the whole exploration process much more accurate and free of map errors. According to the feedback, our system is also good for demonstration. The visualization provides intuitive interface for understanding the trajectory data with many unique attributes. Utilizing the uncertainty lenses, users can gain insight into the speed, time, and direction attributes of the data which are very useful to detect the anomaly of the traffic, study drivers' driving patterns, and investigate the passengers' distribution. Our visual analytics cases are very convincing for domain experts. The analysis of the traffic patterns and trajectory reconstruction is considered very valuable to guide the road traffic capacity design and driving pattern study. The discovery of the map errors in Fig. 6 greatly attracts the experts' attention. The detected map errors are later confirmed by their manual inspection on and comparison of different versions of the maps. The experts also feel excited that the ambiguity analysis and demonstration of our system shows a data exploration process that is more promising than traditional methods to retrieve driving patterns.

6 Conclusion

In this paper we have presented a comprehensive visual analytics system for revealing and resolving the uncertainty in trajectory data. Two novel encoding schemes are designed. The experiments with real data, three case studies, and the feedback from the domain experts, have demonstrated the effectiveness of our system.

Acknowledgments. This project is partially supported by HKUST grant SRFI11EG15 and FSGRF12EG40. Siyuan Liu's research is supported by the Singapore National Research Foundation under its International Research Centre @ Singapore Funding Initiative and administered by the IDM Programme Office, Media Development Authority (MDA) and the Pinnacle Lab at Singapore Management University.

References

1. Collins, C., Penn, G., Carpendale, S.: Bubble sets: Revealing set relations with isocontours over existing visualizations. IEEE Transactions on Visualization and Computer Graphics 15(6), 1009–1015 (2009)
2. Davis, T.J., Keller, C.P.: Modelling and visualizing multiple spatial uncertainties. Computers and Geosciences 23(4), 397–408 (1997)
3. Fisher, D.: Incremental, approximate database queries and uncertainty for exploratory visualization. In: IEEE Symposium on Large Data Analysis and Visualization, pp. 73–80 (2011)
4. Greenfeld, J., Joshua, S.: Matching gps observations to locations on a digital map. Environmental Engineering 1(3), 1–13 (2002)
5. Hengl, T., Toomanian, N.: Maps are not what they seem: representing uncertainty in soil-property maps. In: International Symposium on Spatial Accuracy Assessment in Natural Resources and Environmental Sciences, pp. 805–813 (2006)
6. Hummel, B., Tischler, K.: Robust, gps-only map matching: exploiting vehicle position history, driving restriction information and road network topology in a statistical framework. In: The GIS Research UK Conference, pp. 68–77 (2005)
7. Kraak, M.J.: The space-time cube revisited from a geovisualization perspective. In: International Cartographic Conference, pp. 1988–1995 (2003)
8. Liu, S., Liu, Y., Ni, L.M., Fan, J., Li, M.: Towards mobility-based clustering. In: ACM SIGKDD Conference on Knowledge Discovery and Data Mining, pp. 919–928 (2010)
9. Liu, S., Pu, J., Luo, Q., Qu, H., Ni, L.M., Krishnan, R.: Vait: A visual analytics system for metropolitan transportation, pp. 1586–1596 (2013)
10. Lou, Y., Zhang, C., Zheng, Y., Xie, X., Wang, W., Huang, Y.: Map-matching for low-sampling-rate gps trajectories. In: Geographic Information Systems, pp. 352–361 (2009)
11. Pang, A.T., Wittenbrink, C.M., Lodha, S.K.: Approaches to uncertainty visualization. The Visual Computer 13(8), 370–390 (1997)
12. Pfoser, D., Jensen, C.S.: Capturing the uncertainty of moving-object representations. In: Güting, R.H., Papadias, D., Lochovsky, F.H. (eds.) SSD 1999. LNCS, vol. 1651, pp. 111–131. Springer, Heidelberg (1999)
13. Pink, O., Hummel, B.: A statistical approach to map matching using road network geometry, topology and vehicular motion constraints. In: Intelligent Transportation Systems, pp. 862–867 (2008)
14. Quddus, M.A., Ochieng, W.Y., Noland, R.B.: Current map-matching algorithms for transport applications: state-of-the art and future research directions. Transportation Research 15(5), 312–328 (2007)
15. Taylor, G., Brunsdon, C., Li, J., Olden, A., Steup, D., Winter, M.: Gps accuracy estimation using map matching techniques: Applied to vehicle positioning and odometer calibration. Computers, Environment and Urban Systems 30(6), 757–772 (2006)
16. Trajcevski, G., Wolfson, O., Hinrichs, K., Chamberlain, S.: Managing uncertainty in moving objects databases. ACM Transaction Database System 29(3), 463–507 (2004)

Activity Recognition Using a Few Label Samples

Heidar Davoudi[1], Xiao-Li Li[2], Nguyen Minh Nhut[2], and Shonali Priyadarsini Krishnaswamy[2]

[1] School of Computer Engineering, Nanyang Technological University, Singapore
[2] Data Analytics Department, Institute for Infocomm Research, A*Star, Singapore

Abstract. Sensor-based human activity recognition aims to automatically identify human activities from a series of sensor observations, which is a crucial task for supporting wide range applications. Typically, given sufficient training examples for all activities (or activity classes), supervised learning techniques have been applied to build a classification model using sufficient training samples for differentiating various activities. However, it is often impractical to manually label large amounts of training data for each individual activities. As such, semi-supervised learning techniques sound promising alternatives as they have been designed to utilize a small training set L, enhanced by a large unlabeled set U. However, we observe that directly applying semi-supervised learning techniques may not produce accurate classification. In this paper, we have designed a novel *dynamic temporal extension* technique to extend L into a bigger training set, and then build a final semi-supervised learning model for more accurate classification. Extensive experiments demonstrate that our proposed technique outperforms existing 7 state-of-the-art supervised learning and semi-supervised learning techniques.

Keywords: Activity Recognition, Semi-Supervised Learning, Dynamic Temporal Extension.

1 Introduction

Sensor-based human activity recognition has received considerable attention due to its diverse applications such as healthcare systems [1], pervasive and mobile computing [5], and smart homes [10] etc. For example, in the healthcare applications, understanding elderly people's activities can not only provide real-time useful information to caregivers, but also facilitate a monitoring system to take proper actions (e.g. alert clinicians) in emergency cases (e.g. falling down).

While privacy and complexity are major issues in video-based activity recognition [11], different *supervised* learning approaches have been applied to activity recognition based on body-worn sensors. For example, Naïve Bayes (NB) [23], Hidden Markov Model (HMM) [12,27], Conditional Random Field (CRF) [27] and Support Vector Machine (SVM) [23] etc. These approaches require *sufficient* labeled samples to train accurate classifiers. However, sample labeling is the most labor intensive and time consuming process in activity recognition [11] as different classes of related activities could have nondeterministic natures —

V.S. Tseng et al. (Eds.): PAKDD 2014, Part I, LNAI 8443, pp. 521–532, 2014.

some steps of the activities can be performed in arbitrary order, and the starting and ending points of activities are difficult to define due to the overlaps among different activities, leading to activities could be concurrent or even interwoven [11]. As such, it would be extremely hard to get *sufficient* training samples.

Semi-supervised learning approaches, on the other hand, have been designed to build a classification model using a small training set L and a large unlabeled set U. Particularly, an initial classifier can be built from L by using NB, SVM or another classifier. Then the classifier can be used to classify U to assign a (probabilistic) class label to each sample in U [14, 15], which can be in turn used to refine the classifier, using Expectation-Maximization (EM) [22] or other classifiers (like SVM) iteratively [13, 16–19]. The combination of NB and EM, and transductive Support Vector Machine (TSVM) [8] are two well known semi-supervised learning methods. In addition, Guan *et al* [7] designed En-Co-training algorithm for activity recognition based on co-training algorithm [2] which trains two independent classifiers based on two views of data. Stikic *et al* [26] compared two semi-supervised techniques, namely self-training [4] and co-training, and showed co-training can be applied to activity recognition by using sensors with different modalities, i.e. accelerometers and infra-red sensors. While a few semi-supervised learning algorithms have been proposed, they still need sizable labeled samples for training to achieve accurate results. The reason is that these methods will encounter a cold-start issue when very few labeled samples are present for learning a decent initial classifier. Its poor performance subsequently affects the iterative model refinement process. As such, directly applying semi-supervised learning techniques may not always produce accurate prediction results.

In this paper, we address the challenging issue by proposing a novel *Dynamic Temporal Extension* (DTE) algorithm. Particularly, based on a few labeled samples, we automatically infer some *reliable labeled samples* from U in temporal space [21]. For each labeled sample, we extend it to include its near neighbors along its time axis, if these neighbors are predicted mostly to have a same label with the labeled sample. In effect, our DTE algorithm tries to fill the gap between the *limited* available labeled samples and *sizable* labeled samples required for semi-supervised learning. With our inferred reliable labeled samples, we can thus build a more robust and accurate semi-supervised learning model. Extensive experiments demonstrate that our proposed technique outperforms existing 7 state-of-the-art supervised learning and semi-supervised learning techniques.

2 The Proposed Technique

In this section, we present our proposed technique. First, section 2.1 provides our overall algorithm. Then, we introduce how to build a classifier using our designed semi-supervised learning approach in section 2.2. Finally, section 2.3 elaborates how to extend limited labeled samples using our proposed DTE technique.

2.1 Overall Algorithm

Fig. 1 shows an overall algorithm of our proposed technique. Step 1 is a preprocessing that deals with missing value problem as well as faulty sensors. Missing

---

Input: L, U ▷ Set of Initial Labeled Samples and Unlabeled Samples respectively
Output: M ▷ Classification Model

1: Handle missing values and faulty sensors in a given activity recognition data set
2: Build an initial semi-supervised classifier IC using L and U
3: Apply IC to predict the labels of samples in U and perform smoothing
4: Compute continuous same-label segment size SS of activities using predicted labels in Step 3
5: Extend labeled samples into set E using DTE with the estimated segment size SS
6: Build a final semi-supervised learning model M based on initial labeled set L, the inferred labeled set E, as well as the remaining unlabeled set $U - E$
7: Classify the test samples using the model M

Fig. 1. Overall DTE algorithm for activity recognition using few labeled samples

value problem can be caused by some faulty sensors or unreliable communication channels. We first eliminate sensors which have more than 30% missing values as they can not provide useful and sufficient information for accurate activity recognition. We then apply cubical spline interpolation for replacing those missing values in our data sets [3] and eliminate samples with no activity label.

In order to prepare for labeled sample extension, we want to estimate the median size of continuous same-label activity sequences in $L + U$, as this is useful information that indicates how far we should extend each labeled sample. Since we do not have the label information for all the samples in U, we first design a semi-supervised learning method to build an initial classifier IC in step 2 — we provide its detailed description in section 2.2. Then, we apply IC to classify the unlabeled samples in U in step 3. Given that sensor sampling rates are usually significantly higher than the rates of change between different human activities, we thus perform *smoothing* step to improve the prediction accuracy by removing impulse noises. Our smoothing process moves a sliding window on predicted labels of U and chooses the labels with majority as the label of all the samples in that window. Step 4 computes the median continuous same-label segment size SS of activities based on the predicted labels in U. Step 5 extends all the labeled samples into the inferred reliable labeled set E based on our designed DTE techniques using the prediction results from IC as well as the estimated segment size SS. We have provided the details of step 5 in section 2.3.

Finally, we build the final classification model M using our semi-supervised learning method, based on labeled set L, the extended set E, and the remaining unlabeled set $U - E$ in step 6, and perform final classification in step 7. Note the same method is used for both initial classifier IC and final classifier although the final one is more accurate due to additional reliable labeled sample E.

2.2 Classifier Building Based on Semi-supervised Learning

First, labeled set L and unlabeled set U are defined as follows:

$$L = \{(A_k, C_k)|k = 1, 2, \ldots, |L|\} \tag{1}$$

$$U = \{A_k|k = 1, 2, \ldots, |U|\} \tag{2}$$

$A_k = (a_{k1}, a_{k2}, ..., a_{kn})$ is a n-dimensional feature vector where $a_{ki}(i = 1, 2, \ldots, n)$ is the i'th feature of A_k and $C_k \in C(C = \{C_1, C_2, ..., C_{|C|}\})$ is the associated activity/class label of A_k. The objective of this research is to learn a classification model M that can be used to classify any test sample.

We adopt a mixture model to generate each activity A_k as mixture models are very useful to characterize heterogeneous data or multiple types of activities.

$$P(A_k|\theta) = \sum_{j=1}^{|C|} P(C_j|\theta)P(A_k|C_j;\theta) \tag{3}$$

Where θ parameterized a mixture model. Given any class C_j, assuming that the probabilities of the features are independent as well as normal distribution (based on the central limit theorem) for each attribute in C_j, we have

$$P(A_k|C_j;\theta) = P(A_k|C_j;\mu_j,\sigma_j) = \prod_{i=1}^{n} P(a_{ki}|C_j;\mu_{ij},\sigma_{ij}) \tag{4}$$

and

$$P(a_{ki}|C_j;\mu_{ij},\sigma_{ij}) = \frac{1}{\sqrt{2\pi}\sigma_{ij}} e^{-\frac{(a_{ki}-\mu_{ij})^2}{\sigma_{ij}^2}} \tag{5}$$

Note $\mu_j = (\mu_{1j}, \mu_{2j}, ..., \mu_{nj})$ and $\sigma_j = (\sigma_{1j}, \sigma_{2j}, ..., \sigma_{nj})$ where μ_{ij} and σ_{ij} are the mean and standard deviation of feature i in class j respectively.

In order to perform classification, we compute the posterior probability of class C_j for a given example A_k:

$$P(C_j|A_k;\mu,\sigma) = \frac{P(C_j|\mu,\sigma)\prod_{i=1}^{n} P(a_{ki}|C_j;\mu_{ij},\sigma_{ij})}{P(A_k|\mu,\sigma)} \tag{6}$$

Where $\mu = [\mu_1, \mu_2, ..., \mu_{|C|}]$ and $\sigma = [\sigma_1, \sigma_2, ..., \sigma_{|C|}]$. The class C_k with $k = \arg\max_j P(C_j|A_k;\mu,\sigma)(j = 1, \ldots, |C|)$ will be assigned to the example A_k as its predicted class label.Note in formula (6), the denominator $P(A_k|\mu,\sigma)$ is a constant, i.e. same for all classes. We now elaborate how to calculate $P(C_j|\mu,\sigma)$. Based on formula (3), the probability of both labeled and unlabeled training set $(S = L \cup U)$ can be written as:

$$P(S|\mu,\sigma) = \prod_{A_k \in U} \sum_{j=1}^{|C|} P(C_j|\mu,\sigma)P(A_k|C_j;\mu,\sigma) \times \prod_{(A_k,C_k)\in L} P(C_k|\mu,\sigma)P(A_k|C_k;\mu,\sigma) \tag{7}$$

Local maximum of log likelihood indicated by (7) can be found using Expected Maximization (EM) in an iterative manner [6]. It can be shown that parameters in M-step of EM algorithm can be computed as follow:

$$\mu_{ij}^{t+1} = \frac{\sum_k a_{ki} P(C_j|A_k;\mu^t,\sigma^t)}{\sum_k P(C_j|A_k;\mu^t,\sigma^t)} \tag{8}$$

$$\sigma_{ij}^{2\ t+1} = \frac{\sum_k (a_{ki} - \mu_{ij}^t)^2 P(C_j|A_k;\mu^t,\sigma^t)}{\sum_k P(C_j|A_k;\mu^t,\sigma^t)} \tag{9}$$

Input: L, U ▷ Set of Initial Labeled Samples and Unlabeled Samples respectively
Output: E ▷ Set of Extended Samples

```
1: E ← ∅
2: CfTable[ ][ ] ← 0
3: for all s ∈ L do
4:     C_j = label(s)
5:     ExtS ← EXTENDSAMPLE(s, C_j)
6:     for all p ∈ ExtS do
7:         if p ∉ E then
8:             E ← E ∪ {p}
9:             Add p into CfTable
10:            CfTable[p][C_j] ← 1
11:        else
12:            CfTable[p][C_j] ← CfTable[p][C_j] + 1
13:        end if
14:    end for
15: end for
16: for all p ∈ E do
17:     label(p) ← arg max_j CfTable[p][j]
18: end for
```

Fig. 2. DTE algorithm on labeled example extension

Finally, the class prior probability can be calculated:

$$P(C_j|\mu^{t+1}, \sigma^{t+1}) = \frac{\sum_k P(C_j|A_k; \mu^{t+1}, \sigma^{t+1})}{|L| + |U|} \tag{10}$$

Note that we have modified the original EM algorithm by setting the posterior class probabilities of initially labeled samples in L as their original labels, at the start of each iteration. This guarantees our classification model is built using *correct* labeled examples during its iterative training process.

2.3 Dynamic Temporal Extension for Labeled Samples

In section 2.2, we have proposed a semi-supervised method to learn a classification model from a small labeled set L and a large unlabeled set U. In our case, since L is very small, say a few training examples for each class, the semi-supervised learning method will not capture the characteristics for each class. As such, we have proposed a novel *dynamic temporal extension* algorithm DTE to extend the initial labeled samples into an extended reliable labeled set E by exploring the dependencies between samples in time domain to effectively boost the performance of semi-supervised learning.

The proposed DTE algorithm for labeled example extension is shown in Fig. 2. After we initialize the extended set E into an empty set in step 1, a confidence table $CfTable[\,][\,]$ is created and initialized in step 2 which records the support of an unlabeled example to a given class, e.g. $CfTable[p][C_j]$ indicates how likely and the degree of confidence that an example p belongs to a class C_j. From steps 3–15, we extend each labeled sample one by one and compute the confidence table correspondingly. Particularly, for each seed labeled sample s with class label C_j, we extend it to include its neighbor samples to form $ExtS$

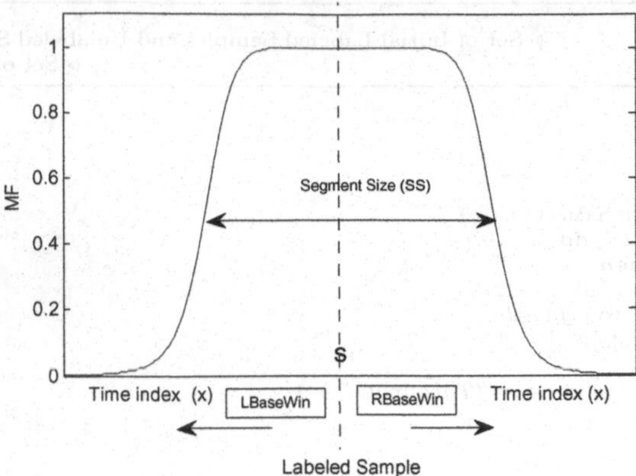

Fig. 3. Membership Function (MF)

along a time line (left/ right hand sides of s) if we judge these samples in $ExtS$ belong to the same class, i.e. C_j, by calling a function EXTENDSAMPLE() that will be described in Fig 4 later.

Steps 6 to 14 is a loop, which computes the support for each extended sample in $ExtS$ by updating the confidence table $CfTable[p][C_j]$. When selected sample p does not exist in the table, it is inserted to the set E (step 8) and *CfTable* with the initial value 1 (steps 9 to 10); otherwise our algorithm increases its respective support value by 1. Finally, steps 16 to 18 assign each sample in E with a label with a maximal class support score wrt to its associated classes in *CfTable*.

Now, we are ready to introduce the function EXTENDSAMPLE() that is used in Fig. 2. We first illustrate the key intuition behind it in Fig. 3. As the sampling frequency of sensors is much higher than the change frequency between different human activities, given an initial labeled sample s at time point t_0, the neighbor samples at its near left time points and at its near right time points are highly likely to share the same label with s, as long as these left/right time points are not too far away from t_0. However, one challenging problem is how long/far we should extend s to its left/right hand side.

In this paper, we use a sliding window to estimate right and left label-consistent neighbor samples of s. Two base sliding windows (left/right sliding window LBaseWin/RBaseWin) are moved forward from both sides of s individually and our algorithm decides whether the extension should be continued or stopped based on the analysis of the neighbor samples' labels inside both sliding windows (Fig. 3). Two important information sources have been used as a stop criteria, namely, 1) neighbor label consistency with s where the labels of s's neighbors are predicted by our initial classification model IC, and 2) the distance of neighbor samples to s. We design a membership function MF based on a distance measure to s as follows:

$$MF(x, s, SS) = \frac{1}{1 + |\frac{x-s}{0.5*SS}|^{SS}} \tag{11}$$

where x is a current neighbor sample inside in a left/right window; s and SS are the center and width of membership function respectively (Fig 3) where SS is estimated in our overall algorithm by computing the median size of continuous same-label activity sequences. The membership function defined in (11) is a bell shaped fuzzy membership function. As sliding windows move forward from s to left/right side, membership value decreases corrspondingly. The bigger this distance $x - s$, the less likely the neighbor sample x should be included. We also take the label-consistency into consideration by computing a support score $Support\_Score_{C_j}$ for all the predicted class labels $C_j \in C$ in current base window (LBaseWin/RBaseWin):

$$Support\_Score_{C_j} = \sum_{\substack{x \in RBaseWin \\ label(x)=C_j}} MF(x, s, SS) \qquad (12)$$

Score $Support\_Score_{C_j}$ defined in (12) basically calculates the label support inside current (right) base window based two information sources mentioned. We will continue our extending process if these labels in the base window are consistent with s's label and they are not too far away from s. Depending on the two factors, we extend s *dynamically*, and the numbers of extended examples are different for each initial labeled sample as well as for the left and right extension.

The detailed description of function EXTENDSAMPLE() is shown in Fig 4. Note we only show how to extend right samples from s as extending left samples is the same except the extension direction. After initializing the extended sample set $ExtS$ as empty set in step 2, we perform a loop from steps 3 to 18. Particularly, steps 5 to 9 sum up all the class membership scores for each example in the current base window. Steps 10 to 11 get the *Winner score* and *Winner Label* from all the classes. Steps 12 to 17 decide if we want to repeat our loop. If the *Winner Label* is consistent with s and the *Winner score*>=50%, then the samples inside the current window are reliable, and we thus add them into $ExtS$ and shift the right window to continue our extension process; Otherwise, we stop the loop from here. Finally, step 19 returns our extension results.

3 Empirical Evaluation

In this section, we evaluate our proposed DTE technique. We compare it with 7 existing state-of-the-art techniques, including 3 *supervised learning techniques*, namely, 1NN (which performs best for time series data) [9], SVM [23], NB (Naïve Bayes) [23], as well as 4 *semi-supervised learning techniques*, namely, Transductive Support Vector Machine (TSVM) [8], NB+EM (NB+Expected Maximization) [22], Self-training [20] and En-Co-training [7] etc.

3.1 Datasets

For evaluation, we used 3 datasets from Opportunity Challenge [24,25]. Each dataset represents 1 subject/person who performed various activities with attached on-body sensors. Particularly, each subject performed one *Drill* session which is 20 predefined atomic activities and 5 *Activities Daily Life* (ADLs) sessions where the subject performed high level activities with more freedom on sequence of atomic activities. Datasets contain 4 locomotion activities/classes,

```
Input: s, C_j        ▷ s and C_j are a labeled sample to extend and its associated class respectively
Output: ExtS                  ▷ A set of extended samples for the labeled sample s

function EXTENDSAMPLE(s, C_j)
    ExtS ← ∅
    Acceptable ← true
    repeat                                                      ▷ Extension towards right
        for i ← 1 to |C| do
            for all x located inside right based window RBaseWin do
                Support_Score_i ← Σ_{label(x)==i} MF(x, s, SS)
            end for
        end for
        WinnerScore ← max Support_Score_i(i = 1, 2, ..., |C|)
        WinnerLabel ← arg max_k Support_Score_i(i = 1, 2, ..., |C|)
        if WinnerLabel==C_j and WinnerScore ≥ 50% then
            ExtS ← ExtS∪ {the samples inside RBaseWin}
            RBaseWin ← ShiftRight(RBaseWin)
        else
            Acceptable ← false
        end if
    until Acceptable == false
    Return ExtS
end function
```

Fig. 4. EXTENDSAMPLE procedure of DTE algorithm

namely, *stand, walk, sit,* and *lie.* All the experiments are performed across 3 different datasets. Drill and first 3 ADLs (ADL 1, ADL2, ADL3) are used for training while the last 2 ADLs (ADL4, ADL5) are used for testing.

Table 1. Opportunity Challenge Data Sets After Preprocessing

	Dataset 1	Dataset 2	Dataset 3
Number of Features	107	110	110
Training Set	139427	126595	141263
Test Set	47291	48486	45774

Table 1 shows the details of the 3 datasets that we used in our experiments after the preprocessing step introduced in section 2.1. To simulate the challenging scenario of learning from limited training samples, for each dataset we randomly select only 3 samples from Training Set for each of the 4 classes, i.e. 12 (3× 4) samples to serve as the labeled set L. We then randomly select 1000 training samples as unlabeled set U which are used by all the semi-supervised learning methods after ignoring their labels. We have also tested how the sizes of U affect the final classification performance. In addition, the base window size (RBaseWin) is set to 10 (Fig 4) and we will report its sensitivity study results. Finally, to make the evaluation reliable, all the experiments are repeated for 10 times and the performances of all the 8 techniques are evaluated on the same test sets in terms of average *Accuracy* and *F-measure*, which are typically used for evaluating the performance of classification models. Note that locomotion activities/classes is moderately imbalanced [3] so we use *Accuracy* and *F-measure* as performance measurements.

Table 2. Comparison of different supervised and semi-supervised learning methods

Methods	Dataset 1		Dataset 2		Dataset 3	
	Accuracy	*F-Measure*	*Accuracy*	*F-Measure*	*Accuracy*	*F-Measure*
1NN	75.5	78.8	58.0	65.1	59.9	67.7
SVM	9.1	24.0	16.0	27.9	15.5	26.7
NB	64.3	76.9	66.0	71.8	50.5	58.2
TSVM	61.4	69.8	54.6	57.4	40.4	49.0
NB+EM	77.7	82.9	75.9	77.6	57.0	56.1
Self-Training	41.8	16.8	36.8	13.5	42.5	14.9
En-Co-Training	59.1	65.8	47.4	64.5	56.6	64.3
DTE	79.2	83.6	77.0	78.4	68.2	68.6
DTE+smoothing	80.2	84.3	77.6	78.9	68.8	69.0

3.2 Experimental Results

The experimental results are shown in Table 2. The first 3 rows show the performance of 3 supervised classification models, i.e. 1NN, SVM, NB, where only initial labeled set (L) is used for training. We observe that SVM performs badly compared to 1NN and NB as SVM can not build accurate hyperplanes based on 12 labeled samples for 4-class classification problem.

The next 4 rows illustrate the experimental results of 4 semi-supervised methods, which have used unlabeled set U to boost their performance. Compared with supervised learning methods, we observe that TSVM is much better than SVM for all datasets, and NB+EM outperforms NB for datasets 1 and 2, while it is slightly worse in dataset 3 according to *F-measure* only. In general, NB+EM works better than TSVM, Self-training and En-Co-Training most of times.

The last two rows in Table 2 show the performance of DTE method and the effect of smoothing for test results. We observe DTE method consistently works better than all the other 7 techniques across all the datasets. Especially for dataset 3, DTE method achieves 11.2% and 12.5% better results than the second best technique NB+EM in terms of average Accuracy and F-measure respectively. We also perform the post-processing by applying smoothing (choose majority labels inside a sliding window) on our predicted test results to remove the impulse prediction errors, which further improve our method in all cases.

Recall that our proposed DTE method has dynamically extended initial labeled samples to left/right windows to include those neighbor samples that have consistent-label and are not too far away from the labeled samples. For comparison, we have also used a fixed extension size (FES) which is obtained by averaging the segments sizes of activities based on the actual labels in U. We observe that our proposed DTE perform much better than those semi-supervised methods (e.g. TSVM and NB+EM) that use the FES to include neighbor samples, indicating our method is very effective to automatically extend initial labeled samples dynamically to include those reliable labeled samples so that it can boost the performance of semi-supervised learning significantly.

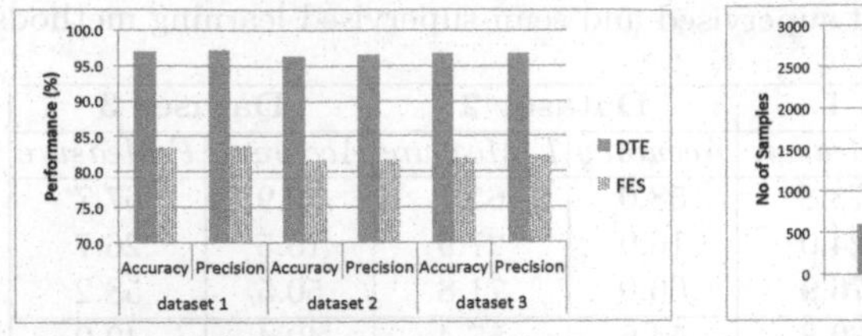

(a) Sample extension performance

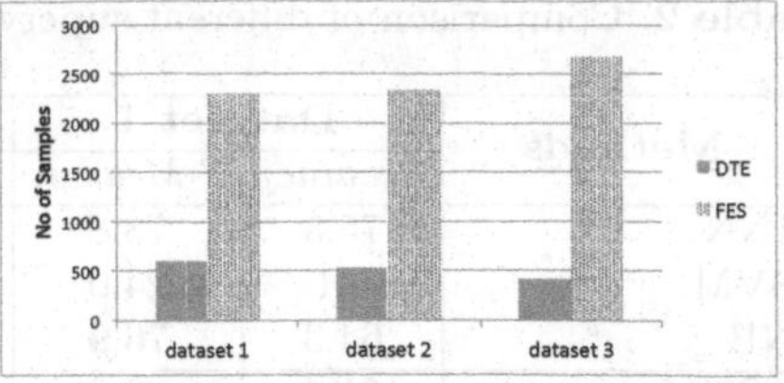

(b) Number of exteneded samples

Fig. 5. No/Performance of different sample extension methods

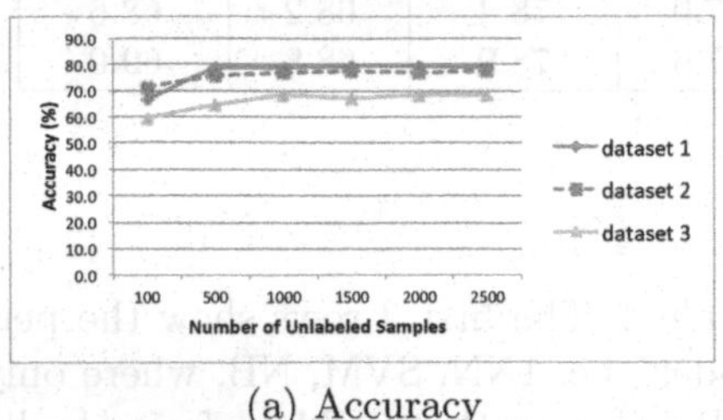

(a) Accuracy

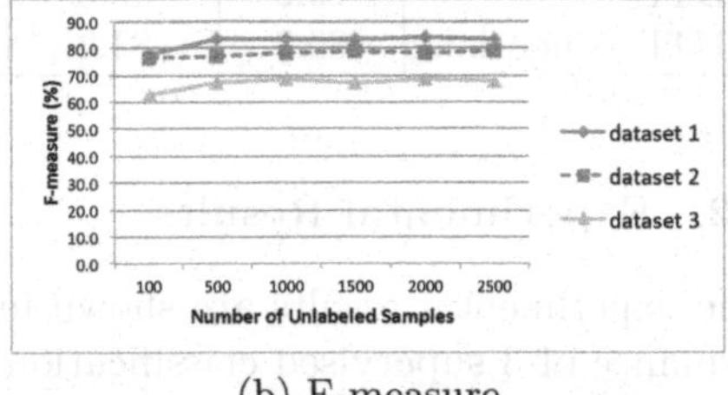

(b) F-measure

Fig. 6. Performance of DTE using diffrent unlabeled samples

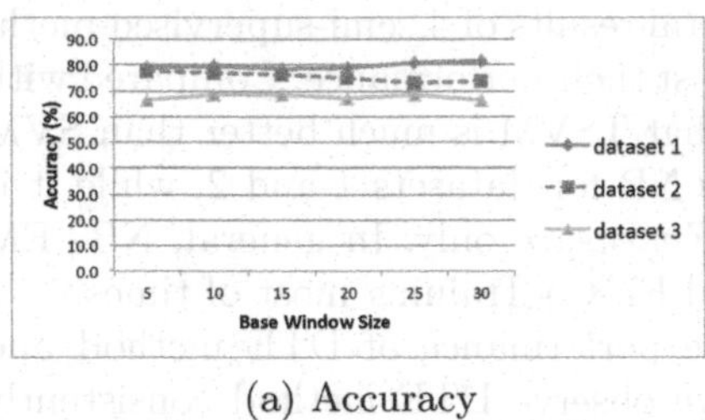

(a) Accuracy

(b) F-measure

Fig. 7. Performance of DTE using diffrent base window size

Fig. 5 shows the comparison between the performance of proposed DTE sample extension and FES sample extension. From Fig. 5(a), the proposed DTE algorithm can produce much more accurate labeled samples than FES, i.e. more than 15% better in terms of both Accuracy and Precision. While FES produces more samples (Fig. 5(b)), they are not accurate and thus bring noisy labels for the subsequent training process, leading to inferior classification results.

Fig. 6 illustrates the performance of proposed DTE method for different datasets in terms of the number of unlabeled samples used. From Fig. 6, with the increase of unlabeled samples, the performance of classification increases in the beginning and it becomes stable, with at least 1000 unlabelled samples, indicating our method is not sensitive to the number of unlabelled samples used.

Finally, Fig. 7 shows the effect of changing base window size on DTE approach performance (Accuracy/F-measure). As we can be observed, by increasing the base window size, the performance of our approach only changes slightly across all the datasets in terms of both average Accuracy and F-measure, indicating that our method is not sensitive to the base window size.

4 Conclusions

Learning from a few labeled samples becomes crucial for real-world activity recognition applications as it is labor-intensive and time-consuming to manually label large numbers of training examples. While semi-supervised learning techniques have been proposed to enhance supervised learning by incorporating the unlabelled samples into classifier building, it still can not perform very well. In this paper, we propose a novel dynamic temporal extension technique to extend the limited training examples into a larger training set which can further boost the performance of semi-supervised learning. Extensive experimental results show that our proposed technique significantly outperforms existing 7 state-of-the-art supervised learning and semi-supervised learning techniques. For our future work, we will investigate how to select the initial labeled samples intelligently so that they can benefit our classification models more effectively.

Acknowledgements. This work is supported in part by a grant awarded by a Singapore MOE AcRF Tier 2 Grant (ARC30/12).

References

1. Avci, A., Bosch, S., Marin-Perianu, M., Marin-Perianu, R., Havinga, P.: Activity recognition using inertial sensing for healthcare, wellbeing and sports applications: A survey. In: The 23rd International Conference on Architecture of Computing Systems (ARCS), pp. 1–10 (2010)
2. Blum, A., Mitchell, T.: Combining labeled and unlabeled data with co-training. In: Proceedings of the 11th Annual Conference on Computational Learning Theory, pp. 92–100 (1998)
3. Cao, H., Nguyen, M.N., Phua, C., Krishnaswamy, S., Li, X.: An integrated framework for human activity classification. In: Ubicomp (2012)
4. Chapelle, O., Schölkopf, B., Zien, A.: Semi-supervised learning, vol. 2. MIT Press, Cambridge (2006)
5. Choudhury, T., Consolvo, S., Harrison, B., Hightower, J., LaMarca, A., LeGrand, L., Rahimi, A., Rea, A., Bordello, G., Hemingway, B., et al.: The mobile sensing platform: An embedded activity recognition system. IEEE Pervasive Magazine, Spec. Issue on Activity-Based Computing 7(2), 32–41 (2008)
6. Dempster, A.P., Laird, N.M., Rubin, D.B.: Maximum likelihood from incomplete data via the em algorithm. Journal of the Royal Statistical Society. Series B (Methodological), 1–38 (1977)
7. Guan, D., Yuan, W., Lee, Y.K., Gavrilov, A., Lee, S.: Activity recognition based on semi-supervised learning. In: 13th IEEE International Conference on Embedded and Real-Time Computing Systems and Applications, pp. 469–475 (2007)
8. Joachims, T.: Making large-scale svm learning practical. In: Schölkopf, B., Burges, C., Smola, A. (eds.) Advances in Kernel Methods-support Vector Learning (1999)
9. Keogh, E., Kasetty, S.: On the need for time series data mining benchmarks: a survey and empirical demonstration

10. Kidd, C.D., Orr, R., Abowd, G.D., Atkeson, C.G., Essa, I.A., MacIntyre, B., Mynatt, E., Starner, T.E., Newstetter, W.: The aware home: A living laboratory for ubiquitous computing research. In: CoBuild, pp. 191–198 (1999)
11. Lara, O., Labrador, M.: A survey on human activity recognition using wearable sensors. IEEE Communications Surveys Tutorials PP(99), 1–18 (2002)
12. Lester, J., Choudhury, T., Borriello, G.: A practical approach to recognizing physical activities. In: Fishkin, K.P., Schiele, B., Nixon, P., Quigley, A. (eds.) PERVASIVE 2006. LNCS, vol. 3968, pp. 1–16. Springer, Heidelberg (2006)
13. Li, X., Liu, B.: Learning to classify texts using positive and unlabeled data. In: IJCAI, pp. 587–592 (2003)
14. Li, X.-L., Liu, B.: Learning from positive and unlabeled examples with different data distributions. In: Gama, J., Camacho, R., Brazdil, P.B., Jorge, A.M., Torgo, L. (eds.) ECML 2005. LNCS (LNAI), vol. 3720, pp. 218–229. Springer, Heidelberg (2005)
15. Li, X.-L., Liu, B., Ng, S.-K.: Learning to classify documents with only a small positive training set. In: Kok, J.N., Koronacki, J., Lopez de Mantaras, R., Matwin, S., Mladenič, D., Skowron, A. (eds.) ECML 2007. LNCS (LNAI), vol. 4701, pp. 201–213. Springer, Heidelberg (2007)
16. Li, X., Liu, B., Ng, S.: Learning to identify unexpected instances in the test set. In: IJCAI, pp. 2802–2807 (2007)
17. Li, X., Liu, B., Ng, S.: Negative training data can be harmful to text classification. In: EMNLP, pp. 218–228 (2010)
18. Liu, B., Lee, W., Yu, P., Li, X.: Partially supervised classification of text documents. In: ICML, pp. 387–394 (2002)
19. Liu, B., Yang, D., Li, X., Lee, W., Yu, P.: Building text classifiers using positive and unlabeled examples. In: ICDM, pp. 179–186 (2003)
20. Longstaff, B., Reddy, S., Estrin, D.: Improving activity classification for health applications on mobile devices using active and semi-supervised learning. In: 4th International Conference on Pervasive Computing Technologies for Healthcare, pp. 1–7 (2010)
21. Nguyen, M., Li, X., Ng, S.: Positive unlabeled learning for time series classification. In: IJCAI, pp. 1421–1426 (2011)
22. Nigam, K., McCallum, A., Thrun, S., Mitchell, T.: Learning to classify text from labeled and unlabeled documents. In: AAAI, pp. 792–799 (1998)
23. Ravi, N., Dandekar, N., Mysore, P., Littman, M.L.: Activity recognition from accelerometer data. In: AAAI, vol. 20, p. 1541. AAAI Press, MIT Press, Menlo Park, Cambridge (1999, 2005)
24. Roggen, D., Calatroni, A., Rossi, M., Holleczek, T., Forster, K., Troster, G., Lukowicz, P., Bannach, D., Pirkl, G., Ferscha, A., et al.: Collecting complex activity datasets in highly rich networked sensor environments. In: 7'th International Conference on Networked Sensing Systems, pp. 233–240 (2010)
25. Sagha, H., Digumarti, S.T., del Millan, J., Chavarriaga, R., Calatroni, A., Roggen, D., Troster, G.: Benchmarking classification techniques using the opportunity human activity dataset. In: IEEE International Conference on Systems, Man, and Cybernetics, pp. 36–40 (2011)
26. Stikic, M., Van Laerhoven, K., Schiele, B.: Exploring semi-supervised and active learning for activity recognition. In: 12th IEEE International Symposium on Wearable Computers, pp. 81–88 (2008)
27. Vail, D.L., Veloso, M.M., Lafferty, J.D.: Conditional random fields for activity recognition. In: Proceedings of 6th International Joint Conference on Autonomous Agents and Multiagent Systems, p. 235 (2007)

A Framework for Large-Scale Train Trip Record Analysis and Its Application to Passengers' Flow Prediction after Train Accidents

Daisaku Yokoyama[1], Masahiko Itoh[1], Masashi Toyoda[1], Yoshimitsu Tomita[2], Satoshi Kawamura[2,1], and Masaru Kitsuregawa[3,1]

[1] Institute of Industrial Science, The University of Tokyo
{yokoyama,imash,toyoda,kitsure}@tkl.iis.u-tokyo.ac.jp
[2] Tokyo Metro Co. Ltd.
{y.tomita,s.kawamura}@tokyometro.jp
[3] National Institute of Informatics

Abstract. We have constructed a framework for analyzing passenger behaviors in public transportation systems as understanding these variables is a key to improving the efficiency of public transportation. It uses a large-scale dataset of trip records created from smart card data to estimate passenger flows in a complex metro network. Its interactive flow visualization function enables various unusual phenomena to be observed. We propose a predictive model of passenger behavior after a train accident. Evaluation showed that it can accurately predict passenger flows after a major train accident. The proposed framework is the first step towards real-time observation and prediction for public transportation systems.

Keywords: Smart card data, Spatio-Temporal analysis, Train transportation, Passenger behavior.

1 Introduction

Public transportation systems play an important role in urban areas, and efficient and comfortable transportation is highly demanded, especially in megacities such as Tokyo. Tokyo has one of the most complex train systems in the world, so a major disruptive event can cause congestion and disruption over a wide area. The effect of such events is hard to predict, even for the operating companies.

Our goal is to implement the ability to analyze and predict passenger behaviors in a complex transportation system. In particular, we want to implement a system for understanding daily passenger flows and event-driven passenger behaviors, suggesting itineraries, and preparing for events.

Understanding Daily Passenger Flows. Operating companies want to know how many passengers are using their stations, lines, and trains. An understanding of the spatio-temporal demands of their passengers would help them

V.S. Tseng et al. (Eds.): PAKDD 2014, Part I, LNAI 8443, pp. 533–544, 2014.

adapt their train operations to the demands. An understanding of the demands would also enable the passengers to avoid crowded trains by enabling them to change to a less-crowded route or a more favorable departure time.

Understanding Event-Driven Passenger Behaviors. Events, such as natural disasters, public gatherings, and accidents, can create unusual passenger behaviors. To help them recover from disruptive events, operating companies need to know where congestion has occurred and how many passengers are present when an event occurs. Since various types of events occur repeatedly, understanding the changes in traffic that occurred with previous events would help them prepare for the next occurrence. Passengers could also use such information to avoid congestion after a disruptive event.

Suggesting Itineraries. If passenger flows could be observed in real time and if future flows could be predicted, operating companies could recommend itineraries to their passengers that would make their trips more convenient and comfortable. Since the current IT infrastructure for train systems in Japan does not support such real-time monitoring, we used a large-scale dataset of train trip records created from smart card data and examined the feasibility of flow prediction. Such recommendations could make the transportation system more efficient.

Preparing for Disruptive Events. Although operating companies already have some knowledge about the effects of major disruptive events, they still do not know what would happen if two or more of them occurred simultaneously. Such knowledge would help them allocate sufficient staff, trains, and other resources. It would also help them educate their staff by emulating arbitrary events.

Our contribution of this paper is as follows:

1. We propose a framework for analyzing large-scale train trip records as the first step to our goal. While there are various information sources, such as the number of passengers at stations, and train operation logs, we used smart card data for passengers using the Tokyo Metro subway system as such data reflects actual passenger demand. Within this framework, we developed a method for deriving passenger flows from the origin-destination records created from the smart card data and a function for visualizing unusual phenomena.
2. We propose a method to predict how passengers behave after an accident. Our method is based on two models: passenger demand and passenger behavior. The demand model is constructed using the passenger flows derived from the origin-destination records. The behavior model is used to predict passenger behavior after an accident. It uses the Abandonment Rate as a parameter. Our prediction method is based not simply on the results of simulation but on actual trip records.
3. We evaluate the accuracy and effectiveness of the proposed method by comparing the trip records after two major accidents. We find that the prediction accuracy could be improved by using an appropriate Abandonment Rate determined from historical post-accident data.

We describe related work in Section 2 and explain our goal of passenger flow analysis and our developed framework for large-scale trip record analysis in Section 3. Our method for predicting passenger behavior after an accident is explained in Section 4. In Section 5, we evaluate the accuracy and usefulness of the proposed method on the basis of our analysis of smart card data related to two major accidents. Section 6 summarizes the key points and mentions future work.

2 Related Work

Transportation log data, including data from operation logs, smart card logs, and equipment monitoring logs has been analyzed in various studies. Ushida et al. used train operation data to visualize and identify delay events and created chromatic diagrams for one line in the Tokyo Metro subway system with the goal of generating a more robust (delay-resistant) timetable [7]. Smart card data has been used as a data source to analyze the operation of a public transportation system [4]. Trépanier et al. used an alighting point estimation model to analyze passenger behavior from incomplete trip records created from smart card data[6]. Ceapa et al. used oyster (smart card) data to clarify passenger flow congestion patterns at several stations in the London Tube system for use in reducing congestion [1]. Their spatio-temporal analysis revealed highly regular congestion patterns during weekdays with large spikes occurring in short time intervals. Sun et al. provided a model for predicting the spatio-temporal density of passengers and applied it to one line in the Singapore railway system [5].

Previous work using smart card data focused only on a single line or a few stations, mainly because the trip records created from the data did not include transfer station information. We overcame this limitation by determining the most probable transfer station(s) for each trip record created from the origin and destination information and were thus better able to analyze how the effects of a disruptive event in a metro network propagate.

3 Framework for Analyzing Trip Records

The system we are constructing for analyzing train trip records currently uses data collected each night rather than in real time. Our immediate aim is to evaluate the effectiveness of our approach by analyzing historical data.

3.1 System Overview

Entrance and exit information is obtained each time a passenger uses a smart card to enter or exit a station gate (wicket). This information is aggregated on a central server each night. Our system can be used to analyze this information.

The system creates a trip route (start point, transfer points, and end point) for each record and estimates the passenger flows – how many passengers traveled a certain section at a certain time (as explained in detail in Section 3.3).

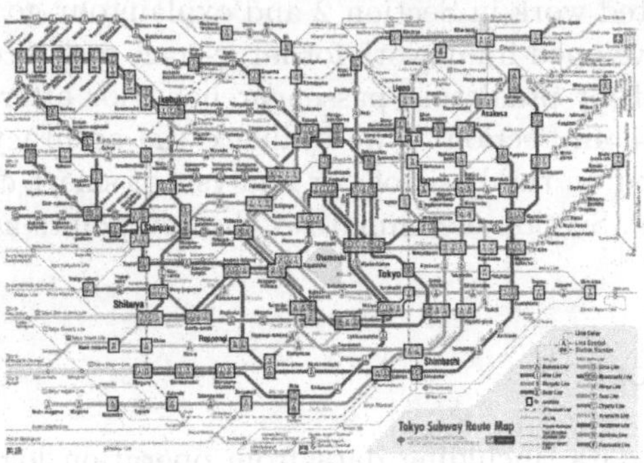

Fig. 1. Tokyo subway map[1]

By analyzing these flows, we can identify disruptive events resulting from various causes. Our interactive visualization framework can be used to identify and understand such events (as explained in Section 3.4).

Our method for predicting passenger behavior after an accident uses a passenger demand model and a passenger behavior model, which describes the effects of an accident (as described in Sections 4.1 and 4.2). Using these models and current traffic condition information, we can predict short-term passenger behavior.

We plan to make a passenger flow simulator that uses the behavior model. We also plan to combine real-time observation data with simulation data to make the predictions more accurate.

3.2 Smart Card Data

We used two years' worth of trip records for the Tokyo Metro subway system created from smart card data. As shown in Figure 1, the train system in Tokyo has a complicated route structure, consisting of lines of various railway companies including Tokyo Metro, Toei Subway, Japan Railway (JR), and many private railroads. We analyzed the Tokyo Metro trip records for almost all of the Tokyo business area, covering 28 lines, 540 stations, and about 300 million trips. The records included lines and stations besides Tokyo Metro ones if passengers used lines of other railway companies for transfers.

In our experiments, we use passengers log data from anonymous smartcards without personal identity information, such as, name, address, age, and gender. From each record, card ID is eliminated. Each record consisted of the origin, destination, and exit time. Since transfer information was not included, we estimated the probable route for each trip (as explained in Section 3.3).

During the week, the trains are mainly used by people going to or returning home from work while they are used more generally on weekends. We thus

[1] `http://www.tokyometro.jp/en/subwaymap/index.html`

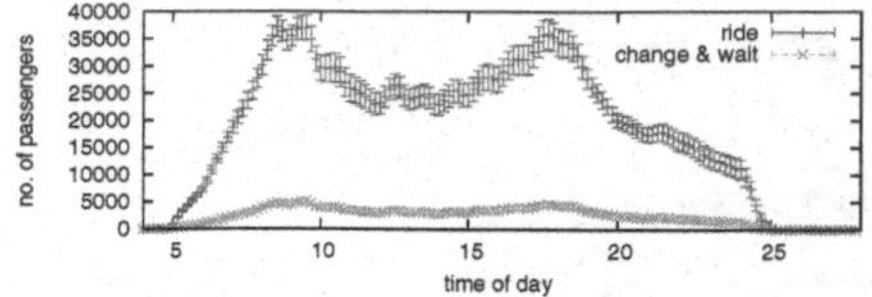

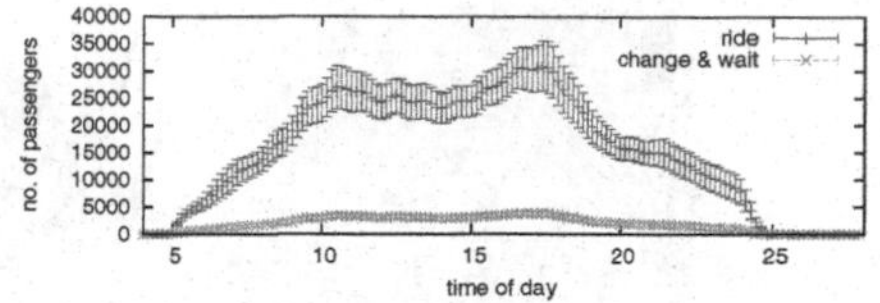

Fig. 2. Average and standard deviation of number of passengers over one year (Apr. 2012 to Mar. 2013) for weekdays (left) and weekends and holidays (right); "ride" represents number of passengers riding on a train, "change & wait" represents number of passengers waiting to change trains, error bars indicate standard deviation.

separated the data between weekdays and weekends and analyzed the two sets independently. National holidays and several days during vacation seasons were treated as weekend days.

Passenger behavior was assumed to follow periodic patterns, especially daily ones, so we statistically analyzed these data to identify the patterns. Figure 2 shows the average and standard deviation of the number of passengers over one year. The plot points were obtained by estimating the total trip time for each trip record (as described in Section 3.3) and then determining the number of passengers who were travelling during each 10-minute time period.

The weekday and weekend demand patterns are clearly different. In the weekday one, there are two distinct peaks corresponding to the morning and evening rush hours. In the weekend and holiday one, there is only one distinct peak, around 5:20 p.m. The deviations in the weekday pattern are smaller than those in the weekend one, indicating that most passengers in the weekday behave in a periodic manner. Therefore, we may be able to detect disruptive events by comparing the differences in the average number of passengers for each section of a line.

3.3 Extraction of Passenger Flows

From the trip records, we can determine how many passengers used a certain station. However, since the data did not include the entrance time, we could not determine how many passengers there were within a certain time period. Moreover, the origin-destination pair information was not enough for estimating the crowdedness of each train or the effects of a disruptive incident at a certain location. We need the trip route for each passenger.

There are usually several possible routes for traveling from an origin station to a destination station. A smart card log contains information about where a passenger touched in and where and when he/she touched out at a station gate. It does not include the entrance time and transfer station information. We therefore assumed that the most probable route for each trip (origin and destination pair) was the one with the shortest total trip time.

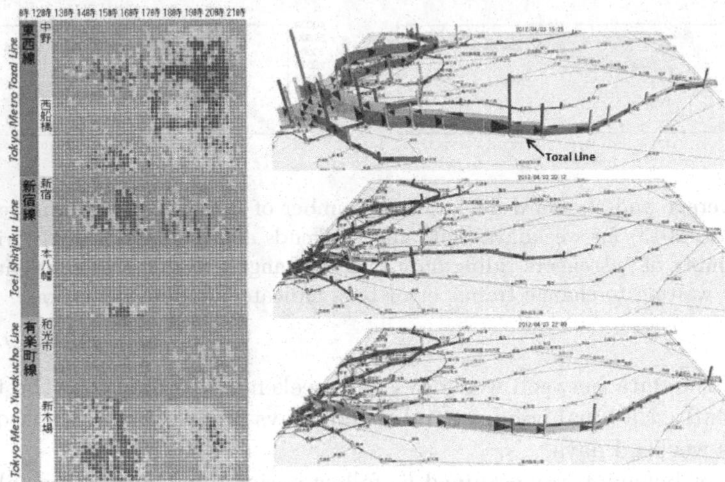

Fig. 3. HeatMap view (left) shows passenger crowdedness during rainstorm in April 2012; RouteMap view (right) shows animated changes in passenger flows and propagation of crowdedness on route map

We defined total trip time $t = T + C + W$, where

- T is the time spent riding, as defined by the timetable,
- C is the walking time when transferring, as determined by the layout of each station and roughly defined using information provided by the train company, and
- W is the time waiting for a train to arrive, as defined by the timetable (average train interval / 2).

Using this definition, we calculated the estimated time for every possible trip route for each origin-destination pair. We then used the Dijkstra algorithm [2] to find the fastest route.

To identify phenomena that differed from the usual cyclical patterns, we first estimated in which section of a line each passenger was during a particular time on the basis of the fastest route and exit time. We then calculated the number of passengers who were in a certain section during a certain (10-minute or one-hour) time period and the average number and standard deviation. The weekday and weekend data sets were independently analyzed as explained in Section 3.2. The average and standard deviation were used to detect unusual patterns, especially in the weekday cyclical patterns.

3.4 Visualization of Passenger Flows

We used a visualization technique [3] to investigate passenger behavior. Our framework provides two visualization views, the HeatMap view and the RouteMap view, for exploring passenger flows and spatio-temporal propagation of crowdedness extracted in Section 3.3.

The HeatMap view (Figure 3, left) provides an overview of the spatio-temporal crowdedness of sections of the lines in the route map. The RouteMap view (Figure 3, right) shows the animated temporal changes in the number of passengers and the crowdedness of each section. The two views are coordinated – selection of lines and time stamps in the HeatMap view causes the RouteMap view to start showing animated changes in the values for the selected lines and time stamps.

The combination of these views enables users to detect unusual events. Two HeatMap thresholds (high and low) are defined for determining unusually crowded areas and unusually empty areas (for a certain train section and for a certain time period). These unusual geo-temporal areas are concatenated into several chunks separated by normal areas. The size and density of each chunk reflect the largeness of affected area, and the severity of the effect. Many of the large extracted chunks corresponded to unusual events, such as natural disasters, public gathering, and accidents. The example shown in Figure 3 illustrates the effects of a spring storm. The large red and blue chunks in HeatMap view indicate that passenger flows changed drastically during that event. The RouteMap reveals corresponding events by showing the spatio-temporal propagation of the unusual passenger behaviors.

4 Prediction Method

Passenger flows after train accidents are predicted using the demand and behavior models. To predict the number of passengers at the time of an accident without using real-time data, we estimate how many passengers will want to travel during or after the time of the accident by using the demand model. We then estimate how these passengers will change their route by using the behavior model.

Our prediction method requires two kinds of information as input: average passenger flow in the past and the expected time to recover from the accident. The proposed method can predict passenger flows after an accident from this information, without real-time data.

4.1 Passenger Demand Model

We estimated each passenger's entrance time by using the trip time estimation method described in Section 3.3. We then calculated the number of passengers starting a trip, for every origin-destination pair, for every 10 minutes. The results reflected the passenger demand during a certain time period. We calculated the average demand for each time slot over one year (Apr. 1, 2012, to Mar. 31, 2013). As mentioned, data for weekdays and weekends were treated independently.

We used this average demand as the passenger demand model.

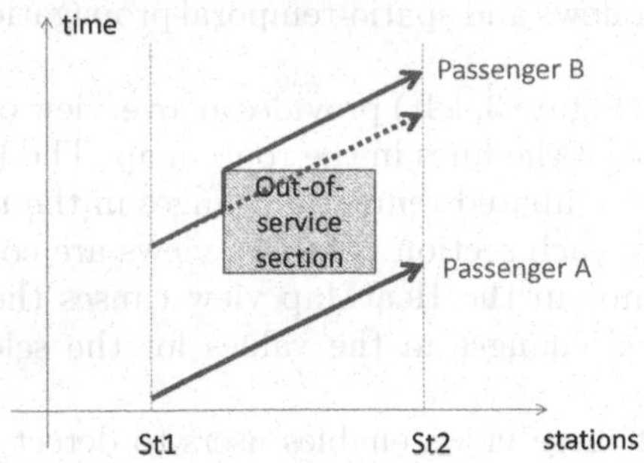

Fig. 4. Model of passenger behavior following an unusual event; red arrows indicate passenger trajectories

4.2 Passenger Behavior Model

In the passenger behavior model, each passenger has an origin station and a destination, and the route taken is calculated as explained in Section 3.3. Example spatio-temporal behaviors of two passengers when service on a section of a line is suspended is illustrated in Figure 4. Two passengers (A and B) want to travel from Station 1 (St1) to St2. We assume that we know about the service suspension and the begin and end times of the suspension (represented in the figure as "out-of-service section"). Passenger A completes passage through that section before service is suspended and is thus not affected. Passenger B is affected and thus must change his route or time of travel:

- If there is another route from St1 to St2 that does not use the out-of-service section, he can change his route, which will likely change the arrival time.
- If there is no such route, he must wait until service is restored (as illustrated in Figure 4).

Since passengers often change their travel method in such situations, such as by switching to travel by bus or taxi, we use the "Abandonment Rate" to capture this behavior. When N passengers are affected by the out-of-service section, we assume only $N \times (1 - \mathit{AbandonmentRate})$ passengers continue to use the train system.

The Abandonment Rate should be based on historical data. The estimation of this parameter is described in Section 5.

5 Evaluation

5.1 Target Data

We found disruptive events for the Tokyo Metro subway system by using the transport information webpage of a third-party company [2]. We checked this page every hour for one year and extracted the train operating condition information.

[2] `http://transit.goo.ne.jp/unkou/kantou.html` (in Japanese).

Here we focus on one line (the "Tozai Line") in order to set the parameter of our model under the simplest conditions. We found that six disruptive events occurred on the Tozai Line during the year. They caused service suspensions on several sections, and it took several hours to restore service.

5.2 Evaluation Metric

Since each trip record contained each passenger's arrival time and not the departure time, we could only determine how many passengers had left from a certain station at a certain time (*Exitnum*). We evaluated the passenger behavior model by comparing these values. We estimated how many passengers left from each station by using our passenger demand and behavior models.

5.3 Test Case 1 — Accident on Nov. 6, 2012

An accident occurred at 17:11 on Nov. 6, 2012. A failure of railway power supply system was found and several sections of the line between Toyocho Station and Kasai Station were closed until 18:34. The accident happened during rush hour, so it affected many passengers.

Figure 5 shows the difference in *Exitnum* between the estimation obtained using the demand model and the actual trip record data. The estimation was made without considering the accident information. Colors are used in the figure to represent the difference in *Exitnum* normalized by the value in the trip record data. Blue represents the situation in which the actual number of passengers was lower than the estimated one, and red represents the opposite case. The large blue area indicates that many passengers were unable to travel as they normally did.

We also estimated *Exitnum* for all the other stations in the metro system. Since the difference was at the highest for stations on the Tozai Line, we focused on that line in our evaluation.

The estimation results when the out-of-service information was considered, with Abandonment Rate = 0, are shown in Figure 6. Most of the blue area (labeled i) disappeared. However, a new large blue area (labeled ii) appeared around 18:50, after the accident.

We adjusted the Abandonment Rate to find the most appropriate setting. We computed the sum of the absolute number of difference in this sections and timespan to estimate the correctness of our model. Figure 8 shows this value normalized by the average number of passengers who used the Tozai Line at this time of the day. The difference for Abandonment Rate = 0 was larger than the case of no information about the service suspension (green line). We can see that the best setting of the Abandonment Rate was 0.9. With the out-of-service information, the difference improved by about 9%. Figure 7 shows the best case, with Abandonment Rate = 0.9. The large problematic blue area (ii) disappeared with this setting.

Even in the best case (Figure 7), we can see several chunks of blue area (iii and iv). We obtained train operation information for that date from the railway

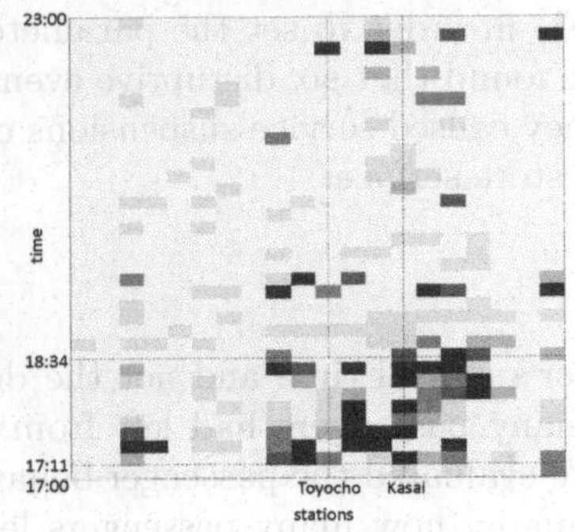

Fig. 5. Difference in number of passengers after accident on Nov. 6, with no out-of-service information

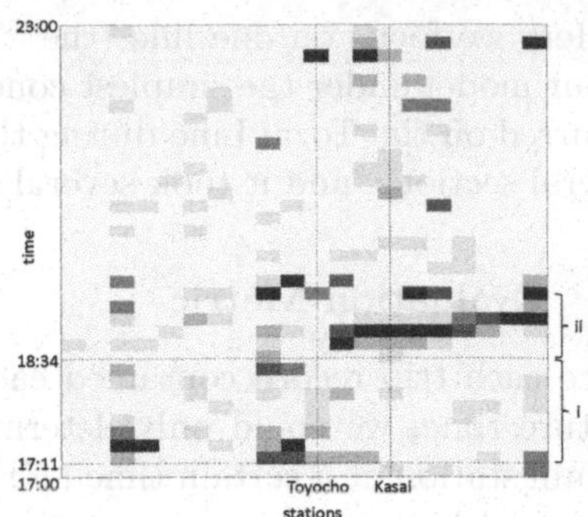

Fig. 6. Difference in number of passengers after accident on Nov. 6, with out-of-service information and Abandonment Rate (AR) = 0.0

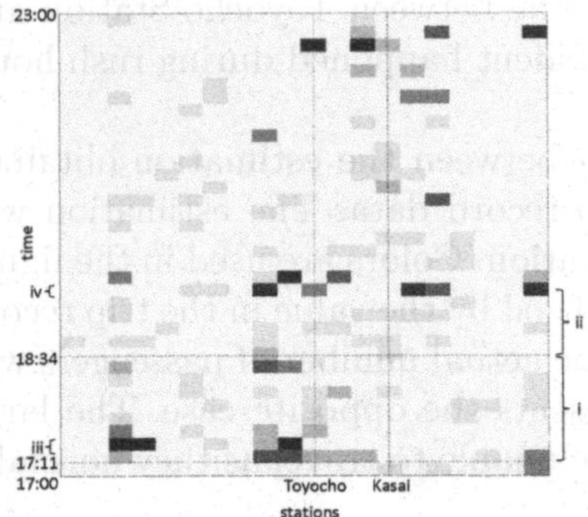

Fig. 7. Difference in number of passengers after accident on Nov. 6, with out-of-service information and $AR = 0.9$

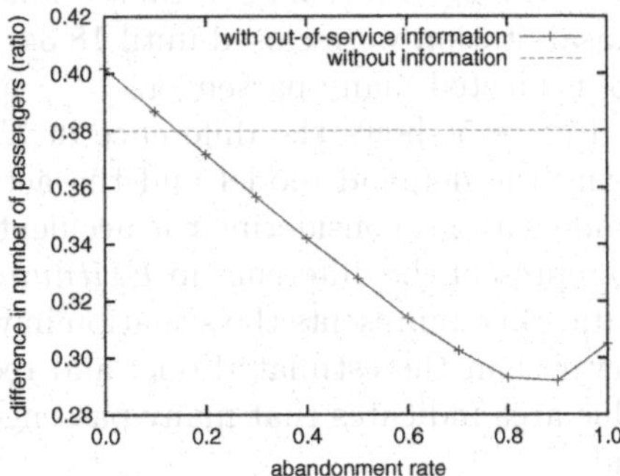

Fig. 8. Difference in number of passengers after accident on Nov. 6

company and determined that most of the decrease in the number of passengers is understandable: the entire Tozai Line was stopped twice for a short time, from 17:11 to 17:28 (iii), and from 19:19 to 19:29 (iv). The blue chunk in Figure 7 is reasonably explained by this information.

5.4 Test Case 2 — Accident on Oct. 10, 2012

Another accident occurred at 11:32 on Oct. 10, 2012. A passenger was injured and sections between Kasai Station and Myouden Station were closed until 13:09. Note that the closed sections and time of day differ from those in the November case (Section 5.3).

Figure 9 shows the difference in *Exitnum* without the out-of-service information. The difference is smaller than in the November case.

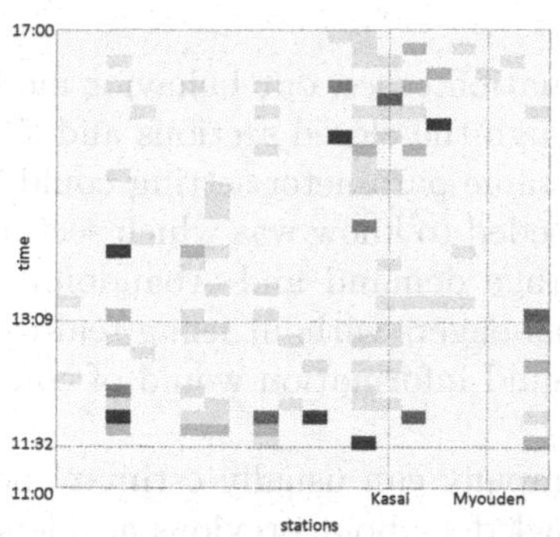

Fig. 9. Difference in number of passengers after accident on Oct. 10, with no out-of-service information

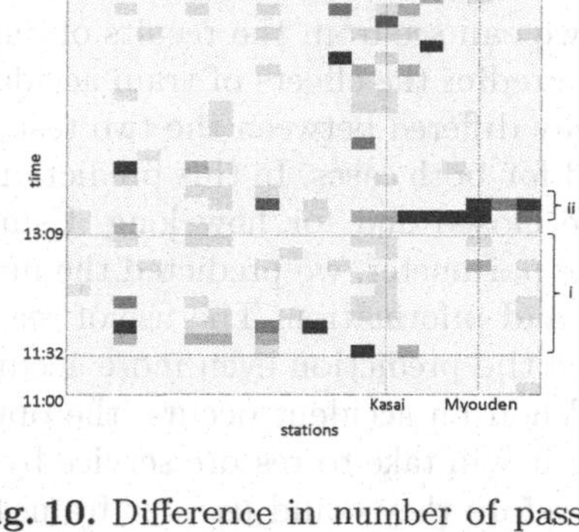

Fig. 10. Difference in number of passengers after accident on Oct. 10, with out-of-service information and Abandonment Rate $(AR) = 0.0$

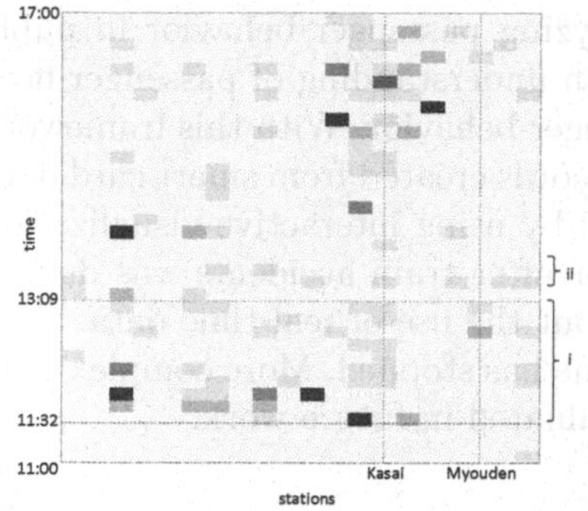

Fig. 11. Difference in number of passengers after accident on Oct. 10, with out-of-service information and $AR = 0.9$

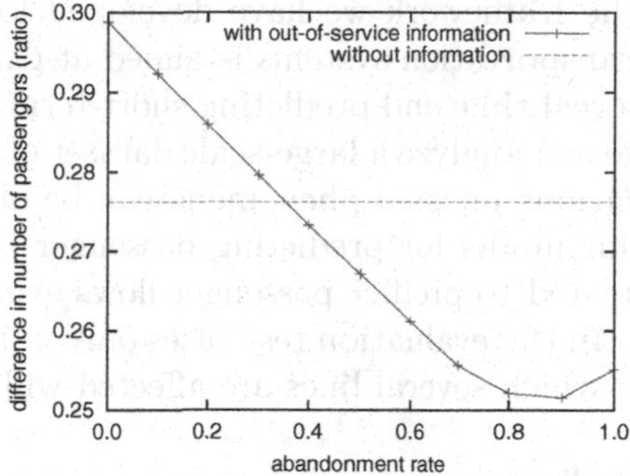

Fig. 12. Difference in number of passengers after accident on Oct. 10

Figure 10 shows the results with the out-of-service information and Abandonment Rate = 0. The difference during the time period of the accident (i) is smaller, but a new blue area (ii) appeared after the accident.

We again adjusted the Abandonment Rate to find the most appropriate setting. Figure 12 shows the total difference in our estimation normalized by the average number of passengers who used the Tozai Line at this time of the day. The best result was achieved when we set the Abandonment Rate to 0.9, the same value as in the November case. This shows that our behavior model can predict user behavior more precisely when the out-of-service information is used.

The best setting (Abandonment Rate = 0.9) results are shown in Figure 11. The large blue chunk in Figure 10-(ii) has disappeared.

5.5 Discussion

As we can see from the results of our two evaluation cases, our behavior model can predict the effects of train accidents. Although the closed sections and time of day differed between the two test cases, the same parameter setting could be used for both cases. In this prediction, all we needed to know was which sections were closed and for how long. Using the average demand and Abandonment Rate parameter, we predicted the number of passengers without using real-time demand information. The use of real-time demand information would of course make the prediction even more accurate.

When an accident occurs, the operating company can usually estimate how long it will take to restore service by using knowledge about previous accidents. Therefore they could use our framework and predict the effects of an accident in real time.

6 Conclusion

The framework we have developed for analyzing passenger behavior in public transportation systems is aimed at gaining an understanding of passenger flows in real time and predicting short-term passenger behavior. With this framework, we can analyze a large-scale dataset of trip records created from smart card data. Various unusual phenomena can be observed by using interactive visualization. Our model for predicting passenger behavior after train accidents was demonstrated to predict passenger flows even without the use of real-time data.

In the evaluation test cases only a single line has stopped. More complex cases in which several lines are affected will be evaluated in future work.

References

1. Ceapa, I., Smith, C., Capra, L.: Avoiding the Crowds: Understanding Tube Station Congestion Patterns from Trip Data. In: Proc. UrbComp 2012, pp. 134–141 (2012)
2. Dijkstra, E.: A note on two problems in connexion with graphs. Numerische Mathematik 1(1), 269–271 (1959), `http://dx.doi.org/10.1007/BF01386390`
3. Itoh, M., Yokoyama, D., Toyoda, M., Tomita, Y., Kawamura, S., Kitsuregawa, M.: Visualization of Passenger Flows on Metro. In: IEEE Conference on Visual Analytics Science and Technology, VAST 2013 (2013) (poster)
4. Pelletier, M.P., Trépanier, M., Morency, C.: Smart Card Data Use in Public Transit: A Literature Review. Transportation Research Part C: Emerging Technologies 19(4), 557–568 (2011)
5. Sun, L., Lee, D.H., Erath, A., Huang, X.: Using Smart Card Data to Extract Passenger's Spatio-temporal Density and Train's Trajectory of MRT System. In: Proc. UrbComp 2012, pp. 142–148 (2012)
6. Trépanier, M., Tranchant, N., Chapleau, R.: Individual Trip Destination Estimation in a Transit Smart Card Automated Fare Collection System. Journal of Intelligent Transportation Systems 11(1), 1–14 (2007)
7. Ushida, K., Makino, S., Tomii, N.: Increasing Robustness of Dense Timetables by Visualization of Train Traffic Record Data and Monte Carlo Simulation. In: Proc. WCRR 2011 (2011)

Patent Evaluation Based on Technological Trajectory Revealed in Relevant Prior Patents

Sooyoung Oh[1], Zhen Lei[2], Wang-Chien Lee[1], and John Yen[3]

[1] Department of Computer Science and Engineering
[2] Department of Energy and Mineral Engineering
[3] College of Information Sciences and Technology,
The Pennsylvania State University, University Park, PA 16802, USA
{sooh,wlee}@cse.psu.edu, zlei@psu.edu, jyen@ist.psu.edu

Abstract. It is a challenging task for firms to assess the importance of a patent and identify valuable patents as early as possible. Counting the number of citations received is a widely used method to assess the value of a patent. However, recently granted patents have few citations received, which makes the use of citation counts infeasible. In this paper, we propose a novel idea to evaluate the value of new or recently granted patents using *recommended relevant prior patents*. Our approach is to exploit trends in temporal patterns of relevant prior patents, which are highly related to patent values. We evaluate the proposed approach using two patent value evaluation tasks with a large-scale collection of U.S. patents. Experimental results show that the models created based on our idea significantly enhance those using the baseline features or patent backward citations.

Keywords: patent, evaluation, ranking.

1 Introduction

Patent evaluation, including predicting a patent's future value, comparing the value of patents in a given patent set, and identifying influential patents in a field or within a company, is a challenging but important task for technology and innovation management in a firm. The *forward citations* of a patent (i.e., citations to the patent made by other patents granted in later/*forward* times), in combination with other patent information such as the number of claims, etc, have been widely used as a measure to assess the patent economic value [1, 2, 3, 4]. As highly cited patents imply a number of successful lines of innovation (that is why they are highly cited), their inventions are likely to be technologically significant and economically valuable. However, it often takes years for a patent to receive sufficient information of forward citations in order to make meaningful assessment of its value.

This paper addresses the challenging problem of evaluating patents at the early stage of their patent life when there is little information about forward citations. To highlight this challenge, we ask the following question: *Can we*

V.S. Tseng et al. (Eds.): PAKDD 2014, Part I, LNAI 8443, pp. 545–556, 2014.

evaluate patents immediately after they are granted when there is no forward citation information at all? This is not merely a theoretically interesting question stressing the limitation of forward citation-based approaches for patent evaluation, but one with practical importance. A *patent grant*, to a large extent, secures the patent protection and ascertains the scope of the right for an invention. This is often the point of time when the patent owner decides what to do with the invention, e.g., developing/incorporating it into products in house or licensing to someone else. This is also the point of time when the deal of licensing/selling the invention, if any, is made. Therefore, it is crucial to predict the value of a patent *right after* its grant.

In the absence of forward citation information, one naturally turns to the *backward citations* of the patent under evaluation, i.e., the references made by the focal patent to prior patents (which are granted in past/*backward* times), besides other useful information such as the number of claims, the number of figures, etc. Indeed, previous studies [1, 2, 3, 4] have shown that the information of patent backward citations can be used to measure the *novelty* of a patent, and thus useful for evaluating its technological and economic value of a patent as well. Patent backward citations provide information about technologically relevant prior patents for the focal patent, and we can use it to estimate how novel and in what stage of the technology trajectory the focal patent is. However, the idea of using backward citations for patent evaluation also has limitations. First, backward citations are to a large extent missing for one or two years prior to the patent grant year. The reason is simple. Applicants and examiners reference prior patents either at patent filing or during the early stage of patent examination, which tend to be one or two years earlier than patent grant. Therefore, backward citations in first one or two lagged years (presumably the most important piece of information regarding the focal patent's technology stage or novelty) are often missing. Second, patent backward citations, unlike paper references that focus on completeness, are often parsimonious and incomplete. The average number of citations is usually less than 20 and many patents have no backward citations at all.

In this paper, aiming to evaluate a newly granted patent with no forward citations and incomplete (and often quite sparse) backward citation information, we develop a novel approach to assess value of a patent by exploring *technologically relevant prior patents* as a supplement to the backward citations. The idea behind our approach is simple yet innovative. Since the key in a backward citation-based patent evaluation approach is to find relevant prior patents of a focal patent via its backward citations, we shall also identify other good (or better) technologically relevant prior patents through other means. We test this idea by identifying a set of technologically relevant prior patents based on *content similarity*, and use them to construct features representing a focal patent's novelty and stage in the technology trajectory. Identification of relevant prior patents based on content similarity seems simple, yet it works well. This is exactly the beauty of our proposed approach: simple, intuitive and working well.

Moreover, to measure the novelty of a patent, its stage of technology trajectory, or other properties related to patent value, we focus on the dynamics of relevant prior patents over past years (i.e., in form of temporal distribution). We have studied a series of effective features exhibiting discriminative temporal behaviors from diverse aspects, and identified six temporal patterns and features. We propose two prediction models and evaluate the performance of the models using features from patent backward citations, relevant prior patents, or both together with baseline features. Our experimental results show that relevant prior patents complement patent backward citations, and significantly enhance the evaluation of patent value.

Our major contributions in this work are as follows:

- **Problem of Patent Evaluation on Newly Granted Patents**: We point out the limitation in using forward citations and backward citations to evaluate newly or recently issued patents. This is a very important and practical problem for technology management in firms.

- **New Patent Evaluation Approach**: Without relying on forward patent citation information, our patent evaluation approach utilizes a set of technologically relevant prior patents identified based on content similarity to supplement information derived from backward citations.

- **Features based on Temporal Trending of Relevant Prior Patents**: We propose six sets of features measuring trending and temporal patterns in multiple subsets of technologically relevant prior patents and backward citations, which are highly related to the value of a patent.

2 Related Work

The study on patent quality evaluation using patent mining techniques has received growing interests. Hasan et al. [5] proposed a patent ranking model, called COA (Claim Originality Analysis), which evaluated the value of a patent based on the novelty (recency) and impact (influence) of the important phrases in Claim section of a patent. They used the number of patent citations received, patent maintenance status, and their own confidential ratings to evaluate the proposed model with patents related to software technology or business process. Jin et al. [6] introduced a model recommending patent maintenance decision. They proposed diverse features, which could measure the value of a patent, and built a ranking model to predict whether a patent should be maintained or abandoned when a renewal decision was made. Liu et al. [7] proposed a latent graphical model to assess patent quality using quality-related features, such as citation quality, technology relevance, claim originality, etc. They used the number of citation received, court decisions (ruled as valid or invalid), and reexamination records as patent quality measurements. Even though the court decisions may be strongly correlated with patent quality, it is hard to get enough samples to train the model. Oh et al. [8] introduced a weighted citation method. They distinguished different types of citations to rank patents by their importance. Hu et al. [9] proposed a topic-based temporal mining approach to assess the novelty and influence of a patent to discover core patents. They extracted topics from a patent document and quantified a patent's novelty and influence by analyzing the topic activeness variations along the timeline. Compared to those previous works, our study is focused on augmenting the values from patent citations and relevant prior arts. Instead of extracting diverse features from patent documents or using text analysis, our features are mainly extracted from temporal patterns and trends of patent citations or relevant patents, which make the model simpler and easier to build.

3 Patent Value Evaluation Approach

In this section, we define the patent value evaluation problem, and present our approaches to solve this problem.

Research Goal: Let $D = \langle D_1, D_2, \cdots, D_N \rangle$ be a set of newly granted N patents, which have no forward citation yet. Our goal is to evaluate these new patents, using only information available when they are granted. As discussed earlier, patent evaluation may include predicting a patent's future value, comparing the value of patents in a given patent set, and identifying influential patents in a field or within a company.

Relevant Prior Patent Based Patent Evaluation: To evaluate the value of a newly granted patent, we use information on technologically relevant prior patents. In particular, we focus on the temporal patterns and trending of the relevant prior patents that reflect the stage of the technology trajectory and/or novelty of the focal patent. For example, if the focal patent is associated with a lot of technologically relevant prior patents in the years immediately prior to its grant, its technology was likely to be new and on the rise, which would in turn suggest the focal patent likely to be novel and valuable. Inspired by existing patent backward citation based patent evaluation [1, 2, 3, 4], our first proposal is to identify the set of relevant prior patents for the focal patent based on its backward citations. However, as we pointed out earlier, patent backward citations are often incomplete overall and seriously missing in the most recent 1-2 years before patent grant year. Alternatively, we propose to identify a comprehensive set of technologically relevant prior patents based on content similarity. We then construct features that reflect the temporal patterns and trends of this set of relevant prior patents, which should be related to the value of the focal patent.

Features of Temporal Trending: Another innovative idea in this paper is that we propose to exploit the trending and the stage of a focal patent among relevant patents. We aim to reveal the technology stage, novelty and value of the focal patent from different aspects. For example, if the focal patent is associated with a lot of recent technologically relevant prior patents that are assigned to the same firm (or invented by the same inventor), this might indicate that the firm (or the inventor) has become very interested in the field of the focal patent and devoted quite a lot of resources in R&D related to the focal patent. Therefore, the focal patent and its technology field might be important to the firm (or the inventor), thus more likely to be valuable.

Patent Evaluation Models: There is no public available gold standard or benchmarks that clearly define the value of a patent in monetary terms. We consider two patent evaluation cases: i) ranking a set of patents, ii) identifying top-ranked patents. We are particularly interested in ranking patents because the patent ranking can provide a relative comparison of patent values. In these two cases, we use patent forward citations and patent maintenance status as indicators of patent values. Corresponding to the two cases, we learn prediction models to evaluate patent values. The first model is a ranking model that ranks patents according to their values. Often firms are more interested in understanding the relative values among a set of patents, rather than predicting their absolute value. For instance, when a firm decides to renew only half of its patents, or two firms decides whether a deal of cross-licensing is worthwhile to pursue, all matters is the relative ranking of the patents involved. In this model, we use the number of forward citations that a patent receives in a long time window (e.g., 12 years after grant) as an indicator of the patent's true value. The second model identifies high-valued patents. To make their technology and patent management decisions, firms often need to know which are the most valuable inventions/patents in their patent portfolio. For example, given limited financial resource, a firm may only afford to maintain the top 10% of its patents. We learn a binary classifier to identify top 10% or 20% most valuable patents based on 12-year forward citations.

4 Feature Extraction

We conjecture that the temporal distribution or trending of technologically relevant prior patents, in combination of patent backward citations, is highly related to the value of a patent. Moreover, the subsets of relevant prior patents in various relations to the focal patent might reflect patent value from different angles. In this section, we introduce the feature sets describing the temporal patterns in technologically relevant prior patents and backward citations of a newly granted patent in a variety of angles.

4.1 Feature Sources and Types

We describe main feature sources for the patent evaluation models in this work as follows: Given a patent, D_i, $\{Q_i, C_i, R_i\}$ denotes the three main feature sources, where $Q_i = \langle f_1, f_2, \cdots, f_l \rangle$ is l patent characteristic features, $C_i = \langle p_1, p_2, \cdots, p_n \rangle$ is n patent backward citations, and $R_i = \langle p_1, p_2, \cdots, p_m \rangle$ is m technologically relevant prior patents. First, we choose some features capturing the characteristics of a patent as our baseline features. These patent characteristics, including the number of claims, figures and assignees, are often used in previous studies. We also include binary indicator variables to identify the technology fields of the focal patent. These variables help to control the variance on patent value among patents in different technology fields. The patent characteristic features that we use are the number of claims, figures, inventors, assignees, foreign references, other references, USPC codes, and IPC codes.

In addition to these baseline features, we propose a variety of features based on temporal distribution pattern and trending of the subsets of relevant prior patents, such as the same assignee set, the same inventor set, or the same technological class set, etc. These features hopefully reflect or are related to the value of a focal patent, and would be used in our three patent evaluation models. The temporal distribution of relevant prior patents is defined as $T(D_i) = f_Y = \langle f_{Y_1}, f_{Y_2}, \cdots, f_{Y_n} \rangle$, where f_{Y_1} is the number (frequency) of prior patents in the set of retrieved relevant prior patents that were granted in the first year prior to the patent D_i, f_{Y_2} is the number of relevant prior patents granted in the second year prior to the patent D_i, and so on. We call the citing year gap, $Y_1, Y_2, \cdots$, as "citation lag" in this paper.

4.2 Features for Temporal Patterns

Our approach is to capture the attractiveness of the technology and the novelty of a new patent using temporal trends in the sets of relevant prior patents and backward citations. Moreover, we look at temporal patterns in the subsets of relevant prior patents or backward citations that share the same assignees, the same inventors, or the same technology class, with the focal patent. These temporal trends, revealed by the activities of inventors and assignees, or the popularity of the technology field in years prior to the grant of the focal patent, could contain further information about the novelty and the technology stage of the focal patent, from the perspectives of the assignees, inventors and technology area of the focal patent. In addition, prior patents owned by other assignees or in other technology fields might also be important to provide a comprehensive understanding of the technology position and the novelty of invention, which could be related to the value of the patent. Thus, we also construct features that characterize the temporal patterns and trending in relevant prior patents or backward citations, owned by other assignees, or in other technology fields. In total, we introduce six sets of features to characterize the temporal patterns and trending in relevant prior patents or backward citations from various perspectives.

We capture the temporal trends based on the number (frequency) of technologically relevant prior patents or backward citations in each of the 20 years prior to the grant of a new patent. The temporal patterns between patent citations and relevant prior patents tend to be quite different. For example, backward citations in most recent one or two years prior to the grant year of the focal patent are very few, due to the fact that most of backward citations are cited by applicants or examiners at the time of filing or shortly after the filing (during the process of prior art search), which is one or two years earlier than patent grant. By contrast, the number of relevant prior patents in the most recent years prior to the grant of the focal patent tends to be greatest, relative to those in other prior years. This is particularly true for patents with higher value. These patterns suggest that information about relevant prior patents, in particular those in most recent years, could be very useful for patent evaluation.

To capture a trend, we use Gaussian filters to measure diverse distribution patterns of relevant prior patents or backward citations according to their grant year. We use this Gaussian filtering technique to construct features capturing temporal trends. A feature, F_y, which characterizes the temporal trend of relevant prior patents (or backward citation) for the last N years regarding the yth citation lag year, is $F_y = \sum_{i=1}^{N} f_{Y_i} e^{-\alpha(i-y)^2}, \alpha > 0$, where f_{Y_i} is the frequency of relevant prior patents (or backward citations) in the Y_i^{th} lag year relative to the grant year of the focal patent. Then, we can use K features, $F_1, F_2, \cdots, F_K$, to evaluate the value of a patent according to their temporal trends. The six sets of features for temporal patterns (trends) in technologically relevant prior patents (or backward citations) are defined as follows:

Temporal Distribution of Backward Citations and Relevant Prior Patents (C1): We first measure the novelty and technology stage of a newly granted patent using the temporal trend of all the technologically relevant prior patents (or backward citations). To capture those temporal patterns, we construct features based on Gaussian filters. In addition, we add the frequency for the most recent prior year (one year lag from the grant of the focal patent), expecting to capture patent value well, especially with those most recent relevant prior patents. Thus, the features for C1 are $C1(D_1) = \langle F_1, F_2, \cdots, F_K, f_{Y_1}, f_{Y_2}, f_{Y_3} \rangle$.

Temporal Distribution of Backward Citations to and Relevant Prior Patents in the Same Assignee (C2): This set of temporal trend features focus on the subsets of relevant prior patents and backward citations assigned to the same assignees as the focal patent. If there are many recent relevant prior patents filed by the same firm, this might suggest that the firm has a strong interest in research and innovation related to the focal patent, thus the focal patent might be important and valuable. However, the number of backward citations in most recent years prior to the grant of the focal patent is quite small, and cannot be used to evaluate patents. Thus again the information on relevant prior patents, in particular in the most recent prior years, could be very useful in patent evaluation. We construct the same features as defined in C1, but only based on temporal patterns in the subsets of relevant prior patents and backward citations that share the same assignees as the focal patent.

Temporal Distribution of Backward Citations to and Relevant Prior Patents in the Same Inventor (C3): We also look into the temporal patterns in relevant prior patents or backward citations filed by the same inventors of the focal patent. It is intuitive if there are a large number of recent relevant prior patents filed by the same inventors, this would suggest that the inventors have been devoted themselves to this line of research, which in turn suggest that the technology is important and valuable. Again here, the number of most recent backward citations that were invented by the same inventors are small and not useful to capture this intuition. However, we can use the recent relevant prior patents by the same inventors. We construct the same features as defined in C1, but only applied to the subsets of relevant prior patents and backward citations that share the same inventors as the focal patent.

Temporal Distribution of Backward Citations to and Relevant Prior Patents in the Same Technology Class (C4): The temporal patterns in relevant prior patents in the same technology field could gauge the popularity of the technology field of the focal patent. We construct the same features as defined in C1, but only based on temporal patterns in the subsets of relevant prior patents and backward citations that share the same technology class (i.e., the primary U.S. patent class) as the focal patent.

Temporal Distribution of Backward Citations to and Relevant Prior Patents in the Different Assignees (C5): The relation of a focal patent to relevant prior patents filed by other assignees (i.e., other firms) can be used to measure the attractiveness and thus the value of the patent because it reflects the interest in the associated technology by other firms. Thus, we use the assignee diversity in relevant prior patents as another feature set in predicting patent value. To capture the diversity of assignees in relevant prior patents, we use the *entropy* of the distribution in different assignees in relevant prior patents in each year prior to the grant of the focal patent. High entropy means that a focal patent is related to other previous inventions filed by a larger number of other firms. We use the entropy values in each prior year to construct this set of features. The features of C5 are defined as $C5(D_i) = \langle E_1, E_2, \cdots \rangle$, where E_y is the entropy in y^{th} citation lag year, $E_y = -\sum_{k=1}^{K} Pr(a_k)log(Pr(a_k))$, where a_k is the kth assignee, K is the number of different assignees in relevant prior patents (or backward citations) for the last y years prior to the grant of the focal patent. $Pr(a_k)$ is the probability of the assignee a_k appeared in relevant prior patents (or backward citations) in that year.

Temporal Distribution of Backward Citations to and Relevant Prior Patents in the Different Classes (C6): We also look into the diversity in technology classes in relevant prior patents (or backward citations), i.e., how many different technology classes in relevant prior patents (or backward citations). This information could be useful in patent evaluation as it indicates how much the focal patent might be related to a diversity of technologies. However, the pattern in relevant prior patents is different from citations. We use the same method in C5 to construct this set of features, C6, to capture the temporal pattern in entropies in technology fields of either relevant prior patents or backward citations.

5 Experiments

We perform an empirical evaluation to validate our proposed ideas. In this section, we describe our experimental setup and discuss the results.

5.1 Experimental Setup

Data Set: We evaluate the proposed approaches using 4 million U.S. patent documents granted since 1980 until 2012. We use 14,000 patents granted on

January 2001 as our evaluation set because those patents are the most recent patents that have information about their 4th, 8th and 12th year renewal status and enough forward citations which we use as the indicator of the patent value. About 2 million patents granted between 1980 and 2000 are used as the pool of prior patents from which we retrieve technologically relevant prior patents for a given focal patent. Backward citations are also restricted to patents granted between 1980 through 2000. Patents granted between 2001 and 2012 are used to count the 12-year forward citations for the focal patent.

Content-Similarity Based Retrieval of Relevant Prior Patents: We build a prior patent retrieval/recommendation engine based on content similarity using information retrieval techniques. A patent document consists of multiple sections such as Title, Abstract, Claims, Description, etc. We extract search query terms from each section of a patent document. Xue and Croft [10] proposed a method to transform a query patent into a search query. Our approach to generate the best query terms is similar with their method. We use Indri [11] as our retrieval model. Given a search query, it returns ranked relevant prior patents with retrieval scores (relevance scores). Indri supports a weighted-term query. Xue and Croft showed a better performance when a log-scaled *TF*(Term Frequency) is used as a weight on a query term. We observed the same and thus use the a log-scaled *TF* as the weight on a query term.

Feature Set for Evaluation: We prepare four feature sets to evaluate our approaches. The first feature set (FS1) is the baseline features, involving only 8 patent characteristics features and 6 HJT-6 technological class indicators. The second feature set (FS2) uses temporal pattern and trend features extracted from backward citations, together with baseline features. FS1, FS2 is the benchmark, to which we compare our proposed approach using technologically relevant prior patents. The third feature set (FS3) uses temporal patterns and trending features extracted from relevant prior patents, together with FS1. The last feature set (FS4) combines temporal trend features based on both relevant prior patents and backward citations, together with FS1. Comparing FS4 to FS2, we can see the incremental improvement in performance from augmenting information on relevant prior patents to backward citations.

Evaluation Metrics: We use the Spearman's rank correlation coefficients to evaluate the patent ranking model. The performance of the top-ranked patent classification and the least valuable patent classification are evaluated by Precision, Recall, F-score and AUC (Area Under Curve).

5.2 Analysis of Experimental Results

We conduct experiments on the two patent evaluation tasks detailed in Section 3 for each feature set. Then, we investigate the relative significance of those six temporal pattern features we propose.

Table 1. Predicting Patent Ranks: * denotes a significant difference (p-value<0.05) from FS2. 2,3,5,10-class are divided with the same size. 5,10-class(FC) use floor(log(# of forward citations+1) as ranking values).

	FS1		FS2		FS3		FS4	
Rank	LR	SVR	LR	SVR	LR	SVR	LR	SVR
2-class	0.3095	0.3115	0.3711	0.3699	0.3593*	0.3875*	0.3974*	0.4112*
3-class	0.3494	0.3690	0.4176	0.4282	0.4075*	0.4418	0.4497*	0.4704*
5-class	0.3682	0.3890	0.4396	0.4522	0.4272*	0.4622	0.4718*	0.4948*
10-class	0.3759	0.3929	0.4495	0.4618	0.4375*	0.4709	0.4831*	0.5070*
5-class(FC)	0.3667	0.3878	0.4378	0.4504	0.4243*	0.4626*	0.4686*	0.4936*
10-class(FC)	0.3764	0.3988	0.4505	0.4651	0.4376*	0.4762*	0.4838*	0.5105*
# of FC	0.3795	0.4049	0.4534	0.4688	0.4407*	0.4794*	0.4870*	0.5121*

Predicting the Ranks of Patents: In this experiment, we evaluate the ranks of patents. Using patent ranks, we compare the relative value among patents or assess overall value for a patent portfolio. To test diverse ranking scenarios, we use several ranking approaches. We first divide the ordered evaluation set based on their 12-year forward citations (ground truth), into 2, 3, 5, or 10 classes of the same size. Thus, we prepare four types of the uniformly distributed ranked values. We also directly use the log-scaled number of forward citations. To build prediction models, we use two regression models, Linear Regression (LR) and Support Vector Regression (SVR). In all these cases the Spearman's rank correlation coefficients are used to evaluate the ranking performance. Table 1 show the performance of predicting patent ranks with diverse ranking values using LR and SVR. The results in Table 1 show that SVR is better than LR in overall performance. According to the SVR results, the feature set based on relevant prior patents (FS3) is better than that based on backward citations (FS2). Overall, the combined feature set (FS4) achieves the best performance which is significantly better than FS1 with more than 27% improvement in all these cases.

Identifying the Top-Ranked Patents: We conduct experiments to predict the top-ranked patents. We prepare two top-ranked patent data sets based on their forward citations. One is the top-10% ranked patents, and the other is the top-20% ranked patents. We build the patent evaluation model using two binary classifier, Random Forest (RF) and Support Vector Machine (SVM). Table 2 shows the results of classifying the top-10% ranked and top-20% ranked patents using RF and SVM. In both cases, the relevant patents (FS3) shows better performance than patent citations (FS2), and again, we can get the best results when we use both patent citations and relevant patents together (FS4). In general, SVM shows better results than RF. The overall results shows our approaches works for predicting the top-ranked patents.

Feature Analysis: Finally, we investigate the relative importance of the six temporal patterns and the extracted trending features discussed in Section 4.2. Table 3 shows the rank prediction results when we add or remove one of these six temporal pattern features in the models. According to the results, the temporal distribution of relevant prior patents and backward citations (C1) is the key factor among six

Table 2. Classification of Top-Ranked Patents: * denotes the significant difference (p-value<0.05) from FS2

(a) Top-10% ranked patent classification

RF					SVM				
features	Precision	Recall	F-Score	AUC	features	Precision	Recall	F-Score	AUC
FS1	0.1738	0.6410	0.2731	0.6462	FS1	0.1912	0.6541	0.2956	0.6683
FS2	0.2110	0.7186	0.3259	0.7057	FS2	0.2208	0.7167	0.3374	0.7135
FS3	0.2251*	0.7430*	0.3452*	0.7250*	FS3	0.2307	0.7730*	0.3551*	0.7387*
FS4	0.2354*	0.7613*	0.3920*	0.7389*	FS4	0.2426*	0.7725*	0.3691*	0.7482*

(b) Top-20% ranked patent classification

RF					SVM				
features	Precision	Recall	F-Score	AUC	features	Precision	Recall	F-Score	AUC
FS1	0.2976	0.6205	0.4018	0.6184	FS1	0.3292	0.6337	0.4329	0.6483
FS2	0.3440	0.6630	0.4526	0.6664	FS2	0.3627	0.6669	0.4695	0.6806
FS3	0.3608*	0.6864	0.4728*	0.6846*	FS3	0.3663	0.7024*	0.4813*	0.6927*
FS4	0.3753*	0.7167*	0.4925*	0.7027*	FS4	0.3819*	0.7110*	0.4966*	0.7055*

Table 3. Feature Analysis: * denotes a significant difference (p-value<0.05) from the baseline

	(a) Adding features			(b) Removing features		
Featues	FS2	FS3	FS4	FS2	FS3	FS4
Base	0.4128	0.4128	0.4128	0.4721	0.4908	0.5207
C1	17.83%*	15.06%*	25.33%*	-0.99%	-3.00%*	-2.21%*
C2	1.16%	2.67%*	3.81%*	0.21%	0.01%	0.31%
C3	0.73%	1.37%	1.90%	0.13%	-0.18%	-0.09%
C4	10.02%*	6.52%*	12.57%*	-0.04%	-0.67%	-0.11%
C5	14.08%*	9.62%*	19.30%*	-0.06%	-1.66%*	-1.35%*
C6	12.58%*	10.21%*	18.40%*	0.04%	-0.50%	-0.07%

temporal patterns. It shows the significant improvement (or decrease) when we add into (or remove it from) the feature sets. Other important features are the measure of assignee or technological class diversity in relevant prior patents or backward citations (C5 and C6). When we add those features, the performance is significantly improved. C2 and C3 do not show the significance, possibly because the number of relevant prior patents and backward citations assigned to the same assignee or invented by the same inventors is too small to reflect useful information regarding patent value.

6 Conclusion

In this study, we propose a novel approach to evaluate newly granted patents, using only information available at the time of patent grant. Our approach is to use the temporal patterns and trends of relevant prior patents that reflect the novelty and technology state of a patent or the attractiveness of the technology associated with a patent. The experimental results show that our approach can achieve significantly better evaluation performance by augmenting information on relevant prior patents to backward citations. The feature analysis results show that temporal patterns based on the distribution of relevant prior patents and backward citations are important features in our proposed patent evaluation models. Moreover, the measures on the dynamics regarding the diversity of other assignees and other technology classes in relevant prior patents and backward citations are highly related to patent value. Our approach based on temporal trends of relevant prior patents is the first of this kind, as an effort to assess the value of a new granted patent. Compared to previous works relying on rigorous feature extraction or text analysis, our approach is much simpler and easier to build patent evaluation models, and quite flexible to expand.

Acknowledgement. This work is supported by U.S. National Science Foundation Grant 5MA-1064194.

References

[1] Trajtenberg, M.: A penny for your quotes: patent citations and the value of innovations. The Rand Journal of Economics, 172–187 (1990)
[2] Harhoff, D., Narin, F., Scherer, F.M., Vopel, K.: Citation frequency and the value of patented inventions. Review of Economics and Statistics 81(3), 511–515 (1999)
[3] Hall, B.H., Jaffe, A.B., Trajtenberg, M.: The nber patent citation data file: Lessons, insights and methodological tools. Technical report, National Bureau of Economic Research (2001)
[4] Hall, B.H., Jaffe, A., Trajtenberg, M.: Market value and patent citations. RAND Journal of Economics, 16–38 (2005)
[5] Hasan, M.A., Spangler, W.S., Griffin, T., Alba, A.: Coa: Finding novel patents through text analysis. In: CIKM, pp. 1175–1184. ACM (2009)
[6] Jin, X., Spangler, S., Chen, Y., Cai, K., Ma, R., Zhang, L., Wu, X., Han, J.: Patent maintenance recommendation with patent information network model. In: ICDM, pp. 280–289. IEEE (2011)
[7] Liu, Y., Hseuh, P.Y., Lawrence, R., Meliksetian, S., Perlich, C., Veen, A.: Latent graphical models for quantifying and predicting patent quality. In: Proceedings of the 17th ACM KDD. ACM (2011)
[8] Oh, S., Lei, Z., Mitra, P., Yen, J.: Evaluating and ranking patents using weighted citations. In: Proceedings of the 12th ACM/IEEE-CS Joint Conference on Digital Libraries. ACM (2012)
[9] Hu, P., Huang, M., Xu, P., Li, W., Usadi, A.K., Zhu, X.: Finding nuggets in ip portfolios: core patent mining through textual temporal analysis. In: Proceedings of the 21st ACM CIKM. ACM (2012)
[10] Xue, X., Croft, W.B.: Automatic query generation for patent search. In: Proceedings of the 18th ACM CIKM, pp. 2037–2040. ACM (2009)
[11] Strohman, T., Metzler, D., Turtle, H., Croft, W.B.: Indri: A language model-based search engine for complex queries. In: Proceedings of the International Conference on Intelligent Analysis, vol. 2(6) (2005)

Deferentially Private Tagging Recommendation Based on Topic Model

Tianqing Zhu, Gang Li*, Wanlei Zhou, Ping Xiong, and Cao Yuan

School of Information Technology, Deakin University, Australia
School of Information, Zhongnan University of Economics and Law, China
School of Mathematics and Computer, Wuhan Polytechnic University, China
tianqing.e.zhu@gmail.com, {gang.li,wanlei.zhou}@deakin.edu.au, pingxiong@znufe.edu.cn, yc@whpu.edu.cn

Abstract. *Tagging recommender system* allows Internet users to annotate resources with personalized tags and provides users the freedom to obtain recommendations. However, It is usually confronted with serious privacy concerns, because adversaries may re-identify a user and her/his sensitive tags with only a little background information. This paper proposes a privacy preserving tagging release algorithm, *PriTop*, which is designed to protect users under the notion of *differential privacy*. The proposed *PriTop* algorithm includes three privacy preserving operations: *Private Topic Model Generation* structures the uncontrolled tags, *Private Weight Perturbation* adds *Laplace* noise into the weights to hide the numbers of tags; while *Private Tag Selection* finally finds the most suitable replacement tags for the original tags. We present extensive experimental results on four real world datasets and results suggest the proposed *PriTop* algorithm can successfully retain the utility of the datasets while preserving privacy.

Keywords: Privacy Preserving, Differential Privacy, Recommendation, Tagging.

1 Introduction

The widespread success of social network web sites introduces a new concept called the *tagging recommender system* [9]. These social network web sites usually enable users to annotate resources with customized tags, which in turn facilitates the recommendation of resources. But the issue of privacy in the recommender process has generally been overlooked [11]. An adversary with background information may re-identify a particular user in a tagging dataset and obtain the user's historical tagging records [8]. How to preserve privacy in tagging recommender systems is an emerging issue that needs to be addressed.

Over the last decade, a variety of privacy preserving approaches have been proposed for traditional recommender systems [11]. For example, *cryptography* is used in the rating data for multi-party data sharing [15]. *Perturbation* adds noise

* Corresponding author.

V.S. Tseng et al. (Eds.): PAKDD 2014, Part I, LNAI 8443, pp. 557–568, 2014.

to the users' ratings before rating prediction [12], and *obfuscation* replaces a certain percentage of ratings with random values [1]. However, these approaches can hardly be applied in tagging recommender systems due to the semantic property of tags. To overcome the deficiency, the *tag suppression* method has recently been proposed to protect a user's privacy by modeling users' profiles and eliminating selected sensitive tags [11]. However, this method only releases an incomplete dataset that significantly affects the recommendation performance. Moreover, most existing approaches suffer from one common weakness: the privacy notions are weak and hard to prove theoretically, thus impairing the credibility of the final results. Accordingly, a more rigid privacy notion is needed.

Recently, *differential privacy* provides a strict privacy guarantee for individuals [6]. It applies a randomized mechanism suitable for both numeric and non-numeric values and has been proven effective in recommender systems [16,17]. This paper introduces *differential privacy* into tagging recommender systems, with the aim of preventing re-identification of users and avoiding the association of sensitive tags (e.g., healthcare tags) with a particular user. However, although these characteristics make *differential privacy* a promising method for tagging recommendation, there remain some barriers: First, the naive *differential privacy* mechanism only focuses on releasing statistical information that can barely retain the structure of the tagging dataset. this naive mechanism lists all the tags, counts the number and adds noise to the statistical output, but ignores the relationship among users, resources and tags. This simple statistical information is inadequate for recommendations; Second, Differential privacy introduces a large amount of noise due to the sparsity of the tagging dataset. For a dataset with millions of tags, the mechanism will result in a large magnitude of noise.

Both barriers imply the naive *differential privacy* mechanism can not be simply applied in tagging recommender systems. To overcome the first barrier, we generate a synthetic dataset retaining the relationship among tags, resources and users rather than releasing statistical information. The second barrier can be addressed by shrinking the randomized domain, because the noise will decrease when the randomized range is limited. The topic model method is a possible way to structure tags into groups and limit the randomized domain. Therefore, we propose a tailored *differential privacy* mechanism that optimizes the performance of recommendation with a fixed level of privacy. The contributions can be summarized as follows:

- We maintain an acceptable utility of the tagging dataset by designing a practical private tagging release algorithm, *PriTop*, with a rigid privacy guarantee.
- In *PriTop*, a novel private topic model-based method is proposed to structure the tags and to shrink the randomized domain. The effectiveness of the proposed method is verified by extensive experiments on real-world datasets.
- With *differential privacy* composition properties, a theoretical privacy and utility analysis confirms an improved trade-off between privacy and utility.

2 Preliminaries and Related Work

Let G be a dataset to be protected, two datasets G and G' are *neighboring datasets* if they differ in only one record. *Differential privacy* provides a randomization mechanism $\mathcal{M}$ to mask the difference between the *neighboring datasets* [6]. We use $\widehat{G}$ to represent the synthetic dataset after applying $\mathcal{M}$.

In tagging recommendation, dataset G contains users $U = \{u_1, u_2, \cdots\}$, resources $R = \{r_1, r_2, \cdots\}$ and tags $T = \{t_1, t_2, \cdots\}$. For a particular user $u_a \in U$ and a resource $r_b \in R$, $T(u_a, r_b)$ represents all tags flagged by the u_a on r_b, and use $T(u_a)$ to denote all tags utilized by u_a. The recommended tags for u_a on a given resource r are represented by $T^p(u_a, r)$. A user u_a's profile $P(u_a) = < T(u_a), W(u_a) >$ is usually modeled by his tagging records, including tag's names $T(u_a) = \{t_1, ..., t_{|T(u_a)|}\}$ and weights $W(u_a) = \{w_1, ..., w_{|T(u_a)|}\}$ [11].

Differential privacy acquires the intuition that releasing an aggregated report should not reveal too much information about any individual in the dataset [6].

Definition 1 (*ϵ-Differential Privacy*). *A randomized mechanism $\mathcal{M}$ gives* ϵ-differential privacy *if for every set of outcomes Ω, $\mathcal{M}$ satisfies: $Pr[\mathcal{M}(G) \in \Omega] \leq \exp(\epsilon) \cdot Pr[\mathcal{M}(G') \in \Omega]$, where ϵ is the privacy budget.*

$\mathcal{M}$ is associated with the *sensitivity* [7], which measures the maximal change of the query f when removing one record from G.

Definition 2 (*Sensitivity*). *For $f : G \rightarrow \mathbb{R}$, the* sensitivity *of f is defined as $\Delta f = \max_{G,G'} ||f(G) - f(G')||_1$,*

Two mechanisms are utilized in *differential privacy*: the *Laplace* mechanism and the *Exponential* mechanism. The *Laplace* mechanism is suitable for numeric output and adds controlled noise to the outcome of a query [7]:

Definition 3 (*Laplace* Mechanism). *Given a function $f : G \rightarrow \mathbb{R}^d$, the mechanism, $\mathcal{M}(R) = f(R) + Laplace(\frac{\Delta f}{\epsilon})^d$, provides the* ϵ-differential privacy.

The *Exponential* mechanism focuses on non-numeric queries and pairs with an application dependent *score function* $q(G, \psi)$, which represents how good an output scheme ψ is for dataset G:

Definition 4 (*Exponential* Mechanism). *[10] An Exponential mechanism $\mathcal{M}$ is* ϵ-differential privacy *if: $\mathcal{M}(G) = \{return\ \psi \propto \exp(\frac{\epsilon q(G,\psi)}{2\Delta q})\}$.*

Privacy violations in recommender systems have been well studied since 2001. The first study concerned with this issue was undertaken by Ramakrishnan et al. [13]. They claimed users who rated items across disjointed domains could face a privacy risk through statistical database queries. Recently, Calandrino et al. [4] presented a more serious privacy violation. By observing temporal changes in the public outputs of a recommender system, they inferred a particular user's historical rating and behavior with background information.

Several traditional privacy preserving methods have been employed in CF, including *cryptographic* [5,15], *perturbation* [12] and *obfuscation* [1]. *Cryptographic* is suitable for multiple parties but induces extra computational cost [5,15]; *Obfuscation* is easy to understand and to implement, but the utility will decrease significantly [1]. *Perturbation* preserves high level of privacy by adding noise to the original dataset, but the magnitude of noise is hard to control [12].

The privacy in tagging recommendation systems is more complicated due to its unique structure and semantic content. Parra-Arnau et al. [11] proposed the *tag suppression* by eliminating sensitive tags from users' profile. They applied a clustering method to structure tags and suppressed the less represented ones. This approach only releases an incomplete dataset, and sensitive tags are subjective. When publicly sharing the dataset, users still have the potential to be identified. The privacy issue in tagging recommender systems remains largely unexplored, and we attempt to fill this void in this paper.

3 Private Tagging Release

In a tagging dataset G, users' profile is represented as $\mathbf{P} = \{P(u_1), ..., P(u_{|U|})\}$. *Differential privacy* assumes all tags have probabilities to appear in $P(u_a)$. Suppose tags in $P(u_a)$ are represented by $T(u) = \{t_1, ..., t_{|T|}\}$ and weights are denoted as $W(u_a) = \{w_1, ..., w_{|T|}\}$, where $w_i = 0$ indicates that t_i is unused. *Differential privacy* will add noise to the weight $W(u_a)$ and release a noisy profile $\widehat{P}(u_a) =< T(u_a), \widehat{W}(u_a) >$. However, this process will introduce large noise because lots of weights will change from zero to a positive value. To reduce the noise is to shrink the randomized domain, which refers to the diminished number of zero weights in the profile. Accordingly, we structure the tags into K *topics* $Z = \{z_1, ...z_K\}$ and define a *topic-based* profile $P_z(u_a) =< T_z(u_a), W_z(u_a) >$, where $T_z(u_a) = \{T_{z_1}(u_a), ..., T_{z_K}(u_a)\}$ represents tags in each topic and $W_z(u_a) = \{w_{z_1}(u_a), ..., w_{z_K}(u_a)\}$ is the frequency of tags. Compared to $W(u_a)$, $W_z(u_a)$ is less sparse. The total noise added will significantly diminish.

In this section, we propose a ***Pri**vate **Top**ic-based Tagging Release* (PriTop) algorithm to publish users' profiles by masking their exact tags and weights under *differential privacy.* As described in Alg. 1, three private operations are involved: *Private Topic Model Generation* creates multiple private topics by masking the topic distribution on tags. *Topic Weight Perturbation* masks the weights of tags to prevent inferring how many tags a user has annotated on a topic. *Private Tag Selection* uses privately selected tags replace the original tags.

3.1 Private Topic Model Generation

This operation categorizes tags into topics to eliminate the randomization domain. We introduce *differential privacy* to *Latent Dirichlet Allocation* (LDA) [2] to generate a private LDA model, which is constructed in three steps: *LDA Model Construction*, *Private Model Generation* and *Topic-based Profile Generation.*

LDA Model Construction. The first step constructs the LDA model by *Gibbs Sampling* [14]. In this model, a resource is considered as a document and a tag is

Algorithm 1. *Private Topic-based Tagging Release (PriTop)* Algorithm

Require: G, privacy parameter ϵ, K.
Ensure: $\widehat{G}$

1. Divided privacy budget into $\epsilon/2$, $\epsilon/4$ and $\epsilon/4$;
2. *Private Topic Generation*: create topic-based user profiles $P(u_a)$ based on the private topic model with $\epsilon/2$ privacy budget;

for each user u_a **do**

3. *Topic Weight Perturbation*: add *Laplace* noise to the weights with $\epsilon/4$;

for each topic z_k in $P(u_a)$ **do**

4. *Private Tag Selection*: Select tags according to the $\widehat{W}(u_a)$ with $\epsilon/4$;

end for

end for

5. Output $\widehat{G}$ for tagging recommendations;

interpreted as a word. Let $Z = \{z_1, ...z_K\}$ be a group of topics, Eq. 1 represents a standard LDA model to specify the distribution over tag t.

$$Pr(t|r) = \sum_{l=1}^{K} Pr(t|z_l)Pr(z_l|r) \tag{1}$$

where $Pr(t|z_l)$ is the probability of tag t under a topic z_l and $Pr(z_l|r)$ is the probability of sampling a tag from topic z in the resource r.

To estimate topic-tag distribution $Pr(t|z)$ and the resource-topic distribution $Pr(z|r)$ in Eq. 1, *Gibbs Sampling* iterates multiple times over each tag t of resource r and samples the new topic z for the tag based on the posterior probability $Pr(z|t_i, r, Z_{-i})$ by Eq. 2 until the model converges.

$$Pr(z|t_i, r, Z_{-i}) \propto \frac{C_{tK}^{TK} + \beta}{\sum_{t_i}^{|T|} C_{t_iK}^{TK} + |T|\beta} \frac{C_{rk}^{RK} + \alpha}{\sum_{K=1}^{K} C_{r_iK}^{RK} + K\alpha} \tag{2}$$

where C^{TK} is the count of topic-tag assignments and C^{RK} counts the resource-topic assignments. Z_{-i} represents topic-tag assignment and resource-topic assignment except the current z for t_i. α and β are parameters of Dirichlet priors. Simultaneously, the evaluation on $Pr(t|z)$ and $Pr(z|r)$ is formulated as follows:

$$Pr(t|z) = \frac{C_{tk}^{TK} + \beta}{\sum_{t_i}^{|T|} C_{t_iK}^{TK} + |T|\beta}, Pr(z|r) = \frac{C_{rK}^{RK} + \alpha}{\sum_{k=1}^{K} C_{r_iK}^{RK} + K\alpha}$$

After converging, the LDA model is generated by $Pr(z|t, r)$, $P(t|z)$ and $P(z|r)$.

Private Model Generation. The second step adds *Laplace* noise to the final counts in the LDA model. There are four difference counts in Eq. 2. If we changed the topic assignment on current t_i, the C_{tK}^{TK} will decrease by 1 and $\sum_{t_i}^{|T|} C_{t_iK}^{TK}$ will increase by one. Similarly, if the C_{rK}^{RK} decreases by 1, the $\sum_{K=1}^{K} C_{r_iK}^{RK}$ will increase by 1 accordingly. So we sample two groups of *Laplace* noise and add them to four count parameters. The new $\widehat{Pr}(z|t,r)$ is evaluated by Eq. 3:

$$\widehat{Pr}(z|t,r) \propto \frac{C_{tK}^{TK} + \eta_1 + \beta}{\sum_{t_i}^{|T|} C_{t_i K}^{TK} - \eta_1 + |T|\beta} \frac{C_{rK}^{RK} + \eta_2 + \alpha}{\sum_{K=1}^{K} C_{r_i k}^{RK} - \eta_2 + K\alpha} \tag{3}$$

where η_1 and η_2 are both sampled from $Laplace(\frac{2}{\epsilon})$ with the *sensitivity* as 1.

Topic-based Profile Generation. The third step creates topic-based user profiles. For each user with tags $T(u_a) = \{t_1, ..., t_{|T(u_a)|}\}$ and related resources $R(u_a) = \{r_1, ..., r_{|R(u_a)|}\}$, each tag can be assigned to a particular topic $z_l \in Z$ according to the $\widehat{Pr}(z|t,r)$. So the user profile can be represented by a topic-based $P_z(u_a) =< T_z(u_a), W_z(u_a) >$ with the weight $W_z(u_a) = \{w_1(u_a), ..., w_K(u_a)\}$.

3.2 Topic Weight Perturbation

After generating $P_z(u_a)$, we will add *Laplace* noise to mask the weights of tags in each topic: $\widehat{W}_z(u_a) = W_z(u_a) + Laplace(\frac{4}{\epsilon})^K$. Noise implies the revision of the list $T_z(u_a)$. Positive noise indicates new tags being added, while negative one indicates tags being deleted from the list. For positive noise in the topic z_l, the operation will choose the tags with the highest probability in the current topic z_j according to the $Pr(t|z)$. For negative noise, the operation will delete the tag with the lowest probability in the current topic z_l.

$$\widetilde{T}_{z_l}(u_a) = T_{z_l}(u_a) + t_{new}, \widetilde{T}_{z_l}(u_a) = T_{z_l}(u_a) - t_{delete} \tag{4}$$

where $t_{new} = \max_{i=1}^{|T|} Pr(t_i|z_l)$ and $t_{delete} = \min_{i=1}^{|T|} Pr(t_i|z_l)$.

After perturbation, we use $\widetilde{P}_z(u_a) =< \widetilde{T}_z(u_a), \widehat{W}_z(u_a) >$ to represent the noisy topic-based user profile. However, the $\widetilde{P}_z(u_a)$ still has the high probability to be re-identified because it retains a major part of the original tags. The next operation will replace all tags in $\widetilde{T}(u_a)$ to preserve privacy.

3.3 Private Tag Selection

Private Tag Selection adopts the *Exponential* mechanism to privately select tags from a list of candidates. Specifically, for a particular tag t_i, the operation first locates the topic z_l to which it belongs and all tags in $\hat{T}_{z_l}(u_a)$ are then included in a candidate list I. Each tag in I is associated with a probability based on a *score function* and the *sensitivity* of the function. The selection of tags is performed based on the allocated probabilities.

The *score function* is defined as the *Jensen-shannon* (JS) divergence between $Pr(z|t_i = t_i)$ and $Pr(z|t_i = t_j)$. Because the JS divergence is bounded by 1, the score function q for a target tag t_i is defined as $q_i(I, t_j) = (1 - D_{JS}(Pr_i||Pr_j))$, where $t_j \in I$ are the candidate tags for replacement and D_{JS} refers to JS divergence. The *sensitivity* for q is measured by the maximal distance of two tags, which is 1. Based on the *score function* and *sensitivity*, the probability arranged to each tags t_j is computed by Eq. 5 with the privacy budget $\frac{\epsilon}{4}$.

$$Pr_{t_j \in I}(t_j) = \exp\left(\frac{\epsilon \cdot q_i(I, t_j)}{8}\right) / \sum_{j \in z_l} \exp\left(\frac{\epsilon \cdot q_i(I, t_j)}{8}\right). \quad (5)$$

where z_l is the topic in which t_j belongs to.

4 Algorithm Analysis

4.1 Privacy Analysis

To analyze the privacy guarantee, we apply two composition properties of differential privacy [10]. The *sequential composition* accumulates ϵ of each step when a series of private analysis is performed *sequentially* on a dataset. The *parallel composition* ensures the maximal ϵ when each private step is applied on disjointed subsets of the dataset. The *PriTop* algorithm contains three private operations and the ϵ is consequently divided into three pieces: $\frac{\epsilon}{2}$, $\frac{\epsilon}{4}$ and $\frac{\epsilon}{4}$, respectively.

- *Private Topic Model Generation* is performed on the whole dataset with the $\frac{\epsilon}{2}$. According to *sequential composition*, it preserves $\frac{\epsilon}{2}$-*differential privacy.*
- *Topic Weight Perturbation* preserves $\frac{\epsilon}{4}-$ *differential privacy* for each user. As a user's profile is independent, according to *parallel composition*, it preserves $\frac{\epsilon}{4}$-*differential privacy.*
- *Private Tag Selection* processes the *Exponential* mechanism successively. For a user u, each selection is performed on the individual tags, according to *sequential composition*, each user guarantees $\frac{\epsilon}{4}$-*differential privacy.* Similar to the *Topic Weight Perturbation*, every user can be considered as subsets. Thus, the *Private Tag Selection* guarantees $\frac{\epsilon}{4}$-*differential privacy.*

Consequently, the proposed *PriTop* algorithm preserves ϵ-*differential privacy.*

4.2 Utility Analysis

Given a target user u_a, the utility level of the proposed *PriTop* algorithm is highly dependent on the distance between $P(u_a)$ and $\widehat{P}(u_a)$, which is referred to as *semantic loss* [11]: $SLoss = \frac{1}{|U|} \sum_{u \in U} (\frac{\sum_{t \in P(u_a)} d(t, \widehat{t})}{\max d \cdot |T(u)|})$, where $\widehat{t}$ is the new tag replacing the tag t. If we consider each private step as a query f, we then apply a utility definition in *differential privacy* suggested by Blum et al [3]. Accordingly, we demonstrate the *SLoss* is bounded by a certain value α with a high probability.

Definition 5 ((α,δ)-usefulness). *A mechanism $\mathcal{M}$ is (α,δ)-useful for a set of query F, if with probability $1-\delta$, for every query $f \in F$ and every dataset G, for $\widehat{G} = \mathcal{M}(G)$, we have $\max_{f \in F} |f(\widehat{G}) - f(G)| \leq \alpha$, where F is a group of queries.*

Theorem 41 *For any user $u \in U$, for all $\delta > 0$, with probability at least $1-\delta$, the $SLoss_1$ of the user in the perturbation is less than α. When $|T(u)| \geq \frac{K \cdot \exp(\frac{-\epsilon\alpha_a}{4})}{\delta}$, the perturbation operation is satisfied with (α, δ)-useful.*

Proof. The perturbation adds *Laplace* noise with $\epsilon/4$ to the weight. According to the property of *Laplace*(b): $Pr(|\gamma| > t) = \exp(-\frac{t}{b})$, we have $Pr(SLoss_1 > \alpha_a) = \frac{K \cdot d(t_{ai}, \widehat{t}_{ai})}{\max d|T(u_a)|} \exp(-\frac{\epsilon\alpha_a}{4})$. As the perturbation step adds new tags or delete tags, the $d(t_{ai}, \widehat{t}_{ai})$ will be less than 1, we obtain the evaluation on the $SLoss_1$: $Pr(SLoss_1 < \alpha_a) \leq 1 - \frac{K \cdot \exp(-\frac{\epsilon\alpha_a}{4})}{|T(u_a)|}$. Let $1 - \frac{K \cdot \exp(-\frac{\epsilon\alpha_a}{4})}{|T(u_a)|} \geq 1 - \delta$, Thus $|T(u_a)| \geq \frac{K \cdot \exp(\frac{-\epsilon\alpha_a}{4})}{\delta}$. The average semantic loss for all the users is less than the maximal value, $\alpha = \max_{u_a \in U} \alpha_a$, we have $|T(u)| \geq \frac{K \cdot \exp(\frac{-\epsilon\alpha}{4})}{\delta}$.

The theorem 41 reveals the *semantic loss* of perturbation depends on the number of tags a user has. More tags results in a lower *semantic loss.*

Theorem 42 *For any user $u \in U$, for all $\delta > 0$, with probability at least $1 - \delta$, the $SLoss_2$ of the user in the private selection is less than α. When $Q \leq \frac{\exp(\frac{\epsilon}{8})}{1-\delta\alpha}$, where Q is the normalization factor that depends on the topic that $t \in T(u)$ belongs to, the private selection operation is satisfied with (α, δ)-useful.*

Proof. According to Marlkov's inequality, we get $Pr(SLoss_2 > \alpha_a) \leq \frac{E(SLoss_2)}{\alpha_a}$. For each tag t_{ai} in $\widehat{P}_a$, the probability of 'unchange' in the private selection is proportional to $\frac{\exp(\frac{\epsilon}{8})}{Q_i}$, where Q_i is the normalization factor depending on the topic t_{ai} belongs to. We then obtain $E(SLoss_2) = \sum_{t_i \in T(u_a)} \frac{d(t_{ai}, \widehat{t}_{ai})}{\max d|T(u_a)|}(1 - \frac{\exp(\frac{\epsilon}{8})}{Q_i})$ and estimate $SLoss_2$ as $Pr(SLoss_2 > \alpha_a) \leq \frac{\sum_{t_i \in T(u_a)} d(t_{ai}, \widehat{t}_{ai})(1 - \frac{\exp(\frac{\epsilon}{8})}{Q_i})}{|T(u_a)|\alpha_a}$. When $d(t_{ai}, \widehat{t}_{ai}) = 1$ and $Q = \max Q_i$, it is simplified as $Pr(SLoss_2 \leq \alpha_a) \geq 1 - \frac{1 - \frac{1}{Q}\exp(\frac{\epsilon}{8})}{\alpha_a}$. Let $1 - \frac{1 - \frac{1}{Q}\exp(\frac{\epsilon}{8})}{\alpha_a} \geq 1 - \delta$, $Q \leq \frac{\exp(\frac{\epsilon}{8})}{1 - \delta\alpha_a}$. Similar to the proof of theorem 41, We obtain $Q \leq \frac{\exp(\frac{\epsilon}{8})}{1-\delta\alpha}$, where $Q_i = \sum_{j \in z_l} \exp\left(\frac{\epsilon \cdot d(t_i, t_j)}{8}\right)$.

The theorem 42 shows the *semantic loss* of private selection mainly depends on the ϵ and Q_i, which measures by the total distance inside topic z to which t_i belongs. The shorter distance leads to a smaller Q_i and less *semantic loss.*

5 Experiment and Analysis

We conduct experiment on three datasets: *Del.icio.us*, *MovieLens* and *Last.fm*. *Del.icio.us* dataset was retrieved from the *Del.icio.us* website by the *Distributed Artificial Intelligence Laboratory* (DAI-Labor), and includes around 132 million resources and 950,000 users. We extracted a subset with 3,000 users, 34,212 bookmarks and 12,183 tags. *MovieLens* and *Last.fm* datasets were obtained from *HetRec 2011*. All datasets are structured as triples (*user*, *resource*, *tag*), and filtered by removing added tags like "imported", "public", etc.

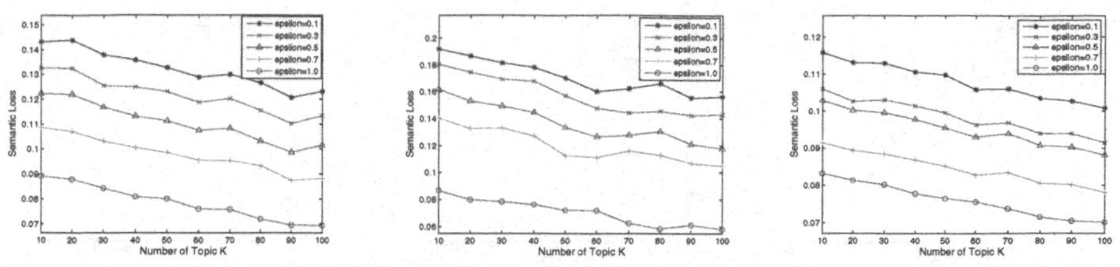

(a) *Semantic Loss* in *Del.icio.us* (b) *Semantic Loss* in *MovieLens* (c) *Semantic Loss* in *Last.fm*

Fig. 1. *Semantic Loss* on Different Datasets

5.1 Semantic Loss Analysis

To maintain consistency with previous research, we compare the *semantic loss* of *PriTop* with *tag suppression* [11]. For the *PriTop* algorithm, we selected $\epsilon = 0.1, 0.3, 0.5, 0.7$ and 1.0 to represent different privacy levels and the number of topic K varies from 10 to 100 with a step of 10. In *tag suppression* [11], the *semantic loss* exhibits a linear relationship with the *eliminate parameter* σ. When we choose the representative value $\sigma = 0.8$, the *sematic loss* is 0.2.

It can be observed from Fig. 1 that the *semantic loss* of the *PriTop* algorithm in a variety of datasets was less than 0.2 with different privacy budgets, which indicates that *PriTop* outperforms *tag suppression* on all configurations. Specifically, the *PriTop* obtains a considerably lower *semantic loss* when $\epsilon = 1$. For example, in Fig. 1a, when $K = 90$ and $\epsilon = 1$, the *semantic loss* is 0.0767, which is 62% lower than *tag suppression* with $SLoss = 0.2$. This trend is retained when K equals other values and in other figures, such as Fig. 1b and 1c. All figures show that *PriTop* obtains a stable *semantic loss* at a lower level, and retains more utility than *tag suppression*. This is because *PriTop* retains the relationship between tags and resources, and makes the profiles of users meaningful.

5.2 Performance of Tagging Recommendation

To investigates the effectiveness of *PriTop* in the context of tagging recommendations, we apply a state-of-the-art tagging recommender system, *FolkRank* [9], to measure the degradation of privacy preserving recommendations. We use *Recall* to quantify the performance and N is the number of recommended tags. The following experiments compare the *PriTop* with *tag suppression* with N varies from 1 to 10. For *PriTop*, we chose $K = 100$, and test the performance when $\epsilon = 1$ and 0.5. For *tag suppression*, we fix $\sigma = 0.8$ and 0.6, corresponding to suppression rates of 0.2 and 0.4, respectively.

Fig. 2 presents the recall of recommendation results. It is observed that the proposed *PriTop* algorithm significantly outperforms the *tag suppression* method on both privacy budgets. Specifically, as shown in Fig. 2a, when $N = 1$, *PriTop* achieves a *recall* at 0.0704 with the $\epsilon = 1$ which outperforms the result from the *tag suppression* with $\sigma = 0.6$, 0.0407, by 42.19%. This trend is retained

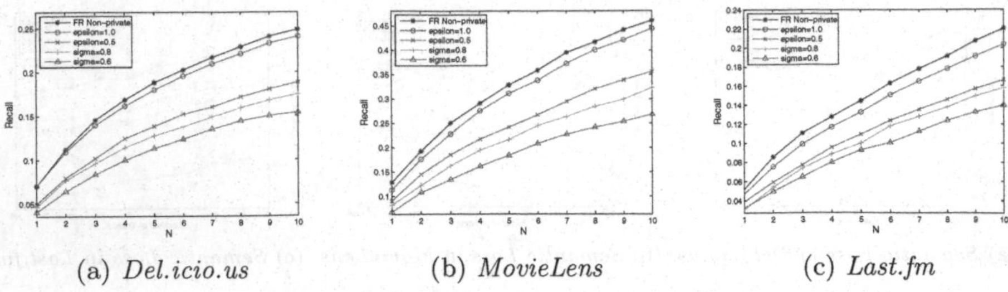

(a) *Del.icio.us* (b) *MovieLens* (c) *Last.fm*

Fig. 2. *FolkRank* Recall Result

as the increasing of N. When $N = 5$, *PriTop* achieves a *recall* at 0.1799 with the $\epsilon = 1$ which outperforms the result from the *tag suppression* by 37.19% when $\sigma = 0.6$, 0.113. When N reaches 10, the *PriTop* still retains 36.09% higher on *recall* than *tag suppression.* Even we choose the lower privacy budget with $\epsilon = 0.5$ and a higher *eliminate* parameter *sigma* $= 0.8$, the improvement is still significant. The *PriTop* has a *recall* of 0.1382, which is also 7.67% higher than *tag suppression* with a *recall* of 0.1276. The improvement of *PriTop* is more obvious when $N = 10$. It achieves *recalls* of 0.1882 and 0.2408 when $\epsilon = 1$ and $\epsilon = 0.5$, respectively. But *tag suppression* only achieves *recalls* of 0.1538 and 0.1881 with $\sigma = 0.6$ and $\sigma = 0.8$. Similar trends can also be observed in Fig. 2b and 2c. In the *MovieLens* dataset, when $N = 10$ and $\epsilon = 1.0$, the recall of *PriTop* is 0.4445, which is 27.33% higher than *tag suppression* with $\sigma = 0.8$. With the same configuration, *PriTop* is 22.43% and 25.22% higher than *tag suppression* in *Last.fm* and *Bibsonomy* datasets. The experimental results show the *PriTop* algorithm outperforms *tag suppression* in variety of N, which implies that *PriTop* can retain more useful information for recommendations than simply deleting the tags. In addition, the performance of *PriTop* is very close to the *non-private* baseline. For example in Fig. 2a, when $\epsilon = 1$, the *recall* of the *De.licio.us* dataset is 0.2408, which is only 3.00% lower than the non-private recommender result. Other datasets show the same trend. As shown in Fig. 2b and 2c, with the same configuration, the *PriTop* result is 3.62% lower than the non-private result on the *MovieLens* dataset, and 7.58% lower on the *Last.fm* dataset. The results indicate that *PriTop* algorithm achieves the privacy preserving objective while retaining a high accuracy of recommendations.

To show the statistical effectiveness of *PriTop*, we apply a paired t test (with a 95% confidence) to examine the difference on the performance of *PriTop* with $\epsilon = 1.0$ and *tag suppression* with $\sigma = 0.2$. The statistics for results are shown in Table 1. All t values are greater than 6 and all p values are less than 0.0001, thus indicating improvement on *recall* are statistically significant.

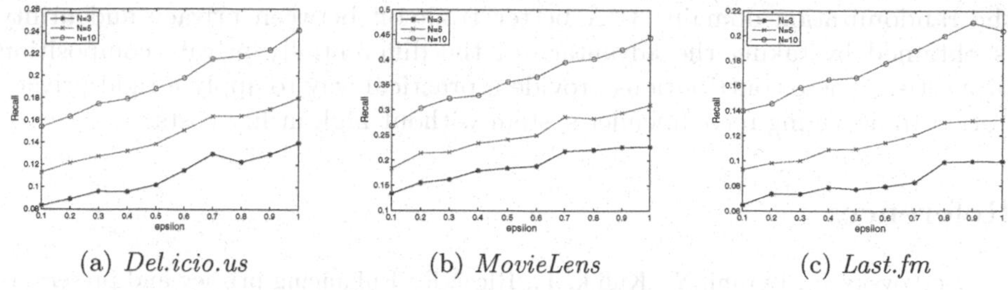

(a) *Del.icio.us* (b) *MovieLens* (c) *Last.fm*

Fig. 3. Impact of Privacy Budget in *FolkRank* Recall Result

Table 1. Paired-t-test betwee *PriTop* and *Tag Suppression*

		df	*t*	*p*-value
De.licio.us	Recall	9	11.3276	< 0.0001
MovieLens	Recall	9	9.0957	< 0.0001
Last.fm	Recall	9	10.7546	< 0.0001

5.3 Impact of Privacy Budget

In the context of *differential privacy*, the lower ϵ represents a higher privacy level. To achieve a comprehensive examination of *PriTop*, we evaluate the performance of recommendation under diverse privacy levels.

Fig. 3 shows the *recall* on the three datasets. It presents the recommendation performance achieved by *PriTop* when the privacy budget ϵ varies from 0.1 to 1 with a step of 0.1. It is clear the *recall* of tag recommendations is significantly affected by the required privacy budget. The *recall* increases as ϵ increases. For example, as plotted in Fig. 3a on the *Del.icio.us* dataset, when $N = 10$, *PriTop* achieves a *recall* at 0.1538 with $\epsilon = 0.1$ and 0.2408 with $\epsilon = 1$. The reason is the privacy and utility issues are two opposite components of the datasets. We have to sacrifice the utility to obtain the privacy, therefore our purpose is to obtain an optimal utility when fixing the privacy at an acceptable level.

As a summary, results on a real tagging recommender system confirm the practical effectiveness of the *PriTop* algorithm.

6 Conclusions

Privacy preserving is one of the most important aspects in recommender systems. However, when we introduce the *differential privacy*, the solution fails to retain the relationship among users, resources and tags; and introduces a large volume of noise, which significantly affects the recommendation performance.

This paper proposes an effective privacy tagging release algorithm *PriTop* with the following contributions: 1) We propose a private tagging release algorithm to protect users from being re-identified in a tagging dataset. 2) A private topic model is designed to reduce the magnitude of noise by shrinking the randomization domain. 3) A better trade-off between privacy and utility is obtained by taking the advantage of the differentially private composition properties. These contributions provide a practical way to apply a rigid privacy notion to a tagging recommender system without high utility costs.

References

1. Berkovsky, S., Eytani, Y., Kuflik, T., Ricci, F.: Enhancing privacy and preserving accuracy of a distributed collaborative filtering. In: RecSys (2007)
2. Blei, D.M., Ng, A.Y., Jordan, M.I.: Latent dirichlet allocation. The Journal of Machine Learning Research 3, 993–1022 (2003)
3. Blum, A., Ligett, K., Roth, A.: A learning theory approach to non-interactive database privacy. In: STOC, pp. 609–618 (2008)
4. Calandrino, J.A., Kilzer, A., Narayanan, A., Felten, E.W., Shmatikov, V.: "You might also like:" privacy risks of collaborative filtering. In: SP (2011)
5. Canny, J.: Collaborative filtering with privacy. In: S&P 2002, pp. 45–57. IEEE (2002)
6. Dwork, C.: A firm foundation for private data analysis. Commun. ACM 54(1), 86–95 (2011)
7. Dwork, C., McSherry, F., Nissim, K., Smith, A.: Calibrating noise to sensitivity in private data analysis. In: Halevi, S., Rabin, T. (eds.) TCC 2006. LNCS, vol. 3876, pp. 265–284. Springer, Heidelberg (2006)
8. Fung, B.C.M., Wang, K., Chen, R., Yu, P.S.: Privacy-preserving data publishing: A survey of recent developments. ACM Comput. Surv. (2010)
9. Jäschke, R., Marinho, L., Hotho, A., Schmidt-Thieme, L., Stumme, G.: Tag recommendations in folksonomies. In: Kok, J.N., Koronacki, J., Lopez de Mantaras, R., Matwin, S., Mladenič, D., Skowron, A. (eds.) PKDD 2007. LNCS (LNAI), vol. 4702, pp. 506–514. Springer, Heidelberg (2007)
10. McSherry, F., Talwar, K.: Mechanism design via differential privacy. In: FOCS 2007, pp. 94–103 (2007)
11. Parra-Arnau, J., Perego, A., Ferrari, E., Forne, J., Rebollo-Monedero, D.: Privacy-preserving enhanced collaborative tagging. IEEE Transactions on Knowledge and Data Engineering 99(PrePrints), 1 (2013)
12. Polat, H., Du, W.: ICDM, pp. 625–628 (November 2003)
13. Ramakrishnan, N., Keller, B.J., Mirza, B.J., Grama, A.Y., Karypis, G.: Privacy risks in recommender systems. IEEE Internet Computing 5(6), 54–62 (2001)
14. Steyvers, M., Griffiths, T.: Probabilistic topic models. In: Handbook of Latent Semantic Analysis, vol. 427(7), pp. 424–440 (2007)
15. Zhan, J., Hsieh, C.-L., Wang, I.-C., Hsu, T.S., Liau, C.-J., Wang, D.-W.: Privacy-preserving collaborative recommender systems. IEEE Transactions on Systems, Man, and Cybernetics, Part C 40(4), 472–476 (2010)
16. Zhu, T., Li, G., Ren, Y., Zhou, W., Xiong, P.: Differential privacy for neighborhood-based collaborative filtering. In: ASONAM (2013)
17. Zhu, T., Li, G., Ren, Y., Zhou, W., Xiong, P.: Privacy preserving for tagging recommender systems. In: The 2013 IEEE/WIC/ACM International Conference on Web Intelligence (2013)

Matrix Factorization without User Data Retention

David Vallet, Arik Friedman, and Shlomo Berkovsky

NICTA, Australia

{david.vallet,arik.friedman,shlomo.berkovsky}@nicta.com.au

Abstract. Recommender systems often rely on a centralized storage of user data and there is a growing concern about the misuse of the data. As recent studies have demonstrated, sensitive information could be inferred from user ratings stored by recommender systems. This work presents a novel semi-decentralized variant of the widely-used Matrix Factorization (MF) technique. The proposed approach eliminates the need for retaining user data, such that neither user ratings nor latent user vectors are stored by the recommender. Experimental evaluation indicates that the performance of the proposed approach is close to that of standard MF, and that the gap between the two diminishes as more user data becomes available. Our work paves the way to a new type of MF recommenders, which avoid user data retention, yet are capable of achieving accuracy similar to that of the state-of-the-art recommenders.

Keywords: data retention, matrix factorization, recommender systems.

1 Introduction

An increasing number of users are nowadays facing the information deluge, which hinders the discovery of content and services of interest. Recommender systems address this problem through the provision of personalized recommendations to users. Personalization, however, entails another problem: potential leak of sensitive user data [10]. Indeed, accurate recommendations inherently require a large volume of personal user data, which may concern some users to the extent that they refrain from using the recommender despite its benefits [1].

While encryption technologies can mitigate the risk of eavesdropping when data is transferred to the recommender system, the retention of the data by the recommender exposes the users to additional privacy risks. Even seemingly innocuous preference information, like that typically stored by recommender systems, introduces privacy risks, as demonstrated in scenarios such as 1) inference of undisclosed sensitive personal information, e.g., relationship status or sexual orientation, through access to public user data [14,20,4,9]; 2) trading or brokering of personal data to an untrusted third party;[1] or 3) disclosure of sensitive personal information to law enforcement agencies, even without a warrant [6].

[1] `http://gizmodo.com/5991070/big-data-brokers-they-know-everything-about-you-and-sell-it-to-the-highest-bidder`

V.S. Tseng et al. (Eds.): PAKDD 2014, Part I, LNAI 8443, pp. 569–580, 2014.

These issues fuel a growing concern that motivates keeping personal user data away from a centralized storage, thereby mitigating data disclosure risks.

In this work, we leverage MF techniques [8] to investigate the use of a semi-decentralized recommender, which retains neither the user ratings nor the latent user vectors. Instead, user ratings are stored on the user side, e.g., on a device owned and carried by the user or on a secure cloud-based service with a personal account. When a user rates a new item or a recommendation is required, the user's ratings are provided to the MF recommender, which derives the latent user vector and updates the latent item matrix. Then, item ratings are predicted and the recommendations are generated. When the interaction with the user is concluded, the recommender discards both the ratings and the latent user vector. This semi-decentralized setting mitigates the above risks by not retaining permanently any user-specific data that could be exploited for the inference of sensitive personal information, while allowing service providers to retain content-specific data, crucial for the provision of recommendations.

We carried out an experimental evaluation, considering scenarios in which the recommender transitions from the centralized to the semi-decentralized model, as well as using the latter from the outset. The results demonstrate that the semi-decentralized model achieves accuracy that is close to that of the centralized recommender, and the gap between the two diminishes over time, as more ratings become available to the system. To summarize, the main contributions of this paper are two-fold. Firstly, we propose a novel semi-decentralized variant of MF without user data retention. Secondly, we evaluate several variants of the algorithm and perform a time-based simulation that demonstrates the applicability of the algorithm to a large-scale real-life MF recommender.

2 Related Work

Several studies focused on the privacy implications of access to user ratings. In the Netflix Prize competition,[2] data anonymization was used to protect user ratings. However, Narayanan and Shmatikov [14] demonstrated that it was possible to de-anonymize the ratings with a little background knowledge, and that sensitive and private information could be inferred from these ratings. In a later work, Calandrino et al. [2] demonstrated that it was possible to infer which items had been rated by a user by sampling changes to aggregated publicly available item recommendations. Weinsberg et al. [20] showed that demographic data could be derived from user ratings. The experiments showed that gender and other sensitive private information (age, ethnicity, political orientation) could be accurately inferred. Chaabane et al. [4] demonstrated a similar problem in information pertaining to users' interests. By accessing supposedly harmless music interests of users, the authors were able to infer sensitive undisclosed information about user gender, age, country, and relationship status. Similarly, Kosinski et al. [9] were able to uncover sensitive information referring to sexual orientation, political alignment, religion, and race, based on Facebook "likes."

[2] `http://www.netflixprize.com/`

Decentralization of personal user data was previously studied in the context of online advertising, which entails gathering browsing and behavioral data. Adnostic [19] and Privad [7] are two privacy preserving systems, which offer advertisement services while storing private user data on the user side. This is achieved by pushing parts of the advertisement selection process to the client. However, this solution is inapplicable to MF and Collaborative Filtering recommenders, which are too computationally intensive to run on the client. Even if that was feasible, many recommenders would need to push to the user side additional information, such as item descriptions, similarity scores, or latent factors. Not only may this information be deemed too sensitive by service providers, it could also be used maliciously, e.g., for inferring other users' data [2].

Our semi-decentralized variant of MF, which retains no user data, shares similarities with prior works on online MF [16,15,18,5,12]. These works leveraged incremental re-training of user/item vectors (e.g., through *folding-in* [16]) when new ratings become available, to avoid the cost incurred by running the full factorization process, and improve system scalability and efficiency. In contrast to these approaches, we present a novel solution that relies neither on a centralized storage nor on the availability of past user ratings.

3 Semi-decentralized MF Recommender

In this section we present how a MF-based recommender can be modified to operate without retaining user ratings, profiles, or latent vectors.

3.1 Preliminaries and Notation

Consider a set U of n users, who assign ratings to a set I of m items. We denote by S the set of available ratings, where each element is a triplet (u, i, r_{ui}), and r_{ui} denotes the rating that user u assigned to item i. We denote by $S_v = \{(u, i, r_{ui}) \in S | u = v\}$ the set of known ratings of user v. The recommendations are generated by predicting the values of unknown ratings. MF achieves this by learning two low-rank matrices with d latent factors: $P_{n\times d}$ for users and $Q_{m\times d}$ for items, where a row p_u in P pertains to user u and a row q_i in Q pertains to item i. The predicted rating of user u for item i is then computed by $\hat{r}_{ui} = p_u q_i^\intercal$. Given the known ratings in S, the latent matrices P and Q are obtained by solving the optimization problem

$$\arg\min_{P,Q} J_S(P, Q), \quad (1)$$

where the loss function $J_S(P, Q)$ with regularization parameter γ is given by

$$J_S(P, Q) := \sum_{r_{ui}\in S} [(r_{ui} - p_u q_i^\intercal)^2 + \gamma(\|p_u\|^2 + \|q_i\|^2)] \ . \quad (2)$$

We consider the Stochastic Gradient Descent (SGD) technique for empirical risk minimization, with a learning rate λ [8]. In this process, P and Q are

Algorithm 1. GetRecommendations(S_u, Q)

Input:
S_u – ratings of user u
Q – the latent item matrix maintained by the recommender

User u sends S_u to the recommender
Given S_u and Q, the recommender solves $\arg\min_{p_u} J_{S_u}(p_u, Q)$ to derive p_u
The recommender predicts ratings $\hat{r}_{ui} = p_u q_i^{\intercal}$ and generates recommendations
The recommender discards S_u and p_u

modified for each $r_{ui} \in S$ using the following update rules, which are applied iteratively until both P and Q converge:

$$p_u := p_u + 2\lambda[(r_{ui} - p_u q_i^{\intercal})q_i - \gamma p_u] \tag{3}$$

$$q_i := q_i + 2\lambda[(r_{ui} - p_u q_i^{\intercal})p_u - \gamma q_i] \tag{4}$$

3.2 Maintaining Decentralized User Profiles

One way to avoid centralized retention of user profiles by a MF recommender is to off-load them to the user side. Predicting unknown scores $\hat{r}_{ui}$ for all unrated items of user u requires the latent user vector p_u and the latent item matrix Q. In a semi-decentralized setting, u can send the user vector p_u to the system, which holds the item matrix Q. The system then computes $\hat{r}_{ui}$ and provides the recommendations to u. Once the interaction with u is concluded, the system discards p_u, as it is not needed for recommendations for other users. This eliminates the need for a permanent centralized retention of all ratings of all users, and mitigates privacy concerns related to misuse of personal data "at rest".

One drawback of this setting is that the user's p_u may not reflect recent ratings provided by other users. In MF, p_u is affected indirectly by ratings of other users, as new ratings gathered by the recommender affect Q and, in turn, affect vectors of users, who previously rated those items. Fixing latent user vectors and keeping them on the user side hinders these updates of p_u, and can cripple the accuracy of the system. Folding-in techniques [16] circumvent this drawback, using an up-to-date matrix Q to re-evaluate the user vector p_u by solving

$$\arg\min_{p_u} J_{S_u}(p_u, Q) \ . \tag{5}$$

This optimization problem can be solved with the SGD technique by fixing Q and only applying update rule (3). Algorithm 1 shows the modified semi-decentralized recommendation process that recomputes p_u from S_u.

3.3 Maintaining the Item Matrix

In addition to generating recommendations, the system needs to update the latent item matrix Q based on new ratings. In a standard MF setting, the user

Algorithm 2. UpdateItem(S_u, r_{ui}, Q, γ, λ)

Input:
S_u – past ratings of user u
r_{ui} – new rating from user u
Q – the item-factor matrix maintained by the recommender
γ – regularizer
λ – learning rate

1: User u sends r_{ui} and S_u to the recommender
2: $S_u := S_u \cup r_{ui}$
3: Given S_u and Q, the recommender solves $\arg\min_{p_u} J_{S_u}(p_u, Q)$ to derive p_u.
4a: $q_i := q_i + 2\lambda[(r_{ui} - p_u q_i^\intercal)p_u - \gamma q_i]$ **OR** 4b: **for** *each* $r_{uj} \in S_u$ **do**
$q_j := q_j + 2\lambda[(r_{uj} - p_u q_j^\intercal)p_u - \gamma q_j]$
5: The system discards p_u and S_u.

and item vectors are typically updated simultaneously. But this cannot be done in the semi-decentralized setting, as no user ratings are retained. Alternatively, following the approaches outlined in [15,16], folding-in can be applied to update the item vectors by fixing the user matrix P, and retraining an item vector q_i through applying SGD and update rule (4) to all the available ratings. While effective, this iterative update of item factors is also inapplicable unless all the ratings for i are available, and another solution is needed.

Algorithm 2 presents a modified process for updating Q that leverages the fact that for each new rating r_{ui}, the update of q_i requires access only to the vector p_u of the user who provided the rating. First, the user u sends r_{ui} along with previous ratings in S_u to the system, which reconstructs p_u. Then, update rule (4) is applied to the newly added rating and, finally, the system keeps the updated latent item vector and discards both p_u and S_u.

We consider two variants of this scheme. In the first variant, denoted by *new*, we perform an update only with the newly added rating r_{ui}, i.e., apply the update in instruction 4a of Algorithm 2. In the second variant, denoted by *rated*, we take advantage of the availability of the entire set of ratings S_u and apply an iterative update rule in instruction 4b of Algorithm 2 to the item vectors of all the items that were rated by u. We note that unlike the *new* variant, which updates one latent vector for every new rating, the *rated* variant updates the vectors of all previously rated items, which may bias the latent item matrix Q towards ratings of highly active users, who provided many ratings. However, this bias can be mitigated, e.g., by weighting the update according to the number of ratings that a user provided.

Note that the semi-decentralized setting can be extended to support more complex variants of MF. For example, a common practice is to incorporate user bias $b_u = \sum_{r_{ui} \in S_u} r_{ui}/|S_u|$ to normalize user ratings. The user bias is discounted from the gathered ratings prior to computing P and Q, and re-introduced when the predictions are computed: $\hat{r_{u,i}} = b_u + p_u q_i^\intercal$. This bias can be stored at the user side, or be directly recomputed by the system from the gathered ratings, similarly to the user latent vector p_u.

4 Experimental Evaluation

In order to evaluate the proposed semi-decentralized variants of MF, we consider the following scenario. The recommender collects user ratings until a cut-off time t_s, after which it transitions to the semi-decentralized model. All the data collected before t_s is considered as the initial *static* data. Then, the recommender retains the latent item matrix Q_s derived at t_s and discards the latent user matrix P_s. As discussed in Section 3, from time t_s and onwards the recommender retains neither the user ratings nor the latent user vectors.

The item matrix Q, initialized to Q_s, is updated as outlined in Algorithm 2, when new ratings become available. We posit that the resulting Q will deviate from the hypothetical Q^* that could be derived if the recommender retained user ratings as in the centralized MF. That is, the proposed approach can only approximate Q^*, since not all user ratings are available when Q is updated. A "good" approach for updating Q will be one that generates predictions as close as possible to those of the centralized MF that continuously updates P and Q.

We denote by $\mathcal{R}_s$ the ratio between the number of ratings collected before time t_s, i.e., ratings used to derive Q_s, and the overall number of ratings processed in the evaluation. $\mathcal{R}_s$ represents the relative volume of the static initialization data. A special case of $\mathcal{R}_s = 0$ reflects a scenario in which the semi-decentralized recommender has no prior user data and starts from the outset with a randomly initialized Q.

In the evaluation, we use two public datasets: MovieLens and Netflix. Since the temporal information of MovieLens ratings was deemed unreliable (many ratings had the same timestamp), we use MovieLens to produce a random split of data and Netflix to produce a time-based split. Specifically, we use the Movielens 10M dataset to perform a 10-fold cross validation for randomly split non-overlapping training and test sets. For the Netflix dataset, we randomly sample 10% of users to obtain a comparable dataset of 10M ratings. For the time-based split we select the most recent 10% of ratings given by each user as the test set. We use RMSE to assess the performance of the proposed approach and the Wilcoxon's signed-rank test to validate statistical significance [17].

We tune the MF parameters in an offline evaluation that is omitted due to space limitation. We set the number of factors to $d = 10$. When training with static data, we use SGD with regularization factor $\gamma = 0.1$, learning rate $\lambda = 0.01$, and $n = 200$ iterations. We use the user and item biases as presented in Section 3.3. When deriving p_u (Equation 5), we run $n = 10$ iterations of SGD. When updating Q (algorithm 2), the learning rate is set to $\lambda = 0.005$ and $\lambda = 0.05$ for the *rated* and *new* approaches, respectively.

4.1 Overall Performance

In this section, we assess which of the proposed variants of the semi-decentralized MF performs better and how they compare to the state-of-the-art MF that retains user data.

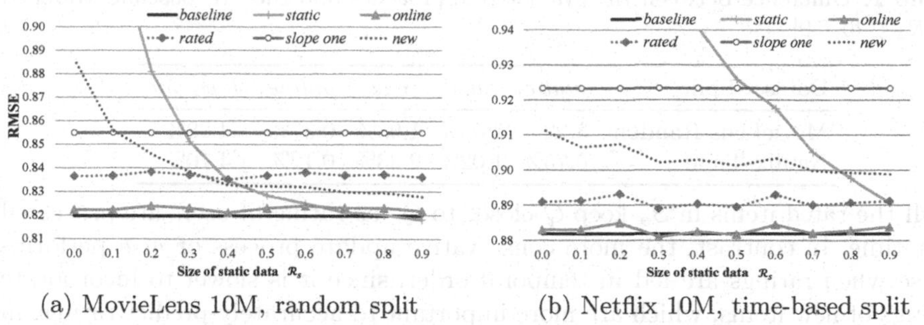

(a) MovieLens 10M, random split (b) Netflix 10M, time-based split

Fig. 1. Comparison between the semi-decentralized MF approach and the state-of-the-art. The differences are significant at $p < 0.01$.

Figures 1a and 1b depict the accuracy achieved by the recommenders for the random and time-based split of training and test data, respectively. We compare the two proposed variants of the semi-decentralized MF, which update the latent item vectors of either all the rated items (*rated*) or of the recently rated items only (*new*), to four baseline approaches:

Baseline Standard centralized implementation of MF, using SGD and retaining all the available ratings [8].

Static Similar to baseline, but using only the static ratings available at t_s, i.e., only Q_s and P_s, to generate recommendations. This mimics a "lazy" approach, where the recommender disregards new ratings and does not update P and Q after t_s.

Online Centralized online MF proposed by Rendle and Schmidt-Thieme in [15], which allows an incremental update of Q by iterating over all the stored ratings. We tuned the parameters using an offline evaluation to $d = 10$, $\lambda = 0.01$, and $n_{iter} = 200$.

Slope-one A simplified item-based Collaborative Filtering (CF) that assumes a linear relationship between the items [11]. This algorithm provides a realistic lower bound for acceptable performance of a personalized recommender. This approach also retains all the available user ratings.

We observe that for the random split (Figure 1a), the *new* variant outperforms the *rated* variant for $\mathcal{R}_s \geq 0.4$, but its error is substantially higher for lower ratios of static data. This is due to the fact that *new* introduces fewer updates to the reasonably stable Q_s than *rated*, and, just like the *static* approach that does not update Q at all, yields good performance for high values of $\mathcal{R}_s$. Conversely, the *rated* variant, which entails more updates, results in a stronger deviation of Q from Q_s that has already stabilized for high $\mathcal{R}_s$, and this leads to less accurate predictions. The performance of the *rated* variant is, however, superior to that of *new* and *static* for lower ratios of $\mathcal{R}_s$, due to the repetitive vector updates of the rated items.

In the more realistic time-based split scenario portrayed in Figure 1b, *rated* outperforms *new* across the board: the repetitive updates of the latent vectors

Table 1. Difference between the evaluated approaches and the MF baseline averaged for $\mathcal{R}_s \in [0, 0.9]$

Dataset-Split	*static*	*rated*	*new*	*online*	*slope one*
MovieLens-Random	3.22%	2.09%	1.98%	0.37%	4.28%
Netflix-Time	6.75%	1.02%	2.33%	0.12%	4.70%

of all the rated items in S_u keep Q closer to Q^* as rating of users are processed over time. In contrast, the more conservative update process of *new* performs worse when ratings are fed in temporal order, since it is slower to incorporate ratings of new items, which are more important to accurately predict ratings in this setup (this is also shown in the *static* approach).[3] We conclude that, when taking a realistically behaving temporal factor of ratings into consideration, the *rated* variant is the best performing semi-decentralized MF approach, and its performance is stable across various values of $\mathcal{R}_s$.

As expected, standard centralized MF achieves the highest accuracy. However, MF is not sufficiently scalable and cannot be deployed in a practical Web-based recommender, since it requires the latent vectors to be re-trained for every new rating. In the following analysis, we consider the RMSE of standard MF as the lower bound for error achievable by other approaches and refer to its accuracy as the baseline.

Table 1 summarizes the performance of the evaluated approaches with respect to the centralized MF baseline. The performance is quantified by averaging the errors obtained for ten values of $\mathcal{R}_s$, ranging from 0 to 0.9. As can be seen, the average error of the *online* approach is only 0.12% higher than that of the MF baseline for the realistic time-based split. The *rated* and *new* semi-decentralized variants are inferior to MF by 1.02% and 2.33%, respectively. Finally, *slope one* and *static* demonstrate substantially higher error rates than MF.

Out of the two semi-decentralized MF variants, *rated* again outperforms *new* and comes closer to the baseline MF. *Online* is very close to MF and is consistently superior to our best performing *rated* variant. However, it should be noted that *online* benefits from access to all the available ratings and can re-train the latent factors, which explains its performance. Placing *rated* on the range between the upper bound of accuracy set by the baseline MF and the lower bound set by *slope one*, we highlight that *rated* is much closer to MF than to *slope one*. We conclude that *rated* offers a reasonable compromise, as it performs online updates of the latent vectors, without retaining user data.

4.2 Impact of the Availability of User Ratings

In this section we investigate how the number of available user ratings affects the performance of the proposed approach. We do this by measuring the fluctuations in the accuracy of the semi-decentralized *rated* variant, as a function of the volume of incrementally added training data. We fix the static data ratio to

[3] Random split of Netflix yielded results similar to those in Figure 1a, highlighting the importance of the temporal data. Due to space limitations we excluded this chart.

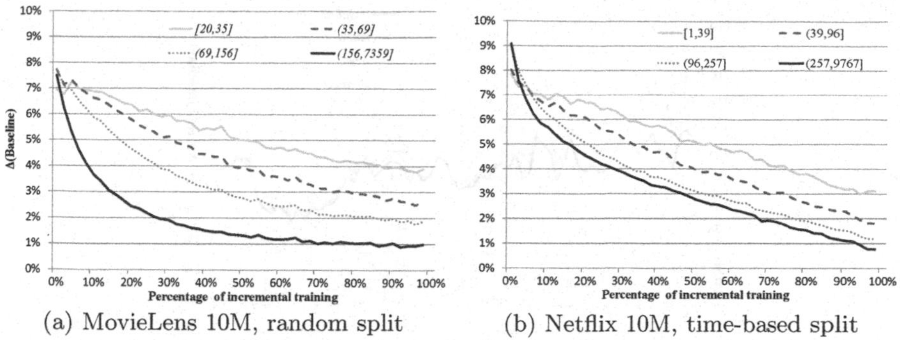

(a) MovieLens 10M, random split (b) Netflix 10M, time-based split

Fig. 2. Relative difference between the *rated* variant and the MF baseline as a function of the volume of the incremental data. Each line pertains to a set of users with a given number of ratings. The differences are significant at $p < 0.01$.

$\mathcal{R}_s = 0.3$, and then gradually increase the number of ratings added on top of the static data from 0% to 100% of the remaining training ratings. We split the users into four equal-sized buckets according to the number of their ratings, and average the user-based accuracy of the *rated* variant across the buckets.

Figures 2a and 2b show the relative difference between the RMSE of the *rated* variant and that of the baseline MF, respectively for the random split and the time-based split, both averaged on a bucket basis for an increasing number of incrementally added ratings. Generally, the performance of *rated* converges to MF as more ratings become available. For the random split, there is an evident difference between the buckets, and the difference with respect to MF gets smaller with the number of user ratings. This is expected, as predictions based on many ratings are generally more accurate than those based on a few ratings. The convergence rate observed in the temporal split evaluation is slower than that observed in the random split evaluation, since the temporal split emphasizes prediction of ratings for new items, and in the early stages of the evaluation there is no sufficient training data pertaining to new items to perform such predictions accurately. The difference between the buckets is less pronounced for the time-based split, and the behavior of the top two buckets, corresponding to users with more than 96 ratings, is similar.

Hence, we conclude that when users have a moderate amount of ratings, the prediction accuracy converges to the baseline MF, while users with fewer ratings will suffer a slight impact on performance.

4.3 Time-Based Simulation

Finally, we investigate the performance of the proposed approach in a realistic large-scale time-aware scenario. We perform this experiment using the complete Netflix 100M ratings dataset, to mimic the evolution of a Web-scale recommender system that transitions to the semi-decentralized MF model. We compare the accuracy of the two semi-decentralized variants that do not retain user data

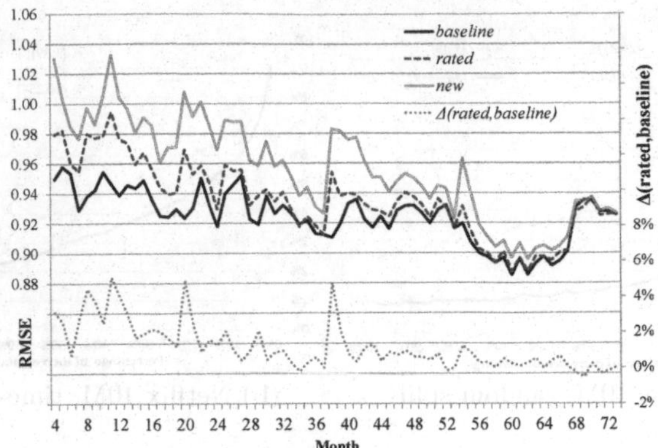

Fig. 3. Difference between the *rated* and *new* variants and the MF baseline. The bottom line shows the difference between *rated* and MF.

with the baseline MF. In this experiment, we sort all the available Netflix ratings according to their timestamps.[4] We then use the increasing-time window evaluation methodology with a window size of 1 month [3]. That is, we use the data of the first $n-1$ months as the training set, predict the ratings of the n-th month, and compute the RMSE for that month. Then, we add the data of the n-th month to the training set and predict the ratings of the $n+1$-th month, and so on. The overall span of Netflix ratings is 72 months and we initialize the training set with the first four months of data.

Figure 3 shows the RMSE of the *new*, *rated*, and MF recommenders (values on the left axis), as well as the relative difference between the *rated* variant and the MF baseline (right axis). Initially (months 4 to 24), the difference hovers around the 2% mark, and it steadily diminishes later on, as the training window size increases. Eventually, the difference becomes smaller than 1% around month 55 and virtually disappears from month 68 onward.[5] We note some lose of accuracy (spikes in the relative difference) up to months 38-40. These spikes are due to the introduction of a large number of new items into the system paired with a lower number of users in the system. This makes our algorithms more prone to error, as their update process is slower than that of the baseline (the *new* approach has a higher spike due to an even slower update cycle).

This large-scale simulation clearly demonstrates that the performance of the proposed approaches comes very close to that of the baseline MF, as more ratings become available to the recommender.[6] This result is encouraging, as it shows that although avoiding user data retention comes at the cost of accuracy,

[4] As the timestamps of Netflix rating are truncated to days, we randomize all the ratings with the same timestamp.

[5] Similar results are obtained in an experiment using the increasing dataset size methodology with quanta of 1M ratings (ratings are sorted, and chunks of 1M ratings are predicted and then added). Due to space limitations we exclude this chart.

[6] The processing of ratings (re-computing the factors and the recommendations) was done at a rate of 900 ratings/second, using one core of AMD Opteron 4184.

the cost diminishes over time. Also, while *rated* outperforms *new* at most stages of the simulation, their performance converges at the later stages, as more ratings are included, implying that *new* could also be viable when enough training data is available. Overall, this simulation demonstrates that the proposed semi-decentralized MF variant can be deployed in a practical Web-scale recommender.

5 Conclusions and Discussion

In this work we present a novel semi-decentralized variant of MF that retains neither user ratings nor user latent factors. This way, we mitigate the risk of undesired access to personal data by malicious third parties and hinder the use of the data for purposes other than the recommendations. Our evaluation indicates that the proposed approach performs on a par with the state-of-the-art MF approaches, thus offering a viable alternative and moving a step forward towards privacy-aware recommender systems. At the same time, our approach allows service providers to keep control of their domain knowledge, encapsulated in the latent item factors matrix.

Note that even if the service providers avoid user data retention, this does not prevent possible leakages of item-related data, which is retained by the service providers. While item-related data does not contribute directly to user privacy risks, it could still be abused to derive personal information, as implied by Calandrino et al. [2], and addressing this risk would require the use of complementing privacy mechanisms, such as differential privacy [13]. However, when compared to accessing readily-available user data, such attacks require more effort, depend on access to auxiliary data, and are harder to execute at scale.

The mitigation of privacy risks relies on the service provider discarding user data. One could argue that no privacy gain is achieved, as user ratings are still accessible to service providers when the recommendations are generated and the users have no means to verify that their data is discarded. Furthermore, since the recommendations requires access to all the user's ratings, service providers could still mine personal information when the ratings are sent to the system. However, such misconduct may be detrimental to user trust and to the reputation of the service. While service providers need to balance user privacy with the benefits of acquiring personal data, it is worth to note the increased awareness of privacy, such that privacy is often advocated as a competitive advantage of services.[7] This incentivizes service providers to conform to privacy regulations and to embrace privacy preserving algorithms. We also note that our approach does not preclude law enforcement agencies from activating "wiretaps" that record the data of certain users. In fact, as made evident in the mandatory data retention initiative,[8] and the recent revelations about the NSA and the PRISM program [6], service providers may be forced to put such mechanisms in place. However, we posit that requiring an explicit wiretap activation, as opposed to

[7] `http://techcrunch.com/2013/04/22/microsoft-launches-new-online-privacy-awareness-campaign/`

[8] `https://www.eff.org/issues/mandatory-data-retention`

grabbing data that is already stored in the system, raises the bar for user privacy, and reduces the risk of users being subject to passive surveillance. In addition, cryptographic techniques could be leveraged for maintaining the confidentiality of user input, and pose an interesting future research direction.

References

1. Berkovsky, S., Eytani, Y., Kuflik, T., Ricci, F.: Enhancing privacy and preserving accuracy of a distributed collaborative filtering. In: RecSys, pp. 9–16 (2007)
2. Calandrino, J.A., Kilzer, A., Narayanan, A., Felten, E.W., Shmatikov, V.: "You might also like:" privacy risks of collaborative filtering. In: IEEE Symposium on Security and Privacy, pp. 231–246 (2011)
3. Campos, P., Díez, F., Cantador, I.: Time-aware recommender systems: a comprehensive survey and analysis of existing evaluation protocols. In: UMUAI (2013)
4. Chaabane, A., Acs, G., Kaafar, M.: You are what you like! information leakage through users' interests. In: Proc. NDSS (2012)
5. Cremonesi, P., Koren, Y., Turrin, R.: Performance of recommender algorithms on top-n recommendation tasks. In: RecSys, pp. 39–46 (2010)
6. Greenwald, G., MacAskill, E.: NSA Prism program taps in to user data of Apple, Google and others. In: The Guardian (June 2013)
7. Guha, S., Reznichenko, A., Tang, K., Haddadi, H., Francis, P.: Serving ads from localhost for performance, privacy, and profit. In: HotNets (2009)
8. Koren, Y., Bell, R., Volinsky, C.: Matrix factorization techniques for recommender systems. IEEE Computer 42(8), 30–37 (2009)
9. Kosinski, M., Stillwell, D., Graepel, T.: Private traits and attributes are predictable from digital records of human behavior. Proceedings of the National Academy of Sciences (March 2013)
10. Lam, S.K., Frankowski, D., Riedl, J.: Do you trust your recommendations? An exploration of security and privacy issues in recommender systems. In: Müller, G. (ed.) ETRICS 2006. LNCS, vol. 3995, pp. 14–29. Springer, Heidelberg (2006)
11. Lemire, D., Maclachlan, A.: Slope one predictors for online rating-based collaborative filtering. Society for Industrial Mathematics 5, 471–480 (2005)
12. Luo, X., Xia, Y., Zhu, Q.: Incremental collaborative filtering recommender based on regularized matrix factorization. Know. -Based Syst. 27, 271–280 (2012)
13. McSherry, F., Mironov, I.: Differentially private recommender systems: Building privacy into the netflix prize contenders. In: KDD, pp. 627–636 (2009)
14. Narayanan, A., Shmatikov, V.: Robust de-anonymization of large sparse datasets. In: IEEE Symposium on Security and Privacy, pp. 111–125 (2008)
15. Rendle, S., Schmidt-Thieme, L.: Online-updating regularized kernel matrix factorization models for large-scale recommender systems. In: RecSys (2008)
16. Sarwar, B., Karypis, G., Konstan, J., Riedl, J.: Incremental singular value decomposition algorithms for highly scalable recommender systems. In: Fifth International Conference on Computer and Information Science, pp. 27–28 (2002)
17. Shani, G., Gunawardana, A.: Evaluating recommendation systems. In: Recommender Systems Handbook, pp. 257–297. Springer (2011)
18. Takács, G., Pilászy, I., Németh, B., Tikk, D.: Scalable collaborative filtering approaches for large recommender systems. JMLR 10, 623–656 (2009)
19. Toubiana, V., Narayanan, A., Boneh, D., Nissenbaum, H., Barocas, S.: Adnostic: Privacy preserving targeted advertising. In: NDSS (2010)
20. Weinsberg, U., Bhagat, S., Ioannidis, S., Taft, N.: BlurMe: inferring and obfuscating user gender based on ratings. In: RecSys, pp. 195–202 (2012)

Privacy-Preserving Collaborative Anomaly Detection for Participatory Sensing

Sarah M. Erfani[1], Yee Wei Law[2], Shanika Karunasekera[1], Christopher A. Leckie[1], and Marimuthu Palaniswami[2]

[1] NICTA Victoria Research Laboratory,
Department of Computing and Information Systems
[2] Department of Electrical and Electronic Engineering,
The University of Melbourne, Australia
{sarah.erfani,ywlaw,karusg,caleckie,palani}@unimelb.edu.au

Abstract. In collaborative anomaly detection, multiple data sources submit their data to an on-line service, in order to detect anomalies with respect to the wider population. A major challenge is how to achieve reasonable detection accuracy without disclosing the actual values of the participants' data. We propose a lightweight and scalable privacy-preserving collaborative anomaly detection scheme called Random Multiparty Perturbation (RMP), which uses a combination of nonlinear and participant-specific linear perturbation. Each participant uses an individually perturbed uniformly distributed random matrix, in contrast to existing approaches that use a common random matrix. A privacy analysis is given for Bayesian Estimation and Independent Component Analysis attacks. Experimental results on real and synthetic datasets using an auto-encoder show that RMP yields comparable results to non-privacy preserving anomaly detection.

Keywords: Privacy-preserving data mining, Anomaly detection, Collaborative learning, Participatory sensing, Horizontally partitioned data.

1 Introduction

Anomaly detection (also known as outlier detection) plays a key role in data mining for detecting unusual patterns or events in an unsupervised manner. In particular, there is growing interest in *collaborative anomaly detection* [1,2,3], where multiple data sources submit their data to an on-line service, in order to detect anomalies with respect to the wider population. For example, in *participatory sensing networks* (PSNs) [4], participants collect and upload their data to a central service to detect unusual events, such as the emergence of a source of pollution in environmental sensing, or disease outbreaks in public health monitoring. A major challenge for collaborative anomaly detection in this context is how to maintain the trust of participants in terms of both the accuracy of the anomaly detection service as well as the privacy of the participants' data. In this paper, we propose a random perturbation scheme for privacy-preserving

V.S. Tseng et al. (Eds.): PAKDD 2014, Part I, LNAI 8443, pp. 581–593, 2014.

anomaly detection, which is resilient to a variety of privacy attacks while achieving comparable accuracy to anomaly detection on the original unperturbed data.

There have been several studies on collaborative anomaly detection, where a number of participants want to build a global model from their local records, while none of the participants are willing to disclose their private data. Most existing work relies on the use of Secure Multiparty Computation (SMC) to generate the global model for anomaly detection [1,2]. While this approach can achieve high levels of privacy and accuracy, SMC incurs a high communication and computational overhead. Moreover, SMC based methods require the simultaneous coordination of all participants during the entire training process, which limits the number of participants. Thus, an open research challenge is how to improve scalability while achieving high levels of accuracy and privacy.

To address this challenge, we propose a privacy-preserving scheme for anomaly detection called *Random Multiparty Perturbation* (RMP). RMP supports the scenario where participants contribute their local data to a public service that trains an anomaly detection model from the combined data. This model can then be distributed to end-users who want to test for anomalies in their local data. In this paper, we focus on the use of an auto-associative neural network (AANN, also known as an autoencoder) as our anomaly detection model, although our scheme is also applicable to other types of anomaly detectors.

In order for the participants of RMP to maintain the privacy of their data, we propose a form of random perturbation to be used by each participant. Previous approaches to random perturbation in this context [5,6,7,3] require all participants to perturb their data in the same way, which makes this scheme potentially vulnerable to breaches of privacy if collusion occurs between rogue participants and the server. In contrast, RMP proposes a scheme in which each participant first perturbs their contributed data using a unique, private random perturbation matrix. In addition, any end-user can apply the resulting anomaly detection model to their own local data by using a public perturbation matrix. This provides a scalable collaborative approach to anomaly detection, which ensures a high level of privacy while still achieving a high level of accuracy to the end-users.

The main contributions of this paper are as follows: (i) We propose a privacy-preserving model of collaborative anomaly detection based on random perturbation, such that each participant in training the anomaly detector can have their own unique perturbation matrix, while the resulting anomaly detection model can be used by any number of users for testing. (ii) In contrast to previous methods, we give the first privacy-preserving scheme for auto-associator anomaly detectors that does not require computationally intensive cryptographic techniques. (iii) We show analytically the resilience of our scheme to two types of attacks—Bayesian Estimation and Independent Component Analysis (ICA). (iv) We demonstrate that the accuracy of our approach is comparable to non-privacy preserving anomaly detection on various datasets.

2 Related Work

In general, privacy-preserving data mining (PPDM) schemes can be classified as *syntactic* or *semantic*. Syntactic approaches aim to prevent syntactic attacks, such as the table linkage attack. Semantic approaches aim to satisfy semantic privacy criteria, which are concerned with minimising the difference between adversarial prior knowledge and adversarial posterior knowledge about the individuals represented in the database. While both approaches have the same goal, i.e., hiding the real values from the data miner, syntactic approaches typically take the entire database as input, whereas semantic approaches do not. Since in participatory sensing the participants are responsible for masking their own data before outsouring them to the data miner, semantic approaches are the only option here.

In the following we review existing work on privacy-preserving back-propagation and anomaly detection in the context of collaborative learning, and highlight the major differences with our work. The majority of semantic PPDM approaches, either in the context of privacy-preserving back-propagation [8,9] or privacy-preserving anomaly detection [1,2], use Secure Multiparty Computation (SMC). SMC approaches suffer from high computational complexity. Moreover, they require the cooperation of all participants throughout the whole process of building the data mining models, and hence suffer a lack of scalability.

Another approach to semantic PPDM is data *randomisation*, which refers to the randomised distortion or perturbation of data prior to analysis. Perturbing data elements, for example, by introducing additive or multiplicative noise, allows the server to extract the statistical properties of records without requiring the participants to allocate substantial resources or coordinate with the server during the training process. We now outline the main types of data perturbation methods: additive, multiplicative, and nonlinear transformation.

Additive perturbation adds noise to the data [10,11]. Since independent random noise can be filtered out [12,13], the general principle is that the additive noise should be correlated with the data [14,15].

Multiplicative perturbation multiplies the data with noise. A typical approach is to use a zero-mean Gaussian random matrix as the noise for a distance-preserving transformation [5]. However, if the original data follows a multivariate Gaussian distribution, a large portion of the data can be reconstructed via attacks to distance-preserving transformations as proposed in [16,17].

In general, distance-preserving transformations are good for accuracy but are susceptible to attacks that exploit distance relationships. In comparison, the *random transformation* approach [7], where the noise matrix is a random matrix with elements uniformly distributed between 0 and 1, is not distance-preserving and hence not susceptible to these attacks.

Nonlinear transformations can be used to break the distance-preserving effect of a distance-preserving transformation on some data points. For example, the function tanh approximately preserves the distance between normal data points, but collapses the distance between outliers [3]—it is hence suitable for anomaly detection. In RMP, we use the *double logistic* function for this purpose and for conditioning the probability density function (pdf) of the transformed data.

The aforementioned works satisfy privacy requirements in the case where all participants are assumed to be semi-honest (i.e., are not colluding with other parties) and agree on using the same perturbation matrix. As discussed in Section 3.1, this assumption is restrictive in applications such as participatory sensing. A few works have tried to extend current randomisation approaches to a collaborative learning architecture [6,18]. For example, Chen et al.'s framework [6] is a multiparty collaborative mining scheme using a combination of additive and multiplicative perturbations. This scheme requires the participants to stay in touch for an extended period to generate the perturbation matrix, which is impractical for large-scale participatory sensing. Liu et al. [18] build a framework on top of [5], and allow each participant to perturb the training data with a distinct perturbation matrix. This scheme then regresses these mathematical relationships between training samples in a unique way, thereby preserving classification accuracy. Although participants use private matrices, the underlying transformation scheme is still vulnerable to distance inference attacks [16,17].

In summary, data randomisation is the most promising approach to PPDM in the paradigm of PSN. Unlike distance-preserving transformations, random transformation does not preserve Euclidean distances and inner products between data distances, hence it thwarts distance inference attacks. However, if this method is applied to collaborative learning, all the participants must agree on the same perturbation matrix, and then collusion attacks may succeed. Furthermore, the mining models generated from the perturbed records are specific to the participants. For these reasons our RMP scheme uses a random transformation and provides each participant with a private perturbation matrix. The perturbation matrices are tailored so that both data privacy and accuracy of the mining models are achieved. In addition, RMP generates models that can be adopted by any user, i.e, data contributors and non-contributors.

3 RMP Privacy-Preserving Anomaly Detection Scheme

In this section we present our Random Multiparty Perturbation (RMP) scheme, which considers a general participatory sensing architecture comprising three types of parties, namely, participants, the data mining server and end-users.

3.1 Problem Statement and Design Principles

We consider the case of a participatory sensing network that comprises a set of participants, $\mathcal{C} = \{c_i | i = 1, \ldots, q\}$, a mining server $\mathcal{S}$, and an arbitrary number of end-users $\mathcal{U}$, as can be seen in Fig. 1. Each participant c_i is an individual who captures data records for the same set of attributes, i.e., a horizontal partition, and contributes the sampled records to $\mathcal{S}$ for training purposes. The server $\mathcal{S}$ is a third-party providing a data mining service to the participants. After receiving the contributed data records, the server trains an anomaly detector that generates a global classification model $\mathbf{M}$ from the locally collected data. The end-users, $\mathcal{U}$, could be the participants themselves or third parties such

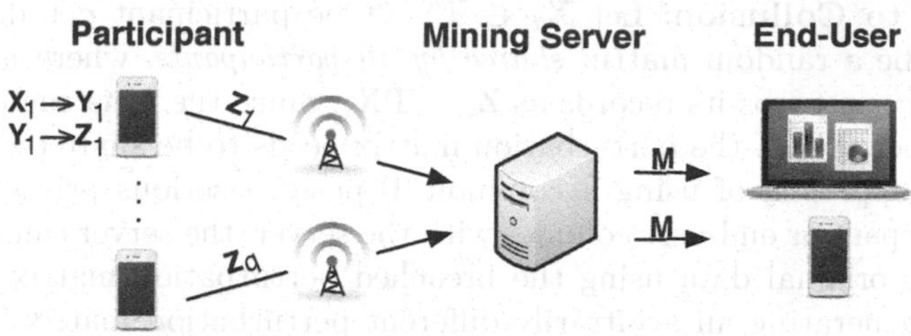

Fig. 1. The RMP architecture

as analysts trying to learn about the monitored phenomena. We share a similar underlying assumption with [4], in which the computational demands on the participants should be minimised, while the anomaly detection model should be in a form that can be disseminated and used by an arbitrary number of end-users, and not limited to the original participants or customised for a small number of end-users. In addition, we adopt a well established assumption in the literature that regards all parties as semi-honest entities, i.e., they follow the protocol. To make the scenario realistic, we assume that a small subset of the participants might collude with the server to infer other participants' submitted records. Designing a collaborative data mining scheme that fulfills these requirements with low communication and computation costs while preserving the participants' privacy is an open problem.

For privacy, we need to ensure that the attribute values in the participants' contributed records are properly masked: given the masked values, the server cannot infer the original values. However, this must be done without over-sacrificing accuracy, i.e., the results of anomaly detection based on the masked data should be close to the corresponding result using the original data. Multiplicative perturbation projects data to a lower dimensional space. The perturbed data matrix has a lower rank than the original data matrix, thereby forcing the attacker to solve an undetermined system of linear equations. However, this is not enough, and the following design principles are pertinent.

Resilience to Distance Inference Attacks: The review of existing multiplicative perturbation schemes in Section 2 reveals that distance-preserving transformations are susceptible to distance inference attacks [16,19]. The challenge is to find a non-distance preserving transform that is suitable for certain data mining tasks. *Random transformation* [7] qualifies as such a transform in that it does not preserve the dot product or Euclidean distance among transformed data points, yet it is suitable for anomaly detection.

Resilience to Bayesian Estimation Attacks: Bayesian Estimation is a general attack that exploits the pdf of the original data. Gaussian data is particularly exploitable because it reduces a maximum a posteriori estimation problem into a simple convex optimization problem [17]. A suitable defense is to prevent this reduction by conditioning the pdf through a nonlinear transformation.

Resilience to Collusion: Let $\mathbf{X}_i \in \mathbb{R}^{n \times m_i}$ be participant c_i's dataset, and $\mathbf{T} \in \mathbb{R}^{w \times n}$ be a random matrix *shared by all participants*, where $w < n$. The participant c_i perturbs its records as $\mathbf{Z}_i = \mathbf{T}\mathbf{X}_i$. Since the constructed anomaly detector is perturbed, the perturbation matrix needs to be shared with all end-users. This approach of using a common $\mathbf{T}$ poses a serious privacy risk. If a rogue participant or end-user colludes with the server, the server can recover any participant's original data using the breached perturbation matrix. The naive solution of generating an arbitrarily different perturbation matrix $\mathbf{T}_i$ for each participant c_i does not work, because building an accurate mining model requires consistency among the perturbation matrices. To overcome this challenge, RMP generates participant-specific perturbation matrices by perturbing $\mathbf{T}$.

3.2 The Scheme

RMP is designed to address the design principles in the previous subsection. Let $\mathbf{T}$ be a $w \times n$ matrix ($w < n$) with $U(0, 1)$-distributed elements. Each participant c_i generates a unique perturbation matrix

$$\tilde{\mathbf{T}}_i = \mathbf{T} + \Delta_i, \tag{1}$$

where each element in Δ_i is drawn from $U(-\alpha, \alpha)$, and $0 < \alpha < 1$. Experimental results show that for small values of α, the accuracy loss in anomaly detection is small. Next, we describe RMP in full detail.

In RMP, a participant transforms $\mathbf{X}_i$ to $\mathbf{Z}_i$ in two stages:

Stage 1: The participant transforms $\mathbf{X}_i$ to $\mathbf{Y}_i$, by applying a *double logistic function* to $\mathbf{X}_i$ element-wise:

$$y_{k,l} \stackrel{\text{def}}{=} \text{sgn}(x_{k,l})[1 - \exp(-\beta x_{k,l}^2)], \tag{2}$$

for $k = 1, \ldots, n$, $l = 1, \ldots, m_i$, where sgn is the signum function. For suitable values of β and normalised values of $x_{k,l}$ (i.e., $|x_{k,l}| \leq 1$), the double logistic function approximates the identity function $y_{k,l} = x_{k,l}$ well. To maximise this approximation, we equate the optimal value of β to the value that minimises the integral of the squared difference between the two functions:

$$\beta^{\star} = \arg\min_{\beta} \int_0^1 [1 - \exp(-\beta x^2) - x]^2 dx \approx 2.81.$$

The role of this nonlinear transformation is to condition the pdf of $\mathbf{Y}_i$ to thwart Bayesian Estimation attacks, as detailed in Section 5.

Stage 2: Using $\tilde{\mathbf{T}}_i$ generated earlier, the participant transforms $\mathbf{Y}_i$ to $\mathbf{Z}_i$:

$$\mathbf{Z}_i \stackrel{\text{def}}{=} \tilde{\mathbf{T}}_i \mathbf{Y}_i. \tag{3}$$

The participant then sends $\mathbf{Z}_i$ to the mining server $\mathcal{S}$, which will receive the following from all the participants:

$$\mathbf{Z}_{\text{all}} \stackrel{\text{def}}{=} [\mathbf{Z}_1 | \mathbf{Z}_2 | \cdots | \mathbf{Z}_q] .$$

The role of $\mathcal{S}$ is to learn an anomaly detection model encoding the underlying distribution of $\mathbf{Z}_{\text{all}}$. End-users given access to the model and $\mathbf{T}$ can then detect anomalies in their data with respect to the model. RMP is independent of the anomaly detection algorithm used, but the auto-associative neural network is used for our study and is discussed next.

4 Anomaly Detection Service

In our collaborative framework, the role of the server $\mathcal{S}$ is to learn an anomaly detection function, which can then be disseminated for use by the end-users. In particular, the learned anomaly detection function should encode a model of the underlying distribution of the (perturbed) training data provided by the participants. A given data record can then be tested using this model for anomalies.

While our general framework can potentially use a wide variety of anomaly detection models, in this paper we use an auto-associative neural network (AANN) [20], also known as an auto-encoder, as the basis for our anomaly detection function. We choose to use an AANN for our anomaly detection function because: (i) it can be trained in an unsupervised manner on either normal data alone, or a mixture of normal data with a small but unspecified proportion of anomalous data; (ii) it is capable of learning a wide variety of underlying distributions of training data; and (iii) the resulting anomaly detection function is compact and computationally efficient—hence practical for dissemination to end-users.

An AANN is a multi-layer feed-forward neural network that has the same number of output nodes as input nodes. Between the input and output layers, a hidden "bottleneck" layer of a smaller dimension than the input/output layers captures significant features in the inputs through compression. Training the AANN means adjusting the weights of the network for each training record, so that the outputs closely match the inputs. In this way, the AANN learns a nonlinear manifold that represents the underlying distribution of the data. In a trained AANN, the reconstruction error (integrated squared error between the inputs and outputs) should be high for anomalies, but low for normal data. Let e_i be the reconstruction error for training dataset X_i. If the reconstruction error for a test record is larger than the threshold $\theta = \mu(e_i) + 3\sigma(e_i)$, where $i = 1, \ldots, q$, the record is identified as an anomaly. Due to space constraints, the interested reader is referred to [20] for the details of the AANN training algorithm.

5 Privacy Analysis

In the *statistical disclosure control* and *privacy-preserving data publishing* literature, the most popular semantic privacy criterion is *differential privacy*. Differential privacy is for *answering queries* to a database containing private data of *multiple individuals*. For the participatory sensing scenario where participants are data owners who *publish data* (instead of answering queries) about *themselves alone*, differential privacy is not a good fit. Below, we propose an alternative (informal) privacy criterion.

Linear multiplicative perturbation schemes aim to project a data matrix to a lower dimensional space so that an attacker has only an ill-posed problem in the form of an underdetermined system of linear equations $\tilde{\mathbf{T}}\boldsymbol{x} = \boldsymbol{y}$ to work with, where $\boldsymbol{y}$ is a projection of data vector $\boldsymbol{x}$ and the projection matrix $\tilde{\mathbf{T}}$ is assumed known in the worst case. An underdetermined system cannot be solved for $\boldsymbol{x}$ exactly, but given sufficient prior information about $\boldsymbol{x}$, an approximation of the true $\boldsymbol{x}$ might be attainable. In a *known input-output attack*, the attacker has some input samples (i.e., some samples of $\boldsymbol{x}$) and all output samples (i.e., all samples of $\boldsymbol{y}$), and knows which input sample corresponds to which output sample [21,22,19]. In the collaborative learning scenario where the data miner may collude with one or more participants to unravel other participants' data, the known input-output attack is an immediate concern. In the following, our privacy analysis is conducted with respect to two known input-output attacks, one based on Bayesian estimation, and one based on Independent Component Analysis (ICA). Suppose the attacker is targeting a particular participant by trying to solve $\mathbf{Z} = \tilde{\mathbf{T}}\mathbf{Y} = (\mathbf{T} + \Delta)\mathbf{Y}$ for $\mathbf{Y}$. In the analysis below, let $\boldsymbol{z} \sim \mathcal{Z}$ represent a column of $\mathbf{Z}$, and $\boldsymbol{y} \sim \mathcal{Y}$ represent a column of $\mathbf{Y}$.

5.1 Attacks Based on Bayesian Estimation

We consider two scenarios where $\tilde{\mathbf{T}}$ is known and where $\tilde{\mathbf{T}}$ is unknown.

Scenario where $\tilde{\mathbf{T}}$ is Known: For a worst-case analysis, we assume the attacker somehow knows $\tilde{\mathbf{T}}$ exactly but not $\mathbf{X}$. In a Bayesian formulation, the *maximum a posteriori* (MAP) estimate of $\boldsymbol{y}$, given $\tilde{\mathbf{T}}$ and $\boldsymbol{z}$, is

$$\hat{\boldsymbol{y}} = \arg\max_{\boldsymbol{y}} p_{\mathcal{Y}}(\boldsymbol{y}|\boldsymbol{z}, \tilde{\mathbf{T}}) = \arg\max_{\boldsymbol{y} \in \mathcal{Y}} p_{\mathcal{Y}}(\boldsymbol{y}), \tag{4}$$

where $\mathcal{Y} = \{\boldsymbol{y} : \boldsymbol{z} = \tilde{\mathbf{T}}\boldsymbol{y}\}$ [17]. Note that MAP estimation is a more general approach than maximum likelihood estimation because the former takes a prior distribution (which in our case is $p_{\mathcal{Y}}$) into account. If $p_{\mathcal{Y}}$ is an n-variate Gaussian with a positive definite covariance matrix, then (4) becomes a quadratic programming problem with solution [17, Theorem 1]:

$$\hat{\boldsymbol{y}} = \bar{\boldsymbol{y}} + \boldsymbol{\Sigma}_{\mathcal{Y}}\tilde{\mathbf{T}}'\boldsymbol{\Sigma}_{\mathcal{Z}}^{-1}(\boldsymbol{z} - \tilde{\mathbf{T}}\bar{\boldsymbol{y}}), \tag{5}$$

where $\bar{\boldsymbol{y}}$ and $\boldsymbol{\Sigma}_{\mathcal{Y}}$ are the sample mean vector and sample covariance matrix of $\mathcal{Y}$ respectively, and $\boldsymbol{\Sigma}_{\mathcal{Z}}$ is the sample covariance matrix of $\mathcal{Z}$. Note that $\boldsymbol{\Sigma}_{\mathcal{Y}}$ *is* positive definite, provided the covariance matrix is full rank and there are more samples of $\mathcal{Y}$ than dimensions of $\mathcal{Y}$ [23]. In this case, we can write $\boldsymbol{\Sigma}_{\mathcal{Y}} = \mathbf{Q}\mathbf{Q}'$, where $\mathbf{Q}$ is a nonsingular matrix. Furthermore, $\boldsymbol{\Sigma}_{\mathcal{Z}} = \tilde{\mathbf{T}}\boldsymbol{\Sigma}_{\mathcal{Y}}\tilde{\mathbf{T}}' = \tilde{\mathbf{T}}\mathbf{Q}\mathbf{Q}'\tilde{\mathbf{T}}'$. Since $\tilde{\mathbf{T}}$ is nonsingular, $\boldsymbol{\Sigma}_{\mathcal{Z}}^{-1}$ and therefore the solution (5) exists.

The analysis above suggests that to thwart MAP estimation, we cannot hope to generate $\tilde{\mathbf{T}}$ such that $\boldsymbol{\Sigma}_{\mathcal{Z}}$ is singular. Instead, it is more productive to prevent MAP estimation from being reducible to a quadratic programming problem solvable by (5) in the first place. RMP achieves this by nonlinearly transforming the original data. If X is a standard Gaussian random variable, then based on [24, Section 4.7], the pdf of $Y = \text{sgn}(X)[1 - \exp(-\beta X^2)]$ is

$$p_Y(y) = \frac{1}{\sqrt{8\pi\beta \ln\left(\frac{1}{1-|y|}\right)}} \left(\frac{1}{1-|y|}\right)^{1-\frac{1}{2\beta}}, |y| < 1, y \neq 0.$$

Unlike a standard Gaussian which is continuous everywhere and has a global maximum at $x = 0$, $p_Y \to \infty$ at $y = -1, 0, 1$. Using p_Y or $\ln p_Y$ in (4) renders the optimisation problem non-convex, and numerically unstable near $y = -1, 0, 1$, which is problematic for numerical methods such as hill climbing and simulated annealing. Applying the same nonlinear transformation to Laplace-distributed data (sparse data) has the same effect. Therefore, RMP's nonlinear transformation converts a potentially Gaussian (Laplace) data distribution to a non-Gaussian (non-Laplace) one that hampers the attacker's solution of (4).

The double logistic function is better than tanh, which is used in [3], in terms of thwarting MAP estimation attacks. If X is a standard Gaussian random variable, then the pdf of $Y = \tanh(X)$ is $p_Y(y) = \frac{1}{(1-y^2)\sqrt{2\pi}} \exp(-\frac{\text{artanh}(y)^2}{2}), |y| < 1$. For $|y| \leq 0.9$, the pdf above is convex. This means the attacker can solve (4) as a convex optimisation problem, when the data are perturbed using tanh.

Scenario Where $\tilde{\mathbf{T}}$ is Unknown: In the previous scenario, the attacker knows $\mathbf{T}$ and the relationship between $\mathbf{T}$ and $\tilde{\mathbf{T}}$ (see Equation 1). We also consider the scenario where the attacker does not know $\tilde{\mathbf{T}}$. Note that even with precise knowledge of $\mathbf{T}$ and α, without further information, any matrix value between $\mathbf{T} - \alpha\mathbf{1}$ and $\mathbf{T} + \alpha\mathbf{1}$ can be an estimate of the victim's matrix $\tilde{\mathbf{T}}$. According to Lemma 1, for every element of $\tilde{\mathbf{T}}$, there is a 50% chance of guessing its value wrong by at least $(2 - \sqrt{2})\alpha$.

Lemma 1. *Let D be the difference between two $U(-\alpha, \alpha)$-distributed random variables. Then for $0 \leq d \leq 2\alpha$, $\Pr[|D| \geq d] = (2\alpha - d)^2/(4\alpha^2)$.*

Proof. Let A and B be two $U(-\alpha, \alpha)$-distributed random variables. Then the pdf of $D = A - B$ is given by the convolution

$$p_D(d) = \frac{1}{2\alpha} \int_{-\alpha}^{\alpha} p_{U(-\alpha,\alpha)}(d + b)\mathrm{d}b = \begin{cases} \frac{1}{2\alpha}\left(1 + \frac{d}{2\alpha}\right) & -2\alpha \leq x < 0, \\ \frac{1}{2\alpha}\left(1 - \frac{d}{2\alpha}\right) & 0 \leq x < 2\alpha, \\ 0 & \text{elsewhere.} \end{cases}$$

To get $\Pr[|D| \geq d]$ where $0 \leq d \leq 2\alpha$, we integrate the expression above from -2α to $-d$, and from d to 2α. With a little algebra we get $\Pr[|D| \geq d] = \frac{(2\alpha - d)^2}{4\alpha^2}$.

MAP estimation can be used to estimate both $\mathbf{Y}$ and $\tilde{\mathbf{T}}$. The MAP estimates of $\mathbf{Y}$ and $\tilde{\mathbf{T}}$, given $\mathbf{Z}$, are the matrix values that maximise $p_{\tilde{\mathcal{T}},\mathcal{Y}|\mathcal{Z}}(\tilde{\mathbf{T}}, \mathbf{Y}|\mathbf{Z}) = p_{\tilde{\mathcal{T}}}(\tilde{\mathbf{T}}|\mathbf{Z}) p_{\mathcal{Y}}(\mathbf{Y}|\mathbf{Z})$. The optimisation problem can be written as

$$\max_{\tilde{\mathbf{T}},\mathbf{Y}} p_{\tilde{\mathcal{T}}}(\tilde{\mathbf{T}})p_{\mathcal{Y}}(\mathbf{Y}) \text{ s.t. } \mathbf{Z} = \tilde{\mathbf{T}}\mathbf{Y};$$

or the following when $p_{\tilde{\mathcal{T}}}$ is substituted with $U_{w\times n}(\mathbf{T} - \alpha\mathbf{1}, \mathbf{T} + \alpha\mathbf{1})$ and $p_{\mathcal{Y}}$ is assumed to be zero-mean Gaussian:

$$\min_{\tilde{\mathbf{T}},\mathbf{Y}} \sum_{j=1}^{m} \boldsymbol{y}_j' \Sigma_{\mathcal{Y}}^{-1} \boldsymbol{y}_j \text{ s.t. } \mathbf{Z} = \tilde{\mathbf{T}}\mathbf{Y}, \mathbf{T} - \alpha\mathbf{1} \preceq \tilde{\mathbf{T}} \preceq \mathbf{T} + \alpha\mathbf{1}. \quad (6)$$

In (6), $\boldsymbol{y}_j$ $(j = 1, \ldots, m)$ are columns of $\mathbf{Y}$. Note that in the equality constraint, both $\tilde{\mathbf{T}}$ and $\mathbf{Y}$ are optimisation variables, so even the Gaussian assumption does not reduce (6) to a convex problem. As previously explained, RMP's nonlinear transformation converts a potentially Gaussian (Laplace) data distribution to a non-Gaussian (non-Laplace) one. This hampers the attacker's solution of not only (4) but also (6), which is a harder problem than (4).

5.2 Attacks Based on Independent Component Analysis (ICA)

ICA can be used to estimate $\tilde{\mathbf{T}} \in \mathbb{R}^{w\times n}$ and $\mathbf{Y} \in \mathbb{R}^{n\times m}$, knowing only their product $\mathbf{Z} = \tilde{\mathbf{T}}\mathbf{Y}$, provided (i) $w = n$ (the even-determined case) or $w > n$ (the over-determined case); (ii) the attributes (rows of $\mathbf{Y}$) are pairwise independent; (iii) at most one of the attributes are Gaussian; (iv) $\tilde{\mathbf{T}}$ has full column rank.

However, RMP enforces $w < n$. When $w < n$, the problem of ICA becomes *overcomplete ICA* (or underdetermined ICA). In this case, the mixing matrix $\tilde{\mathbf{T}}$ is identifiable but the independent components are not [22]. Furthermore, when $w \leq (n+1)/2$, no linear filter can separate the observed mixture $\mathbf{Z}$ into two or more disjoint groups, i.e., recover any row of $\mathbf{Y}$ [5].

6 Empirical Results

In this section we evaluate the quality of our proposed privacy-preserving anomaly detection scheme RMP when used with an AANN. The main objective of our experiment is to measure the trade-off in accuracy of our AANN anomaly detection algorithm as a result of maintaining the participants' privacy. Note that Lemma 1 provides a theoretical relationship between α and the privacy that can be achieved in terms of an attacker's ability to estimate the value of a target victim's perturbation matrix $\tilde{\mathbf{T}}$. Thus, in our empirical evaluation, we consider the effect of different levels of privacy in terms of α on the overall accuracy of anomaly detection. We use the Receiver Operating Characteristic (ROC) curve and the corresponding Area Under the Curve (AUC) to measure the performance of RMP. The effectiveness and change in accuracy of RMP are evaluated by comparing against a non-privacy-preserving neural network, in which the raw data records are fed to the AANN for training and testing.

Experiments are conducted on four real datasets from the UCI Machine Learning Repository (all collected from sensor networks except the fourth): (i) Human Activity Recognition using Smartphones (HARS), (ii) Opportunity activity

Table 1. Comparing AUC values of RMP against non-privacy anomaly detection

Dataset	n	Raw	α				
			0.01	0.1	0.2	0.4	0.6
Abalone	8	1	0.95	0.92	0.87	N/A	N/A
Banana	8	1	0.98	0.92	0.81	N/A	N/A
Gas	168	0.99	0.98	0.98	0.98	0.95	0.92
HARS	561	0.99	0.98	0.98	0.97	0.96	0.95
Opportunity	242	0.99	0.99	0.98	0.94	0.90	0.87

Note: "Raw" indicates the non-privacy preserving anomaly detection on the non-perturbed data, and α values indicate the level of imposed noise in RMP anomaly detection.

recognition (Opportunity), (iii) Gas Sensor Array Drift (Gas), (iv) Abalone. We also use the Banana synthetic dataset, generated from a mixture of overlapping Gaussians to resemble a pair of bananas in any two dimensions. We ran the experiment on the first 1000 records of each dataset. Feature values in each dataset are normalised between $[0, 1]$ and merged with 5% anomalous records, which are randomly drawn from $U(0, 1)$. In each experiment a random subset of the dataset is partitioned horizontally among the participants in batches of 30 records and submitted to the server for training.

We deploy a three-layer AANN with the same number of input and output units, i.e., the number of units is set corresponding to the dataset attributes n. The number of hidden units for each dataset is set to about half of the input units (empirically we found that increasing the number of hidden units causes overfitting). All weights and biases in the neural network are initialised randomly in the range of $[-0.1, 0.1]$. The learning rate and momentum are set to 0.25 and 0.85, respectively, and the number of training epochs range from 300 to 1000.

Table 1 compares the results using an AANN anomaly detector on the unperturbed data records ("Raw") along with the corresponding results using the privacy preserving scheme of RMP. Accuracy in RMP is affected by its two stages of transformation, i.e., applying the double logistic function and the random transformation, and the level of added noise α. As can be seen from the table, when data are perturbed with a marginal level of noise $\alpha = 0.01$, the accuracy decreases slightly (about 1%). Hence, it shows that the transformations do not exert a significant impact on the accuracy of anomaly detection. In the reported

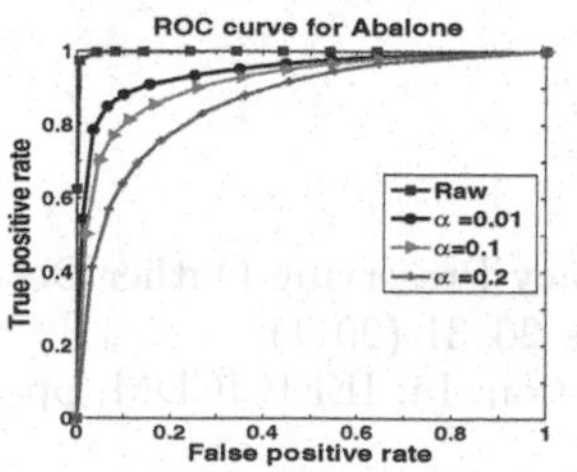

Fig. 2. Abalone Dataset

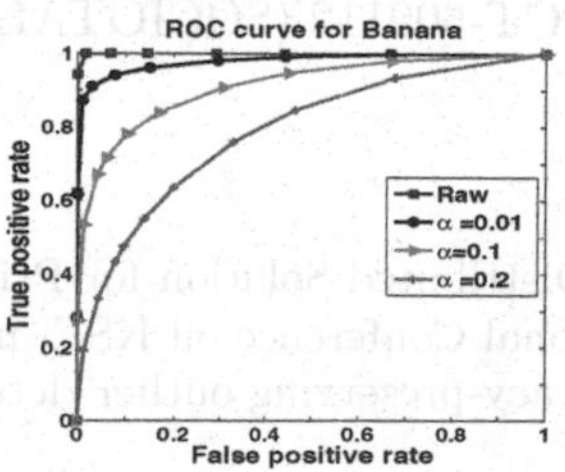

Fig. 3. Banana Dataset

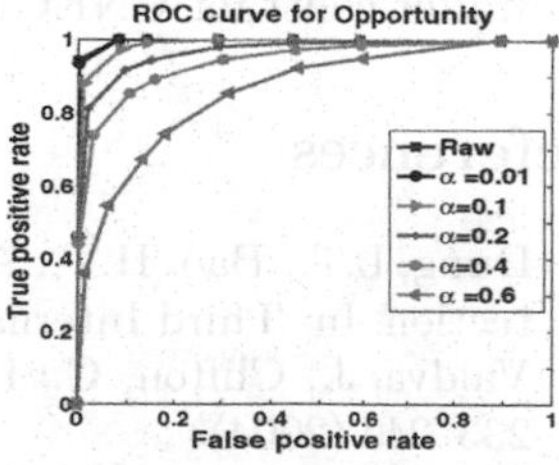

Fig. 4. Opportunity Dataset

results in Table 1, the dimensionality of the datasets is reduced by $r = 1$, where $r = n - w$. Our empirical experiments show that the accuracy on datasets with a larger number of attributes n are less affected by an increase of r, e.g., reducing n by 40% only decreases accuracy by about 1%. However, that is not the case for datasets with small n, e.g., the Abalone and Banana datasets, where reducing the data dimensionality by half might result in a 10% reduction in accuracy.

The level of added noise to the perturbation matrices has a major influence on RMP accuracy. As α increases, so does the loss in accuracy, especially in datasets with smaller numbers of attributes. Since the accuracy loss is generally small, RMP is a highly effective approach for privacy-preserving anomaly detection.

7 Conclusion and Future Work

In a typical participatory sensing scenario, participants send data to a data mining server; the server builds a model of the data; and end-users download the model for their own analyses. Collaborative anomaly detection refers to the case where the model is for anomaly detection. RMP is a privacy-preserving collaborative learning scheme that masks the participants' data using a combination of nonlinear and linear perturbations, while maintaining detection accuracy. RMP protects the private data of participants using individual perturbation matrices, imposes minimal communication and computation overhead on the participants, and scales for an arbitrary number of participants or end-users. We show analytically how RMP is resilient to two common types of attacks: Bayesian Estimation and ICA. Our experiments show that RMP yields comparable results to non-privacy preserving anomaly detection using AANN on a variety of real and synthetic benchmark datasets. As follow-up to this preliminary work, we are in the process of establishing a mathematical framework that relates α to accuracy loss and privacy level—this will allow us to determine α based on the intended trade-off between accuracy and privacy. We are also investigating the resilience of RMP, in conjunction with other supervised or unsupervised learning algorithms, to attacks exploiting sparse datasets, such as overcomplete ICA.

Acknowledgments. NICTA is funded by the Australian Government as represented by the Department of Broadband, Communications and the Digital Economy and the Australian Research Council through the ICT Centre of Excellence program. Yee Wei Law is partly supported by ARC DP1095452 and the EC under contract CNECT-ICT-609112 (SOCIOTAL).

References

1. Dung, L.T., Bao, H.T.: A Distributed Solution for Privacy Preserving Outlier Detection. In: Third International Conference on KSE, pp. 26–31 (2011)
2. Vaidya, J., Clifton, C.: Privacy-preserving outlier detection. In: IEEE ICDM, pp. 233–240 (2004)
3. Bhaduri, K., Stefanski, M.D., Srivastava, A.N.: Privacy-preserving outlier detection through random nonlinear data distortion. IEEE TSMC, Part B 41, 260–272 (2011)

4. Burke, J., Estrin, D., Hansen, M., Parker, A., Ramanathan, N., Reddy, S., Srivastava, M.B.: Participatory sensing. In: ACM SenSys 1st Workshop on World-Sensor-Web: Mobile Device Centric Sensor Networks and Applications (2006)
5. Liu, K., Kargupta, H., Ryan, J.: Random projection-based multiplicative data perturbation for privacy preserving distributed data mining. IEEE TKDE 18(1), 92–106 (2006)
6. Chen, K., Liu, L.: Privacy-preserving multiparty collaborative mining with geometric data perturbation. IEEE TPDS 20(12), 1764–1776 (2009)
7. Mangasarian, O.L., Wild, E.W.: Privacy-preserving classification of horizontally partitioned data via random kernels. In: Proceedings of DMIN (2007)
8. Bansal, A., Chen, T., Zhong, S.: Privacy preserving Back-propagation neural network learning over arbitrarily partitioned data. Neural Computing and Applications 20, 143–150 (2010)
9. Zhang, Y., Zhong, S.: A privacy-preserving algorithm for distributed training of neural network ensembles. Neural Computing and Applications 22, 269–282 (2012)
10. Agrawal, R., Srikant, R.: Privacy-preserving data mining. ACM SIGMOD Record 29(2), 439–450 (2000)
11. Agrawal, D., Aggarwal, C.C.: On the design and quantification of privacy preserving data mining algorithms. In: Proceedings of ACM PODS, pp. 247–255 (2001)
12. Kargupta, H., Datta, S., Wang, Q., Sivakumar, K.: On the privacy preserving properties of random data perturbation technique. In: IEEE ICDM, pp. 99–106 (2003)
13. Huang, Z., Du, W., Chen, B.: Deriving private information from randomized data. In: Proceedings of ACM SIGMOD, pp. 37–48 (2005)
14. Papadimitriou, S., Li, F., Kollios, G., Yu, P.S.: Time series compressibility and privacy. In: Proceedings of VLDB, pp. 459–470 (2007)
15. Ganti, R.K., Pham, N., Tsai, Y.E., Abdelzaher, T.F.: Poolview: stream privacy for grassroots participatory sensing. In: Proceedings of ACM SenSys, pp. 281–294 (2008)
16. Liu, K., Giannella, C., Kargupta, H.: A survey of attack techniques on privacy-preserving data perturbation methods. In: Privacy-Preserving Data Mining. Springer (2008)
17. Sang, Y., Shen, H., Tian, H.: Effective reconstruction of data perturbed by random projections. IEEE Transactions on Computers 61(1), 101–117 (2012)
18. Liu, B., Jiang, Y., Sha, F., Govindan, R.: Cloud-enabled privacy-preserving collaborative learning for mobile sensing. In: Proceedings of SenSys, pp. 57–70 (2012)
19. Giannella, C.R., Liu, K., Kargupta, H.: Breaching euclidean distance-preserving data perturbation using few known inputs. Data & Knowledge Engineering (2012)
20. Rumelhart, D.E., Hinton, G.E., Williams, R.J.: Learning representations by back-propagating errors. Nature 323(6088), 533–536 (1986)
21. Liu, K., Giannella, C., Kargupta, H.: An attacker's view of distance preserving maps for privacy preserving data mining. In: Fürnkranz, J., Scheffer, T., Spiliopoulou, M. (eds.) PKDD 2006. LNCS (LNAI), vol. 4213, pp. 297–308. Springer, Heidelberg (2006)
22. Guo, S., Wu, X.: Deriving private information from arbitrarily projected data. In: Zhou, Z.-H., Li, H., Yang, Q. (eds.) PAKDD 2007. LNCS (LNAI), vol. 4426, pp. 84–95. Springer, Heidelberg (2007)
23. Dykstra, R.L.: Establishing the positive definiteness of the sample covariance matrix. The Annals of Mathematical Statistics 41(6), 2153–2154 (1970)
24. Grimmett, G.R., Stirzaker, D.R.: Probability and Random Processes, 3rd edn. Oxford University Press (2001)

Privacy Preserving Publication of Locations Based on Delaunay Triangulation*

Jun Luo[1,2], Jinfei Liu[3], and Li Xiong[3]

[1] Shenzhen Institutes of Advanced Technology, Chinese Academy of Sciences, China
[2] Huawei Noah's Ark Laboratory, Hong Kong, China
[3] Department of Mathematics & Computer Science, Emory University, Atlanta, USA
jun.luo@siat.ac.cn, jinfei.liu@emory.edu,
lxiong@mathcs.emory.edu

Abstract. The pervasive usage of LBS (Location Based Services) has caused serious risk of personal privacy. In order to preserve the privacy of locations, only the anonymized or perturbated data are published. At the same time, the data mining results for the perturbated data should keep as close as possible to the data mining results for the original data. In this paper, we propose a novel perturbation method such that the Delaunay triangulation of the perturbated data is the same as that of the original data. Theoretically, the Delaunay triangulation of point data set presents the neighborhood relationships of points. Many data mining algorithms strongly depend on the neighborhood relationships of points. Our method is proved to be effective and efficient by performing several popular data mining algorithms such as KNN, K-means, DBSCAN.

Keywords: Privacy Preserving, Location, Delaunay Triangulation, Data Mining.

1 Introduction

Due to the rapid development of location sensing technology such as GPS, WiFi, GSM and so on, huge amount of location data through GPS and mobile devices are produced every day. The flood of location data provides the numerous opportunities for data mining applications and geo-social networking applications. For example, mining the trajectories of floating car data (FCD) in a city could help predict the traffic congestion and improve urban planning. For individual driver, we can analyze his/her driving behavior to improve his/her profit. Moreover, many popular mobile device applications are based on locations such as FourSquare[1], Google Latitude[2], *etc.*

However, exposing location data of individuals could cause serious risk of personal privacy. For example, Buchin *et al.* [1] provide an algorithm to find the commuting pattern for an individual by clustering his/her daily trajectories. With some extra information such as timestamps, we can easily identify his/her home and working place.

* This research has been partially supported by NSF of China under project 11271351 and AFOSR under grant FA9550-12-1-0240.

[1] https://foursquare.com/

[2] http://www.google.com/latitude

V.S. Tseng et al. (Eds.): PAKDD 2014, Part I, LNAI 8443, pp. 594–605, 2014.

Another real world example about the risk of revealing location information in social networking application is that thieves planned home invasions according to user's location information and his/her planned activity published in Facebook[3].

The goals of data mining and privacy protection are often conflicting with each other and can not be satisfied simultaneously. On the one hand, we want to mine meaningful results from the dataset which requires the dataset to be precise. On the other hand, in order to protect privacy, we usually need to modify the original dataset. There are two widely adopted ways for modifying location data:

- cloaking, which hides a user's location in a larger region.
- perturbation, which is accomplished by transforming one original value to another new value.

A widely used privacy principle for cloaking is k-anonymity. A dataset is said to satisfy k-anonymity [14] [13] if each record is indistinguishable from at least $k-1$ other records with respect to certain identifying attributes. In the context of location cloaking, k-anonymity guarantees that given a query, an attack based on the query location cannot identify the query source with probability larger than $1/k$ among other $k-1$ users. The location anonymizer removes the ID of the user and transforms his location by a k-anonymizing spatial region (k-ASR or ASR), which is an area that encloses the user that issued the query, and at least $k-1$ other users [7]. The cloaking based on k-anonymity has the following shortcomings:

1. The location point is cloaked to a region. This causes trouble for point based data mining algorithms such as k-means [9], Density Based Spatial Clustering of Applications with Noise (DBSCAN [4]), K-Nearest Neighbor (KNN [2]) and so on.
2. The result of cloaking based on k-anonymity could be a very large region even for a rather small k if some points are in a sparse region, which could sacrifice the accuracy of data mining results.
3. k-anonymity only protects the anonymity of the users, but may in fact disclose the location of the user, if the cloaked area is small.

Perturbation [11] [15] has a long history in statistical disclosure control due to its simplicity, efficiency and ability to preserve statistical information. The method proposed in this paper for privacy preserving data mining of location data falls in the category of perturbation. Possible transformation functions include scaling, rotating, translation, and their combinations [8]. All previous works on perturbation used the same transformation function for all points. This is not reasonable for nonuniform distributed points (almost all real life location data are nonuniform distribution). For example, one way is to perturb each point p by randomly moving it to another point p' inside a disk with radius r and center p. In this way, if the distance between two points is less than $2r$, after perturbation, those two points could change topological relationship (left point becomes right point and right point becomes left point). This could cause serious problem for point based data mining algorithms since the results of most of those algorithms depend on the relative topological relationships.

[3] http://www.wmur.com/Police-Thieves-Robbed-Homes-Based-On-Facebook-Social-Media-Sites/-/9858568/11861116/-/139tmu4/-/index.html

Delaunay triangulation has been widely used in computer graphics, GIS, motion planning and so on. The Delaunay triangulation of a discrete point set P corresponds to the dual graph of the Voronoi diagram for P [3]. The Voronoi diagram of n points is the partition of plane into n subdivisions (called cells) such that each cell contains one point (called site) and the closest site for all point in one cell is the site in that cell. The relationship between Delaunay triangulation and Voronoi diagram is shown in Figure 1: there is an edge in Delaunay triangulation between two sites if and only if the two cells corresponding to those two sites have common edge in Voronoi diagram (in other words, two cells are neighbors). Since the Voronoi diagram captures the topological relationships of points, the Delaunay triangulation also presents the neighborhood relationships of points. Intuitively, if we can keep the Delaunay triangulation unchanged after perturbation, the results of data mining algorithms based on neighborhood relationships of points will not change or have a very small change. That is the key idea for our perturbation method. On the other hand, we want to maximize the size of the perturbation region, i.e. the uncertainty of the user's exact location, for maximum privacy protection given the constraint of maintaining Delaunay triangulation. The formal definition of the problem we want to solve is as follows:

Problem 1. Given a set of n points $S = \{p_1, p_2, ..., p_n\}$, we want to compute a continuous region R_i (perturbation region) as large as possible for each point such that for any point $p'_i \in R_i$, the topology structure of the Delaunay triangulation of the new point set $S' = \{p'_1, p'_2, ..., p'_n\}$, denoted as $DT(S')$, is the same as that of the original point set S, denoted as $DT(S)$. We use $DT(S) \sim DT(S')$ to denote the topology structure of $DT(S')$ is the same as the topology structure of $DT(S)$.

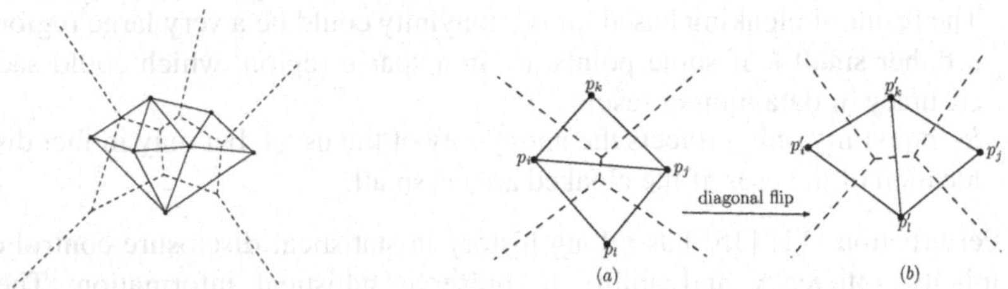

Fig. 1. Delaunay triangulation (solid lines) is the dual graph of Voronoi diagram (dashed lines)

Fig. 2. Illustration of diagonal flip from (a) to (b)

The contributions of our paper are three-fold:

1. We present a novel perturbation method that guarantees the Delaunay triangulation of the point set does not change, which means we can guarantee the utility of privacy preserving data ming algorithms.
2. In our method, the perturbation region for all points are not uniform. Each point has its own distinctive perturbation region that depends on the surrounding situation of that point. Basically, if a point is located in a dense area, its perturbation region will be much smaller than the point in a sparse area. This is different from existing perturbation method that applies a uniform perturbation to each point.

3. Since our method is point perturbation, it can be used in any point based data mining algorithms.

The rest of this paper is organized as follows. Section 2 presents the algorithms of our perturbation method and the analysis of the algorithms. Section 3 covers comprehensive experiment evaluations. Finally, section 4 concludes and points out future directions for research.

2 Algorithms for Computing Perturbation Area R_i and Perturbated Point p_i'.

In this section, we explain how to compute the perturbation area R_i for each point p_i and how to pertubate p_i to p_i'. First we solve Problem 1 for the simple case of four points in section 2.1 and then give the solution for n points case in section 2.2. In section 2.3, we show how to compute p_i' based on R_i.

2.1 A Simple Case: Four Points

The reason why we choose four points is that four points is the simplest case that could have diagonal flip[4]. The analysis of $n > 4$ cases are based on the analysis of the simplest case.

Suppose we only have four points $S = \{p_i, p_j, p_k, p_l\}$ and they are in convex position. Without loss of generality, we assume $\overline{p_ip_j}$ is the edge (diagonal) of Delaunay triangulation (see figure 2(a)). Now the problem is to find a value r such that if $d(p_i, p_i') \leq r$, $d(p_j, p_j') \leq r$, $d(p_k, p_k') \leq r$ and $d(p_l, p_l') \leq r$, then $DT(S) \sim DT(S')$ where $S' = \{p_i', p_j', p_k', p_l'\}$ and $d(.,.)$ is the Euclidean distance between two points. Note the continuous region R_i for each point p_i is the disk with the center p_i and radius r.

In order to let $DT(S) \sim DT(S')$, we need to prevent the diagonal flip case (see figure 2). That means we need to compute the largest r such that there is no diagonal flip (or to compute the smallest r that causes diagonal flip). Therefore, for two adjacent triangular faces $p_ip_jp_k$ and $p_ip_jp_l$, we want to compute the smallest r that makes the four points cocircular. This is equivalent to compute the circle that minimizes the maximum distance to the four points, or to compute the annulus of minimum width containing the points. It has been shown [5] [12], that the annulus of minimum width containing a set of points has two points on the inner circle and two points on the outer circle, interlacing angle-wise as seen from the center of the annulus (see figure 3). Therefore, the center will be the intersection of the bisectors of $\overline{p_ip_j}$ and $\overline{p_kp_l}$. Obviously the value of r can be computed in $O(1)$ time.

2.2 n Points

For n points set S, there is simple solution based on the above algorithm. First we can compute the $DT(S)$. For each pair of adjacent triangle faces, we compute one radius as

[4] Diagonal flip means the topology structure or neighborhood relationship is changed (see figure 2). For more details, please refer to the computation geometry book [3].

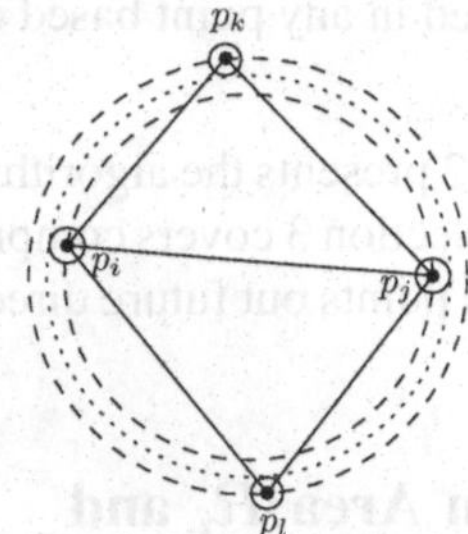

Fig. 3. The annulus of minimum width containing the four points

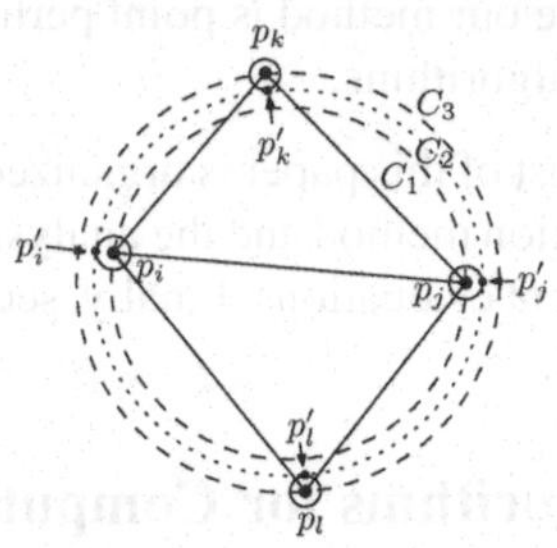

Fig. 4. The worst case scenario for four points

above such that no diagonal flip happens if four points are perturbated within the circle with that radius. Then for each point, there will be several radii. After computing all radii for all pairs of adjacent triangle faces, we get the smallest radius r. Since this is the smallest radius among all radii, no diagonal flip happens if all points are perturbated within the circle with radius r. In this way, all points have the uniform perturbation radius. But actually, except for those four points which produces the smallest radius r, all other points could be perturbated in a larger area. Notice for a point $p \in S$ with m incident edges in $DT(S)$, there are at most $2m$ ways such that p is a vertex of one pair of adjacent triangle faces, which means there are at most $2m$ radii for p. We can get the smallest radius from those $2m$ radii for the radius of p. In this way, we can still guarantee there is no diagonal flip.

Furthermore, we can improve the perturbation area for each point because in above analysis, we assume the worst case scenario. For example, in Figure 4, suppose the inner circle is C_1 and the outer circle is C_3 for the minimum width annulus containing four points p_i, p_j, p_k, p_l. Let the median axis of the annulus be the circle C_2. Let four disks with radius r_1 and center points p_i, p_j, p_k, p_l be $D_i^{r_1}, D_j^{r_1}, D_k^{r_1}, D_l^{r_1}$ and the tangent points between $D_i^{r_1}, D_j^{r_1}, D_k^{r_1}, D_l^{r_1}$ and C_2 be p'_i, p'_j, p'_k, p'_l. Then only when p_i, p_j, p_k, p_l move to p'_i, p'_j, p'_k, p'_l respectively, there is diagonal flip. In other words, p_i, p_j can not move out of C_2 and p_k, p_l can not move into C_2. Now if there is another disk $D_k^{r_2}$ for point p_k where $r_2 > r_1$ which is produced by other pair of adjacent triangles, then we know p_k can move inside $D_k^{r_2} \setminus D_2 = D_k^{r_2} \cap \overline{D_2}$ such that there is no diagonal flip (see Figure 5), where D_2 is the disk corresponding to C_2 and $\overline{D_2}$ is the area outside of D_2. Similarly, we can compute the larger perturbation area for p_i, p_j, p_l. There is only one small difference for p_i, p_j since p_i, p_j are in D_2 while p_k, p_l are outside of D_2. Therefore we need to use $D_i^{r_3} \cap D_2$ instead of $D_i^{r_3} \cap \overline{D_2}$ where r_3 is the larger radius than r_1 for p_i which is produced by other pair of adjacent triangles related to p_i.

For a point p, if there are m cocentric disks with center p and radii $r_1, r_2, ..., r_m$ and their corresponding media axis disks are $D_1, D_2, ..., D_m$, suppose $r_1 \leq r_2 \leq ... \leq r_m$ and the largest disk with radius r_m and center p is D, then the perturbation area R for p is

$$R = D \cap D_1(or\ \overline{D_1}) \cap D_2(or\ \overline{D_2}) \cap ... \cap D_m(or\ \overline{D_m})$$

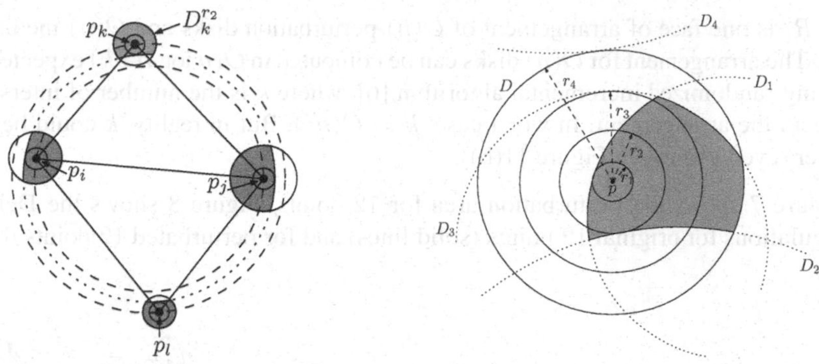

Fig. 5. The larger perturbation area (shaded area)

Fig. 6. The perturbation area (shaded area) for p

If $p \in D_i$, then we use D_i in above formula, otherwise $\overline{D_i}$ is used. See Figure 6 for example. The perturbation area is much larger than the smallest disk with radius R.

The algorithm for computing R_i is shown in Algorithm 1.

Algorithm 1. Algorithm for computing R_i

Require:
The original point set $S = p_1, p_2, ..., p_n$;
Ensure:
The cloaking regions $R_1, R_2, ..., R_n$;

Compute DT(S);
FOR each pair of adjacent triangles in DT(S) with vertices p_i, p_j, p_k, p_l and diagonal $\overline{p_i p_j}$
  Compute the minimum width annulus containing p_i, p_j, p_k, p_l and the median axis disk D_{median};
 Compute corresponding perturbation disks $D_i^r, D_j^r, D_k^r, D_l^r$ with the same radius r;
ENDFOR
FOR each point p_i
  Sort the m cocentric perturbation disks around p_i according their radius and get the largest radius disk D;
  Get the m median axis disks $D_1, D_2, ..., D_m$ corresponding to m cocentric perturbation disks;
  Compute $R_i = D \cap D_1(or\ \overline{D_1}) \cap D_2(or\ \overline{D_2}) \cap ... \cap D_m(or\ \overline{D_m})$;
ENDFOR

Theorem 1. *For any point p, the perturbation area R is larger than* 0.

Proof. Since $R = D \cap D_1(or\ \overline{D_1}) \cap D_2(or\ \overline{D_2}) \cap ... \cap D_m(or\ \overline{D_m})$ $(i = 1, ..., m)$, and $p \in D$ and $p \in D_i(or\ \overline{D_i})$, then at least $p \in R$. Actually, R at least includes the smallest disk with radius r_1. □

Theorem 2. *The perturbation areas* $R_1, R_2, ..., R_n$ *for a set of* n *points* $S = \{p_1, p_2, ..., p_n\}$ *can be computed in* $O(n^2)$ *time.*

Proof. The Delaunay triangulation of S can be computed in $O(n \log n)$ time [3]. For each pair of adjacent faces in $DT(S)$, the perturbation disks and corresponding one medial axis disk for four points can be computed in $O(1)$ time. Since there are $O(n)$ pairs of adjacent faces, there are $O(n)$ perturbation disks and $O(n)$ medial axis disks.

Each R_i is one face of arrangement of $O(n)$ perturbation disks and $O(n)$ medial axis disks. The arrangement for $O(n)$ disks can be computed in $O(n \log n + k)$ expected time by using randomized incremental algorithm [6], where k is the number of intersection points in the arrangement. In worst case, $k = O(n^2)$. But in reality, k could be much smaller (even linear, see Figure 11(b)). □

Figure 7 shows the perturbation area for 12 points. Figure 8 shows the Delaunay triangulations for original 12 points (solid lines) and for perturbated 12 points (dashed lines).

Fig. 7. The perturbation area for 12 points

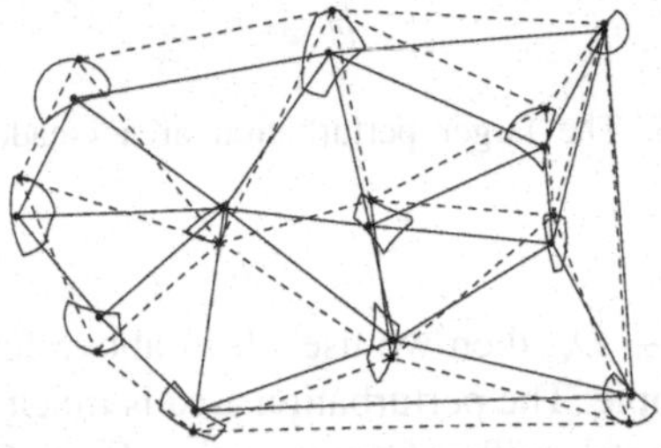

Fig. 8. The Delaunay triangulations for original 12 points (solid lines) and for perturbated 12 points (dashed lines)

2.3 Point Perturbation

After get perturbation region R_i, we need to perturbate p_i inside R_i for publication. In order to achieve maximum privacy protection, we perturbate p_i to the boundary of R_i as follows: choose a random angle $\theta \in [0, 2\pi]$, shoot a ray with angle θ and find the intersection point p_i' of this ray with boundary of R_i, and let p_i' be the perturbated point (see Figure 9).

3 Experimental Evaluation

3.1 Data Set

We use three datasets from University of Eastern Finland's clustering datasets[5]: Flame (240 points), R15 (600 points) and Jain (373 points) for testing our method's utility on KNN, K-means and DBSCAN respectively. The reason we choose those three algorithms is that KNN is the most popular neighborhood querying algorithm in database, K-means and DBSCAN are two popular clustering algorithms in data mining. The other important reason is that the results of those three algorithms all depends on the neighborhood relationship between points. We also generate 29 synthetic datasets with the number of points ranging from 100 to 100000 for testing the performance of our perturbation algorithm.

[5] `http://cs.joensuu.fi/sipu/datasets/`

3.2 Utility Results

The definition of utility varies for different privacy preserving data mining algorithms. In this paper, we make use of the two well-known criteria, i.e., the precision and recall rates [10]. We will give the definitions of precision and recall for privacy preserving KNN, K-means, DBSCAN in the following three subsections and also give the experimental results respectively.

Utility of Privacy Preserving KNN. For KNN, the precision and recall rate is defined as follows:

$$Precision_{KNN} = \frac{||Q(p_i) \cap Q(p_{i'})||}{||Q(p_{i'})||}$$

$$Recall_{KNN} = \frac{||Q(p_i) \cap Q(p_{i'})||}{||Q(p_i)||}$$

where $Q(p_i)$ is the set of points that are the K nearest neighbors of p_i, $p_{i'}$ is the perturbated point of p_i. Note $Precision_{KNN} = Recall_{KNN}$ since $||Q(p_{i'})|| = ||Q(p_i)||$ for fixed K. Therefore we only report precision for KNN. We perform 100 rounds of perturbations and query KNN on each point and get the average value of precision:

$$AP_{KNN} = \frac{\sum_{j=1}^{m} \sum_{i=1}^{n} \frac{||Q(p_i) \cap Q(p_{i'}^j)||}{||Q(p_{i'}^j)||}}{m \times n}$$

where n is the number of query times (in this paper, we perform query on each point, therefore $n = 240$ for Flame dataset) and m is the number of rounds of perturbation ($m = 100$ in this paper), $p_{i'}^j$ is the perturbated point of p_i after the jth round of perturbation. Therefore AP_{KNN} is the average precision rate per query. We test $Precision_{KNN}$ for $K = 1, ..., 100$. For comparison we also perform the **uniform** purturbation with uniform distance r_{max} where

$$r_{max} = \max_i \{d(p_i, p_i') | p_i' \in \partial R_i\}$$

Figure 10 shows the precision rate of two perturbation methods for KNN algorithm. We can see precision rate based on Delaunay triangulation AP_{KNN}^{DT} is always better than that of uniform perturbation $AP_{KNN}^{Uniform}$. When $K < 10$, AP_{KNN}^{DT} is 20 percent higher than $AP_{KNN}^{Uniform}$. When K becomes larger, $AP_{KNN}^{DT} - AP_{KNN}^{Uniform}$ becomes smaller. $AP_{KNN}^{DT} - AP_{KNN}^{Uniform} = 3.93\%$ when $K = 100$. Also when K become larger, both $AP_{KNN}^{Uniform}$ and AP_{KNN}^{DT} seem to converge. This is reasonable since the K nearest neighbors may only change for the far away points (for example, the Kth nearest point) no matter how the points are perturbated.

Utility of Privacy Preserving K-means. For K-means, we adopt B-CUBED algorithm to compute the precision and recall rate of each point for clustering results:

$$Precision_{K-means}^{p_i} = \frac{||C_i \cap C_{i'}||}{||C_{i'}||}$$

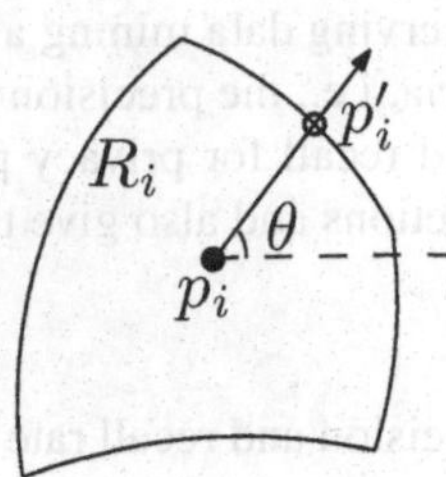

Fig. 9. Illustration of perturbation from p_i to p'_i

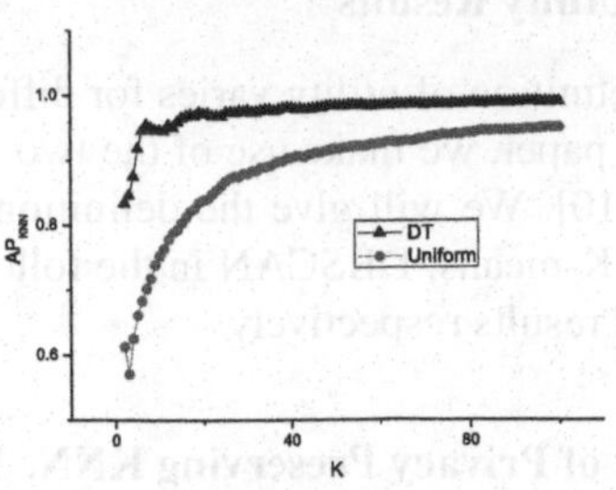

Fig. 10. The utility of privacy preserving KNN. x axis is K and y axis is AP_{KNN}. Red circle curve is for uniform perturbation and black triangle curve is for Delaunay triangulation based perturbation.

$$Recall^{p_i}_{K-means} = \frac{||C_i \cap C_{i'}||}{||C_i||}$$

where $p_i \in C_i, p'_i \in C_{i'}$ and p'_i is the perturbated point of p_i. 100 rounds of perturbations are performed. Therefore we can get the average value of precision and recall rates per point:

$$AP_{K-means} = \frac{\sum_{j=1}^{m} \sum_{i=1}^{n} \frac{||C_i \cap C^j_{i'}||}{||C^j_{i'}||}}{m \times n}$$

$$AR_{K-means} = \frac{\sum_{j=1}^{m} \sum_{i=1}^{n} \frac{||C_i \cap C^j_{i'}||}{||C_i||}}{m \times n}$$

where n is the number of points and m is the number of rounds of perturbation, $p_i \in C_i, p^j_{i'} \in C^j_{i'}$ and $p^j_{i'}$ is the perturbated point of p_i after the jth round of perturbation. In this paper, we use R15 dataset for K-means algorithms. Since we already know there are 15 clusters, we set $k = 15$ and get the results as shown in Table 1.

Table 1. Average precision and recall rate for 15-means clustering on R15 dataset

	DT	Uniform
AP	0.99951371	0.99419382
AR	0.99951316	0.9942013

We can see precision and recall rate of privacy preserving 15-means for DT based perturbation are better than those for uniform perturbation with uniform radius r_{max}. Since the precision and recall rate for uniform perturbation are already very high, it's very difficult to achieve higher precision and recall rate for DT based perturbation.

Utility of Privacy Preserving DBSCAN. For DBSCAN, the utility $Precision_{DBSCAN}$ is defined as the same as $Precision_{K-means}$ since they are both clustering algorithms. The difference is that we don't need to set the number of cluster K in DBSCAN. For Jain dataset, there are two clusters for original dataset. DBSCAN requires two parameters: ε and the minimum number of points required to form a cluster ($minPts$). In our experiments, we set $\varepsilon = 2.4$ and $minPts = 20$ such that DBSCAN algorithm produces two clusters over original Jain dataset. The results for the utility of uniform perturbation and DT based perturbation are shown in Table 2.

Table 2. Average precision and recall rate for DBSCAN clustering on Jain dataset

	DT	Uniform
AP	1	0.76515634
AR	1	0.9349843

We can see the points in each cluster keep exactly same after DT based perturbation while that could change a lot after uniform perturbation with uniform radius r_{max}.

3.3 Privacy

We define the privacy for DT based perturbation as follows: $P = \frac{\sum_{j=1}^{m}\sum_{i=1}^{n} R_i^j}{m \times n}$ where R_i^j means the perturbation area for ith point after jth round of perturbation. Therefore P means the average perturbation area for one point. However, P depends on the whole area of point sets. In order to cancel out this effect, we define the privacy ratio P_r as follows: $P_r = P/Area(CH(S))$ where $CH(S)$ is the convex hull of point set S and $Area(CH(S))$ is the area of the convex hull of S. The experimental results of privacy for three datasets $R15$, $Jain$, $Flame$ are listed in Table 3.

Table 3. Privacy for DT based perturbation

	Jain	Flame	R15
P	0.0235319	0.0259528	0.000893764
$Area(CH(S))$	639.819	132.049	138.938
P_r	3.6779E-05	1.96539E-04	6.43283E-06

We can see the average perturbation areas for Jain and Flame datasets are similar while the average perturbation area for R15 dataset are much smaller which could be caused by the higher average neighboring point density for each point in R15 dataset. The other possible reason is there could be many degenerate cases in R15 dataset such that many four points groups are cocircular (or almost cocircular).

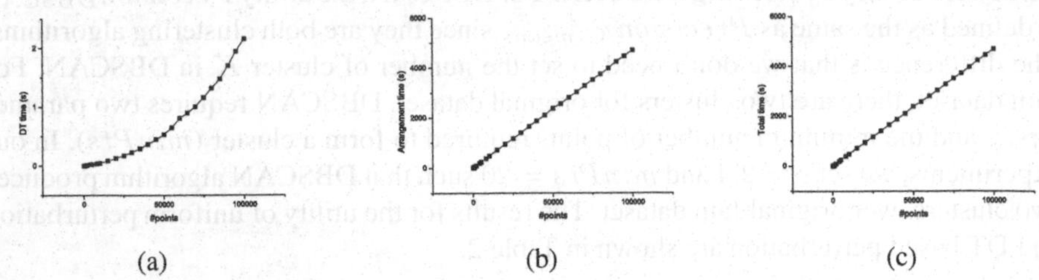

Fig. 11. x axis is the number of points and y axis is the running time of computing (a) Delaunay triangulations, (b) arrangement of perturbation areas, (c) total running time

3.4 Performance of Our Perturbation Algorithm

To test the performance of our perturbation algorithm, the 29 synthetic datasets are used with various number of points from 100 to 100000. We perform all experiments on our Thinkpad laptop with the following configurations: Ubuntu 13.04 operating system, 2 Gbyte of memory, 2.6GHz intel core i3 CPU. Since our perturbation algorithm mainly consists of two parts: computing Delaunay triangulation and computing the arrangement of perturbation areas, we give the results of running time for both parts and the total running time as well.

Figure 11(a) shows the running time for Delaunay triangulation. It almost fits the theoretical bound of $O(n \log n)$ running time.

Figure 11(b) shows the running time for computing the arrangement of perturbation areas. Although, theoretically the running time is $O(n^2)$ for the worst case, in practice, the running time is almost linear.

Figure 11(c) shows the total running time. Since the computation of perturbation areas is the dominant factor (for example, for 100000 points dataset, computing Delaunay triangulation only takes only around 2 seconds while computing the arrangement of perturbation areas takes 4690 seconds), the total running time is also linear. Therefore, in practice, our algorithm is very efficient.

4 Conclusions and Future Work

In this paper, we present a novel method to address the problem of location privacy. Our method achieves both high assurance privacy and good performance. Specifically, our method perturbates each point distinctively according to its own environment instead of uniform perturbation. In this way, the attackers are more difficult to guess original location. Furthermore our perturbation method guarantees that the Delaunay triangulation of point set does not change, which means we can guarantee the utility of privacy preserving data ming algorithms. We also develop a privacy model to analyze the degree of privacy protection and evaluate the utilities of our approach through three popular data mining algorithms. Extensive experiments show our method is effective and efficient for location based privacy preserving data mining.

Several promising directions for future work exist. First of all, the perturbation region or the perturbation distance for each point could be larger. We need to prove it theoretically and find the optimal values in terms of perturbation distance and area. Also we can apply our method on other data mining algorithms to see whether our method is a universal method for location based privacy preserving data mining algorithms. This is significant since many current methods are only good for one kind of data mining algorithm.

References

1. Buchin, K., Buchin, M., Gudmundsson, J., Löffler, M., Luo, J.: Detecting commuting patterns by clustering subtrajectories. Int. J. Comput. Geometry Appl. 21(3), 253–282 (2011)
2. Cover, T., Hart, P.: Nearest neighbor pattern classification. IEEE Transactions on Information Theory 13(1), 21–27 (1967)
3. De Berg, M., Cheong, O., Van Kreveld, M., Overmars, M.: Computational geometry: algorithms and applications. Springer-Verlag New York Incorporated (2008)
4. Ester, M., Kriegel, H.-P., Sander, J., Xu, X.: A density-based algorithm for discovering clusters in large spatial databases with noise. In: KDD, pp. 226–231 (1996)
5. Garcia-Lopez, J., Ramos, P.A.: Fitting a set of points by a circle. In: Symposium on Computational Geometry, pp. 139–146 (1997)
6. Goodman, J., O'Rourke, J.: Handbook of discrete and computational geometry. Chapman & Hall/CRC (2004)
7. Kalnis, P., Ghinita, G., Mouratidis, K., Papadias, D.: Preventing location-based identity inference in anonymous spatial queries. IEEE Trans. Knowl. Data Eng. 19(12), 1719–1733 (2007)
8. Lin, D., Bertino, E., Cheng, R., Prabhakar, S.: Location privacy in moving-object environments. Transactions on Data Privacy 2(1), 21–46 (2009)
9. MacQueen, J., et al.: Some methods for classification and analysis of multivariate observations. In: Proceedings of the Fifth Berkeley Symposium on Mathematical Statistics and Probability, California, USA, vol. 1, p. 14 (1967)
10. Manning, C.D., Raghavan, P., Schütze, H.: Introduction to Information Retrieval. Cambridge University Press, New York (2008)
11. Rastogi, V., Hong, S., Suciu, D.: The boundary between privacy and utility in data publishing. In: VLDB, pp. 531–542 (2007)
12. Rivlin, T.: Approximation by circles. Computing 21(2), 93–104 (1979)
13. Sweeney, L.: Achieving k-anonymity privacy protection using generalization and suppression. International Journal of Uncertainty, Fuzziness and Knowledge-Based Systems 10(5), 571–588 (2002)
14. Sweeney, L.: k-anonymity: A model for protecting privacy. International Journal of Uncertainty, Fuzziness and Knowledge-Based Systems 10(5), 557–570 (2002)
15. Xiao, X., Tao, Y., Chen, M.: Optimal random perturbation at multiple privacy levels. PVLDB 2(1), 814–825 (2009)

A Fast Secure Dot Product Protocol with Application to Privacy Preserving Association Rule Mining

Changyu Dong[1] and Liqun Chen[2]

[1] Department of Computer and Information Sciences,University of Strathclyde, Glasgow, UK
changyu.dong@strath.ac.uk
[2] Hewlett-Packard Laboratories, Bristol, UK
liqun.chen@hp.com

Abstract. Data mining often causes privacy concerns. To ease the concerns, various privacy preserving data mining techniques have been proposed. However, those techniques are often too computationally intensive to be deployed in practice. Efficiency becomes a major challenge in privacy preserving data mining. In this paper we present an efficient secure dot product protocol and show its application in privacy preserving association rule mining, one of the most widely used data mining techniques. The protocol is orders of magnitude faster than previous protocols because it employs mostly cheap cryptographic operations, e.g. hashing and modular multiplication. The performance has been further improved by parallelization. We implemented the protocol and tested the performance. The test result shows that on moderate commodity hardware, the dot product of two vectors of size 1 million can be computed within 1 minute. As a comparison, the currently most widely used protocol needs about 1 hour and 23 minutes.

1 Introduction

Data mining, which generally means the process of discovering interesting knowledge from large amounts of data, has become an indispensable part of scientific research, business analytics, and government decision making. Privacy has always been a concern over data mining. Regulations on data privacy become tighter and tighter. Legislation includes various US privacy laws (HIPAA, COPPA, GLB, FRC, etc.), European Union Data Protection Directive, and more specific national privacy regulations. Since the groundbreaking work [4,21], there have been a lot of research in privacy preserving data mining (PPDM). However, privacy is not cost free. The overhead of privacy preserving protocols is often too high when a large amount of data needs to be protected. Therefore, how to make protocols secure and also efficient becomes a major challenge to PPDM.

The focus of this paper is association rule mining [2], which discovers association patterns from different sets of data and this is one of the most popular and powerful data mining methods. There have been several privacy preserving solutions for association rule mining. According to [27], those solutions can be divided into two approaches: randomization-based and cryptography-based. In the randomization-based approach [10,24,32] , there is a centralized server who aggregates data from many clients and discovers association rules. To protect privacy, data owners (the clients) use some statistical methods to distort the data before sending it to the server. In the distortion procedure,

V.S. Tseng et al. (Eds.): PAKDD 2014, Part I, LNAI 8443, pp. 606–617, 2014.

noise is added to the data in order to mask the values of the records while still allow the distribution of the aggregation data to be recovered. The goal is to prevent the server from learning the original data while still be able to discover the rules. However, it has been shown in [20,17] that in many cases, the original data can be accurately estimated from the randomized data using techniques based on Spectral Filtering, Principal Component Analysis and Bayes estimate. In recent years, the cryptography-based approach becomes very popular. This is because cryptography provides well-defined models for privacy and also a large toolset. In this approach, data is usually split among multiple parties horizontally [19,25] or vertically [28,29]. In general, associate rule mining on vertically partitioned data is much more computationally intensive than on horizontally partitioned data because little data summarization can be done locally. Therefore performance is a more acute problem when dealing with vertically partitioned data.

As we will see in section 2, the core of the algorithm for association rule mining on vertically partitioned data is a secure dot product protocol. Therefore the key to make privacy preserving association rule mining practical is to improve the efficiency of the underlying dot product protocol. Many secure dot product protocols have been developed in the past [8,9,28,18,1,13,11,29,5,6,26]. However, [8,28,18] have been shown to be insecure or incorrect, [9] requires an additional third party, and the others require at least $O(n)$ modular exponentiation operations, where n is the size of the input vectors. Modular exponentiation is a costly operation, thus those protocols are very inefficient and cannot be used in real applications that handle large datasets.

The main contribution of this paper is a very efficient secure dot product protocol, which can significantly accelerate privacy preserving association rule mining. Our analysis and implementation show that our dot product protocol is orders of magnitude faster than previous ones. The efficiency comes from the fact that the protocol relies mostly on cheap cryptographic operations, i.e. hashing, modular multiplication and bit operations. The efficiency can be further increased by parallelization. It is worth mentioning that in addition to association rule mining, secure dot product has also been shown to be an important building block in many other PPDM algorithms such as naive Bayes classification [30], finding K-Nearest Neighbors (KNN) [31] and building decision trees [9]. Therefore our protocol can also be used to boost the performance of those PPDM tasks. We also provide a informal security analysis. We have proved the protocol to be secure in the semi-honest model in terms of multiparty secure computation [14]. For space reason, the proof is omitted here and will appear in the full version.

The paper is organized as follows: in Section 2, we briefly review the privacy preserving Apriori algorithm for association rule mining on vertically partitioned data; in Section 3, we describe our secure dot product protocol and the underlying cryptographic building blocks; in Section 4, we report the performance measurements based on the prototype we have implemented, and compare it against the most widely used secure dot product protocol; in section 5, we conclude the paper and discuss possible future work.

2 Privacy Preserving Association Rule Mining

Association rules show attribute values that appear frequently together in a given dataset. The problem can be formalized as follows [2]: let $I = \{i_1, i_2, \ldots, i_m\}$ be a set of items.

Algorithm 1. Privacy Preserving Apriori Algorithm [28]

L_1 ={large 1-itemsets};
for $k = 2; L_{k-1} \neq \emptyset; k++$ **do**
    C_k =apriori-gen(L_{k-1});
    for each *candidate* $c \in C_k$ **do**
        if *all the attributes in c are entirely at* P_1 *or* P_2 **then**
            that party independently calculates *c.count*;
        else
            let P_1 have attributes 1 to l and P_2 have the remaining m attributes
            construct $\boldsymbol{A}$ on P_1's side and $\boldsymbol{B}$ on P_2's side where $\boldsymbol{A} = \prod_{i=1}^{l} \boldsymbol{X}_i$ and $\boldsymbol{B} = \prod_{i=l+1}^{l+m} \boldsymbol{X}_i$;
            compute $c.count = \boldsymbol{A} \cdot \boldsymbol{B}$
        end
        $L_k = L_k \cup c | c.count \geq minsup$;
    end
end
Answer=$\cup_k L_k$;

Let D be a set of transactions and T be a transaction. Each transaction T is a set of items such that $T \subseteq I$. We say T contains X if $X \subseteq T$. An association rule is of the form $X \Rightarrow Y$ where $X \subset I, Y \subset I$ and $X \cap Y = \emptyset$. A rule $X \Rightarrow Y$ holds in the transaction set D with confidence c if $c\%$ of transactions in D that contain X also contain Y. A rule $X \Rightarrow Y$ has support s in the transaction set D if $s\%$ transactions in D contain $X \cup Y$. The goal of association rule mining is to find all rules having high support and confidence. One classic algorithm for association rule mining is the Apriori algorithm [3]. The Apriori algorithm can find all frequent itemsets, i.e. itemsets that appear in the transaction set at least $minsup$ times, where $minsup$ is a threashold. After all frequent itemsets have been found, association rules can be generated straightforwardly.

The computation model we consider is two parties each holds part of the transaction set that is vertically partitioned, i.e. one party holds some attributes and the other party holds the rest attributes. In [28], a secure algorithm for finding association rules on vertically partitioned data was proposed. The algorithm (see Algorithm 1) is a straightforward extension of the Apriori algorithm. For each attribute, the party holding it can create a binary vector whose length is the size of the transaction set: the absence or presence of the attribute can be represented as 0 and 1. As shown in [28], we can reduce privacy preserving association rule mining to securely computing the dot products of the binary vectors. Candidate itemsets can be generated as in the Apriori algorithm.In the simplest case where the candidate itemset has only two attributes, for example an attribute X_1 held by P_1 and X_2 held by P_2, the two parties can compute the dot product of the corresponding vectors $\boldsymbol{X}_1 \cdot \boldsymbol{X}_2 = \sum_{i=1}^{n} \boldsymbol{X_1}[i] \cdot \boldsymbol{X_2}[i]$. Then the dot product is the support count, i.e. how many times the itemset appears in the transaction set, and is tested against a predefined threshold $minsup$. If the dot product is greater or equal to the threashold, then the candidate itemset is a frequent itemset. This approach can be easily extended to itemsets that have more attributes. Most of the steps in Algorithm 1 can be done locally. The only step that needs to be computed securely between the two parties is step 10, where a dot product needs to be calculated. Being the only cryptographic step, step 10 is the most time consuming part of the algorithm, and its running time dominates the total running time of the algorithm. Therefore improving the performance of the secure dot product protocol is key to improving the performance of the entire mining algorithm.

3 The Secure Dot Product Protocol

To make privacy preserving association rule mining efficient, we need an efficient secure dot product protocol. In this section, we present such a protocol and analyze its security. The protocol is built on two well-defined cryptographic primitives: the Goldwasser–Micali Encryption and the Oblivious Bloom Intersection.

3.1 Cryptographic Building Blocks

Goldwasser–Micali Encryption. The Goldwasser–Micali (GM) encryption is a semantically secure encryption scheme [15]. The algorithms are as follows:

- Key generation algorithm $\mathcal{G}$: it takes a security parameter k, and generates two large prime numbers p and q, computes $n = pq$ and a quadratic non-residue x for which the Jacobi symbol is 1. The public key pk is (x, n), and the secret key sk is (p, q).
- Encryption algorithm $\mathcal{E}$: to encrypt a bit $b \in \{0, 1\}$, it takes b and the public key (x, n) as input, and outputs the ciphertext $c = y^2 x^b \bmod n$, where y is randomly chosen from Z_n^*.
- Decryption algorithm $\mathcal{D}$: it takes a ciphertext c and the private key as input, and outputs the message b: $b = 0$ if the Legendre symbol $(\frac{c}{p}) = 1$, $b = 1$ otherwise.

There are two reasons why we use the GM encryption in our protocol: it is efficient for our purpose and it is homomorphic. The inputs to our protocols are bit vectors and they must be encrypted bit-by-bit in the protocol. Notice that the GM encryption and decryption operations involve only modular multiplications and Legendre symbol computation. Both are very efficient and the computational costs are on the same order of symmetric key operations. This is in contrast to the public key encryption schemes required by the other secure dot product protocols. Those encryption schemes require modular exponentiations, which is usually thousands of time slower. Therefore using the GM encryption makes our protocol much more efficient. The GM encryption is known to be homomorphic and allows computation to be carried out on ciphertexts. More specifically, for two ciphertexts $c_1 = \mathcal{E}(pk, b_1)$ and $c_2 = \mathcal{E}(pk, b_2)$, their product $c_1 c_2$ is a ciphertext of $b_1 \oplus b_2$, where $\oplus$ is the bitwise exclusive or (XOR) operator. This property will be used in our protocol to allow a party to blindly randomize ciphertexts generated by the other party.

Oblivious Bloom Intersection. Oblivious Bloom Intersection (OBI) [7] is an efficient and scalable private set intersection protocol. A private set intersection protocol is a protocol between two parties, a server and a client. Each party has a private set as input. The goal of the protocol is that the client learns the intersection of the two input sets, but nothing more about the server's set, and the server learns nothing. In Section 3.3, we will see how to convert the computation of the dot product of two binary vectors into a set intersection problem. Previously, the private set intersection protocols are equally, or even more, costly as secure dot product protocols. Therefore PSI based secure dot product protocols have no advantage in terms of performance. This situation is changed by the recently proposed OBI protocol. The OBI protocol adapts a very different approach for computing set intersections. It mainly bases on efficient hash operations. Therefore it is significantly faster than previous private set intersection protocols. In addition, the protocol can also be parallelized easily, which means performance can be further improved by parallelization. The protocol is secure in the semi-honest model and an enhanced version is secure in the malicious model.

Briefly, the semi-honest OBI protocol works as follows: the client has a set C and the server has a set S. Without loss of generality, we assume the two sets have the same size, i.e. $|C| = |S| = w$. The protocol uses two data structures: the Bloom filter and the garbled Bloom filter. Both data structures can encode sets and allow membership queries. The two parties agree on a security parameter k and choose k independent uniform hash functions. The hash functions are used here to build and query the filters. The client encodes its set into a Bloom filter, which is a bit vector. The server encodes its set into a garbled Bloom filter, which is a vector of k-bit strings. The size of the Bloom filter (and also the garbled Bloom filter) depends on the security parameter k and the set cardinality w. More precisely, the filter size is $k \cdot w \cdot \log_2 e$. The client then runs an oblivious transfer protocol with the server. The oblivious transfer protocol can be implemented efficiently with hash functions. The number of hash operations required by the oblivious transfer protocol is linear in the filter size. The result of the oblivious transfer protocol is that the server learns nothing and the client obtains another garbled Bloom filter that encodes the set intersection $C \cap S$. The client can query all elements in C against this garbled Bloom filter to find the intersection.

We refer the readers to [7] for more details regarding OBI. We will show in Section 3.3 that by combining OBI with the GM encryption, we can get a much more efficient secure dot product protocol.

3.2 Security Model

We study the problem within the general framework of Secure Multiparty Computation. Briefly, there are multiple parties each has a private input, they engage in a distributed computation of a function such that in the end of the computation, no more information is revealed to a participant in the computation than what can be inferred from that participants input and output [14]. Several security models have been defined in this framework. In this paper, we consider the semi-honest model [14]. In this model, adversaries are honest-but-curious, i.e. they will following the protocol specification but try to get more information about honest party's input. Although this is a weak model, it is appropriate in many real world scenarios where the parties are not totally untrusted. Besides, semi-honest protocols can be upgraded to full security against malicious adversaries using a generic technique [14].

3.3 The Protocol

P_1's input: A binary vector $\boldsymbol{X_1}$ of size n and density at most d.
P_2's input: A binary vector $\boldsymbol{X_2}$ of size n and density at most d.
Other input: A security parameter k, the GM Encryption scheme $(\mathcal{G}, \mathcal{E}, \mathcal{D})$.
Phase 1: Sets initialization

1.1 P_1 generates a random key pair (pk, sk) for the GM encryption, and sends the public key to P_2.
1.2 P_1 then encrypts $\boldsymbol{X_1}$ bit by bit using the public key, and sends the ciphertext $(\mathcal{E}(pk, \boldsymbol{X_1}[1]), ..., \mathcal{E}(pk, \boldsymbol{X_1}[n]))$ as an ordered list to P_2.
1.3 P_2 generates a vector of n distinct random numbers $\boldsymbol{R}$, then creates an empty set T_2, for $1 \leq j \leq n$, if $X_2[j] = 1$, $T_2 = T_2 \cup \boldsymbol{R}[j]$.
1.4 P_2 generates a set C that has n elements $\{c_1, ..., c_n\}$. For $1 \leq j \leq n$, $c_j = (\boldsymbol{R}[j], \mathcal{E}(pk, X_1[j]) \cdot \mathcal{E}(pk, 0))$.
1.5 P_2 chooses a random permutation θ, and sends $C' = \theta(C)$ to P_1.
1.6 P_1 creates an empty set T_1, then for each $1 \leq i \leq n$ and $c_i' = (u_i, v_i)$ in C', if $\mathcal{D}(sk, v_i) = 1$, then $T_1 = T_1 \cup u_i$.

Phase 2: Output Dot Product

2.1 P_1 uses the set T_1 and P_2 uses the set T_2 as inputs. They engage in an execution of the OBI protocol. P_1 plays the role of the OBI client and gets $T_1 \cap T_2$, P_2 plays the role of the OBI server and gets nothing.
2.2 P_1 counts the element in the intersection and outputs $d = |T_1 \cap T_2|$, which is the dot product $\boldsymbol{X_1} \cdot \boldsymbol{X_2}$.

Fig. 1. The Secure Dot Product Protocol

A method [16] that converts the binary vector dot product problem into computing the cardinality of set intersection is as follows: given two binary vectors $\boldsymbol{X_1}$ and $\boldsymbol{X_2}$ both of cardinality n, we can construct two sets S_i $(i \in \{1, 2\})$ such that j is an element of S_i if the $\boldsymbol{X_i}[j] = 1$, i.e. $S_i = \{j \mid \boldsymbol{X_i}[j] = 1\}$. Then the dot product $\boldsymbol{X_1} \cdot \boldsymbol{X_2} = |S_1 \cap S_2|$. In the light of this, in [16] the authors proposed an approximate secure dot product protocol that works by estimating the cardinality of the set intersection.

Our protocol is also based on this set intersection cardinality idea, but can compute the exact intersection cardinality rather than just an approximation. The protocol is shown in Fig. 1. We now explain the rationale behind the design and discuss the security informally. Our insight is that we can build a secure binary vector dot product protocol on top of a private set intersection protocol. It seems trivial: let the two parties convert their private vectors into sets as described above, then run a private set intersection protocol between them, the protocol outputs the set intersection $S_1 \cap S_2$ thus $|S_1 \cap S_2|$ can be obtained as a by-product. This solution, however is flawed. The main problem is that it leaks more information than desired. In fact, the party who obtains the intersection as the output of the private set intersection protocol learns not only the dot product, but also some bits in the other party's vector: for any j in the intersection, the jth-bit of the other party's vector must be 1. To prevent this, in our protocol, we do not use the sets S_1 and S_2 directly, but map them into T_1 and T_2 such that the following two properties hold: (1) $|T_1 \cap T_2| = |S_1 \cap S_2|$, (2) at the end of the protocol P_1 obtains $T_1 \cap T_2$, from which it can obtain $|S_1 \cap S_2|$, but no other information.

The first property ensures the correctness of our protocol. To see why this property holds, observe that in step 1.3, P_2 maps each $1 \leq j \leq n$ to a random number $\boldsymbol{R}[j]$. The mapping is used by both parties to map S_i to T_i: for each $j \in S_i$, or equivalently for each j such that $\boldsymbol{X_i}[j] = 1$, its counterpart $\boldsymbol{R}[j]$ is in T_i. Thus $|T_1 \cap T_2| = |S_1 \cap S_2|$ is guaranteed. The difficulty here is how to let P_1 correctly generate T_1. For P_2, it can easily generate T_2 because it knows the mapping. However, to maintain privacy, P_1 is not allowed to know the mapping. In order to allow P_1 to generate T_1, P_2 creates a labelled permutation of the vector $\boldsymbol{R}$ (step 1.4, 1.5). For each $\boldsymbol{R}[j]$, the label assigned to it is a ciphertext of $\boldsymbol{X}_1[j]$. In step 1.6, P_1 can decrypt the label, if it encrypts 1, then P_1 knows the corresponding vector element should be put into T_1.

The second property ensures the privacy of our protocol. It is easy to see why P_2 gets no more information: the only message P_2 gets in phase 1 is the encrypted version of $\boldsymbol{X_1}$, then by the security of the GM encryption, P_2 gets no information about $\boldsymbol{X_1}$. Phase 2 is essentially a PSI execution and P_2 plays the role of the PSI server, thus by the security of the PSI protocol, it learns nothing. For P_1, it receives C' in step 1.6 and each c'_i is a tuple $(u_i, v_i) = (\boldsymbol{R}[j], \mathcal{E}(pk, X_1[j]) \cdot \mathcal{E}(pk, 0))$. C' allows P_1 to generate T_1 but no more than that. Since $C' = \theta(C)$ and θ is a random permutation, the order of an element in the set i does not leak any information. Also $u_i = \boldsymbol{R}[j]$ for some j is a random number generated independently of j, and the mapping from j to $\boldsymbol{R}[j]$ is known only by P_2, thus u_i leaks no information. On the other hand $v_i = \mathcal{E}(pk, X_1[j]) \cdot \mathcal{E}(pk, 0)$, so v_i incorporates randomness generated by P_2, therefore P_1 cannot link it back to the ciphertext $\mathcal{E}(pk, X_1[j])$ generated by itself. $\mathcal{D}(sk, v_i)$ gives only 1 bit information, i.e. there exists a j such that $\boldsymbol{X_i}[j] = 1$. That is exactly what P_1 should know. At the end of phase 1, P_1 holds a set T_1 which it knows is a equivalent of S_1 but cannot link the elements in T_1 back to elements in S_1: any element in T_1 can be any element in S_1 with equal likelihood. A consequence is that the output in step 2.1 $T_1 \cap T_2$ gives P_1 no more information about $S_1 \cap S_2$ other than the cardinality of the intersection.

The protocol in Fig. 1 has only one output: P_1 gets the dot product and P_2 gets nothing. In some applications, it is preferable to have two outputs such that P_1 gets a number a and P_2 gets a number b and $a + b$ is the dot product. This can be done by executing the protocol in Fig. 1 twice and let the parties switch roles in the two executions. Suppose Alice has a vector $\boldsymbol{Y_1}$ and Bob has a vector $\boldsymbol{Y_2}$, both are of size n. Bob first splits $\boldsymbol{Y_2}$ randomly into two n-bit vectors $\boldsymbol{Y_3}$ and $\boldsymbol{Y_4}$, such that $\boldsymbol{Y_2} = \boldsymbol{Y_3} + \boldsymbol{Y_4}$. In the first round, Alice plays the role of P_1 and uses $\boldsymbol{Y_1}$ as her input. Bob plays the role of P_2 and uses $\boldsymbol{Y_3}$ as input. Alice gets an output a after the execution. Then in the second round, Alice plays the role of P_2 and still uses $\boldsymbol{Y_1}$ as her input. Bob plays the role of P_1 and uses $\boldsymbol{Y_4}$ as his input. Then Bob receives an output b. The outputs satisfy that $a + b = \boldsymbol{Y_1} \cdot \boldsymbol{Y_2}$. To see that, observe that $a = \boldsymbol{Y_1} \cdot \boldsymbol{Y_3}$ and $b = \boldsymbol{Y_1} \cdot \boldsymbol{Y_4}$, so $a + b = \boldsymbol{Y_1} \cdot (\boldsymbol{Y_3} + \boldsymbol{Y_4}) = \boldsymbol{Y_1} \cdot \boldsymbol{Y_2}$.

3.4 Efficiency

From the description of the protocol in Fig. 1 we can see that the computational and communicational complexities of phase 1 are both $O(n)$, where n is the size of the vector. Phase 2 is a single execution of the OBI protocol. Therefore the computational and communicational complexities of this phase are also $O(n)$.

More specifically: in phase 1 the total computational cost is $3n$ modular multiplications plus computing n Legendre symbols, in phase 2 the total computational cost is $2d \cdot k \cdot n(1 + \log_2 e)$ hash operations, where d is the density of the vectors, k is the security parameter (e.g. 80 or 128) and e is the natural logarithm constant. As a comparison, the protocol in [13], which is based on Paillier homomorphic encryption, requires n modular exponentiations and n modular multiplications. At first glance it seems that our protocol is not as efficient as the protocol in [13]. However a closer analysis shows the opposite. This is because modular exponentiation, required by [13], is much more expensive than the operations requires by our protocol. To better illustrate the difference, we show the running time of the cryptographic operations at 80-bit security in Table 1. The modulus used is 1024-bit. The time was measured on a Mac Pro with 2.4 GHz Intel Xeon CPUs. For the cheap operations, the time shown is the average of 1,000,000 executions and for modular exponentiation, the time is the average of 1,000 executions. We can see the difference is 3 orders of magnitude. At higher security levels, the difference is even bigger.

Table 1. Average running time of cryptographic operations at 80-bit security (in seconds)

sha-1 hash	mod mul.	Legendre symbol	mod exp.
0.2×10^{-6}	0.8×10^{-6}	5.4×10^{-6}	3.7×10^{-3}

Communication wise, our protocol in phase 1 transfers $2n$ element in group Z_N^* and in phase 2 transfers $2 \cdot d \cdot k \cdot n \cdot \log_2 e$ k-bit strings. On the other hand, the communication cost of the protocol in [13] is n element in group $Z_{N^2}^*$. Our protocol consumes more bandwidth than the protocol in [13]. The bandwidth consumption of our protocol depends on d, the density of the vector. In the worst case where $d = 1$, the bandwidth consumed by our protocol is about 10 times as much as the protocol in [13], but when $d = 0.1$, the bandwidth consumption of our protocol is only twice that of the protocol in [13]. In real applications the density is often small thus the difference is not significant.

4 Implementation and Performance

We implemented our protocol in Figure 1 and measured the performance. The protocol was implemented in C. The implementation uses OpenSSL [22] and GMP [12] for the cryptographic operations. We used the Oblivious Bloom Intersection as the underlying PSI protocol. We obtained the source code of the OBI protocol, which is also in C, and incorporated it in our implementation. As a reference, we also implemented Goethals et al's protocol [13] in C. Goethals et al's protocol relies on additive homomorphic encryption and we use the Paillier public key scheme [23] as the building block of the protocol. We tested the protocols with 80 bits symmetric keys and1024 bits public keys. The experiments were conducted on two Mac computers. P_1 ran on a Macbook Pro laptop with an Intel 2720QM quad-core 2.2 GHz CPU and 16 GB RAM. P_2 ran on a Mac Pro with 2 Intel E5645 6-core 2.4GHz CPUs and 32 GB RAM. The two computers were connected by 1000M Ethernet. The two parties communicate through TCP socket. In all experiments, we use randomly generated bit vectors as inputs. We use synthesized data rather than real data because of the following two reasons: firstly the performance is not affected by the nature of the input data; secondly the performance of our protocol varies with the density of the vectors and it is hard to demonstrate the worst case performance with real data.

We first show the overall running time of the protocols with different vector sizes. The performance of our protocol depends on density of the vectors. In this experiment we measured the worst case performance by setting the density of the vectors used in our protocol to 1. In the experiment, each party uses one thread for computation and another one for network communication. The result is shown in table 2. As we can see in the table, our protocol is more than 20 times faster than Goethals et al's protocol.

Table 2. Total Running Time Comparison (in Seconds)

Vector size	1,000	10,000	100,000	1,000,000
Our Protocol	0.27	2.25	22.01	238.37
Goethals et al	5.06	49.58	509	5039

In real world applications, the density of the vectors is less than 1 and our protocol can be more efficient. The performance of our protocol has been further improved by exploiting parallelization. Oblivious Bloom Intersection, the underlying PSI protocol, is highly parallel. The implementation of the Oblivious Bloom Intersection protocol has a parallel mode which allows the program to utilize all available CPU cores and distribute the computation among them. The total running time can be significantly shortened if the program is running on multi-core systems. In the next experiment, we measured the performance of our protocol running in non-parallel and parallel modes with different densities. In the experiment, we set the vector size $n = 1,000,000$ and varied the density from 0.1 to 1. The result is shown in Figure 2. As we can see in the figure, the parallel mode does increase the performance significantly. The total running time in the non-parallel mode is $1.8\times - 4.2\times$ of that in the parallel mode. On the other hand, the total running time in each mode increases linearly with the vector density. When the density is 0.1, the total running time is 18.9 and 34.1 seconds in the parallel and non-parallel modes respectively, while when the density is 1 the numbers become 57 and 238 seconds. To compare, we also plot the total running time of Goethals et al's protocol when the vector size is set to $5,000$ and $50,000$. The running time of Goethals et al's protocol is not affected by the vector density. As we can see, the total running time of our protocol in all cases is less than that of the Goethals et al's protocol with $n = 50,000$. The total running time of Goethals et al's protocol is linear in n.

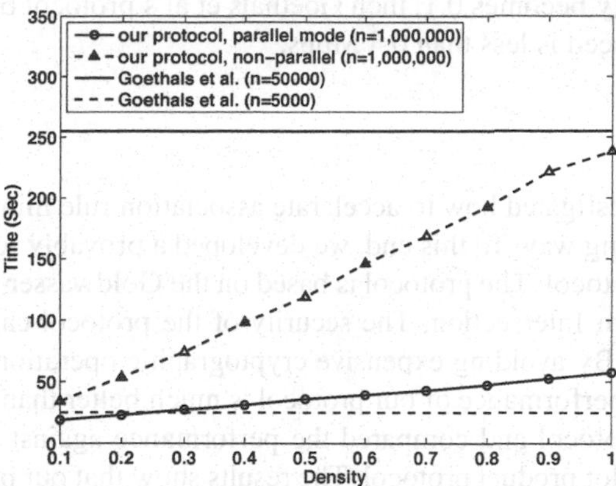

Fig. 2. Running Time in Non-Parallel and Parallel Modes with Various Vector Density

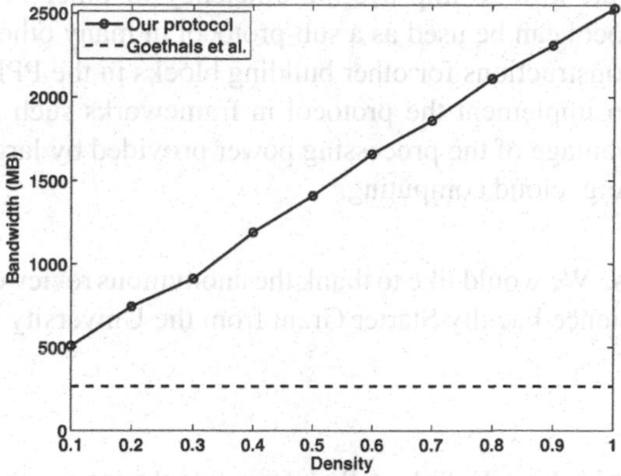

Fig. 3. Bandwidth Consumption

That means our protocol is at least 20 times faster. Our protocol is more than100 times faster than Goethals et al's when running in the non-parallel mode and when the vector density is 0.1, and is 200 times faster when running in the parallel mode and with a vector density below 0.2.

We also measured the bandwidth consumption. In the experiment, we set the vector size $n = 1,000,000$ and varied the density. The bandwidth consumption of Goethals et al's protocol with $n = 1,000,000$ was measured for comparison. The result is shown in Figure 3. Goethals et al's protocol consumes about 266 MB bandwidth. The bandwidth consumption of our protocol is about 1.9× and 9.5× of that when the density is 0.1 and 1 respectively. The number is quite close to our estimation in section 3.4. If the network connection is very slow, then Goethals et al's protocol can be faster than ours because the bottleneck is the network speed. Given the above measurements, we can roughly estimate when to switch. For example, if the vector size is 1,000,000 and density is 1, then when the network speed is less than 3.8 Mbps, Goethals et al's protocol should be used; if the density becomes 0.1, then Goethals et al's protocol becomes faster only when the network speed is less than 0.4 Mbps.

5 Conclusion

In this paper we investigated how to accelerate association rule mining on big datasets in a privacy preserving way. To this end, we developed a provably secure and very efficient dot product protocol. The protocol is based on the Goldwasser–Micali Encryption and Oblivious Bloom Intersection. The security of the protocol can be proved in the semi-honest model. By avoiding expensive cryptographic operations such as modular exponentiation, the performance of our protocol is much better than previous ones. We implemented the protocol and compared the performance against the currently most widely used secure dot product protocol. The results show that our protocol is orders of magnitude faster.

As future work, we would like to extend the protocol to multiple parties. We would also like to investigate how to improve the efficiency of other PPDM tasks. As we mentioned, our protocol can be used as a sub-protocol in many other PPDM tasks. We need also efficient constructions for other building blocks in the PPDM tasks. Another future direction is to implement the protocol in frameworks such as MapReduce, so that we can take advantage of the processing power provided by large scale distributed parallel computing, e.g. cloud computing.

Acknowledgements. We would like to thank the anonymous reviewers. Changyu Dong is supported by a Science Faculty Starter Grant from the University of Strathclyde.

References

1. Agrawal, R., Evfimievski, A.V., Srikant, R.: Information sharing across private databases. In: SIGMOD Conference, pp. 86–97 (2003)
2. Agrawal, R., Imielinski, T., Swami, A.N.: Mining association rules between sets of items in large databases. In: SIGMOD Conference, pp. 207–216 (1993)
3. Agrawal, R., Srikant, R.: Fast algorithms for mining association rules in large databases. In: VLDB, pp. 487–499 (1994)
4. Agrawal, R., Srikant, R.: Privacy-preserving data mining. In: SIGMOD Conference, pp. 439–450 (2000)
5. Amirbekyan, A., Estivill-Castro, V.: A new efficient privacy-preserving scalar product protocol. In: AusDM, pp. 209–214 (2007)
6. De Cristofaro, E., Gasti, P., Tsudik, G.: Fast and private computation of cardinality of set intersection and union. In: Pieprzyk, J., Sadeghi, A.-R., Manulis, M. (eds.) CANS 2012. LNCS, vol. 7712, pp. 218–231. Springer, Heidelberg (2012)
7. Dong, C., Chen, L., Wen, Z.: When private set intersection meets big data: An efficient and scalable protocol. In: ACM Conference on Computer and Communications Security (2013)
8. Du, W., Atallah, M.J.: Privacy-preserving cooperative statistical analysis. In: ACSAC, pp. 102–110 (2001)
9. Du, W., Zhan, Z.: Building decision tree classifier on private data. In: Proceedings of the IEEE International Conference on Privacy, Security and Data Mining, CRPIT '14, Darlinghurst, Australia, vol. 14, pp. 1–8. Australian Computer Society, Inc. (2002)

10. Evfimievski, A.V., Srikant, R., Agrawal, R., Gehrke, J.: Privacy preserving mining of association rules. In: KDD, pp. 217–228 (2002)
11. Freedman, M.J., Nissim, K., Pinkas, B.: Efficient private matching and set intersection. In: Cachin, C., Camenisch, J.L. (eds.) EUROCRYPT 2004. LNCS, vol. 3027, pp. 1–19. Springer, Heidelberg (2004)
12. GMP, `http://gmplib.org/`
13. Goethals, B., Laur, S., Lipmaa, H., Mielikäinen, T.: On private scalar product computation for privacy-preserving data mining. In: Park, C., Chee, S. (eds.) ICISC 2004. LNCS, vol. 3506, pp. 104–120. Springer, Heidelberg (2005)
14. Goldreich, O.: The Foundations of Cryptography, Basic Applications, vol. 2. Cambridge University Press (2004)
15. Goldwasser, S., Micali, S.: Probabilistic encryption and how to play mental poker keeping secret all partial information. In: STOC, pp. 365–377 (1982)
16. He, X., Vaidya, J., Shafiq, B., Adam, N., Terzi, E., Grandison, T.: Efficient privacy-preserving link discovery. In: Theeramunkong, T., Kijsirikul, B., Cercone, N., Ho, T.-B. (eds.) PAKDD 2009. LNCS, vol. 5476, pp. 16–27. Springer, Heidelberg (2009)
17. Huang, Z., Du, W., Chen, B.: Deriving private information from randomized data. In: SIGMOD Conference, pp. 37–48 (2005)
18. Ioannidis, I., Grama, A., Atallah, M.J.: A secure protocol for computing dot-products in clustered and distributed environments. In: ICPP, pp. 379–384 (2002)
19. Kantarcioglu, M., Clifton, C.: Privacy-preserving distributed mining of association rules on horizontally partitioned data. In: DMKD (2002)
20. Kargupta, H., Datta, S., Wang, Q., Sivakumar, K.: On the privacy preserving properties of random data perturbation techniques. In: ICDM, pp. 99–106 (2003)
21. Lindell, Y., Pinkas, B.: Privacy preserving data mining. In: Bellare, M. (ed.) CRYPTO 2000. LNCS, vol. 1880, pp. 36–54. Springer, Heidelberg (2000)
22. OpenSSL, `http://www.openssl.org/`
23. Paillier, P.: Public-key cryptosystems based on composite degree residuosity classes. In: Stern, J. (ed.) EUROCRYPT 1999. LNCS, vol. 1592, pp. 223–238. Springer, Heidelberg (1999)
24. Rizvi, S., Haritsa, J.R.: Maintaining data privacy in association rule mining. In: VLDB, pp. 682–693 (2002)
25. Tassa, T.: Secure mining of association rules in horizontally distributed databases. IEEE Transactions on Knowledge and Data Engineering 99(PrePrints), 1 (2013)
26. Tassa, T., Jarrous, A., Ben-Ya'akov, Y.: Oblivious evaluation of multivariate polynomials. J. Mathematical Cryptology 7(1), 1–29 (2013)
27. Vaidya, J., Clifton, C., Zhu, Y.: Privacy Preserving Data Mining. Advances in Information Security. Springer (2006)
28. Vaidya, J., Clifton, C.: Privacy preserving association rule mining in vertically partitioned data. In: KDD, pp. 639–644 (2002)
29. Vaidya, J., Clifton, C.: Secure set intersection cardinality with application to association rule mining. Journal of Computer Security 13(4), 593–622 (2005)
30. Vaidya, J., Kantarcioglu, M., Clifton, C.: Privacy-preserving naïve bayes classification. VLDB J. 17(4), 879–898 (2008)
31. Wong, W.K., Cheung, D.W.L., Kao, B., Mamoulis, N.: Secure knn computation on encrypted databases. In: SIGMOD Conference, pp. 139–152 (2009)
32. Zhang, N., Wang, S., Zhao, W.: A new scheme on privacy preserving association rule mining. In: Boulicaut, J.-F., Esposito, F., Giannotti, F., Pedreschi, D. (eds.) PKDD 2004. LNCS (LNAI), vol. 3202, pp. 484–495. Springer, Heidelberg (2004)

Author Index